Handbook of Transparent Conductors

David S. Ginley
Editor

Hideo Hosono • David C. Paine
Associate Editors

Handbook of Transparent Conductors

Springer

Editor
Dr. David S. Ginley
NREL
Photovoltaics & Electronic Materials
Center & Basic Sciences Ctr.
Cole Blvd. 1617
80401-3393 Golden Colorado
USA
david_ginley@nrel.gov

Associate Editors
Dr. Hideo Hosono
Tokyo Institute of Technology
Materials & Structures Lab.
Nagatsuta 4259
226-8503 Yokohama
Midori-ku
Japan
hosono@msl.titech.ac.jp

Prof. David C. Paine
Brown University
Division Engineering
610 Barus & Holley
Hope Street 182
02912 Providence Rhode Island
USA
David_Paine@Brown.EDU

ISBN 978-1-4419-1637-2 e-ISBN 978-1-4419-1638-9
DOI 10.1007/978-1-4419-1638-9
Springer New York Heidelberg Dordrecht London

Library of Congress Control Number: 2010935196

Printed on acid-free paper

Springer is part of Springer Science+Business Media (www.springer.com)

Preface

Transparent Conducting Oxides (TCOs) are a unique class of materials that exhibit both transparency and electronic conductivity simultaneously. These materials have found wide spread use in displays, photovoltaics, low-e windows, and flexible electronics. In many of these applications, the TCO's, are enabling in their role as transparent contacts. However, increasingly, the demands required extend beyond the combination of conductivity and transparency, where indeed higher performance is needed, but now include work function, morphology, processing and patterning requirements, long term stability, lower cost and elemental abundance/ green materials. As these needs have begun to emerge over the last 5 years they have stimulated a dramatic resurgence of research in the field leading to many new materials and processes. Overall it is the purpose of this book to provide both a snapshot of the new and enabling work in the field and to provide some indications of what might be coming next. We note that now the field of Transparent Conductors (TC's) includes not only conventional TCOs but also metal and carbon nano-composites, grapheme and polymer based TC materials. While the book primarily focuses on the TCOs some comparisons are made to the newer materials. To do this we have assembled a group of authors representing most of the leading groups in the field.

Historically, TCOs were limited primarily to tin oxide with fluorine doping, zinc oxide with aluminum doping and Indium tin oxide. Over the past 5–10 years the field has exploded to include a vastly increased number of n-type materials and to add in a class of new p-type materials. In addition, the historically held view that crystalline materials have superior properties, has been challenged by an emergence of new amorphous TCOs that have properties as good as or better than their crystalline counterparts. These materials have led to the development of amorphous oxide transistors which offer the advantage of low temperature processing and the promise of flexible electronics on polymer substrates. In their role as a channel material in thin film transistor structures, TCO's with controlled carrier densities are often termed transparent oxide semiconductors (TOS) since their key properties may lie in the limited to non-conductive regime. To capture this diversity of materials, processing and applications, we have organized the book as follows.

Chapter 1 introduces TCOs and covers the historic materials and their properties and uses this background to put some of the newly emergent materials into a technological context. *Chapter 2* presents a detailed discussion of the basic electronic structures of TCO materials emphasizing the key properties which give them their unique properties. *Chapter 3* then provides an overview of methods for the measurement and interpretation of transport properties in TCOs based of the Drude model with a focus on the method of four coefficients for the determination of critical parameters such as carrier type, mobility and scattering mechanisms in multinary oxides. *Chapter 4* covers the basic physics of, and practical tools for, the characterization of important TCO parameters including atomic structure, optical properties, electrical transport, work function and other properties that must be better understood as TCO's become used in novel applications such as thin film transistors. *Chapter 5* presents a picture of the current In based TCOs covering both the traditional $InSnO_x$ materials which have been the gold standard of TCOs and the emerging amorphous materials. *Chapter 6* presents an overview of the tin oxide based TCO materials. While historically these materials have been produced in exceptionally large areas new work has begun to improve their properties. *Chapter 7* reviews the state of the art for ZnO. This material, due to its natural abundance and the ease with which it can be deposited via both physical and chemical routes, has important applications both as a traditional transparent contact and great potential as an active optoelectronic material. To realize this potential, a great deal of work has been done to identify new approaches to both n and p-type doping. *Chapter 8* looks at the rapidly expanding class of multi-cation TCO materials. Recent work shows that much higher performance can be achieved in some TCO materials by the addition of elements that serve to modify defect and electronic band structure. This ability to create multi-component TCO materials without significantly degrading key transport parameters (e.g., carrier mobility) is a characteristic of the TCO class of materials. *Chapter 9* looks at the theoretical framework used to describe the band structures of both n- and p-type oxide materials and includes a discussion of emerging non-oxide based transparent conductors. This fundamental background provides the basis for a discussion on considerations for the discovery of new high performance transparent conducting materials. *Chapter 10* considers new materials that have emerged in the transparent conductor field over the last few years. Historically, the set of elements whose oxides provide useful TCO properties have been constrained to single or mixed oxides of In, Ga, Zn, Sn, and Cd. This chapter discusses how the pallet of useful elements for TCO applications has grown to open whole new classes of materials.

The second half of the book begins to address the applications of TCOs and how new materials can significantly change the paradigm for a technology or be enabling for another. *Chapter 11* discusses the application of TCO materials for solar energy and energy efficiency applications. In fact, though a key focus is the active devices like PV, the reality is that in terms of energy efficiency, the use of TCO's in energy conservation applications are greater in the near term than production. In any case it needs to be looked in an integrated way which is the theme of the chapter. *Chapter 12* considers the idea that TCOs need not be planar films but that in many cases the films can be enabling or integrated into a more complex hybrid (organic/inorganic for example) device by having a nanostructured morphology. Enabling this is a broad set of solution and PVD approaches to creating controlled nanostructures in TCO materials from texture to nano-rods etc. *Chapter 13* explores the application of amorphous TCOs and their semiconducting/insulating TOS counterparts to develop new flexible and transparent electronics for displays and more. The demonstration of TOS materials as a channel materials in thin film transistor applications has dramatically altered the potential for amorphous oxides in an increasingly diverse set of technologies. *Chapter 14* considers the potential for making true oxide based p/n junctions to realize active devices that are entirely based on TCO/TOS materials. The ability to make such junctions expands the potential for oxide based electronics including transparent electronics, oxide based solar cells and LED/lasers. Finally, *Chap. 15* discusses the scaling of TCO materials to large area industrial processing. This is a key issue as it addresses some of the critical properties dependence on process parameters.

We note that there is increasing interest in solution processed transparent conductors consisting of nanostructures of carbon (nanotubes), oxides (nanorods i.e., ZnO) and metals (such as Ag nanorods). However, thus far although they are very interesting, these materials still have conductivities approximately an order of magnitude below those for high performance TCOs. Over the next few years we expect these materials will become increasingly important perhaps in combination with TCO materials. Their inclusion in this volume at present is, however, beyond the intended scope of this publication.

This book presents a picture of an important class of materials that has, in recent years, drawn increasing interest for applications in active devices and as a critical component in any structure that requires both electrical connectivity and optical transparency. Despite their technological importance and relatively long history of use, our understanding of the existing set of TCO materials are only now receiving the kind of combined fundamental/experimental materials research attention that will inevitably lead to new materials discoveries and novel applications. Overall, it is clear that transparent conductive oxides and transparent conductors are a vibrant field that is advancing rapidly across an ever broadening spectrum of applications.

We hope this book will provide a valuable reference for those interested in the topic and stimulate additional development of new TCO materials and their applications.

David Ginley (Editor)
Hideo Hosono and David Paine
(Associate Editors)

Contents

Contributors

Joseph Berry National Renewable Energy Laboratory, Mail-stop 3211, 1617 Cole Blvd, Golden, CO 80401, USA

David P. Cann Associate Professor of Materials Science, Department of Mechanical Engineering, 303D Dearborn Hall, Oregon State University, Corvallis, OR 97331, USA, cann@engr.orst.edu

Dr. Timoth J. Coutts Research Fellow Emertus, National Renewable Energy Laboratory, 1617 Cole Blvd. Golden, CO 80401, USA

Dr. Klaus Ellmer Dept. solar fuels, Helmholtz-Zentrum für Materialien und Energie Berlin GmbH, Hahn-Meitner-Platz 1, 14109, Berlin, Germany, ellmer@ helmholtz-berlin.de

Barbara Falabretti Department of Engineering, University of Cambridge, Trumpington Street, Cambridge, CB2 1PZ, UK, bf222@eng.cam.ac.uk

Dr. Timoth A. Gessert Group Manager Thin Film Photovoltaics, NREL National Center For Photovoltaics, 1617 Cole Blvd, Golden, CO 80401, USA, Tim. Gessert@nrel.gov

David S. Ginley Research Fellow Group Manager Process Technology and Advanced Concepts, NREL SERF W102, 15313 Denver West Pkwy, Golden, CO 80401, USA, david_ginley@nrel.gov

Claes G. Granqvist Professor Solid State Physics, Department of Engineering Sciences, The Angstrom Laboratory Uppsala University, Uppsala, SE-75121, Sweden, claes-goran.granqvist@angstrom.uu.se

Steven P. Harvey Institute of Physical Chemistry, RWTH Aachen University, Landoltweg 2, Aachen D-52056, Germany

Hideo Hosono Professor at Frontier Research Center & Materials and Structures Laboratory, Tokyo Institute of Technology, 4259 Nagatsuta, Midori-ku, Yokohama 226-8503, Japan, hosono@msl.titech.ac.jp

Robert Kykyneshi Department of Materials Science, Oregon State University, Corvallis, OR 97331, USA

Thomas O. Mason Department of Materials Science and Engineering Northwestern University, Materials Research Science and Engineering Center, Evanston, IL 60208, USA, t-mason@northwestern.edu

Dr. Mamoru Mizuhashi School of Science and Engineering, Aoyama Gakuin University, 5-10-1 Fuchinobe, Sagamihara, Kanagawa 229-8558, Japan, dsatelmn@ss.iij4u.or.jp

Dr. Hiromichi Ohta Associate Professor Department of Molecular, Design & Engineering, Graduate School of Engineering Nagoya University, Furo-cyo, Chikusa-ku, Nagoya 464-8603, Japan

Dana C. Olson National Renewable Energy Laboratory, Mail Stop 3211 National Center for Photovoltaics, 1617 Cole Blvd. Golden, CO 80401-3393, USA, dana.olson@nrel.gov

David C. Paine Professor of Engineering Brown University, Division of Engineering, Box D, Providence, RI 02912, USA, David_Paine@brown.edu

Dr. John D. Perkins National Renewable Energy Laboratory, Mail Stop 3211 National Center for Photovoltaics, 1617 Cole Blvd. Golden, CO 80401, USA

Kenneth R. Poppelmeier Professor of Chemistry Northwestern University, Room: Tech GG35 Clark Street, Evanston, IL 60208, USA, krp@northwestern.edu

John Robertson Department of Engineering University of Cambridge, Trumpington Street, Cambridge, CB2 1PZ, UK

Yuzo Shigesato Professor Graduate School of Science and Engineering Aoyama Gakuin University, 5-10-1 Fuchinobe, Sagamihara, Kanagawa 229-8558, Japan, yuzo@chem.aoyama.ac.jp

Art Sleight Chemistry Department Oregon State University, 339 Weniger Hall, Corvallis, OR 97331, USA, arthur.sleight@oregonstate.edu

Burag Yaglioglu Plastic Logic Limited, 296 Cambridge Science Park, Milton Road, Cambridge, CB4 0WD, UK, buragy@gmail.com

David L. Young Senior Scientist National Renewable Energy Laboratory National Center for Photovoltaics, Silicon Materials and Devices, 1617 Cole Blvd. M.S. 3219, Golden, CO 80401, USA, david_young@nrel.gov

Jin Zeng Materials Science Oregon State University, Corvallis, OR 97331, USA, zengj@onid.orst.edu

Chapter 1
Transparent Conductors

David S. Ginley and John D. Perkins

1.1 Basics

Over the last 6 years the field of transparent conducting oxides has had a dramatic increase in interest with a huge influx in the number of active groups and the diversity of materials and approaches. Why? There are a number of primary motivators for this, some of the most compelling are the increase in portable electronics, displays, flexible electronics, multi-functional windows, solar cells and, most recently, transistors. The diverse nature of the materials integrated into these devices, including semiconductors, molecular and polymer organics, ceramics, glass, metal and plastic, have necessitated the need for TCO materials with new performance, processibility and even morphology. The remarkable applications dependent on these materials have continued to make sweeping strides. These include the advent of larger flat-screen high-definition televisions (HDTVs including LCD, Plasma and OLED based displays), larger and higher-resolution flat screens for portable computers, the increasing importance of energy-efficient low-emittance ("low-e"), solar control and electrochromic windows, a dramatic increase in the manufacturing of thin film photovoltaics (PV), the advent of oxide based transistors and transparent electronics as well as a plethora of new hand-held, flexible and smart devices, all with smart displays. Driven by the increased importance and potential opportunities for TCO materials in these and other applications, there has been increasing activity in the science of these materials. This has resulted in new n-type materials, the synthesis of p-type materials and novel composite TCO materials as well as an increased set of theoretical and modeling tools for understanding and predicting the behavior of TCOs. Considering that, over the last 20 years, much of the materials work on TCOs has been empirical with a focus on minor variants of ZnO, In_2O_3 and SnO_2, it is quite remarkable how dramatically this field has grown recently in both basic and

D.S. Ginley (✉)
National Renewable Energy Laboratory, 1617 Cole Blvd., Golden, CO 80401, USA
e-mail: david_ginley@nrel.gov

D.S. Ginley et al. (eds.), *Handbook of Transparent Conductors*,
DOI 10.1007/978-1-4419-1638-9_1, © Springer Science+Business Media, LLC 2010

applied science. This is reflected in the thousands of papers published over the last 5 years. This may be a function of not only the need to achieve higher performance levels for these devices, but also of the increasing importance of transition-metal-based oxides in devices in a broader sense including ferroelectric, piezoelectric, thermoelectric, gas sensing superconducting and other materials applications. An important realization is that despite a very long history in the application of TCOs, there is still not a complete theoretical understanding of the materials nor an ability to reliably predict the properties of new materials. This has been emphasized recently by the emergence of amorphous mixed metal oxide TCO materials, typified by amorphous In-Zn-O [1, 2], where even the basic transport physics is not understood. Broadly, this book will summarize the current state of the art across a broad range of TCO science including materials, theory, thin film deposition and applications.

At this point, a brief summary of the relevant opto-electronic properties of conventional TCO materials will provide a useful baseline for comparison and discussion. The left panel in Fig. 1.1 shows optical reflection, transmission and absorption spectra for a typical commercial ZnO TCO on glass which, collectively, show the key spectral features of a TCO material. First, the material is quite transparent, ~80%, in the visible portion of the spectrum, 400–700 nm. Across this spectral region where the sample is transparent, oscillations due to thin film interference effects can be seen in both the transmission and reflection spectra. The short wavelength cut off in the transmission at ~300 nm is due to the fundamental band gap excitation from the valence band to the conduction band as depicted in the right panel of Fig. 1.1. The gradual long wavelength decrease in the transmission starting at ~1,000 nm and the corresponding increase in the reflection starting at ~1,500 nm are due to collective oscillations of conduction band electrons known as plasma oscillations or plasmons for short. There can also be substantial absorption due to these plasma oscillations as is the case for this particular sample with the maximum absorption occurring at the characteristic plasma wavelength, λ_P, as shown in the figure. As the number of electrons in the

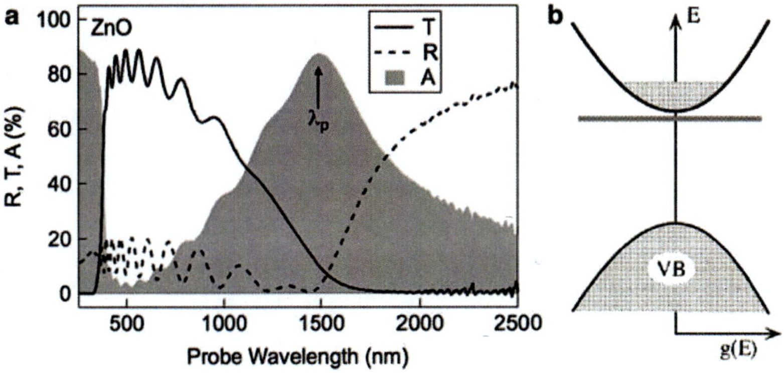

Fig. 1.1 Optical spectra of typical (ZnO) transparent conductor (*left side*) and schematic electronic structure of conventional TCO materials (*right side*)

conduction band, N, is increased, such as by substitutional doping, the plasma wavelength shifts to shorter wavelengths as $\lambda_P \propto 1/\sqrt{N}$ which creates a fundamental tradeoff between conductivity and the long wavelength transparency limit. At very high electron concentrations, this can even decrease the visible wavelength transparency. The left panel in Fig. 1.2 shows how the infrared transparency increases for SnO_2 TCOs as the sheet resistance is increased from 5 to 100 Ω/sq. Even though both of these SnO_2 samples have similar visible wavelength transparency, the 5 Ω/sq. sample would be unusable as a transparent conductor for telecom applications at 1,500 nm or for giving a high solar throughput. The right panel in Fig. 1.2 shows how the plasma wavelength varies with the dopant level in Ti doped In_2O_3 and hence how TCO properties can be tuned [3]. Collectively, the examples shown in Figs. 1.1 and 1.2 should make it clear that there is no such thing as a single "best" TCO and that TCOs must be tailored to the constraints of the specific application.

The current use of TCOs in industry is dominated by just a few materials. We will present an overview of the current state of the field in order to help the reader develop an appreciation for the size and demands of the industry as well as the need for new materials. At present, the dominant markets for TCOs are in architectural window applications and flat panel displays (FPDs), followed closely by the rapidly growing photovoltaics industry.

The architectural use of TCOs is predominately for energy efficient windows. Fluorine-doped tin oxide (SnO_2:F), deposited by a chemical vapor deposition (CVD) process, is the TCO most often used in this application [4, 5]. Metal-oxide/Ag/Metal-oxide stacks such as ZnO/Ag/ZnO are also common [6, 7]. Windows with tin oxide coatings are efficient in preventing radiative heat loss, due to their low thermal emittance, ~0.15, compared to ~0.84 for uncoated glass [8]. Such "low-e" windows are ideal for use in cold or moderate climates. In addition, pyrolytic tin oxide is also used for heated glass freezer doors in commercial use. In this application, the doors can be defrosted by passing a small current through the slightly

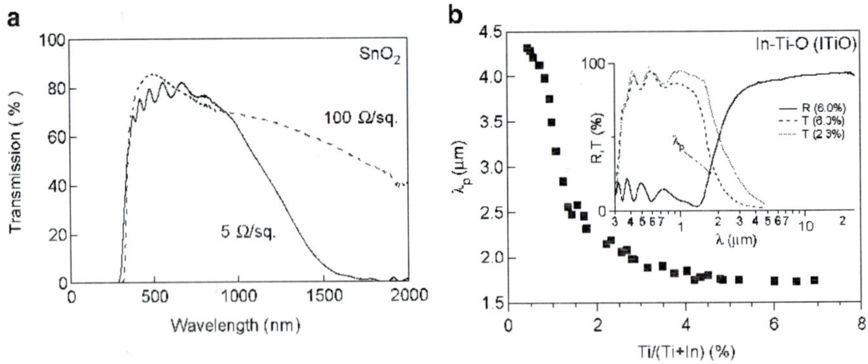

Fig. 1.2 Optical spectra of TCO materials: SnO_2 (*left side*) and Ti-doped In_2O_3 (*right side*) from van Hest et al. [3]

resistive TCO coating. In 2007, the annual demand for low-e coated glass in Europe was 60×10^6 m^2 (~23 square miles) and this is projected to increase to about 100×10^6 m^2 in a few years [9]. Rapid growth in China is also increasing the demand for low-e glass with a projected demand of 97×10^6 m^2 in 2010 and a projected domestic production capacity of 50×10^6 m^2 in 2010, up significantly from 3×10^6 m^2 in 2004 [10]. Added to these demands for low-e coated glass are increasing amounts of TCO coated glass for solar control applications [5, 9] as well as the increasing amounts used in displays and PVs [11]. Multilayer stacks as referred to above represent an increasing market for TCOs in the conventional low-e applications but also in an expanding automotive and specialty market.

Pyrolytic tin oxide is also used in PV modules, touch screens, and plasma displays. However, for the majority of flat panel display (FPD) applications, crystalline tin doped indium oxide (indium-tin-oxide, ITO) and, more recently, amorphous In-Zn-O (IZO) are the TCOs used most often in these higher value added products. In FPDs, the primary function of the TCO is as a transparent electrode. However, often, the TCO will also have additional functions such as an antistatic electromagnetic interference shields or as an electric heater.

The annual volume of FPDs produced, and hence the volume of TCO (ITO) coatings required, continues to grow rapidly. New analysis from Frost & Sullivan (http://www.electronics.frost.com) World Flat Panel Display Markets, reveals that the FPD market earned revenues of $65.25 billion in 2005 and estimates this to roughly double to $125.32 billion in 2012 as shown in Fig. 1.3. This by far exceeds the initial market projections and arises from the rapid worldwide adoption of flat panel displays in place of conventional vacuum tube display.

Recently, a significant fraction of the FPD industry has begun to use amorphous indium-zinc-oxide (IZO) in place of ITO as the main TCO. Amorphous IZO has the advantages of being amorphous along with room temperature deposition, easy patterning and improved thermal stability relative to ITO. While it is currently not known exactly what fraction of the display industry uses IZO, it is estimated to be between 30 and 40%.

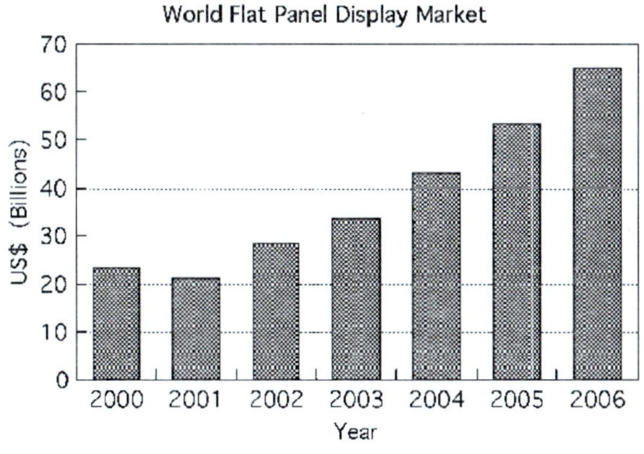

Fig. 1.3 TCO markets vs. year from http://www.electronics.frost.com

Currently, the third, and fastest growing segment, of the TCO market is for photovoltaic (PV) cells, predominately driven by crystalline and polycrystalline silicon solar cells which represent more than 93% of the PV market at present. Even for PV cells based on bulk Si material TCOs are important. For example, the two-sided Sanyo HIT cell (Heterojunction with Intrinsic Thin-layer) shown in Fig. 1.4 actually uses TCO layers on both the front and back. In addition, thin film photovoltaics based on amorphous-Si (a-Si), CdTe and Cu(In,Ga)Se_2 (CIGS) absorber layers all depend on one or more layers of a high performance TCO as shown in Figs. 1.4 and 1.5. This thin film PV application represents a growing and important market for TCO materials. Figure 1.6 shows the projected market growth of the PV industry based on various growth rates ranging from 15 to 30% per year. In actuality, the PV market is currently (2007) growing at over 50% per year with no sign of slowing down.

There are potentially two other areas where there could be a rapid increase in TCO use on a large scale. These are electrochromic windows and oxide based thin film transistors (TFTs). In the former, this technology is either all metal oxide based or organic/inorganic based. In either case, there is a key reliance on TCOs as the transparent electrodes. The second area, oxide based TFTs, is embryonic at present, but could rapidly mature [12]. There has been a longstanding search for higher mobility transistors for displays and flexible electronics and recently amorphous metal oxide based transistors have emerged as a promising alternative to the conventional a-Si TFTs. Here, the key driver is higher electron mobility, $\mu \approx 10$–$40\ cm^2/V\ s$ whereas $\mu < 1\ cm^2/V\ s$ for a-Si as well as processibility, easy integration into flexible electronics due to low temperature (room temperature) processing, and the potential mechanical advantages of an amorphous material. In addition to these major applications, there is also use of TCOs in the emerging areas of opto-electronic components, other electrochromic devices (automobile windows, mirrors, sun roofs, displays etc.) and flexible electronics. Together, all these needs

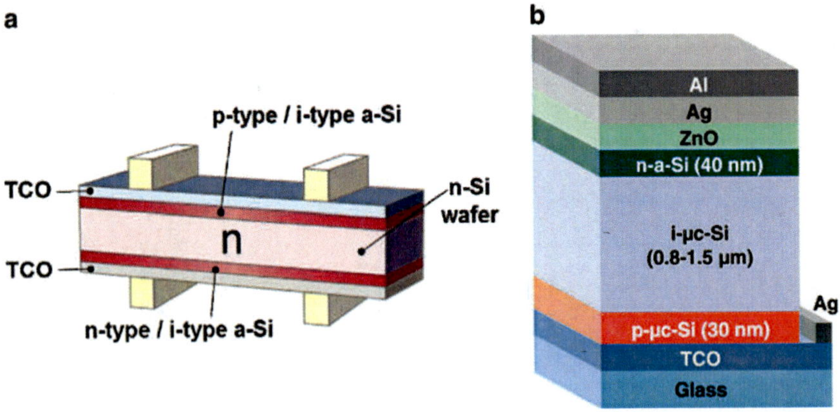

Fig. 1.4 Typical configuration for (**a**) the Sanyo HIT cell and (**b**) for an amorphous-Si nip cell

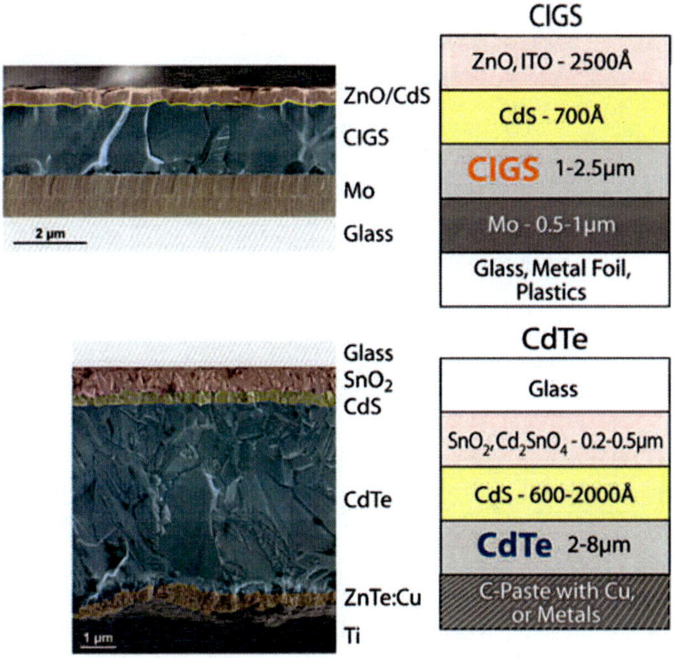

Fig. 1.5 CIGS (*top*) and CdTe (*bottom*) PV structures seen in cross-section using SEM (*left side* of *panel*) and viewed schematically (*right side* of *panel*)

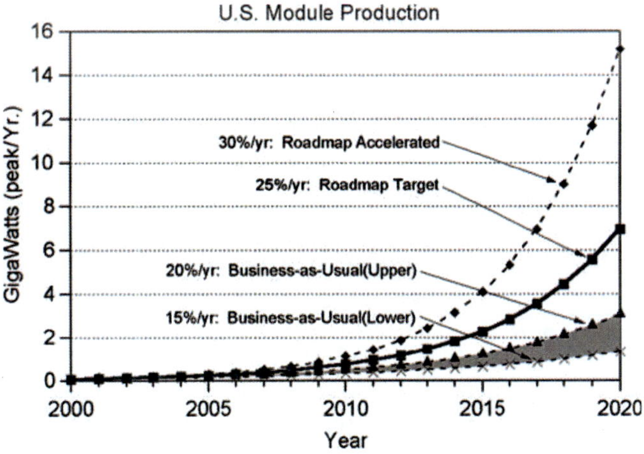

Fig. 1.6 Projected growth of PV markets

provide the driving impetus for the increasing importance of TCO materials both technologically and economically across a wide variety of applications.

Cleary, the driving forces for new TCOs are quite diverse and are being driven by a large number of concerns. Some of these are summarized in Table 1.1. All or

Table 1.1 Properties relevant to TCO materials and applications

General criteria	Opto-electronic criteria	Processing criteria
Green materials	Visible transparency	Deposition temperatures and conditions
Green processing	Conductivity	Annealing stability
Cost	Carrier concentration	Compatibility with vacuum or non-vacuum processing
Availability	Mobility	Chemical stability
Ease of application	Infrared transparency	Etchability, patterning and electrical contacts
	High mobility	Interfacial chemistry and surface states
	High mobility with low carrier concentration	Ionic diffusion properties
	Suitability to flexible electronics	Temperature sensitivity
	Work function	Atmospheric sensitivity

Table 1.2 TCO materials for various applications[a]

Property application	Material		
	Simple	Binary	Ternary
Highest transparency	$ZnO{:}F$	Cd_2SnO_4	
Highest conductivity	$In_2O_3{:}Sn$		
Highest plasma frequency	$In_2O_3{:}Sn$		
Highest work function	$SnO_2{:}F$	$ZnSnO_3$	$Zn_{0.45}In_{0.88}Sn_{0.66}O_3$
Lowest work function	$ZnO{:}F$		
Best thermal stability	$SnO_2{:}F$	Cd_2SnO_4	
Best mechanical durability	$SnO_2{:}F$		
Best chemical durability	$SnO_2{:}F$		
Easiest to etch	$ZnO{:}F$		
Best resistance to H plasmas	$ZnO{:}F$		
Lowest deposition temperature	$In_2O_3{:}Sn$ $ZnO{:}B$ a-InZnO		
Least toxic	$ZnO{:}F$, $SnO_2{:}F$		
Lowest cost	$SnO_2{:}F$		
TFT channel layer	ZnO	a-InZnO, a-ZnSnO	$InGaO_3(ZnO)_5$, a-InGaZnO
Highest mobility	CdO, $In_2O_3{:}Ti$ $In_2O_3{:}Mo$		
Resistance to water	$SnO_2{:}F$		

[a]Adapted from Gordon [4]

any of these criteria may be important for a particular TCO application. This diverse matrix of needs may also necessitate new individual TCO materials and/or the development of new multilayer stacks incorporating TCOs. Table 1.2 gives a partial list of some of the properties and some of the representative TCOs based on the conventional materials as well as a column showing the role of some of the new materials. This, in part, is driving the exploration of new and improved materials as well as the increasing emphasis on improved processibility and environmental properties.

Thus, in the last few years, there has been an increasing realization that the conventional TCO material set of substitutionally doped crystalline SnO_2, ZnO and In_2O_3 materials are no longer sufficient to meet the needs of all TCO applications. As is the case in many technological areas, this is a consequence of the acknowledgment of the limitations of the existing materials as well as a realization that new materials can open the way to new and improved devices. Amplifying this is the need for TCO materials with certain specific properties other than just high transparency and conductivity as applications are emerging where work function, surface roughness, nano-structure, thermal and chemical reactivity/diffusivity or ease of patterning are critical TCO functionalities.

1.2 History of Transparent Conducting Oxides

The list of TCO materials in Table 1.3, though not fully inclusive, clearly shows the wide diversity of current TCO materials. As one can see, there was a dramatic and on-going increase in the number of TCO materials starting after 1995. This rate of materials discovery has continued with more new materials every year. Furthermore, transparent conductors now also include thin metal films, sulphides, selenides, nitrides, nanotube composites, graphenes and polymers in additional to the traditional metal oxide based TCOs.

As we have seen, there are numerous technological drivers for the development of new and improved TCOs and we have also seen that worldwide the field has expanded dramatically in the last 10 years with the number of researchers and their efforts increasing substantially every year. There are also global societal drivers for the development of improved TCOs due to their critical role in the development of various energy related technologies. For example, Fig. 1.7, which shows the world energy consumption by region, makes very clear the rapidly increasing energy use worldwide [13]. Furthermore, as the undeveloped world rapidly becomes more technological with the associated increasing energy needs and vehicular traffic, the total global energy consumption will continue to rise rapidly. One clear consequence of this is that global atmospheric CO_2 levels which are a major cause of global warming are increasing dramatically. In Fig. 1.8, it is clear that the present on-going rapid increase in CO_2, which appears instantaneous on the 45,000 year time span of the graph, is significantly beyond any previous short term event in the history spanned by the figure [14–18]. These two interrelated facts are increasingly leading to a view that society must drive towards a truly sustainable life style. To achieve this, sustainability must be a consideration in all aspects of a technology including design, processing, delivery, application and, finally, end of service life and recycling. So how do TCOs relate to this global problem? First off, they are key elements in a number of "green" technologies. In particular, they are critical to low-e and solar control windows, photovoltaics, OLEDs for indoor lighting and vehicle heat management. Collectively, this combination of technologies which depend on TCO materials has the potential to significantly change the energy use

Table 1.3 Selected historical TCO references

Material	Year	Process	Reference
Cd-O			
CdO	1907	Thermally Oxidation	K. Badeker, Ann. Phys. (Leipzig) 22, 749 (1907)
Cd-O	1952	Sputtering	G. Helwig, Z. Physik, 132, 621 (1952)
Sn-O			
SnO_2:Cl	1947	Spray pyrolysis	H.A. McMaster, U.S. Patent 2,429,420
SnO_2:Sb	1947	Spray pyrolysis	J.M. Mochel, U.S. Patent 2,564,706
SnO_2:F	1951	Spray pyrolysis	W.O. Lytle and A.E. Junge
SnO_2:Sb	1967	CVD	H.F. Dates and J.K. Davis, USP 3,331,702
Zn-O			
ZnO:Al	1971		T. Hada, Thin Solid Films 7, 135 (1971)
In-O			
In_2O_3:Sn	1947		M.J. Zunick, U.S. Patent 2,516,663
In_2O_3:Sn	1951	Spray pyrolysis	J.M. Mochel, U.S. Patent 2,564,707 (1951)
In_2O_3:Sn	1955	Sputtering	L. Holland and G. Siddall, Vacuum III
In_2O_3:Sn	1966	Spray	R. Groth, Phys. Stat. Sol. 14, 69 (1969)
Ti-O			
TiO_2:Nb	2005	PLD	Furubayashi et al., Appl. Phys. Lett. 86, 252101 (2005)
Zn-Sn-O			
Zn_2SnO_4	1992	Sputtering	Enoki et al., Phys. Stat. Solid A 129, 181 (1992)
$ZnSnO_3$	1994	Sputtering	Minami et al., Jap. J. Appl. Phys. 2, 33, L1693 (1994)
a-ZnSnO	2004	Sputtering	Moriga et al., J. Vac. Sci. & Tech. A 22, 1705 (2004)
Cd-Sn-O			
Cd_2SnO_4	1974	Sputtering	A.J. Nozik, Phys. Rev. B, 6, 453 (1972)
a-CdSnO	1981	Sputtering	F.T.J. Smith and S.L. Lyu, J. Electrochem. Soc. 128, 1083 (1981)
In-Zn-O			
$Zn_2In_2O_5$	1995	Sputtering	Minami et al., Jap. J. Appl. Phys. P2 34, L971 (1995)
a-InZnO			
In-Ga-Zn-O			
$InGaZnO_4$	1995	Sintering	Orita et al., Jap. J. Appl. Phys. P2. 34, 1550 (1995)
a-InGaZnO	2001	PLD	Orita et al., Phil. Mag. B 81, 501 (2001)

CVD chemical vapor deposition; *PLD* pulsed laser deposition

balance by both enabling new energy generation technologies and improving energy efficiency technologies. Again, this provides further motivation to move to new TCO materials for less environmental impact, lower cost, sustainability and efficiency improvements in important devices.

1.3 Diversity of Transparent Conductors

As stated previously, TCOs have historically been dominated by a small set of oxide materials including predominately SnO_2, In_2O_3, InSnO and ZnO. However, stimulated by the concerns enumerated above, the field has more recently expanded not just into a broader spectrum of oxides, but also into other materials as well. The traditional TCO oxide composition space is nominally focused on oxides and

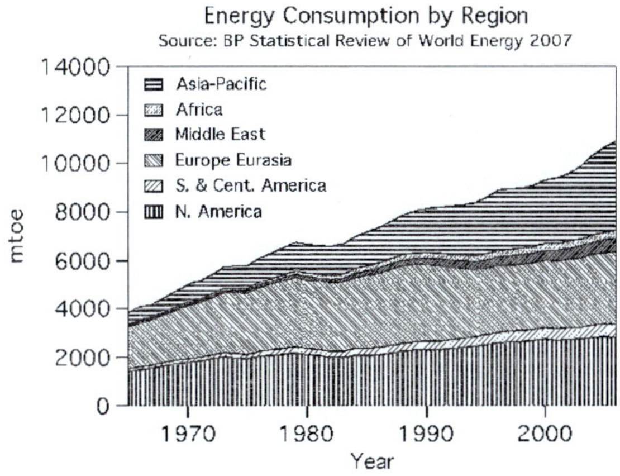

Fig. 1.7 Energy consumption vs. year by global region (mtoe: millions of tons oil equivalent) [13]

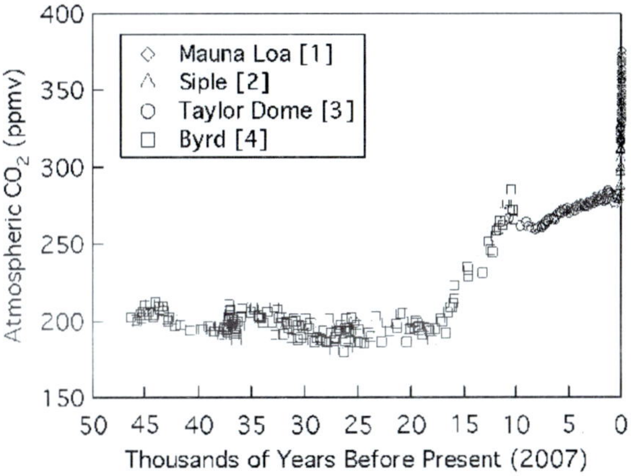

Fig 1.8 Atmospheric CO_2 vs. year. Data compiled from refs. [14–17]

combinations of oxides of Zn, Sn, In, Ga, and Cd. These main TCO materials tend to fall in to groups that can be defined by structure type as shown in Table 1.4. The building blocks in the first two rows of Table 1.4 can then be combined as shown in Table 1.5 to form most of the known TCO materials. This basic composition space is depicted pictorially in Fig. 1.9.

This has created a very diverse set of crystalline and, more recently, amorphous transparent conducting materials. There has been considerable modeling and speculation of the range of crystalline oxide TCO materials with both empirical and first principles models having been developed recently [19–23]. This has led both to a much better understanding of the n-type materials and to an emerging understanding of the limits of p-type materials. Thus far however, the theory has not been applied extensively to the amorphous mixed-metal-oxide n-type systems such as the prototypical In-Zn-O and Zn-Sn-O systems. This is largely due to the difficulty in applying electronic structure calculation approaches to non-period amorphous materials but some initial work has been done on In-Ga-Zn-O [24].

Table 1.4 Cation coordination and carrier type of TCO materials[a]

Structural feature	Carrier type	Examples
Tetrahedrally-coordinated cations	n-Type	ZnO
Octahedrally-coordinated cations	n-Type	CdO, In_2O_3, SnO_2, $CdIn_2O_4$, Cd_2SnO_4, etc.
Linearly-coordinated cations	p-Type	$CuAlO_2$, $SrCu_2O_2$, etc.
Cage framework	n-Type	12CaO-7Al_2O_3

[a]From Inger et al. Journal of Electroceramics 13, 167 (2004)

Table 1.5 Doping of TCO materials[a]

Material	Dopant or compound
SnO_2	Sb, F, As, Nb, Ta
In_2O_3	Sn, Ge, Mo, F, Ti, Zr, Mo, Hf, Nb, Ta, W, Te
ZnO	Al, Ga, B, In, Y, Sc, F, V, S, Ge, Ti, Zr, Hf
CdO	In, Sn
Ga_2O_3	
ZnO-SnO_2	Compounds Zn_2SnO_4, $ZnSnO_3$
ZnO-In_2O_3	$Zn_2In_2O_5$, $Zn_3In_2O_6$
In_2O_3-SnO_2	$In_4Sn_3O_{12}$
CdO-SnO_2	Cd_2SnO_4, $CdSnO_3$
CdO-In_2O_3	$CdIn_2O_4$
$MgIn_2O_4$	
$GaInO_3$, (Ga, In)$_2O_3$	Sn, Ge
$CdSb_2O_6$	Y
Zn-In_2O_3-SnO_2	$Zn_2In_2O_5$-$In_4Sn_3O_{12}$
CdO-In_2O_3-SnO_2	$CdIn_2O_4$-Cd_2SnO_4
ZnO-CdO-In_2O_3-SnO_2	

[a]From Minami, Semiconductor Science and Technology 20, S35 (2005)

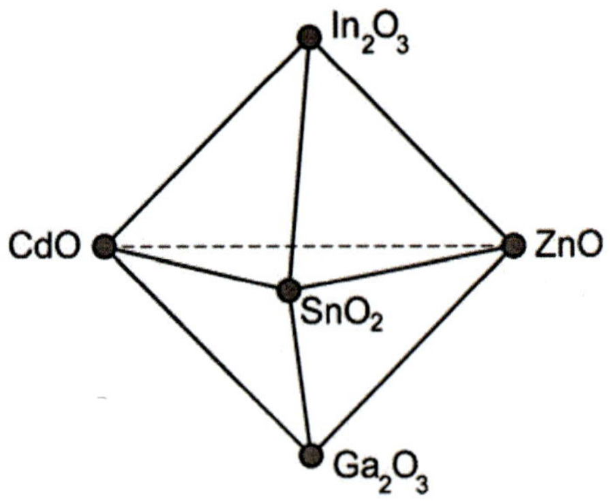

Fig. 1.9 Composition space for conventional TCO materials

The generality of the amorphous TCOs is further illustrated by the ternary cation InGaZnO [24, 25] and the binary cation CdSnO [26] systems. In all of these amorphous materials, the electronic transport mechanism appears to be complex but nevertheless, the performance is very good, especially the electron mobilities which can be as high as 50 cm^2/V s [2], better than many commercial crystalline TCOs. These new amorphous TCO materials are amorphous mixtures of the composition phase space shown in Fig. 1.9 in which all the single metal oxide basis members have filled d-shells and the conduction band states come mostly from the empty metal atom s-states. Hosono et al. [27, 28] have proposed that the high electron mobility in these amorphous TCO materials is due to direct overlap of these large and non-directional metal atom s-states as depicted in Fig. 1.10. Candidate metallic elements which form oxides which satisfy these basic criteria are highlighted in Fig. 1.10. However, materials synthesis work over the past decade has made clear that the specific metal element mixtures selected from this candidate set can make an order of magnitude or more difference in the maximum achievable conductivity.

Recently, attention has also been directed to Nb and Ta doped anatase TiO_2 as a potential new TCO material. In this material, which falls outside the conventional TCO materials shown in Fig. 1.9, the conduction band is formed largely from Ti 3d states instead of metal atom s-states as discussed above. Highly conducting films have been deposited onto single crystal $SrTiO_3$ by both pulsed laser deposition (PLD) (σ = 4,300 S/cm) [29] and sputtering (σ = 3,000 S/cm) [30]. For films on glass, a conductivity of σ = 2,200 S/cm has been obtained for films deposited by PLD and then subsequently annealed in H_2 [31]. Another novel class of unconventional oxide based transparent conductors system is the $12CaO{\cdot}7Al_2O_3$ ("C12A7"

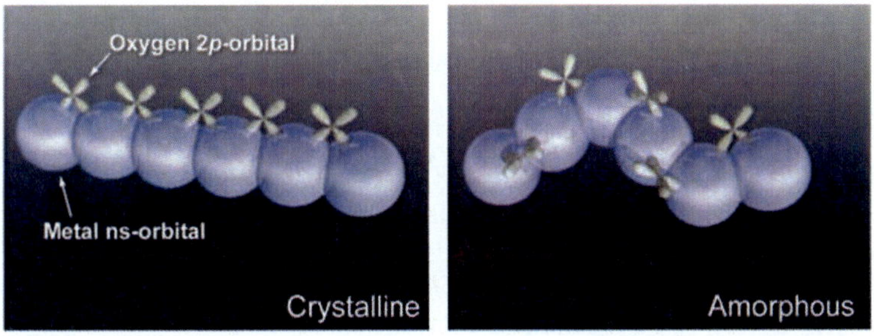

1 H																	2 He
3 Li	4 Be											5 B	6 C	7 N	8 O	9 F	10 Ne
11 Na	12 Mg											13 Al	14 Si	15 P	16 S	17 Cl	18 Ar
19 K	20 Ca	21 Sc	22 Ti	23 V	24 Cr	25 Mn	26 Fe	27 Co	28 Ni	**29 Cu**	**30 Zn**	**31 Ga**	**32 Ge**	**33 As**	34 Se	35 Br	36 Kr
37 Rb	38 Sr	39 Y	40 Zr	41 Nb	42 Mo	43 Tc	44 Ru	45 Rh	46 Pd	**47 Ag**	**48 Cd**	**49 In**	**50 Sn**	**51 Sb**	52 Te	53 I	54 Xe
55 Cs	56 Ba	57 La	72 Hf	73 Ta	74 W	75 Re	76 Os	77 Ir	78 Pt	**79 Au**	**80 Hg**	**81 Tl**	**82 Pb**	**83 Bi**	84 Po	85 At	86 Rn

Fig 1.10 Effect of disorder on electron orbital overlap in metal oxide semiconductors (*top*) and candidate metal ions for amorphous mixed metal oxide transparent conductors (*bottom*) from Hosono [82]

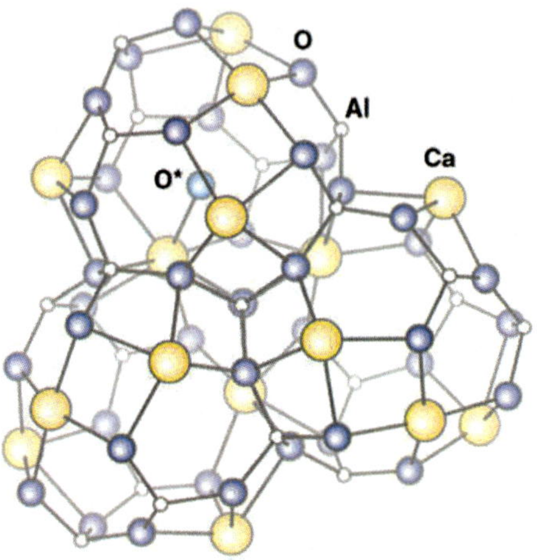

Fig. 1.11 Structure of $12CaO{\cdot}7Al_2O_3$ (C12A7) from Medvedeva et al. [32]

or $Ca_{12}Al_{14}O_{33}$) cage compounds based on electride materials [32–34] as shown in Fig. 1.11. The conductivity in these materials must be activated. In the initial photo-activated materials, conductivities of order 1 S/cm were obtained [33]. Recently, using a multi-step film growth, crystallization and in-situ annealing process, a conductivity of ~800 S/cm has been obtained [35].

Generally TCOs are predominately n-type because of the ease of forming oxygen vacancies or cation interstitials in the oxides such as those depicted in Fig. 1.9 [36]. However one of the current major research challenges for TCO materials is the development of p-type materials with comparable conductivities to their n-type counterparts, i.e., of order 10^3 S/cm. The on-going search for p-type TCOs was pushed to a much higher level by the work of Kowazoe and Hosono in the 1990s on the Cu based materials such as $CuAlO_2$ [37] and $SrCu_2O_2$ [38]. These materials were clearly p-type, but the doping levels and mobilities were low, typically $N \leq 10^{18}/cm^3$ and $\mu < 1$ cm^2/V s. To date these problems have not been solved in these copper based metal oxides.

Starting in 1999 [39, 40], there was a great flurry of activity trying to make ZnO p-type which would have yielded a very versatile opto-electronic material similar to GaAs. To date, while p-type materials have been observed, they do not seem to be stable and reproducible [34, 41–47]. Many groups are still working on this area because of the promise of ZnO which can have the band gap increased or decreased by the addition of Mg or Cd respectively and can be made a spintronic material by the addition of Co. This, along with a viable p-type material, would make a remarkably versatile optoelectronic system. Towards this end, Tsukazaki et al. have developed a laser MBE deposition with in-situ repeated temperature modulation during growth that can consistently make p-type nitrogen doped ZnO [48] as well as ZnO/MgZnO quantum well structures of sufficient quality to enable observation of the quantum hall effect in a ZnO based heterostructure [49].

The search for p-type materials has expanded beyond oxides to copper based sulfides and selenides such as LaCuOS [50], BaCuSF [51] and related materials

[52–54]. Several precious metal based transparent oxides have also shown p-type conduction including crystalline ZnM_2O_4 (M = Ir, Rh, Co) [55–57] as well as amorphous Zn-Rh-O [58, 59].

Outside of the range of the basic oxide materials and primarily driven by the OLED and nanomaterials communities is an emerging interest in organic based transparent conductors. The focus has been on intrinsically conducting polymers [60], charge transfer polymers like PEDOT:PSS [61–64] and, more recently, on carbon nanotube composites [65–68] and graphenes [69–73]. All of these materials of are of great interest for the OLED, flexible electronics and polymer/thin film photovoltaics communities because of the potential to write, print, or spin on coatings at, or near, room temperature from liquid based precursors at atmospheric pressure. Typically, the conductivities are around 2–1,200 S/cm, about an order of magnitude less than for ITO. Still, the materials may be of use as an intermediate contact layer to a conventional TCO in order to provide a better electronic interface to a polymer or molecular electronic device.

Overall, as transparent conductors are used in an increasingly broad array of applications, each with their own particular needs, nearly everything in the transparent conductors materials tool box will likely become important.

1.4 Emerging Applications

In addition to the conventional TCO applications discussed above, there are a number of emerging applications that have the potential to significantly impact the use of TCOs and, in some cases, clearly require very different TCO performance characteristics. Several of these currently emerging applications are briefly described here.

1.4.1 Transistors and Flexible Transparent Electronics

One high profile and rapidly emerging area is that of oxide based thin film transistors (TFTs) including transparent thin film transistors (TTFTs). Conducting TCOs can be combined with amorphous semi-insulating high mobility TCO materials to create a viable alternative to the conventional amorphous Si TFTs currently used in flat panel displays. This has the potential to create a whole new class of transparent electronics. At present, most of the schemes employ an amorphous semi-insulating transparent oxide semiconductor (TOS) such as indium zinc oxide (IZO) for the channel layer as shown in Fig. 1.12 [74]. This simple back-gated test structure designed for channel layer materials studies is not transparent due to both the Si substrate and the Ti gate. However, transparent top-gated structures utilizing TCOs for the source, drain and gate electrodes have been demonstrated on both glass and flexible plastic substrates as shown in Fig. 1.13 [12]. The semiconducting

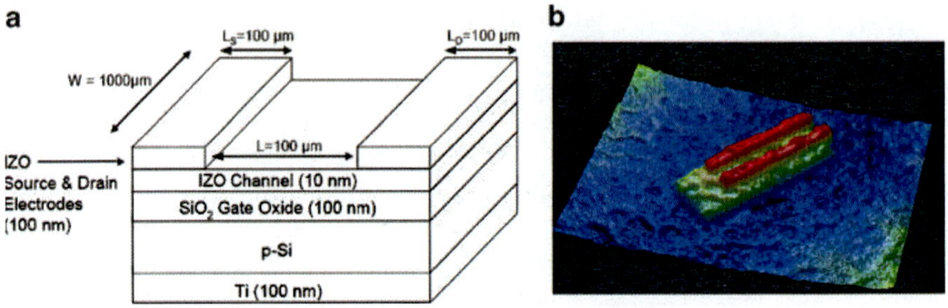

Fig. 1.12 Schematic (*left side*) and top-down image (*right side*) of IZO based TFT structure on Si substrate. Schematic image from Yaglioglu et al. [74]

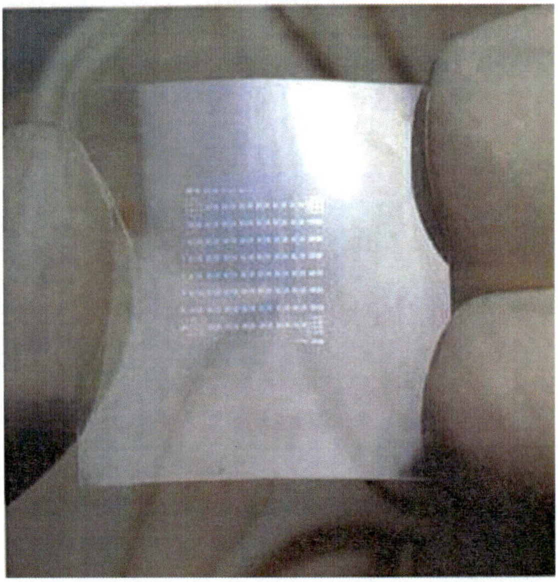

Fig. 1.13 Transparent TFT on flexible PET substrate from Nomura et al. [12]

IZO films used as the channel layer can be sputtered from the same target as conducting IZO simply by adding oxygen to the sputtering gas, about 10% O_2 in Ar is typical [2]. These amorphous TOS layers can have a high electron mobility, $\mu \sim 10\text{–}50\,\mathrm{cm^2/V\,s}$, across a wide range of doping levels compared to $\mu < 1\,\mathrm{cm^2/V\,s}$ in typical amorphous Si TFTs [75]. So, while the process technology for oxide TFT fabrication is not yet up to the reliability and device density of a-Si, it is a rapidly developing area with the potential for not just faster TFTs to replace a-Si TFTs, but transparent flexible electronics are possible. This will be covered in more detail in Chaps. 13 and 14.

1.4.2 Electrochromic Windows

Electrochromic devices have been emerging as a commercial reality over the last 10 years first finding application in the automotive arena as self dimming and then smart rear view mirrors as well as in smart windows with electronically adjustable transmission for building applications [7, 76, 77]. For these electrochromic applications either one TCO layer is needed for reflective applications and two are needed for transmissive applications such as the window illustrated in Fig. 1.14 [78].

1.4.3 Optical Arrays

The operation of liquid crystal displays is based on the change in the liquid crystal index of refraction when a voltage is applied and as the index of refraction changes, so does the effective optical path length. Based on this effect, electronic optical beam steerers can be made by using pixelated contacts to apply a spatially varying voltage across a liquid crystal resulting in an electronically reconfigurable optical wedge, an optical phased array in short. Such devices require a transparent contact at the operating wavelength, typically 1,500 nm for free space laser based telecom which requires high mobility, low-carrier concentration TCO materials to obtain sufficient IR transparency [79] such as the Ti-doped In_2O_3 shown in Fig. 1.2 [3].

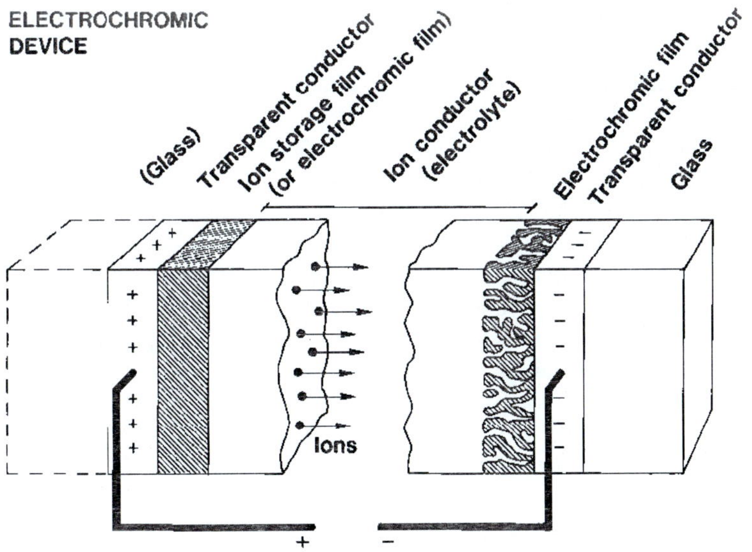

Fig. 1.14 Schematic structure of an electrochromic window showing front and back TCO contacts from Granqvist [78]

1.5 Chapter Topics

The structure of the book is summarized in the following section and ties directly to the themes discussed above.

Based on the discussion thus far, we have hopefully set the stage for the book explaining the size of the current TCO industry and its limitations with respect to existing technologies and emerging new technologies. The rest of the book will present a background for TCO science, a discussion of current and emerging materials and a discussion of the emerging roles of TCO as active electronic elements.

Chapter 2 will begin with the basics and will present a discussion of the electronic structure of transparent conducting oxides including a discussion of the role of different modeling approaches, both first principal and empirical, to TCOs and the ability of theory to understand and predict the conductivity, mobility, band gap, etc. This will be discussed in the context of both n-type and p-type TCO materials. Theoretical results will be compared with current experimental data which will form the basis to discuss the evolution of "design rules" for TCOs.

Chapter 3 will start with the basic empirical models from Chap. 2 and expand them to a discussion of the empirical modeling of basic optical and electronic properties of TCs. This is a critical area as these design tools are necessary to the development and optimization of device applications especially for multilayer stacks and complex heterostructures. It will include for the first time a complete description of the method of four coefficients which is a key tool for characterizing the new generation of TC materials.

Chapter 4 will take the modeling work of the previous two chapters and discuss the actual approaches to the measurement of TC films. Thus, the purpose of this chapter is to illustrate methods for the characterization of the microstructural, chemical, thermal, electrical and optical properties of transparent conducting oxides. Examples taken from the TCO literature will illustrate the use of both established and novel characterization tools. As the needs of TCOs expand well beyond simple transparency and conductivity so must the associated characterization approaches. This chapter is meant to provide a basis for understanding and interpreting the characterization data provided in Appendix 1. This chapter is organized in three sections: (1) Electrical/optical characterization, (2) microstructural and crystallographic characterization, and (3) Novel methods and approaches.

Chapter 5 discusses, in detail, the structure, electrical properties and structural characteristics on the indium oxide (In_2O_3) based TCO films.

One component of all flat-panel display devices – from simple calculator liquid-crystal displays to large-area, active-matrix, liquid-crystal color screens for top-of-the-line portable computer monitors or televisions – are the electrodes that control the orientation of the liquid-crystal molecules and, consequently, the on-off state with respect to the passage of light. At least one of the electrodes in a flat-panel display must be transparent since either ambient or transmitted (i.e., back light) light must pass through the device and reach the viewer's eye. The transparent conductors of choice is currently tin-doped In_2O_3 (ITO) although other transparent conductors are available and include SnO_2 and ZnO. To date, the indium based materials are favored for these optoelectronic applications and are the main topic for Chap. 5. ITO is favored because, within this class of materials, it offers the highest available transmissivity for visible light combined with the lowest electrical resistivity.

Tin-doped indium oxide is actually an n-type, highly degenerate, wide band gap semiconductor that owes its relatively low electrical resistivity to its high free carrier density. The free carrier density of ITO can be increased through appropriate processing to a level of order $10^{21}/cm^3$. Free carriers are contributed to the material by two different kinds of electron donor sites: substitutional four-valent tin atoms and oxygen vacancies. The electrically active tin atoms are positioned in substitutional sites in the 80 atom unit cell bixbyite In_2O_3 crystal structure. The highest quality thin films of ITO that can be routinely produced via physical vapor deposition (PVD) processes have a resistivity of $\rho = 1\text{–}2 \times 10^{-4}\ \Omega\,cm$ ($\sigma = 5\text{–}10 \times 10^3$ S/cm) when deposited at a substrate temperature of 300–400°C by conventional e-beam evaporation or magnetron sputtering. The resistivity of a material is inversely proportional to the product of the carrier mobility and the carrier density in the material; in the case of ITO, both of these parameters depend on the material microstructure.

Chapter 6 extends the discussion on the primary TCO materials to tin oxide (SnO_2). This material probably represents, by area, the largest application for TCOs due to its extensive use in windows for low-e and heat management applications. This system is generally deposited by CVD during the float glass process, unlike most of the other TCOs which are deposited by conventional PVD approaches. The chapter will cover the general structure of SnO_2, the defect structure and approaches to doping primarily with F and Sb. It will also discuss the chemistry and physical properties of doped tin oxide films. Interestingly though deposited essentially by chemical vapor deposition process, the resultant film is chemically resistant which although it makes it relatively difficult to pattern makes it environmentally stable. This characteristic is, in many cases, important enough to live with a slightly reduced conductivity (2×) compared to ITO. It is also substantially cheaper than ITO but may not be well suited to low temperature processing.

Chapter 7 focuses on zinc oxide (ZnO) based TCOs. Despite being one of the oldest TCOs, it is also one of the newest. It has the potential to be the next major opto-electronic material, competing with GaN and the like. Al:ZnO usually deposited by sputtering is a key TCO for many optical window, solar cell and display applications. The Ga doped analogue is recently attracting considerable attention due to its resistance to phase segregation and ease of deposition by solution and sol gel based approaches. But the real interest in ZnO has been stimulated by a number of non-conventional results. One of the holy grails for TCO materials is demonstration of a comparable p-type material to the n-type materials. In the last 5 years there has been a significant debate on the ability and stability of doping ZnO p-type this will be discussed. It is also important that ZnO can be substitutionally doped with

Mg or Cd to either increase or decrease the band gap and that substitution of Co etc. can convert it into a spintronic material. This combination of characteristics makes ZnO uniquely suited to a diverse set of transparent electronics with the potential for technological impact in a wide number of areas. However, ZnO is also chemically not very stable and is sensitive to the environment. Thus, this simple material is actually quite complex and potentially very versatile.

Chapter 8 undertakes the very difficult task of addressing the rest of n-type phase space including ternary and higher cation mixtures. In the classical TCO composition space of the oxides of Zn, Sn, Cd, In, and Ga (see Fig. 1.9) there is still a large amount of unexplored terrain. Some of the binaries, CdO for example, have demonstrated superb properties ($\mu \approx 600$ cm^2/V s for CdO [80]) but are not practical materials. Cd_2SnO_4 is a much more manufacturable TCO and, in fact, a combination of Cd and Zn stannates were used in the world record CdTe solar cell [81]. Thus the functionality and the potential utility of TCOs may be substantially enhanced by moving to compositionally more complex systems. The difficulty is clearly in how to develop design rules or experimental methodologies suitable to optimizing materials in this phase space. A number of sets of complex systems as examples include: In_2O_3:Sn,O_i, In_2O_3 co-doped with either Cd,Sn or Zn,Sn, $CdIn_2O_4$-Cd_2SnO_4 spinels, $Ga_3In_6Sn_2O_{16}$, $(ZnO)_kIn_2O_3$ and Ga-substituted layered phases, and the delafossite systems which have demonstrated both n and p-type components. The summary of the chapter tries to establish some basic design rules for developing new materials.

Chapter 9 extends the work of the previous chapter to look at the potential to use band edge engineering as another tool for the design of TCO properties. In particular, controlling holes in the oxygen 2p band is discussed in terms of the basic defect structure of model and TCO systems. This is then specifically applied to 3D, 4D and 5S or 6S based oxides. The substitution of sulfur or selenium for oxygen and the effects on the resultant defect structure is also discussed.

Chapter 10 continues this discussion of newly developed materials. In particular, this includes looking a very high performance Ga_2O_3, the newly developed LaCuOCh (Ch = S and Se) which are p-type TCOs with the potential for higher doping levels than existing metal oxide only materials and, finally, the $12CaO{\cdot}7Al_2O_3$ (C12A7) system which is a unique persistent photoconductor based on electride materials. These emerging materials are extending the horizon of the applications of TCOs.

Chapter 11 then takes the many properties of transparent conductors as enumerated in the previous chapters and undertakes a survey of some of their conventional and emerging applications in general. The chapter looks at the broad picture starting with a discussion of the radiation in our natural surroundings: thermal (blackbody) radiation, solar irradiance on the atmospheric envelope, atmospheric transmittance and then couples in the luminous efficiency of the human eye, spectral efficiency of photosynthesis for example as determining the "useful spectral ranges" for certain applications/phenomena. Spectral selectivity, angular dependence, and temporal variability are singled out as the distinctive features, and the desired properties of materials for solar energy and energy efficiency are then specified.

Chapter 12 presents the new generation of multifunctional applications for transparent conductors. Many of these include the use of nanostructured materials in conjunction with a optoelectronic heterostructure device. These include new building applications using nanostructured materials such as self cleaning and multifunctional windows, organic photovoltaics and the Gratzel cell. These devices all not only rely on the typical TCO characteristics but also depend on a number of other factors that control the interfaces between the heterostructure components. Some of these factors include the materials nanostructure both in a film or 3D, the work function, surface chemistry i.e. the wetting and contact to an organic, nature of surface states, etchability/chemical and physical stability, surface/bulk photochemistry and transport in nanostructured materials. The context of this will be illustrated by the use of ZnO nanofibriles for organic photovoltaics, the use of nanostructured TiO_2 for the Gratzel cell, they uses of superhydrophylic TiO_2 and TCOs based on carbon nanotube composites. The application of TCOs for everything from contacts for OLEDs to fuel cells is being investigated.

Chapter 13 discusses how the combination of low temperature deposition, flexible substrates and amorphous oxides is leading rapidly to a whole new field of TCO applications. It will discuss the application of low temperature deposition and its suitability especially for amorphous TCO materials and then extend this to applications on flexible substrates and to transparent TFTs. The higher mobility in these amorphous oxides (10–40 $cm^2/V\ s$) over that of the conventional amorphous Si (1 $cm^2/V\ s$) has the potential to dramatically improve displays. Coupled to their transparency this could create a whole new generation of applications. Integrated with this is the potential for low temperature printable materials that could lead to printable opto-electronics.

Chapter 14 continues the theme of Chap. 13 by way of extending the TFT work to transparent all oxide electronics. This would require a good p-type TCO and the design rules and prospects for this are discussed. In addition there is more discussion of the construction of all oxide transistors and TFTs, electron emitters and the prospects for future all transparent devices.

Chapter 15 then presents a current snapshot of existing and emerging process technology and industrial processes for TCs. Several different kinds of deposition methods have been used for the growth of thin-film TCO on suitable substrates. The current trend toward higher quality flat-panel devices has led to new display technologies with a much thinner electrode width and consequent demands for further optimization of TCO materials properties and processing. These new and demanding technologies require even lower film resistivity, and this must be achieved at lower substrate temperatures during deposition. The requirement for lower deposition temperatures derives from device designs that call for the deposition of ITO films on polymer color filters and other polymer films that cannot survive vacuum processing temperatures above 100–150°C. Meeting this demand requires the application of new PVD techniques that include various plasma and energetic ion processes, because under conventional sputtering or reactive evaporation deposition conditions at substrate temperatures below 150°C the deposited films are typically amorphous and have undesirable properties.

Utilization of plasma-enhanced processes provides a route to the formation of high-quality crystalline materials at low substrate deposition temperatures.

Examples of the successful implementation of these include modification of low-voltage, direct current (DC), magnetron sputtering to yield an enhanced plasma density or activated EB evaporation using either tungsten electron emitters or arc plasma generators. One consequence of utilizing plasma-enhanced techniques is that the growth surface during deposition is bombarded with energetic ions. This interaction between the growing film and the plasma during deposition has a dramatic effect on the microstructure of the resulting films.

Finally, we have a set of appendixes provide summary tables of useful TC properties and deposition parameters, etc.

Acknowledgments The National Center for Photovoltaics at the National Renewable Energy Laboratory (NREL) funded this work. NREL is a U.S. Department of Energy laboratory operated by Midwest Research Institute-Battelle-Bechtel under contract No. DE-AC36-99-GO10337.

References

1. "Highly transparent and conductive ZnO-In2O3 thin films prepared by dc magnetron sputtering", T. Minami, T. Kakumu, Y. Takeda and S. Takata, Thin Solid Films **291**, 1–5 (1996).
2. "General mobility and carrier concentration relationship in transparent amorphous indium zinc oxide films", A.J. Leenheer, J.D. Perkins, M. Van Hest, J.J. Berry, R.P. O'hayre and D.S. Ginley, Physical Review B **77**, 115215 (2008).
3. "Titanium-doped indium oxide: A high-mobility transparent conductor", M. Van Hest, M.S. Dabney, J.D. Perkins, D.S. Ginley and M.P. Taylor, Applied Physics Letters **87**, 032111 (2005).
4. "Criteria for choosing transparent conductors", R.G. Gordon, MRS Bulletin **25**, 52 (2000).
5. "Transparent conductors as solar energy materials: A panoramic review", C.G. Granqvist, Solar Energy Materials and Solar Cells **91**, 1529–1598 (2007).
6. "History of the development and industrial production of low thermal emissivity coatings for high heat insulating glass units", H.J. Glaser, Applied Optics **47**, C193-C199 (2008).
7. "Nanomaterials for benign indoor environments: Electrochromics for "smart windows", sensors for air quality, and photo-catalysts for air cleaning", C.G. Granqvist, A. Azens, P. Heszler, L.B. Kish and L. Osterlund, Solar Energy Materials and Solar Cells **91**, 355–365 (2007).
8. "Annual energy window performance vs. glazing thermal emittance - the relevance of very low emittance values", J. Karlsson and A. Roos, Thin Solid Films **392**, 345–348 (2001).
9. "Pilkington and the flat glass industry 2008", Pilkington Group Limited, St. Helens, United Kingdom WA10 3TT, www.pilkington.com.
10. "China Low-E Glass Market Report, 2007–2008", Research and Markets, Guinness Centre, Taylors Lane, Dublin 8, Ireland, www.researchandmarkets.com/reports/c88669.
11. "Global market and technology trends on coated glass for architectural, automotive and display applications", H. Ohsaki and Y. Kokubu, Thin Solid Films **351**, 1–7 (1999).
12. "Room-temperature fabrication of transparent flexible thin-film transistors using amorphous oxide semiconductors", K. Nomura, H. Ohta, A. Takagi, T. Kamiya, M. Hirano and H. Hosono, Nature **432**, 488–492 (2004).

13. "BP Statistical Review of World Energy June 2007", *http://www.bp.com/statisticalreview* (2007).
14. "Asynchrony of Antarctic and Greenland climate change during the last glacial period", T. Blunier, J. Chappellaz, J. Schwander, A. Dallenbach, B. Stauffer, T.F. Stocker, D. Raynaud, J. Jouzel, H.B. Clausen, C.U. Hammer and S.J. Johnsen, Nature **394**, 739–743 (1998).
15. "Holocene carbon-cycle dynamics based on CO2 trapped in ice at Taylor Dome, Antarctica", A. Indermuhle, T.F. Stocker, F. Joos, H. Fischer, H.J. Smith, M. Wahlen, B. Deck, D. Mastroianni, J. Tschumi, T. Blunier, R. Meyer and B. Stauffer, Nature **398**, 121–126 (1999).
16. "Historical CO2 record from the Siple Station ice core. In Trends: A Compendium of Data on Global Change", Carbon Dioxide Information Analysis Center, Oak Ridge National Laboratory, U.S. Department of Energy, Oak Ridge, TN, USA, http://cdiac.ornl.gov/trends/co2/siple.html (1994).
17. "Atmospheric carbon dioxide dry air mole fractions from quasi-continuous measurements at Barrow, Alaska; Mauna Loa, Hawaii; American Samoa; and South Pole, 1973–2006, Version: 200710-01", Version: 2007-10-01, ftp://ftp.cmdl.noaa.gov/ccg/co2/in-situ/ (2007).
18. "Proxy-based reconstructions of hemispheric and global surface temperature variations over the past two millennia", M.E. Mann, Z.H. Zhang, M.K. Hughes, R.S. Bradley, S.K. Miller, S. Rutherford and F.B. Ni, Proceedings of the National Academy of Sciences of the United States of America **105**, 13252–13257 (2008).
19. "Combining high conductivity with complete optical transparency: a band structure approach", J.E. Medvedeva and A.J. Freeman, Europhysics Letters **69**, 583–587 (2005).
20. "Combining high conductivity with complete optical transparency: a band-structure approach", J.E. Medvedeva and A.J. Freeman, Los Alamos National Laboratory, Preprint Archive, Condensed Matter 1–3, arXiv:cond-mat/0405317 (2004).
21. "Bipolar doping and band-gap anomalies in delafossite transparent conductive oxides", X. Nie, S.-H. Wei and S.B. Zhang, Physical Review Letters **88**, 066405/1–066405/4 (2002).
22. "Indium tin oxide alternatives - high work function transparent conducting oxides as anodes for the organic light-emitting diodes", J. Cui, A. Wang, N.L. Edleman, J. Ni, P. Lee, N.R. Armstrong and T.J. Marks, Advanced Materials (Weinheim, Germany) **13**, 1476–1480 (2001).
23. "Chemical and thin-film strategies for new transparent conducting oxides", A.J. Freeman, K.R. Poeppelmeier, T.O. Mason, R.P.H. Chang and T.J. Marks, MRS Bulletin **25**, 45–51 (2000).
24. "Local coordination structure and electronic structure of the large electron mobility amorphous oxide semiconductor In-Ga-Zn-O: Experiment and ab initio calculations", K. Nomura, T. Kamiya, H. Ohta, T. Uruga, M. Hirano and H. Hosono, Physical Review B **75**, 035212 (2007).
25. "Amorphous transparent conductive oxide InGaO3(ZnO)m ($m \leq 4$): a Zn 4s conductor", M. Orita, H. Ohta, M. Hirano, S. Narushima and H. Hosono, Philosophical Magazine B: Physics of Condensed Matter: Statistical Mechanics, Electronic, Optical and Magnetic Properties **81**, 501–515 (2001).
26. "Amorphous transparent conductors: the cadmium oxide-tin dioxide system", F.T.J. Smith and S.L. Lyu, Journal of the Electrochemical Society **128**, 1083–1088 (1981).
27. "Working hypothesis to explore novel wide band gap electrically conducting amorphous oxides and examples", H. Hosono, N. Kikuchi, N. Ueda and H. Kawazoe, Journal of Non-Crystalline Solids **200**, 165–169 (1996).
28. "New amorphous semiconductor: 2CdO center dot PbOx", H. Hosono, Y. Yamashita, N. Ueda, H. Kawazoe and K. Shimidzu, Applied Physics Letters **68**, 661–663 (1996).
29. "A transparent metal: Nb-doped anatase TiO2", Y. Furubayashi, T. Hitosugi, Y. Yamamoto, K. Inaba, G. Kinoda, Y. Hirose, T. Shimada and T. Hasegawa, Applied Physics Letters **86**, 252101/1–252101/3 (2005).

30. "rf magnetron sputter deposition of transparent conducting Nb-doped TiO2 films on SrTiO3", M.A. Gillispie, M. Van Hest, M.S. Dabney, J.D. Perkins and D.S. Ginley, Journal of Applied Physics **101**, 033125 (2007).
31. "Fabrication of highly conductive Ti1-xNbxO2 polycrystalline films on glass substrates via crystallization of amorphous phase grown by pulsed laser deposition", T. Hitosugi, A. Ueda, S. Nakao, N. Yamada, Y. Furubayashi, Y. Hirose, T. Shimada and T. Hasegawa, Applied Physics Letters **90**, 212106 (2007).
32. "Electronic structure and light-induced conductivity of a transparent refractory oxide", J.E. Medvedeva, A.J. Freeman, M.I. Bertoni and T.O. Mason, Physical Review Letters **93**, 016408 (2004).
33. "Light-induced conversion of an insulating refractory oxide into a persistent electronic conductor", K. Hayashi, S. Matsuishi, T. Kamiya, M. Hirano and H. Hosono, Nature **419**, 462–465 (2002).
34. "Function cultivation of transparent oxides utilizing built-in nanostructure", H. Hosono, T. Kamiya and M. Hirano, Bulletin of the Chemical Society of Japan **79**, 1–24 (2006).
35. "High electron doping to a wide band gap semiconductor 12CaO center dot 7Al(2)O(3) thin film", M. Miyakawa, M. Hirano, T. Kamiya and H. Hosono, Applied Physics Letters **90**, 182105 (2007)
36. "Dopability, intrinsic conductivity, and nonstoichiometry of transparent conducting oxides", S. Lany and A. Zunger, Physical Review Letters **98**, 045501 (2007).
37. "P-Type electrical conduction in transparent thin films of CuAlO2", H. Kawazoe, M. Yasukawa, H. Hyodo, M. Kurita, H. Yanagi and H. Hosono, Nature **389**, 939–942 (1997).
38. "SrCu2O2: A p-type conductive oxide with wide band gap", A. Kudo, H. Yanagi, H. Hosono and H. Kawazoe, Applied Physics Letters **73**, 220–222 (1998).
39. "p-Type electrical conduction in ZnO thin films by Ga and N codoping", M. Joseph, H. Tabata and T. Kawai, Japanese Journal of Applied Physics Part 2-Letters **38**, L1205–L1207 (1999).
40. "Solution using a codoping method to unipolarity for the fabrication of p-type ZnO", T. Yamamoto and H. Katayama-Yoshida, Japanese Journal of Applied Physics Part 2-Letters **38**, L166–L169 (1999).
41. "Studies of minority carrier diffusion length increase in p-type ZnO:Sb", O. Lopatiuk-Tirpak, L. Chernyak, F.X. Xiu, J.L. Liu, S. Jang, F. Ren, S.J. Pearton, K. Gartsman, Y. Feldman, A. Osinsky and P. Chow, Journal of Applied Physics **100**, 086101/1–086101/3 (2006).
42. "Fabrication of p-type Li-doped ZnO films by pulsed laser deposition", B. Xiao, Z. Ye, Y. Zhang, Y. Zeng, L. Zhu and B. Zhao, Applied Surface Science **253**, 895–897 (2006).
43. "Progress in ZnO materials and devices", D.C. Look, Journal of Electronic Materials **35**, 1299–1305 (2006).
44. "p-Type behavior in In-N codoped ZnO thin films", L.L. Chen, J.G. Lu, Z.Z. Ye, Y.M. Lin, B. H. Zhao, Y.M. Ye, J.S. Li and L.P. Zhu, Applied Physics Letters **87**, 252106/1–252106/3 (2005).
45. "Diffusion of phosphorus and arsenic using ampoule-tube method on undoped ZnO thin films and electrical and optical properties of P-type ZnO thin films", S.-J. So and C.-B. Park, Journal of Crystal Growth **285**, 606–612 (2005).
46. "Pulsed-laser-deposited p-type ZnO films with phosphorus doping", V. Vaithianathan, B.-T. Lee and S.S. Kim, Journal of Applied Physics **98**, 043519/1–043519/4 (2005).
47. "p-Type ZnO thin films grown by MOCVD", X. Li, S.E. Asher, B.M. Keyes, H.R. Moutinho, J. Luther and T.J. Coutts, Conference Record of the IEEE Photovoltaic Specialists Conference 31st, 152–154 (2005).
48. "Repeated temperature modulation epitaxy for p-type doping and light-emitting diode based on ZnO", A. Tsukazaki, A. Ohtomo, T. Onuma, M. Ohtani, T. Makino, M. Sumiya, K. Ohtani, S.F. Chichibu, S. Fuke, Y. Segawa, H. Ohno, H. Koinuma and M. Kawasaki, Nature Materials **4**, 42–46 (2005).
49. "Quantum Hall effect in polar oxide heterostructures", A. Tsukazaki, A. Ohtomo, T. Kita, Y. Ohno, H. Ohno and M. Kawasaki, Science **315**, 1388–1391 (2007).

50. "Transparent p-type semiconductor: LaCuOS layered oxysulfide", K. Ueda, S. Inoue, S. Hirose, H. Kawazoe and H. Hosono, Applied Physics Letters **77**, 2701–2703 (2000).
51. "p-Type conductivity in wide-band-gap BaCuQF (Q=S,Se)", H. Yanagi, J. Tate, S. Park, C.H. Park and D.A. Keszler, Applied Physics Letters **82**, 2814–2816 (2003).
52. "p-Type transparent conducting oxides", S. Sheng, G. Fang, C. Li, S. Xu and X. Zhao, Physica Status Solidi A: Applications and Materials Science **203**, 1891–1900 (2006).
53. "Frontier of transparent conductive oxides", H. Ohta, M. Orita, H. Hiramatsu, K. Nomura, M. Miyakawa, K. Ueda, M. Hirano and H. Hosono, Advances in Science and Technology (Faenza, Italy) **33**, 983–994 (2003).
54. "Non-vacuum and PLD growth of next generation TCO materials", D. Ginley, B. Roy, A. Ode, C. Warmsingh, Y. Yoshida, P. Parilla, C. Teplin, T. Kaydanova, A. Miedaner, C. Curtis, A. Martinson, T. Coutts, D. Readey, H. Hosono and J. Perkins, Thin Solid Films **445**, 193–198 (2003).
55. "Fabrication and characterization of heteroepitaxial p-n junction diode composed of wide-gap oxide semiconductors p-ZnRh2O4/n-ZnO", H. Ohta, H. Mizoguchi, M. Hirano, S. Narushima, T. Kamiya and H. Hosono, Applied Physics Letters **82**, 823–825 (2003).
56. "ZnIr2O4, a p-type transparent oxide semiconductor in the class of spinel zinc-d(6)-transition metal oxide", M. Dekkers, G. Rijnders and D.H.A. Blank, Applied Physics Letters **90**, 021903 (2007).
57. "Electrical and magnetic properties of spinel-type magnetic semiconductor ZnCo2O4 grown by reactive magnetron sputtering", H.J. Kim, I.C. Song, J.H. Sim, H. Kim, D. Kim, Y.E. Ihm and W.K. Choo, Journal of Applied Physics **95**, 7387–7389 (2004).
58. "A p-type amorphous oxide semiconductor and room temperature fabrication of amorphous oxide p-n heterojunction diodes", S. Narushima, H. Mizoguchi, K. Shimizu, K. Ueda, H. Ohta, M. Hirano, T. Kamiya and H. Hosono, Advanced Materials **15**, 1409–1413 (2003).
59. "Electrical properties and structure of p-type amorphous oxide semiconductor xZnO center dot Rh2O3", T. Kamiya, S. Narushima, H. Mizoguchi, K. Shimizu, K. Ueda, H. Ohta, M. Hirano and H. Hosono, Advanced Functional Materials **15**, 968–974 (2005).
60. "Progress in optically transparent conducting polymers", V. Seshadri and G.A. Sotzing, Optical Science and Engineering **99**, 495–527 (2005).
61. "Functional conductive 3,4-polyethylendioxythiophene layers", F. Jonas and S. Kirchmeyer, Jahrbuch Oberflaechentechnik **62**, 162–169 (2006).
62. "Conducting and transparent polymer electrodes", F. Zhang and O. Inganaes, Optical Science and Engineering **99**, 479–494 (2005).
63. "Technical trend in conductive polymers", M. Takeuchi, Idemitsu Giho **48**, 323–327 (2005).
64. "PEDOT/PSS patterning for plastic electronics", H. Okuzaki, Kobunshi Kako **54**, 367–372 (2005).
65. "Transparent, Conductive, and Flexible Carbon Nanotube Films and Their Application in Organic Light-Emitting Diodes", D. Zhang, K. Ryu, X. Liu, E. Polikarpov, J. Ly, M.E. Tompson and C. Zhou, Nano Letters **6**, 1880–1886 (2006).
66. "Metallic single-walled carbon nanotubes for transparent conductive films", Y.-P. Sun, Y. Lin, K.A.S. Fernando, X. Wang and Y. Liu, Abstracts of Papers, 231st ACS National Meeting, Atlanta, GA, United States, March 26–30, 2006 IEC-197 (2006).
67. "Transparent conductor coating films containing carbon nanotubes prepared by contacting carbon-containing compounds to impregnated metals", Y. Maeda and Y. Ozeki, Application: JP (Toray Industries, Inc., Japan). 15 pp (2006).
68. "Transparent and conductive coatings with carbon nanotubes", M. Kaempgen, Jahrbuch Oberflaechentechnik **61**, 88–92 (2005).
69. "Transparent and conducting electrodes for organic electronics from reduced graphene oxide", G. Eda, Y.Y. Lin, S. Miller, C.W. Chen, W.F. Su and M. Chhowalla, Applied Physics Letters **92**, 233305 (2008).
70. "The rise of graphene", A.K. Geim and K.S. Novoselov, Nature Materials **6**, 183–191 (2007).

71. "Electric field effect in atomically thin carbon films", K.S. Novoselov, A.K. Geim, S.V. Morozov, D. Jiang, Y. Zhang, S.V. Dubonos, I.V. Grigorieva and A.A. Firsov, Science **306**, 666–669 (2004).
72. "Electronic and structural properties of graphene-based transparent and conductive thin film electrodes", A. Vollmer, X.L. Feng, X. Wang, L.J. Zhi, K. Mullen, N. Koch and J.P. Rabe, Applied Physics a-Materials Science & Processing **94**, 1–4 (2009).
73. "Transparent, conductive graphene electrodes for dye-sensitized solar cells", X. Wang, L.J. Zhi and K. Mullen, Nano Letters **8**, 323–327 (2008).
74. "High-mobility amorphous In2O3-10 wt %ZnO thin film transistors", B. Yaglioglu, H.Y. Yeom, R. Beresford and D.C. Paine, Applied Physics Letters **89**, 062103 (2006).
75. "Novel oxide amorphous semiconductors: Transparent conducting amorphous oxides", H. Hosono, M. Yasukawa and H. Kawazoe, Journal of Non-Crystalline Solids **203**, 334–344 (1996).
76. "Electrochromic tungsten oxide films: Review of progress 1993-1998", C.G. Granqvist, Solar Energy Materials and Solar Cells **60**, 201–262 (2000).
77. "Stand-alone photovoltaic-powered electrochromic smart window", S.K. Deb, S.H. Lee, C.E. Tracy, J.R. Pitts, B.A. Gregg and H.M. Branz, Electrochimica Acta **46**, 2125–2130 (2001).
78. C.G. Granqvist (1995). *Handbook of Inorganic Electrochromic Materials*. Amsterdam, Elsevier.
79. "Optical analysis of thin film combinatorial libraries", J.D. Perkins, C.W. Teplin, M. Van Hest, J.L. Alleman, X. Li, M.S. Dabney, B.M. Keyes, L.M. Gedvilas, D.S. Ginley, Y. Lin and Y. Lu, Applied Surface Science **223**, 124–132 (2004).
80. "Highly conductive epitaxial CdO thin films prepared by pulsed laser deposition", M. Yan, M. Lane, C.R. Kannewurf and R.P.H. Chang, Applied Physics Letters **78**, 2342 (2001).
81. "High-efficiency polycrystalline CdTe thin-film solar cells", X.Z. Wu, Solar Energy **77**, 803–814 (2004).
82. "Recent progress in transparent oxide semiconductors: Materials and device application", H. Hosono, Thin Solid Films **515**, 6000–6014 (2007).

Chapter 2
Electronic Structure of Transparent Conducting Oxides

J. Robertson and B. Falabretti

2.1 Introduction

Metallic oxides are a materials class showing one of the greatest range of properties – superconducting, ferroelectric, ferromagnetic [1], multiferroic, magneto-resistive, dielectric, or conducting. Of particular interest are the so-called transparent conducting oxides (TCOs) and amorphous semiconducting oxides (ASOs). The TCOs are heavily used for flat panel displays, photovoltaic cells, low emissivity windows, electrochromic devices, sensors and transparent electronics [2–4]. Oxides are of particular interest because the metal-oxide bond is strong so that the oxides have a combination of a high heat of formation and a wide band gap, compared to any similar compound.

This chapter describes the basic electronic structure of oxides that allows this to occur, why they can be doped, what controls the polarity of the doping, and the effect of disorder on their properties. The majority of the TCOs are n-type electron conductors. A few p-type hole conductors have been discovered following the break through of Kawazoe et al. [5].

2.2 Band Structures of n-Type Oxides

There are numerous n-type TCOs. We will focus here on the electronic structure of a subset of them, SnO_2, In_2O_3, ZnO, Ga_2O_3 and CdO, which illustrate their main properties. These all are oxides of group IIB-IVB metals. They have smaller ions and are not as electropositive as the corresponding alkaline earth metals of groups

J. Robertson (✉)
Department of Engineering, Cambridge University, Cambridge CB2 1PZ, UK
e-mail: jr@eng.cam.ac.uk

D.S. Ginley (ed.), *Handbook of Transparent Conductors*,
DOI 10.1007/978-1-4419-1638-9_2, © Springer Science+Business Media, LLC 2010

IIA. They are predominantly ionic bonded except for ZnO. Their crystal structures are summarised in Fig. 2.1.

SnO_2 is perhaps the simplest of the TCOs. It has the rutile structure, in which each tin atom is surrounded by six oxygens in an octahedral array, and each oxygen is surrounded by three tin atoms in a planar array. Figure 2.2 shows the band

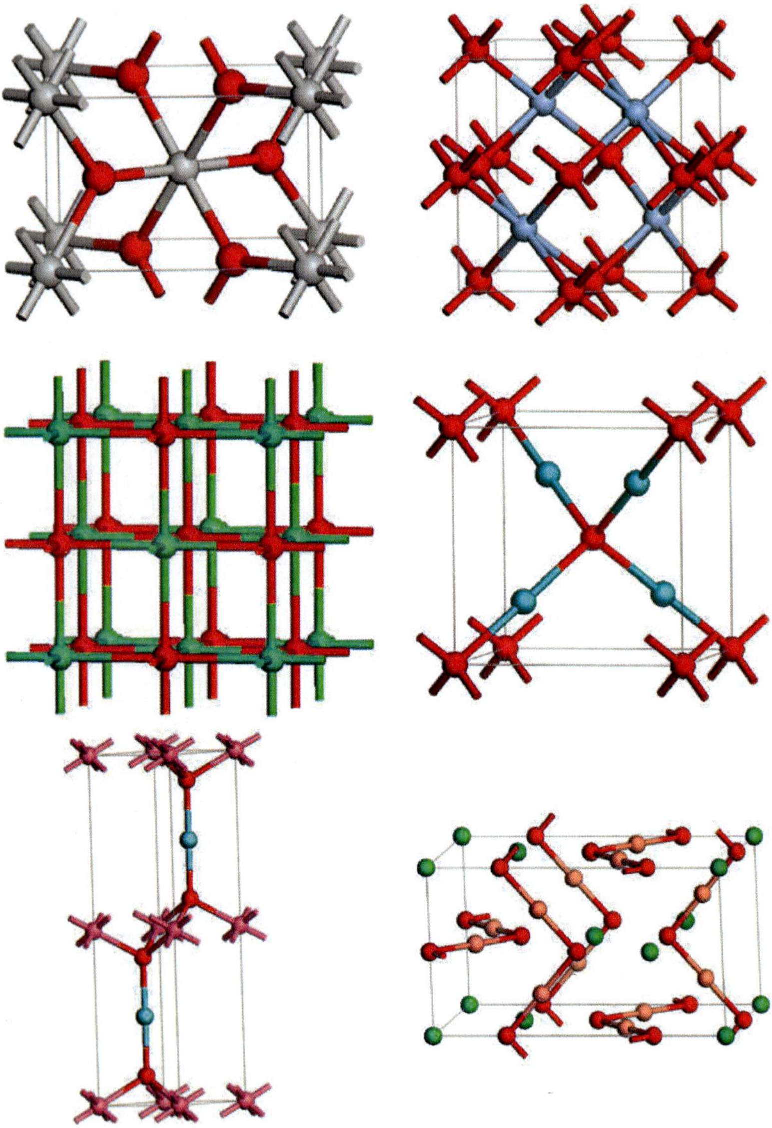

Fig. 2.1 Crystal structures of SnO_2, In_2O_3, CdO, 2H-$CuAlO_2$ and $SrCu_2O_2$ (oxygen = *darker balls*)

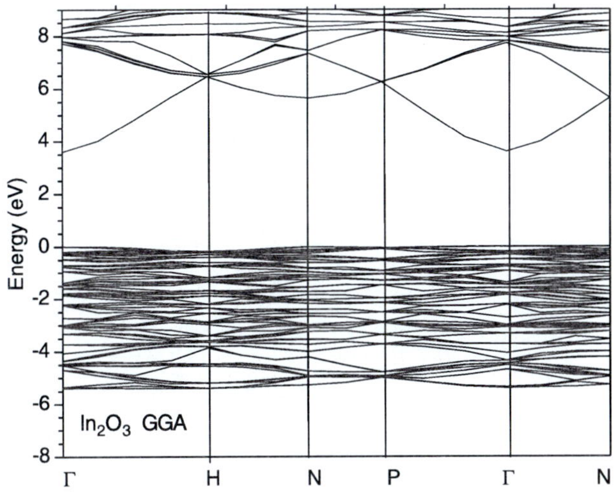

Fig. 2.2 Band structure of SnO_2. Band gap fitted

structure of SnO_2. The band gap is 3.6 eV and direct. The band structure shown here was calculated by the plane wave pseudopotential method, using the generalised gradient approximation (GGA) of the local density formalism (LDF). The GGA functional represents the exchange-correlation energy of the electron gas. These LDF and GGA methods under-estimate the band gap. This error has been corrected in the band structure shown by the "scissors operator," a rigid upward shift of the conduction bands. The band structure of SnO_2 was first calculated correctly by Robertson [6], followed by numerous calculations using improved methods [7–9].

The most obvious feature in Fig. 2.2 is the free-electron-like conduction band minimum at the zone centre Γ. It is noticeable that this state from 3.6 eV upwards is a single minimum, without any subsidiary minima leading to indirect gaps. This main minimum is formed out of Sn 5s states [6]. In a tight-binding description, it consists of 96% of Sn s states. The band gap is direct. The electron effective mass is 0.23–0.3, reasonably small, but not 0.1 like for example Si.

The upper valence band from 0 eV down to −8.1 eV consists mainly of O 2p states, mixed with some Sn s and p states. The ionicity of SnO_2 is about 60%, so this sets the Sn content of the valence band, averaged over the Brillouin zone, as rather low. Finally, at −16 eV, there are O 2s states which do not contribute to the bonding. Any Sn 4d states lie below this, and can be ignored.

The upper valence band in SnO_2 is typical of many oxides. It is relatively flat, and thus has a large effective mass, which does not favour conduction by holes. The valence band maximum has a Γ_2^- symmetry, which leads to a direct forbidden band gap [10]. The valence band is consistent with the experimental ultraviolet photoemission spectra [11].

The next most important TCO, In_2O_3, has the bixbyite structure, in which the oxygens form a close packed lattice and the In ions lie at sixfold and fourfold

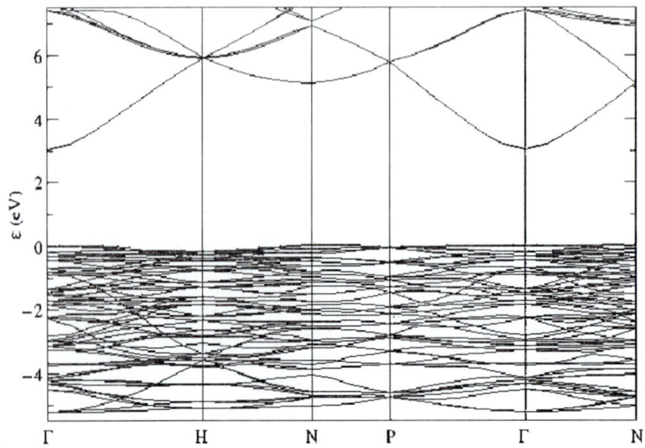

Fig. 2.3 Band structure of In_2O_3 by screened exchange

interstices. The In sites are sixfold coordinated by oxygen. The overall symmetry is cubic, but the unit cell is large, 40 atoms. Figure 2.2 shows our calculated band structure of In_2O_3. The band gap was originally believed to be 3.7 eV. The first good band calculation of In_2O_3 was by the Shigesato et al. [12], followed by Mi et al. [8], and Mryasov and Freeman [13].

We see again in Fig. 2.3 that In_2O_3 has a single free-electron-like conduction band minimum, formed from In s states. The minimum band gap is direct. The effective mass is 0.3 m. Thus, the conduction band minimum of In_2O_3 has the same nature as that of SnO_2. The main valence band is 5.2 eV wide, less than that of SnO_2, indicating that In_2O_3 is more ionic than SnO_2. This is consistent with the ultra violet photoemission spectra of Christou et al. [14] and Klein [15]. Below the O 2p states come In 4d states, then O 2s states.

The band gap of In_2O_3 has recently been re-appraised. A smaller indirect gap was once proposed [16]. However, there can be no indirect gap due to the parabolic nature of the conduction band. It is now realised by Walsh et al. [17–19] that the minimum band gap is 2.9 eV, direct and forbidden. The screened exchange band structure in Fig. 2.3 gives this value. The upper valence bands all have the wrong symmetry for allowed optical transitions to the conduction band, as in SnO_2. The first allowed optical transition is 0.8 eV below the valence band top [17]. This now gives a consistent view of the band structure, optical gap and surface band bending of In_2O_3.

A third important TCO, especially for photovoltaic applications, is ZnO. Interest in ZnO is wider than just as a TCO, because it is a prototype direct-gap, wide band gap optoelectronic semiconductor in competition with GaN [20]. It has been an important phosphor. It is also easy for make as "nanorods." ZnO typically has the hexagonal wurzite structure in which each Zn or O atom is surrounded by four neighbours of the other type. There is also a hypothetical zinblende polymorph of ZnO, with the same bonding. The band structure of this zincblende phase is shown in Fig. 2.4, as the gaps and band widths are the same. It band gap is 3.35 eV and direct, and the conduction band minimum is again a single broad minimum formed from Zn s states [8]. There are numerous band calculations of ZnO [21, 22], some at high levels of accuracy such as GW [23]. Note that the uncorrected GGA band gap of ZnO is only ~0.9 eV, very small compared to experiment. Donors such as interstitial Zn or substitutional Al are shallow in ZnO, but other defects like the O vacancy are deep.

Ga_2O_3 is the least studied of the binary oxides. It has more complex crystal structure such as the β-Ga_2O_3 structure. In this, the Ga sites are both fourfold or sixfold coordinated. Its band structure is shown in Fig. 2.5. The band structure has a minimum direct band gap of 4.52–4.9 eV [24, 25]. The broad conduction band

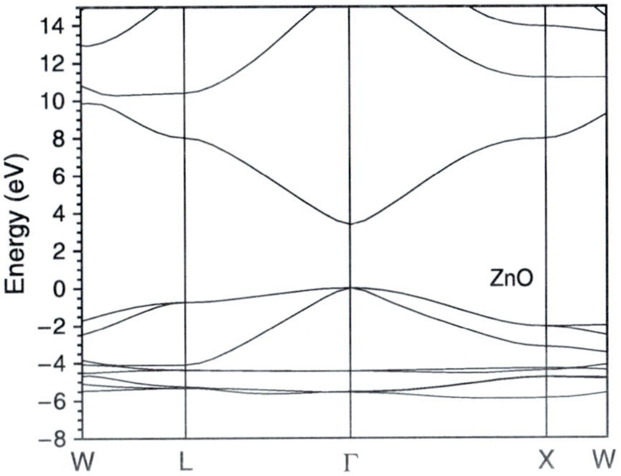

Fig. 2.4 Band structure of Ga_2O_3

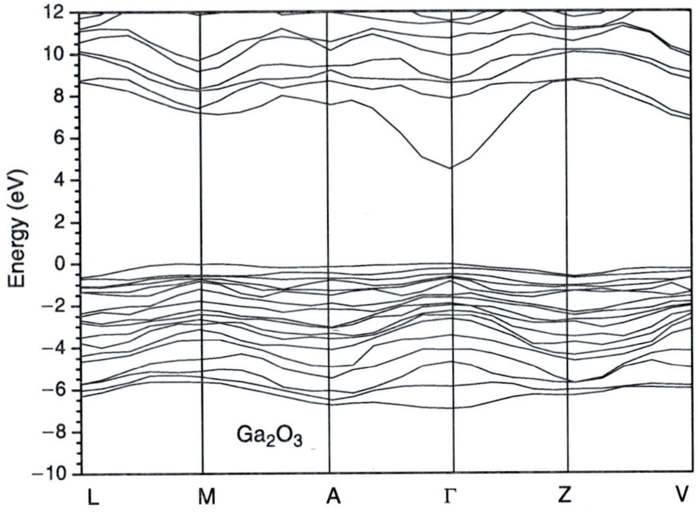

Fig. 2.5 Band structure of zincblende (*cubic*) ZnO

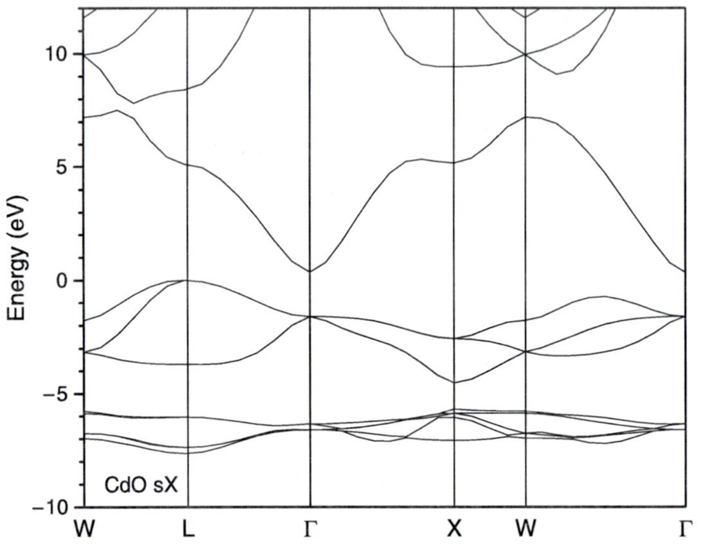

Fig. 2.6 Band structure of CdO using screened exchange

minimum at Γ is formed of Ga s states. The valence band maximum is very flat, a slightly indirect gap and a maximum at M 0.05 eV above Γ.

The last n-type oxide considered is CdO. This has the rock-salt structure, in which each Cd or O ion is surrounded by 6 neighbours. Its band structure is shown in Fig. 2.6. This is similar to that found by others [22, 26, 27]. In contrast to the other TCOs, the minimum gap of CdO is indirect, at 0.8 eV [28, 29]. The conduction band minimum is at Γ, free-electron-like, and is formed from Cd s states. However, the valence band maximum is not at Γ. It is displaced to the zone boundary at the L point and along ΓW, due to a repulsion of the O p states in the upper valence band by Cd d states lying at −7 eV. The three upper valence bands in CdO consist mainly of O 2p states. However, instead of Cd p states leading to a downward repulsion of these states away from Γ, the upward repulsion of Cd d states is stronger. This leads to a calculated minimum indirect gap of 0.6 eV and a minimum direct gap of 2.1 eV, compared to experimental values of 0.8 and 2.3 eV, respectively. The valence band width is consistent with that seen experimentally by photoemission [30].

When these various oxides are doped with donors, the free electrons lie in the lowest conduction band. The next available empty state after this is not the conduction band minimum itself, but higher unoccupied conduction states. This increases the energy of the lowest optical transition, the optical gap. This is the Moss-Burstein shift, and varies inversely with the effective mass [3].

An aspect that is not so important for the overall understanding of TCOs, but nevertheless relevant to this chapter, is the question of the calculated band gaps. The standard method of calculating electronic structure uses the local density functionals such as the GGA to represent the electron's exchange-correlation energy. These so-called ab initio methods are able to give the structure, bond lengths and total energies quite well. However, they under-estimate the band gap of semiconductors and insulators. It is generally found that this under-estimate is of order 30%, as is the case of Si. However, for the oxides of interest here, the under-estimate is very large, typically 70%. For SnO_2 the GGA band gap is 1.2 eV compared to the experimental value of 3.6 eV. The largest problem is for CdO, where LDF gives a negative band gap of −0.5 eV.

In Figs. 2.2 and 2.7–2.11 we have corrected the GGA band gaps by the scissors operator, rigidly shifting the conduction bands upwards. The conduction band dispersions in GGA are correct. Thus, once corrected for the error, the band structures shown are correct.

There are a number of methods beyond LDF which do give better band gaps. The best known of these is the GW method [31], but this is computationally very expensive. Other popular methods are the self-interaction correction (SIC) [32] and the B3LYP functional [33–35]. B3LYP is a hybrid functional, that is a LDF-type functional of the exchange-correlation energy containing a fraction of the Hartree-Fock function, which can give the correct total energy and reasonable eigenvalues. Other hybrid functionals are PBE0, HSE and screened exchange.

A hybrid functional which is valuable is the method of screened exchange. This was first implemented by Kleinman [36]. It was then taken up by Freeman et al. [37, 38] and applied to various oxides of interest here [26]. We have also used screened exchange for TCOs [39, 40]. The CdO band structure shown in Fig. 2.6 is that calculated by the screened exchange method. It gives a reasonable band gap compared to experiment. In other cases, the sX or WDA band structures can be used to verify that the conduction bands formed by the scissors operator are indeed correct, and that gaps do have the character as shown in Figs. 2.2–2.10.

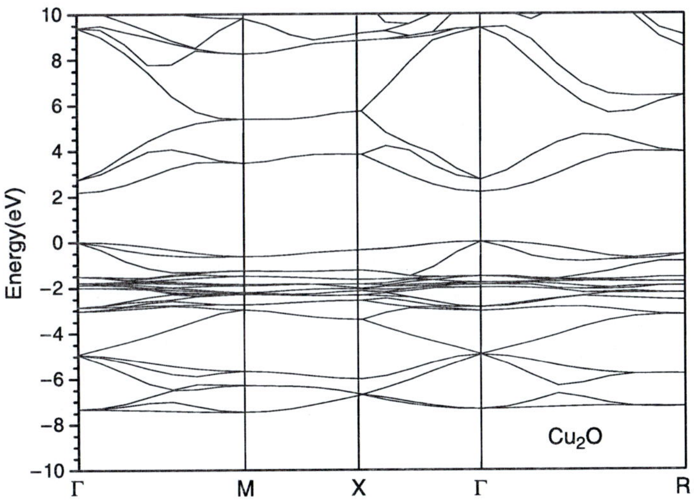

Fig. 2.7 Band structure of Cu_2O

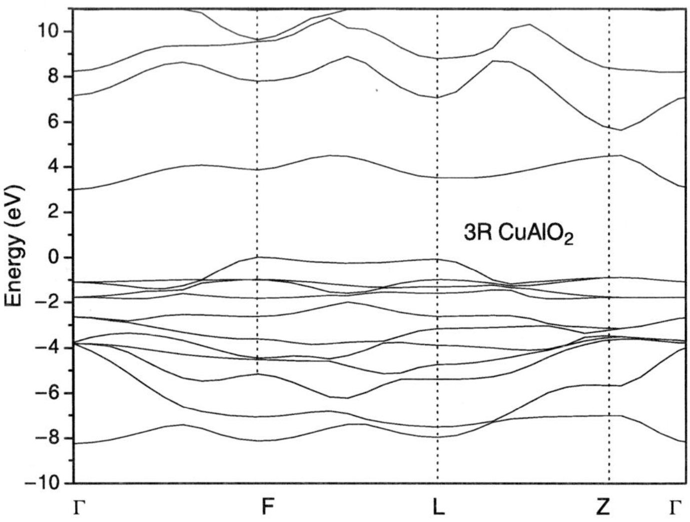

Fig. 2.8 Band structure of 3R-$CuAlO_2$

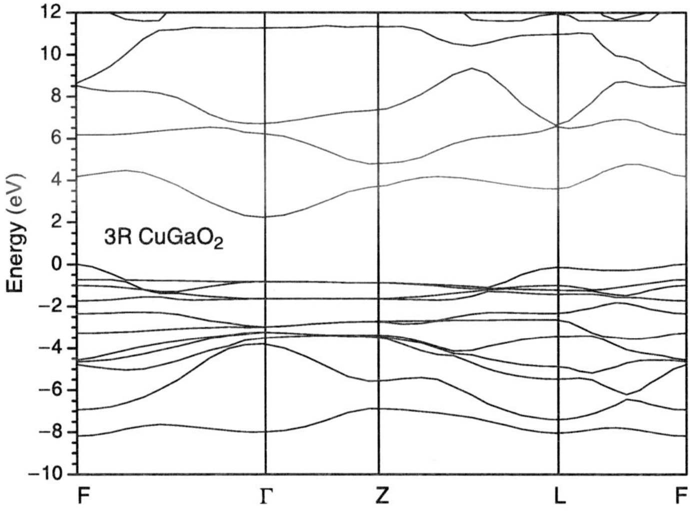

Fig. 2.9 Band structure of 3R-$CuGaO_2$

2.3 Band Structures of p-Type Oxides and Other Cu Oxides

For many years, the only apparent p-type conducting oxide was Cu_2O. This was used since early days in Cu-Cu_2O rectifiers. However, its band gap is only 2.17 eV. Its band structure calculated in the GGA method is shown in Fig. 2.7. The Cu_2O

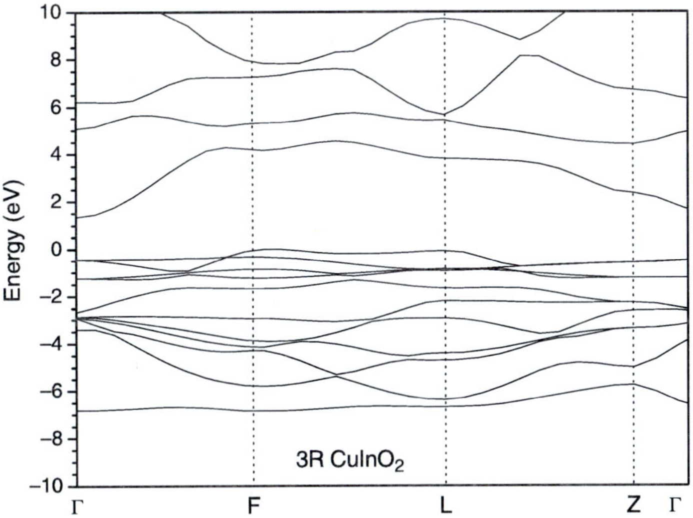

Fig. 2.10 Band structure of 3R-$CuInO_2$

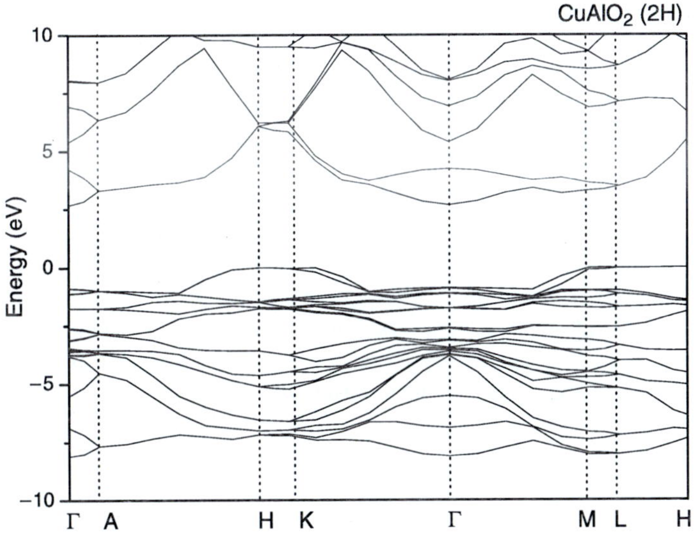

Fig. 2.11 Band structure of 2H-$CuAlO_2$

crystal structure is unusual, the oxygens are fourfold coordinated, and the Cu atoms are only twofold coordinated, in a linear configuration (Fig. 2.1). In the band structure, there is again a broad conduction band minimum at Γ due to Cu s states [41–45]. The Cu 3d states are all filled. They lie as a mass of narrow bands at around

-1 to -3 eV. The Cu d states mix with the O 2p states and form a continuous band, which extends down to -7 eV.

Kawazoe et al. [5] noted that the main problem with oxides is the high effective mass of their holes. Thus acceptors would not ionise. The high hole effective mass arises in SiO_2 or SnO_2 because the valence band maximum states are nonbonding pπ states, directed perpendicular to the bonding direction. Thus, they have a small dispersion. ZnO is one case without nonbonding p states, but it still has a rather large hole mass. Kawazoe et al. [5] noted that we should try to increase the dispersion of the valence band top states, and the way to do this is to hybridise (mix) them with d states of a cation at a similar energy. The Cu d states are the best case of this. Thus, the interaction of Cu d and O 2p states should reduce the effective hole mass in Cu_2O.

However, the problem with Cu_2O itself is that its 2.17 eV band gap is too narrow to be transparent across the optical spectrum. This arises because its conduction band is too broad, and its conduction band minimum falls too low. The dispersion of the Cu s band is mainly due to Cu-Cu interactions. The width of the Cu s band is proportional to the number of Cu neighbours. By using $CuAlO_2$ in the defossalite structure, we can reduce the Cu-Cu coordination from 12 to 3, and this should reduce the conduction band dispersion and increase the band gap [5]. This indeed occurs [46]. Of course, the largest interest in Cu based oxides arose from high temperature superconductivity, so the role of holes in Cu-O layers is well known.

$CuAlO_2$ has a layered structure in which the Cu ions a linearly coordinated to two O atoms, as in Cu_2O. The Al ions are surrounded by six oxygens. These AlO_6 units form a layer of hexagonal symmetry. The AlO_6 and Cu layers can be stacked in various patterns. A two-layer repeat gives the 2H (P63/mmc) form and a three-layer repeat gives the 3R (R-3m) rhombohedral form. Despite having more layers, the primitive cell of the 3R form has the fewer atoms. Figure 2.8 shows the band structure of 3R-$CuAlO_2$ calculated using the GGA functional. The indirect band gap has been adjusted to the experimental value. The bands are very similar to those given by Ingram et al. [47]. The other important calculations of $CuAlO_2$ are by Yanagi [48] and Zhang [49]. We see that 3R-$CuAlO_2$ is an indirect gap semiconductor. The conduction band minimum is at Γ and consists of Cu s states, while the valence band maximum is at F. The upper valence band from 0 eV down to -8 eV consists of a mixture of Cu d and O 2p states, with the Cu d states tending to lie higher. There are O 2s states at -20 eV.

The minimum indirect gap is 3.0 eV, and the minimum direct gap is 3.5 eV at L. Ingram et al. [47] and Zhang et al. [49] have discussed the nature of the optical transitions. We see that the conduction band has subsidiary minima at F and L, so it is not as simple as those in Cu_2O or SnO_2.

In Fig. 2.8, when using the scissors operator, we fitted the main direct gap at 3.5 eV, not the minimum gap, which is not well known experimentally. The resulting band gap is nevertheless consistent with that found by the screened exchange method or weighted density approximation [39, 40].

Figures 2.9 and 2.10 show the band structures of 3R-$CuGaO_2$ and 3R-$CuInO_2$ calculated using the GGA functional. In each case, the main direct gap was fitted [49]. We see that the conduction band minimum of $CuInO_2$ is mainly at Γ, and there are no other subsidiary minima as was found in $CuAlO_2$. The valence band of $CuInO_2$ is reasonably similar to that of $CuAlO_2$. The band width is slightly smaller and the band dispersions flatter, so the hole effective mass is larger. There is also a mass of flat bands at −14 eV due to In d states.

The minimum indirect band gap of $CuInO_2$ is now 1.4 eV, and the minimum direct gap is 3.9 eV at L. Thus, the indirect gap decreases from $CuAlO_2$ to $CuInO_2$, but the direct gap increases, which is unusual [47, 49]. This gives the appearance that the optical gap increases when changing from $CuAlO_2$ to $CuInO_2$, which is against the trend in ionic radii.

$CuCrO_2$ is an oxide with the defossalite structure which also shows p-type behaviour [50]. Now the band gap appears in the middle of the Cr d states. This is consistent with the insulating property of Cr_2O_3. The alloy $CuAl_{1-x}Cr_xO_2$ can be doped p-type [51] by substitutional Mg_{Al}.

Figure 2.11 shows the band structures of the 2H polymorph of $CuAlO_2$ [35]. It confirms that the stacking does not alter the main band gaps so much.

The final oxide considered is $SrCu_2O_2$. This is an amipolar oxide [52–54]. $SrCu_2O_2$ has the body-centred tetragonal $D_{4h}{}^{10}$, 4/mmm space group. Its structure consists of O-Cu-O zig-zag chains in the x and y direction, separated by SrO_6 octahedra. Cu is again twofold coordinated by O. There are four Cu atoms in the primitive unit cell. The band structure of $SrCu_2O_2$ is shown in Fig. 2.12. The conduction band has a minimum at Γ and the valence band has a weak maximum at Γ. The bands have been calculated by Ohta et al. [53], Robertson et al. [35] and

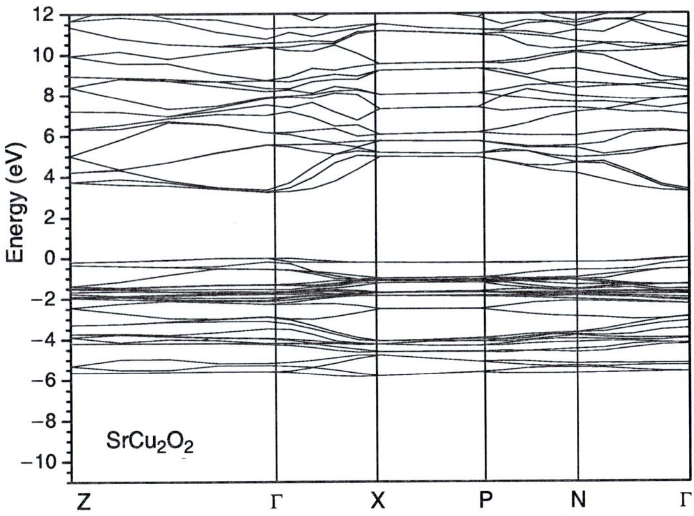

Fig. 2.12 Band structure of $SrCu_2O_2$

Nie et al. [44]. The different appearance of some of the published bands arises because not all authors used the primitive cell. There are a number of other ternary oxides of interest which are covered elsewhere [47, 48].

2.4 Band Line-ups and Work Functions

An important application of TCOs is as electrodes on semiconductor devices, such as solar cells or organic light emitting diodes. A key criterion is that these electrodes form ohmic rather than injecting contacts. This depends on the band alignment of the conducting oxides to the semiconductor, the Schottky barrier height. The barrier height can be measured by photoemission, internal photoemission or by electrical means. It can be estimated by a variety of theoretical means, which we now discuss.

An often-used method is to compare the work function of the conductor ϕ_M with the electron affinity (EA) of the semiconductor (for the n-type case), χ, each measured from the vacuum level. The barrier height is then taken as the difference,

$$\phi_n = \phi_M - \chi \tag{2.1}$$

This approximation is the so-called electron affinity rule. It can work for very wide gap semiconductors, or those with van der Waals bonding such as organic semiconductors.

Figure 2.13 plots the work functions of the various TCOs, data taken from Minami et al. [56]. It is interesting that the n-type TCOs have very high work functions, their conduction band minima lying well below the vacuum level. The work functions of the oxide films are often maximised by ensuring an oxygen-rich outer surface, by processing. It is interesting in Fig. 2.13 that the conduction band minima of SnO_2, and In_2O_3 lie *deeper* than the Fermi level of the *parent metal*. This behaviour distinguishes the best n-type TCOs from normal oxides.

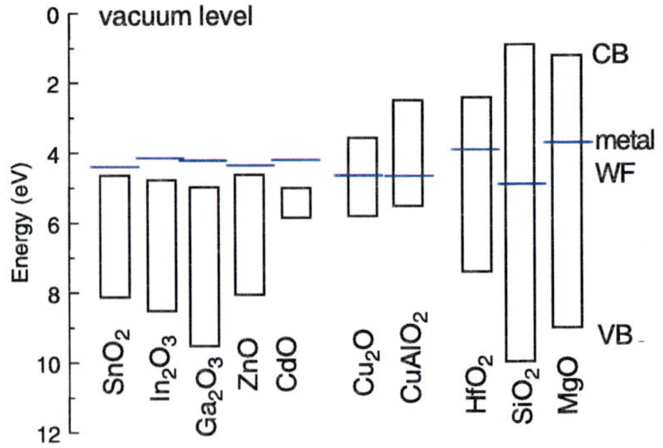

Fig. 2.13 Work functions and electron affinities of various oxides, compared to the work functions of their parent metal. Note how the metal WF of n-type TCO lies above the oxide conduction band

In fact, the electron affinity rule does not work for metals on typical semiconductors [57–59]. This is because the semiconductor interfaces possess mid gap states which tend to pin the metal Fermi level from changes in barrier height. The semiconductor's mid gap states on the neutral surface are filled up to some energy, which we will call the charge neutrality level (CNL). The effect of these states is to try to pin the metal work function towards this CNL. The degree of pinning depends on the density of these states and their extent into the semiconductor.

The net effect is that the Schottky barrier heights tend to follow the equation

$$\phi_n = S(\Phi_M - \Phi_S) + (\Phi_S - \chi_s) \tag{2.2}$$

Here Φ_S is the CNL energy measured from the vacuum level. S is the Schottky barrier pinning factor. S = 1 for the strongly pinned case, a narrow gap semiconductor, and S = 0 for the unpinned case, like SiO_2, as in the electron affinity rule.

Monch [57] found that S follows an empirical dependence on the electronic dielectric constant, ε_∞,

$$S = \frac{1}{1 + 0.1(\varepsilon_\infty - 1)^2} \tag{2.3}$$

This model can also be applied to the interfaces between two semiconductors [60], to derive their band offsets, where the electron barrier ϕ_n or conduction band (CB) offset between semiconductors *a* and *b* is given by

$$\phi_n = (\chi_a - \Phi_{S,a}) - (\chi_b - \Phi_{S,b}) + S(\Phi_{S,a} - \Phi_{S,b}) \tag{2.4}$$

There are more detailed methods of calculating the band offsets, as for example given by van de Walle et al. [61], or by Zunger et al. [62], based on calculations of explicit interface structures.

The CNL can be calculated from the oxide band structure as the energy at which the simple Greens function is zero;

$$G(E) = \int_{BZ} \int_{-\infty}^{\infty} \frac{N(E')dE'}{E - E'} = 0 \tag{2.5}$$

The CNL can be calculated from the bands calculated by the local density approximation/pseudopotential method, after adjusting the band gap to the experimental value.

The CNL normally lies near the centre of the band gap. For most ionic oxides, the CNL energy tends to vary with the metal valence, because the large number of oxygen-related valence states repels the CNL up in the gap [60, 63]. The transparent oxides are different. The CNL is effectively the mid point of the *average* gap over the Brillouin zone. But the s-band oxides have a broad CB minimum, which comes

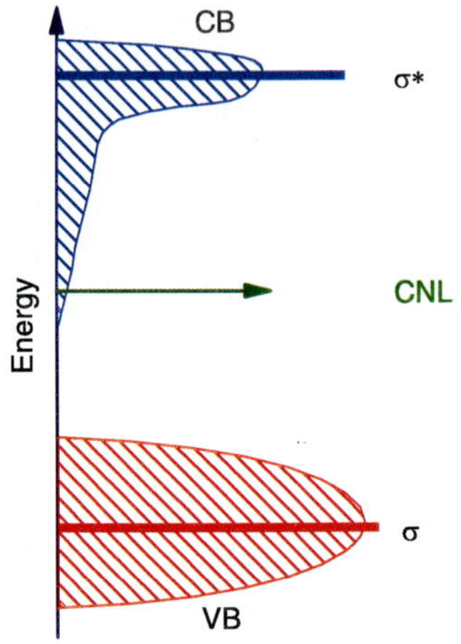

Fig. 2.14 Schematic of the charge neutrality level in an oxide with a deep s-like conduction band minimum of low density of states

well below the average CB energy. This causes the CNL to lie either close to the CB, as in ZnO, or even above the CB minimum, as in SnO_2 and In_2O_3 (Fig. 2.14). This is an unusual situation.

Parameters are listed in [64]. The S value is calculated from their experimental refractive index ($\varepsilon_\infty = n^2$). The electron affinities are taken from experiment for poly-crystalline oxide films; those for the dielectrics are tabulated previously [60]. The work function of doped SnO_2 is large, 4.5 eV or more [50]. This is partly because the surface is treated to be O-rich, to maximise the work function.

2.5 Ability to Dope

Substitutional doping is a key requirement for a semiconductor to be used in practical devices. It is often stated that doping of both polarities is required. This not strictly accurate; thin film transistors of amorphous hydrogenated silicon (a-Si:H) are the main-stay of the flat panel display industry and only use n-type doping.

There are three requirements for successful doping of a given polarity;

- Solubility of the dopant in the lattice
- Shallowness of the dopant level
- Lack of compensation of the dopant by an intrinsic defect

When designing the p-type TCOs, Kawazoe et al. [5] only considered point 2. Point 1 is usually easily satisfied, it is only a problem in cases like diamond where there is a large mismatch between the atomic radii of a shallow donor (e.g. Sb) and the small diamond lattice.

Compensation, point 3, is the most interesting case [65–68]. In wide gap semiconductors, a donor electron can lower its energy if it falls into an empty intrinsic defect state (such as a vacancy state) at the bottom of the gap. Indeed the energy gain can be so much that this energy gain exceeds the cost of creating the defect. In that case, the donor action is completely compensated by the defect, if there is thermal equilibrium.

Another way to express this is that the creation energy of the intrinsic defect (say the vacancy) depends on the Fermi energy, E_F as [66].

$$\Delta H(E_f) = qE_f + \Delta E$$

This is shown schematically in Fig. 2.15. It means that there will be some Fermi energy at which the cost of the vacancy is zero. If a dopant would move the Fermi level to this energy, called the dopant pinning energy, then vacancies will be spontaneously created at no cost. There will be two pinning energies, one for n- and one for p-type doping. Thus, if equilibrium holds, it will be impossible to shift the Fermi level beyond these two pinning energies $E_{pin,\,n}$ and $E_{pin,\,p}$. Practical doping will only occur if these pinning energies lie in the conduction or valence band, respectively, and not in the gap. If for example, $E_{pin,\,p}$ lies above the valence band edge E_v, then there will be no p-type doping, because it will not be possible to shift E_F to the valence band edge. Note that the type of defect doing the compensation will differ for n- and p-type doping.

The NREL group have applied these ideas to study the limits to doping of the tetrahedral semiconductors such as III-Vs, ZnO and the chalcopyrites [66–68]. The bands of the various semiconductors are aligned with each other using the calculated or observed band offsets, and the pinning energies are found to lie at a roughly constant energy across the series, Fig. 2.16. Zunger [68] prefers to reference band offsets to the vacuum level E_{vac}. In their picture, a semiconductor cannot easily be doped n-type if its conduction band energy lies too high towards E_{vac}, i.e. its electron affinity is too small. AlN would be an example. On the other hand, a semiconductor cannot easily be doped p-type if its valence band lies too far below E_{vac}, if its photoelectric threshold is too large. ZnO is a good example.

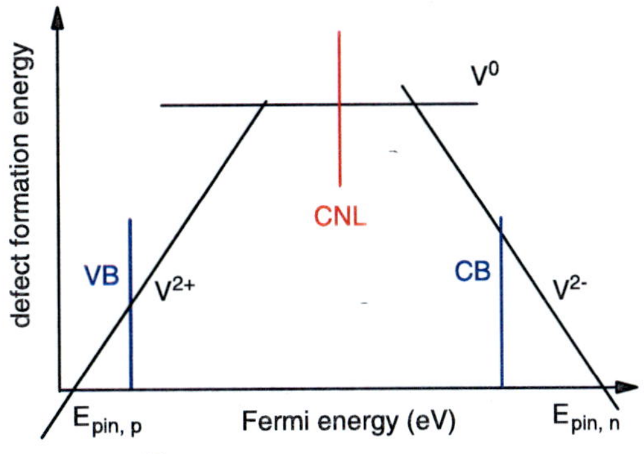

Fig. 2.15 Concept of the doping pinning levels for n- and p-type dopants, on the band diagram. If the pinning level lies inside the bands, then the semiconductor can be doped; if it falls in the gap the semiconductor cannot easily be doped to that polarity

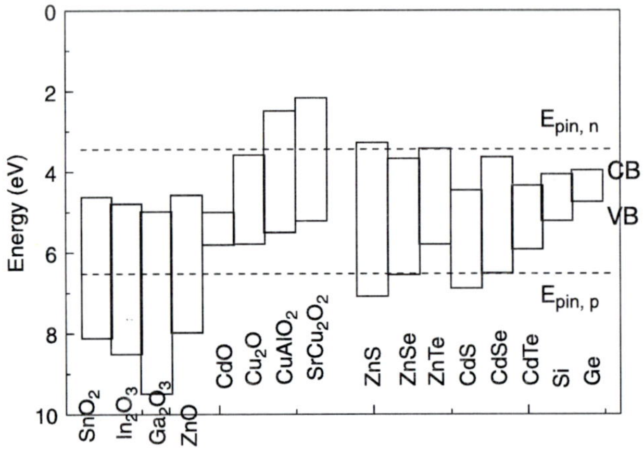

Fig. 2.16 Band diagrams of various oxides and comparative semiconductors aligned according to the vacuum level, with dopant pinning levels indicated

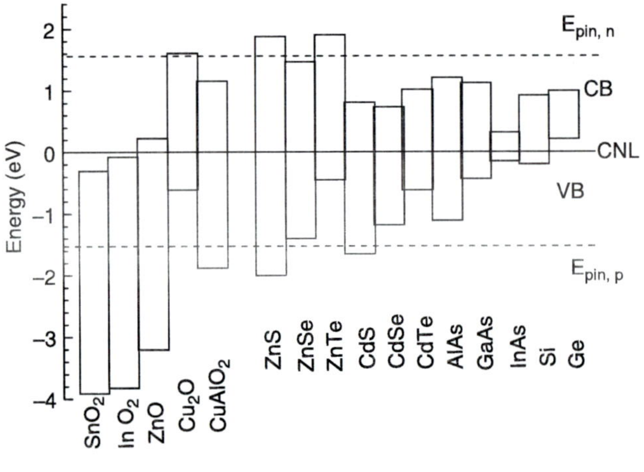

Fig. 2.17 Band diagrams of various oxides and comparative semiconductors aligned according to their charge neutrality levels, with dopant pinning levels indicated

This method gives a good view of doping possibilities in TCOs. The n-type TCOs SnO_2, In_2O_3, ZnO stand out as having very large work functions (when n-type). As they have the same band gap, their valence band maxima are very deep below E_{vac}. In this model, p-type ZnO is only possible by inhibiting thermal equilibrium occurring.

There is a second, related view of doping limits, which is more consistent with band offset models. The semiconductor oxides of interest here have $\varepsilon_{\infty} \sim 3.7$–4.0, giving $S \sim 0.5$. Thus, the bands should be aligned [55] using CNLs and (2.3) not just electron affinities. This is done in Fig. 2.17.

In addition, van de Walle and Neugebauer [69] note that for AB compounds the CNL tends to equal the average of the dangling bond energies of the cation and anion species. This arises because defect levels ultimately depend on bulk band structures. Turning to AB_n compounds, the CNL will be a weighted average of cation and anion site defect levels. Thus, in our view [70] shown in Fig. 2.17, each semiconductor is aligned according to the band offset, using primarily the CNL, with no reference to the vacuum level. The doping pinning levels then lie at some energy above and below the CNL. (This energy is not necessarily constant.) In this case, n-type doping is difficult if the conduction band edge lies too far above the CNL, and p-type doping becomes difficult if the valence band edge lies too deep below the CNL.

This is consistent with experiment. SnO_2, In_2O_3 are very unusual in that their CNLs actually lie *above* the conduction band edge [18, 70], rather than in the gap as normal. In ZnO, the CNL lies close to the conduction band edge. This accounts for their ease of n-type doping. On the other hand, that the CNL is so far above the valence band edge in SnO_2, In_2O_3 and ZnO accounts for why these oxides are difficult to dope p-type.

In the case of $CuAlO_2$ and Cu_2O, the CNL is calculated to lie closer to the valence band, and this is consistent with their p-type behaviour. In contrast, in $CuInO_2$, the CNL lies higher in the gap [64], and now this oxide can be doped both n- and p-type. $SrCu_2O_2$ also has a CNL near midgap, allowing ambipolar doping.

What controls the CNL energy? From (2.5), the CNL lies between the density of states (DOS) of valence band (VB) and conduction band (CB). A larger DOS repels the CNL away [60, 64]. A large VB DOS repels the CNL to the upper gap. A high valence and large O content gives a large VB DOS and a high CNL energy. But a second factor is the nature of the conduction band minimum. The CNL lies midway in the *indirect* gap. A deep CB minimum with only a small DOS as in SnO_2 makes only a small contribution to the integral in (2.5) and has little effect on the CNL. The CNL lies high in SnO_2, In_2O_3 and ZnO because the DOS in their CB minima is small. On the other hand, in $CuAlO_2$ its multiple CB minima push the CNL down.

A complicating factor which is not discussed here is the "self" doping of TCOs, in which vacancies are generated in multi-component oxides such as Cd_xSnO_3.

2.6 Effect of Disorder–Disorder in Amorphous Semiconductors

We now turn to a different question, why n-type TCOs work well even when they are amorphous [71, 72]. The ineffectiveness of disorder scattering has been noted for some time [73, 74]. The n-type TCOs consist of the oxides of post-transition metals, Sn, In, Ga, Zn, Cd, etc. As we noted above, the conduction band minima of these oxides are free-electron-like states, localised on the metal s states. Unlike the alkaline earth metal oxides, because of their lower ionicity, these metal oxides can be made amorphous [71].

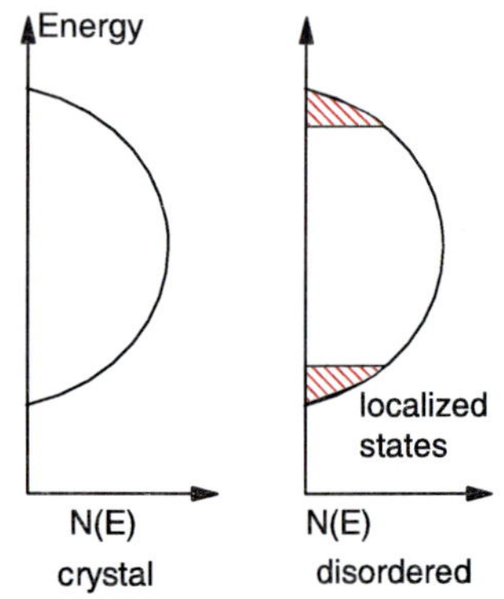

Fig. 2.18 Effect of disorder on an s band

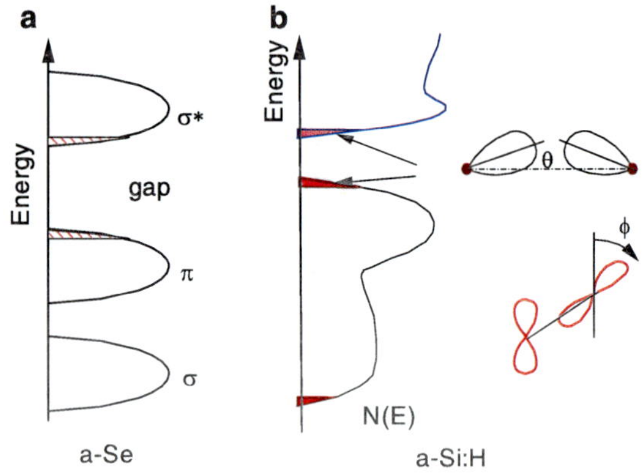

Fig. 2.19 (**a**) Schematic density of states of a-Se, (**b**) a-Si, and the effect of bond angle and dihedral angle disorder

Since the work of Anderson [75], it is known that disorder will cause a localization of electron states in the band structure. Mott [76, 77] then showed that disorder first localizes states at the band edges, and that the extended states and localized tail states are separated by an energy called the mobility edge. With increasing disorder, the mobility edges move further into the bands, as in Fig. 2.18. Eventually, the whole band becomes localized. These results were worked out for s states, which are spherically symmetric.

The first amorphous semiconductors to be studied in depth experimentally were the amorphous chalcogenides, such as amorphous Se (a-Se). The chalcogenides are in fact p bonded [78], and their simplified band diagram is shown in Fig. 2.19a. Disorder introduces localized band tail states.

The next and most important amorphous semiconductors are a-Si and hydrogenated amorphous silicon (a-Si:H). Its bonds are sp^3 states, but the states around its band gap are p states [78]. Figure 2.19b shows a schematic of its density of states. The valence band maximum consists of pure p states, whereas the conduction band minimum consists of mixed s,p states. The effect of disorder has been considered in

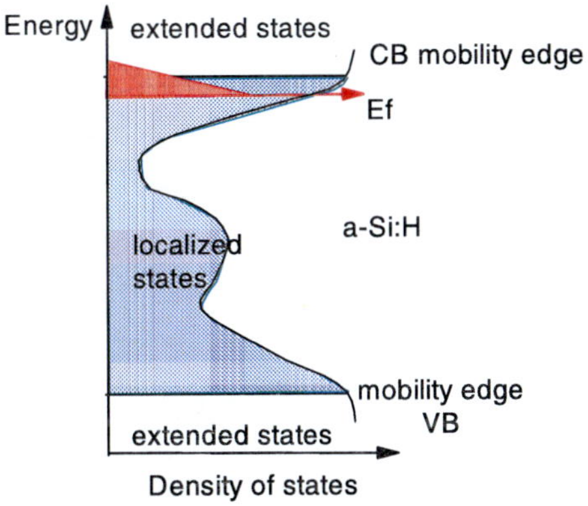

Fig. 2.20 Schematic density of states and mobility edges of a-Si:H

more detail in a-Si than in most other amorphous semiconductors [78–83]. The Si-Si bond length is relatively fixed in a-Si:H. On the other hand, the bond angle θ varies by 10° and the dihedral angle ϕ varies by 180°. The nearest neighbour V (ppπ) interaction equals 0.7 eV and it varies directly with ϕ, Fig. 2.19b, and therefore this is a strong source of disorder in the valence band edge [79]. This causes a strong tailing of the valence band edge, giving a characteristic tail width of at least 60 meV. The conduction band is less affected by dihedral angle disorder, but is affected by bond angle disorder. This also gives quite strong tailing, but less than for the valence band edge [80–82]. This gives rise to density of states in the gap, as shown schematically in Fig. 2.20.

The overall result of this strong effect of angular disorder is that the electron mobility of a-Si:H is quite low, ~ 1 cm^2/V s. The hole mobility is very low, 10^{-3} cm^2/V s.

A second limitation of a-Si:H is that its substitutional doping is very inefficient. Whereas every substitutional atom in c-Si produces a carrier, in a-Si:H there is an unusual self-compensation mechanism which limits the doping efficiency [78, 83]. The net effect of this is that the Fermi level can never move up to the donor level, or above the mobility edge. Thus, E_F is stuck in localised states. Conduction is in extended states, but only after thermal excitation of the carriers.

2.7 Disorder in Oxide Semiconductors

The amorphous n-type TCOs are in fact the first practical examples of disorder in an s band. The effect of disorder on s states is rather weak, compared to p states. As the conduction band minimum state is 90–95% localized on metal s states, then its energy depends mainly on the V(ss) interaction between second neighbour metal sites, and not much on the V(sp) interaction between metal s and oxygen p states, Fig. 2.21. A two-centre Slater-Koster interaction V(l,m) between orbitals

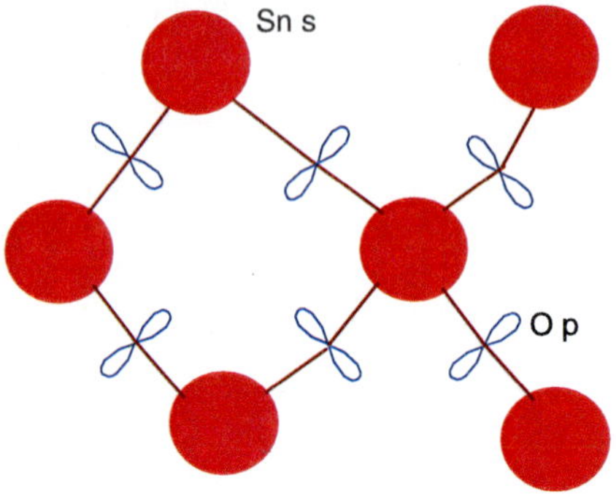

Fig. 2.21 s-Like atomic orbitals of CB states disordered SnO_2

on atoms l and m [84] would normally depend on their distance (r), the angles θ between the orbitals and the separation vector r, and their dihedral angle (ϕ),

$$V(l, m) = V(r, \ \theta, \phi)$$

However, for an interaction between two s states, this reduces to

$$V(ss) = V(r) \tag{2.6}$$

because of their spherical symmetry. Thus, the only source of disorder is the variation of the metal-metal distance, and any angular disorder has no effect on s states.

The first known effect of this was on the conduction band edge on amorphous SiO_2 (silica) [85]. SiO_2 has a wide band gap, 9 eV, and low screening. The conduction band minimum of SiO_2 has an effective mass of 0.5 m and is formed from Si s states and O 3s states [86]. The effect of disorder theoretically is small because the O-O distance is relatively fixed due to the small disorder of the O-Si-O bond angle. Holes form polarons and have a very low mobility. In contrast, unexpectedly, free electrons have a high mobility, indicating an absence of disorder effects and localized states. (Of course, there are not many free electrons, due to the band gap).

The same effect occurs in SnO_2 and related TCOs [87]. The large metal ion radius means that ion packing keeps the metal-metal distance rather constant, s states mean that the angular disorder has no effect, so the effects of disorder on the s-like conduction band minimum is very small. This is the case for the pure oxide. The density of states is summarised in Fig. 2.22. There is no mobility edge.

Experimentally, these s-like TCOs differ very strongly from a-Si:H. The electron mobility is large, of order 10–40 cm^2/V s. The Fermi level can be moved far into the conduction band without any problem, creating large free carrier concentrations [71]. Thin film transistors of n-type oxides show high field effect mobilities, 10–40 cm^2/V s [87–95], much higher than a-Si:H.

It is interesting that the recent oxide-based TFTs use mixed oxides of Ga, In and Zn [87–89] to control the off-current and vacancy concentration. The carrier

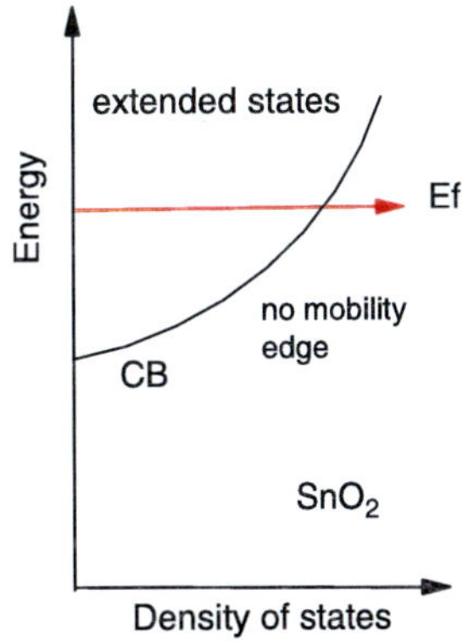

Fig. 2.22 Schematic density of states of disordered SnO_2, with no mobility edge, and the Fermi level able to enter the extended conduction band states

mobility is still high. Clearly the conduction band edge is still very delocalized and it is not even affected by compositional disorder. The reason for this is that Al, Ga or In form shallow donor states in ZnO [96], while ZnO forms resonant states in Ga or In oxide. Ga forms a shallow bound state below the ZnO CB edge, but at high concentrations, this forms a continuous band with the ZnO states in the alloy. Similarly, in SnO_2, substitutional Sb gives a shallow state [97]. The absence of deep states due to aliovalent impurities means that there are no localized states at the conduction band edge, and no effects that break the delocalisation of the CB states.

The absence of localized states due to aliovalent dopants also means that disorder does not introduce localized tail states below E_c. Thus, in a TFT, the field effect mobility is given essentially by the free carrier mobility or Hall mobility. The free carrier mobility in s states is much higher than in p states, which partly accounts for the higher FE mobility. A second factor is that the Fermi wavevector is large. This contrast with a-Si where the Hall coefficient has the opposite sign to the carrier.

A very narrow conduction band tail of order 0.1 eV was found in a calculation of the alloy $InGaZnO_x$ [98]. Note that the Urbach energy of these disordered oxides is ~0.2 eV [99], because of valence band tailing. Extended X-ray Fine Structure shows the local bonding [100].

2.8 Summary

The band structures of the various transparent conducting oxides are given, and discussed in terms of their band edge properties. The reasons for the ability to dope them n- or p-type given. Finally, the oxides are shown to be able to support a higher electron mobility than amorphous silicon due to their s-like conduction band minima.

Acknowledgements We thank Dr. S. J. Clark for many band calculations.

References

1. C H Ahn, J M Triscone, J Mannhart, Nature 424 (2003) 1015
2. D S Ginley, C Bright (eds) MRS Bull 25 (2000)
3. I Hamberg, C G Gramquist, J App Phys 60 (1986) R123
4. M Batzill, U Diebold, Prof Surf Sci 79 (2005) 47
5. H Kawazoe, N Yasukawa, H Hyodo, M Kurita, H Yanagi, H Hosono, Nature 389 (1997) 939
6. J Robertson, J Phys C 12 (1979) 4767
7. K C Mishra, K H Johnson, P C Schmidt, Phys Rev B 51 (1995) 13972
8. Y Mi, H Odaka, S Iwata, Jpn J App Phys 38 (1999) 3453
9. C Kilic, A Zunger, Phys Rev Lett 88 (2002) 095501
10. V T Agekyan, Phys Stat Solidi A 43 (1977) 11
11. R G Egdell, W R Flavell, P Tavener, J Solid State Chem 51 (1984) 345
12. H Odaka, S Iwata, N Taga, S Ohnishi, Y Kaneta, Y Shigesato, Jpn J App Phys 36 (1997) 5551
13. O N Mryasov, A J Freeman, Phys Rev B 64 (2001) 233111
14. R L Weiher, R P Ley, J App Phys 37 (1966) 299
15. V Christou, M Etchells, O Renault, P J Dobson, OV Salata, J App Phys 88 (2000) 5180
16. A Klein, App Phys Lett 77 (2000) 2009
17. A Walsh, J L F DaSilva, S H Wei, C Korber, A Klein, L F J Piper, A DeMast, K E Smith, G Panaccione, P Torelli, D J Payne, A Bourlange, R G Egdell, Phys Rev Lett 100 (2008) 167402
18. P D C King, T D Veal, D J Payne, A Bourlange, R G Egdell, C F McConville, Phys Rev Lett 101 (2008) 116808
19. A Bourlange et al, App Phys Lett 92 (2008) 092117
20. U Ozgur, Y I Alikov, C Liu, A Teke, M A Reshchikov, H Morkoc, J App Phys 98 (2005) 041301
21. A F Kohan, G Ceder, D Morgan, C G van de Walle, Phys Rev B 61 (2000) 15019
22. A Schleife, F Fuchs, J Furthmuller, F Bechstedt, Phys Rev B 73 (2006) 245212
23. M Usuda et al, Phys Rev B 66 (2002) 125101
24. N Ueda, H Hosono, R Waseda, H Kawazoe, App Phys Lett 71 (1997) 933
25. M Orita, H Ohta, M Hirano, H Hosono, App Phys Lett 77 (2000) 4166
26. R Asahi, A Wang, J R Babcock, N L Edleman, A W Metz, M A Lane, V P Dravid, C R Kannewurf, T J Marks, Thin Solid Films 411 (2002) 101
27. J E Medvedeva, A J Freeman, Europhys Lett 69 (2005) 583
28. P Koffyberg, Phys Rev B 13 (1976) 4470
29. R C Whited, C J Flaten, W C Walker, Solid State Commun 13 (1973) 1903
30. Y Dou, R G Egdell, D S L Law, N M Harrison, B G Searle, J Phys Cond Mat 10 (1998) 8447
31. M S Hybertsen, S G Louie, Phys Rev B 34 (1986) 5390
32. A Filippetti, N A Spaldin, Phys Rev B 67 (2003) 125109
33. A D Becke, J Chem Phys 98 (1993) 5648
34. J Muscat, A Wander, N M Harrison, Chem Phys Lett 342 (2001) 397
35. J Robertson, P W Peacock, M D Towler, R Needs, Thin Solid Films 411 (2002) 96
36. B M Bylander, L Kleinman, Phys Rev B 41 (1990) 7868
37. R Asahi, W Mannstadt, A J Freeman, Phys Rev B 59 (1999) 7486
38. C B Geller, W Wolf, S Picozzi, A Continenza, R Ashi, W Mannstadt, A J Freeman, E Wimmer, App Phys Lett 79 (2001) 368
39. J Robertson, K Xiong, S J Clark, Thin Solid Films 496 (2006) 1
40. J Robertson, K Xiong, S J Clark, Phys Stat Solidi B 243 (2006) 2054
41. L Kleinman, K Mednick, Phys Rev B 21 (1980) 1549
42. J Robertson, Phys Rev B 28 (1983) 3378
43. W Y Ching, Y N Xu, K Wong, Phys Rev B 40 (1989) 7684

44. X Nie, S H Wei, S B Zhang, Phys Rev B 65 (2002) 075111
45. M Nolan, S R Elliott, Phys Chem Chem Phys 8 (2006) 5350
46. A Buljan, P Alemany, E Ruiz, J Phys Chem B 103 (1999) 8060
47. B J Ingram, T O Mason, R Asahi, K T Park, A J Freeman, Phys Rev B 64 (2001) 155114
48. H Yanagi, S Inoue, K Ueda, H Kawazoe, N Hamada, J App Phys 88 (2000) 4159
49. X Nie, S H Wei, S B Zhang, Phys Rev Lett 88 (2002) 066405
50. R Nagarajan, A D Draeseke, A W Sleight, J Tate, J App Phys 89 (2001) 8022
51. D O Scanlon, A Walsh, B J Morgan, G W Watson, D J Payne, R G Egdell, Phys Rev B 79 (2009) 035101
52. A Kudo, H Yanagi, H Hosono, H Kawazoe, App Phys Lett 73 (1998) 220
53. H Ohta, M Orita, M Hirano, I Yagi, K Ueda, H Hosono, J App Phys 91 (2002) 3074
54. D Segev, S H Wei, Phys Rev B 71 (2005) 125129
55. S B Zhang, S H Wei, App Phys Lett 80 (2002) 1376
56. T Minami, T Miyaia, T Yamamoto, Surf Coating Technol 108 (1998) 583
57. W Mönch, Phys Rev Lett 58 (1987) 1260
58. W Mönch, Surface Sci 300 (1994) 928
59. J Tersoff, Phys Rev B 30 (1984) 4874
60. J Robertson, J Vac Sci Technol B 18 (2000) 1785
61. C G Walle, J Neugebauer App Phys Lett 70 (1997) 2577
62. S H Wei, A Zunger, App Phys Lett 72 (1998) 2011
63. P W Peacock, J Robertson, J App Phys 92 (2002) 4712
64. J Robertson, B Falabretti, J App Phys 100 (2006) 014111
65. W Walukiewicz, Physica B 302 (2001) 123
66. S B Zhang, S H Wei, A Zunger, J App Phys 83 (1998) 3192
67. S B Zhang, S H Wei, A Zunger, Phys Rev Lett 84 (2000) 1232
68. A Zunger, App Phys Lett 83 (2003) 57
69. C G van de Walle, J Neugebauer, Nature 423 (2003) 626
70. B Falabretti, J Robertson, J App Phys 102 (2007) 123703
71. H Hosono, J Non-Cryst Solids 352 (2006) 851
72. J Robertson, Phys Stat Solidi B 245 (2008) 1026
73. J R Bellingham, W A Phillips, C J Adkins, J Phys Condens Mat 2 (1990) 6207
74. J R Bellingham, W A Phillips, C J Adkins, J Mats Sci Lett 11 (1992) 263
75. P W Anderson, Phys Rev 109 (1958) 1492
76. N F Mott, Phil Mag 19 (1969) 835
77. N F Mott, Phil Mag 22 (1970) 7
78. J Robertson, Adv Phys 32 (1983) 361
79. J Singh, Phys Rev B 23 (1981) 4156
80. T Tiedje, J M Cebulka, D L Morel, B Abeles, Phys Rev Lett 46 (1981) 1425; Solid State Commun 47 (1983) 493
81. G Allan, C Delerue, M Lannoo, Phys Rev B 57 (1998) 6933
82. R Atta-Fynn, P Biswas, P Ordejon, D A Drabold, Phys Rev B 69 (2004) 085207
83. R A Street, Phys Rev Lett 49 (1982) 1187
84. W A Harrison, Electronic structure, W A Freeman, San Francisco, 1979, p 480
85. N F Mott Adv Phys 58 (1977) 363; Phil Mag B 58 (1988) 369
86. J R Chelikowksy, M Schluter, Phys Rev B 15 (1977) 4020
87. K Nomura, T Kamiya, H Ohta, K Ueda, M Hirano, H Hosono, App Phys Lett 85 (2004) 1993
88. K Nomura, H Ohta, K Ueda, T Kamiya, M Hirano, H Hosono, Science 300 (2003) 1269
89. K Nomura, H Ohta, A Takagi, T Kamiya, M Hirano, H Hosono, Nature 432 (2004) 488
90. H Q Chiang, J F Wager, R L Hofmann, J Jeong, A Keszler, App Phys Lett 86 (2005) 013503
91. W B Jackson, R L Hoffman, G S Herman, App Phys Lett 87 (2005) 193503
92. E Fortunato, A Pimentel, A Goncalves, A Marques, R Martins, Thin Solid Films 502 (2006) 704
93. B Yaglioglu, H Y Yeon, R Beresford, D C Paine, App Phys Lett 89 (2006) 062103

94. P F Garcia, R S McLean, M H Reilly, G Nunes, App Phys Lett 82 (2003) 1117
95. Navamathavan R et al, J Electrochem Soc 153 (2006) G385
96. S B Zhang, S H Wei, A Zunger, Phys Rev B 63 (2001) 075205
97. J Robertson, Phys Rev B 30 (1984) 3520
98. K Nomura, T Kamiya, H Ohta, T Uruga, M Hirano, H Hosono, Phys Rev B 75 (2007) 035212
99. K Nomura, T Kamiya, H Yanagi, H Hosono, App Phys Lett 92 (2008) 202117
100. D Y Cho, J Song, K D Na, App Phys Lett 94 (2009) 112112

Chapter 3
Modeling, Characterization, and Properties of Transparent Conducting Oxides

Timothy J. Coutts, David L. Young, and Timothy A. Gessert

3.1 Introduction

Other authors in this book have discussed at length the applications and synthesis of transparent conducting oxides (TCOs). Our purpose in this chapter is to discuss some elementary aspects of TCO properties, which can be explained surprisingly well using the Drude free-electron theory [1]. Although this theory explains the electrical properties and fits the optical data so well, many have questioned whether any fundamental understanding of TCOs can be gained from its use. We believe that much can be learned about the properties of the conduction electrons in some, but not all, TCOs. The conduction electrons are important because they dominate the optical properties of the materials in the visible and near-infrared (NIR) wavelengths. The functional form of the free-electron theory often accounts for measurable properties of TCOs such as transmittance and reflectance, and their relationship to extrinsically controllable properties (e.g., carrier concentration and relaxation time) and intrinsic, uncontrollable, properties (e.g., crystal lattice and effective mass,). In different ranges of wavelengths, it is reasonable to expect that the optical properties are dominated by mechanisms not involving the conduction electrons. Nevertheless, in the range of wavelengths of relevance to applications in which TCOs are used (e.g., flat-panel displays, solar cells, heat reflectors), the free electrons dominate the electrical and optical properties. We find that electrical properties such as effective mass and relaxation time, which may be derived from the optical characteristics, are essentially identical to those measured using the electrical characterization technique discussed in the second section of this chapter. We discuss several standard techniques used for characterizing TCOs, as well as more advanced approaches. These enable us to determine fundamental properties of relevance and we provide examples of this for several TCOs. The use of optical techniques is discussed in the first part of this chapter. Specifically, we consider the

T. Coutts (✉)
National Renewable Energy Laboratory, Golden, CO 80401, USA

D.S. Ginley et al. (eds.), *Handbook of Transparent Conductors*,
DOI 10.1007/978-1-4419-1638-9_3,

properties of the conduction electrons and do not consider band-to-band transitions described by, for example, the Burstein-Moss effect.

In the second section, we go on to discuss electrical measurements using an advanced characterization technique that provides fundamental information about the properties of the TCOs (or semiconductors, in general) such as the effective mass and relaxation time of the electrons, the position of the Fermi level (regardless of the shape of the Fermi surface), and a good indication of the dominant scattering mechanism of the electrons. A comparison of the values of the effective mass and relaxation time with the values determined using optical techniques provides a probe of the band structure of the material.

The most successful, and commonly used, TCO is indium tin oxide (ITO), but indium is likely to become a commodity in short supply [2]. Vendors of sputtering targets already require their customers to return spent targets. It is also economically volatile. Consequently, there is great incentive to develop new TCOs that do not contain scarce or toxic metals, but which have equivalent electrical and optical properties to ITO [2]. In Sect. 3.3, we discuss our efforts to control the deposition parameters carefully and thereby improve conventional materials such as tin oxide. We also discuss the use of a novel dopant in a conventional material; this appears to be a particularly promising topic. We focus exclusively on *n*-type materials in this chapter even though *p*-type TCOs have been demonstrated and have considerable potential for device applications [3].

3.1.1 A Semi-quantitative Assessment of Grain-Boundary Scattering in TCOs

In this section, we develop approaches that enable us to estimate fundamental properties using optical probes. To establish whether or not these optical measurements provide realistic estimates that compare usefully with electrical measurements, we first need to clarify the role of grain-boundary scattering. Grain-boundary scattering is often assumed to be the dominant mechanism in polycrystalline TCO thin films because of the rather small electron mobility compared with that of single-crystal samples. Although, many researchers [4–6] have claimed that grain-boundary scattering is the limiting scattering mechanism for TCO films, others [7–10] have argued against it.

Typically, single-crystalline samples have a much higher mobility than their polycrystalline counterparts, and it is often assumed that this is due to the absence of grain-boundary scattering in the single-crystal material. We do not believe that this is necessarily true. The techniques used to grow single crystals inevitably produce higher-quality material, with fewer point and extended defects, and other sources of scattering than exist in polycrystalline material. Even if this material were polycrystalline, we believe it would probably have greater mobilities than highly defective polycrystalline thin films. Some of the highest mobilities found in TCOs are for films that are best described as "amorphous" as characterized using

X-ray diffraction (XRD) [11–13]. This has been found for indium oxide, cadmium stannate, zinc indium oxide, and perhaps other TCOs. If grain barriers were truly responsible for lowering the mobility, then their influence would be seen most obviously in these materials.

Firstly, it is instructive to examine the various length scales that apply to degenerate TCO films. The de Broglie wavelength is given by

$$\lambda_{db} = 2\pi/\left(3\pi^2 n\right)^{1/3}, \tag{3.1}$$

whereas the classical mean free path is

$$\lambda_{mfp} = \frac{\hbar\mu}{e}\left(3\pi^2 n\right)^{1/3}. \tag{3.2}$$

The ratio of these two quantities is

$$\frac{\lambda_{mfp}}{\lambda_{db}} = \frac{\hbar\mu}{2\pi e}\left(3\pi^2 n\right)^{2/3}. \tag{3.3}$$

In these equations, μ is the electron mobility, n is the carrier concentration, e is the electronic charge, and $\hbar = h/2\pi$. Plotting (3.3) against the carrier concentration, n, reveals that, for a realistic value of μ of 50 $cm^2\ V^{-1}\ s^{-1}$, the classical mean free path concept is valid for values of $n > 1$–$2 \times 10^{19}\ cm^{-3}$. The carrier concentration for typical TCOs is at least an order of magnitude greater than this critical value, and we therefore conclude that the mean free path construct may legitimately be used for practical TCOs.

At optical and NIR frequencies, the amplitude of oscillation of an electron, under the influence of an electromagnetic field is on the order of 10^{-7} nm for each volt per meter of electric field strength,[1] i.e., much less than the average grain size (~50–100 nm) of typical TCO films. Therefore, optical mobilities should be a good probe of the intragrain mobility because the oscillating electrons do not typically encounter the grain boundaries.

Likewise, direct current (d.c.) Hall mobilities should be a good indication of *intergrain* properties and should include the influence of many grain boundaries. If grain boundaries are a significant source of scattering, then we should expect the Hall mobility to be significantly less than the optical mobility. For films with

[1]This can be shown from the second of Newton's equations of motion, $s = 1/2at^2$. We put $\bar{a} = e\bar{E}/m^*$ and $t = 1/2f$, where $f = \omega/2\pi$. For a wavelength of 1.5 μm, an effective mass of 0.35 m_e and an electric field strength of 1 V m^{-1}, the amplitude, s, is about 2×10^{-7} nm. Even though the electric field strength is assumed to be constant, this approach is sufficient to make an order of magnitude estimate. If we include scattering of the electrons, then an estimate of the amplitude may be made from (3.7). With a mobility of 50 $cm^2\ V^{-1}\ s^{-1}$, and the same conditions as above, the amplitude is about 4×10^{-8} nm.

a carrier concentration of 5×10^{20} cm^{-3} and a mobility of 50 cm^2 V^{-1} s^{-1}, the mean free path of electrons is about 8 nm, whereas the typical grain sizes are on the order of 50–100 nm. Therefore, we expect that an electron moving under the influence of an applied d.c. voltage will be scattered many times within a grain before reaching a grain boundary. The influence of the alternating electromagnetic field is insignificant, as we have shown above.

For an electron attempting to pass from one grain to another, the issue is whether or not a potential barrier exists at the grain boundary. If a barrier is present, the electron must either tunnel through the barrier or be thermionically emitted over it [14]. In both cases, the issues are how high (in eV) and how wide (in nm) the barrier is because these are the factors that govern the transmission probability.

There are two opposing theories for the nature of grain boundaries [15]. The first suggests that impurity segregation makes grain boundaries an impurity atom sink. Intragrain densities of impurity atoms are less than at the boundaries, which leads to lower carrier concentrations overall and higher resistivity within the grains. The other model suggests that defects at the grain boundaries form electron traps. This reduces the number of carriers in the film, at least near the grain boundaries, and decreases the conductivity. Trapped carriers charge the grain boundaries and create a potential energy barrier to mobile carriers moving from grain to grain.

To test whether or not grain-boundary scattering is a plausible mechanism for limiting the transport of carriers in ZTO films, the theory developed by Seto [15] and extended by Zhang and Ma [10] was applied to the temperature-dependent data of Sect. 3.2. For non-degenerate semiconductors (Maxwell-Boltzmann statistics), the grain-boundary limited mobility is expected to follow

$$\mu_{eff} = Le\left(\frac{1}{2\pi m^* k_B T}\right)^{1/2} Exp\left(\frac{-E_B}{k_B T}\right), \tag{3.4}$$

where L is the size of a grain, E_B is the height of the grain-boundary potential barrier, and k_B is Boltzmann's constant. For degenerate semiconductors (Fermi-Dirac statistics), the equivalent equation is

$$\mu_{eff} = BT^{-1} Exp\left(\frac{-E_a}{k_B T}\right), \tag{3.5}$$

with $E_a = E_B - (E_F - E_c)$ being the activation energy needed to surmount the grain-boundary potential energy barrier and B is a constant. $(E_F - E_c)$ is the difference in energy between the Fermi energy and the bottom of the conduction band. Plots of $\ln(\mu T^{1/2})$ and $\ln(\mu T)$ versus $1/k_B T$ for zinc stannate and tin oxide films showed that the height of the potential barrier was typically less than $k_B T$. Thus, grain-boundary scattering does not appear to be a major scattering mechanism for these particular degenerate ZTO films. We conclude that the Fermi electrons, which lie at an energy of about 0.5 eV above the top of the grain-boundary potential barrier, are unlikely to be impeded as they move from grain to grain. Finally, and most importantly, as we show later in this section, our estimates of the effective mass and relaxation time obtained using optical and electrical measurements are in good agreement, exactly as this analysis of grain-boundary scattering implies they should be. We therefore believe that it is valid to use the optical characterization techniques described in this section.

3.1.2 The Free-Electron Theory

In the Drude theory, electrons are assumed to oscillate in response to the electric-field component of the electromagnetic field, i.e., the light [16, 17]. The so-called Lorentz oscillator equation of motion is used to describe the position, velocity, and acceleration of an average electron as functions of time and of the angular frequency of the electromagnetic field. The differential equation of motion of the oscillating electrons, in one dimension, is

$$m * \ddot{\vec{x}} + \frac{m*}{\tau}\dot{\vec{x}} + Kx = e\vec{E}(t). \tag{3.6}$$

$m*$ and τ are the effective mass and relaxation time of the electrons, respectively; K is the restoring force per unit displacement between the electrons and their host ion cores; and $\vec{E}$ is the electric field strength. The relaxation time, τ, can be thought of in several different ways. When the system of electrons is perturbed from equilibrium, τ is the characteristic time for the system to return to equilibrium when the perturbing force is removed. It can also be thought of as the interval of time between successive randomizing collisions of the electrons. In general, it is a function of electron energy but, as we shall see later, this is not the case for TCOs. For the moment, we shall assume that the electrons are not totally free and that the restoring force is finite. The solution of (3.6) gives the time-dependent position, velocity, and acceleration of the average electron. The velocity, in particular, may then be used to calculate the current density and the frequency-dependent conductivity. The latter is substituted into the expression for the complex permittivity derived from Maxwell's equations.

The solutions of (3.6) for the position and velocity of the electron is

$$\vec{x} = \frac{e\vec{E}/m*}{\left(\omega_0^2 - \omega^2\right) - i\omega/\tau} \tag{3.7}$$

and

$$\vec{v} = \dot{\vec{x}} = \frac{-i\omega e\vec{E}/m*}{\left(\omega_0^2 - \omega^2\right) - i\omega/\tau}, \tag{3.8}$$

in which $\omega_0 = \sqrt{\frac{K}{m*}}$.

From (3.8), we can derive the frequency-dependent current density, viz.,

$$\vec{J} = nev = \frac{-i\omega ne^2\vec{E}/m^*}{(\omega_0^2 - \omega^2) - i\omega/\tau}, \tag{3.9}$$

in which n is the carrier concentration. This leads directly to the frequency-dependent optical conductivity

$$\sigma = \frac{\vec{J}}{E} = \frac{-i\omega ne^2/m^*}{(\omega_0^2 - \omega^2) - i\omega/\tau} = \frac{\omega ne^2}{m^*}\left\{\frac{-i(\omega_0^2 - \omega^2) + \omega/\tau}{(\omega_0^2 - \omega^2)^2 + (\omega/\tau)^2}\right\}. \tag{3.10}$$

For a TCO, the electrons are generally regarded as free and the restoring force to the ion core is assumed to be zero. The natural frequency of oscillation of the electrons about the ion cores is, therefore, very small compared with frequencies of interest (visible and near-infrared). At much lower frequencies, a different mechanism occurs and the electromagnetic radiation interacts with the ion cores themselves and causes them to oscillate about their equilibrium positions in the lattice. However, because the ions are much heavier than the electrons, this occurs in the terahertz range, and it has been observed for heavy fermion materials and for ZnO using terahertz spectroscopy [18, 19]. This is not, however, relevant to the free-electron effects under discussion. For the free electrons in a TCO, we may set $\omega_0 \simeq 0$. Hence, (3.10) may be simplified to

$$\sigma(\omega,\tau) = \frac{\omega ne^2}{m^*}\left\{\frac{i\omega^2 + \omega/\tau}{\omega^4 + \omega^2/\tau^2}\right\} = \frac{ne^2}{m^*}\left\{\frac{i\omega\tau^2 + \tau}{1 + \omega^2\tau^2}\right\}. \tag{3.11}$$

This consists of real and imaginary parts, and we may write the complex conductivity as

$$\sigma^* = \sigma_1 + i\sigma_2. \tag{3.12}$$

Therefore, from (3.11) and (3.12), the real and imaginary parts of the conductivity are, respectively,

$$\sigma_1(\omega,\tau) = \frac{ne^2\tau}{m^*(1 + \omega^2\tau^2)} = \frac{\sigma_0}{(1 + \omega^2\tau^2)} \tag{3.13}$$

and

$$\sigma_2(\omega,\tau) = \frac{ne^2\omega\tau^2}{m^*(1 + \omega^2\tau^2)} = \frac{\sigma_0\omega\tau}{(1 + \omega^2\tau^2)}, \tag{3.14}$$

where σ_0 is the d.c. conductivity, given by

$$\sigma_0 = \frac{ne^2\tau}{m^*}. \tag{3.15}$$

Equations (3.13) and (3.14) govern the free-carrier absorptance in a TCO (or a simple metal), and they are plotted in Fig. 3.1 as a function of wavelength and angular frequency, for a relaxation time, τ, of 5×10^{-15} s and a d.c. conductivity of $5.6 \times 10^3\ \Omega^{-1}\ \text{cm}^{-1}$. The d.c. conductivity, given by (3.15), is also shown by the horizontal straight line in Fig. 3.1. The maximum of (3.14) occurs when $\omega_{\max}\tau = 1$, i.e., an angular frequency of 2×10^{14} Hz, at which the value of σ_2 is half the d.c. conductivity. It is also straightforward to show that the width of the imaginary part of the conductivity curve, at $1/\sqrt{2}$ times the peak height, is given by $\Delta\omega = 2/\tau$. The frequency at which the maximum occurs, divided by the width, gives us the relationship $Q = \omega_{\max}/\Delta\omega = 0.5$. This is extremely lossy compared with very large Q factors obtained for very high-quality optical or electronic filters. However, it corresponds to a critically damped system, meaning that the system returns to equilibrium in the minimum possible time, without oscillating [20]. From Maxwell's equations, we have $\varepsilon * = \varepsilon_\infty + \frac{i\sigma}{\varepsilon_0\omega}$, and substitution of (3.11) into this expression leads to the real and imaginary parts of the permittivity, through the definition [21],

$$\varepsilon * = \varepsilon_1 + i\varepsilon_2. \tag{3.16}$$

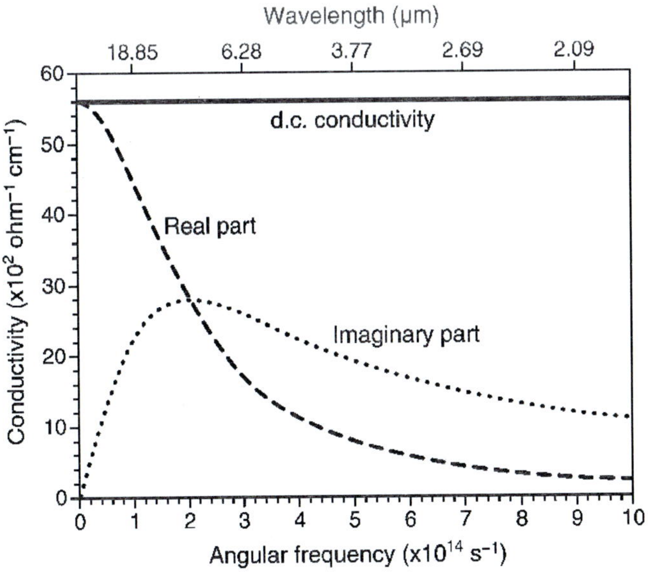

Fig. 3.1 Modeled real and imaginary parts of the conductivity for a relaxation time of 5×10^{-15} s and a d.c. conductivity of $5.6 \times 10^3\ \Omega^{-1}\ \text{cm}^{-1}$. The d.c. conductivity is also shown

The real and imaginary parts of the frequency-dependent dielectric permittivity are

$$\varepsilon_1 = \varepsilon_\infty - \frac{\sigma_0 \tau}{\varepsilon_0}\left\{\frac{1}{1+\omega^2\tau^2}\right\} \tag{3.17}$$

and

$$\varepsilon_2 = \frac{\sigma_0}{\varepsilon_0 \omega}\left\{\frac{1}{1+\omega^2\tau^2}\right\}. \tag{3.18}$$

If we take the ratio of (3.14) and (3.18), we obtain the expression

$$\sigma_2 = \varepsilon_0 \varepsilon_2 \omega^2 \tau. \tag{3.19}$$

Consequently, we can reconstruct the dependence of the alternating current (a.c.) conductivity on frequency by plotting $\omega^2\varepsilon_2$ against ω. We have tested this approach using a film of gallium-doped zinc oxide (ZnO:Ga). This was deposited using pulsed laser deposition onto a silicon substrate at a temperature of about 190°C. The thickness of the substrate was approximately 400 μm and that of the film was estimated to be about 200 nm. The carrier concentration had been measured as 9×10^{20} cm^{-3}. The reflectance was measured at an angle of incidence of approximately 12°, between 1.5 and 25 μm using a Nicolet Magna-IR 550 Spectrometer with a Pike Technologies MappIR accessory. The data were imported into TFCalc and fitted to a Drude model. The agreement between the measured and fitted reflectances was excellent (root-mean-square [r.m.s.] deviation of 0.6% for nearly 1,000 points), and we were able to extract the optical constants as functions of wavelength. This enabled us to test the approach described immediately above. The result is plotted in Fig. 3.2b. The position of the maximum in the conductivity function gave the relaxation time as 5.2 fs, and the width of this function, at $1/\sqrt{2}$ times the height, confirmed this value. With the value of the relaxation time now known, it is straightforward to plot the real part of the a.c. conductivity, as was shown in Fig. 3.1. This, however, does not lead to any additional useful information. Note that we can convert the function representing the imaginary part of the conductivity into the actual imaginary conductivity by multiplying by the relaxation time and the permittivity of free space. The peak occurs at a value of half the d.c. conductivity. Hence, the latter corresponds to a resistivity of about 3×10^{-4} Ω cm. The carrier concentration had previously been measured as 9×10^{20} cm^{-3}, and the mobility must therefore be about 22.6 cm^2 V^{-1} s^{-1}. In addition, we estimate that the effective mass is about 0.4 m_e and the high-frequency permittivity is 3.66. So far as we are aware, this is the first time that the a.c. conductivity has been reconstructed in this frequency range using spectroscopic measurements.

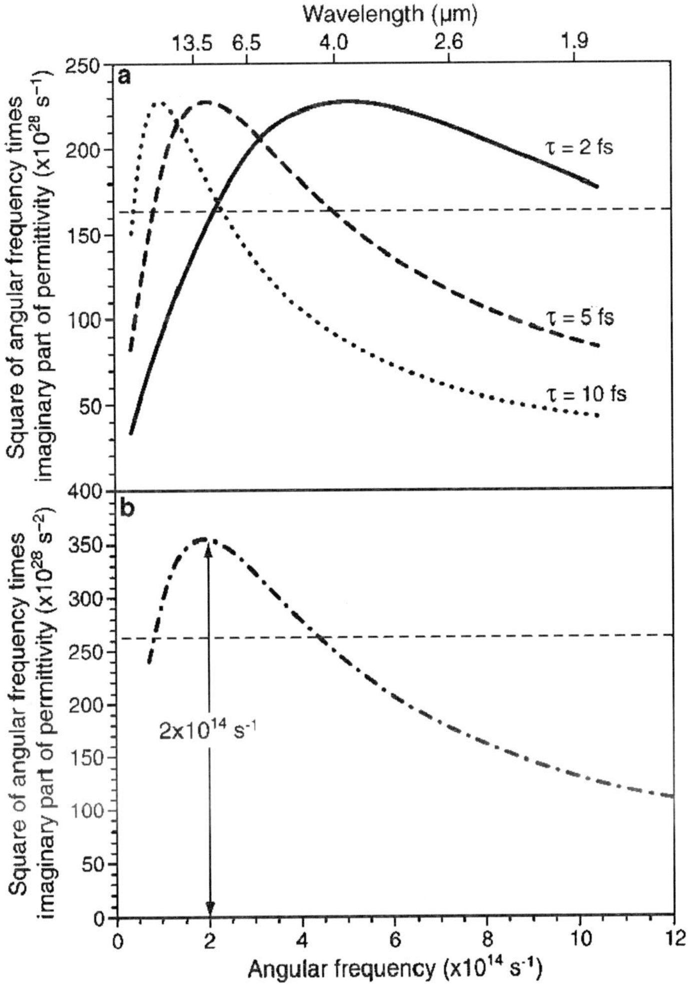

Fig. 3.2 (**a**) Modeled data for the angular frequency squared times the imaginary part of the a.c. conductivity plotted against the angular frequency. The three curves correspond to three different relaxation times. The frequency corresponding to the peak of the curve is equal to the reciprocal of the relaxation time, and the width at $1/\sqrt{2}$ times the peak height (the *horizontal dashed line*) gives twice the reciprocal of the relaxation time. (**b**) Same function plotted for a ZnO film doped with gallium. The real and imaginary parts of the permittivity were derived from measurements of reflectance in the wavelength range 1.5–25 μm

If this approach could be used in situ, as the film is being deposited, with the relaxation time being followed as growth proceeded, it would provide an excellent method for monitoring and optimizing the growth conditions for a TCO. However, it is rather time consuming at present, and it is not clear how in situ implementation could be performed.

Notice that (3.17) is the difference of two terms, which means that it is zero at some combination of values of the parameters and angular frequency. The plasma frequency, ω_p, determined from (3.17) and assuming that $\omega_p{}^2\tau^2>>1$,[2] is

$$\omega_p = \sqrt{\frac{\sigma_0}{\varepsilon_0\varepsilon_\infty\tau}} = \sqrt{\frac{ne^2}{\varepsilon_0\varepsilon_\infty m*}}. \tag{3.20}$$

Equation (3.20) shows that the plasma frequency depends on the carrier concentration, but does not, to a very good approximation, depend on the relaxation time, τ.

The free-carrier concentration may be estimated from (3.20), using the observed value of the plasma frequency, because this corresponds, approximately, to the peak of the absorptance curve, and the predetermined values of m^* and ε_∞. Taylor et al. have demonstrated how the optical properties may be used to estimate the carrier concentration of various TCOs, and we shall discuss this further later [13].

From (3.17) and (3.18), and using the inequality $\omega_p{}^2\tau^2>>1$, we see that

$$\varepsilon_1 = \varepsilon_\infty - \frac{ne^2}{\varepsilon_0 m * \omega^2} \tag{3.21}$$

and

$$\varepsilon_2 = \frac{\sigma_0}{\varepsilon_0\omega^3\tau^2} = \frac{ne^2}{\varepsilon_0 m * \omega^3\tau}, \tag{3.22}$$

the latter showing that the imaginary part of the permittivity is inversely proportional to the relaxation time.

ε_1 and ε_2 are related to the optical constants by the following two equations:

$$\varepsilon_1 = N^2 - k^2 \tag{3.23}$$

and

$$\varepsilon_2 = 2Nk. \tag{3.24}$$

Equations (3.23) and (3.24) may be used to obtain expressions for the refractive index and the extinction coefficient (using the Fresnel equations[3]), viz.,

[2]For a typical plasma wavelength of 1.5 μm, an effective mass of 0.35 m_e, and a relaxation time of about 5×10^{-15} s, corresponding to a mobility of about 25 $cm^2\ V^{-1}\ s^{-1}$, this inequality is obeyed to within about 2.5%. It is obeyed to better than 10% over the full range of frequencies shown in Fig. 3.2.

[3]The Fresnel reflection coefficient is given by $R = \frac{(N-1)^2+k^2}{(N+1)^2+k^2}$. Hence, at low frequency, where both N and k are relatively large, the reflection coefficient is also large.

$$N = \sqrt{\frac{1}{2}\left(\varepsilon_1{}^2 + \varepsilon_2{}^2\right)^{1/2} + \frac{\varepsilon_1}{2}} \tag{3.25}$$

and

$$k = \sqrt{\frac{1}{2}\left(\varepsilon_1{}^2 + \varepsilon_2{}^2\right)^{1/2} - \frac{\varepsilon_1}{2}}. \tag{3.26}$$

Hence, the optical constants both decrease in the visible range of frequencies as the relaxation time increases because ε_2 decreases. They are not affected by changes in ε_1, at least within the range of frequencies for which $\omega\tau >> 1$, because ε_1 is not a strong function of relaxation time in this range. The absorption coefficient is defined as

$$\alpha = \frac{4\pi k}{\lambda}. \tag{3.27}$$

As the frequency increases, in the vicinity of the plasma wavelength, there is a phase change. At higher frequencies, the electrons lag the electric-field vector. At lower frequencies, they lead the electric field. Significant changes in the optical properties occur in the vicinity of the plasma frequency. The modeled real and imaginary parts of the optical constants and the measurable optical properties are shown in Fig. 3.3a and b, respectively. In Fig. 3.3a, we also show the phase change on reflection of the *s*-polarized ray at normal incidence. These figures were modeled using a carrier concentration of 5×10^{20} cm^{-3} and a mobility of 50 cm^2 V^{-1} s^{-1}; the high-frequency permittivity and the effective mass were taken as 4 and 0.35 m_e, respectively.

Figure 3.3 illustrates several important features that emerge from the theory. The phase change (indicated by the ordinate λ_{PC}) occurs at a significantly shorter wavelength than the plasma wavelength, indicated by the ordinate shown as λ_{PW} that passes through the intersection of *N* and *k*. However, this is not necessarily a general observation, but may be specific to this particular set of parameters.

Figure 3.3b shows the modeled reflectance, transmittance, and absorptance of a 500-nm-thick film of the modeled material. In this film, there are three additional spectral features of possible relevance. The first corresponds to the wavelength at which the reflectance and transmittance curves intersect, λ_{RT}. For this set of modeling parameters, $\lambda_{RT} \approx \lambda_{PW}$. However, there is not an exact correspondence. The second feature is the maximum absorptance that occurs at a wavelength λ_A, and the third is the minimum in reflectance that occurs immediately before the rapid increase in reflectance, at a wavelength λ_M. None of these wavelengths is equal to any other and, more particularly, none of them may be expressed in analytical form, as is the case for λ_{PW}. In addition, different material parameters can lead to a different sequence of the various features in Fig. 3.3. Ideally, we wish to use (3.20) and the observed plasma wavelength (or frequency) to infer the carrier

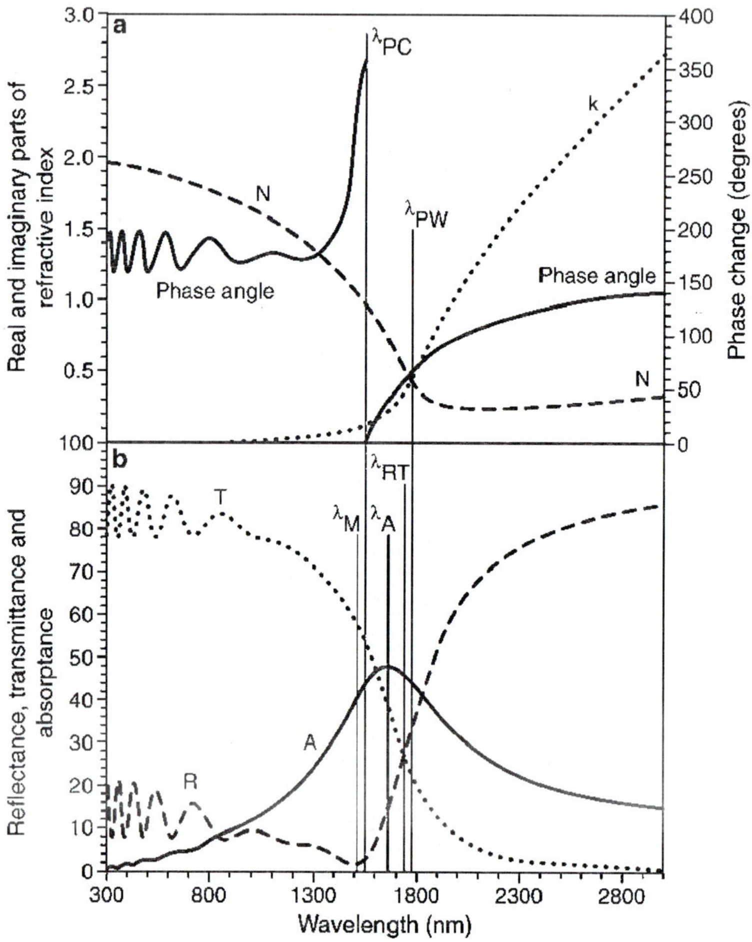

Fig. 3.3 **(a)** The modeled real and imaginary parts of the optical constants and the phase change on reflection. **(b)** The modeled reflectance, transmittance, and absorptance. The carrier concentration was 5×10^{20} cm^{-3}, effective mass was 0.4 m_e, high-frequency permittivity was 4, and mobility was 50 cm^2 V^{-1} s^{-1}. The film thickness was taken as 500 nm

concentration. This discussion has shown that to do so may not necessarily be straightforward because it is impossible to relate λ_{PW} to one of the readily observable features in the measurable quantities. We shall return to this later.

If $\varepsilon_2 = 0$, $N = \sqrt{\varepsilon_1}$ and $k = 0$, i.e., the material does not absorb light. This is either because there are no free electrons to absorb electromagnetic energy ($n = 0$), or they have a very large relaxation time. As the angular frequency increases, (3.21) and (3.22) show that ε_1 increases and ε_2 decreases. These changes cause N to increase and k and $\alpha$to decrease. In this range, the electrons are unable to respond sufficiently quickly to the applied electric field and are therefore unable to absorb radiation. At lower frequencies (say, in the NIR), ε_1 decreases and becomes negative, while in this range, ε_2, N, and k all increase strongly, as does the reflectance. As discussed above, the plasma frequency occurs between the two regions and defines the transition from high transmittance and low reflectance to high reflectance and low transmittance.

In the vicinity of the plasma frequency, light is absorbed to an extent that depends inversely on the relaxation time. This is illustrated in Fig. 3.4, which shows the modeled absorptance of 500-nm-thick films that have a carrier concentration of 5×10^{20} cm^{-3}, high-frequency permittivity of 4, and effective mass of 0.35 m_e. The maximum absorptance and the half-width both decrease with increasing relaxation time, and the position of the maximum remains fixed. However, the frequency at which the peak absorptance occurs is less than that of the plasma edge, although the magnitude of the difference is unpredictable.

In contrast to the effect of the relaxation time, we show the absorptance as a function of carrier concentration time in Fig. 3.5. In this figure, the relaxation time was taken as 10^{-14} s, the effective mass was 0.4 m_e, and the high-frequency permittivity was 4. Again, several effects are noteworthy. Firstly, the peak of the absorptance curves moves to shorter wavelengths as the carrier concentration increases, in accordance with (3.20) and (3.21). Secondly, the height of the absorptance band increases with carrier concentration, in accordance with (3.22) and (3.26). Thirdly, the half-width of the absorptance band decreases with increasing carrier concentration. Based on Figs. 3.4 and 3.5, there are clearly great incentives to develop materials with long relaxation times and high mobilities.

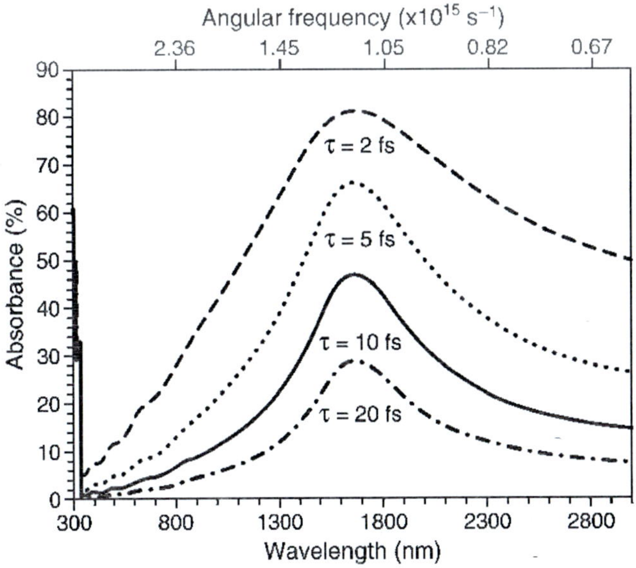

Fig. 3.4 Modeled absorptance as a function of wavelength and angular frequency for various relaxation times. The carrier concentration was 5×10^{20} cm^{-3}, effective mass was 0.4 m_e, and high-frequency permittivity was 4

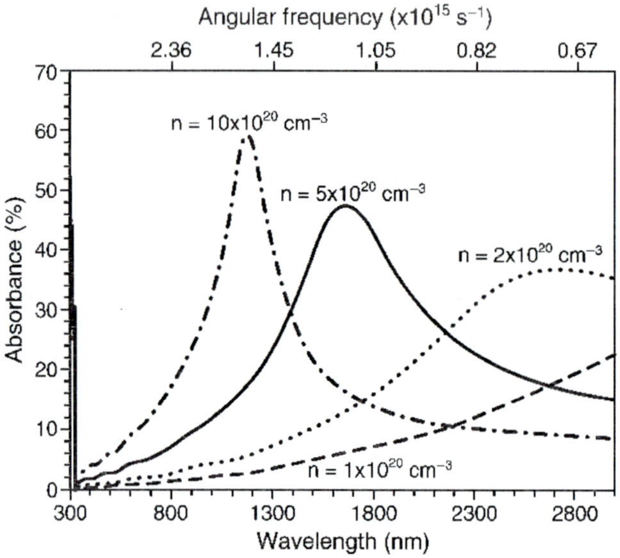

Fig. 3.5 Modeled absorptance as a function of wavelength and angular frequency, with carrier concentration being the variable. The relaxation time was 10^{-14} s, effective mass was 0.4 m_e, and high-frequency permittivity was 4

Given that a TCO is a "compromise" material because the requirements of high transmittance and conductivity are conflicting, a figure of merit (FOM) is sometimes used to compare TCOs [22]. Various FOMs have been used, but the most useful, which is thickness independent, is

$$\psi = \frac{\sigma}{\alpha}. \tag{3.28}$$

As we have seen, conductivity and absorption coefficient are both functions of frequency, but we shall assume for simplicity that the d.c. conductivity will suffice. From (3.22), (3.23), and (3.27), we can show that

$$\alpha = \frac{1}{Nc}\left(\frac{ne^2}{\varepsilon_0 m*}\right)\frac{1}{\omega^2 c}. \tag{3.29}$$

Equation (3.29) may now be substituted into (3.28) and, with (3.15), we find

$$\psi = \sqrt{\varepsilon_1} c \varepsilon_0 \tau^2 \omega^2. \tag{3.30}$$

Equation (3.30) is important because it makes two useful points. Firstly, the FOM is proportional to the square of the relaxation time, which emphasizes the importance of this quantity. Secondly, it is proportional to $\sqrt{\varepsilon_1}$. The latter quantity depends directly on the high-frequency permittivity, suggesting that this may offer an additional way of adjusting the properties. We shall return to this in Sect. 3.3.

3.1.3 Application of the Drude Theory

3.1.3.1 Determination of the Carrier Concentration

The plasma wavelength, or frequency, is defined as the wavelength, at which the real part of the permittivity equals zero, i.e., $\varepsilon_1 = 0$ corresponding to $N = k$. The plasma frequency is expressed by either of (3.17) or (3.20). Clearly, if the plasma frequency is known, then an estimate of the carrier concentration can be made. Although, strictly, the plasma wavelength (or frequency) is a function of relaxation time, it has a relatively weak dependence and, for $\omega\tau \gg 1$, there is essentially no dependence. As mentioned earlier, we wish to relate the plasma wavelength to a particular spectral feature. Taylor et al. defined the plasma wavelength as the minimum in reflectance that occurs immediately before the plasma edge [23]. Here, we estimate the magnitude of the uncertainty introduced by calculating the carrier concentration from another spectral feature, viz., the wavelength at which the maximum absorptance occurs. This approach leads to an estimate of the plasma wavelength that is longer than the true value, i.e., a value of carrier concentration that is lower than the correct value. This, however, is not known to be generally true. The reverse is true for the approach used by Taylor et al., which leads to too high a carrier concentration.

Firstly, we solve (3.17) numerically for $\varepsilon_1 = 0$ with typical values of $m^* = 0.35m_e$ and $\varepsilon_\infty = 4$ over a range of carrier concentration 10^{20}–10^{21} m^{-3}. The procedure was repeated for $\varepsilon_\infty = 3.5$ and for values of relaxation time between 2 and 10 fs. The results of these calculations are shown in Fig. 3.6a ($\varepsilon_\infty = 4$) and b ($\varepsilon_\infty = 3.5$).

These two figures can be viewed as "calibration curves" in the sense that they can be used to make an estimate of the carrier concentration using any of the spectral features identified in Fig. 3.3a and b. In the present case, we consider the wavelength at which the maximum absorptance occurs for a particular sample. We have used the data shown in Fig. 3.5 to generate quasi-experimental values of wavelength at which the three maxima occur in the range of wavelengths up to 3 μm. These are then plotted as the crosses in Fig. 3.6a and b. To calculate the absorptance curves in Fig. 3.5, a value of $\varepsilon_\infty = 4$ was used, and it is clear from Fig. 3.6a and b that the crosses are, as expected, nearer the set of curves in Fig. 3.5a. The relaxation time was taken as 10 fs to calculate Fig. 3.5, and it can be seen in Fig. 3.5a that the crosses are near the bottom of the set of curves, where the relaxation time is largest. Consequently, we can be confident that an accurate estimate can be made of the carrier concentration provided we have good estimates of the high-frequency permittivity and effective mass. Although m^* changes with carrier concentration, the variation is not strong and it is unlikely to have a large effect. The role of the relaxation time is minimal for all but the shortest value, particularly in the mid-range of carrier concentrations. This is because the inequality $\omega\tau \gg 1$ is more closely obeyed.

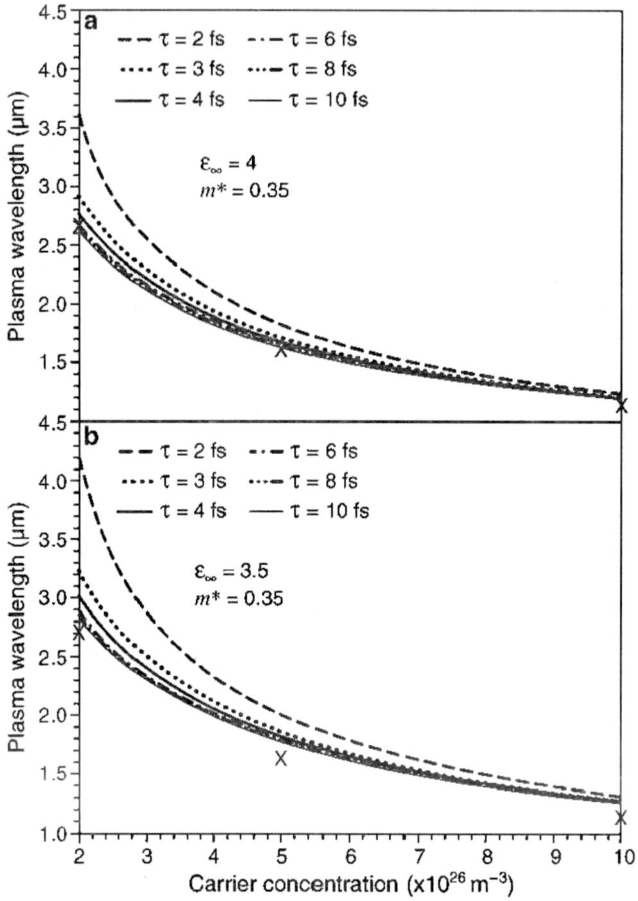

Fig. 3.6 Variation of the plasma wavelength with carrier concentration for various relaxation times. (**a**) A high-frequency permittivity of 4, whereas for (**b**), this quantity is 3.5. The effective mass was taken as 0.35 m_e in both cases

3.1.3.2 Estimation of the Relaxation Time

Fitting of Spectrophotometer Data

Firstly, we use the full spectrophotometric (reflectance and transmittance) data to calculate the electronic properties of TCOs. To do this, we once again use the application TFCalc but, on this occasion, in the visible/NIR range of wavelengths. The latter software package can take the reflectance and transmittance data as functions of wavelength and perform fitting to obtain estimates of the variable we have discussed here. It must be appreciated that fitting the spectrophotometric data necessitates specifying the model to be used. Hence, it does not show that the Drude model is applicable; this is assumed. With this assumption, we are then able to derive estimates of the key parameters. The alternative would be to measure the optical constants by another method – for example, spectroscopic ellipsometry – and then to use TFCalc to fit the data. We used the carrier concentration as an input to the modeling although, as already seen, this is not absolutely necessary because a reasonable estimate may be obtained from either the absorptance or reflectance spectrum. The full data are shown in Fig. 3.7 although, because of the excellent fit between the measured and fitted data, it is not immediately obvious that there are four lines. The data obtained optically and using the method of four coefficients (see Sect. 3.2 of this chapter) are shown in Table 3.1 [24]. In the fitting procedure, the thickness was permitted to be an adjustable variable, and the fact that it is about 30 nm less than the value measured using a Dektak is not of major concern.

To obtain data of this quality, it is necessary to take into account the dispersion of the optical constants of the substrate, to have a film that is very uniform in thickness over an area greater than the diameter of the spectrophotometer beam, and for the film to be uniform in its properties in depth. The free-carrier concentration

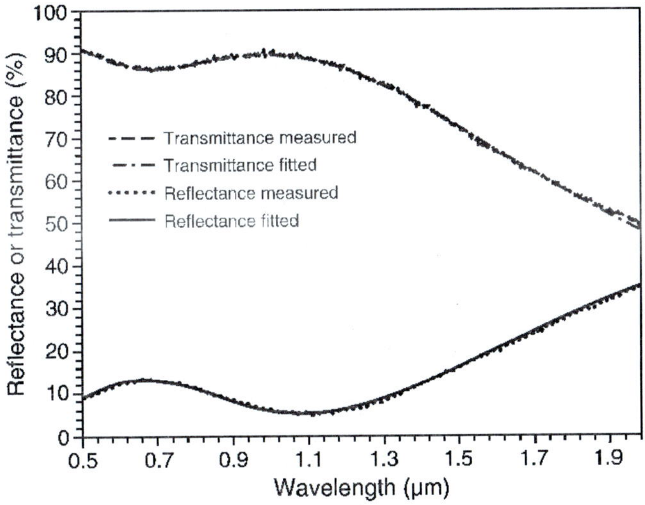

Fig. 3.7 Comparison of measured and fitted transmittance and reflectance for the cadmium stannate film

Table 3.1 Comparison of transport properties of a film of cadmium stannate measured using the *method of four coefficients* and the values from optical fitting

Property	Method of four coefficients	Fitted
Thickness (nm)	165	132
Effective mass ($\times m_e$)	0.31	0.32
Relaxation time (fs)	11.5	10.7
Mobility ($cm^2\ V^{-1}\ s^{-1}$)	52	58
Resistivity ($\times 10^{-4}\ \Omega$ cm)	2.0	1.8
Sheet resistance ($\Omega/\square$)	12.1	13.6

was measured using the *method of four coefficients* (MFC) and it is not included in the table because it was used as an input to calculate the resistivity and sheet resistance.

As with the a.c. conductivity approach described earlier, the full fit, using the transmittance and reflectance data, is very time-consuming and would certainly not be suitable for rapid, possibly in situ, monitoring of the film properties. For this purpose, simpler approaches are needed. It should be relatively straightforward to use an in situ scanning ellipsometer and/or a spectrophotometer to monitor the real and imaginary parts of the permittivity, as well as the absorptance, and to obtain the carrier concentration, relaxation time, and effective mass. These could then be correlated with film composition, possibly during combinatorial synthesis, and deposition conditions to optimize the film properties. However, very rapid data acquisition would be required to implement this approach.

Correlation Between Real and Imaginary Parts of Permittivity

Equations (3.17) and (3.18) lead to the expression

$$\varepsilon_1 = \varepsilon_\infty - \omega\tau\varepsilon_2. \tag{3.31}$$

Hence, plotting ε_1 against $\omega\varepsilon_2$ should yield a straight line with slope equal to τ and intercept equal to ε_∞. This approach of estimating the relaxation time provides an alternative to the more difficult methods of reconstructing the imaginary part of the a.c. conductivity or fitting the ultraviolet (UV)/visible spectrophotometer data. Before applying the technique to a film, the approach was first applied to a modeled film with a relaxation time of 5 fs and carrier concentration of 5×10^{20} m^{-3}, shown in Fig. 3.8. The technique returned precisely the same values of the relaxation time and high-frequency permittivity as were used to generate the original data. Of course, this merely shows that the approximations leading to (3.31) are valid. This technique was then applied to a cadmium stannate (CTO) film made by sputtering. The optical constants, N and k, and the permittivities, ε_1 and ε_2, were obtained from spectrophotometric measurements of transmittance and reflectance and then fitting these to a Drude model, using TFCalc.

The plot described above is shown in Fig. 3.8 over the wavelength range 500–2,000 nm. The short-wavelength limit was imposed by the need to avoid near-band-edge absorptance. The plasma wavelength for this film (i.e., the wavelength at which $N = k$) was found at about 1.45 μm, implying that the carrier concentration should be about 7×10^{20} cm^{-3} (as estimated from Fig. 3.6b). The measured carrier concentration for this film was 6×10^{20} m^{-3}. The relaxation time, determined using this approach, was 9.6 fs and the high-frequency permittivity was 3.53. The equivalent value of the relaxation time, measured using the *method of four coefficients*, was 11.5 fs, in reasonably good agreement with the optical value.

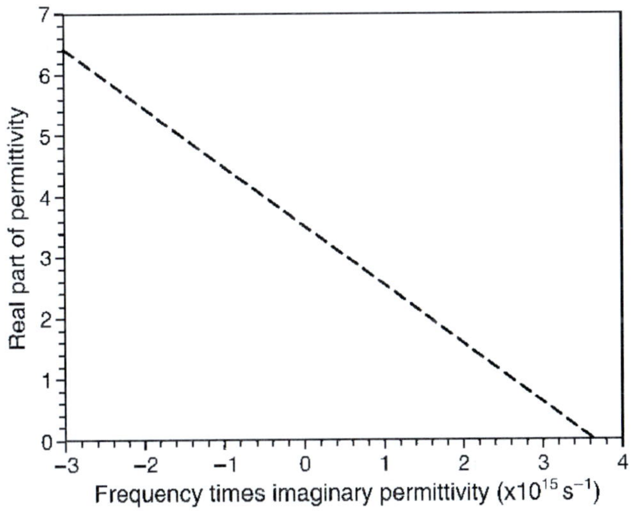

Fig. 3.8 Real part of permittivity plotted against angular frequency times the imaginary part of the permittivity. The data were measured for a cadmium stannate film. In this case, the relaxation time was 9.6 fs

3.1.3.3 Estimation of the Effective Mass

For $\omega\tau \gg 1$, (3.18) can be reduced to (3.22), which shows that the imaginary part of the permittivity is proportional to $1/\omega^3$. Hence, by plotting the imaginary part against $1/\omega^3$, we should expect a straight line with slope proportional to the conductivity. However, as already demonstrated, it is possible to calculate the carrier concentration and the relaxation time, making it simple to estimate the effective mass. We first applied this approach to the modeled film discussed in the last section for which the carrier concentration was 5×10^{20} cm^{-3}, and the relaxation time was 5 fs. An effective mass of 0.35 m_e was used to model the optical properties. The line of $\varepsilon \sim 1/\omega^3$ was linear, as expected, with a correlation coefficient of very near unity. The slope gave the d.c. conductivity, which led to an effective mass of 0.35 m_e, in perfect agreement with the value used for the modeling. Again, this merely shows that the approximations used to derive (3.22), and the assumption that the free-electron model adequately represents the behavior of a TCO, are reasonable. We have used this approach for the CTO film discussed above, the data being shown in Fig. 3.9.

The slope of this line yields an effective mass of 0.39 m_e, compared with the value measured with the MFC of 0.34 m_e. The values of the carrier concentration (estimated from Fig. 3.6) and relaxation time (from Fig. 3.8) were 7×10^{20} cm^{-3} and 9.6 fs, respectively. These values, and the slope of Fig. 3.9, were used to calculate the effective mass.

Equation (3.21) may be used in precisely the same way, except that the real part of the permittivity is proportional to the reciprocal of the square of angular

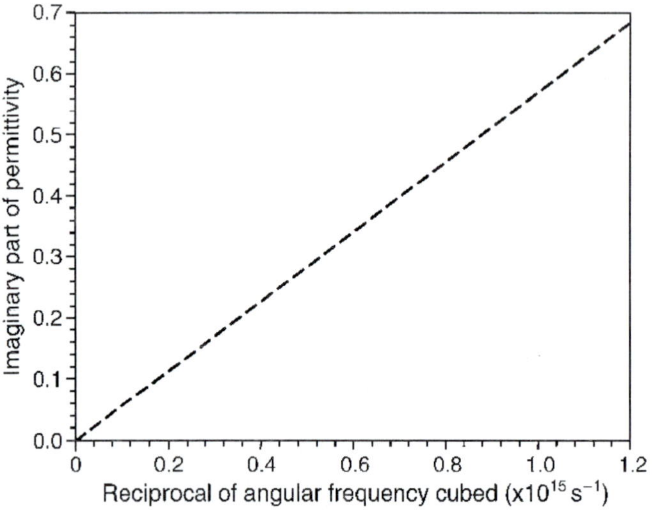

Fig. 3.9 Imaginary part of permittivity as a function of the reciprocal of angular frequency cubed. This was for the same CTO film shown in Figs. 3.7 and 3.8

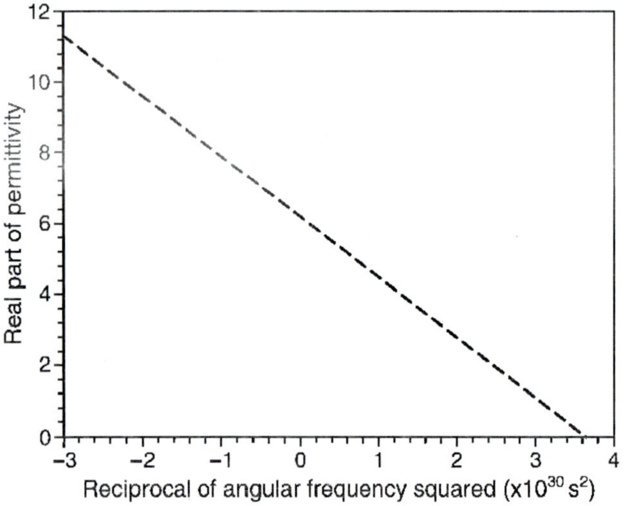

Fig. 3.10 Real part of permittivity as a function of angular frequency squared. The data were obtained for the same CTO film shown in Fig. 3.8

frequency. Figure 3.10 shows this plot, which once again yields excellent linearity over the wavelength range 500–2,000 nm. The effective mass derived from this plot was 0.36 m_e, once again in good agreement with the value derived from Fig. 3.9 and with the value obtained from MFC, shown in Table 3.1.

3.2 Measuring Transport Properties in TCOs

Measuring transport properties in TCO thin films is very similar to that of measuring metals. High carrier concentrations allow simple ohmic contacts to be made that allow current and voltage probes to measure sheet resistance (four-point probe) and mobility (Hall effect) easily. The ability to transmit light, as well as reflect light off a TCO, allows very accurate thickness measurements via ellipsometry or simple interference fringes. Knowing the thickness of the film allows measurement of the resistivity and carrier density. However, because of their relatively low mobilities ($<300\ \mathrm{cm^2\ V^{-1}\ s^{-1}}$), in part due to their high carrier concentrations, compared with other semiconductors (i.e., Si = 1,350 $\mathrm{cm^2\ V^{-1}\ s^{-1}}$, GaAs = 1,200 $\mathrm{cm^2\ V^{-1}\ s^{-1}}$) and short relaxation times ($\sim 10^{-15}$ s), many traditional methods for measuring the effective mass (cyclotron resonance [25], Shubnikov-de Haas [26], and de Haas–van Alphen oscillations) are not applicable to TCO films [17]. Measurement of the effective mass is *the* key parameter for understanding carrier transport in materials. Knowledge of its magnitude and how it varies with Fermi level reveals the electronic band structure of the material, which, in turn, allows *accurate* interpretation of transport data.

Resonance techniques for measuring the effective mass are only valid for high-mobility materials because of the constraint that the product $\mu B \gg 1$ (B is an externally applied magnetic field) so that electrons complete at least one revolution of their helical orbit before they are scattered in time, τ. With typical magnetic fields in the laboratory on the order of one Tesla, the mobility must be in the tens of thousands $\mathrm{cm^2\ V^{-1}\ s^{-1}}$ for $\mu B >> 1$ to be satisfied. Values of effective mass for polycrystalline, thin-film TCOs are usually inferred from optical modeling (plasma frequency) or assumed to be equal to single-crystal bulk values. Studies of single-crystalline TCO materials are rare [25, 27, 28], but suggest that there is a great discrepancy in mobilities for thin-film samples and bulk, single-crystalline samples. At least one study on crystalline SnO_2 was able to use oscillation techniques [25] to measure the effective mass due to the very large mobilities found in single-crystalline samples. Due to the nearly defect-free nature of the single crystals, the samples had very low carrier concentrations and hence low Fermi energies. Single-crystal studies may not reveal the effective mass and scattering mechanisms involved in defect-dominated, highly degenerate thin films. Mobilities in thin-film TCOs may be limited by a variety of scattering mechanisms including: ionized impurities, neutral impurities, phonon scattering, and grain-boundary scattering, whereas single-crystalline samples are ideally limited only by intrinsic lattice phonon scattering. This section describes the theory and experimental techniques available for measuring the relaxation time, effective mass, and scattering mechanism in low-mobility TCO thin films. The section starts with a qualitative overview of several transport phenomena, followed by a detailed solution to Boltzmann's transport equation. Finally, experimental details for measuring transport properties in TCOs and results from several TCOs will be presented.

3.2.1 History of Transport Phenomena

In 1879, E.H. Hall [29], a Fellow at Johns Hopkins University, discovered an electrical phenomenon that seemed to deflect and compress electrical flow to one side of a gold-leaf sample in crossed electric and magnetic fields. This discovery of a "compressible" electrical flow, later coined the Hall effect, led many researchers to explore this phenomenon in detail and with other materials [30]. Ettingshausen [31], prompted by a prediction by Boltzmann [32], found a temperature difference between the electrical Hall probes in bismuth metal. The phenomenon was named the Ettingshausen effect and helped to establish experimental evidence that heat could be carried by electrical flow. Nernst [33], while exploring the Hall effect in a plate of bismuth metal in 1886, discovered a voltage established by a magnetic field perpendicular to heat current flowing in the material. The Nernst effect, as it was called, again added to the idea that heat flow and electrical flow could be tied to the same mechanisms. With the discovery of the electron in 1897 by J. J. Thomson, the conduction mechanism for both electrical currents and heat currents in metals was easily interpreted [17]. Those early experiments and subsequent theoretical explanations led to the field of transport, or motion, of charge carriers in materials. Though many refinements in the subject have been made in the past 120 years, the basic experiments and mathematical modeling of electron transport remain fundamentally the same and are widely applied to simple metals and degenerate semiconductors today.

3.2.2 Qualitative Review of Transport Phenomena

A brief review of the transport phenomena related to TCOs will be helpful to our understanding of the mathematical results presented in this section. Each transport phenomenon manifests itself under different experimental conditions. Figure 3.11 depicts conditions for the Hall, Seebeck, and Nernst phenomena for n-type, isotropic, parabolic energy-band materials, exhibiting ionized impurity scattering. The Hall effect is a measure of the deflection of the drift carriers in a magnetic field. This deflection is measured by the Hall angle, ϕ. The magnitude of ϕ depends on the scattering mechanism through the relaxation time,τ, (see (3.44)), whereas its orientation is sensitive to carrier type. The Seebeck effect is an electric field established under steady-state conditions between diffusion current from a temperature gradient and the counteracting drift current. The direction of the field depends on carrier type. The Nernst effect is analogous to the Hall effect except that the carriers are driven by a temperature gradient, rather than by an applied bias. Carriers originating from the hot end of the sample will, on average, have more energy and have a higher velocity with respect to carriers from the cold end. The magnetic field will deflect the slower-moving carriers into a shorter-radius orbit than the faster-moving carriers and will establish a net charge, and thus, a Nernst

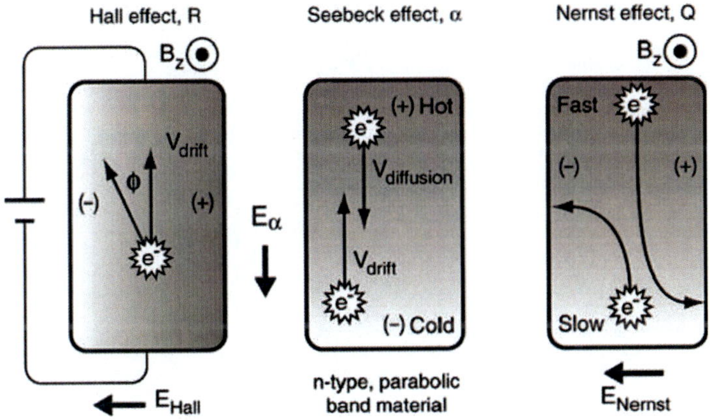

Fig. 3.11 Qualitative depiction of the Hall, Seebeck, and Nernst transport phenomena

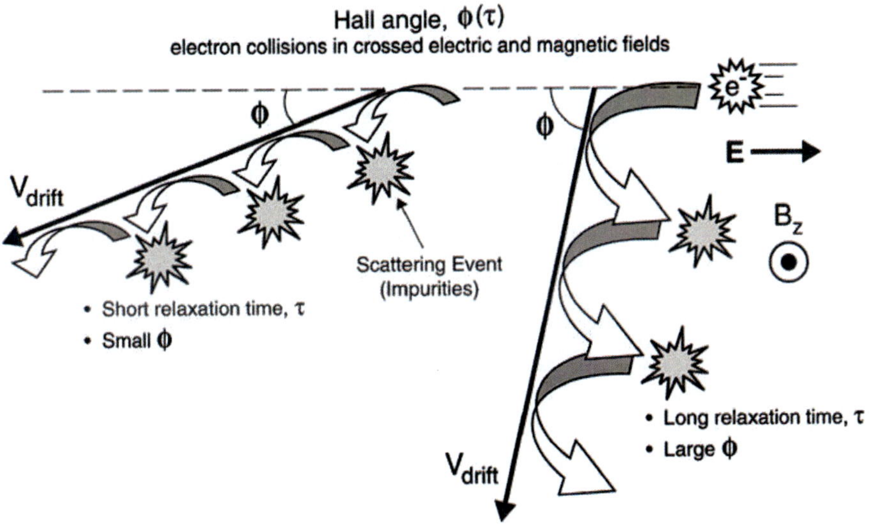

Fig. 3.12 Qualitative depiction of the origin of the Hall angle due to scattering of electrons in crossed electric and magnetic fields

electric field, as shown [34]. The Nernst field is not carrier-type sensitive, but, as will be shown next, is highly dependent on the scattering mechanism.

As will be shown later, the relaxation time of the charge carriers is related to the type of scattering experienced by the carriers through the variable s' (3.44). Figure 3.12 depicts the path of an electron in crossed electric and magnetic fields and relates the relaxation time to scattering events. The net path or drift velocity is shown as a result of scattering events between the electron and the lattice. The Hall angle, ϕ, is shown with respect to the electric field. The time between collisions, τ, is proportional to the arc length of the orbit. If the relaxation time,τ,

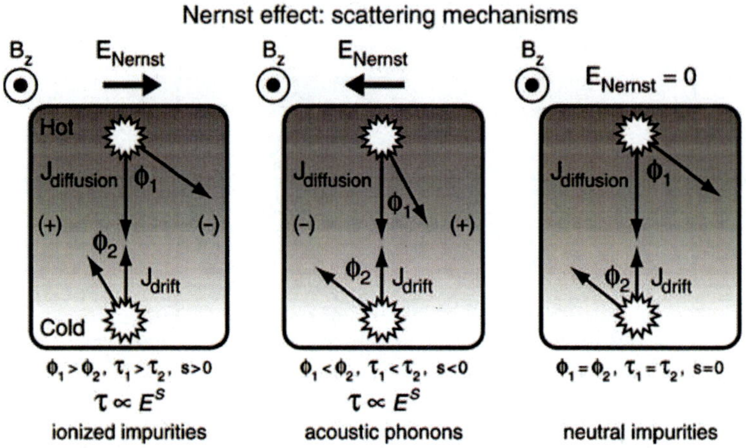

Fig. 3.13 Qualitative depiction of the Nernst effect for three scattering mechanisms

is short for a material, ϕ will be smaller than for a material with a longer relaxation time. Consider Fig. 3.13, which shows the Nernst effect for three different scattering mechanisms for n-type, isotropic, parabolic energy-band materials. The left figure shows the Nernst effect in the presence of ionized impurity scattering. Note that because $s'>0$, the relaxation time, and hence the Hall angle, must be larger for the more energetic (warmer) electrons. The relaxation time and energy of the electrons have a direct relationship (3.44). This condition gives a Nernst electric field directed to the right. Following Gerlach's sign convention [35], this configuration would give a positive Nernst coefficient. The center diagram of Fig. 3.13 depicts scattering by acoustic phonons. For this mechanism, $s'<0$, making the relaxation time and the Hall angle larger for the less energetic (colder) electrons, establishing a Nernst electric field directed to the left and a negative Nernst coefficient. The relaxation time and energy of the electrons have an indirect relationship. The right-most diagram of Fig. 3.13 shows that if scattering is by neutral impurities ($s' = 0$), then both Hall angles are equal and no Nernst electric field is established. The relaxation time is energy independent. Hence, the Nernst effect gives a clear signal for the type of scattering occurring in a parabolic-band material by changing the sign of the Nernst voltage. However, as will be shown, the Nernst coefficient (3.63) is affected by the non-parabolicity of the bands as measured by the λ term (3.64).

3.2.3 Transport Theory

Experimentalists and theorists working in the field of transport phenomena made many significant contributions to the field around the mid 1900s. Seitz's [36] work on mobility in pure insulators, Barrie's [37] understanding of electronic conduction in solids, and Kane's [38] band-structure modeling using the $k{\cdot}p$ perturbation approach influenced others [39, 40] in the field. These early scientists paved the way for modern applications of transport theory.

Transport of carriers may be described in a quasi-classical way by Boltzmann's transport equation, which relates changes in the carrier distribution (occupied energy states) to forces that drive the carriers to a non-equilibrium state. The carriers (electrons for n-type materials) in equilibrium are described by the Fermi-Dirac distribution function, $f_o(\vec{k},\vec{r})$, with $\vec{k}$ the wavevector. When external forces from a temperature gradient, electric or magnetic fields are imposed on a material, the carriers are driven to higher energy (excited) states. The redistribution of electrons over the allowed quantum states from the equilibrium configuration may be described by a non-equilibrium distribution function, $f(\vec{k},\vec{r})$. The non-equilibrium distribution will return to equilibrium upon removing the external fields or after scattering by various mechanisms [41]. The electric and thermal current densities in a material are described by

$$\vec{J} = \frac{e}{4\pi^3} \int \vec{v} f(\vec{k},\vec{r}) d\vec{k} \;\; amps/m^2 \tag{3.32}$$

and

$$\vec{Q}_{heat} = \frac{1}{4\pi^3} \int (E - E_F) \vec{v} f(\vec{k},\vec{r}) d\vec{k} \;\; watts/m^2, \tag{3.33}$$

where $\vec{v}$ is the carrier velocity. To solve these integrals, $f(\vec{k},\vec{r})$ must be determined by solving the steady-state Boltzmann transport equation [42],

$$\left(\frac{\partial f}{\partial t}\right) \approx \left(\frac{\partial f}{\partial t}\right)_{collision} + \frac{e}{\hbar}(\vec{E} + \vec{v} \times \vec{B}) \cdot \vec{\nabla}_k f(\vec{k},\vec{r}) + \vec{v} \cdot \vec{\nabla}_r f(\vec{k},\vec{r}) = 0, \tag{3.34}$$

which relates the rate of change of the non-equilibrium function due to collisions, $(\partial f/\partial t)_{collisions}$, to driving forces acting on the non-equilibrium energy distribution of carriers (right-most terms of (3.34)). Obviously, this is a very difficult equation to solve because it requires knowledge of occupation and transition probabilities between excited and equilibrium states. Traditionally, an approximation, known as the relaxation time, has been made to allow analytical solutions. This approximation assumes a mean time, τ, between collisions, when carriers are forced from an excited to an equilibrium distribution, or vice versa. Additionally, the approximation assumes: (1) the electron distribution after a collision in no way depends on the distribution of the electrons prior to the collision, and (2) the local temperature thermodynamic equilibrium is maintained by the collision events [17].

Using the relaxation-time approximation, the rate of change in the distribution function due to collisions can be approximated by

$$\left(\frac{\partial f}{\partial t}\right)_{collisions} \approx \left(\frac{f_o - f}{\tau}\right) \quad \mathrm{s}^{-1}. \tag{3.35}$$

The relaxation time is often the time between scattering collisions and is a function of energy and possibly temperature [41], not of direction of the carrier motion. The relaxation-time approximation is valid if the scattering events are elastic or semi-inelastic when scattering completely randomizes directions of the velocity vector [9, 43]. This approximation is very good for ionized impurity, dislocation, or acoustic phonon scattering, This approximation may also be valid for optical phonon scattering if the optical-mode energy is comparable to the Fermi energy. The frequency of elastic collisions, $1/\tau(E)$, depends on the density-of-states function, $g(E)$, and the square of the matrix element of the scattering transition probability from one energy state to a final state of the same energy, $w(E)$. For isotropic, parabolic energy bands, the energy of the carriers is described by

$$E = \frac{\hbar^2 k^2}{2m^*}, \tag{3.36}$$

where m^* is the effective mass of the carriers in a periodic potential. The effective-mass concept arises from equating the motion of carriers in a periodic lattice potential to the motion of a carrier in free space when an external force acts on the carrier. Newton's second law of motion has a proportionality constant, m, between force and acceleration. For carriers in a periodic lattice potential, the proportionality constant is different from the free-carrier constant. Thus, the effective mass, m^*, is the proportionality constant that preserves Newtonian mechanics in a periodic lattice potential. For a particle of mass m^* and momentum $\vec{p}$, Newton's second law may be written in two forms:

$$\vec{F}dt = d\vec{p} = \hbar d\vec{k} \Rightarrow \vec{F} = \hbar\frac{d\vec{k}}{dt} \tag{3.37}$$

and

$$\vec{F} = m^*\frac{d\vec{v}_g}{dt}, \tag{3.38}$$

where $\vec{v}_g = \hbar^{-1}\nabla_{\vec{k}}E(\vec{k})$ is the group velocity. Equation (3.37) relates force to a change in momentum, while (3.38) relates force to a change in acceleration. To make (3.37) and (3.38) consistent, m^* must be defined carefully. In one dimension,

$$\frac{dv_g}{dt} = \frac{1}{\hbar}\frac{d}{dt}\left(\frac{dE}{dk}\right) = \frac{1}{\hbar}\frac{d}{dk}\left(\frac{dk}{dt}\frac{dE}{dk}\right). \tag{3.39}$$

Substituting (3.37) into (3.39) gives, in one dimension,

$$F = \left(\frac{\hbar^2}{d^2E/dk^2}\right)\frac{dv_g}{dt}. \tag{3.40}$$

Equating (3.38) and (3.40) reveals that the effective mass must be defined by

$$m^* = \frac{\hbar^2}{d^2E/dk^2}. \tag{3.41}$$

Equation (3.41) describes the *dynamic* effective mass. The *dynamic* term is used to indicate that the effective mass was derived from the group velocity of the electron distribution. In three dimensions, the effective mass becomes a second-rank tensor relating the *i*th component of acceleration with the *j*th component of force and is given by [44]:

$$\frac{1}{m_{ij}^*} = \frac{1}{\hbar^2}\frac{\partial^2 E}{\partial k_i \partial k_j}. \tag{3.42}$$

Equation (3.42) reveals that the reciprocal of the effective mass is proportional to the curvature of the energy band. The smaller the effective mass value, the sharper the curvature of the band and, thus, the fewer the energy states for a given $\vec{k}$range. Figure 3.14a [9] shows the $E(k)$ dispersion relation for two parabolic conduction bands with different effective masses and, thus, curvatures, as shown. The shaded areas represent equal numbers of electrons in filled energy states. It is apparent that the band with the smaller effective mass fills the band to higher energy states because of the lower density of states.

For completeness, the effective-mass concept may also be formulated with momentum, $p = m^*v$. Using $E = p^2/2m^*$ leads to a *momentum* effective mass [9] given by

$$\frac{1}{m^*} = \frac{1}{p}\frac{dE}{dp}. \tag{3.43}$$

The *momentum* and *dynamic* effective masses are equal at the band minimum, but may differ as one moves away from $k = 0$, depending on the shape of the energy band. It is the *momentum* effective mass that is used in transport theory, and it is related to the density-of-states effective mass discussed later in this chapter.

The density of states in a parabolic band is given by $g(E) \propto E^{1/2}$, while w(E) is a power function of energy $w(E) \propto E^{-s} \propto (k^2)^{-s}$ [43]. Thus,

$$\tau^{-1}(E) \propto w(E)g(E) = E^{-s}E^{1/2} = E^{1/2-s}, \tag{3.44}$$

which is often written $\tau \propto E^{s-1/2} \propto E^{s'}$ with $s' = s - 1/2$. For the parabolic band model, s′, or the scattering parameter, has been identified for several scattering mechanisms [42, 43]. Some of these parameters for different scattering mechanisms are listed in Table 3.2.

A more general description of the energy bands allows the bands to be ellipsoidal and non-parabolic. For this case, the energy of the carriers can be described by the function

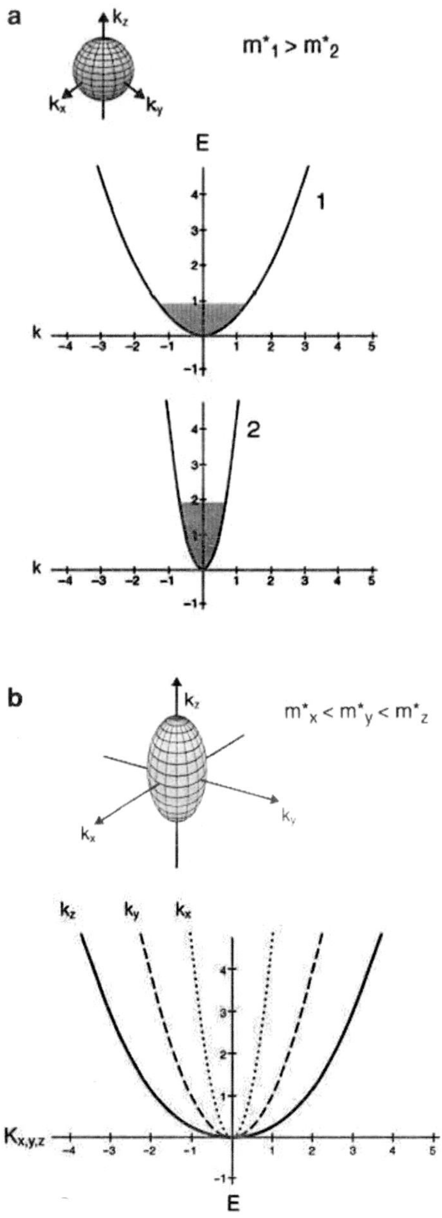

Fig. 3.14 (**a**) Conduction-band curvatures related to effective-mass values. *Dark shading* at the bottom of the bands represents filled energy states. (**b**) Conduction-band curvatures related to effective-mass values for different crystallographic directions

Table 3.2 Parameters for different scattering mechanisms

Scattering mechanism	s'	s
Acoustic phonon	−1/2	0
Optical phonon	1/2	1
Neutral impurity	0	1/2
Ionized impurity	3/2	2

$$\frac{\hbar^2}{2}\sum_{i=1}^{3}\frac{k_i^2}{m^*_{do_i}} = \gamma(E) = E + \frac{E^2}{E_1} + \frac{E^3}{E_2^2} + \cdots \,, \tag{3.45}$$

where E_1 and E_2 are constants that preserve the same transition-matrix element dependence on k^2 as in the parabolic case. The summation index, i, represents the three directions of the principal axes of the ellipsoid in k-space. A parabolic band is represented by $\gamma(E) = E$, and only the first term of the series is required. $\gamma(E)$, then, describes the state of non-parabolicity for the energy bands in the material. Figure 3.14b illustrates the ellipsoidal energy-band concept. Using (3.45), the non-parabolic density-of-states function,

$$g_{N.P.}(E) \propto \gamma^{1/2}\left(\frac{d\gamma}{dE}\right), \tag{3.46}$$

leads to

$$\tau^{-1}_{N.P.} = w(E)_{N.P.}g(E)_{N.P.} \propto \gamma^{-s}\gamma^{1/2}\left(\frac{d\gamma}{dE}\right), \tag{3.47}$$

and finally,

$$\tau \propto \gamma^{s-1/2}\left(\frac{d\gamma}{dE}\right)^{-1} \propto \gamma^{s'}\left(\frac{d\gamma}{dE}\right)^{-1}. \tag{3.48}$$

Clearly, the s' values for different scattering mechanisms in the non-parabolic case are the same as in the parabolic case.

Many texts [34, 41–43] outline the solution of Boltzmann's equation for $f(\vec{k}, \vec{r})$ using the relaxation-time approximation and further solve (3.32) and (3.33) for electric and thermal current densities. In general, (3.32) and (3.33) have the following form [41]:

$$J_\alpha = \frac{e}{4\pi^3}\int v_\alpha^2 \tau G_\alpha(E)\frac{\partial f_0}{\partial E}dk \tag{3.49}$$

and

$$Q_{heat\alpha} = \frac{1}{4\pi^3}\int (E - E_F)v_\alpha^2 \tau G_\alpha(E)\frac{\partial f_0}{\partial E}dk, \tag{3.50}$$

where $G_\alpha(E)$ is a function of temperature gradients and electric and magnetic forces, and α represents different crystallographic directions. Equations (3.49) and (3.50) may be related to the phenomenological equations for carrier transport,

$$J_\alpha = \sigma_{\alpha\beta}E_\beta - \theta_{\alpha\beta}\frac{dT}{dx_\beta} \tag{3.51}$$

and

$$Q_{heat\alpha} = \chi_{\alpha\beta} E_\beta - \xi_{\alpha\beta} \frac{dT}{dx_\beta}, \tag{3.52}$$

to identify specific transport coefficients. In (3.51) and (3.52), E_β is an electric field, dT/dx_β is a temperature gradient across a sample, and $\sigma_{\alpha\beta}, \theta_{\alpha\beta}, \chi_{\alpha\beta},$ and $\xi_{\alpha\beta}$ are fundamental transport tensors that depend on the orientation and strength of a magnetic field [45]. All transport phenomena may be expressed by these tensors [45, 46]. From the Onsager relation [47, 48], the relationship

$$\chi_{\alpha\beta}(H) = T\theta_{\beta\alpha}(-H) \tag{3.53}$$

holds. Casting (3.51) and (3.52) into more experimentally controllable variables, we get

$$E_\gamma = \rho_{\gamma\alpha} J_\alpha + \alpha_{\gamma\alpha} \frac{dT}{dx_\alpha} \tag{3.54}$$

and

$$Q_{heat\gamma} = \pi_{\gamma\alpha} J_\alpha + \mathrm{X}_{\gamma\alpha} \frac{dT}{dx_\alpha}, \tag{3.55}$$

where the new tensors are defined as:

$$\begin{aligned} \rho_{\gamma\alpha} &= \sigma_{\gamma\alpha}^{-1} && \textit{resistivity tensor} \\ \alpha_{\gamma\alpha} &= \sigma_{\gamma\beta}^{-1}\theta_{\beta\alpha} && \textit{thermoelectric tensor} \\ \pi_{\gamma\alpha} &= \chi_{\gamma\beta}\sigma_{\beta\alpha} && \textit{Peltier tensor} \\ \mathrm{X}_{\gamma\alpha} &= \xi_{\gamma\alpha} - \chi_{\gamma\beta}\alpha_{\beta\alpha} && \textit{thermal conductivity tensor.} \end{aligned}$$

The Onsager relation (3.53) makes the connection between the thermoelectric tensor and Peltier tensor by

$$\pi_{\alpha\gamma}(H) = T\alpha_{\gamma\alpha}(-H). \tag{3.56}$$

The fundamental transport tensors have been calculated [46] for energy bands given by (3.45). To match experimental setups, temperature gradients and electric fields are placed in the x, y plane, whereas magnetic fields are parallel to the z-axis. This arrangement leads to analytical solutions for conductivity (σ), Hall (R_H), Seebeck (α), and Nernst (Q) coefficients. Integral equations for the transport coefficients for isotropic material are [42]

$$\text{conductivity } \sigma = \frac{e^2 n}{m_c^*} \langle \tau \rangle, \tag{3.57}$$

$$\text{Hall}\, R_H = \frac{1}{ne}\frac{\langle \tau^2 \rangle}{\langle \tau \rangle^2}, \tag{3.58}$$

$$\text{seeback}\, \alpha = \frac{1}{eT}\left[\frac{\langle \tau E \rangle}{\langle \tau \rangle} - E_F\right], \tag{3.59}$$

$$\text{Nernst}\, Q = \frac{-\mu_H}{eT}\left[\frac{\langle \tau^2 E \rangle}{\langle \tau \rangle^2} - \frac{\langle \tau^2 \rangle \langle \tau E \rangle}{\langle \tau \rangle^3}\right], \tag{3.60}$$

where

$$\langle \tau^m E^n \rangle = \frac{\int_0^\infty (\tau^m E^n) E g(E) \frac{\partial f_0}{\partial E} dE}{\int_0^\infty E g(E) \frac{\partial f_0}{\partial E} dE} \tag{3.61}$$

and $g(E)$ is a density-of-states function. m_c^* is the conductivity effective mass and n is the carrier concentration. μ_H is the Hall mobility defined by $\mu_H = R_{Hall}\sigma$, e is the fundamental charge and is positive for holes and negative for electrons. Note that the sign of the Nernst coefficient does not depend on the sign of the carrier due to the cancellation of q with mobility.

Kolodziejczak and Zukotynski [45] first solved (3.57)–(3.60) for materials with ellipsoidal energy surfaces in a series of landmark papers [39, 45, 46, 49–51]. For highly degenerate materials, analytic expressions for (3.57)–(3.60) are obtained using the method of Bethe-Sommerfeld [17] to approximate the Fermi integrals given by (3.61). Keeping only lowest-order, non-vanishing terms, the conductivity and Hall coefficients require only first-order terms in the Sommerfeld expansion, whereas the Seebeck and Nernst coefficients require retention of second-order terms to first-order in temperature. The results [45] for the Seebeck and Nernst coefficients for low magnetic fields $\mu B<1$ and isotropic materials of high degeneracy are, respectively,

$$\alpha = \frac{2}{3}\alpha_\infty\{s + 1 - \lambda\} \tag{3.62}$$

$$Q = \mu\frac{2}{3}\alpha_\infty\left\{s - \frac{1}{2} - \lambda\right\} \tag{3.63}$$

where

$$\lambda = 2\gamma\left(\frac{d\gamma}{d\varepsilon}\right)^{-2}\frac{d^2\gamma}{d\varepsilon^2} = 3\frac{n}{m_d^*}\frac{dm_d^*}{dn}, \tag{3.64}$$

where s is the scattering parameter, andμ is the mobility. The term

$$\alpha_\infty = \frac{k_B^2 T}{e\hbar^2}\left(\frac{\pi}{3n}\right)^{2/3} m_d^* \tag{3.65}$$

is the Seebeck coefficient under high magnetic field ($\mu B > 1$), and $m_d^* = m_{do}^* \frac{d\gamma}{d\varepsilon}$ is the energy-dependent density-of-states (DOS) effective mass, with m_{do}^* representing the DOS effective mass at the conduction-band minimum. The term DOS effective mass was explained earlier. Notice that the last term within the braces for (3.62) and (3.63), λ, depends on the $E(\vec{k})$ dependence in the band and vanishes for parabolic-band materials when $\gamma(E) = E$ (3.45). If the λ term is large (large non-parabolicity in the energy band) or if the scattering parameter is large and of the appropriate sign, the Seebeck and Nernst coefficient values could possibly change signs. The Nernst coefficient was seen to change sign due to this effect in InSb [50], where non-parabolicity in the conduction band is large. Kolodziejczak and Sosnowski [50] noticed that the high-field Seebeck effect (3.65) could be used to probe the energy band of a material without knowledge of the scattering mechanism or the shape of the bands. Dubrovskaya and Ravich [52] investigated single-crystal PbTe in this manner. Unfortunately, this technique is not applicable for TCOs, again, because of their low mobilities and the requirement for high-field measurements to be made subject to the criterion that $\mu B >> 1$.

Kaidanov et al. [51, 53, 54] were the first to recognize that (3.62) and (3.63) could be combined to give equations for the effective DOS mass at the Fermi level,

$$m_d^* = \left(\frac{3n}{\pi}\right)^{2/3} \frac{e\hbar^2}{k_B^2 T}\left(\alpha - \frac{Q}{\mu}\right), \tag{3.66}$$

and s, the scattering parameter,

$$s = \frac{3}{2}\frac{(Q/\mu)}{(\alpha - Q/\mu)} + \frac{1}{2} + \lambda. \tag{3.67}$$

Substituting into (3.66), the expressions $\mu = |R_H|\sigma$ and $n \approx 1/R_H e$, we obtain:

$$m_d^* = \left(\frac{3}{|R_H| e\pi}\right)^{2/3} \frac{e\hbar^2}{k_B^2 T}\left(\alpha - \frac{Q}{|R_H|\sigma}\right), \tag{3.68}$$

which illustrates that the DOS effective mass may be experimentally determined by measuring the conductivity, Hall, Seebeck, and Nernst coefficients of a sample. This technique has been coined *the method of four coefficients* [9] (MFC) and was originally applied to n-type PbTe single crystals [53]. Young et al. has applied MFC to thin-film TCO samples [55], using a specially designed instrument [56] to measure the four coefficients on the same sample. It should be noted that the conductivity effective mass does appear in the conductivity transport equation and, thus, the validity of (3.68) is questionable. However, the ratio $Q/(|R_H|\sigma)$ that appears in (3.68) can be shown to depend on the DOS effective mass only, for cubic materials.

3.2.4 First-Order Non-parabolicity

In addition to providing the DOS effective mass and scattering parameter values, MFC may also be used to obtain m^*_{do}, the effective mass at the conduction-band minimum, and to measure the Fermi energy level relative to the conduction-band minimum. Equation (3.45) describes the energy function used for solving the transport equations. A first-order, non-parabolicity approximation may be defined by keeping only the first two terms in the energy expansion,

$$\gamma(E) = E + \frac{E^2}{E_1}. \tag{3.69}$$

Differentiating (3.69) with respect to E gives

$$\frac{d\gamma}{dE} = 1 + \frac{2E}{E_1}. \tag{3.70}$$

From the energy relationship of (3.45) and the momentum effective mass of (3.43), letting m^* equal $m^*_d$and E equal $\gamma(E)$,

$$m^*_d = p\frac{dp}{d\gamma(E)}, \tag{3.71}$$

we obtain,

$$m^*_d = m^*_{do}\frac{d\gamma(E)}{dE}. \tag{3.72}$$

Equating the squares of (3.70) and (3.72) and substituting $\gamma(E)$ in (3.69), we obtain

$$\left(\frac{m^*_d}{m^*_{do}}\right)^2 = \left(\frac{d\gamma}{dE}\right)^2 = 1 + \frac{4\gamma(E)}{E_1}. \tag{3.73}$$

The carrier concentration for energy surfaces described by (3.45) is given by (see (3.92))

$$n = \frac{8\pi}{3}\frac{\left(2m^*_{do}\right)^{3/2}}{h^3}\gamma(E)^{3/2}. \tag{3.74}$$

Substituting $\gamma(E)$ from (3.73) into (3.74) gives

$$\left(m_d^*\right)^2 = \left(m_{do}^*\right)^2 + 4\frac{m_{do}^*}{E_1}\left[\left(\frac{3}{8\pi}\right)^{2/3}\frac{h^2}{2}\right]n^{2/3}. \tag{3.75}$$

Equation (3.75) reveals that a plot of $\left(m_d^*\right)^2$versus $n^{2/3}$ gives a y-intercept value of $\left(m_{do}^*\right)^2$. Knowledge of m_{do}^* allows calculation of the Fermi level with respect to the conduction-band minimum:

$$E_{Fermi} = \left(\frac{3}{8\pi}\right)^{2/3}\frac{h^2}{2m_{do}^*}n^{2/3}\frac{1}{(1+\eta)}, \tag{3.76}$$

where

$$\eta = \left(\frac{m_d^*(\gamma)}{2m_{do}^*} - \frac{1}{2}\right). \tag{3.77}$$

$m_d^*(\gamma)$ is the DOS effective mass calculated from (3.68) at a given energy level or carrier concentration. Equation (3.76) was derived by substituting (3.69) into (3.74) while using (3.73).

3.2.5 Density-of-States Effective Mass

At this point in our discussion, we have mentioned two different effective masses: conductivity (optical) and density of states (MFC). It is instructive to distinguish these values to eliminate possible confusion during the presentation of the data. Consider constant energy surfaces in k-space similar to n-Si shown in Fig. 3.15. In this figure, the ratio of parallel to perpendicular effective-mass values, with respect to a principal axis in k-space, is 2:1; however, in n-Si, the ratio is 1:0.2. There are six equivalent ellipsoids of revolution symmetrically located around the gamma point of the first Brillouin zone. The ellipsoids are positioned away from the gamma point, indicative of an indirect-bandgap material if the valence band is centered at the gamma point. These surfaces represent available states within which electrons can move under an applied force. A conductivity tensor may be defined to account for possible anisotropy within the material:

$$\overleftrightarrow{\sigma} = \begin{pmatrix} \sigma_{xx} & \sigma_{xy} & \sigma_{xz} \\ \sigma_{yx} & \sigma_{yy} & \sigma_{yz} \\ \sigma_{zx} & \sigma_{zy} & \sigma_{zz} \end{pmatrix}. \tag{3.78}$$

For materials with similar cubic symmetry, the off-diagonal elements reduce to zero because the ellipsoids lie along the k-vector axes. In addition, for cubic

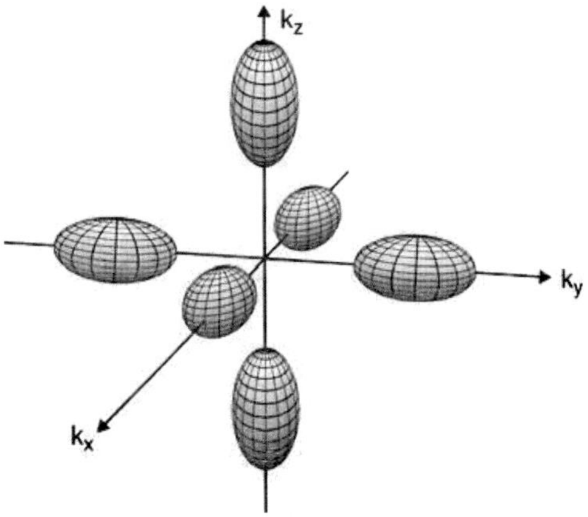

Fig. 3.15 Ellipsoidal Fermi surfaces for silicon

material, all of the diagonal elements are equivalent. These considerations reduce the matrix to

$$\begin{pmatrix} \sigma_{ii} & 0 & 0 \\ 0 & \sigma_{ii} & 0 \\ 0 & 0 & \sigma_{ii} \end{pmatrix} \text{ or } \sigma \begin{pmatrix} 1 & 0 & 0 \\ 0 & 1 & 0 \\ 0 & 0 & 1 \end{pmatrix}. \tag{3.79}$$

The contribution to the conductivity provided by a single ellipsoid can then be calculated by

$$\sigma = e^2 n_{ellip.} \frac{\langle \tau \rangle}{m_{ij}}, \tag{3.80}$$

where $n_{ellips.}$ is the carrier concentration within a single ellipsoid; $\langle \tau \rangle$ is the energy-dependent, average relaxation time; and m_{ij} is the effective mass of the electrons in a given direction in k-space. The total conductivity is the sum of the individual ellipsoid contributions. For the n-Si band, we have two ellipsoids with revolution axes parallel to the [100] direction, two in the [010] direction, and two in the [001] direction. Thus, the conductivity is given by

$$\sigma = \sum_{i=1}^{6} \sigma^i = e^2 n_i \langle \tau \rangle \left(\frac{2}{m_{||}} + \frac{4}{m_{\perp}} \right), \tag{3.81}$$

where i is the sum over the six ellipsoids and $m_{||}$ and $m_{\perp}$ are the effective masses parallel and perpendicular to the k_x direction, respectively. The conductivity reduces to

$$\sigma = e^2 n_{ellips.} \langle \tau \rangle 2 \left(\frac{1}{m_{||}} + \frac{2}{m_{\perp}} \right) \tag{3.82}$$

$$\sigma = \left(6n_{ellips.}\right)e^2\langle\tau\rangle\frac{1}{3}\left(\frac{1}{m_{\|}}+\frac{2}{m_{\perp}}\right). \tag{3.83}$$

The ($6n_{ellips.}$) term is the sum of the carrier concentrations over all six ellipsoids, which equals the total carrier concentration, n_{total}. Finally, the conductivity can be written:

$$\sigma = \frac{e^2 n_{total}\langle\tau\rangle}{m_c^*}, \tag{3.84}$$

where

$$\frac{1}{m_c^*} = \frac{1}{3}\left(\frac{1}{m_{\|}}+\frac{2}{m_{\perp}}\right) \tag{3.85}$$

is the conductivity effective mass. From the above derivation, we see that the conductivity effective mass is an average over all of the effective-mass values in different crystallographic directions.

Now consider the number of states (electrons) available in a filled sphere in k-space for an isotropic band. The total number of states, n, within a spherical volume of radius k_f, is

$$n = \frac{k_f^3}{3\pi^2}, \tag{3.86}$$

with

$$\frac{dn}{dk} = \frac{3n^{2/3}}{\left(3\pi^2\right)^{1/3}}. \tag{3.87}$$

For electron energies given by

$$\frac{\hbar^2 k^2}{2m_o^*} = \gamma(E), \tag{3.88}$$

with

$$\frac{d\gamma}{dk} = \frac{\hbar^2 k}{m_o^*}, \tag{3.89}$$

we can solve for n using the chain rule:

$$\frac{dn}{dk} = \frac{dn}{d\gamma}\frac{d\gamma}{dk}. \tag{3.90}$$

Substituting (3.87) and (3.89) into (3.90) and eliminating k by (3.88), we get

$$\frac{dn}{n^{2/3}} = \frac{3}{(3\pi^2)^{1/3}} \frac{m_o^{*1/2}}{2^{1/2}\hbar} \gamma^{-1/2} d\gamma. \tag{3.91}$$

After integration, we obtain:

$$n = \frac{8\pi}{3} \frac{(2m_o^*)^{3/2}}{h^3} \gamma^{3/2}, \tag{3.92}$$

which is the carrier concentration for a spherical energy surface. For N equivalent energy surfaces, the total carrier concentration is $N \cdot n$. Therefore, the total carrier concentration is

$$n_{total} = N \cdot n = N \frac{8\pi}{3} \frac{(2m_o)^{3/2}}{h^3} \gamma^{3/2}. \tag{3.93}$$

Moving N inside the parentheses gives:

$$n_{total} = \frac{8\pi}{3} \frac{(N^{2/3} 2m_o^*)^{3/2}}{h^3} \gamma^{3/2}. \tag{3.94}$$

The quantity $N^{2/3} m_o^*$ was first defined as the density-of-states effective mass by Kolodziejczak and Zukotynski [45] and given the symbol m_{do}^*. Therefore,

$$n_{total} = \frac{8\pi}{3} \frac{(2m_{do}^*)^{3/2}}{h^3} \gamma^{3/2}. \tag{3.95}$$

If the constant energy surface is instead an ellipsoid, then the energy is described by

$$\gamma(E) = \frac{\hbar^2}{2} \left(\frac{k_1^2}{m_{1o}^*} + \frac{k_2^2}{m_{2o}^*} + \frac{k_3^2}{m_{2o}^*} \right), \tag{3.96}$$

where k_i is the k-vector along principal axis i and m_{io}^* is the effective mass parallel to principal axis i. The number of states within a single ellipsoid can be shown to be equal to [9]

$$n = \frac{8\pi 2^{3/2}}{3} \frac{(m_{1o}^* m_{2o}^* m_{3o}^*)^{1/2}}{h^3} \gamma(E)^{3/2}. \tag{3.97}$$

Summing up the contribution to the number of states from N ellipsoids gives

$$n_{total} = \frac{8\pi 2^{3/2}}{3} \frac{N (m_{1o}^* m_{2o}^* m_{3o}^*)^{1/2}}{h^3} \gamma(E)^{3/2}, \tag{3.98}$$

or

$$n_{total} = \frac{8\pi 2^{3/2}}{3}\frac{\left(m_{do}^{*}\right)^{3/2}}{h^{3}}\gamma(E)^{3/2}, \tag{3.99}$$

where

$$m_{do}^{*} = N^{2/3}\left(m_{1o}^{*}m_{2o}^{*}m_{3o}^{*}\right)^{1/3}. \tag{3.100}$$

Equation (3.100) reveals the averaging nature of the DOS term and the accounting for the number of ellipsoids.

Further connection between the conductivity and DOS effective masses may be made by considering a cubic material with N equivalent ellipsoids of revolution. For cubic materials, like that shown in Fig. 3.15, $m_{2o}^{*} = m_{3o}^{*}$, which reduces (3.100) to

$$m_{d}^{*} = N^{2/3}\left(m_{\|}m_{\perp}^{2}\right)^{1/3}. \tag{3.101}$$

Recalling the conductivity effective-mass definition from (3.85), we see that it applies to our N ellipsoids of revolution material provided that all ellipsoid effective masses may be folded along a principal axis. The conductivity effective mass is given by

$$\frac{1}{m_{c}^{*}} = \frac{1}{3}\left(\frac{1}{m_{\|}} + \frac{2}{m_{\perp}}\right). \tag{3.102}$$

Multiplying (3.101) by (3.102) gives

$$\frac{m_{d}^{*}}{m_{c}^{*}} = N^{2/3}\frac{(2\beta + 1)}{3\beta^{2/3}}. \tag{3.103}$$

where $\beta = \frac{m_{\|}}{m_{\perp}}$ is the ratio of the parallel and perpendicular effective masses with respect to a principal axis in k-space. Equation (3.103) makes the final connection between the DOS and the conductivity effective masses for cubic materials. Knowledge of both quantities reveals the nature of the Fermi surfaces in terms of the number of equivalent energy surfaces, N, and the shape of the ellipsoids, β.

3.2.6 Sample Configuration

Now that we have outlined the basic mathematical concepts behind MFC, we will briefly discuss the experimental techniques involved with measuring transport properties in TCO thin films.

Measurement of the four transport coefficients on thin films has been attempted by other groups [57] and emphasizes the need for a clever pattern in the film to accomplish all of the measurements on the same sample. Young et al. designed an instrument [56] for this purpose that was used for performing MFC on several TCO films. TCOs films are deposited on electrically insulating substrates (glass) and then patterned using standard photolithographic techniques. Photolithographic masks form the shape of the pattern shown in Fig. 3.16. Etching TCO films can be challenging. Tin oxide and ITO are especially resistive to chemical etching and often require reactive-ion etching to form a clearly defined pattern. The pattern used by Young et al. has four contact pads, marked 1–4, with $a<<W$, $b<<L$, and $W<<L$. The first two dimensions provide small-area contacts critical for the van der Pauw technique, whereas the latter dimension ensures a large transverse Nernst signal by maximizing the temperature gradient between contacts 2 and 4 and the Nernst voltage between contacts 1 and 3. This pattern conforms to the specimen shape specified in the ASTM designation F 76-86 [58] for van der Pauw resistivity and Hall measurements. The thickness of the film may be measured by stylus profilometry, ellipsometry, or cross-sectional scanning electron microscopy (SEM). Samples are mounted film-side down across heater blocks A and B, as shown in Fig. 3.17. Good electrical and thermal contact to the heater blocks was made to contact pads 2 and 4 with small pieces of indium metal. A small amount of thermal Apiezon™ N grease is placed between the substrate and the heater blocks along the sample's length to improve thermal contact and temperature uniformity. This grease is easily removed by hexane. Contact pads 1 and 3 are indium-soldered to copper wire 0.0076 cm (0.003 in.) in diameter, which minimizes conductive heat loss from the sample through contacts 1 and 3. These wires are then indium-soldered

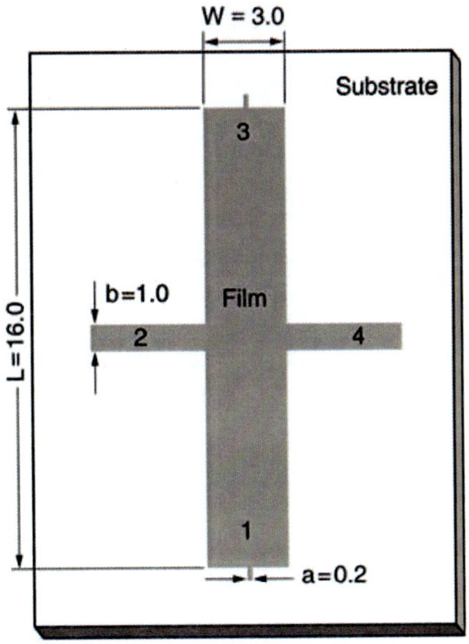

Fig. 3.16 Thin-film pattern on substrate for method of four coefficients technique. All dimensions are in mm

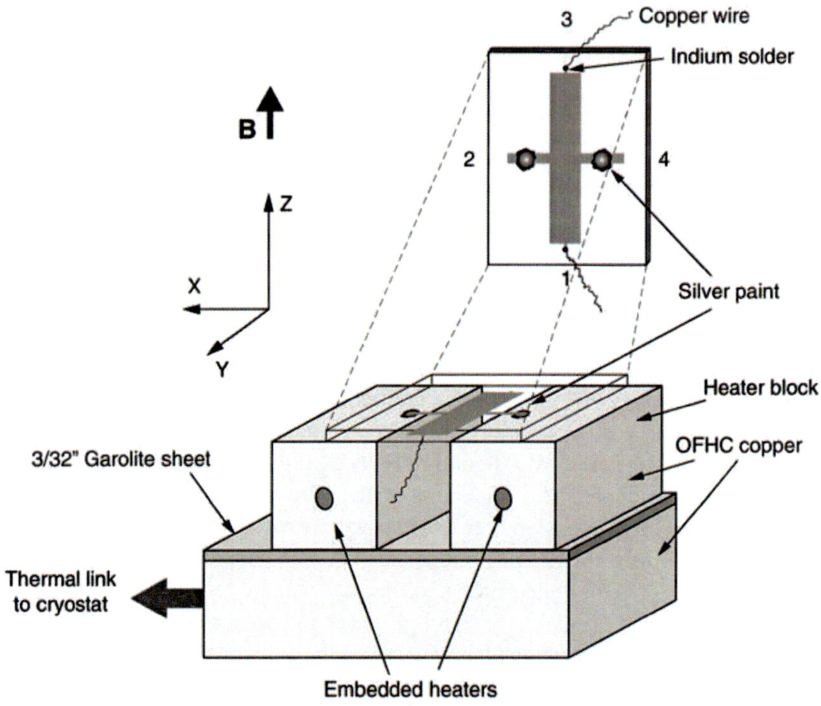

Fig. 3.17 Depiction of sample placement on a sample holder for the method of four coefficients. In the figure, OFHC signifies "oxygen-free, high conductivity"

to copper binding posts located near contacts 1 and 3, which provide continuity to the system's electronics.

With both heater blocks at the same temperature, measurements of the van der Pauw resistivity begin by first zeroing the offset voltages due to stray thermovoltages in the system. Next, contacts 1 and 2 are biased, while the van der Pauw voltage is measured between contacts 3 and 4. This sequence is repeated for reversed current and for the remaining three configurations of contact pairs, as per ASTM designation F 76-86.

The Hall coefficient is measured by first turning on the magnetic field perpendicular to the sample plane (Fig. 3.17). The induction and direction of the magnetic field are measured by a calibrated Hall generator mounted between the poles of the magnet. After a reasonable amount of settling time for the field, contacts 1 and 3 are biased, and the Hall voltage is measured between contacts 2 and 4. Again, this sequence is repeated for reversed current and reversed magnetic field for the remaining configuration of contact pairs, as per ASTM designation F 76-86. Thermal and contact misalignment offset voltages are zeroed, while maintaining isothermal conditions in the sample minimizes errors due to the Ettingshausen effect. Recall that the Ettingshausen effect is a temperature gradient that arises in a magnetic field due to the thermal distribution in the electron velocities. For

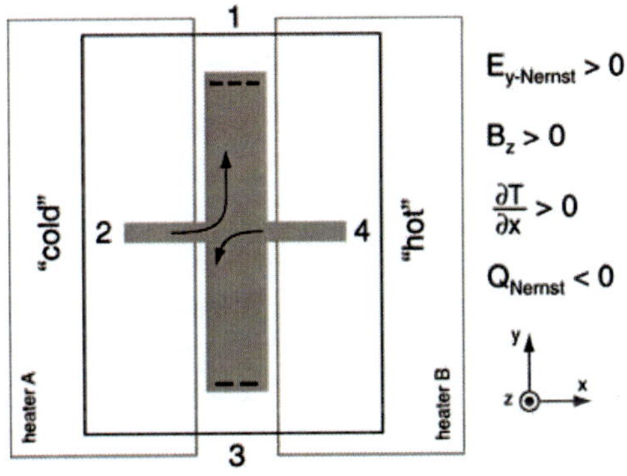

Fig. 3.18 Depiction of electron deflection during the Ettingshausen effect

instance, if electrons flow from contacts 4 to 2 (Fig. 3.18) with a *z*-directed magnetic field, the magnetic field will divert the "hot" (fast) electrons less than the "cold" (slow) electrons [34]. This condition establishes a temperature gradient directed from contact 3 to 1. The temperature gradient will then generate a Seebeck electric field from contact 1 to 3 (*n*-type materials). Note that this Ettingshausen-induced field is superimposed on the Hall field and cannot be subtracted by either reversal of the current or the magnetic field. Thus, isothermal conditions must be maintained within the sample to counteract the Ettingshausen effect. Fortunately, the relatively thick substrate acts as a large thermal mass for the film and hence maintains good isothermal conditions across the film. Similarly, this same substrate thermal mass also acts to decrease the magnitude of the Peltier effect during both the resistivity and the Hall measurements.

Thermoelectric (Seebeck) and thermomagnetic (Nernst) effects are measured with a temperature gradient $\nabla_x T$ across the film by turning on one of the heaters. For the Seebeck coefficient, the instrument turns off the magnetic field and measures the temperature difference between the heater blocks ($\Delta T_{2,4}$) with the differential thermocouple, and the voltage difference ($\Delta V_{2,4}$) between contacts 2 and 4. The Seebeck coefficient can be approximated by

$$\alpha = \frac{\partial V_{2,4}}{\partial T_{2,4}} \xrightarrow[\text{limit } \partial T \to 0]{} \frac{\Delta V_{2,4}}{\Delta T_{2,4}}. \tag{3.104}$$

To eliminate offset voltage errors, $\Delta V_{2,4}$ is measured for several values of $\Delta T_{2,4}$, and the slope of the best-fit line of $\Delta V_{2,4}$ versus $\Delta T_{2,4}$ gives the Seebeck coefficient of the film and the sample-holder couple for a film temperature equal to the average temperature of the two heater blocks. Four $\Delta T_{2,4}$ values are measured: two with $T_2 > T_4$and two with $T_2 < T_4$. The average temperature is then very close to the temperature for which (3.32) is true: $\Delta T_{2,4} \approx 0$.

The absolute Seebeck coefficient for a material is determined with respect to another, known material. When two wires of different material are joined (spot welded) and a temperature difference exists between their junction and the opposite ends of the wires, then a voltage difference between the loose ends will be established. The sum of the voltages in this circuit would equal

$$\alpha_{A,B} = \alpha_A - \alpha_B, \tag{3.105}$$

where the ΔT was eliminated by (3.104). Equation (3.105) indicates that to measure the absolute Seebeck coefficient of a material, one must know the Seebeck coefficient of one material or use a superconducting material that has a Seebeck coefficient of zero. For our system, material B represents the sample-holder contribution to the couple consisting of indium and copper wire. Material A is the absolute Seebeck coefficient for the film in question. Calibration of the total system is required for absolute Seebeck coefficient measurements.

The Nernst coefficient is measured by turning on the magnetic field perpendicular to the plane of the film while maintaining the temperature gradient $\partial T/\partial x$ across the film. The magnetic field and the temperature gradient establish an isothermal, transverse Nernst voltage between contacts 1 and 3. The Nernst field is analogous to the Hall field, except that the charge carriers move due to a temperature gradient, rather than by an applied bias. The isothermal, transverse Nernst coefficient, Q, can be measured according to

$$E_y = -QB_z \frac{\partial T}{\partial x}. \tag{3.106}$$

The quantity $\partial T/\partial x \approx \Delta T/\Delta x$ is valid provided that the complete temperature drop occurs between the two heater blocks and that no temperature drop occurs at the film/Cu contact. This is a valid assumption for the very uniform, low thermal conductivity of glass substrates and excellent thermal conductivity of indium and thermal grease.

Figure 3.18 depicts the generation of a Nernst voltage for a material with a negative Nernst coefficient [34]. By reversing the magnetic field, re-measuring the Nernst voltage, and averaging the difference of the two values, we eliminate offset thermoelectric voltages in the system. Because Nernst voltages tend to be in the nanovolt range for low-mobility, high-degeneracy films, it is advantageous to measure the Nernst coefficient by extracting the slope from plots of $\Delta V_{1,3}$ versus $\Delta T_{2,4}$ or $\Delta V_{1,3}$ versus B_z.

During the transverse Nernst voltage measurement, as in the Hall measurements, one must be careful of multiple transport phenomena occurring simultaneously [59]. Note that because of the temperature gradient and the magnetic field, "hot" and "cold" diffusing carriers are deflected perpendicularly to the temperature gradient in the plane of the film (see Fig. 3.18). For an n-type material, when heater B is warmer than heater A and the magnetic field is in the z-direction, contact 3 becomes warmer than contact 1. This phenomenon is known as the Righi-Leduc effect [34]. The temperature difference will cause a Seebeck voltage to develop between contacts 3 and 1. Analogous to the interference of the Hall effect by the Ettingshausen effect, this unwanted voltage due to the Righi-Leduc effect is superimposed and inseparable from the Nernst voltage. However, the large thermal mass of the substrate and good thermal contact between the substrate and the heater blocks again minimize the unwanted effect. These conditions, and the fact that the natural temperature gradient of the cryostat is parallel to the short length of the sample, should make the temperature gradient between contacts 1 and 3 negligible.

3.2.7 MFC Results on TCOs

3.2.7.1 ZnO (Ionized and Neutral Impurity Scattering)

The first example of MFC data on TCOs will be from radiofrequency (rf) magnetron-sputtered, undoped zinc oxide, and zinc oxide doped with aluminum thin films. These data were previously published in 2000 [55], and they are included in this chapter as an example of the MFC technique and of ionized and neutral impurity scattering in TCO films. After patterning to the MFC shape, film thickness was measured by profilometry. The four relevant transport-phenomena coefficients (conductivity, Hall, Seebeck, and Nernst) were measured by the method and by the instrument described above, and they were then used in (3.68) to calculate the DOS effective mass, m_d^*. Figure 3.19 shows the m_d^* values plotted against the carrier density, n. A best-fit line was drawn through the points, which was then used to calculate λ in (3.64). We first note from Fig. 3.19 that m_d^* is not constant with carrier density and varies from about 0.3 to 0.45 m_e as the carrier density changes from 2×10^{19} to 5×10^{20} cm^{-3}. This trend shows that the conduction energy band for ZnO is non-parabolic in this carrier-density range and will make the energy-band term, λ, non-zero in (3.64). In addition to our m_d^* values for ZnO plotted in Fig. 3.19, conductivity effective-mass values are plotted, as well. These values were taken from the literature [60, 61] on conductivity effective masses calculated from plasma frequency. Note that the conductivity and DOS effective mass are in good agreement with each other, but that the error bars on the DOS effective mass are much smaller. The relationship between the two effective-mass values is given by (3.103), for an isotropic, cubic material and N is the number of equivalent ellipsoids of revolution. Analysis of our data requires that N be one and that the effective-mass values $m_{||}$ and $m_{\perp}$, with respect to crystallographic directions, be approximately equal. This being the case, we may conclude that ZnO has a single-valley minimum in the conduction energy band and that the constant-energy surface must be spherical to within experimental uncertainty. The single-valley minimum of ZnO is predicted by theory [62], but the uncertainty in the plasma-frequency data makes it difficult to be certain of the spherical nature of the energy surface. We note that ZnO is hexagonal and that the films in this study have a (200) orientation. Anisotropy in the effective mass is not found in the literature, nor is it observed in the mobility above 100 K in ZnO [28].

With m_d^* known as a function of n, we solve for λ to determine the degree to which the band shape affects the scattering parameter, s, in (3.67). Figure 3.20a

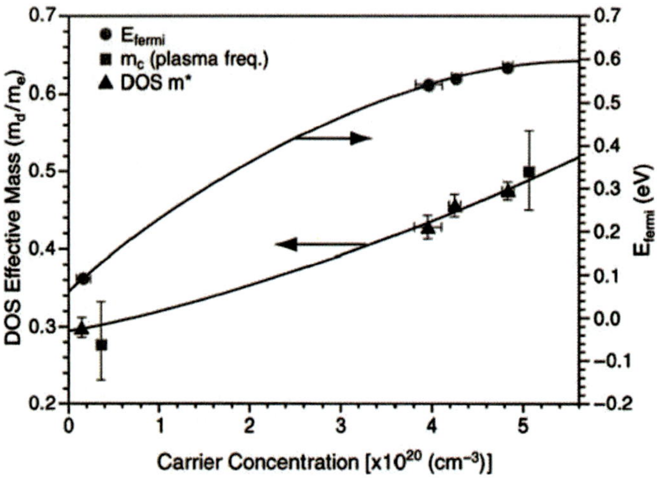

Fig. 3.19 Density-of-states effective mass and Fermi energy level with respect to the conduction-band minimum, and plasma-frequency conductivity effective mass, m_c, versus carrier concentration for ZnO and ZnO:Al

shows the calculated scattering-parameter values from (3.67) for our ZnO samples, along with the predicted trends in the scattering parameter for three different scattering mechanisms. For the ZnO:Al samples, the measured scattering parameter matches well with the trend expected for ionized impurities. This result is expected for these films, where aluminum is added to dope the films *n*-type by contributing an electron to the conduction band, leaving behind an ionized aluminum atom. For the undoped material, the scattering parameter is found to lie most closely with the neutral-impurity trend. This has also been observed by Yang et al. who used Hall effect measurements [63]. The observation raises the interesting issue of the nature of the neutral impurities. Lany and Zunger addressed this issue and concluded that oxygen vacancies are energetically likely to be created in equilibrium conditions but are unlikely to be ionized. Hence, oxygen vacancies may be the source of neutral impurity scattering [64]. For films grown in non-equilibrium conditions, a metastable state of the oxygen vacancy may be ionized and, thus, the source of free carriers. The ratio of these two species may therefore define the nature of the scattering mechanism. To further support our claim that neutral and ionized impurity scattering are the main scattering mechanisms in ZnO and ZnO:Al, respectively, Fig. 3.20b shows our Seebeck measurements plotted against expected Seebeck values (3.62) for three scattering mechanisms. Again, ionized and neutral impurity scattering are the apparent mechanisms.

Another way to check for the type of scattering mechanism occurring in films is to study the mobility as a function of temperature. Mobility varies uniquely as a function of temperature according to the type of scattering mechanism dominant in the films. Table 3.3 [42, 43] shows the mobility versus temperature relationships for several scattering mechanisms.

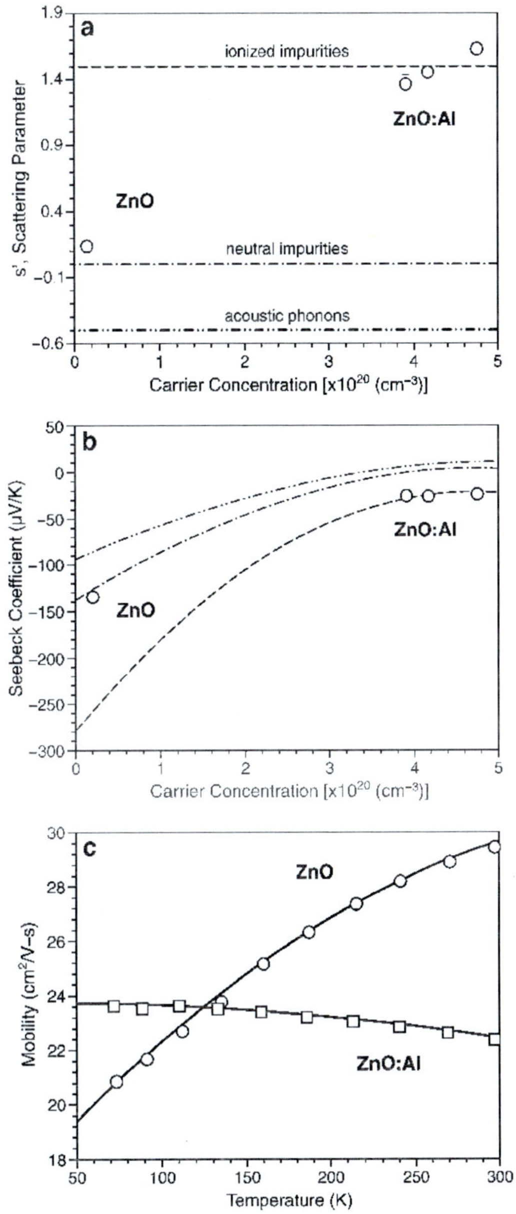

Fig. 3.20 Scattering parameter of ZnO films as a function of carrier concentration ZnO:Al films show ionized impurity scattering, whereas the ZnO films show neutral impurity scattering. (**a**) Scattering parameters, s' versus n. Data points are plotted, along with theoretical values of s', for three different scattering mechanisms. (**b**) Seebeck coefficient versus n. Data points are plotted, along with theoretical curves, for three different scattering mechanisms. (**c**) Mobility, μ versus n

A graph of $ln(\mu)$versus $ln(T)$reveals the temperature power law for mobility and temperature ($\mu \propto T^x$). Mobility versus temperature data for both a ZnO and ZnO:Al film are shown in Fig. 3.20c. The undoped ZnO film shows a clear positive dependence of mobility on temperature, indicative of neutral impurities. As the temperature decreases, the number of neutral impurity centers will increase,

Table 3.3 Mobility vs. temperature relationships for several scattering mechanisms

Scattering mechanism	$\mu(T)\ \alpha$
Point defects	$T^{-1/2}$
Acoustic phonons	$T^{-3/2}$
Optical phonons	T^{-x} $(2.5 < x < 1.5)$
Piezoacoustic phonons	$T^{-1/2}$
Neutral impurities	T^{x} $(0 < x < 1)$
Ionized impurities	T^{x} $(0 < x < 3/2)$ degen. nondegen.

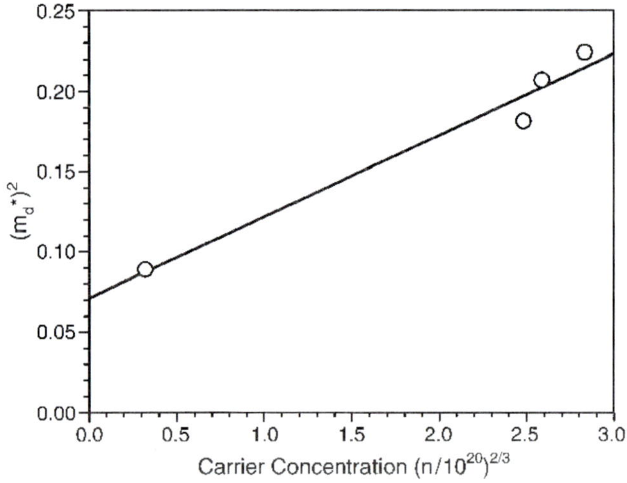

Fig. 3.21 Density-of-states m^{*2} versus $n^{2/3}$. The linear fit of the data justifies our first-order, non-parabolicity approximation. The y-intercept gives the square of the effective mass at the bottom of the conduction band

decreasing the relaxation time between collisions, and thus decreasing the mobility [43]. For the doped material, the mobility is seen to be almost constant with temperature, characteristic of ionized impurity scattering for degenerate materials $(x \sim 0)$ [42].

In addition to providing the DOS effective mass and scattering parameter values, MFC may also be used to extrapolate m^*_{do}, the effective mass at the conduction-band minimum, and to measure the Fermi energy level relative to the conduction band.

Equation (3.75) shows that the y-intercept of a plot of $(m^*_d)^2$ versus $n^{2/3}$ will reveal m^*_{do}, and the slope will define the constant, E_1. Figure 3.21 shows a plot of $(m^*_d)^2$ versus $n^{2/3}$ with an m^*_{d0} value of 0.27 m_e. Note that the plot is linear, indicating that our first-order, non-parabolicity assumption is valid. In addition, the Fermi energy level, with respect to the bottom of the conduction band, can be calculated from (3.76). Figure 3.19 shows the calculated Fermi energy level above the conduction band versus the carrier concentration, revealing a strong degeneracy in the doped samples ($E_{Fermi} - E_{conduction} > 0.5$ eV), but only a slight degeneracy in the undoped sample.

3.2.7.2 Epitaxial ITO and CTO Polycrystalline Films

It is obvious from the depth of the analysis of the transport data on ZnO films that MFC is a powerful technique for understanding fundamental properties of TCO films. We now move to a class of TCO material systems that have exhibited relatively high mobilities: epitaxially grown indium tin oxide (epi-ITO) [65] and cadmium stannate (CTO) [56]. Both materials have nearly perfect intragrain atomic lattice growth, which makes them almost electrically equivalent to single-crystal samples. They also both exhibit phonon-like carrier scattering, which might be expected in single crystals, but is quite rare in thin-film TCOs. This section will show how MFC was used to measure the effective mass and scattering mechanism in CTO and epi-ITO.

Epitaxial ITO

Epitaxial films represent an upper limit of possible electrical quality in TCO films due to their nearly perfect lattice growth and lack of grain boundaries (other than at the surface and substrate interface). Studies were made on epitaxially grown thin films of ITO on YSZ substrates developed at the Tokyo Institute of Technology [65]. The mobilities in these films were as high as 77 $cm^2 V^{-1} s^{-1}$ – nearly twice as high as commercially available TCO films – which attests to the quality of the epitaxy. A series of four films with varying carrier concentrations were measured by MFC so that the effective mass and non-parabolicity of the conduction band were determined. Correcting for the non-parabolicity of the conduction band, the calculated scattering parameter, s', indicated that the carriers are being scattering by optical phonons. To confirm this unusual scattering parameter in thin films, the Seebeck coefficient versus carrier concentration for the epi-ITO films is plotted in Fig. 3.22. The experimental Seebeck data follow quite well the theoretical trend (see Sect. 3.2.7.1 for details of this technique) of optical phonon scattering. Furthermore, mobility versus temperature data increase steadily with decreasing temperature, as shown in Fig. 3.23. This phonon-like trend is added evidence that optical phonon scattering is the limiting scattering mechanism in these epitaxially grown ITO films.

The fact that phonon-like scattering has been observed in epitaxially grown TCO thin films is significant, especially considering that these films are intentionally doped. This intrinsically limiting scattering mechanism would be the ideal for all TCOs and has only been observed in a small number of TCO materials systems [66]. We highlight one such system below: cadmium stannate.

Cadmium Stannate

CTO, Cd_2SnO_4, is grown by rf sputtering onto glass substrates at room temperature and then annealed in a CdS atmosphere to both crystallize the as-deposited

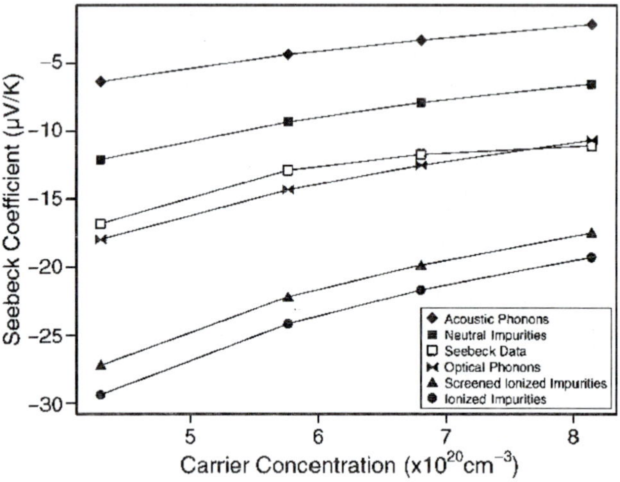

Fig. 3.22 Seebeck coefficient data as a function of carrier concentration for epitaxially grown films of indium oxide doped with tin (ITO). Theoretical Seebeck coefficients for several scattering mechanisms are plotted, as well. The ITO data most closely follow optical phonon scattering

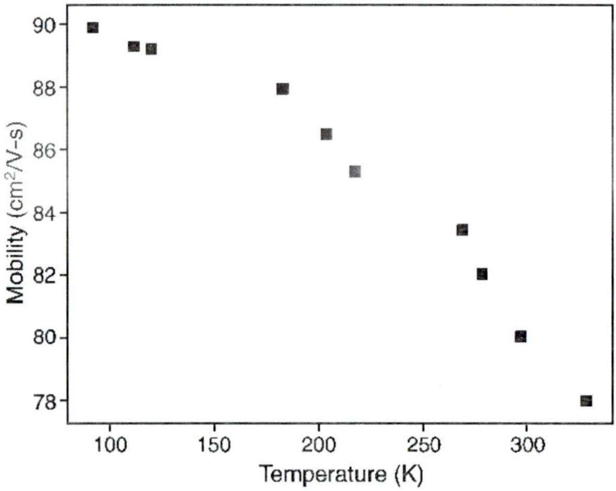

Fig. 3.23 Mobility as a function of temperature for epitaxially grown ITO films. The decrease in mobility with increasing temperature is indicative of phonon scattering

amorphous film and to promote grain growth [67]. A large series of CTO films were grown with varying carrier concentrations for this study. Intrinsic doping in CTO is controlled by growth and annealing conditions; thus, no extrinsic dopants were incorporated into the films. Figure 3.24 shows the effective-mass values calculated by MFC technique versus carrier concentration. Note the positive slope in the effective-mass data, which indicates a slight non-parabolicity of the conduction

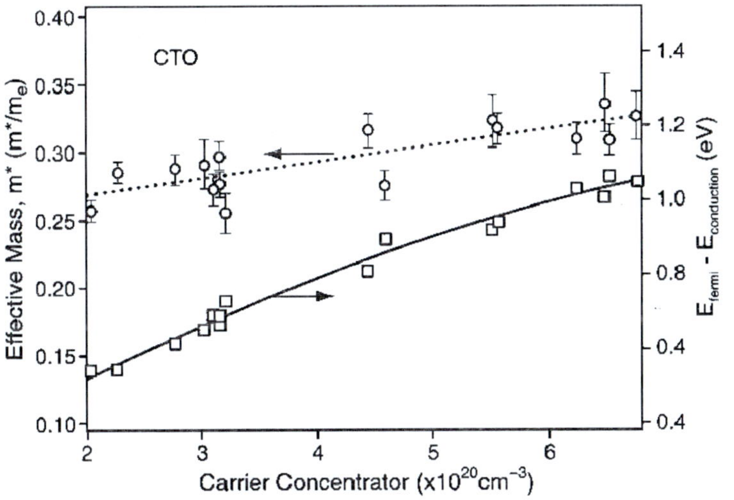

Fig. 3.24 Polycrystalline cadmium stannate (CTO) thin films. *Left axis:* DOS effective mass as a function of carrier concentration. *Right axis:* Fermi energy with respect to the bottom of the conduction band as a function of carrier concentration

band within CTO. As with ZnO, the non-parabolicity term, λ, (3.64) will factor into the complete calculation in the MFC formalism. m_o^*, the effective mass at the bottom of the conduction band, is calculated to be 0.21 m_e. This relatively small effective mass may account for the very large mobility values seen in this material system [67]. The right-hand side of Fig. 3.24 shows the Fermi-energy position above the conduction-band minimum. Even the CTO films with lowest doping concentration are 0.5-eV degenerate. These high Fermi-energy levels are one reason for the absence of grain-boundary scattering in these films.

In an attempt to categorize the main scattering mechanism in CTO, we plot the Seebeck coefficient versus carrier concentration in Fig. 3.25. Along with the experimental data, we show theoretical curves of the Seebeck and Nernst coefficient for different scattering mechanisms. Both sets of data for CTO lie along the theoretical curves for optical phonon scattering, especially for films with high carrier concentration. To confirm our conclusions, we plot mobility versus temperature for two CTO films in Fig. 3.26. Mobility in CTO increases with decreasing temperature, indicative of phonon scattering. As the temperature drops, the lattice phonon density decreases, which allows electrons to move more easily through the lattice. These data are the first indication that a phonon-like scattering mechanism could occur in a polycrystalline thin-film TCO. The assumption has always been that thin-film polycrystalline materials were too defective to ever be limited by phonon scattering. These results for CTO may mean that further increases in mobility are unlikely given the intrinsic nature of phonons. Indeed, these results are remarkable.

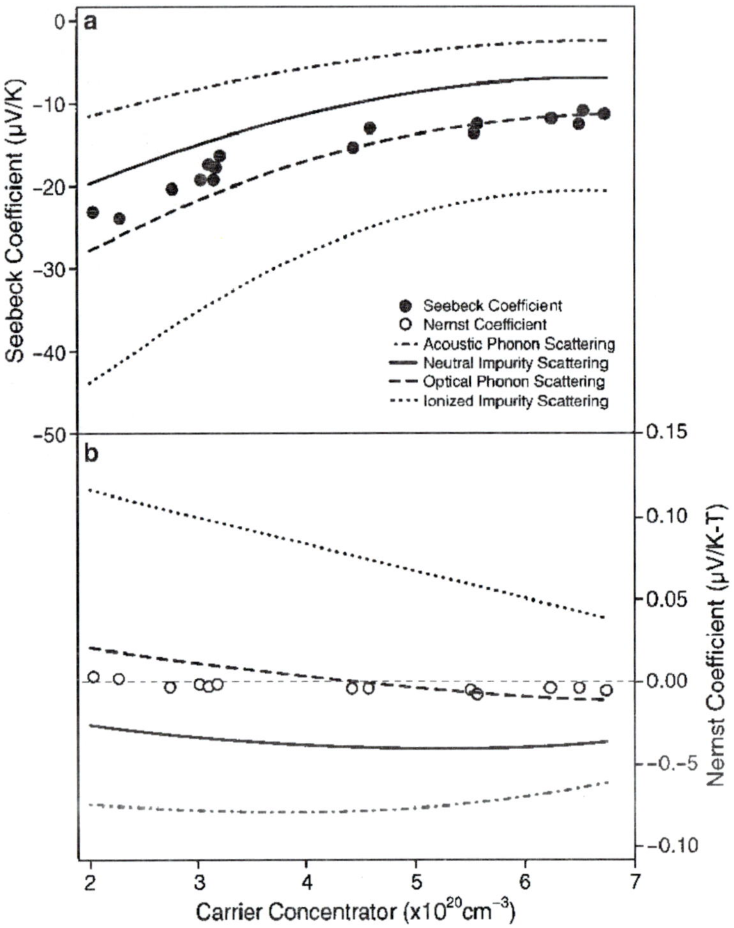

Fig. 3.25 Seebeck and Nernst coefficients for CTO thin films as a function of carrier concentration. The *solid lines* represent the theoretical Seebeck and Nernst variations for different scattering mechanisms. (**a**) Variation of the Seebeck coefficient with carrier concentration. (**b**) Variation of the Nernst coefficient with carrier concentration. The CTO data most closely follow the optical phonon curves in both cases

3.3 Improving TCO Figure of Merit Through Improved Optical Performance

The previous discussions suggest two pathways to increase the figure of merit for TCO materials. The first focuses on improving the electrical properties – specifically, the mobility – of a TCO film without reducing the optical transparency. Research and industrial groups have pursued this pathway extensively [68]. The second pathway, pursued less often, is to improve the optical properties while

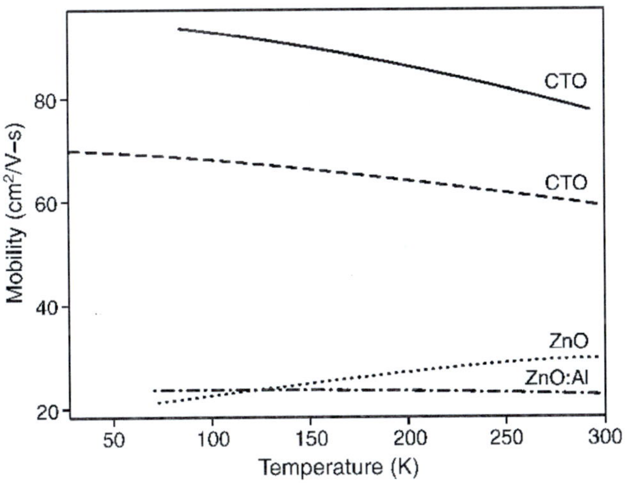

Fig. 3.26 Mobility as a function of temperature for CTO and ZnO thin films

maintaining or improving the electrical properties. Although TCO technologists strive to produce the highest-transparency films possible, the traditional approach is complicated because film transparency and electrical quality are both functions of oxygen availability. Sufficient oxygen must be available during growth or post-deposition treatment to limit formation of reduced or metallic phases, while controlled oxygen deficiency is used to produce high carrier concentration through oxygen-vacancy-related donor defects [69]. In a large-area production deposition system, the correct amount of oxygen can be very difficult to control, especially because oxygen is incorporated from several sources – including the target, process gas, chamber, physisorbed surface species, and unintentional leaks. Further, the amount of oxygen incorporated into the film is affected by deposition parameters (e.g., substrate temperature, substrate bias) that are sometimes used to control other film requirements (e.g., density, wear resistance, and/or adhesion) (Figs. 3.27 and 3.28).

Recently, our studies at the National Renewable Energy Laboratory (NREL) have led us to an alternative pathway to enhance the optical properties of TCO films [70]. In this method, we exploit the fact that the optical absorptance for a TCO film is a function not only of its oxygen content, but also, of material parameters embodied by the real part of the dielectric permittivity (ε_1), as indicated in the figure of merit defined by (3.28). The importance of the high-frequency permittivity (ε_∞) is its impact on ε_1, as shown in (3.17).

These relationships suggest two things: First, if ε_∞ can be increased while maintaining constant values of n and m^*, the FOM should be improved. Second, because increasing n will reduce ε_1, improved TCO optical properties due to increasing ε_∞ may be more readily observed for films with lower carrier concentration.

Using the formalisms described earlier in this chapter, we have mathematically modeled the effect of varying ε_∞ over the range 4–7 on the optical performance of

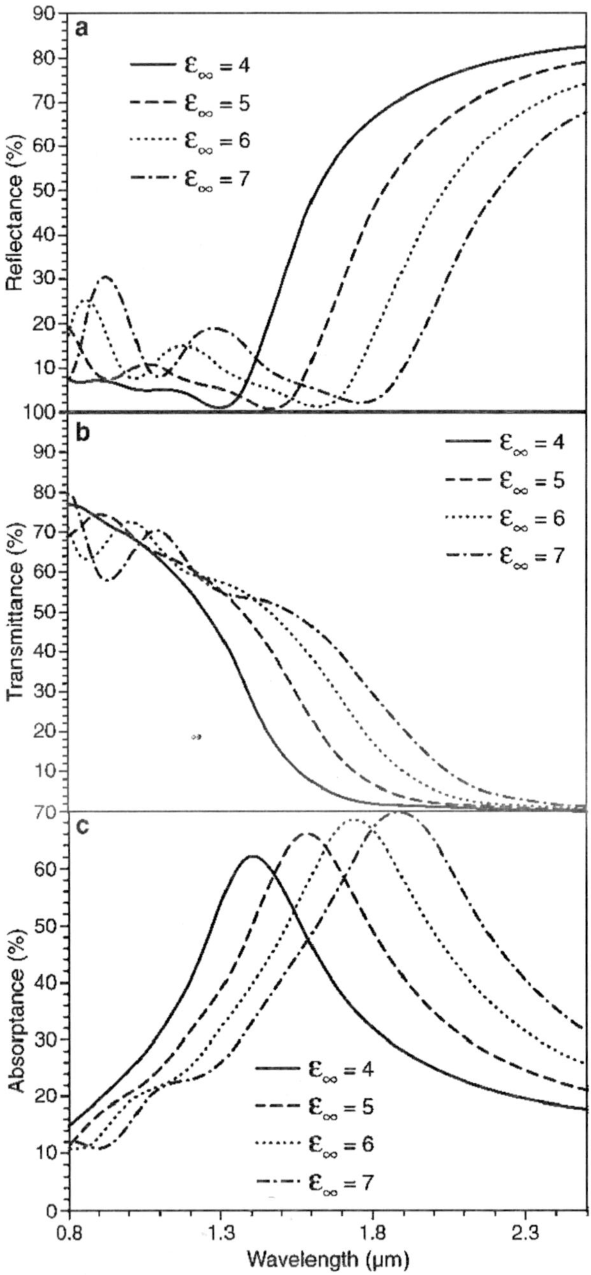

Fig. 3.27 Mathematically modeled effect of varying ε_∞ over the range 4–7 on the optical properties of a typical TCO. TCO film parameters include $n = 7 \times 10^{20}$ cm^{-3}, $\mu = 35$ cm^2 V^{-1} s^{-1}, $m^* = 0.35\ m_e$, high-frequency permittivity = 4, and film thickness = 500 nm. (**a**) Reflectance versus wavelength. (**b**) Transmittance versus wavelength. (**c**) Absorptance versus wavelength

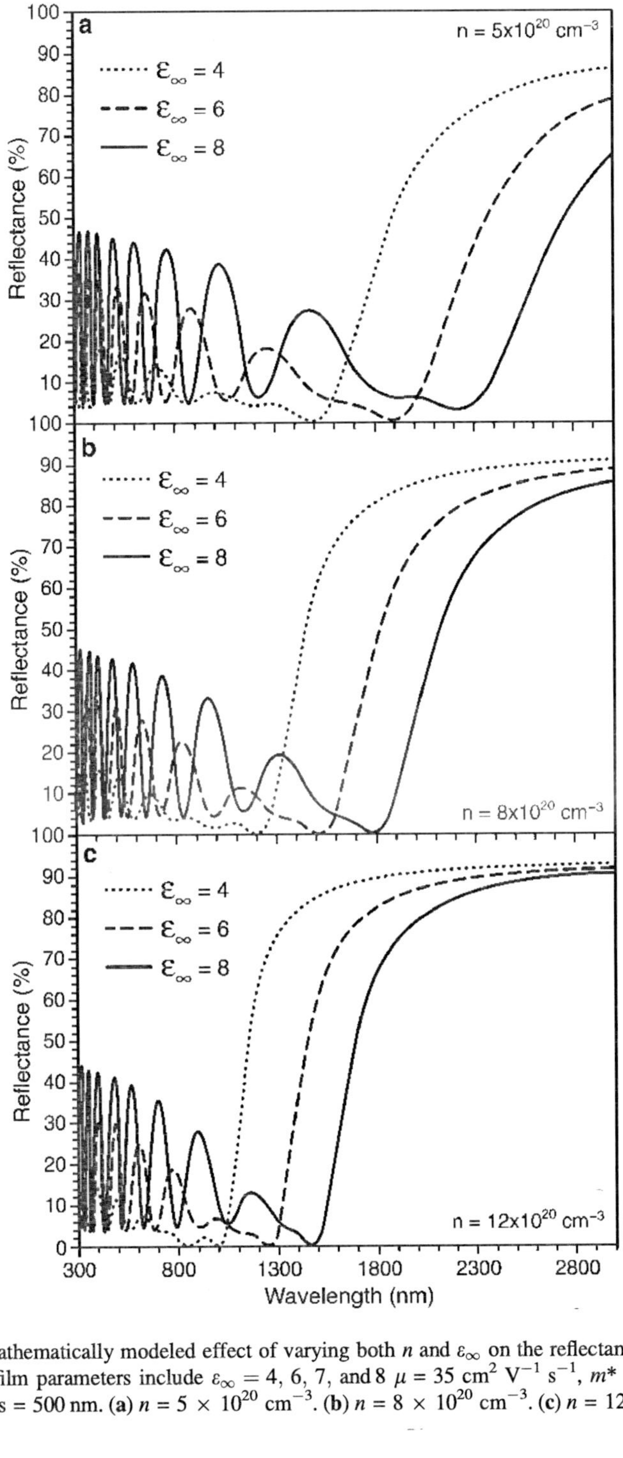

Fig. 3.28 Mathematically modeled effect of varying both n and ε_∞ on the reflectance of a typical TCO. TCO film parameters include $\varepsilon_\infty = 4$, 6, 7, and 8 $\mu = 35\ \text{cm}^2\ \text{V}^{-1}\ \text{s}^{-1}$, $m^* = 0.3\ m_e$, and film thickness = 500 nm. (**a**) $n = 5 \times 10^{20}\ \text{cm}^{-3}$. (**b**) $n = 8 \times 10^{20}\ \text{cm}^{-3}$. (**c**) $n = 12 \times 10^{20}\ \text{cm}^{-3}$

a typical TCO. As will be discussed, we believe it is possible to control this range of ε_∞ in some TCO films. Other modeled TCO film parameters include $n = 7 \times 10^{20}$ cm^{-3}, $\mu = 35\ \text{cm}^2\ \text{V}^{-1}\ \text{s}^{-1}$, $m^* = 0.3\ m_e$, and 500-nm film thickness. Results of these modeled predictions are shown in Figs. 3.29–3.31, and they indicate that this degree of variation of ε_∞ has a significant effect on the NIR transmittance. This enhanced optical transmittance could be advantageously exploited in devices requiring higher NIR transmittance, including certain photovoltaic and photodetector applications [71]. Additionally, the ability to control permittivity could also have significant impact on the design of TCOs for applications where variable work function is desired, such as organic light-emitting diodes [72–74]. The following discussion reviews our efforts to exploit this insight regarding the role

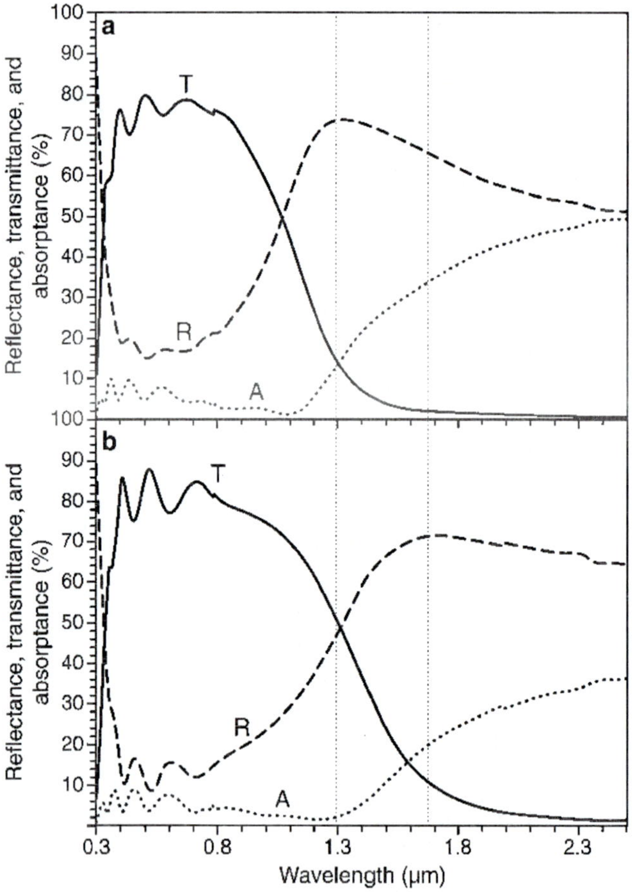

Fig. 3.29 Comparison of transmission, reflection, and absorptance of actual ITO and ITZO films deposited by rf sputtering showing effect of increasing permittivity of a typical TCO material. (**a**) ITO film: thickness = 0.451 μm, $n = 7.29 \times 10^{20}\ \text{cm}^{-3}$, mobility = $36.9\ \text{cm}^2\ \text{V}^{-1}\ \text{s}^{-1}$. (**b**) ITZO film: thickness = 0.448 μm, $n = 7.92 \times 10^{20}\ \text{cm}^{-3}$, mobility = $35.0\ \text{cm}^2\ \text{V}^{-1}\ \text{s}^{-1}$

of ε_∞ in TCOs to produce films with improved optical performance without compromising the electrical quality of the film.

Increasing ε_∞ of any material implies compositional and/or structural modifications that increase the material's ability to be polarized within the frequency range of interest. Polarizability is determined by the combined dipole response from both ion cores and electrons; however, in the frequency range of interest for TCO applications – NIR and optical, 10^{13}–10^{15} s^{-1} – the dipole response of the electrons is expected be more important [75]. Nevertheless, changes in TCO permittivity are not well understood at this time, and so, contributions from ion cores may also be significant.

Insight into modifying the dielectric constant of TCOs can be gained through studies of SiO_2-based films used in metal-oxide-semiconductor (MOS) devices. In this case, new process designs to reduce MOS feature size would benefit significantly from increased permittivity of the oxide layer at the operating frequency of 10^8–10^{10} s^{-1}. These studies have demonstrated that relatively small additions of high-permittivity oxides, such as ZrO_2 and HfO_2, can yield permittivity values significantly higher than expected from a linear extrapolation between SiO_2 and ZrO_2 [76]. In our efforts to increase TCO ε_∞, we have pursued a similar approach and fabricated thin films from a traditional TCO material (ITO) such that it also contains a small amount of a high-ε_∞oxide (ZrO_2). The high-ε_∞ oxide addition was also chosen to maintain electrical quality (i.e., Zr has been shown to demonstrate similar doping in In_2O_3 as does Sn [77, 78]). The goal of this addition is to gain an advantage in the optical properties from the high-ε_∞ material without sacrificing electrical quality.

Test films were produced by rf magnetron sputtering using targets that were custom fabricated to contain 90 wt% In_2O_3 and 9 wt% SnO_2 for the traditional (host) TCO material (ITO) and 1 wt% ZrO_2 for the high-ε_∞ component. The mixture was hot pressed to the approximate target dimensions required, followed by mechanically sizing to more accurate dimensions. The target was then fully re-oxidized using a high-temperature anneal in oxygen ambient. These target fabrication procedures are the same as those used to produce our "standard" ITO sputtering targets. Also, it should be noted that although ZrO_2 and ITO were used for these initial studies, we expect that similar high-ε_∞ oxides (e.g., TiO_2, HfO_2) and/or other traditional host TCO materials (e.g., SnO_2:F, ZnO:Al) will produce similar results if appropriate high-ε_∞ dopant(s) are incorporated into the host TCO material.

Film deposition was conducted in a high-vacuum sputtering system that allows for control of ambient containing both Ar and oxygen. Films were deposited onto Corning 1737 glass substrates at 350°C. We call the rf-sputtered films produced using the ITO/ZrO_2 target "ITZO" films. The study also included fabrication of rf-sputter-deposited ITO films. The ITO sputtering target was produced exactly as indicated above, except that it did not contain ZrO_2 (i.e., the mixture was 9 wt% SnO_2 and 91 wt% In_2O_3). The ITO films were produced during the same time period so that the ITZO films could be used as "control samples" to isolate effects due to the high-ε component (ZrO_2). It should also be noted that the ITO films produced for this comparison are among the highest-quality ITO films ever produced, and so, a favorable comparison with these particular ITO films represent a very high standard. In Figs. 3.32 and 3.33, we compare the ITO and ITZO films that we have produced. In each figure, we have been careful to choose films that demonstrate similar thickness and electrical properties (i.e., carrier concentrations and mobility, measured using room-temperature Hall measurements). This degree of film similarity will minimize effects due to the free-carrier absorptance discussed earlier. The figures show that the ITZO films not only demonstrate significantly higher NIR transmittance, but that the films with the lower value of n, shown in Fig. 3.32, demonstrate a greater difference in optical transmittance due to ZrO_2 addition than do films with higher values of n, shown in Fig. 3.33. This is consistent with the predictions suggested by (3.17) and discussed previously.

3.4 Summary

At frequencies near the free-carrier absorptance band, light is absorbed by electrons in the near-surface region – known as the skin depth – of the TCO. For a typical TCO, the skin depth is about 100 nm and, for thicknesses less than this, the transmittance is enhanced because full absorptance is unable to occur. The effect of reduced thickness may be beneficial in some applications, but there is a corresponding increase in the sheet resistance that may be harmful in others. As explained in the first section of this chapter, the only realistic strategy is to increase the mobility by increasing the relaxation time of the electrons. In this chapter, we have focused on determining the relaxation time from optical measurements, assuming that the Drude theory is applicable, and on using transport measurements to measure it more directly and without assuming any specific model of the electrons. We have also shown that using the free-electron theory helps to interpret the optical properties of a conventional oxide doped with a less familiar impurity.

For a TCO in which free electrons dominate the optical and electrical properties, we have available expressions that relate the fundamental electrical parameters to the optical properties of the material. For some TCO materials, all the essential features of the optical properties predicted by the Drude theory are observed in practice. For these, the free-electron model works surprisingly well and, as discussed, the values of m^* and τ that are derived from optical measurements on some TCOs are very close to those measured using the advanced electrical measurement techniques discussed in the second section of this chapter. This means that if we have reasonably accurate estimates of the fundamental quantities (i.e., m^*, ε_∞, n, and τ), then we can model the optical properties (i.e., transmittance, reflectance, and absorptance) of a film of specified thickness. Conversely, if we have measured optical data, it is possible to determine values of the electrical parameters that best fit the data. However, this is much less straightforward.

As we have seen, the peak of the free-carrier absorptance band is near, but not exactly equal to, the plasma wavelength. In the very low-frequency range, where $\omega\tau << 1$, the real part of the permittivity approaches a constant value, which is independent of frequency: the imaginary part approaches infinity and the reflection coefficient approaches unity. In general, none of the readily identifiable spectral features corresponds precisely to the plasma frequency. However, they are sufficiently close that they provide a good approximation to the carrier concentration.

Several assumptions about the electronic structure of the TCOs are implicit in the approach described above. In the free-electron theory, it is assumed that a single relaxation time is adequate to represent the behavior of all electrons, whereas, in general, the relaxation time is a function of electron energy – to an extent that is governed by the precise scattering mechanism. However, TCOs are generally, if not always, degenerate semiconductors. The Fermi electrons are the only ones that contribute to the conductivity and a single relaxation time appears to be adequate. Stated another way, the Drude theory works well for TCOs simply because they are usually degenerate.

The study of transport properties in TCO films straddles the boundaries between classical and quantum descriptions of solid-state materials. Band-to-band absorptance in the UV demands a quantum description, whereas infrared reflectance is best described by a classical electron sea. Likewise, the method of four coefficients technique mixes both classical and quantum ideas to describe accurately the transport properties of the carriers in TCO films. The four macroscopic, observable, thermomagnetic transport phenomena – Hall, conductivity, Seebeck and Nernst – can all be described qualitatively by a simple classical view of carriers in a material under electric and/or magnetic fields or temperature gradients. Yet, the observed transport phenomena reveal the underlying quantum nature of materials. Even the solution of Boltzmann's transport equation outlined in this chapter is a mixture of classical and quantum theory. From a simplistic relaxation-time approximation for classical billiard-ball-like collisions, to the use of the Fermi-Dirac distribution function, this appears justified. Again, this tie to both the classical and quantum worlds is an adequate description of electronic transport in many TCO films.

We have shown that a series of justified approximations leads to an analytical solution of Boltzmann's transport equation. The main approximation involves simplifying the nature of collisions between carriers and defects within a crystal by assuming that this time is a simple function of energy. This model ensures that all of the relevant collisions involve carriers at the Fermi level. It is the local probing of the transport phenomena at the Fermi level that reveals the nature of the electronic bands in the material under study. Combining the equations that describe the four transport phenomena gives the curvature of either the conduction (n-type) or valence (p-type) band in a material. The curvature is simply related to the effective mass of the carriers. Knowing the effective mass and the mobility allows the relaxation time between carrier-scattering collisions to be calculated; this is the most important transport parameter to know – and the most difficult to measure – in order to understand the optical and electrical properties of TCO materials. Additionally, measuring the effective mass as a function of carrier concentration essentially measures the curvature of the energy band as a function of Fermi energy. Knowing the curvature of the bands reveals their fundamental nature in a material and helps identify more clearly the type of scattering collisions occurring with a given material.

The three experimental material examples given in this section on the use of MFC illustrate the usefulness of the technique in TCO films to reveal practical and fundamental information about the nature of the bands and transport within the bands. We have shown that MFC, complemented by other experimental techniques, clearly identifies dominant scattering mechanisms within TCO films. The data presented on the two high-quality TCO films – epitaxially grown ITO and polycrystalline CTO – show phonon-like scattering. This mechanism represents an upper limit on allowed mobility in films and is rarely seen in TCO materials. For CTO, MFC identified the high mobilities within the material as being due to high relaxation times, rather than to small values of effective mass.

Applying Drude theory to understand electrical quality of TCO films is equally useful to understand and improve the optical quality of these films. Specifically, the theory suggests that fabricating films with either higher or lower permittivity can modify TCO optical quality. For example, modeling shows that the NIR transmittance can be increased significantly by increasing the high-frequency permittivity of a TCO material from typical values of ~4 to values of ~8. The degree of transmittance improvement predicted by this modeling could provide significant benefit to detectors and photovoltaic solar cells that require high NIR transmittance. In contrast, controlled reduction of NIR transmittance through reduced TCO permittivity can have applications in the design of "low-e" glass coatings for energy-efficient architectural windows. Finally, control of TCO permittivity can provide advantages in devices such as organic light-emitting diodes where the work function of the TCO film is a critical device parameter.

As a representative demonstration of this technique to improve the TCO optical properties, small amounts of high-permittivity ZrO_2 have been incorporated into rf-sputtered ITO films. Consistent with the predictions of Drude theory, optical analysis of these Zr-containing ITO films reveal significantly improved NIR transparency. Further, because Zr is an effective dopant in ITO, the electrical quality of the film remains high. The combination of these results produces films with higher figure of merit, especially in the NIR region of the optical spectrum. The ITO:Zr films also reveal improved visible transparency. This indicates an additional advantage of refractory-metal incorporation into TCOs, namely, improved resistance to the formation of low-transparency metal phases. It is expected that these attributes related to high-permittivity additions in TCOs will find increasing applications in future optoelectronic devices.

Acknowledgements This work was performed under U.S. Department of Energy contract number DE-AC36-GO9910337. The authors would like to express their thanks for the input given by Yuki Yoshida (now of Sanyo) and Viktor Kaydanov (formerly of the Colorado School of Mines).

References

1. P. Drude, Annalen der Physik **1, 3,** 566, 369 (1900).
2. B. O'Neill, in *Indium: Markets, Applications and Alternatives*, Lisbon, Portugal (2005).
3. T. J. Coutts, X. Li, T. M. Barnes, B. M. Keyes, C. L. Perkins, S. E. Asher, S. B. Zhang, S. H. Wei, and S. Limpijumnong, in *Zinc Oxide Bulk, Thin Films and Nanostructures:*

Processing, Properties and Applications, edited by C. Jagadish and S. J. Pearton (Elsevier, Amsterdam, 2006), pp. 43–84.
4. S. Ghosh, A. Sarkar, S. Chaudhuri, and A. K. Pal, Thin Solid Films **205,** 64–68 (1991).
5. H. Kim, C. M. Gilmore, J. S. Horwitz, A. Pique, H. Murata, G. P. Kushto, R. Schlaf, Z. H. Kafafi, and D. B. Chrisey, Applied Physics Letters **76,** 259–261 (2000).
6. M. W. J. Prins, D.-O. Grosse-Holz, J. F. M. Cillessen, and L. F. Feiner, Journal of Applied Physics **83,** 888–893 (1998).
7. W. P. Mulligan, Ph.D. Thesis, Colorado School of Mines (1997).
8. T. J. Coutts, D. L. Young, X. Li, W. P. Mulligan, and X. Wu, Journal of Vacuum Science and Technology A **18,** 2646–2660 (2000).
9. V. Kaydanov, Thesis, Colorado School of Mines (1999).
10. D. H. Zhang and H. L. Ma, Applied Physics A **62,** 487–492 (1996).
11. J. R. Bellingham, Thesis, Emmanuel College (1989).
12. X. Li, T. A. Gessert, and T. J. Coutts, Applied Surface Science **223,** 138–143 (2004).
13. M. P. Taylor, Ph.D. Thesis, Colorado School of Mines (2005).
14. R. H. Williams, R. R. Varma, and V. V. Montgomery, Journal of Vacuum Science and Technology **16,** 1418 (1979).
15. J. Y. W. Seto, Journal of Applied Physics **46,** 5247–5254 (1975).
16. M. Born and E. Wolf, *Principles of Optics: Electromagnetic Theory of Propagation, Interference and Diffraction* (Cambridge University Press, New York, NY, 1999).
17. N. W. Ashcroft and N. D. Mermin, *Solid State Physics* (Harcourt Brace College Publishers, Fort Worth, TX, 1976).
18. L. Scheff, M. Dressel, M. Jourdan, and H. Adrian, Nature **438,** 1135 (2005).
19. A. K. Azad, Applied Physics Letters **88,** 021103-1–021103-3 (2006).
20. ?
21. J. I. Pankove, *Optical Processes in Semiconductors* (Dover Publications, New York, NY, 1971).
22. P. A. Iles and S. I. Soclof, in *Design Factors for Transparent Conducting Layers in Solar Cells*, Baton Rouge, LA, (1976) (IEEE), pp. 978–987.
23. M. P. Taylor, D. W. Readey, C. W. Teplin, M. F. A. M. van Hest, J. L. Alleman, M. S. Dabney, L. M. Gedvilas, B. M. Keyes, J. D. Perkins, and D. S. Ginley, Measurement Science and Technology **16,** 90–94 (2005).
24. V. I. Kaidanov and I. A. Chernik, Soviet Physics – Semiconductors **1,** 1159–1163 (1967).
25. K. J. Button, C. G. Fonstad, and W. Dreybrodt, Physical Review B **4,** 4539–4542 (1971).
26. T. Wang, J. Bai, and S. Sakai, Applied Physics Letters **76,** 2737–2739 (2000).
27. J. A. Marley and R. C. Dockerty, Physical Review **140,** A304–A310 (1965).
28. P. Wagner and R. Helbig, The Journal of Physics and Chemistry of Solids **35,** 327–335 (1972).
29. E. H. Hall, American Journal of Mathematics **2,** 287 (1879).
30. L. L. Campbell, *Galvanomagnetic and Thermomagnetic Effects the Hall and Allied Phenomena* (Longmans, Green and Co, New York, NY, 1923).
31. A. V. Ettingshausen, Anzeiger der Akademie der Wissenshaften in Wien 16 (1887).
32. L. Boltzmann, Anzeiger der Akademie der Wissenshaften in Wien **24,** 217 (1886).
33. A. V. Ettingshausen and W. Nernst, Anzeiger der Akademie der Wissenshaften in Wien **23,** 114 (1886).
34. E. Putley, *The Hall Effect and Related Phenomena* (Butterworths, London, 1960).
35. W. Gerlach, *Handbuch der Physik*, Vol. 13 (Springer, Berlin, 1928).
36. F. Seitz, Physical Review **73,** 549–564 (1948).
37. R. Barrie, Proceedings of the Physical Society B **69,** 553–561 (1956).
38. E. O. Kane, The Journal of Physics and Chemistry of Solids **1,** 249–261 (1957).
39. W. Zawadzki and J. Kolodziejczak, Physica Status Solidi **6,** 419–428 (1964).
40. N. V. Kolomoets, Soviet Physics – Solid State **8,** 799–803 (1966).
41. W. F. Leonard and J. T. L. Martin, *Electronic Structure and Transport Properties of Crystals*, First ed. (Robert E. Krieger Publishing Company, Malabar, FL, 1987).
42. S. S. Li, *Semiconductor Physical Electronics* (Plenum, New York, NY, 1993).

43. B. M. Askerov, *Electron Transport Phenomena in Semiconductors* (World Scientific, Singapore, 1994).
44. R. H. Bube, *Electrons in Solids, an Introductory Survey*, 3 ed. (Academic Press, San Diego, CA, 1992).
45. J. Kolodziejczak and S. Zukotynski, Physica Status Solidi **5,** 145–158 (1964).
46. S. Zukotynski and J. Kolodziejczak, Physica Status Solidi **3,** 990–1000 (1963).
47. L. Onsager, Physical Review **37,** 405–426 (1931).
48. L. Onsager, Physical Review **38,** 2265–2279 (1931).
49. J. Kolodziejczak, Acta Physica Polonica **XX,** 289–302 (1961).
50. J. Kolodziejczak and L. Sosnowski, Acta Physica Polonica **21,** 399 (1962).
51. S. Zukotynski and J. Kolodziejczak, Physica Status Solidi **19,** k51–k54 (1967).
52. I. N. Dubrovskaya and Y. I. Ravich, Soviet Physics – Solid State **8,** 1160–1164 (1966).
53. M. K. Zhitinskaya, V. I. Kaidanov, and I. A. Chernik, Soviet Physics – Solid State **8,** 295–297 (1966).
54. I. A. Chernik, V. I. Kaidanov, N. V. Kolomoets, and M. I. Vinogradova, Soviet Physics – Semiconductors **2,** 645–651 (1968).
55. D. L. Young, T. J. Coutts, V. I. Kaydanov, A. S. Gilmore, and W. P. Mulligan, Journal of Vacuum Science and Technology A **18,** 2978–2985 (2000).
56. D. L. Young, T. J. Coutts, and V. I. Kaydanov, Review of Scientific Instruments **71,** 462–466 (2000).
57. W. Jiang, S. N. Mao, X. X. Xi, X. Jiang, J. L. Peng, T. Venkatesan, C. J. Lobb, and R. L. Greene, Physical Review Letters **73,** 1291–1294 (1994).
58. ASTM, in *Annual Book of ASTM Standards* (American Society of Testing and Materials, West Conshohocken, 1996).
59. V. I. Kaidanov and I. S. Lisker, Zavodskaya Laboratoriya **32,** 1091–1095 (1966).
60. S. Brehme, F. Fenske, W. Fuhs, E. Nebauer, M. Poschenrieder, B. Selle, and I. Sieber, Thin Solid Films **342,** 167–173 (1999).
61. *Landolt-Börnstein Numerical Data and Functional Relationships in Science and Technology: Semiconductors*; *Vol. 17*, edited by K. H. Hellwege (Springer, Berlin, 1982)
62. S. Bloom and I. Ortenburger, Physica Status Solidi (b) **58,** 561–566 (1973).
63. X. Yang, C. Xu, and N. Giles, in *Role of Neutral Impurity Scattering in the Analysis of Hall Data from ZnO*, Denver, CO (2007).
64. S. Lany and A. Zunger, Physical Review Letters **98,** 045501-1–045501-4 (2007).
65. H. Ohta, M. Orita, and M. Hirano, Journal of Applied Physics **91,** 3547 (2002).
66. Y. Yoshida, D. M. Wood, T. A. Gessert, and T. J. Coutts, Applied Physics Letters **84,** 2097–2099 (2004).
67. X. Wu, T. J. Coutts, and W. P. Mulligan, Journal of Vacuum Science and Technology A **15,** 1057–1062 (1997).
68. T. J. Coutts, D. L. Young, and X. Li, MRS Bulletin **25,** 58–65 (2000).
69. Y. Yoshida, T. A. Gessert, C. L. Perkins, and T. J. Coutts, Journal of Vacuum Science and Technology A **21,** 1092–1097 (2003).
70. T. A. Gessert, Y. Yoshida, and T. J. Coutts (USA, 2004).
71. J. Zhou, X. Wu, T. A. Gessert, Y. Yan, G. Teeter, and H. R. Moutinho, Materials Research Society Proceedings **865,** 387–392 (2005).
72. T.-H. Chen, Y. Liou, T. J. Wu, and J. Y. Chen, Applied Physics Letters **85,** 2092–2094 (2004).
73. Y.-J. Lin, C.-W. Hsu, Y.-M. Chen, and Y.-C. Wang, Journal of Electronic Materials **34,** L9–L11 (2005).
74. H. Kim, J. S. Horwitz, W. H. Kim, S. B. Qadri, and Z. H. Kafafi, Applied Physics Letters **83,** 3809–3811 (2003).
75. G. D. Wilk, R. M. Wallace, and J. M. Anthony, Applied Physics Letters **89,** 5243–5275 (2001).
76. G. Lucovsky and G. B. Rayner, Applied Physics Letters **77,** 2912–2914 (2000).
77. R. Groth, Physica Status Solidi **14,** 69–75 (1966).
78. T. Koida and M. Mondo, Applied Physics Letters, 082104-1–082104-3 (2006).

Chapter 4
Characterization of TCO Materials

David C. Paine, Burag Yaglioglu, and Joseph Berry

4.1 Introduction

The full array of microstructural, optical, and electrical materials characterization tools are used in the development of materials and processes for TCO applications. Most of these characterization tools are well described in other sources and are widely applied in the semiconductor and oxide fields of research. In this chapter, discussion is limited to a few of the methods used for characterizing the relationship between microstructure and electrical transport properties in transparent conducting materials particularly those that are based on the amorphous and crystalline oxides of In_2O_3, SnO_2, and ZnO. These materials are currently of greatest practical importance for passive transparent electrode applications and they form the basis for the emerging field of oxide thin film transistor electronics.

One of the most interesting properties of TCO materials – at least those that are based on indium oxides – is that electron scattering does not significantly increase in the amorphous phase or as a function of processing. This is shown in Table 4.1 which lists a set of seven indium oxide based thin films all deposited in the same system to the same thickness but under different sputter, substrate temperature, and/or post-deposition processing conditions so that examples of amorphous and crystalline – both bixbyite and rhombohedral phase – materials were produced. Note that in every case listed in Table 4.1, the mobility is moderately high (10–50 $cm^2/V\,s$) regardless of the phase. Of particular interest is the similarity in electron mobility in amorphous and crystalline films which, in In_2O_3-based systems, results from two effects. First, both the amorphous phase and the 80-atom bixbyite In_2O_3 crystal structure are based on the same $In\text{-}O_6$ structural unit. This is readily established by comparing the crystalline indium oxide phase (bixbyite) with the amorphous structure

D.C. Paine (✉)
Division of Engineering, Brown University, Box D, 02912 Providence, RI, USA
e-mail: David_Paine@brown.edu

D.S. Ginley et al. (eds.), *Handbook of Transparent Conductors*,
DOI 10.1007/978-1-4419-1638-9_4, © Springer Science+Business Media, LLC 2010

Table 4.1 Transport properties of seven indium oxide based materials showing remarkably high mobility over a wide range of processing conditions

	Resistivity (Λcm)	Carrier density (/cm^3)	Mobility (cm^2/V s)	Structure
IZO-unheated	5.96E−04	2.02E+20	51.9	Amorphous
ITO-unheated	7.18E−04	2.02E+20	41.4	Amorphous
IO-unheated	7.48E−04	1.88E+20	44.5	Amorphous
IZO-350°C substrate	4.20E−04	4.30E+20	34.8	Bixbyite
ITO-350°C substrate	2.23E−04	6.03E+20	46.4	Bixbyite
IO-350°C substrate	1.73E−03	7.90E+19	46.0	Bixbyite
IZO air anneal 500°C for 150 min	9.80E−02	3.40E+18	18.7	Rhombohedral

using radial distribution function analysis. Both synchrotron radial distribution function (RDF) analysis [1] and electron scattering studies [2], reveal that the In-O nearest neighbor distance is nearly identical and the two characteristic In-In separations in crystalline bixbyite are present in the amorphous state though not in the same proportion. Small structural differences do not lead to increased scattering because, as Nomura [3] points out, the bottom of the conduction band in these highly ionic oxide semiconductors is composed primarily of a spherically symmetric ns (n = principal quantum number) orbital which allows direct ns-orbital overlap between neighboring metal ions. The oxygen 2p-orbitals, though highly directional, are small and do not contribute to conduction. The ns-overlap eliminates the effect of oxygen-metal-oxygen (O-M-O) bond distortions in the amorphous state and allows high carrier mobility even in the degenerately doped state.

In this chapter we discuss two methods for the determination of carrier mobility including conventional Hall Effect measurement and, in highly resistive low carrier density materials, field effect mobility using TFT test structures. The ability to measure carrier mobility as a function of microstructure is, evidently, not necessarily interesting in the case of In-based oxides, but ZnO-based TCO's reveal a profound effect of microstructure on mobility [4]. Due to its relatively simple wurtzite structure, the deposition of ZnO onto unheated substrates always yields polycrystalline material. Such material shows crystallographic surface topography that may limit the use of ZnO in TFT applications to gate-down devices but, more importantly, the grain boundaries in ZnO also contain trap levels that increase threshold voltages in TFT-applications and provide conduction pathways at high fields. Studies of mobility in poly- and single crystal material [4] show that polycrystalline ZnO carrier mobility decreases with decreasing carrier density. This does not occur in single crystal ZnO (or in conventional semiconducting materials) where decreased doping leads to decreased ionized impurity scattering and suggests that grain boundaries in ZnO contain traps that become activated at lower carrier concentration.

The increasing importance of metal oxide channel materials deposited by physical vapor deposition at low temperatures for TFT applications makes the determination of both the fundamental transport properties (mobility and carrier density) and the as-deposited microstructure of candidate TCO materials of great interest. For this reason, we have chosen to combine, in this chapter, a discussion of microstructure characterization using electron microscopy with methods for the determination of carrier mobility including a brief discussion of conventional Hall Effect measurement methods and associated challenges and, at low carrier density, field effect mobility using TFT test structures.

4.2 Electrical/Optical Characterization

4.2.1 Resistivity Measurements

The conductivity σ (S/cm), and its inverse, resistivity, ρ (Ω cm), describe a fundamental property of a material. Generally the conductivity is a tensor $\sigma_{\alpha\beta}$ in the spatial dimensions of system and relates the internal electric field ($\boldsymbol{E}$ or E_α) with the current density $\boldsymbol{j}$ with j_α given as

$$j_\alpha = \sum_\beta \sigma_{\alpha\beta} E_\alpha \tag{4.1}$$

with α and β running over the relevant spatial dimensions. In the case of an isotropic material this α and β may be dropped and σ given as

$$\sigma = e(\mu_n n + \mu_p p) \tag{4.2}$$

where n/p are the number of electrons/holes with μ_n and μ_p their respective mobilities given in the limit of $n \gg p$ and $p \gg n$. The resistivity, ρ, is simple to measure directly and in the case of a uniform sample can be expressed in term of the isotropic electric field (E) and current density (J) or the sample length (L), cross sectional area (A), current (I) and voltage (V).

$$\rho = \frac{E}{J} = \frac{VA}{IL} \tag{4.3}$$

Units for E are typically V/cm and J is most commonly reported in A/cm^2. The resistance R (Ω) is a quantity that removes the length dimension includes sample shape and resistivity and is given by $R = \rho(L/ad)$ where L is the length down which the current flows, a is the width of the sample, and d is the thickness of the sample.

A convenient method for measuring sheet resistance is the four-point in-line probe technique. In this setup, current is supplied through an outer pair of probes, and the voltage is measured using a high input impedance voltmeter across the inner two probes. If the probes are equally spaced (s), as shown in Fig. 4.1, the resistivity

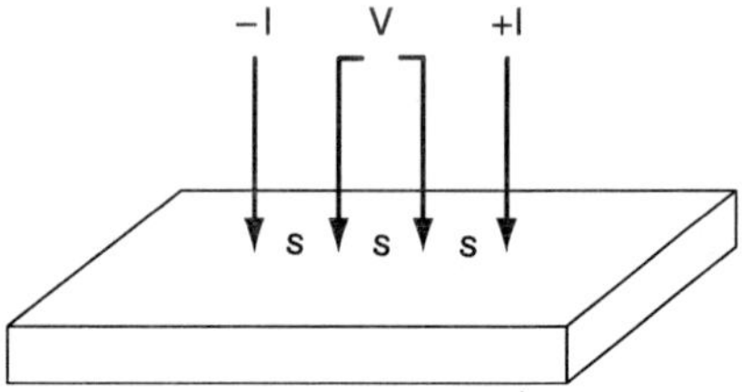

Fig. 4.1 Schematic view of four-point probe configuration

of a thin film sample of thickness, d, and infinite lateral dimension (where s ≪ sample size) is:

$$\rho = \frac{V}{I}\frac{\pi\,\mathrm{d}}{\ln 2} \tag{4.4}$$

Another method that is widely used in both resistivity and Hall measurements on thin film samples of arbitrary shape is the van der Pauw [5] technique. In this method, four small contacts are made at the periphery of an arbitrarily shaped two-dimensional sample (number the contacts counter clockwise 1, 2, 3, and 4). To measure sheet resistance, a DC current is passed from one contact to the adjacent contact (start with $I_{1\text{-}2}$), while the corresponding voltage drop is measured across the two contacts on the opposing side ($V_{4\text{-}3}$). This measurement is repeated twice for each set of contact pairs by switching polarity. Repeated for all four pairs of contacts yields:

$$\begin{aligned} R_{21,34} &= V_{34}/I_{21}, \quad & R_{12,43} &= V_{43}/I_{12} \\ R_{32,41} &= V_{41}/I_{32}, \quad & R_{23,14} &= V_{14}/I_{23} \\ R_{43,12} &= V_{12}/I_{43}, \quad & R_{34,21} &= V_{21}/I_{34} \\ R_{14,23} &= V_{23}/I_{14}, \quad & R_{41,32} &= V_{32}/I_{14} \end{aligned} \tag{4.5}$$

The sheet resistance can then be determined by calculating two characteristic resistances: $R_A = (R_{21,34} + R_{12,43} + R_{43,12} + R_{34,21})/4$ and $R_B = (R_{32,41} + R_{23,14} + R_{14,23} + R_{41,32})/4$ and solving the van der Pauw equation: $e^{-\pi RA/Rs} + e^{-\pi RB/Rs} = 1$. In the usual case of a square shaped sample of homogeneous material with contacts placed at each of the four corners, it can be shown that $R_A = R_B$ and the sheet resistance, $R_s = (\pi/\ln 2)(R_A + R_B)/2$ from which the resistivity is found via $\rho = R_s d$. The materials being tested must be much larger laterally than in the thickness direction.

4.2.2 Thickness Determination: X-Ray Reflectivity

Accurate thickness determination is a key step in the evaluation of resistivity (a fundamental material property) from van der Pauw measurements of sheet resistance. The precision of subsequent carrier density measurements (rather than just sheet density) using the Hall effect also requires an accurate measure of film thickness. Optical thickness measurement techniques such as ellipsometry and mechanical methods such as surface profilometry at a step edge are widely used for this purpose. However, there can be ambiguities associated with model fitting in simple (non-spectroscopic) ellipsometry and profilometry requires the use of a mechanical stylus that can deform the surface of, for example, a soft polymeric substrate which makes the resulting step height unreliable. Furthermore, the mechanical techniques require patterning via lift-off or mask and etch to form a step.

X-ray reflectivity is an alternative method for thickness determination that is less widely used but has proven invaluable for providing a precise and unambiguous measurement of thickness from smooth thin films in the thickness range 1–100 nm on substrates. The attractive feature of this method is that it provides a direct, non-destructive measure of thickness that is dependent, to a first order, only on the wavelength of the incident X-rays and a measure of the spacing between intensity oscillations at low incident angles. Briefly, the determination of thickness by X-ray reflectivity consists of measuring the reflected intensity of X-rays from a sample surface at near-glancing incidence angles ($0.5 < 2\theta < 3.5$ on a conventional $\theta - 2\theta$ two circle diffractometer). Partial reflection of the incident monochromatic X-rays occurs at the interfaces between the regions of different indices of refraction; first at the air/film top interface and then at the film/substrate bottom interface. The reflected X-rays at the bottom surface have an angle-dependent path length and interfere with those X-rays scattered at the top interface to produce intensity oscillations as a function of incidence angle as shown in Fig. 4.2. As the angle

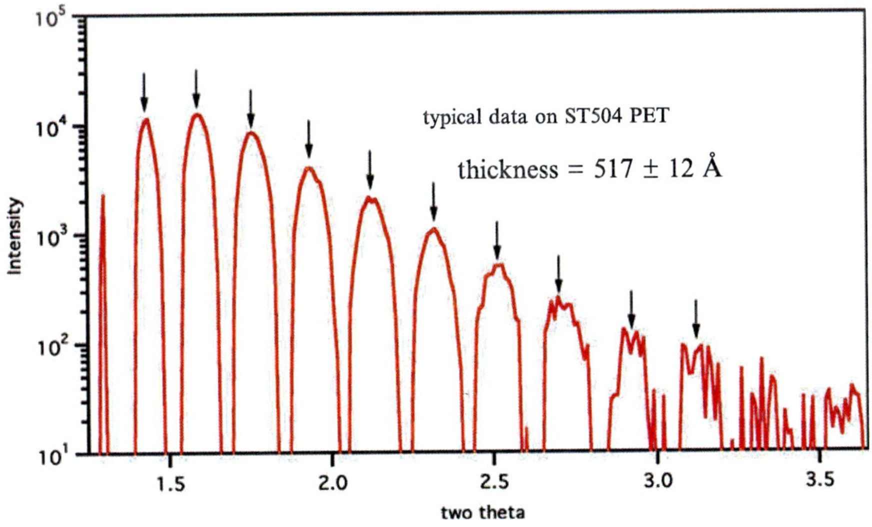

Fig. 4.2 Reflected X-ray intensity obtained using a simple $\theta - 2\theta$ diffractometer with monochromatic Cu radiation. The film thickness is simply found by taking the average difference in 2θ and applying: $d = \lfloor /(2\otimes\theta)$

increases, the path length of the X-rays scattered at bottom interface decreases so that interference with X-rays scattered at the top surface leads to X-ray intensity oscillations at regular uniform 2θ intervals given by: $d = \lambda/2\theta$. Note that the measure of thickness, d, depends only on the wavelength of the radiation (say Cu k_α at 0.154 nm) and the period of the intensity oscillations.

4.2.3 Hall Measurement of Transport Properties: Hall Mobility and Carrier Density

The carrier transport properties of TCO materials can be determined using a variety of magneto-optical and standard transport measurements. The most common measurements are based on the Hall effect. The Hall effect refers to the voltage that results from the deflection of the charged carriers by the Lorenz force that develops in the direction that is defined by the charge on the carrier and the vector cross product of the drift velocity of the carriers (direction set by the current flow) and the magnetic field:

$$\boldsymbol{F}_{Lorentz} = -e(\boldsymbol{v}_{\mathrm{drift}} \times \boldsymbol{H}) = \boldsymbol{I} \times \boldsymbol{H} \tag{4.6}$$

The total force is given by

$$\boldsymbol{F}_{total} = -e(E + \boldsymbol{v}_{\mathrm{drift}} \times \boldsymbol{H}) = \boldsymbol{I} \times \boldsymbol{H} \tag{4.7}$$

In a typical set-up, a square planar van der Pauw sample is created with contacts made at the corners of the square. A current is driven through two diagonally opposing contacts while a magnetic field is applied in the plane-normal direction as shown in Fig. 4.3. Assuming current only along the x direction, the Lorenz force is in the y direction given by $\mathbf{F}_{Lorentz} = -e(\mathbf{v}_{drift} \times \mathbf{H})_y = -e(v_z H_x - v_x H_z) = -ev_x H_{z,}$, however the total force along y, (F_y) must be zero as no current is flowing in this direction. The electric field along y, (E_y) must balance F_y to give

$$F_y = 0 = ev_x H_z = -eE_y \tag{4.8}$$

The voltage associated with E_y is known as the Hall voltage (V_H), for electrons traveling in one in-plane direction, an excess surface charge develops in the transverse direction resulting in the Hall voltage. Explicitly this is then given as

$$V_H = -wv_x H_z \tag{4.9}$$

with the width of the sample given by w. The sign and magnitude of the Hall voltage is therefore dependent on the carrier type (electrons, in the case of most TCO's) and sheet carrier density (carriers/unit area). For the situation shown in Fig. 4.3 with an

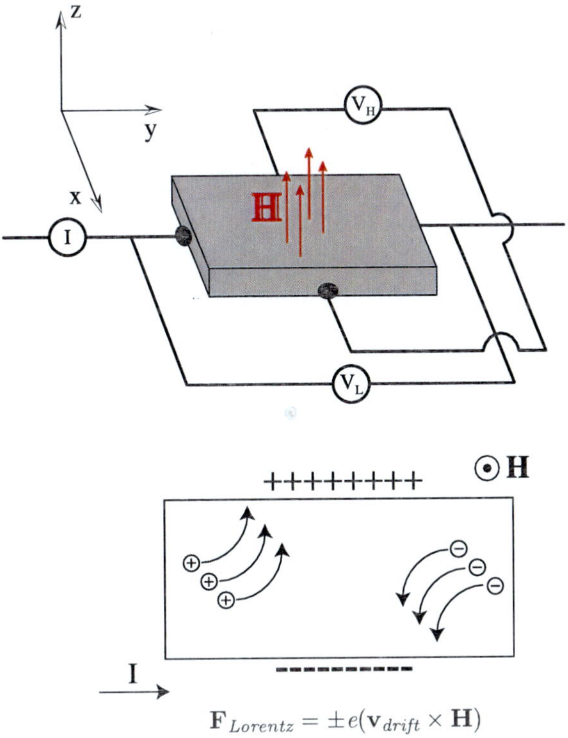

Fig. 4.3 Schematic of the Hall effect in a long, thin bar of semiconductor with four Ohmic contacts

applied magnetic field running in the z-direction (sample plane-normal direction) the electrons travelling in the +x in-plane direction are subjected to a Lorentz force given by $F = ev \times \mathrm{B}$, where v is the drift velocity, and B is the magnetic field strength. This transverse voltage V_H can be written as:

$$V_H = \frac{IH_z}{eNd} \tag{4.10}$$

where I is the current, d is the sample thickness, e is the elementary charge and N is the bulk carrier density. Thus, by measuring the Hall voltage V_H and from the known values of I, H_z, and e, one can determine the areal charge density, $N*d$ and, if the sample thickness is known, the carrier density N. If the sample thickness is known, the van der Pauw sheet resistance can be used to find resistivity from the same sample which then provides the Hall mobility, μ_H, using:

$$\mu_{\mathrm{H}} = 1/(eN\rho) \tag{4.11}$$

Here again we have assumed an isotropic conductivity/resistivity for our sample. The Hall resistance $R_H = V_H/(wIH_z)$ is also a useful quantity, and is defined as the slope of the resistance as function of the applied field.

$$R_H = \pm 1/ne \tag{4.12}$$

Where the ($\pm$) reflects the direction of the Lorenz force and the sign for the carrier types, ($-$) for electrons and ($+$) for the holes. In the case in which both types of carriers are present then this expression is insufficient as both types of carriers may contribute and R_H becomes:

$$R_H = \frac{\left[\left(p - \left(\frac{\mu_n}{\mu_p}\right)^2 n\right) + (\mu_n H_z)^2 (p-n)\right]}{p\left[\left(p + \left(\frac{\mu_n}{\mu_p} n\right)\right)^2 + (\mu_n H_z)^2 (p-n)^2\right]} \tag{4.13}$$

This expression collapses appropriately in the case where the contribution to the conductivity is dominated by the mobility of one carrier being considerably larger than the other to:

$$N = \frac{r}{eR_H} \tag{4.14}$$

Where N is again simply the number of carriers (electrons or holes) and r is the scattering rate which depends on the scattering mechanism which dominates. The value for r can be obtained by examination of the temperature dependence of the μ_H with typical values being between 1 and 2 depending on the dominant scattering mechanism.

Hall measurements, and associated van der Pauw sheet resistance determination, should be carried out using a sample current (I) that minimizes dissipated power (<1 mW is suitable for most materials at room temperature). Low dissipated power is necessary to prevent Joule heating of the sample and subsequent temperature dependent changes in transport characteristics and temperature gradients which create a thermal potential. In highly resistive samples, the current limit may also be practically limited by the applied potential needed to drive the current. Equation (4.10) above shows that decreasing carrier concentration (N) increases the Hall voltage for a given magnetic field, sample thickness, and sample current. This would suggest that the Hall voltage on low carrier density samples should be relatively easy to measure. However, if we set the sample current so that the dissipated power is at a maximum ($P < 1$ mW) the allowed current in these low carrier concentration samples becomes unmanageably small and the voltage that must be applied to the sample to drive this current becomes large. For example, a 100-nm-thick film with a typical TCO mobility of 50 cm^2/V s and a carrier concentration of $10^{15}/cm^3$ (suitable for TFT applications) has a sheet resistance of 12.5 M Ω/square. To keep the dissipated power below the 1 mW maximum, a current no greater than 8.9 μA is allowed and this requires an applied voltage of 112 V. Challenges in testing highly resistive samples include minimizing sample geometric asymmetries with respect to the placement and alignment of contacts, low currents achievable at reasonable bias, sheet resistances that are large relative to the input impedance of the measurement tools. Solutions to these challenges include the use of patterned thin film specimens (Hall spiders) to minimize geometric asymmetries, the use of higher magnetic fields, and the use of high sensitivity AC and/or bridge measurements. In general, however, in the case of moderate mobility in thin layers of low carrier concentration material, carrier surface concentrations (n × thickness) is low and accurate measurement of carrier density and mobility can be challenging. It should also be mentioned that in the case of films which display ferromagnetic behavior the analysis becomes more complicated as an additional contribution to the resistivity/conductivity perpendicular to the current flow may be present. This contribution is known as the anomalous or extraordinary coefficient.

4.2.4 Transistor-Based Transport Measurements

In the TCO literature over the past 40 years, there has been considerable insight developed into the mechanisms of doping and methods for the maximization of carrier density in degenerately doped amorphous and crystalline materials. In conventional passive TCO applications, it is desirable to maximize both mobility and carrier density (at the cost of optical transmissivity in the near IR) in order to minimize resistivity. These materials are, in most cases, doped via the ionization of native point defects and substitutional impurities. With the recent use of TCO-based channel materials for TFT applications there is now interest in developing carrier density control at low carrier densities (10^{14}–10^{16}/cm^3) where, for narrow channel devices, the surface density of carriers (n × d/cm^2) is small, and the sample resistivity is high making Hall/van der Pauw measurements difficult.

Strategies for the control of native defects vary depending on the TCO material. In Indium oxide based materials, for example, doping control amounts to modification of the oxygen stoichiometry since doubly charged oxygen vacancies act as donors (two electrons per ionized vacancy). In these low carrier density moderate mobility materials it is convenient to fabricate thin film transistor test structures to determine field effect mobility and to obtain insight into potential device performance. The discussion below describes a TFT test structure that was used [6] to determine the field effect mobility in an amorphous-IZO channel TFT fabricated on a Si substrate using a thermally grown SiO_2 gate dielectric. It should be recognized that the field effect mobility measured in a TFT is dependent on the condition of the channel/dielectric interface and is not the same as the Hall mobility which, because of the effect of the magnetic field on scattering, is not the same as the bulk carrier mobility.

To evaluate suitable TCO TFT channel materials without the added complications associated with deposited gate dielectric materials, a gate-down test device structure with Si/SiO_2 as the gate metallization/insulator may be used. In one evaluation [6] of a-IZO channel materials, 100-nm-thick silicon oxide was thermally grown in dry oxygen on a p-type Si (24–49 Ω cm) substrate. The backside oxide was etched and a Ti (100 nm thick) contact electrode was deposited using DC magnetron sputtering. On the front side of the wafer, an amorphous IZO channel layer was sputter deposited to a thickness of 10 nm from a commercially available In_2O_3-10 wt% ZnO target (Idemitsu Corp.) and the channel region was defined using a lift-off process. A sputter power of 0.22 W/cm^2 at 280 V, a target-substrate distance of 5 cm and Ar/O_2 (90/10 vol.%) gas mixture at a pressure of 10 mTorr were used. In a second lift-off step, the source and drain regions were defined and IZO was deposited using pure Ar (0 vol.% O_2), while all other sputter parameters were kept the same. Devices with W/L ratios of 1,000 μm/100 μm, 2,000 μm/200 μm, 2,000 μm/400 μm, and 2,000 μm/1,000 μm were fabricated on a single wafer. A schematic view of the full device is shown in Fig. 4.4. In this device the source drain contacts were metalized using conducting (high carrier density) IZO but deposited metal contacts could also be used for test purposes.

The performance and device characteristics of these TFT test structures are revealed by measurement of the transistor transfer and output characteristics. These measurements should be made the dark using a light tight box since it is known that both ZnO and In_2O_3 are sensitive to UV light. The device characteristics are determined using a semiconductor parameter analyzer by measuring the drain current (I_D) as the drain voltage (V_D) is swept from, in the above case, 0–30 V at gate voltages (V_G) that, in this case, ranged in 5 V steps from −5 to 20 V. Figure 4.4 shows drain current – voltage ($I_D - V_D$) curves that are typical of IZO TFT's. At zero gate voltage ($V_G = 0$ V), there is considerable drain current ($I_D = 3 \times 10^{-5}$ A at $V_D =$ 2 V) and the channel is not pinched-off completely until the application of negative

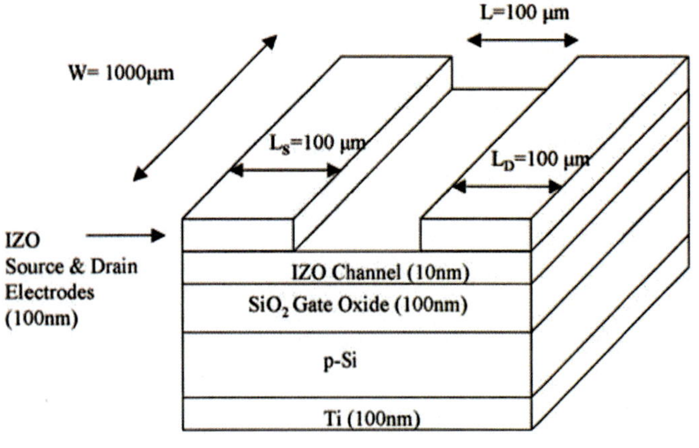

Fig. 4.4 Schematic view of the TFT test structure with W/L ratio of 1,000 m/100 m

gate voltages. Since these are n-type channels, this behavior indicates that the device is operating in depletion mode, and behaves similarly to a buried-channel MOSFET.

The transfer characteristics are investigated in the saturation regime of the $I_D - V_D$ curve. A series of measurements are made in which the gate voltage is swept from, for this example, −15 to +15 V at a fixed drain bias of 20 V. This high drain bias was selected to ensure that the devices were well into the saturation regime. Under these conditions, I_D is a function of the field effect mobility (μ), oxide capacitance (C_{ox}), the aspect ratio of the device (W/L), the gate voltage (V_G) and the threshold voltage (V_T) as shown below [7]

$$I_{D_{sat}} = \mathrm{mCox}\frac{W}{2L}V_G - V_T^2 \qquad (4.15)$$

where the threshold voltage, $V_T = V_{FB} - Q_T/C_{ox}$, reflects the flat band voltage V_{FB} and the magnitude of the channel charge, $Q_T = qN_Tt$. The carrier density in the channel is N_T and the channel thickness is t. Note from (4.7) that a transistor operating in the saturation regime at a fixed drain voltage should show a square root dependence of the drain current with gate voltage. This is shown for the IZO TFT devices in Fig. 4.5 where $(I_D)^{1/2}$ is plotted as a function of gate voltage at a constant drain voltage of $V_D = 20$ V and appears to be linear over the range 0–15 V. As indicated in (4.7), the field-effect mobility can be derived from the slope of the $(I_D)^{1/2}$ vs. V_G curve while the threshold voltage (V_T) of the device can

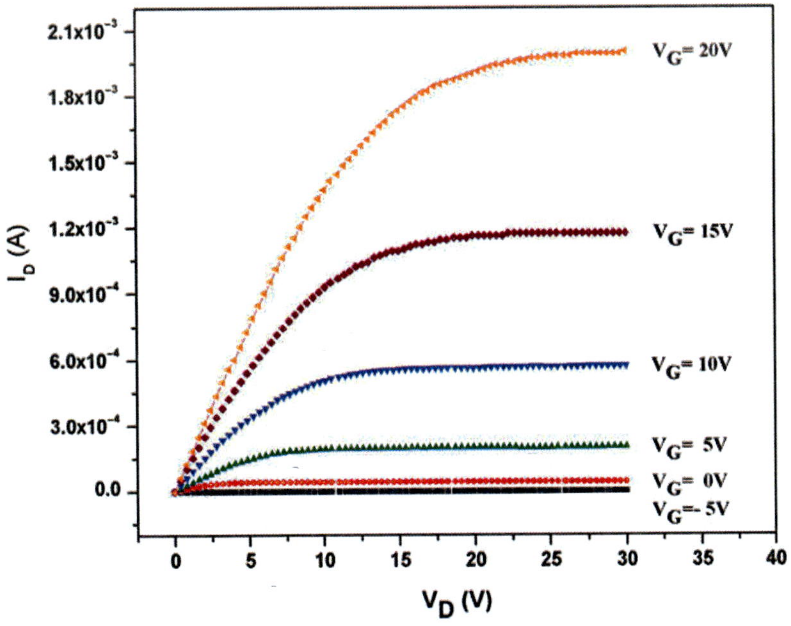

Fig. 4.5 $I_D - V_D$ characteristics at $V_G = 5$ V intervals (W/L:1,000/100) for an amorphous IZO TFT fabricated on a Si substrate with a thermal oxide dielectric

be estimated by extrapolating the linear portion of the curve to zero drain current. In this example, measurements made on 16 devices showed that the threshold voltage and field effect mobility are $V_T = -3.2 \pm 0.9$ V and $\mu = 20.4 \pm 1.9$ cm^2/V s, respectively. In a practical device fabricated using deposited dielectric and channel layers, positive or negative interface charge at the gate dielectric/channel interface due to local non-stoiciometry will result in either a negative or positive shift in the threshold voltage.

For the determination of the field effect mobility at low fields, the drain current at low drain voltages can be examined and the mobility calculated from the slope in the linear regime at different gate voltages. In this example, the output characteristics of four sets of TFT's with the W/L ratios given previously above were determined. These measurements allow an estimation of both the channel conductance and the contact resistance of the devices. This method is referred to as the transmission line method (TLM) and has been used by several research groups in the analysis of amorphous Si and organic TFT's [8, 9]. Equation (4.16) gives the relationship between drain current, I_D and drain voltage, V_D in the linear regime.

$$I_D = \mathrm{mC_{ox}} W L V_G - V_D - I_D R_C \tag{4.16}$$

The channel resistance, which is the reciprocal of the slope of the drain current vs. drain voltage (dI_D/dV_D) of (4.8) is given in (4.9). R_C represents the contact resistance and it adds to channel resistance to calculate total resistance.

$$R = \frac{L}{\mathrm{WmC_{ox}} V_G - V_T} + R_c \tag{4.17}$$

A plot of the variation of the product of resistance, (R) and channel width, (W) with respect to L for different gate voltages, V_G, gives the contact resistance and information about the intrinsic resistance of the channel material. Alternatively, the channel conductance can be plotted as a function of V_G from which the mobility can be found from the slope and the intercept with the V_G-axis gives the threshold voltage.

An important transistor parameter for use in active matrix TFT applications is the device on/off ratio. In this work the on/off ratio was determined by measuring the drain current as a function of gate voltage while holding the drain voltage at 20 V. The left axis of Fig. 4.6 shows I_D (plotted on a log scale) vs. V_G. Note that at a gate voltage of -15 V (chosen arbitrarily as the off state) the drain current is 3.7×10^{-11} A whereas it is 10^{-3} A in the on-state at $V_G = 15$ V. The resulting on/off ratio of $\sim 10^8$, is well above the level (10^6) currently required for active matrix display applications. Also plotted in Fig. 4.6 is the gate leakage current (right-most axis) as a function of V_G, which shows that the gate oxide permits negligibly small gate leakage at all V_G's tested. This is as expected for a thermally grown SiO_2 gate insulator and highlights one of the advantages of the Si-based test

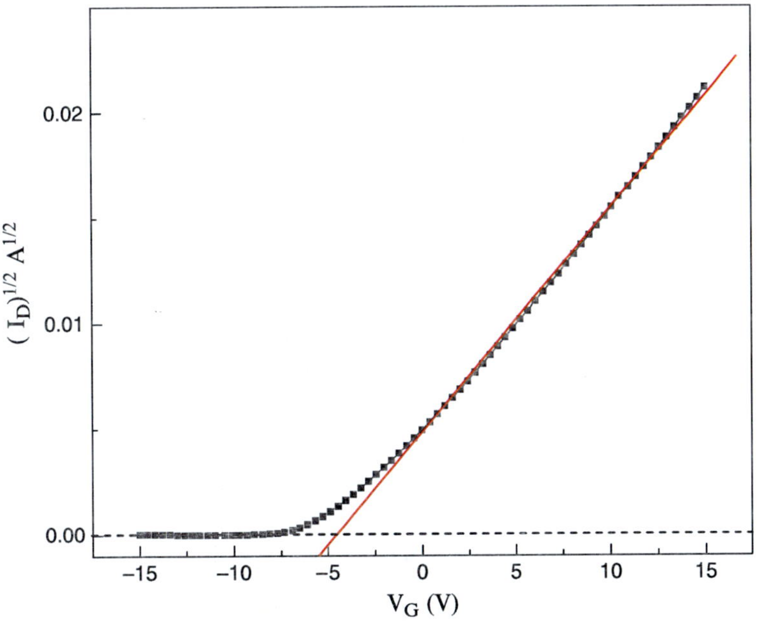

Fig. 4.6 Plot of $(I_D)^{1/2}$ vs. V_G. The threshold voltage, V_T is estimated by extrapolating the linear portion of the curve. The slope yields the field effect mobility

structure: compared to deposited oxide gate dielectric materials, thermal SiO_2 is reproducible, robust, and well understood.

4.2.5 Surface Characterization Via Scanning Probe Microscopy

Scanning probe microscopy techniques are extremely powerful tools for examining a myriad of physical phenomena. In their simplest form they provide basic interfacial/morphological information about materials; atomic force microscopy (AFM) for example, is now a routine method for the characterization of surfaces. AFM is particularly well suited to examining semiconductors or insulators such as mixed metal oxides, which are not easily evaluated at high resolution using scanning electron microscopy (SEM) due to charging and may suffer damage from the electron beam. AFM is utilized extensively for examining materials surfaces after processing to determine process parameters such as etch rate and etch uniformity. The AFM technique consists of a sharp probe whose height is recorded as it is dragged over, or in close proximity to, a surface. Typically the height is monitored via a laser and photodiode as shown in Fig. 4.8 although capacitive or other electronic monitoring systems can be used. The resolution of the measurement is dependent on the size and shape of the probe tip, the sensitivity of the height measurement, and the probe scanning resolution. The scanning motion is accomplished using piezo-electric actuators, which expand and contract in response to an applied voltage. The greater the control of the applied voltage the greater the resolution (aside from noise due to vibration) over the *x,y,z* motion. Variations of this method include oscillating techniques in which the cantilever probe is driven by a separate piezo at the cantilever's characteristic resonant frequency (typically in the kHz range). In these AC-style measurements (also known as tapping mode), the cantilever at the end of the cantilever intermittently contacts the sample surface as it is dragged across the surface. The tapping mode AFM method has higher sensitivity and intrinsic noise filtering typical of AC measurements in general and also permits the phase information of the cantilever response to be measured. By monitoring changes in the phase of the cantilever response the mechanical properties of the surface can be deduced.

Other scanning probe modes that rely on tip/sample interactions and that are most relevant for TCO materials include examination of the surface potential and surface conductivity which can be examined in a correlated fashion to obtain surface conductivity maps for conductive or semiconducting materials. These measurements range in their degree of sophistication but in the simplest form uses a conductive AFM probe in contact mode with the sample surface at fixed bias and with the current monitored. More complex methods use a trace and retrace scheme in which the topography data is taken in contact or tapping mode and then that data is used by the instrument to sweep the tip over the sample at a fixed tip-to-sample separation and the current is then monitored. Other scanning surface potential measurements use this type of approach but add an oscillating tip that is used in conjunction with an applied bias. Measurement of the current as the probe cantilever moves up and down relative to the sample surface will reveal current modulations that are due to the difference in potential of the sample and tip. This is essentially a Kelvin measurement from which a measure of the variation of the surface potential of the sample can be found and, with appropriate referencing, the work function of the surface can be obtained.

Scanning tunneling microscopy (STM) is the grandfather of the scanning probe techniques but its application to TCO materials is difficult in comparison to AFM-based measurements. The key difference between STM and AFM lies in the use of very small and probe/surface sensitive tunneling electrical current from the sample to a sharpened probe tip as feedback to obtain height information in STM. The tips used in STM are atomically sharp spikes but small tip-to-tip differences in this tip sharpness lead to imaging artifacts and difficult interpretation which are worsened when the sample conductivity is anything less than metallic. Despite these difficulties, a considerable amount of work on semiconducting systems has been reported but it is unlikely that STM would be used on a routine characterization basis. Other novel techniques that are based on STM include ballistic electron emission microscopy (BEEM) in which a three terminal configuration is used for the examination of semiconducting samples has show some very interesting results and permits the surface electronic structure of semiconducting samples to be examined [10].

4.2.6 *Optical Characterization*

For a TCO to be of interest for display applications it must transmit freely across the visible spectrum from approximately 0.4–1.5 μm. The short wavelength (UV) cut-off corresponds to the fundamental band gap of the material, while the long wavelength (IR) edge corresponds to the free carrier plasma resonance frequency. These critical optical properties are directly influenced by the carrier density and mobility. In general, for a material to be transparent across the visible spectrum its band gap must be greater than 3 eV ($\lambda \approx 0.4$ μm) and, at the other end of the absorption spectrum, the free carrier plasma resonance must be close to, or above, the near infra-red ($\lambda \approx 1.5$ μm).

An increase in free carrier density decreases the resistivity of the TCO material but also shifts the IR absorption edge to shorter wavelengths, thus narrowing the transmission window. This shift at the IR end of the spectrum is determined by the plasma oscillation of the free carriers that screen the incident electromagnetic wave via intraband transitions within the conduction band. The work of Frank and Kostlin [11] showed that the characteristic frequency for IR reflection in ITO is well predicted by the Drude model. The origin of the transmission drop at the IR end of the spectrum is seen from the frequency-dependent dielectric function ($\varepsilon(\omega)$):

$$\varepsilon(\omega) = \varepsilon_\infty \left(1 - \frac{\omega_p^2}{\omega^2 + i\frac{\omega}{\tau}}\right) \tag{4.18}$$

where τ is the relaxation time associated with the scattering mechanism and ω_p is the plasma resonance frequency. The optical transmission drops sharply near the plasma resonance frequency. The plasma resonance frequency is dependent on the carrier density (N) and inversely dependent on the effective mass (m^*) of the free carriers and is given by

$$\omega_p^2 = \frac{Ne^2}{\varepsilon_0 \varepsilon_\infty m^*} \tag{4.19}$$

Note that any effort to increase the carrier density to reduce resistivity will have the effect of shifting the plasma edge to shorter wavelengths.

The position of the UV edge is also dependent, in part, on the free carrier density in the material. Consider that the UV absorption edge is a result of the band-to-band transfer of electrons across the gap. In degenerate materials, the Fermi level lies in the conduction band and Fermi-statistics indicate that the bottom of this band is effectively filled with carriers. Consequently, the optical band gap (and UV edge) shifts to higher energy (shorter wavelength) with increasing carrier density since absorption by band-to-band transitions requires an electron transition from the top of the valence band to an unoccupied state in the conduction band. This carrier-density-dependent shift is known as the Burnstein-Moss effect [11] and straightforward analysis of the density of states in the conduction band reveals that the UV edge will shift to shorter wavelengths with increasing carrier density (n) since the change in the optical band gap energy increases with carrier density as $n^{3/2}$.

4.2.7 Work Function Measurement

Successful device design requires insight into the electronic structure of the materials used in those devices. These insights are also indispensable for the purpose of tailoring TCO materials to various device applications. The band gap, for instance, is obviously an important property that, among other things, sets the UV end of the absorption spectrum in TCO materials. Similarly, for electronic device applications, the location of the Fermi level is of great importance particularly for heterostructure device applications. For example, TCO's are used as the electrode in organic/inorganic light emitting devices and require careful band matching to enable the efficient injection of holes at the heterostructure interface.

The critical parameter for heterostructure design is the work function which is defined as the total energy needed to remove an electron from the material to vacuum. This can be expressed in terms of the Fermi energy E_F plus any additional energy (W_s) required to overcome surface effects as given in (4.9) and shown schematically in Fig. 4.7.

$$W = W_S + E_F \tag{4.20}$$

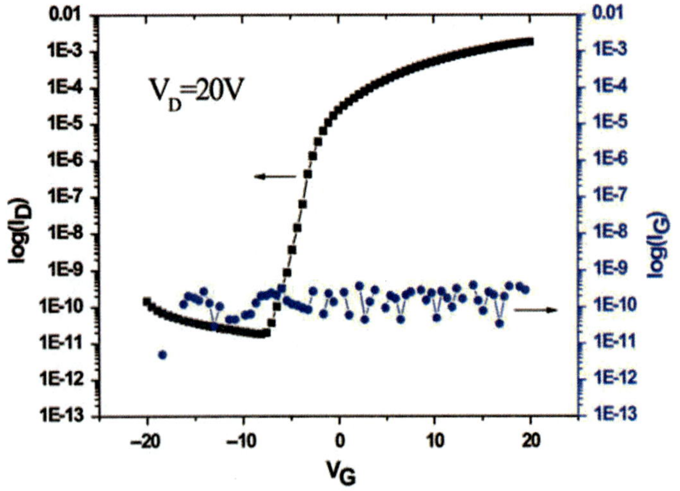

Fig. 4.7 Transfer characteristics found by measuring drain current vs. gate voltage with the drain held at 20 V to ensure saturation. The right hand axis shows the gate leakage current

Measurement techniques to determine the work function are generally of two different classes. The first and most direct measurement techniques are based on the photoelectric effect. Here the sample is irradiated in vacuum under bias and the ejected electrons are analyzed using the equation

$$hv = W + E_k \tag{4.21}$$

where E_k is the kinetic energy of the ejected electron. Due to the wide band gap of TCO materials, high energy photons are needed to liberate carriers from the materials. As a result both X-ray photoemission spectroscopy (XPS) and ultraviolet photoemission spectroscopy (UPS) are used to make a direct and absolute determination of the work function in TCO materials. While XPS is a powerful technique able to probe the surface chemistry of a material in addition to the valance states of the material, UPS is usually preferred for work function measurements as it typically has greater energy resolution. The higher energy photons of XPS make it the favored technique when a more comprehensive chemical analysis is desired. Spatial mapping is possible using photoemission but the lateral resolution is typically limited to 5 μm or greater.

The second class of work function measurements is based on the contact potential generated by the disparity in work function between different materials. If the materials are separated by a distance *d* the situation is similar to that of a capacitor and can be described by the equation.

$$C = \frac{Q}{V} = \frac{\epsilon_0 A}{d} \tag{4.22}$$

If these two materials are connected electrically through the bulk, the potential difference of the surfaces, and the associated voltage, will be nullified by mobile charges. However, because the voltage between the materials (and the number of charges needed to screen it) is dependent on the separation distance, if that separation is altered, current will flow as the charges move to establish a new equilibrium. Changing the distance (d) as a function of time will generate an alternating current as the electric field changes. The voltage V in (4.11) is set by the difference in work function between the sample and the reference. The change of the capacitance and resulting current then can be used to determine V, which then give the work function of the sample if the reference work function is known.

Techniques based on this approach are all essentially variations on the Kelvin probe technique mentioned above in the scanning probe section but there are a number of technical approaches to performing these measurements. The simplest of them use large area probes to reproduce the basic geometry shown in Fig. 4.8. This can provide a good measure of the average work function over a relatively large area. Baikie et al. [12], on the other hand, have demonstrated the use of a very small area oscillating probe which, while requiring a more technically sophisticated approach, allows relatively small areas to be examined. It is also possible to make these measurements using an electrostatic voltmeter or other electrostatic

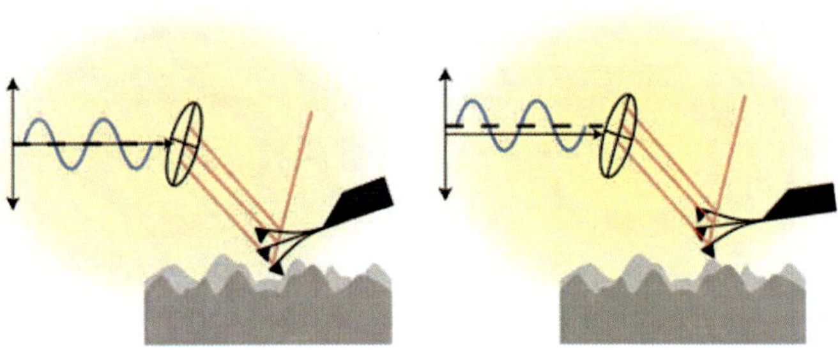

Fig. 4.8 Scanning probe microscopy schematic showing cantilever deflection and resulting change in position of the HeNe laser beam after reflection and onto a four quadrant detector

probe. This approach, however, does not rely on current detection and is therefore not strictly speaking a Kelvin probe measurement but is the voltage analog to it. Since the electrostatic field seen by the probe may be easily modulated without physically moving the probe, it does offer advantages for some experimental geometries.

Kelvin probe or electrostatic measurements are predicated on the ability of charge carriers to move in the material. This means that care must be used in the examination of insulating samples. TCO materials with lower electron mobility present additional challenges as the experimental uncertainty of the work function measured using the Kelvin probe-based techniques will increase as the ability of charge carriers to move in response to the fluctuating field decreases. In addition, because Kelvin probe measurements are fundamentally based on the difference in work function, their ability to make an absolute work function determination is dependent on one or more reference materials of known work function. As photoemission is typically used to establish standards, the stability of the work function of these standards from vacuum to atmosphere is crucial. Another critical aspect in all work function measurements is the effect of buried interfaces where the chemistry may create interfacial dipoles and significantly alter the charge transport at the interface.

4.3 Microstructural/Crystallographic Characterization

TCO thin films are deposited by practically all available physical vapor deposition methods and in each method the details of the microstructure of the deposited film is highly dependent on the process conditions and methods. In sputter deposited films considerations include substrate temperature, kinetic energy of the deposition flux, and composition. In the case of materials that have a relatively simple equilibrium crystal structure, such as ZnO (wurtzite crystal structure with two molecular units per unit cell) or SnO_2 (rutile crystal structure with two molecular units per cell), the deposited film is typically crystalline in the as-deposited state regardless of the deposition conditions. The microstructure of the deposited thin film, however, will range from large grained columnar to nanocrystalline and is often highly textured (particularly in the case of ZnO) with a specific growth plane favored. In TCO's with more complex equilibrium crystal structure such as those based on Indium oxide (bixbyite crystal structure, 16 molecular units per unit cell) typical physical vapor deposition processes such as DC or RF sputtering used to deposit films onto unheated substrates produce an amorphous phase that preserves the short range order (metal – oxygen octahedral structural units) but which do not possess the long range order of the crystalline phase. Mixed In-rich cation oxides containing, for example, Zn, Ga, and/or Sn also may assume an amorphous structure.

The deposition of Indium Oxide rich (ITO or IZO $<$ 10 wt% tin or zinc oxide respectively) thin films onto unheated substrates results in the formation of an amorphous phase. Both pure indium oxide and ITO (10 wt% SnO_2) crystallize from the amorphous state at remarkably low temperatures. For example, a-ITO crystallizes at measurable rates [13] at temperatures as low as 150°C. The addition of zinc oxide to indium oxide thin films deposited at room temperature significantly increases the stability of the amorphous phase with a-IZO remaining amorphous indefinitely at annealing temperatures as high as 500°C. The increased stability of the amorphous phase is likely due to the insolubility [14] of Zn in the bixbyite lattice. This means that the crystallization of the amorphous IZO phase requires either phase separation or the creation of a metastable crystalline phase that can accommodate Zn. The following sections discuss the characterization of one metastable crystalline phase that results when amorphous IZO is heated to cause crystallization.

The optical and electrical performance of TCO's are intimately tied to the fundamental band structure of the material. The band structure, in turn, depends on the periodic distribution of potential in the crystal. This would, at first glance, suggest that amorphous materials where the long-range periodicity of the crystalline structure has been disrupted would yield materials with dramatically different optical and electrical properties. In indium oxide based materials, however, this proves not to be the case and consequently, the amorphous phase is technologically very important. Amorphous ITO and IZO, for example, have desirable isotropic wet etch characteristics combined with optical and electrical performance that nearly matches that of the crystalline phase. In the case of multicomponent oxides (ZITO, for example), the addition of Sn and/or Zn beyond the equilibrium solubility limit is possible in the amorphous phase which allows the displacement of costly indium with Sn or Zn. In this section we describe some techniques for the characterization of the microstructure of TCO materials with an emphasis on the analysis of amorphous near neighbor structure.

The characterization of the microstructure and crystallography is most effectively performed using Transmission electron microscopy – a well known technique which is well described in numerous texts. For work on thin films of multi-component TCO's the value of this technique is that it provides, from the same region of material and at the same time, a structural image and a diffraction record of the local crystallography. A simple comparison of results obtained from simple

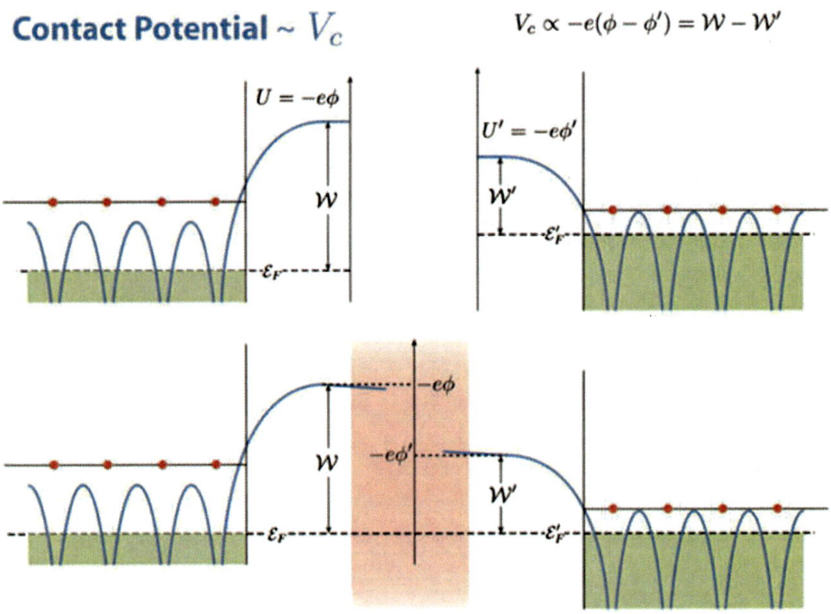

Fig. 4.9 Energy diagram Kelvin probe system. The *shaded area* of width *d* has an induced electric field if the samples on both sides are in electrical contact which is proportional to the difference in the work functions of the two materials

θ–2θ X-ray diffraction of IZO deposited onto a crystalline substrate (NaCl) illustrates the point. In Fig. 4.9a an XRD spectra taken from bixbyite-phase IZO deposited at 350°C is presented and appears to show a single peak. Whether the crystalline film is single crystal or textured polycrystalline such that a single family of planes lie parallel to the substrate surface but randomly rotated in the plane cannot be determined from this data. In contrast, Fig. 4.9b shows a TEM bright field/dark field image pair revealing polycrystalline material and the selected area diffraction (SAD) image (inset) in panel (b) reveals that there is considerable in-plane texturing of the polycrystalline material. The work presented below is intended to further illustrate the use of TEM to characterize microstructure and crystallography in a selected TCO (IZO).

4.3.1 Crystal Structure and Microstructure: Crystalline Indium Oxide

As described elsewhere in this volume, when Indium oxide is deposited under pseudo-equilibrium conditions, it adopts the cubic C-type rare earth oxide structure (rare earth refers to the crystal prototype structure not the element) known as bixbyite (space group, Ia3). The atomic arrangement for the C modification was first determined for the mineral bixbyite $(Fe,Mn)_2O_3$. Bixbyite has an 80-atom unit cell that corresponds to a body centered cubic lattice with 16 molecules and a lattice parameter of 1.012 nm [11]. The bixbyite structure is based on edge-sharing InO_6 octahedra arranged in a $2 \times 2 \times 2$ array. The bixbyite arrangement can be derived from the CaF_2 structure [11] by removing one-quarter of the anions and then slightly repositioning the remaining atoms. In the fluorite structure, however, the metal atoms are coordinated with eight anions while in the bixbyite structure the metal ions have sixfold anion coordination. The 32 indium atoms in bixbyite are distributed in two special symmetry related positions in the Ia3 space group, with 24 in (d) and 8 in (b) while the 48 oxygen atoms are found the general positions. The In atoms in (d) and (b) sites are designated as In (2) and In (1), respectively. Each In (1) is coordinated with six oxygen atoms with a single In-O distance of 2.18 Å. The oxygen atoms lie at the corners of a cube with two body diagonally opposite corners unoccupied. In (2) is also surrounded by six oxygen atoms. In this case two face diagonally opposite corners are unoccupied, and there are three different sets of In-O distances; (2) 2.13 Å, (2) 2.19 Å and (2) 2.23 Å.

At the ZnO-rich end of the In_2O_3-ZnO phase diagram, there are a series of homologous compounds with the chemical formula, $Zn_kIn_2O_{k+3}$, that exist in the temperature range 1,100–1,400°C [15]. These $Zn_kIn_2O_{k+3}$ compounds have two closely related structures consisting of either a hexagonal unit cell (space group $P6_3/mmc$) for $k = 2n$ or, for $k = 2n + 1$, a rhombohedral unit cell (space group $R\bar{3}m$) where n is an integer [16]. In both these structures, the (111) planes of the bixbyite form of In_2O_3 are interleaved in the c-axis direction of the hexagonal wurtzite ZnO structure. In this manner, a range of Zn-rich compositions can be accommodated by modifying the c-axis repeat distance and layer thickness while preserving both the tetrahedral oxygen coordination favored by Zn and the octahedral coordination that is favored by In.

In the examples that are presented below, amorphous IZO films were DC magnetron sputter deposited to a thickness of about 100 and 200 nm from a commercial (Idemitsu Corp) IZO (In_2O_3-10 wt% ZnO) target onto suitable substrates. A sputter power of 0.22 W/cm^2 at 280 V, a target-substrate distance of 10 cm, and pure (0 vol.% O_2) Ar sputter gas at a pressure of 10 mTorr were used. To compare the effect of Zn and Sn on crystallization of the amorphous phase, a set of ITO films (In_2O_3-10 wt% SnO_2) were deposited under identical conditions (at zero reactive oxygen rather than the optimum of 0.1 vol.%) and were annealed in air at 500°C for 150 min. The samples were evaluated before and after annealing using X-ray diffraction (XRD) in a glancing incidence angle thin film system with parallel optics and Cu kα radiation at 40 kV and 30 mA. Plan-view and cross-section TEM samples were prepared using mechanical thinning followed by ion milling.

In the as-deposited state, both the ITO and IZO films deposited at room temperature under the conditions described above are amorphous. After annealing in air at 500°C for 150 min, sharp crystalline peaks become clearly visible in X-ray diffraction spectra taken from both samples and shown in Fig. 4.10. There is also a broad

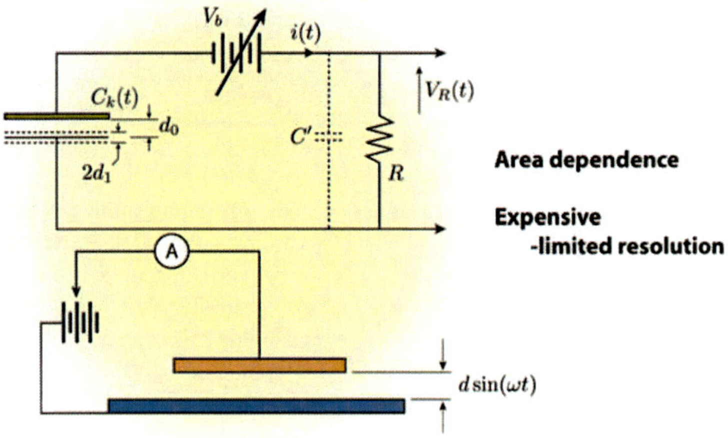

Fig. 4.10 Schematic of Kelvin probe measurement systems. *Lower figure* shows basic setup while *upper* is for a system that uses a back bias to null the induced electric field

peak in the spectra ($15 < 2\theta < 30$), in this X-ray data that is due to the glass substrate. The ITO/glass sample (bottom spectrum) shows peaks that are all readily indexed as belonging to the bixbyite In_2O_3 phase, whereas the peaks in the IZO/glass spectrum (top spectrum) match neither bixbyite nor any of the homologous compounds previously reported in the In_2O_3-ZnO system [15, 16]. Furthermore, this set of peaks does not appear in any of the In-containing phases in the commonly available X-ray databases.

TEM studies were performed [17] on both the single and mixed phase samples to establish the microstructure, composition, and crystallography of the unknown IZO phase. The microstructural difference between bixbyite and the unknown phase is clearly established by comparing the microstructure in both mixed and phase-pure samples. This is shown in the cross-sectional TEM image presented in Fig. 4.11, which is taken from a typical region of the two-phase sample formed after annealing a-IZO on SiO_2 in air at 500°C for 150 min. In this figure, the Si substrate appears in the bottom of the figure, followed by a ~300-nm-thick SiO_2 layer, followed by the 200-nm-thick crystallized IZO layer. Tilting experiments showed strong diffraction contrast from all regions of the sample (establishing that it is fully crystalline) and revealed two distinct crystalline microstructures: (a) a region of lighter intensity that is free of strain contrast, typical of the bixbyite form of In_2O_3 and, (b) a highly defected region that appears in Fig. 4.11 as the darker areas of mottled contrast.

Microdiffraction experiments (including convergent beam techniques) in the transmission electron microscope were used to analyze the crystal structure of the

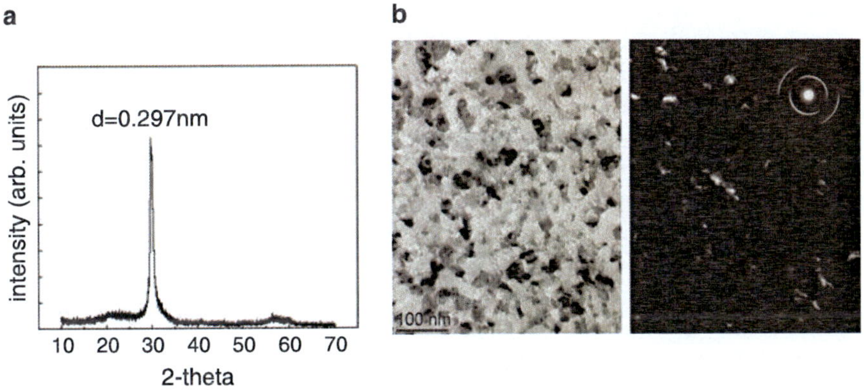

Fig. 4.11 (**a**) XRD of a textured crystallized IZO film, and (**b**) a bright field/dark field TEM image pair showing the polycrystalline film with the selected area diffraction pattern (*inset*) showing a wide arc of in plane orientations

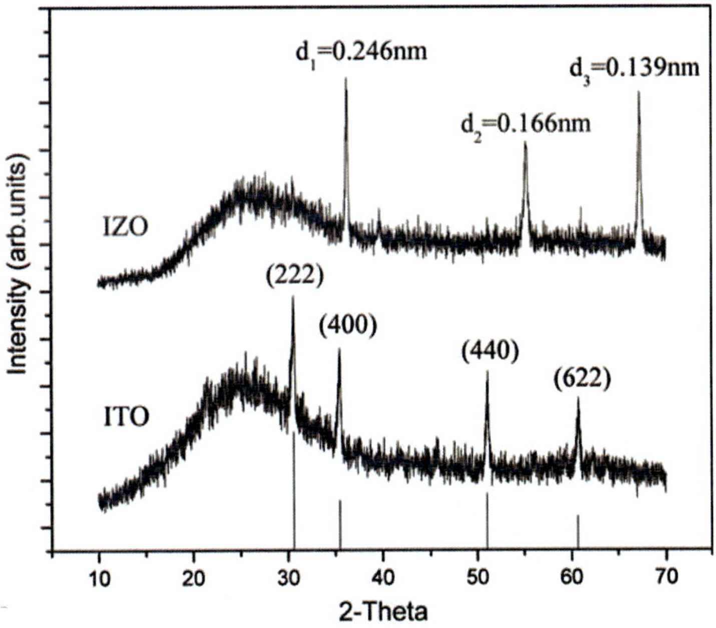

Fig. 4.12 XRD patterns from crystallized ITO (*bottom*) and IZO (*top*)

unknown phase. Figure 4.12 compares microdiffraction patterns collected from bixbyite (left set of images) and the unknown (right set of images). Analysis of these and other electron diffraction patterns from these samples reveals that the unknown phase is based on the high-pressure modification of In_2O_3, which has the corundum structure (space group, R$\bar{3}$*c*). [18] The zone axes presented in Fig. 4.12 are, on the left side, the bixbyite [111], and [001] and, on the right hand side, the corundum-based phase of Indium oxide [0001], [1$\bar{1}$00].

An energy dispersive spectroscopy (EDS) compositional analysis was performed on both the phase-pure unknown and mixed bixbyite/unknown phases to establish any compositional changes that may have occurred during crystallization due to, for example, loss of Zn from the film or by reaction with the substrate. As-deposited, a-IZO was used as a calibration standard for In and Zn by assuming that sputter deposited a-IZO films have the same composition as the sputter target from which they were deposited. Energy dispersive X-ray spectroscopy reveals that the as-deposited a-IZO films, the unknown IZO phase, and the bixbyite IZO phases all have the same In:Zn stoichiometry and no additional elements were detected. Unfortunately, EDS is not a reliable method for establishing oxygen stoiciometry.

Figure 4.13 shows a cross-sectional bright field image of a region of a single IZO corundum grain and an associated selected area diffraction image. The bright field image clearly shows bright/dark modulations in contrast that, although wavy, can be seen to generate satellite reflections in the <0001> crystallographic direction in the diffraction pattern. Modulations of this sort may be due to structural distortions, compositional changes or both. Both the diffraction pattern and the bright field image reveal a periodicity of approximately 2.3 nm. The exact nature of the compositional modulations observed here is not yet known, however, it is noted that the observed periodicity (2.3 nm) is not an integer multiple of the c-repeat distance of the high-pressure In_2O_3 corundum phase where $c = 1.45$ nm. Bright field images show that while the periodicity lies in the <0001> direction, the compositional modulations that we observe are wavy and do not have a well defined (0001) habit plane.

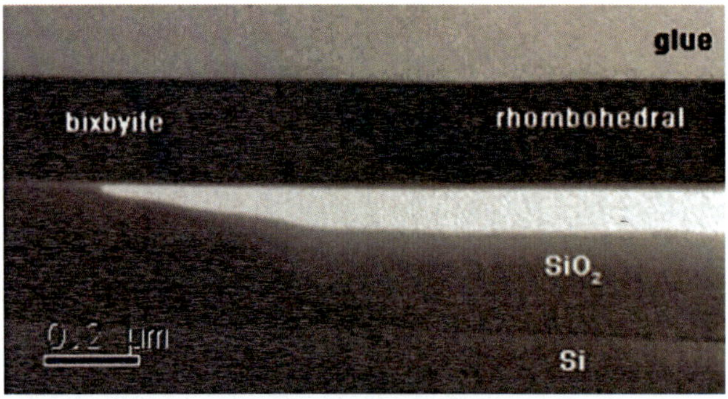

Fig. 4.13 Cross-sectional TEM image showing a crystallized IZO sample that contains both the bixbyite and the rhombohedral phase of IZO

4.3.2 *Electron Diffraction Studies of Amorphous In_2O_3-Based Thin Films*

The amorphous TCO's are of increasing importance in applications where the isotropic etch behavior of the amorphous phase and the advantages of room temperature deposition outweigh the slight increase (vs crystalline ITO) in resistivity. The structural characterization of these amorphous materials is challenging since these materials lack long range periodicity and consequently the well-defined diffraction patterns of crystalline materials are not available. Nevertheless, the scattering of X-ray, neutrons and electrons from amorphous materials is used to perform structural analysis in amorphous materials using a variant of the radial distribution function analysis described below. Electron scattering is of particular interest if small lateral dimensions must be sampled since the interaction cross section of electrons is much higher than alternatives such as X-rays and, hence, very small volumes (thin films and small selected areas) can be sampled (areas <1 μm in width and as thin enough to be electron transparent). The theory of scattering from amorphous materials is briefly reviewed below followed by a brief description of the procedures used in a recent Brown University PhD thesis study [19] of electron scattering from amorphous I(Z)O films.

Atomic arrangements in amorphous materials lack the long range order that is a defining characteristic of crystalline materials. Therefore, instead of a periodic solid, a homogeneous distribution of atom centers with an average density is defined to describe the structure. In this section, the derivation of the radial density functions from scattering information will be explained according to the treatment by Warren [20]. For a system composed of many atoms, the coherent scattering amplitude can be expressed as:

$$I(q) = \sum_j \sum_k f_j f_k \exp[2\pi i q(r_j - r_k)] \tag{4.23}$$

where f_j, f_k are the atomic scattering amplitudes, and q is the scattering vector. Since amorphous solids are isotropic the system can be considered to be spherically symmetrical. Therefore, $r_{jk} = r_j - r_k$ can have all positions with equal probability on the surface of the sphere. The average of the exponential term in (4.23) can then be expressed as:

$$\langle \exp[2\pi i q r_{jk}] \rangle = \frac{1}{4\pi r_{jk}^2} \int_{\phi=0}^{\pi} \exp[i q r_{jk} \cos\phi] 2\pi r_{jk}^2 \sin\phi \, d\phi \tag{4.24}$$

This yields the Debye formula, which describes the intensity of scattering from a random array of atoms:

$$I(q) = \sum_j \sum_k f_j f_k \frac{\sin q r_{jk}}{q r_{jk}} \tag{4.25}$$

If we consider that the material is monatomic $f_j = f_k = f$ and introduce the density function $\rho_j(r_{jk})$ for the origin atom j, we obtain:

$$I(q) = \sum_j f^2 + \sum_j f^2 \int \rho_j(r_{jk}) \frac{\sin qr_{jk}}{qr_{jk}} dV_k \tag{4.26}$$

The microscopic density function averaged over all atoms, j, in the sample can be written as $\rho(r) = \left\langle \rho_j(r_{jk}) \right\rangle$ and considering the macroscopic average density as ρ_0, (4.26) can be reformulated by adding and subtracting ρ_0 as:

$$I(q) = \sum_j f^2 + \sum_j f^2 \int 4\pi r^2 [\rho(r) - \rho_0] \frac{\sin qr}{qr} dr$$
$$+ \sum_j f^2 \int 4\pi r^2 \rho_0 \frac{\sin qr}{qr} dr \tag{4.27}$$

The term, $4\pi r^2 \rho(r)$ in (4.27) represents radial distribution function. The average number of atom centers lying between r and $r + dr$ from the center of an arbitrary origin atom is calculated as $4\pi r^2 \rho(r)dr$. Since amorphous materials don't have long-range order $[\rho(r) - \rho_0]$ tends to zero for distances greater than a few atomic separations. The third term in (4.27) represents small-angle scattering due to finite sample size and can be neglected. Considering that summation over j gives N, the total number of atoms in the sample, (4.27) can be rewritten as:

$$I(q) = Nf^2 + Nf^2 \int_0^\infty 4\pi r^2 [\rho(r) - \rho_0] \frac{\sin qr}{qr} dr \tag{4.28}$$

We can now define the reduced scattering intensity function (interference function), which can be observed experimentally:

$$F(q) = \left[\frac{I(q)/N - f^2}{f^2} \right] \tag{4.29}$$

By combining (4.28) and (4.29) and rearranging, it can be shown that:

$$qF(q) = 4\pi \int_0^\infty r[\rho(r) - \rho_0] \sin qr dr \tag{4.30}$$

Using the relationship in (4.30) we can now invert the interference function by taking the Fourier transform to obtain:

$$r[\rho(r) - \rho_0] = \frac{1}{2\pi^2} \int_0^\infty qF(q) \sin rq dq \tag{4.31}$$

which, with further rearrangement gives the radial density function:

$$G(r) = 4\pi r[\rho(r) - \rho_0] = \frac{2}{\pi}\int_0^{\infty} qF(q)\sin rq\,dq \tag{4.32}$$

G(r) describes the deviation of local atomic density ρ(r) from average atomic density ρ_0, and oscillates around zero. The radial distribution function, RDF, can be calculated by:

$$R(r) = 4\pi r^2\rho(r) = rG(r) + 4\pi r^2\rho_0 \tag{4.33}$$

For a polyatomic system, (4.26) needs to be modified as:

$$\begin{aligned} I(q)/N = &\sum_{i=1}^{n} x_i f_i^2 + \sum_{i=1}^{n}\sum_{j=1}^{n} x_i f_i f_j \int_0^{\infty} 4\pi r^2\rho_{ij}(r)\frac{\sin qr}{qr}dr \\ &- \left[\sum_{i=1}^{n} x_i f_i\right]^2 \int_0^{\infty} 4\pi r^2\rho_0\frac{\sin qr}{qr}dr \end{aligned} \tag{4.34}$$

where x_i is the atomic fraction of element i, f_i is the scattering factor of element i and $\rho_{ij}(r)$ is the average number of j atoms per unit volume at distance r from i atom. Defining the following functions:

$$\langle f\rangle = \sum_{i=1}^{n} x_i f_i \tag{4.35}$$

$$\langle f^2\rangle = \sum_{i=1}^{n} x_i f_i^2 \tag{4.36}$$

$$\tilde{n}(r) = \sum_{i,j} W_{ij}\rho_{ij}(r) \tag{4.37}$$

$$W_{ij} = x_i f_i f_j/\langle f\rangle^2 \tag{4.38}$$

leads to the reformulation of (4.23) as:

$$I(q)/N = \langle f^2\rangle + \langle f\rangle^2\int_0^{\infty} 4\pi r^2[\rho(r) - \rho_0]\frac{\sin qr}{qr}dr \tag{4.39}$$

The reduced interference function is now:

$$F(q) = q\left[\frac{I(q)/N - \langle f^2 \rangle}{\langle f \rangle^2}\right] = \int_0^\infty 4\pi r[\rho(r) - \rho_0] \sin qr dr \qquad (4.40)$$

The analysis above describes coherent scattering of X-rays from amorphous solids. X-rays are scattered by the electrons present in the atoms of the sample, whereas electrons are scattered both by the electrons and nuclei of the atoms [21]. The relationship between form factors for X-rays and electrons is given by Mott's formula:

$$f^e(q) = \frac{m_e e^2}{2h^2}[(Z - f^x(q))/q^2] \qquad (4.41)$$

The principles governed in extracting RDF information from scattering data is the same for X-rays and electrons. However, scattering amplitudes for electrons decrease very rapidly as scattering vectors increase, and the data cannot be collected for large values of scattering vectors [22] Electron diffraction patterns also suffer from a background caused by inelastically scattered electrons. There are different scattering mechanisms that contribute to the inelastic background such as plasmon excitation, secondary electron excitation and Auger electrons [23]. For a quantitative analysis of diffraction data this background has to be removed. Inelastic scattering is eliminated mainly by two approaches: In the first method, the electron beam is tilted from the optical axis up to the maximal scattering angles and after each step the zero-loss filtered intensity is measured using a post-column electron energy loss spectrometer (EELS) [24]. The second approach involves the usage of an energy filter, a magnetostatic so-called omega-filter, between the sample and the detector which eliminates the inelastically diffracted electrons [25]. In the absence of energy filtering, a method was developed by Fujime [26], which assumes that incoherent scattering of an amorphous film is the same as the background intensity of the same region after crystallization. This technique is employed in this study for the analysis of diffraction patterns collected from amorphous In_2O_3-based thin films.

4.3.3 Structural Analysis of Amorphous IZO Films

The procedure used for investigating short range order in amorphous thin films starts with the collection of selected area diffraction patterns. In this example, the experiments were performed on both amorphous IZO and pure In_2O_3 (IO) films in order to analyze structural differences between two materials. This required collection of scattering data from the two materials under identical illumination

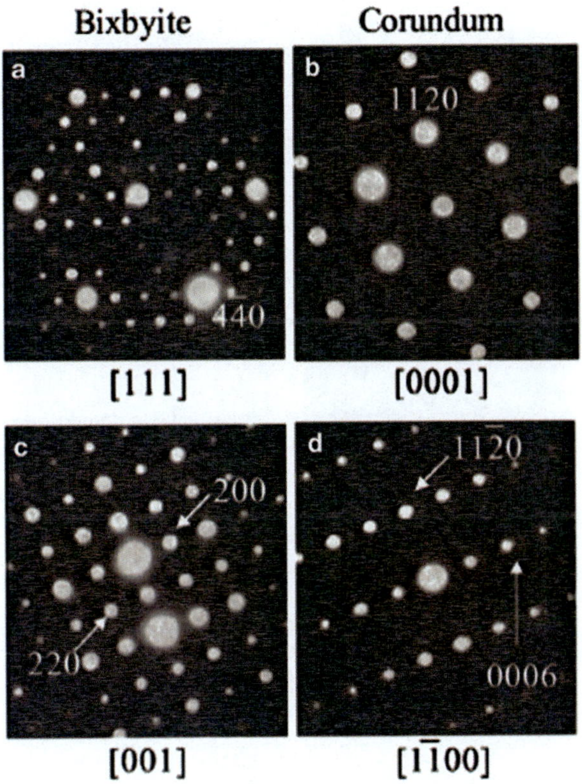

Fig. 4.14 Microdiffraction patterns taken from the (**a**) [111] zone axis of the bixbyite phase, (**b**) the [0001] zone axis of the corundum phase, (**c**) the [001] bixbyite zone and (**d**) the [$1\bar{1}00$ corundum zone axis

conditions. To achieve this, amorphous IZO and IO films were deposited on NaCl substrates to a thickness of 30 nm and then floated off the NaCl crystal onto a single Cu mesh grid. Selected area diffraction (SAD) patterns were obtained from both films without changing any of the illumination conditions by simply translating the stage from the IO to the IZO side of the grid. In situ annealing (in the microscope) at a temperature of approximately 750 crystallized the amorphous IZO (IO crystallized at 200°C) and was used to calibrate the amorphous data. Figure 4.14 shows selected area diffraction patterns of the same IZO film before and after crystallization.

The intensity of the scattering data has contributions from both incoherent and coherent scattering:

$$I_{obs}(q) = I_{coh}(q) + I_{incoh}(q) \tag{4.42}$$

We are interested in the coherent scattering intensity, I_{coh}, and this requires the subtraction of the incoherent contribution, which is obtained by fitting a background curve to the crystalline diffraction profile. Figure 4.15 illustrates line

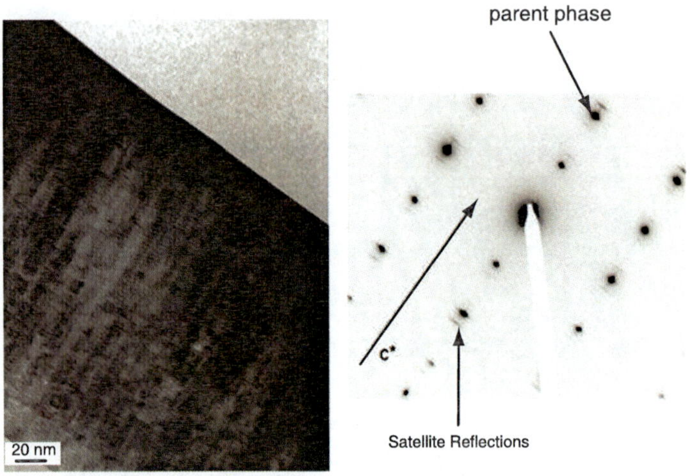

Fig. 4.15 Corundum IZO film oriented in the $[1\bar{1}00]$ zone showing wavy bright/dark contrast modulations with a ~2.3 nm period running in the <001> direction. The period and direction of the modulations are revealed as satellite peaks in the selected area diffraction $[[1\bar{1}00]$ pattern

profiles collected from IZO films before and after crystallization with the background curve superimposed on the crystalline profile ($q = 2\sin\,\theta/\,\lambda$). An exponential decay function given in (4.43) was used in constructing the background curve:

$$I_{bg}(q) = 15.43 + 22437\ \exp(-q/0.025) + 297.7\ \exp(-q/0.46) \qquad (4.43)$$

Equation (4.44) gives the extraction of interference function from observed intensities, I_{obs} by a normalization step:

$$\frac{F(q)}{q} = \frac{I_{obs}(q) - aI_{bg}(q) - b\langle f^2\rangle}{b\langle f\rangle^2} \qquad (4.44)$$

where the constants a, b are normalization parameters. Before the application of the Fourier transform, the interference function was also multiplied with a damping factor, $M(q) = \exp(-mq^2)$. This is necessary to reduce the effects of noise in F(q) at large q because of multiplying factor q and dividing factor b<f>^{2} [3]. It should be noted that the choice of the background curve in (4.43), and the determination of normalization parameters are rather arbitrary. However, a criterion of judging the correctness of estimation for interference function can be established [26].

Equation (4.19) suggests that the radial density function should be linear in r for distances less than the shortest interatomic distance in an amorphous material:

$$G(r) = -4\pi r\ \rho_0 \qquad (4.45)$$

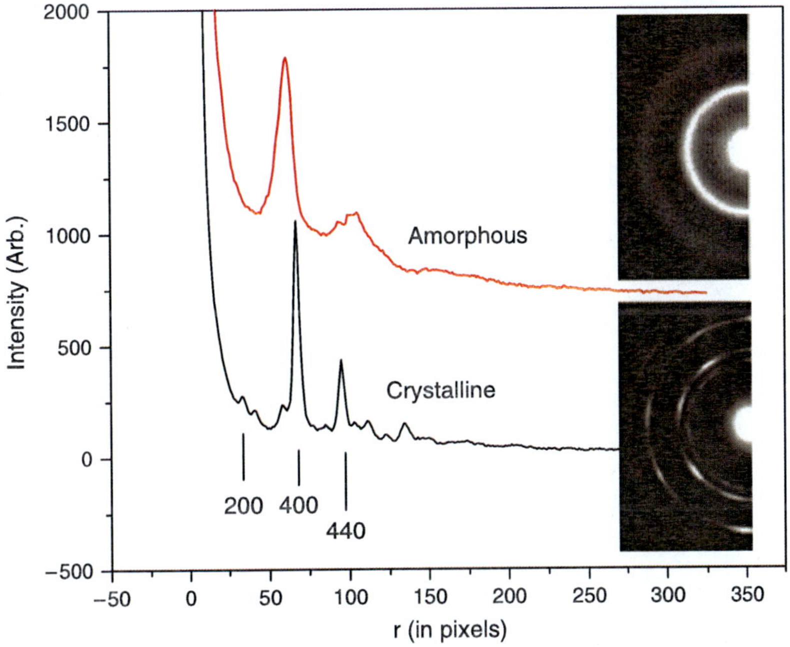

Fig. 4.16 Selected area diffraction patterns (*left side*) and profiles (*right side*) of IZO films before (*top*) and after (*bottom*) crystallization

The parameters in the construction of the interference function were chosen such that G(r) follows the expression given in (4.45) at short distances. This requires the knowledge of average atomic density, ρ_0. For the purposes of this analysis, it is assumed that both amorphous and crystalline In_2O_3 have the same density of 0.077 1/A^3. Figure 4.16 demonstrates the interference function and its Fourier transform, G(r), when a = 1.2, b = 0.1 and m = 2. It is worth noting that the interference function terminates at q ~2 1/Å, which is the limit of data collected.

The first peak at 0.2 nm in Fig. 4.16b gives the In-O correlation, whereas the peak at 0.36 nm is due to In-In correlations – note that only one correlation is present here whereas in the crystalline state there are two distinct In-In correlations corresponding to two different In positions. Different a, b parameters and damping factor, m, can be chosen to obtain the interference function. Although the peak positions are not sensitive to the choice of these parameters the intensities can change considerably making interpretation of coordination density difficult. This brings about errors in the estimation of the number of first, second, third, etc. neighbors about a given atom that are calculated by integrating the area under the peaks in radial distribution function:

$$C = \int_{r-\Delta}^{r+\Delta} 4\pi r^2 \rho dr \tag{4.46}$$

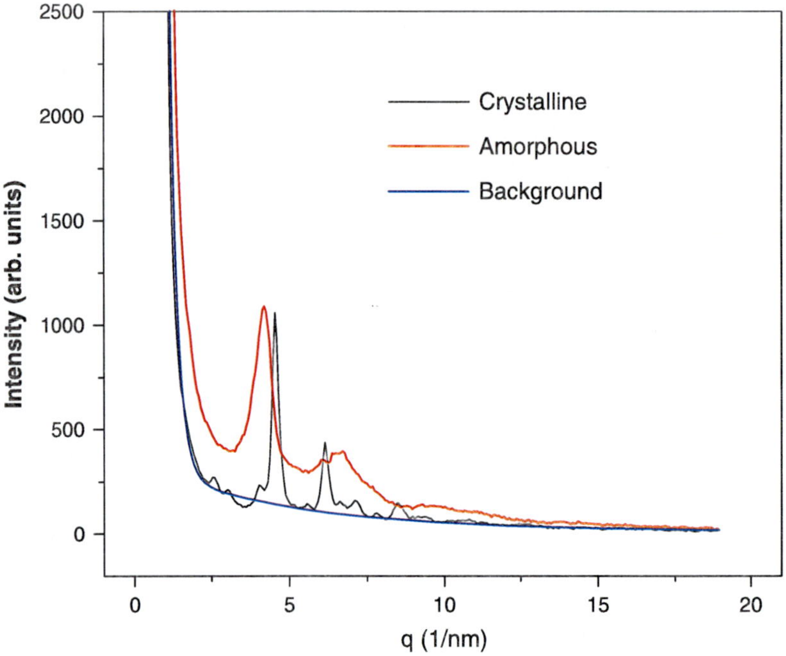

Fig. 4.17 Diffraction profiles of amorphous and crystalline patterns along with the fitted background curve

The pair of curves in Fig. 4.17 shows the RDF and average atomic densities in our amorphous IZO films. The peaks in RDF curves can be represented as Gaussian functions in order to calculate coordination numbers. Using a Gaussian approximation (Fig. 4.17b) to the first peak, the area is estimated to be 6.2 (±0.5). As mentioned before, determination of the coordination numbers are very sensitive to the errors in choosing parameters for normalization. However, if it is assumed that G (r) is $-4\pi r\rho_0$ for small r, a first nearest neighbor number of ~6.2 is found, which is within the error limits of these measurements shows that In atoms in the amorphous state are sixfold coordinated with oxygen as they are in the crystalline state.

4.4 Measurement of Biaxial Stress in TCO Thin Films

Thin films deposited onto rigid substrates are constrained by the substrate in a way that develops biaxial stresses in the film. The origin of these stresses can be understood in terms of all of the processes that would change the volume of the film if it were not constrained by the substrate. Thermal stresses, for example, arise due to differences in the thermal expansion coefficient of the substrate and film and are readily predicted. Intrinsic stresses, however, that develop during the growth of the film depend sensitively on deposition conditions such as substrate temperature, total Ar pressure during deposition, and sputter power and are not well understood. These stresses are of importance since, on thin or deformable substrates, they can lead to curling and/or premature failure. In this section, we briefly describe a measurement approach and show how it is used to determine some of the stresses that can develop as a result of the deposition of thin films of ITO at room temperature at different oxygen partial pressures, the stresses that result when the amorphous phase undergoes crystallization, and finally, the stresses that develop as a result of the creation or elimination of the point defects that contribute carriers in un-doped indium oxide thin films.

Film stress was measured in the as-deposited state and during crystallization by isothermal annealing in air at 250°C using an in situ multibeam optical stress sensor (MOSS) technique. MOSS is a laser-based method for measuring wafer curvature that uses an etalon to create multiple, highly parallel beams from a single low power He-Ne laser beam. The parallel beams are reflected off the substrate surface and into a CCD camera. The change in spacing of the beams incident on the CCD during the transformation yields a measure of the change in substrate curvature from which film stress is obtained [27]. Application of simple elasticity theory yields film stress, film strain and molar volume change.

Changes in carrier density with oxygen content during deposition of pure indium oxide is understood to be due to changes in the concentration of doubly charged oxygen vacancies which are known to contribute two free electrons per defect. Since vacancies are likely to affect the molar volume of the material, changes in vacancy concentration should be accompanied by a change in film stress. It is evident that an increase/decrease in volume of a film bonded to a rigid substrate will result in an overall increase/decrease in compressive film stress. This film stress was measured for all three sets of samples as a function of annealing time and is presented in Fig. 4.18. Note first that in the as-deposited state, all of the samples are compressively stressed with the magnitude of the compressive stress highest (most negative) in the high oxygen case and lowest in the low oxygen case. This trend is in good agreement with results reported by others [28]. Note that it is well known that the sputter film stress is also a function of total process gas pressure but, in this work, a fixed total gas (Ar + O_2) pressure of 10 mTorr was used. Examination of the changes in stress for all three of the samples presented in Fig. 4.18 reveals three regimes. During the first few minutes, while the sample comes to temperature, there is an increase (<0.1 GPa) in compressive stress (i.e. becomes more negative) due to the difference in thermal expansion coefficient between ITO and the glass substrate. This is followed by a marked decrease (0.2 – 0.3 GPa) in the compressive stress and is indicated on Fig. 4.18 for the 2% oxygen samples as the transformation stress. Note that this change occurs during the first 75 min of the heating process while the amorphous oxide undergoes crystallization. Finally, after long times at temperature in air there is a gradual linear decrease in compressive stress as oxygen vacancies are oxidized.

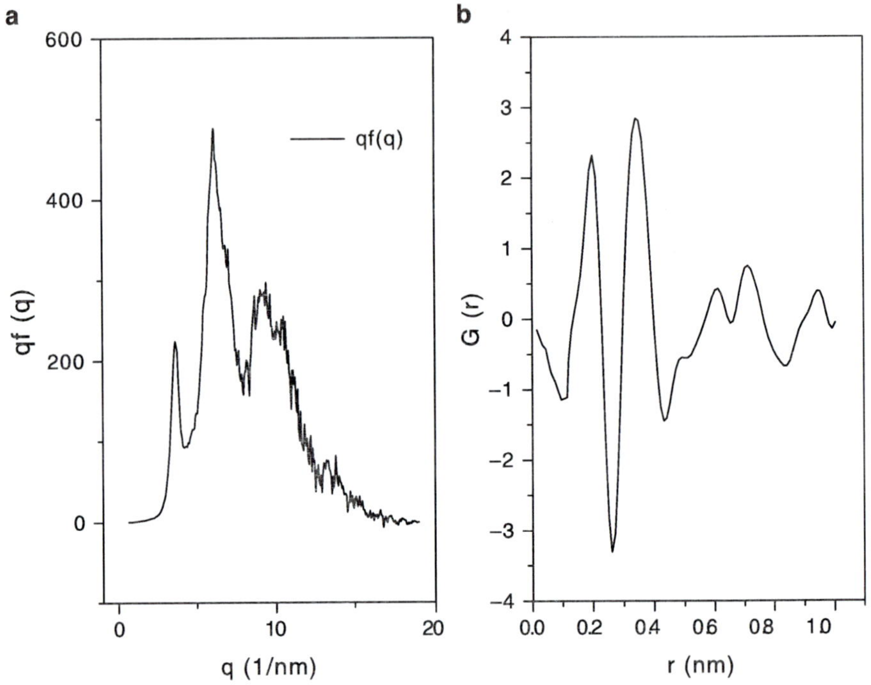

Fig. 4.18 (**a**) The interference function qF(q) after multiplication by the damping factor, M(q). (**b**) The Fourier transform of the interference function, G(r)

After deposition, changes in film stress are due to changes in the molar volume of the material. For instance, the biaxial film stress shown in Fig. 4.18 is readily converted to volume change via:

$$\Delta V/V = e_x + e_y + e_z = (2\sigma/E)(1 - 2v) \tag{4.47}$$

where $e_{x,y,z}$ are strains calculated using the biaxial stress state where σ is the in-plane principal film stress, E and v are the Young's modulus and Poisson's ratio of the ITO. The result shows that the amorphous-to-crystalline transformation results in a molar volume decrease of approximately 0.35% in the film. This decrease in molar volume is due to two effects: (i) the smaller molar volume of the crystalline state vs. the amorphous state and (ii) the effect of the smaller ionic radius of six-coordinated Sn^{IV} (83 pm) placed on In^{III} (94 pm) sites. After long times at 250°C in air, the molar volume of the material continues to gradually decrease. This decrease corresponds to a decrease in carrier density and is remarkable because the oxidation of vacancies is actually increasing the number of atoms in the film while decreasing the volume that those atoms occupy. Assuming that all the changes in carrier density after transformation are due to the oxidation of ionized vacancies we calculate that the consumption of an oxygen vacancy results in a small and negative

(-8.6×10^{-24} cm^3/vacancy) change in the molar volume. Presumably, this decrease in molar volume is due to the bridging of adjacent In atoms with stronger In-O bonding and, possibly, a decrease in Coulombic repulsion between the positively charged oxygen vacancies.

Direct characterization of point defects in oxides is not currently possible. However, changes in point defect density directly affect the film volume and, through elastic deformation, the stress state of any thin film. This effect can be used to examine amorphous (and crystalline reference samples) TCO thin films to attempt to characterize the defects that produce carriers. By annealing samples in a reducing or oxidizing ambient, it is possible to examine both the structural changes associated with the creation or consumption of point defects in the amorphous state and the kinetics of the underlying diffusion processes. The method is based on the determination of film stress with a high degree of precision during annealing in an ambient in which the gas phase is adjusted to control the oxidation state.

An illustration of this approach and the clear relationship between the creation and annihilation of defects and film stress is presented in Fig. 4.19 where the effect of cycling the anneal ambient from reducing (H_2) to oxidizing (O_2) every 120 min on a-IZO film volume and carrier density are shown. In this experiment the goal was to test the ability of this system to measure the small reversible volume changes

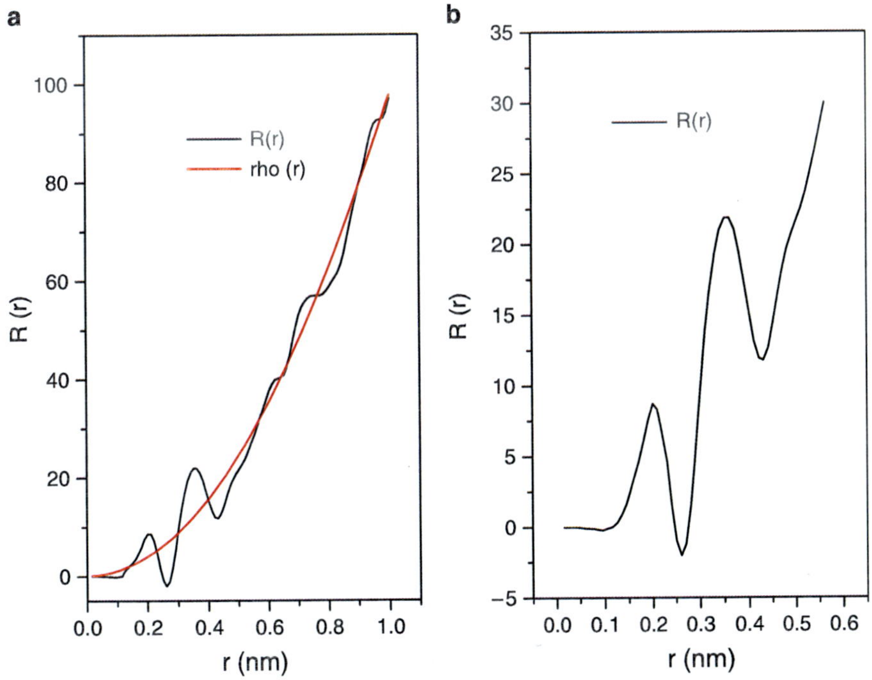

Fig. 4.19 (**a**) Radial distribution function obtained from IZO films. (**b**) A Gaussian approximation to the first peak in the IZO RDF

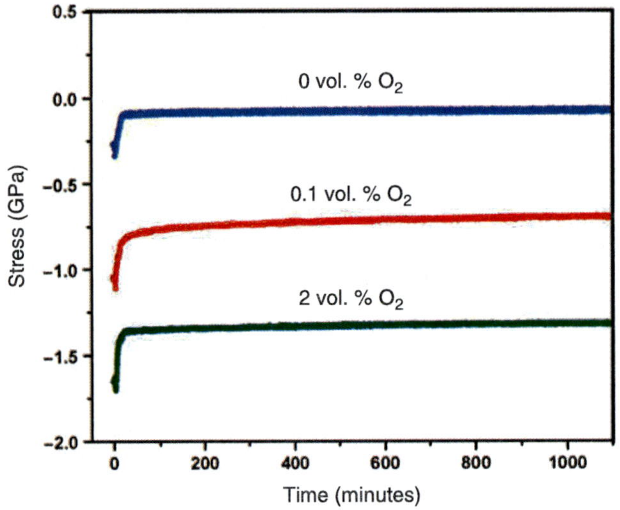

Fig. 4.20 Film stress measured as a function of annealing time at 250°C in air. The as-deposited films are compressively stressed in the as-deposited state, the stress decreases rapidly in the first few minutes as the amorphous phase crystallizes, then the compressive stress continues to relax as oxygen vacancies are consumed

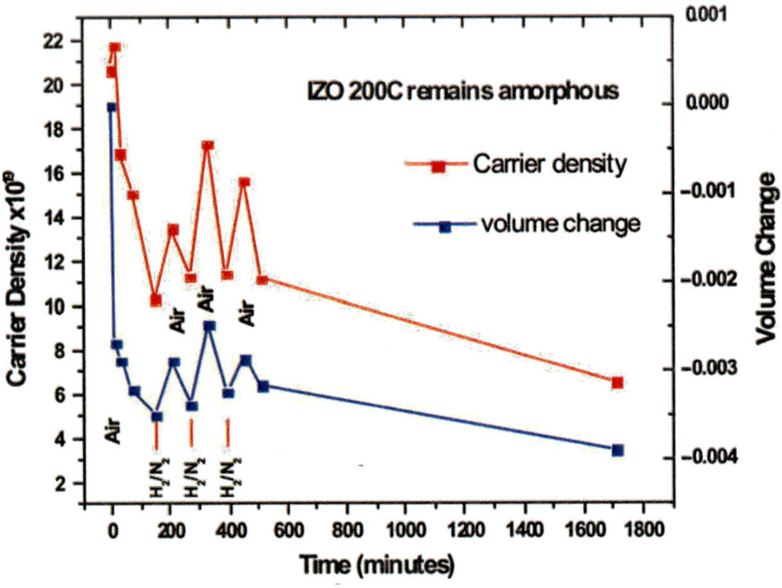

Fig. 4.21 Effect of cycling the reducing and oxidizing ambient on the molar film volume (determined after an initial relaxation on the vacancy concentration) and the carrier density (as measured using the Hall technique)

that accompany the consumption and creation of carriers. The experiment was performed in a stop-and-look mode but both the carrier density change and the reversible film volume changes can be easily observed with each reduction/oxidation cycle. Since we are measuring changes in very large (and known) numbers of defects ($5 \times 10^{19}/cm^3$) the volume per defect is easily found if it is assumed that only one type of defect is being created or consumed Figs. 4.20 and 4.21.

Stress measurements provide insight into practical engineering problems such as debonding or stress-induced curvature of thin or compliant substrates and, as shown above, more fundamental insights into the physical nature of the point defects that contribute carriers to TCO materials.

4.5 Summary

In this chapter a few of the many techniques available to characterize TCO microstructure (primarily TEM) and transport properties (Van der Pauw/Hall and TFT device channel characterization) have been illustrated. Although the examples have been taken from the author's various published works on Indium oxide based TCO materials, these techniques can be easily applied (when applicable) to any inorganic thin film TCO. The connection between microstructure and performance is critical to the optimization of existing TCO's and to the development of new ones.

References

1. Utsuno, F, et al., *Structural study of amorphous In_2O_3 film by GIXS with synchrotron radiation*. Thin Solid Films, 2006. **496**: p. 95–98
2. Yaglioglu, B., *Electrical and structural characterization of IZO thin films for device applications*. Brown University PhD Thesis, 2007.
3. Nomura, K., Ohta, H., Takagi, A., Kamiya, T., Hirano, M., and Hosono, H., *Room-temperature fabrication of transparent flexible thin-film transistors using amorphous oxide semiconductors*. Nature, 2004. **432**: p. 488–491.
4. Ellmer, K., Vollweiler, G., *Electrical transport parameters of heavily-doped zinc oxide and zinc magnesium oxide single and multilayer films heteroepitaxially grown on oxide single crystals*. Thin Solid Films, 2006. **496**: p. 104–111.
5. van der Pauw, L.J., Philips Tech Rev, 1958. **20**: p. 220
6. Yaglioglu, B., Yeom, H.Y., Beresford, R., and Paine, D.C., *High-mobility amorphous In2O3-10 wt% ZnO thin film transistors*. Appl Phys Letts, 2006. **89**: p. 062103.
7. Sze, S.M., *Physics of semiconductor devices*. 2nd ed., 1981, New York: Wiley. xii, 868.
8. Luan, S.W., and Neudeck, G.W., *An experimental-study of the source drain parasitic resistance effects in amorphous-silicon thin-film transistors*. J Appl Phys, 1992. **72**(2): p. 766–772.
9. Horowitz, G., et al., *Extracting parameters from the current-voltage characteristics of field-effect transistors*. Adv Funct Mater, 2004. **14**(11): p. 1069–1074.
10. Yi, W., et al., *Ballistic electron microscopy and spectroscopy of metal and semiconductor nanostructures*. Surf Sci Rep, 2009. **64** (5): p. 169–190

11. Frank, G., and Kostlin H., *Electrical-properties and defect model of tin-doped indium oxide layers*. Appl Phys A: Mater Sci Process, 1982. **27**(4): p. 197–206.
12. Baikie, I.D., Mackenzie, S., Estrup, P.J.Z., Meyer J.A., *Noise and the Kelvin method*, Rev Sci Instrum, 1991. **62**(5): p. 1326.
13. Paine, D.C., Whitson, T., Janiac, D., Beresford, R., Yang, C.O., and Lewis, B., *A study of low temperature crystallization of amorphous thin film indium-tin-oxide*. J Appl Phys, 1999. **85** (12): p. 8445–8450.
14. Moriga, T., Hiruta, K., Fujiwara, A., Nakabayash, I., and Tominaga, K., *Structures and physical properties of films deposited by simultaneous DC sputtering of ZnO and In2O3 or ITO targets*. J Solid State Chem, 2000. **155**: p. 312–319.
15. Moriga, T., Edwards, D.D., Mason, T.O., Poeppelmeier, K.R., Kannewurf, C.R., Nakabayash, I., *Phase relationships and physical properties of homologous compounds in the zinc oxide-indium oxide system*. J Am Ceram Soc, 1998. **81**(5): p. 1310–1316.
16. Kasper, V.H., Z Anorg Allg Chem, 1967. **349**: p. 113.
17. Yaglioglu, B., Yeom, H.Y., and Paine, D.C., *Crystallization of amorphous* In_2O_3*-10 wt% ZnO thin films annealed in air*. Appl Phys Letts, 2005. **86**(26): p. Art. No. 261908.
18. Prewitt, C.T., Shannon, R.D., Rogers, D.B., Sleight, A.W., Inorg Chem, 1969. **8**(9): p. 1985.
19. Yaglioglu, B, *Electrical and structural characterization of IZO thin films for device applications*, Brown University PhD Thesis, May 2007.
20. Warren, B.E., *X-ray diffraction*. 1969, Reading, MA: Addison-Wesley Pub. Co. vii, p. 381.
21. Elliott, S.R., *Physics of amorphous materials*. 1983, London, New York: Longman. xiii, p. 386
22. Moine, P., Pelton, A.R., and Sinclair, R., *Structural determination of small amorphous volumes by electron-diffraction*. J Non-Crystalline Solids, 1988. **101**(2–3): p. 213–222.
23. Ankele, J., et al., *Structure factor of amorphous-germanium by quantitative electron-diffraction*. J Non-Crystalline Solids, 1995. **193**: p. 679–682.
24. Cockayne, D.J.H., and Mckenzie, D.R., *Electron-diffraction analysis of polycrystalline and amorphous thin-films*. Acta Crystallogr Sect A, 1988. **44**: p. 870–878.
25. Sigle, W., et al., *Advances in energy-filtering transmission electron microscopy*. J Electron Microsc, 2001. **50**(6): p. 509–515.
26. Fujime, S., Jpn J Appl Phys., 1966. **5**(9): p. 764.
27. Floro, J.A., et al., *Real-time stress evolution during Si1-xGex heteroepitaxy: Dislocations, islanding, and segregation*. J Electronic Mater, 1997. **26**(9): p. 969–979.
28. Vink, T.J., et al., *On the homogeneity of sputter-deposited ITO films.1. Stress and microstructure*. Thin Solid Films, 1995. **266**(2): p. 145–151.
29. Chason, E., Floro, J.A., *Measurements of stress evolution during thin film deposition*. Mat Res Soc Symp Proc, 1996. **428**: p. 499–504.
30. Yeom, H.Y., Popovich, N., Chason, E., and Paine, D.C., *A study of the effect of process oxygen on stress evolution in d.c. magnetron-deposited tin-doped indium oxide*. Thin Solid Films, 2002. **411**(1): p. 17–22.

Chapter 5
In Based TCOs

Yuzo Shigesato

5.1 Introduction for In Based TCOs

A critical component in all flat-panel display devices (FPDs), from simple calculator liquid-crystal displays to large-area, active-matrix, liquid-crystal color screens, are the electrodes that control the orientation of the liquid-crystal molecules and, consequently, the on-off state for the passage of light. At least one of the electrodes in a flat-panel display must be transparent since either ambient or transmitted (i.e., backlight) light must pass through the device and reach the viewer's eyes. The transparent conductor of choice has been Sn-doped In_2O_3 (ITO) although other transparent conductors are available, for instance SnO_2 and ZnO. ITO is mostly favored because within this class of materials, it offers the highest available transmissivity for visible light combined with the lowest electrical resistivity. While ITO is highly transparent to visible light, it effectively reflects light in the infrared part of the spectrum. For this reason, it is also used for energy-efficient window coatings as well as SnO_2 based TCOs [1].

ITO is an n-type, highly degenerate, wide band gap semiconductor that owes its relatively low electrical resistivity to its high free carrier density. The free carrier density of ITO can be increased through appropriate processing to a level of about $1–2 \times 10^{21}/cm^3$ by the heavy doping of Sn to a level of 5–10 at.%. Free carriers are formed in the material by two different kinds of electron donor sites substitutional four-valent tin ions and oxygen vacancies [1–4]. The electrically active tin atoms are positioned at indium substitutional sites in the 80-atom unit cell of the bixbyite In_2O_3 crystal structure. The highest quality thin films of ITO are routinely produced via magnetron sputtering or activated electron beam (EB) evaporation, and have a resistivity of $1.5–2.0 \times 10^{-4}\ \Omega$ cm. The resistivity of a material is inversely

Y. Shigesato
Graduate School of Science and Engineering, Aoyama Gakuin University, 5-10-1 Fuchinobe, Chuo-ku, Sagamihara, Kanagawa, 252-5258 Japan
e-mail: yuzo@chem.aoyama.ac.jp

D.S. Ginley et al. (eds.), *Handbook of Transparent Conductors*,
DOI 10.1007/978-1-4419-1638-9_5, © Springer Science+Business Media, LLC 2010

proportional to the product of the carrier mobility and the carrier density in the material; in the case of ITO. Both of these parameters depend on the material micro-nano structure [5–8].

The recent trend towards higher quality flat-panel devices has led to new display technologies with a much thinner electrode width and consequently demands for further optimization of ITO material properties and processing conditions. These new and demanding technologies require even lower film resistivities (to 1.0×10^{-4} Ω cm), and this necessitates lower substrate temperatures during deposition. The requirement for lower deposition temperatures derives from newer device designs including flexible displays that call for the deposition of ITO films on polymer films that cannot survive vacuum processing temperatures above 100°C. Meeting this demand requires new approaches for improving conventional deposition processes. Examples of the successful implementation of such new approaches include modification of low-impedance, direct current (dc), magnetron sputtering by enhancement of the magnetic field strength near the cathode or by superimposing radio frequency (rf) discharge, or activated EB evaporation using either tungsten electron emitters or arc plasma generators (high-density plasma assisted EB evaporation; HDPE) [5, 9].

With increasing demand for the large-area and high-quality FPDs, amorphous transparent conductive films including amorphous ITO films have attracted significant attention due to their very flat surfaces, low internal stresses and good etchability for micro-patterning. However, the crystallinity of sputter deposited ITO films have been reported to depend heavily on total gas pressure during the deposition and amorphous ITO films could be deposited only at high total gas pressures in the range of 3–5 Pa, wherein the substrate temperature (Ts) was around RT [10, 11]. Currently In_2O_3-ZnO (IZO) films with a completely amorphous structure can be deposited by dc magnetron sputtering using an oxide target of In_2O_3:ZnO = 89.3:10.7 by wt.%, with high reproducibility under a wide range of deposition conditions such as total gas pressure or substrate temperature up to 300°C [12, 13]. The resistivity of the films deposited with 100% Ar was about 4×10^{-4} Ω cm. These amorphous IZO films have smaller compressive stresses and much smoother surfaces compared to polycrystalline ITO films deposited under the same sputtering conditions [14]. The wet etching properties of the amorphous IZO films with optimized electrical properties were much superior to those of the polycrystalline ITO films. Therefore, the IZO is expected to be a new candidate for transparent electrodes, especially for thin film transistor (TFT)-LCDs or organic LEDs.

5.2 Crystal Structure and Electronic Structure of In_2O_3 and Sn Doped In_2O_3 (ITO)

The crystal structure of In_2O_3 is bixbyite (cubic, space group symmetry: Ia3, lattice constant: 1.0117 nm, JCPDS: 6-0416) as shown in Fig. 5.1. Each unit cell contains 16 formula units of In_2O_3; indium atoms occupy Wyckoff positions of 8b and 24d, while oxygen atoms occupy Wyckoff positions of 48e [15]. The bixbyite structure can be regarded as an imperfect crystal of fluorite structure by extracting 16 oxygen atoms for 1 bixbyite unit cell, which are called empty anion (quasi anion) sites.

Energy band calculation was performed for bixbyite In_2O_3 [16] using the linear muffin-tin orbital method with atomic sphere approximation (LMTO-ASA) [17, 18]. Exchange-correlation interactions were described using the Ceperley and Alder formula [19]. Because the filling ratio of usual muffin-tin sphere to interstitial region was small, combined correction terms and empty spheres were introduced to improve the calculation accuracy [18]. The energy band structure is shown in Fig. 5.2, and the corresponding density of states is shown in Fig. 5.3 together

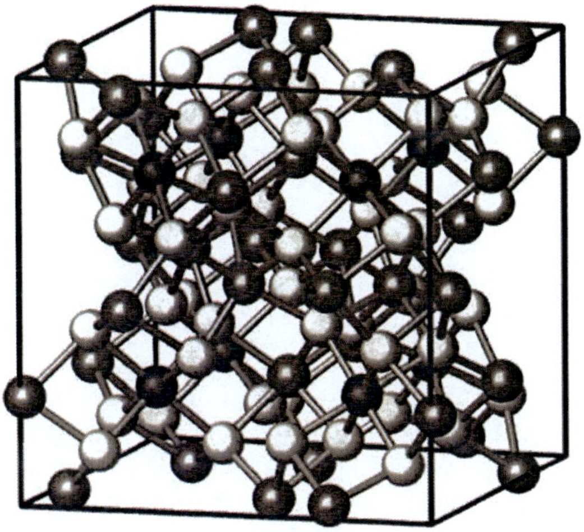

Fig. 5.1 Crystal structure of In_2O_3 (bixbyite). Black, shaded and white spheres represent 8b indium, 24d indium and 48e oxygen atoms, respectively, in a conventional unit cell

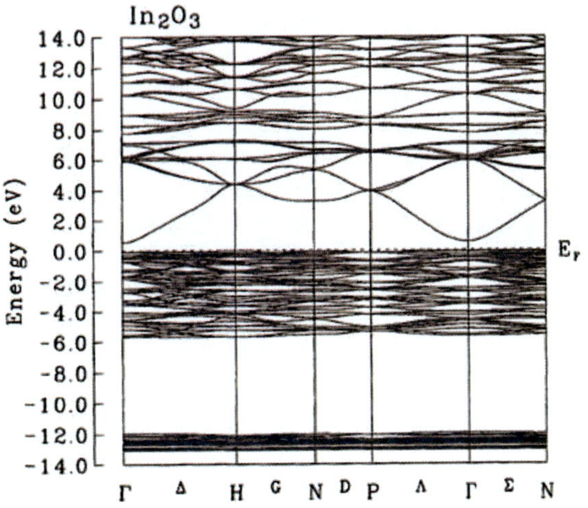

Fig. 5.2 The energy band structure of In_2O_3

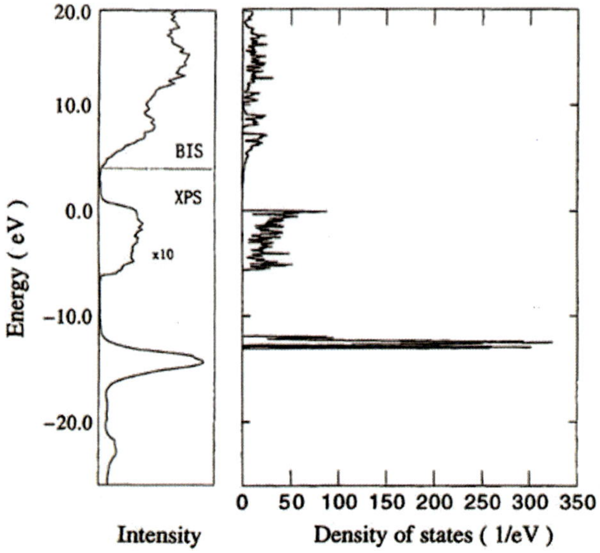

Fig. 5.3 The total density of states (right), and XPS and BIS intensity [20] (left) of In_2O_3

with XPS and BIS spectra [20], wherein the top of the valence band was chosen to be the zero of the energy scale.

The effective mass of electrons in the conduction band was calculated by fitting the dispersion of the conduction band to a parabola: we obtained 0.4182 m_e in the Δ direction, 0.3018 m_e in the Σ direction and 0.3619 m_e in the Λ direction where m_e is the rest mass of an electron. These results agreed well with experiment results [21], wherein the electron mass was estimated to be about 0.3 m_e from the plasma frequency data [22, 23].

For theoretical analysis of Sn-doped In_2O_3 a primitive unit cell similar to that of the In_2O_3 crystal was used [24]. For the supercell calculation, one of the indium atoms at the 8b or 24d site was replaced by an Sn atom, yielding a system with a 2.5 at.% concentration of Sn. Approximately $1 \times 10^{21}/cm^3$ concentration of free electrons should have appeared if all the substituted Sn atoms had donated free electrons. Since ITO films with very low resistivities are known to have about $1 \times 10^{21}/cm^3$ concentration of free electrons by some experimental methods [1, 5, 6], this supercell model seems to be a good approximation of real ITO films with a low resistivity. The calculated density of states (DOS) is shown in Fig. 5.4 [24]. Comparing the total density of states (TDOS) of this doped system with that of an In_2O_3 crystal, it can be seen that the overall features of DOS corresponding to the valence and conduction bands of In_2O_3 are almost conserved except for the appearance of a new band below the valence band. Analysis of the partial density of states (PDOS) shows that this band originates mainly from the bonding states between Sn 5s-like and O 2p-like states. The other two bands composed of Sn 5s-like states appear in the conduction band of In_2O_3 and slightly disturb this band,

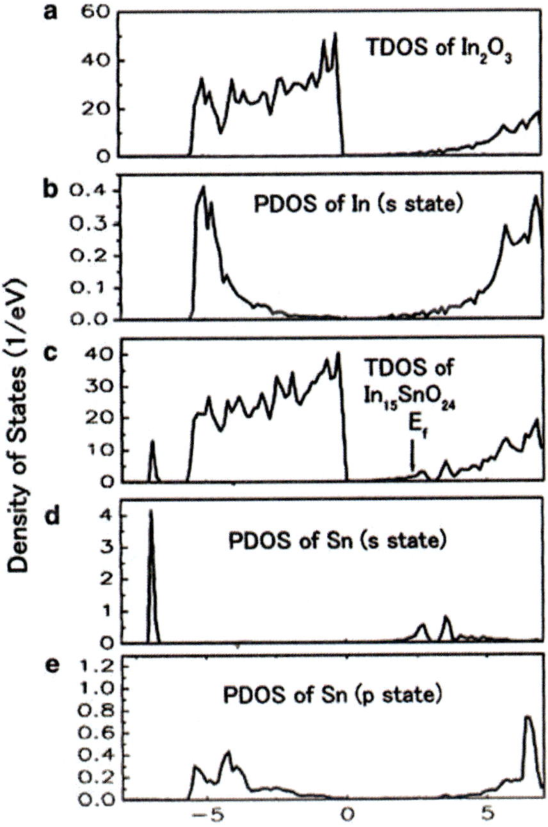

Fig. 5.4 Density states analyses of $In_{15}SnO_{24}$

i.e., two antibonding bands with Sn 5s-like characteristic exist in the conduction band and the Fermi energy is located in the lower band. In summary, the calculated PDOS analyses showed that a Sn atom substituting for an indium atom formed three impurity bands with s-like symmetry, the second band of the three bands overlapped with the conduction band of In_2O_3, and the Fermi energy of ITO was captured in this impurity band. The PDOS analyses also revealed that the substitution of a Sn atom did not significantly destroy the shape of the density of states around the bottom of the conduction band, which gave a physical foundation for the Burstein-Möss shift model to explain the optical gap winding with increasing carrier density [1, 25].

5.3 Electrical and Optical Properties of In_2O_3-SnO_2 Films

ITO films have been deposited by various methods including dry processes such as CVD or PVD, and also some wet processes. Among such deposition methods, dc magnetron sputtering using a ceramic ITO target is by far the preferred method, because of the advantages of controllability of the sputtering conditions and adaptability to large area deposition [26]. It is well known that physical properties of ITO films are strongly affected by the Sn doping concentration. Usually ceramic targets containing 5–10 wt.% SnO_2 have been used to deposit ITO films with a low resistivity of about 10^{-4} Ω cm [5, 26]. In addition, over the last decades, high-density ITO targets have developed to improve electrical properties of ITO films [27–29]. It was previously reported that ITO films with low resistivities of 1.49×10^{-4} Ω cm could be deposited by dc magnetron sputtering on glass heated at 200°C using an ultra high-density target with 99% theoretical density that could contribute to arc-less deposition processes [28].

Recently, considerable efforts have been focused on depositing In_2O_3-SnO_2 films not only with significantly reduced resistivity, but also with excellent wet-etching properties in order to accommodate the increasing technological demand for larger area flat panel displays with higher image quality. ITO films with stable medium range resistivity of 10^{-2}–10^{-3} Ω cm are also required to produce a sufficiently large electric field gradient and/or high transmittance in a limited spectral region [29]. In order to deposit the best-suited In_2O_3-SnO_2 for a particular application, it is important to investigate how the SnO_2 concentration in the high-density target affects the film properties over a wide composition range in the In_2O_3-SnO_2 system and to develop a useful database for the various device designs. Here we report on In_2O_3-SnO_2 films deposited by dc magnetron sputtering using high-density In_2O_3-SnO_2 ceramic targets containing SnO_2 over the whole range of concentrations from 0 to 100 wt.% [30].

High-density ceramic targets (packing density: 99.7%) of the In_2O_3-SnO_2 system containing 0, 5, 10, 15, 20, 44.5 or 100 wt.% SnO_2 were prepared by atmospheric sintering under an oxygen atmosphere [28, 29]. In_2O_3-SnO_2 thin films were deposited by a batch type dc magnetron sputtering system on unheated non-alkali glass substrates (Corning #1737). The total gas pressure and dc power were maintained at 0.5 Pa and 200 W, respectively, for all the depositions. Oxygen flow ratios $[O_2/(O_2 + Ar)]$ were adjusted to 0, 0.1, 0.5, 1, 1.5, 2, 3, 4 or 5%. The maximum horizontal magnetic field strength at the target surface was enhanced to 1,000 Gauss to decrease the plasma impedance and sputtering voltage for all the depositions [5, 31]. The kinetic energy of negative oxygen ions (O−) which were accelerated in the cathode sheath and bomber growing film surfaces could be suppressed by the decreased sputtering voltage. The deposition time was adjusted to obtain a film thickness of 150 ± 5 nm. The film thickness was measured using DEKTAK^3 (Sloan Tech.). All the as-deposited films were postannealed in air, Ar (100%) or Ar (97%) + H_2 (3%) for 1 h at 200°C using a tube-type furnace.

Figure 5.5a–d show the resistivity, Hall mobility and carrier density of the In_2O_3-SnO_2 films with various Sn concentrations for (a) as-deposited, postannealed (b) in air, (c) in Ar (100%) and (d) in Ar (97%) + H_2 (3%) for 1 h at 200°C as a function of the oxygen flow ratio $[O_2/(O_2 + Ar)]$ during the sputter deposition. The minimum resistivities for the as-deposited and postannealed In_2O_3-SnO_2 films were obtained for a certain value of the $O_2/(O_2 + Ar)$ ratio. These represent the optimum $O_2/(O_2 + Ar)$ [29, 32]. For the as-deposited films with 0–20 wt.% SnO_2,

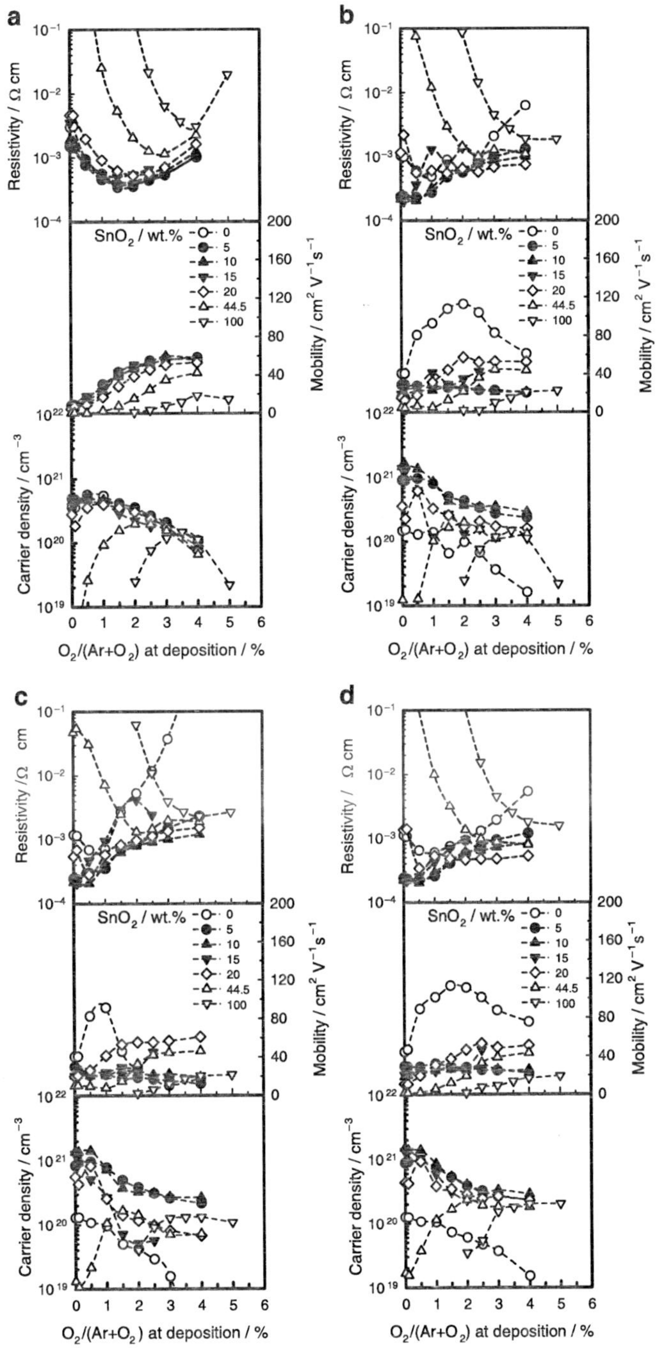

Fig. 5.5 Resistivities, Hall mobilities and carrier densities of (**a**) as-deposited In_2O_3-SnO_2 films, post-annealed films (**b**) in air, (**c**) in Ar (100%) and (**d**) in Ar (97%) + H_2 (3%) for 1 h at 200°C as a function of O_2/Ar flow ratio $O_2/(O_2 + Ar)$ during sputtering deposition

the optimum $O_2/(O_2 + Ar)$ was between 1.5 and 2% where films with 4–5 × 10^{-4} Ω cm were deposited, whereas for the films with 44.5 and 100 wt.% SnO_2, the optimum $O_2/(O_2 + Ar)$ was 4 and 5% with resistivities larger than 1 × 10^{-3} Ω cm. Details on variations of as-deposited films are described in a previous report [5]. On the other hand, variations in electrical properties as a function of $O_2/(O_2 + Ar)$ were almost the same irrespective of the post-annealing treatment in air, Ar (100%) or Ar (97%) + H_2 (3%). For the post-annealed In_2O_3 films, resistivity decreased to the minimum values around 1.0% of $O_2/(O_2 + Ar)$ and then increased with increasing $O_2/(O_2 + Ar)$. Variations in the resistivity strongly depend on the mobility, i.e., the mobility increased to the maximum value around 1% of $O_2/(O_2 + Ar)$ and then decreased with increasing $O_2/(O_2 + Ar)$. For the films with SnO_2 concentration ranging from 5 to 15 wt.%, the resistivity showed minimum values of about 2.0 × 10^{-4} Ω cm between 0 and 0.5% of $O_2/(O_2 + Ar)$ and then gradually increased with increasing $O_2/(O_2 + Ar)$. These variations correspond to those of the carrier density as a function of $O_2/(O_2 + Ar)$, because the carrier density reached maximum value of 1.2 × 10^{21}–1.4 × 10^{21}/cm^3 between 0 and 0.5% of $O_2/(O_2 + Ar)$ while the mobility remained nearly constant for the whole range of SnO_2 concentration. Therefore, the carrier density of these films should be strongly affected by $O_2/(O_2 + Ar)$ in the as-deposited film. For the post-annealed films with 44.5 and 100 wt.% SnO_2, the variations in electrical properties with $O_2/(O_2 + Ar)$ showed similar tendency to those of the as-deposited films. XRD measurements showed that both as-deposited and post-annealed films with 44.5 and 100 wt.% SnO_2 were amorphous. Therefore, these variations corresponded to that the films with 44.5 and 100 wt.% SnO_2 were amorphous before and after the post-annealings.

Figure 5.6a–d show XRD profiles of the In_2O_3-SnO_2 films deposited at the optimum $O_2/(O_2 + Ar)$ using ceramic targets with SnO_2 concentrations ranging

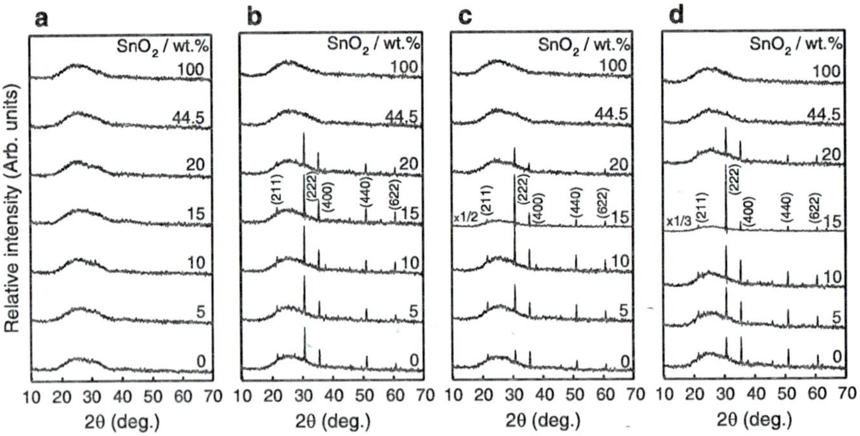

Fig. 5.6 XRD profiles of In_2O_3-SnO_2 films deposited at the optimum $O_2/(O_2 + Ar)$ using the ceramic targets containing SnO_2 from 0 to 100 wt.%. **(a)** As-deposited films, and post-annealed films **(b)** in air, **(c)** in Ar (100%) and **(d)** in Ar (97%) + H_2 (3%) for 1 h at 200°C

from 0.0 to 100 wt.%. Figure 5.6a shows data for the as-deposited films, while Fig. 5.6b–d show data for the films post-annealed in air, Ar (100%) and Ar (97%) + H_2 (3%) for 1 h at 200°C. As seen from Fig. 5.6a, all the as-deposited films were amorphous from 0 to 100 wt.% SnO_2. In Fig. 5.6b–d, all the films deposited using the target with SnO_2 concentrations of 0–20 wt.% showed the bixbyte In_2O_3 polycrystalline structure. The films deposited using the target with SnO_2 concentrations of 44.5 and 100 wt.% remained amorphous after all the post-annealings.

The resistivities, mobilities and carrier densities of the films deposited at the optimum $O_2/(O_2 + Ar)$ value and post-annealed under the three different atmospheres are shown in Fig. 5.7a–d as a function of the SnO_2 concentration of the targets. The values for as-deposited films are also shown for comparison. The resistivity of the as-deposited films gradually increased with increasing SnO_2 concentration because of the decrease in both mobility and carrier density. The minimum resistivity was obtained for pure In_2O_3 (SnO_2: 0 wt.%) films with the highest mobility and carrier density, which can be explained by the fact that the Sn^{4+} do not generate any carriers but act as neutral scattering centers with increasing SnO_2 concentration in the amorphous films [6]. After the post-annealing treatments under all three atmospheres, the resistivity decreased with increasing SnO_2 concentration from 0 to 5 wt.% SnO_2. The resistivity was almost constant around 2.1×10^{-4} Ω cm between 5 and 15 wt.% SnO_2 and then increased at 44.5 and 100 wt.% SnO_2. These variations in resistivity were inversely proportional to those of the carrier density, i.e., the carrier density increased significantly with increasing SnO_2 concentration and reached the highest value ($1.4 \times 10^{21}/cm^3$) at 10 wt.% SnO_2. On the other hand, the carrier density decreased remarkably with increasing SnO_2 concentration to 20 wt.%. The increase in carrier density should be

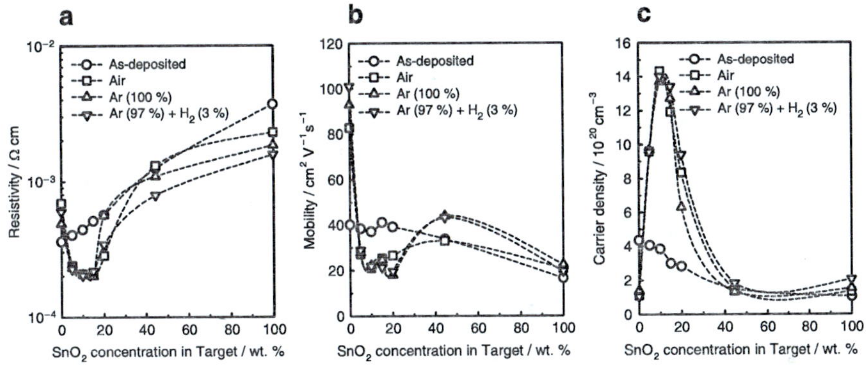

Fig. 5.7 (**a**) Resistivity, (**b**) mobility and (**c**) carrier density of the films post-annealed in three different atmospheres at the optimum $O_2/(O_2 + Ar)$, as a function of SnO_2 concentration of the targets. The *square symbols* represent the 200°C, 1 h air annealing; the *triangles markers* represent the 200°C, 1 h Ar (100%) annealing; the *reverse triangles markers* represent the 200°C, 1 h Ar (97%) + H_2 (3%) annealing. The values for the as-deposited films are also shown for comparison (*open circles*)

caused by an increase in electrically activated substitutional Sn^{4+} at In^{3+} sites, which provide free electrons and increased in proportion to the SnO_2 concentration during crystallization [6, 33]. On the other hand, the mobility of the post-annealed films decreased remarkably from 82, 93, and 101 cm^2/V s for 0 wt.% to around 20 cm^2/V s for 5 wt.% SnO_2 and remained constant between 5 and 20 wt.% SnO_2. This could be attributed to the generation of Sn-oxide based complexes, which should act as neutral scattering centers for the films with the higher SnO_2 concentrations [6, 34]. The solubility limit of Sn in In_2O_3 films was found to be well over 60% [35]. For relatively high Sn concentrations below the solubility limit, Sn^{4+} in solution was not only replaced substitutionally at the In^{3+} sites but also at interstitial sites or by forming SnO_2 complexes. It ha already been reported that an electrically inactive Sn^{4+} increased with increasing Sn concentration above 4 at.% [3, 4, 34].

The optical properties of (a) the as-deposited and the postannealed films (b) in air, (c) in Ar (100%) and (d) in Ar (97%) + H_2 (3%) for 1 h at 200°C are shown in Fig. 5.8a–d. All the films exhibited a high transmittance of over 80% in the region of visible light. On the other hand, the transmittance of the postannealed films decreased with increasing Sn^{4+} concentration, where reflectance increased, in the region of over 1 μm. These variations could be explained by the increase in free electrons in terms of Drude's theory [36].

The measurements of work function and shallow energy levels at the film surface were carried out using photoelectron spectroscopy in air (PESA, AC-2, Riken Keiki). In PESA the electrons emitted by UV irradiation from a solid surface into air were detected by an air-filled counter (open counter) with two grids, one of which is for quenching of the counter discharge using an external circuit and the other is for suppression of positive ion bombardment [37]. Before the measurements, in order to eliminate carbon contamination all the films were cleaned by

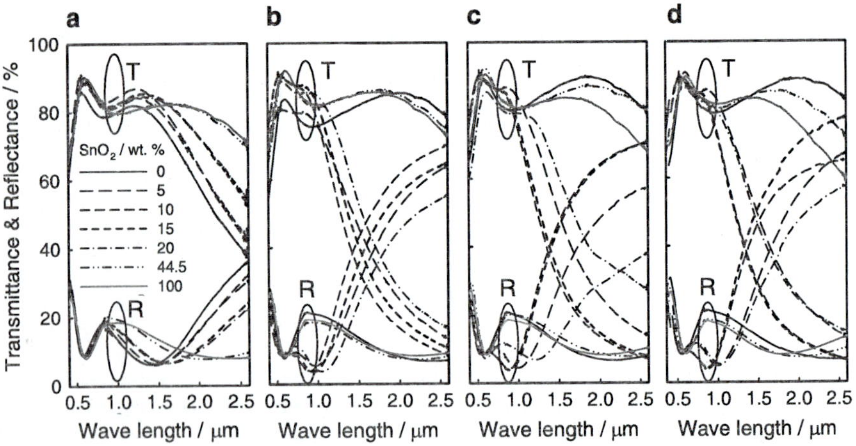

Fig. 5.8 Spectral optical transmission and reflection of (**a**) the as-deposited and the post-annealed films (**b**) in air, (**c**) in Ar (100%) and (**d**) in Ar (97%) + H_2 (3%) for 1 h at 200°C

ultrasonic agitation in deionized water and acetone followed by exposure to high-pressure UV light (Ushio lighting, USH-350D) of 360–370 nm at a distance of 3 cm for 5 min in air. The certified repeatable accuracy for the work function data in this measurement was ± 0.02 eV.

Work functions of the as-deposited and postannealed In_2O_3-SnO_2 films with various SnO_2 concentrations were determined from the low energy thresholds of these photoelectron emission curves analyzed by PESA. Figure 5.9 shows the cubic root of the photoelectron emission intensity curves as a function of irradiated photon energy for the films deposited using the targets of 0–100 wt.% SnO_2 and postannealed in air for 1 h at 200°C as an example. It is observed that the onset energy point of the photoemission intensity, i.e., the work function, was shifted clearly to a lower energy with increasing SnO_2 concentration to 20 wt.%.

Variations in work function for the as-deposited and post-annealed films as a function of SnO_2 concentration in the target are shown in Fig. 5.10. For the

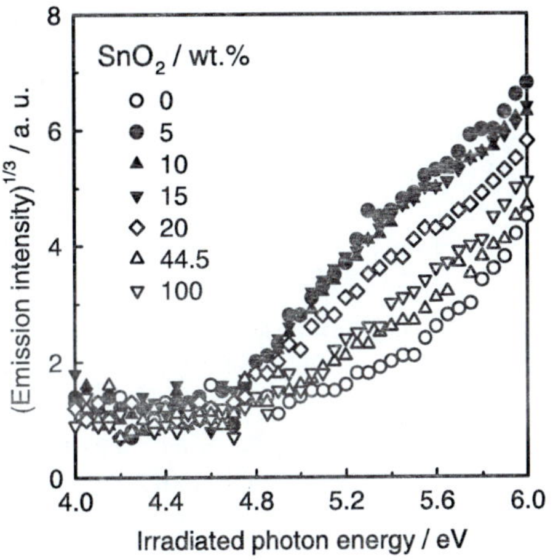

Fig. 5.9 Cubic root of photoelectron emission intensity curves as a function of irradiated photon energy on the film surface analyzed by PESA for the films with 0–100 wt.% SnO_2 post-annealed in air for 1 h at 200°C

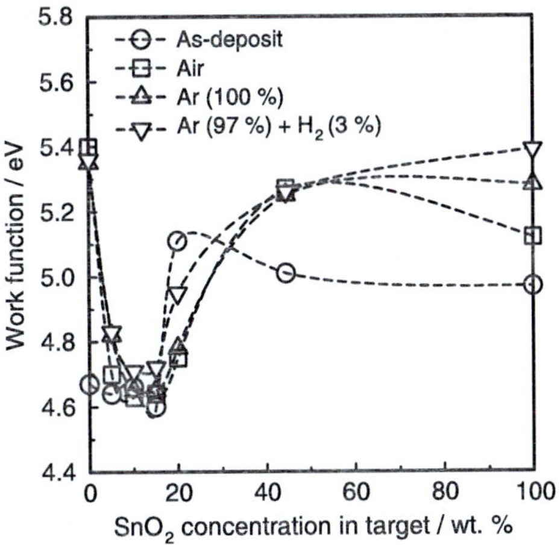

Fig. 5.10 Work functions of the In_2O_3-SnO_2 films as a function of the SnO_2 concentration in the targets. The *squares* represent the 200°C, 1 h air annealing; the *triangles* represent the 200°C, 1 h Ar (100%) annealing; the *reverse triangles* represent the 200°C, 1 h Ar (97%) + H_2 (3%) annealing. The values for the as-deposited films are also shown for comparison (*open circles*)

as-deposited films with SnO_2 concentration from 0 to 15 wt.% and from 20 to 100 wt.%, the work functions were nearly constant around 4.6–4.7 and 5.0–5.1 eV, respectively. On the other hand, for the films postannealed under the three different atmospheres, the work function of the pure In_2O_3 films was in the range of 5.3–5.4 eV. In contrast, the work function of the In_2O_3-SnO_2 films with between 5 and 20 wt.% SnO_2 sharply decreased and remained constant around 4.6–4.8 eV. Above 20 wt.% SnO_2, the work function gradually increased to a value larger than 5.0 eV again. In some previous studies, the work functions of commercial ITO films with relatively low resistivities, analyzed by ultraviolet photoelectron spectroscopy (UPS) after cleaning the surface using UV-O_3 treatment, O_2 plasma or some chemical processes were reported to be 4.4–4.8 eV [38, 39]. These values were almost equal to those of our post-annealed films with 5–20 wt.% SnO_2. The variation in the work function with increasing SnO_2 concentration from 0 to 20 wt.% should correspond to that in the carrier density caused by crystallization and activation of Sn^{4+}.

Furthermore, the work function of the post-annealed In_2O_3-SnO_2 films with 0–20 wt.% SnO_2 is plotted in Fig. 5.11 as a function of the two-thirds power of the carrier density. These films were already confirmed by XRD to be polycrystalline bixbyte In_2O_3. The work functions of these films showed an inverse linear relationship to the two-thirds power of the carrier density. This relationship could be explained in terms of the shift of the Fermi level to a higher energy in the conduction band with increasing carrier density for the highly degenerate semiconductors, which corresponded to the linear relationship between the optical band gap and the two-thirds power of the carrier density [23, 35, 41–43]. The reduced effective mass (m^*_{vc}) of the films was estimated from these relationships to be 0.52 m_0 based on the present carrier density dependence of the work function. This m^*_{vc} value is in good agreement with those for ITO films [m^*_{vc} = (0.5–0.7)] estimated by the carrier density dependence of the optical band gap, which are well-known as the Burstein-Möss shift in previous studies [23, 35, 41–43].

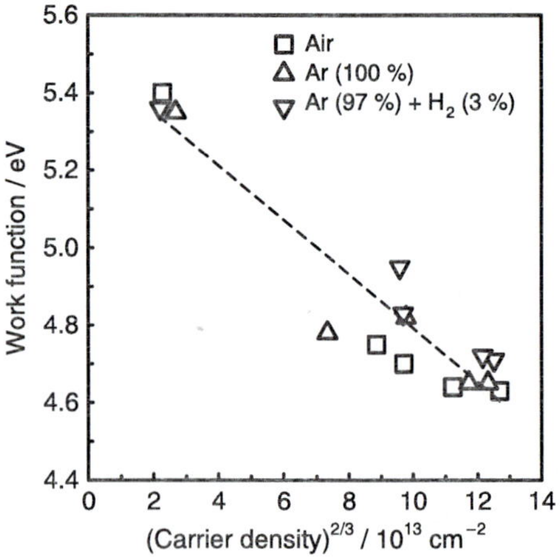

Fig. 5.11 Relationship between work function and two-third power of the carrier density for the In_2O_3-SnO_2 films with 0–20 wt.% SnO_2. The *squares* represent the 200°C, 1 h air annealing; the *triangles* represent the 200°C, 1 h Ar (100%) annealing; the *reverse triangles* represent the 200°C, 1 h Ar (97%) + H_2 (3%) annealing

5.4 Electrical, Optical and Thermal Properties of Amorphous In_2O_3-ZnO Films

With recent demands for large-area and high-quality FPDs, amorphous transparent conductive films including amorphous Sn-doped In_2O_3 (ITO) films have attracted significant attention because of their very flat surfaces, low internal stresses and good etchability for micro-patterning. However, the crystallinity of sputter-deposited ITO films is reported to depend heavily on the total gas pressure during the deposition and fully amorphous films could be deposited only at rather high total gas pressures in the range of 3–5 Pa, when the substrate temperature (Ts) was around RT [10, 11]. K. Inoue et al. reported that In_2O_3-ZnO (IZO) films (In_2O_3:ZnO = 89.3:10.7 by wt%) with an entirely amorphous structure could be obtained with high reproducibility under a the wide range of deposition conditions such as total gas pressures from 0.1 to 5.0 Pa or substrate temperatures up to 300°C, where the resistivity of the films was about 4–5 × 10^{-4} Ω cm [12, 44]. These amorphous IZO films were reported to have smaller compressive stresses and much smoother surfaces compared to polycrystalline ITO films deposited under the same sputtering conditions [45]. The wet etching properties of the amorphous IZO films with optimized electrical properties were also reported to be superior to those of the polycrystalline ITO films since they could be etched easily with weak acid solution such as oxalic acid and could be formatted into the taper-shaped fine pitch patterns [12]. Therefore, IZO is expected to be a new candidate for the transparent electrode material especially for thin film transistor (TFT)-LCDs or organic ELDs [13].

The IZO films were deposited on unheated non-alkali glass (AN100, Asahi Glass) or Si wafer substrates by dc magnetron sputtering using the oxide ceramic In_2O_3-ZnO target (89.3 wt% In_2O_3 and 10.7 wt% ZnO, Idemitsu Kosan Co., Ltd.) [13]. The substrate temperature was confirmed to be lower than 50°C using the thermo-label, which could recognize the highest temperature of the substrate during the deposition. The total gas pressure (P_{tot}) of Ar or Ar + H_2 was maintained at 0.5 Pa for all the depositions. The water partial pressure of the residual gas was maintained at less than 8 × 10^{-4} Pa, which was monitored using a quadrupole mass spectrometer (Transpector XPR2, Inficon), in order to guarantee high reproducibility of the film properties. Control of the electron concentration was accomplished by the following two methods: (a) Impurity dopings, such as with Sn, Al and F, or (b) H_2 gas introduction into the IZO deposition process in order to generate oxygen vacancies or interstitial Zn^{2+} ions by slight reduction. The Sn or Al doping was carried out by placing each oxide pellets on the erosion area of the target surface, whereas F doping was carried out by introducing CF_4 gas during the deposition which was confirmed to be effective for F doping of the polycrystalline In_2O_3 [46]. The thickness of all the films in this study was adjusted to be about 200 nm by controlling the deposition time [13].

Figure 5.12 shows XRD patterns of the IZO films deposited under 100% Ar gas without substrate heating and post-annealed in air at 100–500°C for 1 h. The XRD profiles of the as-deposited film and post-annealed films at 100–400°C showed halo

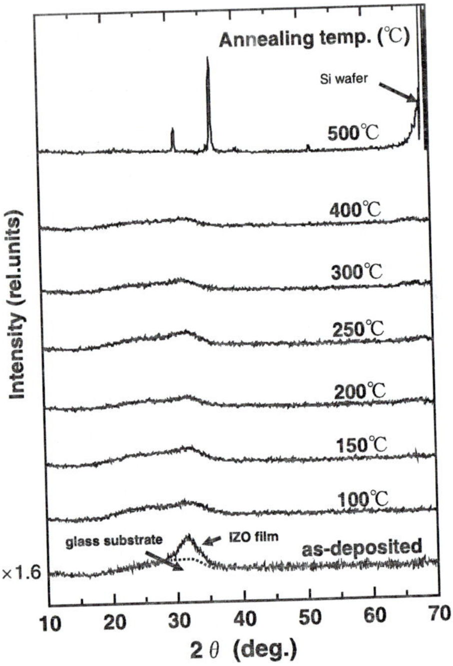

Fig. 5.12 XRD patterns of the IZO film deposited in 100% Ar gas without substrate heating and post-annealed at various temperatures for 1 h in air. The films were deposited on non-alkali glass, except for the 500°C annealing films for which a Si wafer was used as the substrate

pattern around 2θ = 30–35°, implying that these films were entirely amorphous. The film post-annealed at 500°C showed an In_2O_3 polycrystalline structure, indicating that the crystallization temperature of the IZO films was between 400 and 500°C in air. The crystallization temperatures of amorphous In_2O_3 or ITO films have been reported to be 160–170°C or 170–180°C, respectively [47, 48]: it can be thus be stated that the amorphous IZO films are thermally more stable than the amorphous In_2O_3 or ITO films. These IZO films were found to have a very smooth surface with Ra = 0.2 nm as analyzed by an AFM, whereas the polycrystalline ITO and GZO films were reported to have a rougher surface of Ra = 0.8–1.6 nm.

Figure 5.13 shows (a) resistivity, ρ, (b) Hall mobility, μ and (c) carrier density, n of the IZO films as-deposited (open triangle) and post annealed for 1 h at various temperatures in vacuum (6×10^{-4} Pa) (open square) and post-annealed in air (closed square), respectively. After the post-annealing in air, the resistivity increased drastically at temperature higher than 150°C, which can be attributed to a decrease in carrier density. On the other hand, in the case of post-annealing in vacuum, the resistivity remained relatively constant, while the carrier density was almost constant. The decrease in carrier density by the post-annealing in air is likely caused by extinction of oxygen vacancies or interstitial Zn^{2+} ions, implying that carrier density of the amorphous IZO films should be mainly dominated by oxygen vacancies or interstitial Zn^{2+} ions.

Figure 5.14 shows the variation in (a) resistivity, (b) Hall mobility and (c) carrier density of the IZO films deposited with H_2 introduction at various $H_2/(Ar + H_2)$

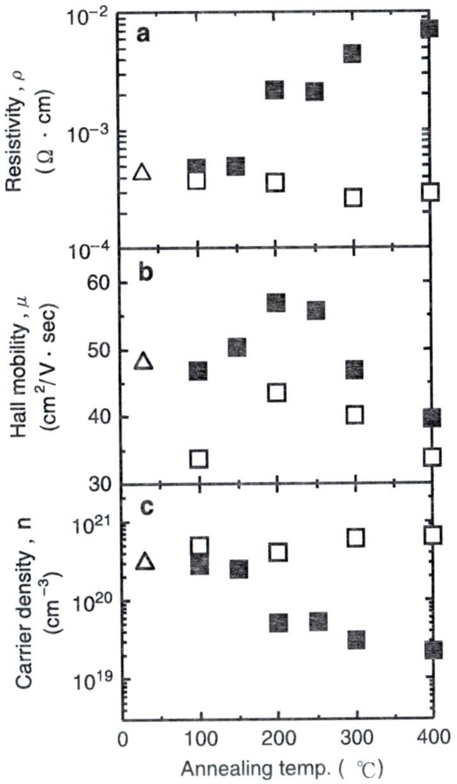

Fig. 5.13 (**a**) Resistivity, ρ, (**b**) Hall mobility, μ and (**c**) carrier density, n of the IZO films as-deposited (*open triangles*) and post-annealed for 1 h at various temperatures in-vacuum (6×10^{-4} Pa) (*open squares*) and post-annealed in-air (*closed squares*), respectively

flow ratios (R_{H2}) in the range of 0–50%. All of these films were also confirmed to show only a halo pattern in XRD and to have an amorphous structure. The systematic increase in carrier density from 3.08×10^{20} to $7.65 \times 10^{20}/cm^3$ with the increase in R_{H2} from 0 to 30% was clearly observed. The lowest resistivity of 3.39×10^{-4} Ω cm was obtained at $R_{H2} = 10\%$. H_2 introduction into the IZO deposition process was confirmed to be effective in order to increase the carrier density. On the other hand, n-type impurity doping of cations or anion, such as Sn^{4+} for In^{3+}, Al^{3+} for Zn^{2+}, or F^- for O^{2-}, by 0.1–3.0% did not lead to an increase in the carrier density in the IZO. These results correspond to previous reports indicating that the doped Sn could not be electrically active in the amorphous ITO films, whereas the doped Sn became electrically active and released free electrons after crystallization [6, 48].

Figure 5.15 shows the transmittance and reflectance from 300 to 2,600 nm for the IZO films deposited with H_2 introduction (R_{H2} = 0, 10 and 30%) at P_{tot} of 0.5 Pa. For the films deposited at larger R_{H2}, transmittance in the near-infrared wavelength region larger than 1,000 nm decreased, where the reflectance increased. This behavior is caused by an increase in the plasma oscillation frequency with increasing carrier density, which could be explained quantitatively by the Drude theory and is well

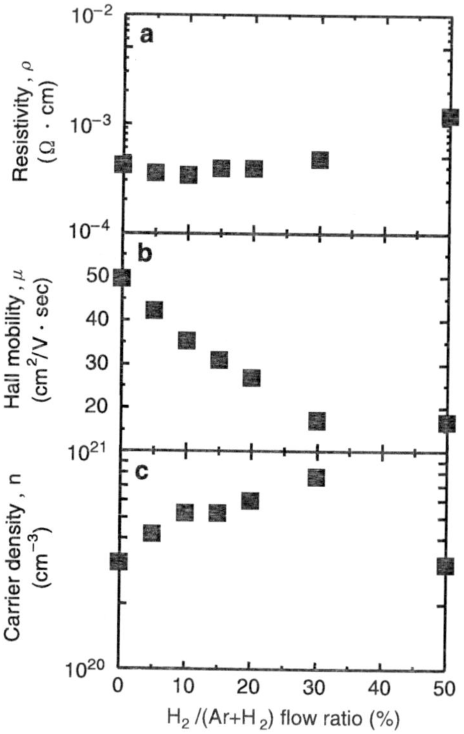

Fig. 5.14 (**a**) Resistivity, (**b**) Hall mobility and (**c**) carrier density of the IZO films deposited at $P_{tot} = 0.5$ Pa with H_2 introduction at various $H_2/(Ar + H_2)$ flow ratios (R_{H2}) of 0–50%

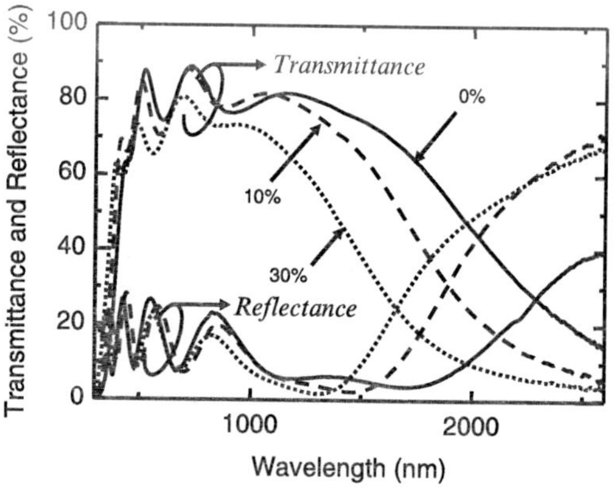

Fig. 5.15 Transmittance and reflectance from 300 to 2,600 nm for the IZO films deposited at $P_{tot} = 0.5$ Pa with H_2 introduction at various $H_2/(Ar + H_2)$ flow ratios (R_{H2}) of 0, 10 and 30%

known in degenerate TCO films [1, 7]. The optical constants of the refractive index (n_{op}) and extinction coefficient (k) were estimated by Newton-Raphson's method from the transmittance and reflectance spectra in the near-ultraviolet region. From the value of k, the absorption coefficient α can be derived as:

$$\alpha = 4\pi k/\lambda.$$

It is commonplace to use the relation

$$\alpha(\hbar\omega - E_g)^{1/2} \quad \text{for} \quad \hbar\omega > E_g$$

to obtain the optical band gap (E_g) for highly degenerate oxide semiconductors based on the assumption of directly allowed transitions to an empty parabolic conduction band [1, 7]. α^2 is plotted as a function of the photon energy ($\hbar\omega$) in Fig. 5.16. E_g is extrapolated to increase from 3.48 to 3.84 eV by increasing R_{H2} from 0 to 30%. This increase in E_g can be explained by the Burstein-Möss (BM) effect in which the lowest states in the conduction band are blocked and transitions can take place only to energies above the Fermi level. Thus the optical constants both in the near-ultraviolet and near-infrared regions imply an increase in n with the introduction of H_2 during the deposition. Consequently, both the electrical and optical analyses indicate that the introduction of H_2 is effective to generate oxygen vacancies or interstitial Zn^{2+} ions, and hence increase the carrier density of the amorphous IZO films.

Figure 5.17 shows the electrical properties of the IZO film deposited with H_2 introduction at $R_{H2} = 10\%$ in the temperature range from 500 to 90 K. The mobility, carrier density and resistivity were almost constant, which corresponded with the typical electrical properties of highly degenerate TCO films [7]. In degenerately-doped polycrystalline semiconductors the following carrier scattering centers have to be taken into account: ionized impurities, neutral impurities, grain boundaries, lattice vibrations, and dislocations. As shown in Fig. 5.17, mobility was almost constant in this temperature range, implying that lattice vibration scattering or dislocation scattering might not be important. In addition, since the IZO films have a completely amorphous structure, electron scattering at grain boundaries should have also little importance. The contribution of ionized scattering centers using the Fermi-Dirac distribution function has been described by the following Brook-Herring-Dingle (BHD) theory [1, 7, 49, 50],

$$\mu_I = \frac{24\pi^3(\varepsilon_0\varepsilon_r)^2\hbar^3 n}{e^3 m^{*2} g(x) Z^2 n_I}, \tag{5.1}$$

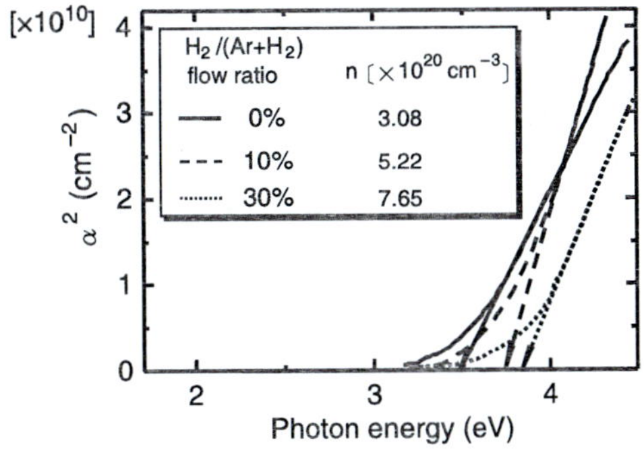

Fig. 5.16 The square of absorption coefficient (α^2) of the IZO films deposited at $P_{tot} = 0.5$ Pa with H_2 introduction at various H_2/($Ar + H_2$) flow ratios (R_{H2}) of 0, 10 and 30%, as a function of the photon energy (eV)

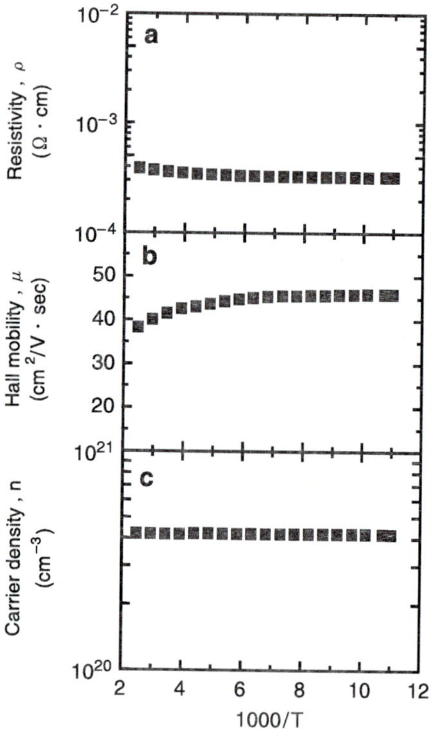

Fig. 5.17 (**a**) Resistivity, (**b**) Hall mobility and (**c**) carrier density of the IZO film deposited at $P_{tot} = 0.5$ Pa with H_2 introduction at $H_2/(Ar + H_2)$ flow ratio (R_{H2}) of 10% as a function of the temperature from 500 to 90 K

where the scattering function g(x) is given by

$$g(x) = \ln\left(1 + \frac{4}{x}\right) - \frac{1}{\left(1 + \frac{x}{4}\right)}$$

and

$$x = \frac{e^2 m^*}{\pi \varepsilon_0 \varepsilon_r \hbar^2 \sqrt[3]{3\pi^5 n}}.$$

Here ε_0 and ε_r are the dielectric constant of free space and the relative permittivity, and Z and n_I are the charge and density of the ionized scattering centers, respectively. The values of the effective mass, m* and dielectric constant, ε_r were selected from the literature on ITO: a value of $0.3m_e$ was used for m* and 9 for ε_r [1, 6, 7]. Calculated values of μ_I using (5.1) are plotted as a function of n in Fig. 5.18 assuming that the carrier density is entirely due to oxygen vacancies (dotted line). The mobility values of all the IZO films obtained in this study are also plotted in Fig. 5.18. In the case of the as-deposited IZO films using only the Ar sputter gas, the

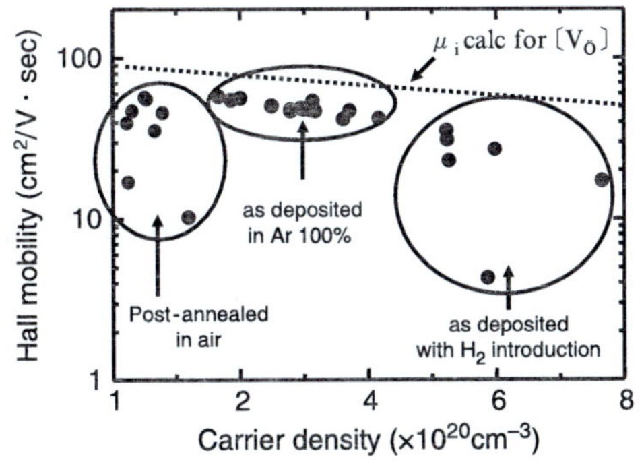

Fig. 5.18 Mobility vs. carrier density of the IZO films prepared under various deposition conditions or post-annealing conditions. The calculated mobility (μ_I) based on the BHD theory with the assumption that all the scattering centers of the free carriers are doubly charged oxygen vacancies (Vö) is also shown by the *dotted line*

mobility showed a slightly smaller value than μ_I implying that electron scattering should be mainly dominated by the ionized impurity. The other IZO films showed roughly a similar tendency, however, some films had much smaller mobilities than μ_I. This might be explained by electron scattering also from neutral scattering centers, similar to those previously reported for ITO films [6, 7]. Leenheer et al. reported that the electron transport in IZO, thermally activated at N smaller than $10^{19}/cm^3$, becomes limited by lattice scattering at N of about $10^{20}/cm^3$ and then limited by ionized impurity scattering at N larger than $5 \times 10^{20}/cm^3$ [51].

Recently, mechanical properties such as internal stress and thermal properties such as thermal diffusion coefficient of ITO or IZO films have also been investigated in detail, and are reported in the references [45, 52], and [53–55], respectively. The thermal diffusivity of the ITO films was $1.5\text{–}2.2 \times 10^{-6}$ m^2/s, depending on the electrical conductivity. The thermal conductivity carried by free electrons could be estimated using the Wiedemann-Franz law. The phonon contribution to the heat transfer in ITO films with various resistivity was found to be almost constant ($\lambda_{ph} = 3.95$ W/m K), which was about twice that for amorphous indium zinc oxide films [54, 55].

References

1. I. Hamberg and C. G. Granqvist, J. Appl. Phys. 60 (1986) R123
2. G. Frank and H. Kostlin, Appl. Phys. A 27 (1982) 197
3. N. Yamada, I. Yasui, Y. Shigesato, H. Li, Y. Ujihira and K. Nomura, Jpn. J. Appl. Phys. 38 (1999) 2856
4. N. Yamada, I. Yasui, Y. Shigesato, H. L. Ujihira and K. Nomura, Jpn. J. Appl. Phys. 39 (2000) 4153
5. Y. Shigesato, S. Takaki and T. Haranoh, J. Appl. Phys. 71(7) (1992) 3356
6. Y. Shigesato and D. C. Paine, Appl. Phys. Lett. 62(11) (1993) 1268
7. Y. Shigesato, D. C. Paine and T. E. Haynes, J. Appl. Phys. 73(8) (1993) 3805
8. Y. Shigesato and D. C. Paine, Thin Solid Films 238 (1994) 44
9. Y. Shigesato, I. Yasui, Y. Hayashi, S. Takaki, T. Oyama and M. Kamei, J. Vac. Sci. Technol. A 13(2) (1995) 268

10. P. K. Song, Y. Shigesato, I. Yasui, C. W. Ow-Yang and D. C. Paine, Jpn. J. Appl. Phys. 37(4) (1998) 1870
11. P. K. Song, Y. Shigesato, M. Kamei and I. Yasui, Jpn. J. Appl. Phys. 38 (1999) 2921
12. A. Kaijyo, K. Inoue S. Matsyzaki and Y. Shigesato, Proceedings for the fourth pacific rim international conference on advanced materials and processing (PRICM-4), 2 (2001) 1787–1790
13. N. Ito, Y. Sato, P. K. Song, A. Kaijio, K. Inoue and Y. Shigesato, Thin Solid Films 496 (2006) 99
14. T. Sasabayashi, N. Ito, M. Kon, P. K. Song, K. Utsumi, A. Kajio and Y. Shigesato, Thin Solid Films 445 (2003) 219
15. M. Marezio, Acta Cryst. 20 (1966) 723
16. H. Odaka, S. Iwata, N. Taga, S. Ohnishi, Y. Kaneta and Y. Shigesato, Jpn. J. Appl. Phys. 36 (1997) 5551
17. K. L. Chopra, S. Major and D. K. Pandya, Thin Solid Films 102 (1983) 1
18. J. L. Yao, S. Hao and J. S. Wilkinson, Thin Solid Films 189 (1990) 227
19. D. M. Ceperley and B. J. Alder, Phys. Rev. Lett. 45 (1980) 566
20. M. Orita, H. Sakai, M. Takeuchi, Y. Yamaguchi, T. Fujimoto, N. Fukumoto and I. Kojima, J. Surf. Sci. Soc. Jpn., Hyomen Kagaku 17 (1996) 440
21. P. Hohenberg and W. Kohn, Phys. Rev. 136 (1964) B864
22. Z. M. Jarzebski, Phys. Status Solidi A 71 (1982) 13
23. Y. Ohhata, F. Shinoki and S. Yoshida, Thin Solid Films 59 (1979) 255
24. H. Odaka, Y. Shigesato, T. Muranaka and S. Iwata, Jpn. J. Appl. Phys. 40 (2001) 3231
25. Y. Sato, R. Tokumaru, K. Utsumi, H. Iigusa, P. K. Song and Y. Shigesato J. Vac. Sci. Technol. A 23(4) (2005) 1167
26. R. Latz, K. Michael and M. Scherer, Jpn. J. Appl. Phys. 30 (1991) L149
27. H. Kimura, H. Watanabe, S. Ishihara, Y. Suzuki and T. Ito, Shinku 30 (1987) 6
28. K. Utsumi, O. Matsunaga and T. Takahata, Thin Solid Films 344 (1998) 30
29. K. Utsumi, H. Iigusa, R. Tokumaru, P. K. Song and Y. Shigesato, Thin Solid Films 445 (2003) 229
30. Y. Sato, R. Tokumaru, K. Utsumi, H. Iigusa, P. K. Song and Y. Shigesato, J. Vac. Sci. Technol. A 23(4) (2005) 1167
31. S. Ishibashi, Y. Higuchi, Y. Ota and K. Nakamura, J. Vac. Sci. Technol. A 8 (1990) 1403
32. H.-Y. Teom, N. Popovich, E. Chason and D. C. Paine, Thin Solid Films 411 (2002) 17
33. Y. Shigesato, Y. Hayashi and T. Haranoh, Appl. Phys. Lett. 61 (1992) 73
34. G. Frank and H. Köstlin, Appl. Phys. A27 (1982) 197
35. H. Köstlin, R. Jost and W. Lems, Phys. Status Solidi A 29 (1975) 87
36. I. Hamberg and C. G. Granqvist, J. Appl. Phys. 60 (1986) 123
37. H. Kirihara and M. Uda, Rev. Sci. Instr. 52 (1981) 68
38. K. Sugiyama, H. Ishii, Y. Ouchi and K. Seki, J. Appl. Phys. 87 (2000) 295
39. Y. Park, V. Choong, Y. Gao, B. R. Hsieh and C. W. Tang, J. Appl. Phys. 68 (1996) 2699
40. T. Kugler, W. R. Salaneck, H. Rothberg and A. B. Holmes, Chem. Phys. Lett. 310 (1999) 391
41. V. M. Vainshtein and V. I. Fistul, Sov. Phys.-Semicond. 4 (1971) 1278
42. W. G. Haines and R. H. Bube, J. Appl. Phys. 49 (1978) 304
43. J. C. Manifacier, L. Szepessy, J. F. Bresse and M. Perotin, Mater. Res. Bull. 14 (1979) 163
44. K. Inoue, Kinouzairyou 19(9) (1999) 39 (in Japanese)
45. T. Sasabayashi, N. Ito, M. Kon, P. K. Song, K. Ustumi, A. Kajio and Y. Shigesato, Thin Solid films 445 (2003) 219
46. Y. Shigesato, N. Shin, M. Kamei, P. K. Song and I. Yasui, Jpn. J. Appl. Phys. 39 (2000) 6422
47. C. W. Ow-Yang, D. Spinner, Y. Shigesato and D. C. Paine, J. Appl. Phys. 83(1) (1998) 145
48. M. Kamei, H. Akao, P. K. Song, I. Yasui and Y. Shigesato, J Korean Ceram. Soc. 6(2) (2000) 107
49. R. B. Dingle, Philos. Mag. 46 (1955) 831
50. R. Clanget, Appl. Phys. 2 (1973) 247

51. A. J. Leenheer, J. D. Perkins, M. F. A. M. Van Hest, J. J. Berry, R. P. O'Hayre and D. S. Ginley, Phys. Rev. B 77 (2008) 115215
52. E. Nishimura, T. Sasabayashi, N. Ito, Y. Sato, K. Utsumi, K. Yano, A. Kaijyo, K. Inoue and Y. Shigesato, Jpn. J. Appl. Phys. 46(12) (2007) 7806
53. T. Yagi, K. Tamano, Y. Sato, N. Taketoshi, T. Baba and Y. Shigesato, J. Vac. Sci. Technol. A 23(4) (2005) 1186
54. T. Ashida, A. Miyamura, Y. Sato, T. Yagi, N. Taketoshi, T. Baba and Y. Shigesato, J. Vac. Sci. Technol. A 25(4) (2007) 1178
55. T. Ashida, A. Miyamura, N. Oka, Y. Sato, T. Yagi, N. Taketoshi, T. Baba and Y. Shigesato, J. Appl. Phys. 105 (2009) 073709

Chapter 6
Transparent Conducting Oxides Based on Tin Oxide

Robert Kykyneshi, Jin Zeng, and David P. Cann

6.1 Introduction

Tin oxide (SnO_2) is an important and widely used wide band-gap semiconductor and is part of a family of binary transparent conducting oxides (TCO), such as Sn- and ZnO-doped In_2O_3 (ITO, ZIO) [1] and ZnO:Al [2], CdO. It is of great interest in corrosive environment applications due to its high stability. This includes applications such as batteries, low emission windows coatings, solar cells, etc. In this chapter we will introduce the structural, electrical and optical properties of undoped and doped tin oxides. In addition, thin films via various deposition methods are discussed.

6.2 Crystal Structure

Tin oxide (SnO_2) occurs in nature as the mineral Cassiterite. It possesses the rutile crystal structure with a tetragonal unit cell ($P4_2/mnm$) and $a = b = 4.738$ Å and $c = 3.188$ Å [3]. Many other metal dioxides share this structure such as TiO_2, PbO_2, TaO_2, TeO_2, and RuO_2. The crystal lattice of SnO_2 is shown in Fig. 6.1a, where the sixfold coordination of the Sn atoms to the oxygen nearest neighbors can be observed.

The formal atom positions in RX_2 are described as:

$$\mathrm{R} : (2\mathrm{a})\, 000;\ {}^1\!/\!_2{}^1\!/\!_2{}^1\!/\!_2$$

$$\mathrm{X} : (4\mathrm{f}) \pm (uu0; u + {}^1\!/\!_2,\ {}^1\!/\!_2 - u,\ {}^1\!/\!_2)$$

R. Kykyneshi (✉)
Department of Materials Science, Oregon State University, Corvallis, OR, 97331, USA
e-mail: kykynesr@onid.orst.edu

D.S. Ginley et al. (eds.), *Handbook of Transparent Conductors*,
DOI 10.1007/978-1-4419-1638-9_6, © Springer Science+Business Media, LLC 2010

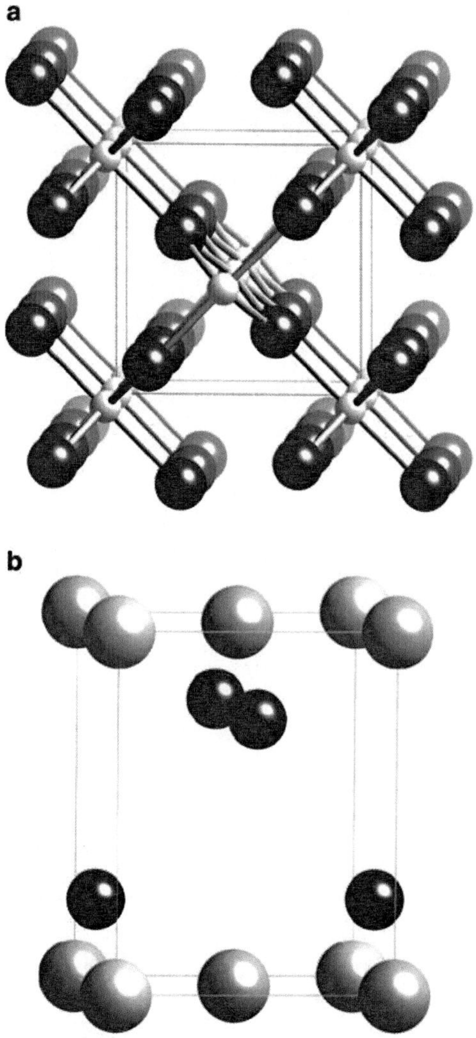

Fig. 6.1 (**a**) Crystal structure of the tetragonal SnO_2 with the cassiterite structure and (**b**) litharge SnO

The variable parameter u is approximately 0.307 in SnO_2. Each oxygen octahedra in SnO_2 is connected to two adjacent octahedra through edge sharing along the c-axis, and connected to other octahedra through corner sharing.

The lower symmetry of the tetragonal crystal structure of SnO_2 leads to an anisotropy in the electronic and optical properties. For example, the smaller distance between Sn atoms along the c-axis, may facilitate a higher atomic orbital overlap and thus a more conductive pathway for electron transport. This has been confirmed in conductivity measurements on single crystals of the isostructural rutile compound TiO_2 with a conductivity five to ten times greater along the c-axis as compared to the <110>.[4]

Although SnO_2 is the most stable and technologically important oxide, other tin oxides (SnO and Sn_3O_4 [5]) also exist, but less is known about them. The crystal structure of SnO is schematically shown in Fig. 6.1b. It possesses the litharge crystal structure with a tetragonal unit cell. These oxides, which contain Sn(II), are less stable and convert to a mixture of SnO_2 and Sn metal at elevated temperatures [6]. Despite the lower stability, realization of p-type conductivity in SnO could be possible, taking advantage of the Sn(II) to Sn(IV) transition upon appropriate doping, similarly to the case of SnS_x ($x = 1, 2$) [7] materials (SnS is an n-type and SnS_2 is a p-type semiconductor). Band structure calculations [8] establish the higher valence band dispersion in SnO compared to SnO_2. Furthermore, there is one report by Ogo et al. of a thin film transistor which incorporates semiconducting SnO [9]. Nonetheless, high mobility p-type wide band-gap semiconductors are much needed for the development of CMOS equivalent transparent electronics.

6.3 Chemistry

Tin oxide is an n-type conductor, in which wide band-gap and high conductivity coexist. As in the case of other binary metal oxides (e.g. ZnO, In_2O_3), oxygen vacancies are responsible for conduction electron-carrier generation in SnO_2. The electron generation can be described by the following equation, using Kröger-Vink notation:

$$O_O^X \rightarrow V_O^{\bullet\bullet} + 2e' + \frac{1}{2}O_2(g)$$

According to this equation, oxygen anions escaping the crystal structure (½O_2) from an occupied oxygen site creates a doubly ionized vacancy site (Vö) and two free electron carriers. Creation of multiple carriers from a single defect is highly desirable, due to minimization of carrier scattering on crystal impurities. The presence of oxygen vacancies in bulk SnO_2 is observed by X-ray diffraction [10] and coulometric titration [11] experiments. This oxygen non-stoichiometry in bulk SnO_{2-x} is in the $x = 0.02$–0.034 [11] over the temperature range 700–990 K. Such a vacancy concentration yields up to $5 \times 10^{20}/cm^3$, assuming all vacancies are doubly ionized. In fact, the most relevant chemical diffusion process in undoped and cation-doped SnO_{2-x} single crystals is due to oxygen anions [12], indicating the low formation energy of oxygen vacancies. This donor level lies about 30 and 140 meV below the conduction band for the singly and doubly ionized states, respectively [13, 14]. The oxygen deficiency can be easily controlled in thin films by adjusting the oxygen partial pressure during sputtering or other PVD deposition [for example 15]. A decrease of the carrier concentration (and therefore electrical conductivity) with higher oxygen partial pressures, indicates a decreasing number of oxygen vacancies in SnO_{2-x} thin films.

First-principles investigation of undoped SnO_2 by Zunger et al. [16], show that although oxygen vacancies are contributing to carrier generation, the majority electron generation takes place due to tin interstitials (Sn_i). The Sn_i defects states appear above the conduction band minimum (CBM) and thus readily create conduction electrons, whereas oxygen vacancies are shallow donors (located somewhat below CBM). It is possible that the high oxygen vacancy concentrations in SnO_2 are tolerated due to the multivalency of Sn.

Apart of oxygen deficiency, a variety of dopants have been employed to induce conductivity in SnO_2. In case of donor doping the most common choices are Sb^{5+} substitution on the Sn^{4+} site and F^{1-} on the O^{2-} site [15, 17]. Some of the less traditional n-type dopants are Mo, Ta, and Cl [15, 18, 19]. In most of these doping scenarios a single electron is produced per substitution. Anion doping by F^{1-} on the threefold coordinated O^{2-} site is the least ambiguous method for electron generation. Moreover, the crystal lattice is affected the least by this substitution, due to the similarity of the oxygen and fluorine ionic radii. Such a size-matched dopant minimizes the lattice distortion, thus preserving the transport properties of the undoped lattice. Doping by the larger Cl^- anion, which also prefers a sixfold coordinated environment, results in lowering carrier mobility and carrier concentration [19]. The incorporation of Ta produces free electrons, but only low doping levels of a few at.% of Ta and Mo can be accomodated in the SnO_2 lattice [15, 18, 20] beyond which the conductivity of doped SnO_2 decreases due to a reduction in mobility.

The size of Sn^{4+} is 0.69 Å which is appreciably larger than the ionic radii of most cation-site donor dopants in the pentavalent oxidation state and sixfold coordination [21]. These cation dopants are often multivalent and thus may be incorporated in a different oxidation state. For example, the inclusion of Sb^{5+} ($r = 0.60$ Å) results in carrier generation, while Sb^{3+} ($r = 0.76$ Å) would result in creation of acceptor states. In fact, the presence of Sb^{3+} is shown in bulk and thin film SnO_2:Sb samples by XPS measurements [22].

Group II and III cation doping, such as In, Ga, Al, Fe, Zn, etc., as well as substitution on the anion site by N^{3-} create acceptor-like states in SnO_2. Incorporation of Zn is reported to decrease the electron carrier concentration [21], whereas SnO_2:Fe gas sensors [23] showed p-type response in an oxygen atmosphere. A few reports on p-type tin oxide use high doping levels (typically >8 at.%) of In [24–26], Ga [27] or Al [28] to produce a low hole carrier concentration. It is typically difficult to induce hole carriers in wide band-gap oxide semiconductors. The reason for such behavior lies in the generation of compensating donor-states. Based on thermoluminescence measurements, modeling [29], and first-principles calculations [15], the number of oxygen vacancies and tin (or trivalent acceptor dopant) interstitials is predicted to increase with a higher acceptor doping level. Therefore only low hole carrier concentrations can be achieved and a changeover to n-type carriers with time may occur, due to slow diffusion processes at room temperature.

Doping studies on SnO are not present in the literature to date. In this case, doping by all group III and IV elements, as well as oxygen deficiency and F-doping are all expected to create donor-like states. Some possibilities for acceptor-doping are excess oxygen, nitrogen incorporation on the anion site, and substritution on the Sn (II) site by group I elements. It has been noted that undoped SnO thin films do exhibit p-type conductivity [30] due to access oxygen present in the film.

6.4 Physical Properties

6.4.1 *Band Structure*

Band structure calculations of undoped and Sb-doped SnO_2 are shown in Fig. 6.2a and b, and undoped SnO in Fig. 6.2c [31–33]. Focusing on the properties of undoped SnO_2 first, the band structure predicts a direct optical band-gap at the Γ point [30], where the conduction band minimum (CBM) and the valence band maximum (VBM) occur. The low estimated 1.08 eV optical band-gap is much smaller than the commonly observed experimental value near 3.9 eV. This is a well-known issue with the local-density approximation within the density functional

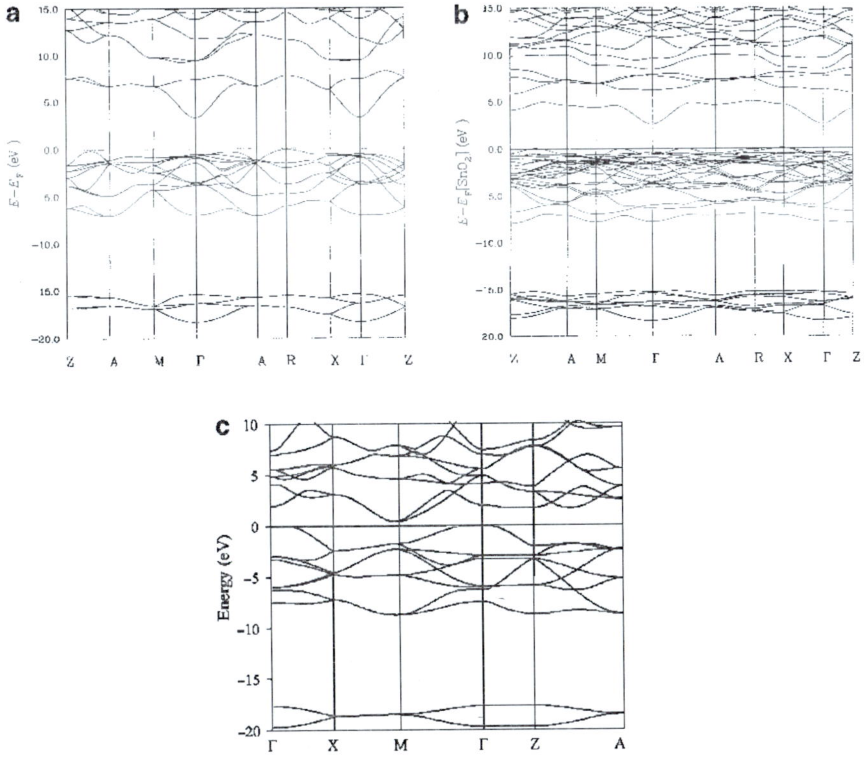

Fig. 6.2 Band structures of (**a**) undoped SnO_2 [32, Fig. 1], (**b**) Sb-doped SnO_2 (or Sn_3SbO_8) [31, Fig. 3], and (**c**) undoped SnO [31, Fig. 5]

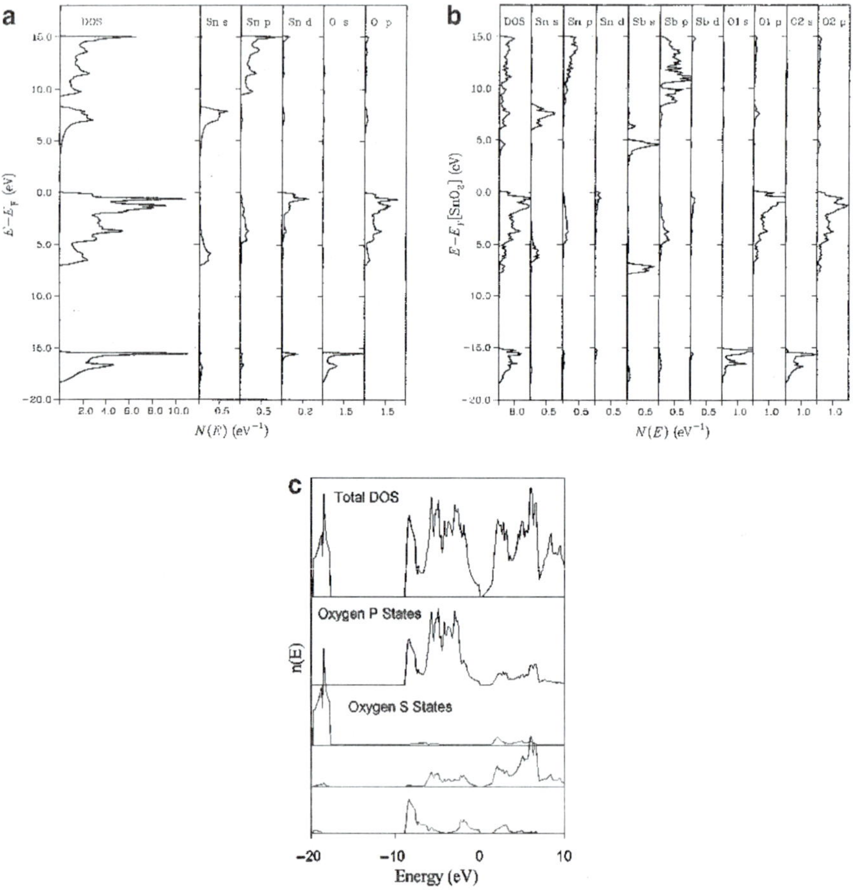

Fig. 6.3 Total and partial density of states of (**a**) undoped SnO_2 [32, Fig. 2], (**b**) Sb-doped SnO_2 (or Sn_3SbO_8) [31, Fig. 4], and (**c**) undoped SnO [31, Fig. 3b]

theory framework. The high dispersion of the density of states near the CBM is indicative of a low conduction electron effective mass or, in other words, a high electron mobility. Calculated electron effective masses are 0.26 and 0.20 m_0 along the $\Gamma - X$ and $\Gamma - Z$ crystallographic directions [34]. This is due to the anisotropy in the tetragonal crystal structure and is in good agreement with experimental values [35].

The total density of states (DOS) and its partial contributions by Sn and O states is illustrated in Fig. 6.3a. The main contribution to the CBM comes from the Sn *5s*, the origin of the high degree of conduction band dispersion, and the O *2s* and *2p* states. The strong *s*-character of the CBM is a common feature of the binary TCO's, resulting in the similarity of the respective band structures [32].

The band structure of SnO_2:Sb was calculated using the Sn_3SbO_8 system (Fig. 6.2b). The addition of Sb introduces a new free-electron-like band, widening the band-gap due to repulsion of the donor and host states. The total density of states (Fig. 6.3b) has a clear Sb *5s* contribution and forms the CBM, while the Sn *4s* states are pushed to higher energies. This band may be partially filled with the extra electrons from the incorporatoin of Sb^{5+} into the lattice, providing the electron density for high conductivity in SnO_2.

The valence band of SnO_2 is composed mainly of O *2p* states and some Sn *5d*, indicating the ionic character of this oxide. As a consequence, the dispersion of the density of states near the VBM is low, and the hole effective mass in p-type SnO_2 is much higher than that of electrons. In fact, modeling of the optical band-gap in degenerate SnO_2 films suggests a hole effective mass of $\sim 1.0\ m_0$ [36].

As for SnO, the band structure shown in Fig. 6.2c suggests a much smaller 0.18 eV indirect band-gap and a lower dispersion of the conduction band. While this makes SnO an inferior n-type TCO compared to SnO_2, it is interesting to note the increase in the valence band dispersion. Now, the top of the valence band is composed of Sn *5s* and O *2p* mixing (Fig. 6.3c), consequently reducing the carrier effective mass (similarly to the case of the SnO_2 conduction band). A detailed study of p-type conductivity in SnO thin films has not yet been carried out.

6.4.2 Transport Properties

A wide range of electrical conductivity from $\sim 10^{-6}$ S/cm [37] up to $0.4\text{–}1.2 \times 10^4$ S/cm [38, 39] can be obtained in SnO_2 as required by application, such as gas-sensor and TCO, respectively. As we have discussed above, there are several ways to modulate the conductivity in tin oxide. Namely by controlling the carrier concentration via oxygen defiency and cation- and/or anion-site doping. In the following discussion the conduction mechanisms of various undoped and doped single-crystals, and epitaxial and polycrystalline thin films is presented.

Transport measurements are best carried out on single crystal samples due to the high crystal quality. Fitting of the temperature dependent Hall mobility reveals the carrier scattering mechanism. Single crystals of SnO_2 with low carrier concentrations ($\sim 10^{16}/cm^3$) exhibit polar mode optical phonon scattering of the carriers and have a room temperature mobilities near 260 cm^2/V s [13] (Fig. 6.4a). Donor doping by Sb to a $10^{18}/cm^3$ carrier concentration results in a nearly degenerate semiconductor, as seen in Fig. 6.4b. This is due to the shallow ~ 12 meV donor level associated with Sb incorporation. A reduction of the room temperature mobility to about 150 cm^2/V s is observed, and the main scattering mechanism at this doping level is due to ionized impurities. Although these Hall measurements do not reveal anisotropy in carrier transport due to the tetragonal structure, the difference in electron effective masses perpendicular 0.30 m_0 and parallel 0.23 m_0 to the c-axis of the tetragonal structure have been measured [35].

Thin films of SnO_2 have additional scattering mechanisms such as line defects and interfaces. The electrical properties of epitaxial films, typically grown on sapphire substrates, exhibit a reduction in carrier concentration and mobility with

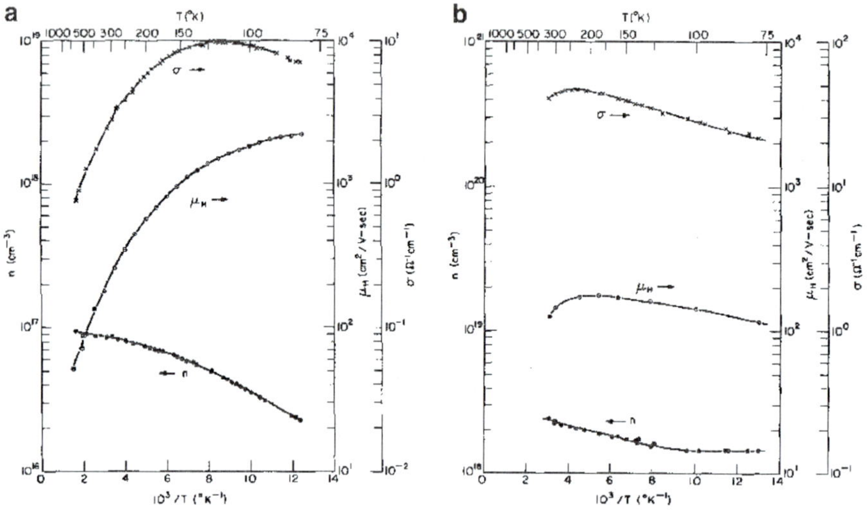

Fig. 6.4 Electrical properties of (**a**) low carrier concentration and (**b**) Sb-doped SnO_2 single crystals [13]

successively lower film thickness [40]. From this work it is concluded that the decrease in mobility is due to scattering on crystal defects, such as antiphase boundaries and partial dislocations, the density of which varies with film thickness. Note, that even in epitaxial films (100 nm thick) the electron mobility at 37 cm^2/V s [40, 41] is much lower than that measured in SnO_2 single crystals. The carrier concentration is speculated to decrease because these same crystal defects may be charged leading to the presence of a space charge region. In the former case, a negatively charged surface would cause the depletion of electrons in the bulk of the film.

Lastly we consider the transport properties of polycrystalline films deposited onto glass or polymer substrates. Apart from the above mentioned scattering processes, electron scattering on charged grain boundaries needs to be considered [14, 17, 42] as an effect similar to the charged surface in epitaxial films. A negative surface charge builds up near the grain boundary due to electron trapping, and conduction electrons predominantly tunnel through the space-charge region [40]. Generally, thin films with larger grain sizes possess higher electron mobility [19, 38, 41, 42], depending on thin film growth conditions and doping. The grain boundary potential can be reduced by annealing in N_2 gas or vacuum, resulting in the recovery of mobility up to ~60 cm^2/V s [13]. Increasing film thickness also helps to obtain higher carrier concentration, mobility, and thus conductivity [43, 44].

Typical mobility values obtained in polycrystalline thin films vary between 1 and 20 cm^2/V s [14, 16, 41]. At carrier concentrations near $10^{21}/cm^3$ the obtained thin film conductivity is of the order of magnitude 10^3 S/cm. The reported highest conductivity of 1.5×10^4 S/cm is obtained by annealing an undoped SnO_2 film at

550°C capped with a thin layer of SiO_2 [45]. This improves the film stucture and helps maintains oxygen stoichiometry. High conductivity and transparency SnO_2 films are successfully grown on flexible substrates [46] for emerging light-weight and foldable device applications.

6.4.3 *Optical Properties*

The interest in SnO_2 coatings is due to the coexistance of high conducivity and transparency in the visible range of the eletromagnetic radiation spectrum. For TCO applications a band-gap of at least 3.1 eV is required. Undoped conductive tin oxide films exhibit a larger band-gap, typically greater than 3.6 eV [15, 17, 47]. The direct nature of the band-edge transition is established from band structure calculations [31, 32]. In addition, a direct gap is indicated by the good linear fit of the $\alpha h\nu$ $(h\nu - E_g)^{1/2}$ relationship shown in Fig. 6.5, where α is the absorption coefficient, $h\nu$ is the incident photon energy, and E_g is the band-gap energy required to promote an electron from the top of the valence band to the first available unoccupied state in the conduction band. With addition of various donor dopants (Sb and F), Fig. 6.5, the energy needed for the transition of an electron increases due to filling of the conduction band with free electron carriers. This effect is known as the Burstein-Moss shift [48], responsible for band-gap widening with an increase in carrier concentration.

Undoped and doped tin oxides are highly transparent in the visible-light wavelength range (400–800 nm) [15, 17, 49], as shown in Fig. 6.6. The high transparency can be explained by a low concentration of mid-gap states, typically responsible for absorption of photons with energies below the band-gap value. Refractive index values between 1.9 and 2 are commonly observed in SnO_2 thin films.

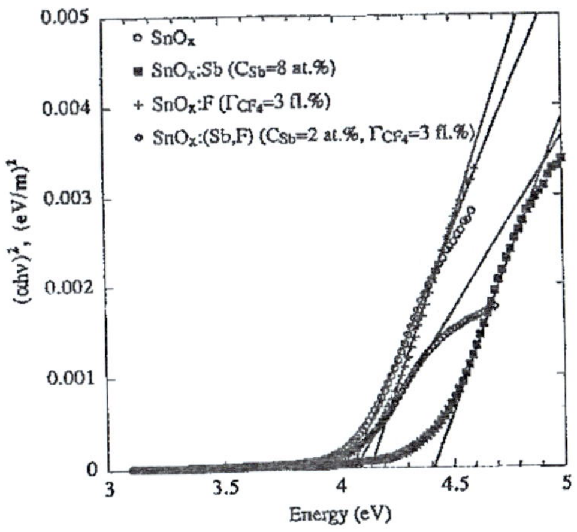

Fig. 6.5 Direct band-gap evolution and Burstein-Moss shift of the band edge with Sb-, F-doping of tin oxide thin films [15, Fig. 14]

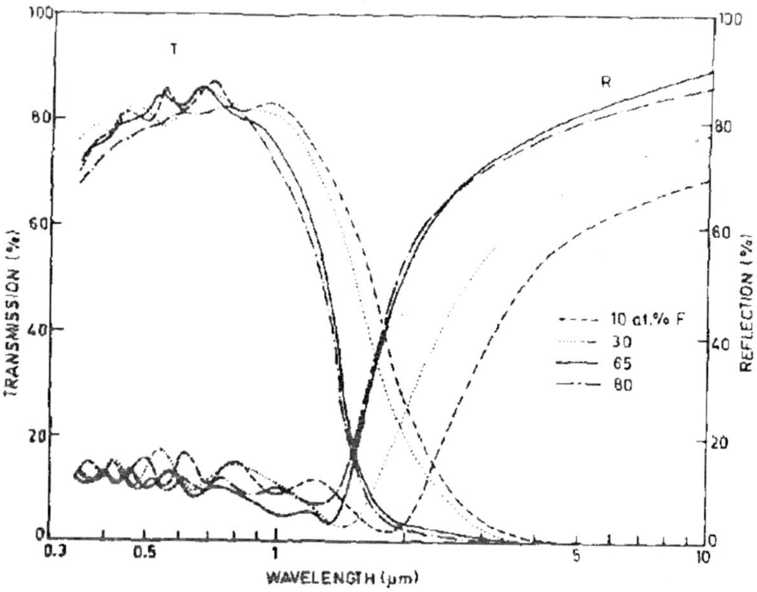

Fig. 6.6 Transmission and reflection spectra of F-doped SnO_2. The crossover between the transmission and reflection curves in the IR-range shows the position of the plasma wavelength due to free carrier reflection [47, Fig. 7]

Optical transmission of the conductive SnO_2 films is limited in the near infra-red (IR) region. This plasma wavelength (or more commonly plasma frequency) depends on the carrier concentration and effective mass. Thus, the observation of the plasma frequency (ω_p) together with the measured carrier concentration (N) allows the determination of the carrier effective mass (m^*), which is expressed from the classical Drude theory as:

$$\omega_p = \frac{4\pi N e^2}{\varepsilon \varepsilon_0 m^*}$$

where e is the electron charge, ε is the high frequency dielectric constant, and ε_0 is the permittivity of free space. The electron effective masses of doped SnO_2 derived from these measurements are in the range of 0.15 – 0.4 m_0 [15, 17, 49]. This is close to the average value of 0.25 m_0 reported for the single crystal [35]. The high reflectivity in the near IR region below the plasma wavelength allows the use of highly doped SnO_2 as low-emissivity window coatings.

The wide band-gap of tin oxide may enable applications in UV LED's. The high exciton binding energy of 130 meV is promising [50], since it is much higher than the reported binding energy value of 60 meV for ZnO [51]. Band-edge and acceptor-donor photoluminescence peaks are observed due to spontaneously forming oxygen vacancies donor levels and intentional acceptor levels from N-doping [52].

The properties of SnO are little studied, but band-gap values in the 2.7–2.9 eV range are reported [53, 54]. The band-gap value below 3.1 eV leads to absorption in the visible range and thus coloration of SnO thin films.

6.4.4 Amorphous SnO_2

Nowadays, amorphous wide band-gap semiconductor thin films are replacing crystalline counterparts in TCO applications, due to lower processing temperatures, reduced cost, and the ability to deposit TCO films on flexible polymer substrates. Successful candidates exhibit optoelectronic properties comparable to those of the crystalline form and high stability at elevated temperatures. A good example of an amorphous TCO is In_2O_3-$(ZnO)_x$ thin films with a conductivity of about 1,500 S/cm and phase stability up to 500°C [1].

Most current TCO materials are based on In, Zn, Sn, and Cd oxides. These elements belong to a group of heavy metal cations of the $(n-1)d^{10}\,ns^0$ electronic configuration [55]. The conduction band of these binary oxides and their mixtures have a strong s-character with large isotropic spatial width due to the spherical symmetry of the s-bands. The short range order is similar in both the amorphous and crystalline phases, except the bond angles vary in the former case, which does not influence the electron conduction. The absence of grain boundaries excludes the associated carrier scattering mechanism, thus higher performance of amorphous materials is expected. Carriers are a typically the result of anion vacancies, and cation doping as a rule does not increase the carrier concentration, but rather stabilizes the amorphous phase.

Tin oxide thin films are amorphous as deposited on to unheated substrates [15, 56, 57]. The disordered phase is maintained upon post-annealing in air up to 300°C [56], and a high mobility of 22 cm^2/V s is achieved comparable to that of the crystalline phase. The highest obtained conductivities are about 500–2,000 S/cm. Amorphous SnO_2 films are about 80% transparent in the visible range with a band-gap near 3.65 eV. The observation of a lower band-gap compared to near 4.0 eV of the crystalline phase of the similar carrier concentration, is typically observed in TCOs possibly due to smearing of the conduction and valence band edges. In the same work [56] the ionized impurity scattering with free electron screening is determined as the main scattering mechanism.

As mentioned above the amorphous phase can be further stabilized by addition of other heavy metal cations, for example Zn. The stabilization of the disordered phase occurs due to the difference in the coordination of Sn in SnO_2 (sixfold) and Zn in ZnO (fourfold). Zinc tin oxide (ZTO) consisting of equal parts of ZnO and SnO_2 shows the highest temperature stability of the amorphous phases [58, 59]. The ZTO thin films are typicaly amorphous when deposited at substrate temperatures below 300°C (via sputtering) and 500°C (via PLD) and retain this structure upon post-annealing at temperatures up to 600°C [60]. Although, ZTO is not highly conductive with a modest conductivity of 54 S/cm, a high carrier mobility is obtained near 30 $cm^2/V\ s$ [59] leading to successful application in thin film transistors as the channel layer [60]. This will be discussed in more detail below.

In the case of indium-tin-oxide mixtures, the low level Sn incorporation into In_2O_3 leads to the well known high conductivity ITO [1] in the amorphous and crystalline forms, not discussed here. The highest conductivity of thin films rich in SnO_2 is exhibited at a In/Sn ratio equal to 1 with a conductivity near 2×10^3 S/cm. Decreasing the In content in thin films results in lowering of the carrier concentration and mobility towards those of undoped amorphous tin oxide thin films [61]. In the same work, the acceptor-like doping of In in crystallized films is shown, as expected from substitution of In^{3+} on the Sn^{4+} site.

Thus amorphous tin oxide films, especially with admixtures are technologically important and find implementation in TCO and TFT applications.

6.4.5 Mechanical Properties and Corrosion Behavior

Tin oxide thin films show good adhesion onto glass substrates. The adhesion of crystalline SnO_2 films is >300 kgF/cm^2, as measured by the pull-off method [62]. It can be further increased with higher substrate temperatures during deposition and with post-deposition annealing.

SnO_2 is found in nature, typically in riverbeds, and therefore has high resistance to atmospheric influences – good for architectural low-emission window coatings. Undoped and Sb-doped SnO_2 films show high stability up to 800°C in air and inert atmospheres [63]. This makes them attractive for applications requiring low variations in electrical properties over a wide temperature range.

6.5 Thin Film Disposition

6.5.1 Substrates

The microstructure of a thin film is strongly dependent on the choice of substrate. In most cases of non-lattice-matched, including amorphous, substrates the resultant thin films are polycrystalline, with the crystallite sizes dependent upon the deposition parameters. Typically, in most applications inexpensive glass substrates are used for polycrystalline film deposition. Undoped and doped SnO_2 polycrstalline films are deposited by chemical and physical vapor deposition methods discussed in the following sections.

Epitaxial SnO_2 films require lattice matched substrates and are successfully deposited on various surfaces of TiO_2 [64–66] and the (0001) and $(10\bar{1}2)$ surfaces of α-Al_2O_3 (sapphire) [40, 67] single crystal substrates. Although the (100) TiO_2

surface provides the lowest lattice mismatch of ~2.6%, reports on the optoelectronic properties of epitaxial SnO_2 films on TiO_2 substrates are not found, possibly due to the coloration and conductive nature of the substrate. The insulating sapphire substrates are perhaps better suited for such studies. Epitaxial SnO_2 thin films are grown on r-cut sapphire $(10\bar{1}2)$ substrates with the epitaxial relationship $SnO_2(101)$ $[010]//Al_2O_3(10\bar{1}2)[\bar{1}2\bar{1}0]$ [40]. The large lattice mismatch of ~12% causes the generation of a large number of crystal defects at the film-substrate interface, including antiphase boundaries and partial dislocations. The carrier concentration and mobility of the epitaxial SnO_2 films are found to depend on film thickness. The electron mobility is reduced in ultra-thin films due to scattering on the antiphase boundaries which tend to terminate inside the film. Also the carrier concentration is lower in thinner films due to charge trapping associated with the defects.

Epitaxial thin films are deposited by a variety of methods including pulsed laser deposition [40, 64], atomic layer deposition [68], reactive sputter deposition [69], molecular beam epitaxy [70], chemical vapor deposition [65] and excimer laser annealed layers grown by spin coating [66].

6.5.2 Spray Pyrolysis

One of the most commonly used and inexpensive large-area deposition techniques for SnO_2 thin films is the spray pyrolysis method. The typical spray mixture contains $SnCl_2$ or $SnCl_4$ in a solvent with the source of oxygen contained in the mixture or the carrier gas. Films are deposited at elevated substrate temperatures, usually in the 350–550°C range at which a substitution reaction of Cl by oxygen takes place. Nominally undoped SnO_{2-x} films of high $\sim 10^3$ S/cm conductivity are obtained by the spray pyrolysis deposition [17, 71] with carrier concentrations in the 10^{19}–10^{20}/cm^3 range. The use of higher substrate temperatures increases the electron carrier concentration, likely due to creation of oxygen vacancies. The increase in electron mobility at higher substrate temperatures is possibly the result of higher film crystallinity.

Doping of tin oxide thin films takes place by the addition of $SbCl_3$ and NH_4F or HF into the spraying solution for Sb-doped and F-doped films, respectively. The donor doping increases the carrier concentration to values near 10^{21}/cm^3 [46, 49, 71, 72], yielding thin film conductivities as high as 4,500 S/cm in SnO_2:F. The carrier mobility of fluorine-doped films is invariably near 20 cm^2/V s [49, 72], while Sb-doped films exhibit a lower mobility near 15 cm^2/V s that decreases with higher Sb content [17, 46]. A preferential (200) crystallite orientaion of Sb and F-doped SnO_2 thin films is observed instead of the (211) orientation of the undoped layers [46, 72]. The Sb incorporation decreases the grain size of the tin oxide films, as shown in Fig. 6.7, resulting in higher grain boundary scattering [46].

Fig. 6.7 SEM images of SnO_2 thin films doped with Sb at (**a**) 0 at.%, (**b**) 1 at.%, (**c**) 3 at.% and (**d**) 5 at.% levels. The average grain size is reduced from ~250 to ~150 nm, yielding a decrease in carrier mobility from 15 to 3 cm^2/V s [42, Fig. 2]

6.5.3 Chemical Vapor Deposition

High conductivity and transparency SnO_2 thin films are successfully prepared by chemical vapor deposition methods (CVD) [40, 63]. In addition, modified CVD techniques such as metal-organic [18, 20, 57] and atmospheric pressure [73] CVD techniques are also effective. Undoped and F-doped SnO_2 thin films deposited by CVD exhibit similar opto-electronic properties to the sprayed films. A notable difference is the ability to prepare amorphous SnO_2 layers by CVD at low substrate temperatures in the range of 200–350°C with a high conductivity of 5,000 S/cm [57].

Another point of interest are the studies conducted on polycrystalline and epitaxial Ta-doped SnO_2 thin films deposited on glass and Al_2O_3 (0001) substrates, respectively [18, 20]. With approximately 4 at.% Ta-doping in polycrystalline SnO_2 on glass substrate, the result is a high conductivity of 5×10^3 S/cm with a carrier concentration of $1.3 \times 10^{21}/cm^3$ and a mobility of 25 cm^2/V s. In epitaxial films, the conductivity of 3.3×10^3 S/cm is the highest for the 1.4 at.% Ta-doped sample. The dopant is found to improve the lattice plane alignment, resulting in an increased mobility of 43 cm^2/V s due to minimizing grain boundary scattering.

6.5.4 *Sputtering*

Thin film deposition by sputtering is another widely used method for industrial large-area thin film coatings. An extensive study of undoped and F-, Sb-, Mo-doped SnO_{2-x} thin films was carried out by Granqvist et al. [15] using the R.F. magnetron sputtering method. The highest conductivity achieved in this study were $\sim 1.1 \times 10^3$ S/cm in the case of (Sb+F) co-doped films, under optimized oxygen background pressure, substrate temperature, and doping concentrations. The lower conductivity was ascribed to the small grain size of 6–30 nm thus reducing the electron mobility to below 20 cm^2/V s.

Deposition of conductive SnO_2:Sb onto flexible substrates for novel applications in light-weight electronics has been demonstrated by same sputtering method [44]. A strong (110) preferential orientation of the grains is observed. An improvement in electrical properites is also oberved with larger film thicknesses possibly due to growth of larger grains.

Epitaxial SnO_2 thin films are grown by the reactive sputter deposition method on single crystal r-cut sapphire Al_2O_3 $(10\bar{1}2)$ [69].

6.5.5 *Other Deposition Methods*

High quality SnO_2 films are deposited onto r-cut Al_2O_3 substrates by pulsed laser deposition (PLD) and pulsed electron deposition (PED) methods [47]. A high oxygen stoichiometry can be achieved by PED in 8 mTorr O_2 deposition gas, resulting in lower 0.01 S/cm conductivity of the undoped film. Epitaxial films grow under similar deposition conditions by PLD which exhibit a conductivity of 50 S/cm at room temperature. The electrical transport properties with respect to crystal defects is studied on epitaxial SnO_2 thin films grown by a femtosecond PLD method [40].

PLD growth of SnO_2:Sb (5 at.%) films on glass substrates show a strong dependence on the oxygen background pressure during the deposition [74]. The films grow preferentially (200) oriented at 4 and 6.7 Pa oxygen pressure. In the case of low 1.1 Pa O_2 pressure, a secondary phase with a SnO stoichiometry appears. The carrier mobility of the Sb-doped tin oxide films is below 10 cm^2/V s, and the grain size and film thickness is not discussed in this contribution.

Spin coating is an emerging thin film deposition method. SnO_2:Sb (7 mol.%) spin coated onto glass substrates require annealing at temperatures as high as 650°C to crystallize the thin films [43]. Thicker films exhibit slightly higher crystallite sizes, and an approximately 1-μm thick film has a room temperature conductivity of 150 S/cm.

Epitaxial thin films of tin oxide for gas sensing applications are also prepared by the atomic layer deposition method [68].

6.6 Applications

6.6.1 *Transparent Thin Film Transistors*

A new field of application has been on the rise which utilizes both the transparency and high carrier mobility of TCOs [75]. The thin film transistor (TFT) structure is that of a metal-insulator-semiconductor field effect transistor (MISFET). In such a device the channel layer, in this case SnO_2, is deposited on top of an insulator (the gate dielectric) and conductor (the gate electrode) stack. This is followed by the deposition of the source/drain electrodes. The structure of this bottom-gate design is shown in Fig. 6.8. The device is transparent in the visible range if the electrodes are TCO materials (e.g. ITO) supported on glass substrates. The SnO_2 channel layers prepared by R.F. magnetron sputtering [76] are enhancement mode devices, in which the device is normally off and a positive voltage is applied to the gate electrode to turn the device on. The source–drain current–voltage characteristics of SnO_2 TTFT at various gate voltages is illustrated in Fig. 6.9. The device exhibits a maximum drain current of 90 μA, and an on/off ratio of 10^5 with a field-effect mobility of about 1 cm^2/V s. The high optical transparency ~80% of the device in the visible range is also shown in Fig. 6.8.

A similar performace TFT device is produced by inkjet printing [77]. The obtained layers are highly porous, with applications in TFT gas sensors and nanostructured solar cells. The performance of SnO_2 is somewhat inferior to other TTFTs [for example 78], possibly due to a low defect control in SnO_2 and its sensitivity to ambient environment.

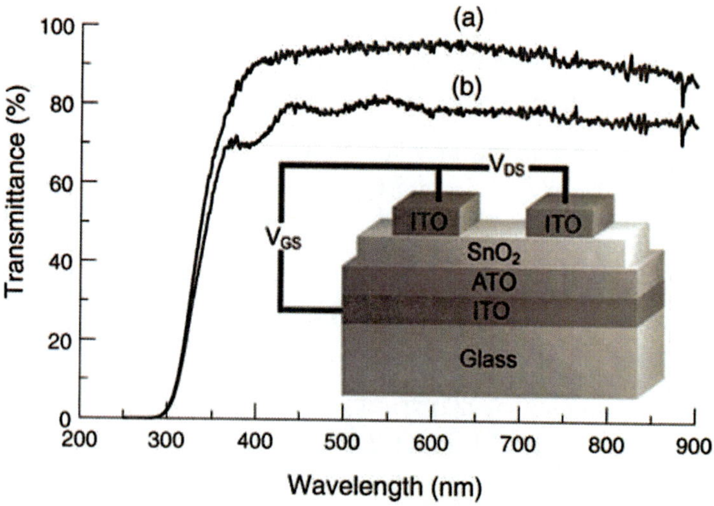

Fig. 6.8 Bottom-gate SnO_2 TTFT structure (*inset*) and optical transmission of the device [73, Fig. 3]

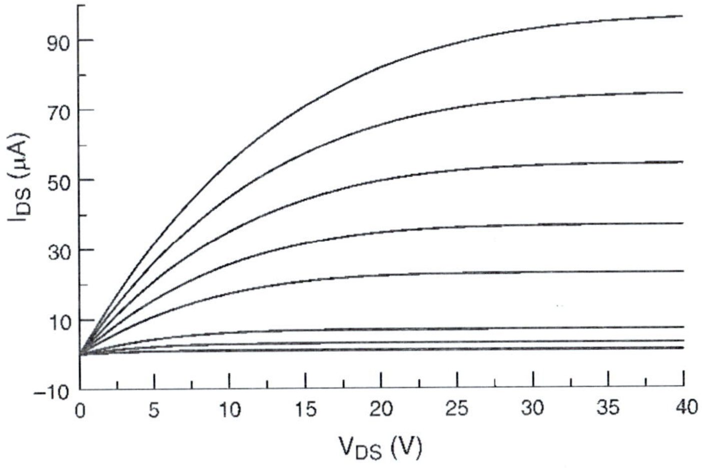

Fig. 6.9 Schematic of the I_{ds}–V_{ds} characteristics of SnO_2-channel based TTFT [73, Fig. 1]

Higher performance devices are produced using amorphous zinc-tin-oxide channel layers [58], exhibiting higher field-effect mobilities up to 50 cm^2/V s and with on/off ratios of close to 10^7. Zinc tin oxide devices have been succesfully applied on flexible substrates [79], which require low temperature processing.

6.6.2 *Gas Sensors*

Gas sensors take advantage of the naturally occuring oxygen non-stoichiometry of many binary oxides. Generally, a measurable change in electrical conductivity is observed in the presence of an oxidizing or reducing atmosphere. Adsorbing species cause a decrease or increase in carrier concentration near the film surface or grain boundaries. Undoped and surface-doped (Pt-, Pd-, Cu-) SnO_2 thin films are good candidates for gas sensing applications due to the high oxygen non-stoichiometry accommodated by the SnO_2 lattice. A full description of gas sensing devices is nicely summarized in reference [80]. For a review on the related surface chemistry of SnO_2 films an excellent review is reference [81].

6.6.3 *Transparent Conductive Oxides*

Transparent conductive oxide (TCO) applications require high conductivity and transparency of thin metal oxide layers. The main application is in flat panel display and thin film photovoltaic technologies such as transparent n-type leads. Another application of SnO_2 is in low-emission window coatings, due to its high stability in various environmental conditions. The possibility to induce a high carrier concentration, typically by F-doping, results in the positioning of the plasma wavelength in the near IR-region, thus reflecting low energy heat radiation.

6.6.4 Magnetic Properties of SnO_2 Semiconductors

Dilute magnetic semiconductors with a high Curie temperature is of interest for spintronics applications. Oxide semiconductors have been the subject of many reports, and there have been a small number of reports about the magnetic properties of doped SnO_2. There are reports on the magnetic properties of epitaxial thin films of Co-doped SnO_2 thin films on sapphire substrates grown by PLD [82] and a more recent study of epitaxial SnO_2:Co films grown by MBE [70]. Another report presents data on Fe-doped tin oxide films [83] and similar results are shown in bulk SnO_2:Fe studies [84]. Low temperature magnetic properties of (Mn+Sb)-doped SnO_2 has also been reported [85].

References

1. Flexible flat panel displays, edited by G. P. Crawford, Wiley, Chichester, England (2005), pp.79–98.
2. A. Sarkar, S. Ghosh, S. Chaudhuri, and A. K. Pal, Thin Solid Films, 204 (1991) 255.
3. W. H. Baur, and A. A. Khan, Acta Crystallogr. B 27 (1971) 2133.
4. O. Byl, and J. T. Yates, J. Phys. Chem. Lett. 110 (2006) 22966–22967.
5. L. Luxman, and R. Dobner, Metall. (Berlin) 34 (1980) 821.
6. J. Guertz, S. Rau, W. Richter, and F. J. Schmitte, SnO films and their oxidation to SnO_2: Raman scattering, IR reflectivity and X-ray diffraction studies, Thin Solid Films 121 (1984) 217.
7. A. Sanchez-Juareza, A. Tiburcio-Silver, and A. Ortizc, Fabrication of SnS_2/SnS heterojunction thin film diodes by plasma-enhanced chemical vapor deposition, Thin Solid Films 480–481 (2005) 452.
8. G. W. Watson, The origin of the electron distribution in SnO, J. Appl. Phys. 114 (2001) 758.
9. Y. Ogo, H. Hiramatsu, K. Nomura, H. Yanagi, T. Kamiya, M. Hirano and H. Hosono, Appl. Phys. Lett. 93 (2008) 032113.
10. P. Kofstad, Non-stoichiometry, diffusion and electrical conductivity of binary metal oxides. Wiley, New York (1972) p. 335.
11. L.-Z. Yang, Z.-T. Sui and C.-Z. Wang, Solid State Ionics 50 (1992) 203.
12. B. Kamp, R. Merkle, R. Lauck, and J. Maier, Chemical diffusion of oxygen in tin oxide: effects of dopants and oxygen partial pressure, J. Solid State Chem. 178 (2005) 3027.
13. C. G. Fornstad, and R. H. Rediker, Electrical properties of high-quality stannic oxide crystals, J. Appl. Phys. 42 (1971) 2911.
14. R. Y. Korotkov, A. J. E. Farran, T. Culp, D. Russo, and C. Roger, Transport properties of undoped and NH_3-doped polycrystalline SnO_2 with low electron background concentrations, J. Appl. Phys. 96 (2004) 6445.
15. B. Stjerna, E. Olsson, and C. G. Granqvist, Optical and electrical properties of radio frequency sputtered tin oxide films doped with oxygen vacancies, F, Sb, or Mo, J. Appl. Phys. 76 (1994) 3797.

16. C. Kilic, and A. Zunger, Origins of coexistance of conductivity and transparency in SnO_2, Phys. Rev. Lett. 88 (2002) 095501.
17. E. Shanti, V. Dutta, A. Banerjee, and K. L. Chopra, Electrical and optical properties of undoped and antimony-doped tin oxide films, J. Appl. Phys. 51 (1980) 6243.
18. S. W. Lee, Y.-W. Kim, and H. Chen, Electrical properties of Ta-doped SnO_2 thin films prepared by the metal-organic chemical-vapor deposition method, Appl. Phys. Lett. 78 (2001) 350.
19. C. Agashe, and S. S. Major, Effect of F, Cl and Br doping on the electrical properties of sprayed SnO_2 films, J. Mater. Sci. Lett. 15 (1996) 497.
20. Y.-W. Kim, S. W. Lee, and H. Chen, Microstructural evolution and electrical property of Ta-doped SnO_2 films grown on Al_2O_3 (0001) by metalorganic chemical vapor deposition, Thin Solid Films 405 (2002) 256.
21. R. D. Shannon, Acta Cryst. A32 (1976) 751.
22. I. Saadeddin, B. Pecquenard, J. P. Manaud, R. Decourt, C. Labrugere, T. Buffeteau, and G. Campet, Synthesis and characterization of single and co-doped SnO_2 thin films for optoelectronic applications, Appl. Surf. Sci. 253 (2007) 5240.
23. K. Galatsis, L. Cukrov, W. Wlodarski, P. McCormick, K. Kalantar-zadeh, E. Comini, C. Sberveglieri, p- and n-type Fe-doped SnO_2 gas sensors fabricated by the mechanochemical processing technique, Sens. Actuators B 93 (2003) 562.
24. Z. Ji, Z. He, Y. Song, K. Liu, and Z. Ye, Fabrication and characterization of indium doped p-type SnO_2 thin films, J. Cryst. Growth 259 (2003) 282.
25. Z. Ji, Z. He, Y. Song, K. Liu, and Y. Xiang, A novel transparent pn^+ juction based on indium tin oxides, Thin Solid Films, 460 (2004) 324.
26. C. Chen, Z. Ji, C. Wang, L. Zhao, and Q. Zhou, P-type tin–indium oxide films prepared by thermal oxidation of metallic InSn alloy films, Mater. Lett. 60 (2006) 3096.
27. Y. Huang, Z. Ji, and C. Chen, Preparation and characterization of p-type transparent conducting tin-gallium oxide films, Appl. Surf. Sci. 253 (2007) 4819.
28. M.-M. Bagheri-Mohagheghi, and M. Shokooh-Saremi, Influence of Al doping on the electrical, optical and structural properties of SnO_2 transparent conducting films deposited by the spray pyrolysis technique, J. Phys. D Appl. Phys. 37 (2004) 1248.
29. L. Poupon, P. Iacconi, and C. Pijolat, J. Eur. Ceram. Soc. 19 (1999) 747.
30. X. Q. Pan, L. Fu, Tin Oxide thin films grown on the (1¯012) sapphire substrate, J. Electroceram. 7 (2001) 35.
31. E. L. Peltzer y Blanca, A. Svane, N. E. Christensen, C. O. Rodriguez, O. M. Cappannini, and M. S. Moreno, Calculated static and dynamic properties of β-Sn and Sn-O compounds, Phys. Rev. B Condens. Matter 48 (1993) 15712.
32. K. C. Mishra, K. H. Johnson, and P. C. Schmidt, Electronic structure of antimony-doped tin oxide, Phys. Rev. B 51 (1995) 13972.
33. G. W. Watson, The origin of the electron distribution in SnO, J. Chem. Phys. 114 (2001) 758.
34. Y. Mi, H. Odaka, and S. Iwata, Electronic structure and optical properties of ZnO, SnO_2 and In_2O_3, Jpn. J. Appl. Phys. 38 (1999) 3453.
35. K. J. Button, C. G. Fonstad, and W. Dreybrodt, Determination of electron masses in stannic oxide by submillimeter cyclotron resonance, Phys. Rev. B 4 (1971) 4539.
36. G. Sanon, R. Rup, A. Mansingh, Band gap narrowing and band structure in degenerate tin oxide (SnO_2) films, Phys. Rev. B 44 (1991) 5672.
37. S.-K. Song, J.-S. Cho, W.-K. Choi, H.-J. Jung, D. Choi, J.-Y. Lee, H.-K. Baik, and S.-K. Koh, Structure and gas-sensing characteristics of undoped tin oxide thin films fabricated by ion-assisted deposition, Sens. Actuators 46 (1998) 42.
38. C. Agashe, B. R. Marathe, M. G. Takwale, and V. G. Bhide, Thin Solid Films 164 (1988) 261.
39. H. S. Randhawa, M. D. Matthews, and R. F. Bunshah, Thin Solid Films 83 (1981) 267.
40. J. E. Dominiguez, L. Fu, and X. Q. Pan, Effect of crystal defects on the electrical properties in epitaxial tin dioxide thin films, Appl. Phys. Lett. 81 (2002) 5168.

41. J. E. Dominiguez, X. Q. Pan, L. Fu, P. A. van Rompay, Z. Zhang, J. A. Nees, and P. P. Pronko, Epitaxial SnO_2 films grown on (-1012) sapphire by femtosecond pulsed laser deposition, J. Appl. Phys. 91 (2001) 1060.
42. M. Kojima, H. Kato, A. Imai, Electronic conduction of tin oxide thin films prepared by chemical vapor deposition, J. Appl. Phys. 64 (1988) 1902.
43. T. R. Giraldi, M. T. Escote, M. I. B. Bernardi, V. Bouquet, E. R. Leite, E. Longo, and J. A. Varela, Effect of thickness on the electrical and optical properties of Sb-doped SnO_2 (ATO) thin films, J. Electroceram. 13 (2004) 159.
44. H.-L. Ma, X.-T. Hao, J. Ma, Y.-G. Yang, J. Huang, D.-H. Zhang, X.-G. Xu, Thickness dependence of properties of SnO_2:Sb films deposited on flexible substrates, Appl. Surf. Sci. 191 (2002) 313.
45. T. Isono, T. Fukuda, K. Nakagawa, R. Usui, R. Satoh, E. Morinaga, and Y. Mihara, High conductivity SnO_2 thin films for flat panel displays, SID Digest (2006) 1874.
46. S.-Y. Lee, and B. O. Park, Structural electrical and optical characterizatics of SnO_2:Sb thin films by ultrasonic spray pyrolisys, Thin Solid Films 510 (2006) 154.
47. R.J. Choudhary, S.B. Ogale, S.R. Shinde, V.N. Kulkani, T. Venkatessan, K.S. Harshavardhan, M. Strikovski, and B. Hannoyer, Pulsed-electron-beam deposition of transparent conducting SnO_2 films and study of their properties, Appl. Phys. Lett. 84 (2004) 1483.
48. E. Burstein, Phys. Rev. 104 (1954) 632.
49. E. Shanthi, A. Banerjee, V. Dutta, and K. L. Chopra, Electrical and optical properties of tin oxide films doped with F and (Sb+F), J. Appl. Phys. 53 (1982) 1615.
50. B. Yu, C. Zhu, and F. Gan, Opt. Mater. Amsterdam, Neth. 7 (1997) 15.
51. Z. K. Tang, G. K. L. Wong, P. Yu, M. Kawasaki, A. Ohtomo, H. Koinuma, and Y. Segawa, Appl. Phys. Lett. 72 (1998) 3270.
52. S. S. Pan, C. Ye, X. M. Teng, L. Li, and G. H. Li, Localized exciton luminescence in nitrogen-incorporated SnO_2 thin films, Appl. Phys. Lett. 89 (2006) 251911.
53. S. S. Fouad, A. Y. Moursy, I. El-Fallal, and M. A. Harith, Optical properties of stannous oxide thin films, Czech. J. Phys. 42(2) (1992) 235.
54. R. Sivaramasubramaniam, M. R. Muhamad, and S. Radhakrishna, Optical properties of annealed tin(II) oxide in different ambients. Physica Status Solidi A Appl. Res. 136(1) (1993) 215.
55. H. Hosono, M. Yasukawa, and H. Kawazoe, Novel oxide amorphous semiconductors: transparent conducting amorphous oxides, J. Non-Cryst. Solids 203 (1996) 334.
56. I. H. Kim, J. H. Ko, D. Kim, K. S. Lee, T. S. Lee, J.-H. Jeong, B. Cheong, Y.-J. Baik, and W. M. Kim, Scattering mechanism of transparent conducting tin oxide films prepared by magnetron sputtering, Thin Solid Films 515 (2006) 2475.
57. S. Shirakata, A. Yokoyama, and S. Isomura, Preparation of SnO_2 thin films by plasma-assisted metaloganic chemical vapor deposition, Jpn. J. Appl. Phys. 35 (1996) L722.
58. T. Minami, H. Sonohara, S. Takata, and H. Sato, Highly transparent and conductive zinc-stannate thin films prepared by rf magnetron sputtering, Jpn. J. Appl. Phys. 33 (1994) L1693.
59. J. D. Perkins, J. A. del Cuerto, J. L. Alleman, C. Warnsingh, B. M. Keyes, L. M. Gedvilas, P. A. Parilla, B. To, D. W. Readey, and D. S. Ginley, Combinatorial studies of Zn-Al-O and Zn-Sn-O transparent conducting oxide thin films, Thin Solid Films 411 (2002) 152.
60. H. Q. Chang, J. F. Wager, R. L. Hoffman, J. Jeong, and D. A. Keszler, High mobility transparent thin film transistors with amorphous zinc tin oxide channel layer, Appl. Phys. Lett. 86 (2005) 013503.
61. T. Minami, T. Kakumu, K. Shimokawa, and S. Takata, New transparent conducting ZnO-In_2O_3-SnO_2 thin films prepared by magnetron sputtering, Thin Solid Films 317 (1998) 318.
62. J. B. Yadav, R. B. Patil, R. K. Puri, and V. Puri, Studies on undoped SnO_2 thin film deposited by chemical reactive evaporation method, Mater. Sci. Eng. B 139 (2007) 69.
63. T. Minami, T. Miyata, and T. Yamamoto, Stability of transparent conducting oxide films for use at high temperatures, J. Vac. Sci. Technol. A 17 (1999) 1822.

64. H. Wakabayashi, T. Suzuki, Y. Iwazaki, and M. Fujimoto, Defect structure of heteroepitaxial SnO_2 thin films grown on TiO_2 substrates, Jpn. J. Appl. Phys. 40 (2001) 6081.
65. N. Nagano, Chemical vapor deposition of SnO_2 on thin films on rutile single crystals, J. Cryst. Growth 67 (1984) 465.
66. T. Tsuchiya, A. Watanabe, T. Kumagai, and S. Mizuta, Epitaxial growth of tin oxide films on (001) TiO_2 substrates by KrF and XeCl excimer laser annealing, Appl. Surf. Sci. 248 (2005) 118.
67. J. E. Dominiguez, L. Fu, and X. Q. Pan, Epitaxial nanocrystalline tin dioxide thin films grown on (0001) sapphire by femtosecond pulsed laser deposition, Appl. Phys. Lett. 79 (2001) 614.
68. A. Rosental, A. Tarre, A. Gerst, J. Sundqvist, A. Harsta, A. Aidla, J. Aarik, V. Sammelselg, and T. Uustare, Gas sensing properties of epitaxial thin films prepared by atomic layer deposition, Sens. Actuators B Chem. 93 (2003) 552.
69. R. E. Cavicchi, S. Semancik, M. D. Antonik, and R. J. Lad, Layer-by-layer growth of epitaxial SnO_2 on sapphire by reactive sputter deposition, Appl. Phys. Lett. 61 (1992) 1921.
70. M. Batzill, J. M. Burst, and U. Diebold, Pure and co-doped SnO_2 (101) films grown by MBE on Al_2O_3, Thin Solid Films 484 (2005) 132.
71. S. Shanti, C. Subramanian, and P. Ramasamy, Preparation and properties of sprayed undoped and fluorine doped tin oxide, Mater. Sci. Eng. B 57 (1999) 127.
72. E. Elangovan, and K. Ramamurthi, Studies on micro-structural and electrical properties os spray-deposited fluorine-doped tin oxide thin films from low cost precursor, Thin Solid Films 476 (2005) 231.
73. S. Suh, Z. Zhang, W.-K. Chu, and D. M. Hoffman, Atmospheric-pressure chemical vapor deposition of fluorine-doped tin oxide thin films, Thin Solid Films 345 (1999) 240.
74. H.-G. Hong, J.-O Song, S.-H. Kim, T. Lee, and T.-Y. Seong, Effects of oxygen partial pressure on the electrical and optical properties of pulsed-laser-deposited Sb-doped SnO_2 films, J. Electrochem. Soc. 153 (2006) G922.
75. J. F. Wager, Transparent electronics, Science 300 (2003) 1245.
76. R. E. Presley, C. L. Munsee, C.-H. Park, D. Hong, J. F. Wager, and D. A. Keszler, Tin oxide transparent thin film transistors, J. Phys. D Appl. Phys. 37 (2004) 2810.
77. D.-H. Lee, Y.-J. Chang, W. Stickle, and C.-H. Chang, Functional porous tin oxide thin films fabricated by inkjet printing process, Electrochem. Solid-State Lett. 10 (2007) K51.
78. R. L. Hoffman, B. J. Norris, and J. F. Wager, ZnO-based transparent thin film transistors, Appl. Phys. Lett. 82 (2003) 733.
79. W. B. Jackson, G. S. Herman, R. L. Hoffman, C. Taussig, S. Braymen, F. Jeffery, and J. Hauschildt, Zinc tin oxide transistors on flexible substrates, J. Non-Cryst. Solids, 352 (2006) 1753.
80. M. J. Madou, and S. R. Morrison, Chemical sensing with solid-state devices, Academic Press, Boston, 1989.
81. M. Batzill, and U. Diebold, The surface and materials science of tin oxide, Prog. Surface Sci. 79 (2005) 47.
82. S. B. Ogale, R. J. Choudhary, J. P. Buban, S. E. Lofland, S. R. Shinde, S. N. Kale, V. N. Kukarni, J. Higgins, C. Lanci, J. R. Simpson, N. D. Browning, S. Das Sarma, H. D. Drew, R. L. Greene, and T. Venkatesan, High temperature ferromagnetism with giant magnetic moment in transparent Co-doped $SnO_{2-\delta}$, Phys. Rev. Lett. 91 (2003) 077205.
83. J. M. D. Coey, A. P. Douvalis, C. B. Fitzgerald, and M. Venkatesan, Ferromagnetism in Fe-doped SnO_2 thin films, Appl. Phys. Lett. 84 (2004) 1332.
84. A. Punnoose, J. Hays, A. Thurber, M. H. Engelhard, R. K. Kukkadapu, C. Wang, W. Shutthanandan, and S. Thevuthasan, Development of high-temperature ferromagnetism in SnO_2 and paramagnetism in SnO by Fe doping, Phys. Rev. B 72 (2005) 054402.
85. H. Kimura, T. Fukumura, M. Kawasaki, K. Inaba, T. Hasegawa, and H. Koinuma, Rutile-type oxide-diluted magnetic semiconductor: Mn-doped SnO_2, Appl. Phys. Lett. 80 (2002) 94.

Chapter 7
Transparent Conductive Zinc Oxide and Its Derivatives

Klaus Ellmer

7.1 Introduction

Zinc oxide (ZnO) is a ceramic material and was used early in history, for instance as a precursor material for brass manufacturing in old china. ZnO occurs naturally as the mineral zincite (Fig. 7.1a). In 1810 it was discovered by Bruce in Franklin (New Jersey, USA) as red oxide of zinc, in 1845 named as zincite by Haidinger [1]. It can be found at different places on earth: Franklin (New Jersey, USA), Sarawezza (Tuscany, Italy), Tsumeb (Namibia), Olkusz (Poland), Spain, Tasmania and Australia. Most of the natural crystals are small, but freely-formed crystals up to 25 mm were reported, which are exceedingly rare [2].

ZnO is composed of abundant elements, the abundances of Zn and O in the earth crust are 132 ppm and 49.4%, respectively. This is important for large-scale applications of ZnO, for instance as transparent electrodes in thin film solar cells or flat panel displays and an advantage compared to indium tin oxide (ITO), the dominant transparent electrode material today, wherein the metal In has an abundance in the earth crust of only 0.1 ppm.

Investigations on the electrical properties of ZnO date back to 1912, when Somerville measured the resistivity of ZnO rods up to a high temperature of about 1,125°C [3]. The first electronic characterization of polycrystalline, bulk ZnO samples, of ZnO thin films and even of the first ZnO single crystals was performed by Fritsch in 1935 [4]. He showed that ZnO exhibits a strong thermal activation of the conductivity typical for a semiconductor. He also performed first Hall measurements on his ZnO samples and obtained Hall mobilities between 7 and

K. Ellmer
Dept. solar fuels, Helmholtz-Zentrum für Materialien und Energie Berlin GmbH, Hahn-Meitner-Platz 1, 14109 Berlin, Germany
e-mail: ellmer@helmholtz-berlin.de

Ed. D. Ginley, H. Hosono, Y. Shigesato "Transparent Conductors: Materials, Processing and Applications (From Transparent Conducting Oxides to Transparent Electronics)".

D.S. Ginley et al. (eds.), *Handbook of Transparent Conductors*,
DOI 10.1007/978-1-4419-1638-9_7,

Fig. 7.1 Photographs of (*left*) natural (red coloured ZnO size) zinc oxide (*zincite*) in calcite (ZnO size: 1.5 × 1.2 cm^2, Franklin mineral museum, New Jersey, USA) and of (*right*) hydrothermally grown ZnO single crystals (Mineral Ltd., Alexandrov, Russia). The diameter of the synthetic single crystal is about 70 mm

30 cm^2/V s and carrier concentrations in the range of 10^{18} cm^{-3}. Both, from the Hall as well as from Seebeck coefficient measurements he found, that ZnO is an n-type semiconductor.

After world war II, with the advent of the semiconductor industry, also ZnO was investigated as one of the first compound semiconductors.

In the 1950s the ZnO research was focused in the United States at Bell Labs (Hutson, Lander, Thomas) and in Germany at the University of Erlangen (Mollwo, Heiland, Helbig). Hutson discovered the large piezoelectric effect of ZnO, which is higher than that of quartz [5]. This lead to one of the first applications of ZnO as a piezoelectric filter using the principle of surface-acoustic waves [6]. Already in 1959 a first review on the properties of ZnO was given by Heiland et al. [7]. In the 1980s two other reviews were published on ZnO single crystal properties by Hirschwald [8] and on thin ZnO films as transparent electrodes by Chopra and Major [9].

During the 1970s ZnO was investigated with much effort as a prospective semiconductor for blue light emitting devices, due to its high bandgap energy of 3.4 eV. These efforts were canceled in the 1980s, caused to the inability to prepare p-type conducting ZnO, a prerequisite for the manufacturing of pn-homo junctions. The comprehensive optical investigations on ZnO and other direct gap semiconductors were reviewed at that time by Klingshirn [10].

A renaissance of the ZnO research started about 1995 as a follow-up of the breakthrough of gallium nitride as a blue light emitting semiconductor [11–13], when new efforts were undertaken to dope ZnO p-type, especially by adding group V elements (N, P, As). In the last years many groups reported p-type conducting ZnO with mobilities around 1 cm^2/V s [14], however, stable, highly conducting p-ZnO still has to be developed.

The recent review by Özgür et al. summarizes especially the search for p-ZnO [15]. Other interesting reviews which appeared recently are those of Nickel and Terukov [16], Jagadish and Pearton [17] and Ellmer, Klein and Rech, the latter focussing on the application of zinc oxide in thin film solar cells [18].

In this chapter we describe structural, electrical and optical properties of doped ZnO and ZnO alloy thin films, especially of degenerately undoped and doped ZnO films which can be used as transparent conductive electrodes for instance in flat panel displays or thin film solar cells.

7.2 Basic Properties of ZnO

7.2.1 Crystallographic Structure

Zinc oxide is a compound semiconductor which crystallizes in the hexagonal wurtzite structure ($P6_3mc$), elucidated by Bragg already in 1914, shortly after the discovery of X-ray diffraction [19] and shown in Fig. 7.2 in two different views. In the hexagonal unit cell of ZnO, which contains two ZnO molecules, the Zn atoms are surrounded tetrahedrally by oxygen atoms. The Zn–O distances in c-direction ($d_{Zn-O}[1] = 0.196$ nm) are a little bit shorter than the distances to the other three oxygen atoms ($d_{Zn-O}[2] = 0.198$ nm). Perpendicular to the c-axis the unit cell consists of double layers composed of planes of zinc and oxygen atoms, respectively. Of the four bonds of each atom three are realized within such a double layer, i.e., these double layers can be viewed as building blocks of the wurtzite lattice [20]. This (nearly) tetrahedral coordination induces a polar symmetry in zinc oxide, which gives rise to some unique properties of ZnO: piezoelectricity, spontaneous polarization, and a strongly anisotropic etching behavior. Some structural data of ZnO are summarized in Table 7.1.

The (001) planes of ZnO are built of Zn atoms only, while the ($00\bar{1}$) surfaces consist only of oxygen atoms (see Fig. 7.2). These two crystal surfaces can be

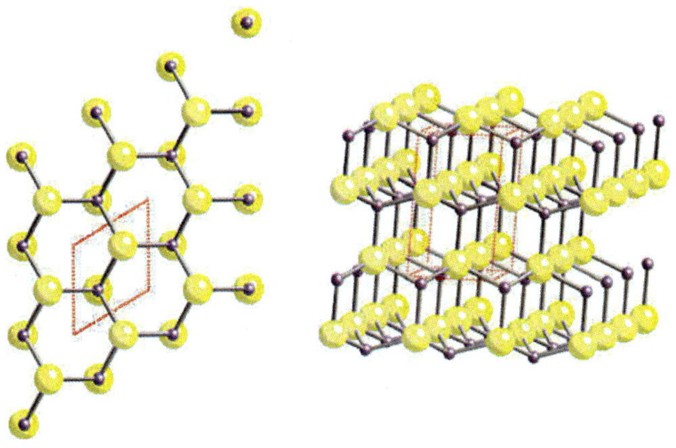

Fig. 7.2 Two views of the crystal structure of zinc oxide (ZnO). *Left*: View along the c-axis on the terminated (00-1) plane. *Right*: perspective view perpendicular to the c-axis. The *upper side* is the oxygen terminated (001) plane, the *bottom plane* is zinc terminated (001). *Large balls*: oxygen, *small balls*: zinc. The unit cell is shown as *dotted lines*

Table 7.1 Structural data of zinc oxide

Lattice (normal conditions)	Hexagonal, wurtzite (B4-type, $P6_3mc$) [22]
Lattice constants	$a = 0.32498$ nm, $c = 0.52066$ nm, $c/a = 1.6021$, $u = 0.3832$ [23]
Density	5.67 g cm^{-3}, 4.2×10^{22} ZnO molecules/cm^3
Hardness	4–5 (Mohs), 5 GPa [24]
Bulk modulus [GPa]	142.4 [23]
Lattice (high pressure)	Cubic, NaCl (B1-type, Fm3m) [25]

differentiated by X-ray diffraction analysis [20]. Also, the etching behavior is different for these two surfaces. While the oxygen surface is strongly etched in slightly diluted acids (HCl, H_2SO_4, HNO_3, H_3PO_4), the zinc-terminated surface is quite resistant against these acids [20, 21].

At high pressure (9 GPa) ZnO transforms to the cubic NaCl phase which is stable down to about 2 GPa at room temperature, but not at normal pressure [23, 25]. An overview on experimental and theoretical structural data of the different ZnO modifications was given recently by Serrano et al. [26].

7.2.2 *Band Structure*

Zinc oxide is a direct semiconductor with a band gap energy of 3.4 eV [27]. The band structure of ZnO has been calculated by density-functional theory (DFT) models, see for instance [28–31]. Depending on the DFT method used, the calculated band gaps are in the range of 0.4–3.77 eV. Especially, the local density (LDA) and the generalized-gradient approximations (GGA) underestimate the band gap energy significantly [28, 31]. The calculated band structures (see Fig. 7.3a, c) confirm the experiment in that the valence band maximum and the conduction band minimum both occur at the Γ point k = 0, which means that ZnO is a direct semiconductor. The low lying valence bands ($E–E_V \approx -8.5$ eV) belong to Zn *3d* states, while the upper valence bands (−5 to 0 eV) are composed of O *2p* bonding levels. A schematic band structure of ZnO is displayed in Fig. 7.3b which shows the attribution of the different bands to the atomic orbitals as well as the bonding and antibonding *sp*-bands.

The lowest conduction bands are strongly localized at Zn *3s* levels. This is comparable to the band structure of indium oxide, where the conduction band is also composed of s-like states of cation, i.e., indium atoms [32–36].

Robertson et al. performed comprehensive band structure calculations of different oxides as well as of the position of the hydrogen defect level H^0 in these oxides [37]. The positions of the neutral H^0 defect level relative to the positions of valence and conduction bands are shown in Fig. 7.4. It can be seen that ZnO belongs to the few oxides, in which the H^0 levels is situated in the conduction band, i.e., it constitutes a donor in ZnO under all conditions (see Sect. 7.4).

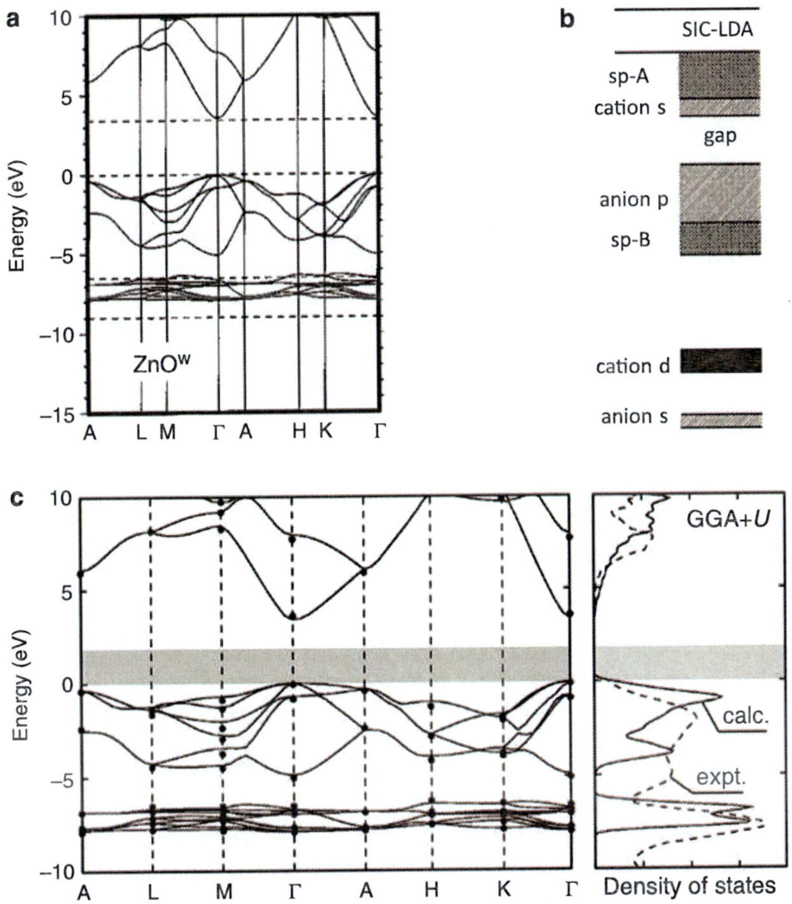

Fig. 7.3 Calculated band structures of zinc oxide (wurtzite structure). **(a)** shows a local-density approximation with self-interaction correction (SIC-LDA) calculation of Vogel et al. [28], **(b)** the schematic band structure of ZnO (and other II–VI semiconductors), derived from **(a)** with the attribution of the atomic orbitals to the bands [28]. **(c)** displays a generalized-gradient approximation (GGA) calculation of Erhart et al. [31]. The GGA calculation yields a band gap energy, which is too small (*gray bar*). Therefore, the conduction bands have been shifted upwards to the experimental band gap of $E_g = 3.4$ eV. The right-hand side of Fig. 7.3c depicts the calculated (*solid line*) and the experimental (*dashed line*) densities of states

Another fact is revealed from this figure: the positions of the valence and conduction band edges and of the band gap energies vary significantly from oxide to oxide. The position of the conduction band edge is important for heterojunctions of ZnO with other semiconductors, for instance in thin film solar cells or light emitting devices [38]. A tailoring of the band edge alignment can be done by alloying of ZnO with other oxides, for instance MgO or CdO [39, 40].

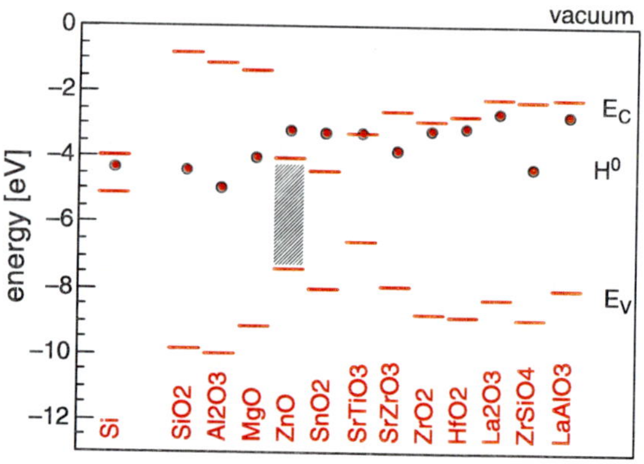

Fig. 7.4 Positions of the defect level H^0 (dots) of neutral hydrogen in different oxides and in silicon relative to the valence and conduction band edges as calculated by Robertson et al. [37] (taken from [41]). Zinc oxide is marked by the hatched band gap area

7.2.3 Thermodynamic Properties

Zinc oxide has a high melting point of nearly 2,000°C, i.e., it is a ceramic material. This is comparable to the other TCO materials SnO_2 and In_2O_3. However, all three oxides decompose into the elements below their melting points if the oxygen partial pressure is too low. The vapor pressure is high already at about 1,400°C [42], which makes it difficult to grow single crystals from its own melt. This is important for the crystal growth of ZnO (and of the other TCOs) which requires a correct adjustment of the growth atmosphere [43]. The melting point of zinc is quite low which implies a high metal vapor pressure at typical substrate temperatures (<600 K). Though no phase diagram of the Zn–O system was reported in literature, Wrieth suggested a binary oxygen–zinc phase diagram [44], which shows a very clear structure. Above 200°C only the binary compound ZnO exists. Below 200°C also zinc peroxide ZnO_2 is stable, which can be prepared by chemical synthesis [48]. Besides the hexagonal phase a cubic phase with the rocksalt structure is known, which is formed at a hydrostatic pressure of 9 GPa [49]. In difference to MgO and CdO, two other oxides of metals from the second group of the periodic table, which crystallize in the zincblende structure, the zincblende modification of ZnO is metastable [50]. According to density-functional calculations of the total energy of the different ZnO modifications, the zincblende phase of ZnO is only about 50 meV less stable than ZnO–wurtzite [51, 52]. Owing to the higher lattice symmetry of ZnO–zincblende, such a material is expected to have advantages for semiconductor devices, such as higher mobility, higher doping efficiency and the possibility for epitaxial growth on lattice-type and -constant matched 3C-SiC. Though bulk crystals of ZnO–zincblende have not been reported, heteroepitaxial ZnO films were grown on substrates like GaAs(001) or on sapphire, in both cases by using a ZnS (zincblende structure) as a buffer layer, the details of which were reviewed recently by Ashrafi and Jagadish [50]. The linear thermal expansion coefficient at room temperature is $2.9 \times 10^{-6}\,K^{-1}$ ($c_{\parallel}$) and $4.75 \times 10^{-6}\,K^{-1}$ ($c_{\perp}$), which is only slightly

Table 7.2 Thermodynamic properties of zinc oxide

Property (unit)	Value
Thermal expansion coefficient α (300 K) [10^{-6} K^{-1}]	‖c: 2.92 ⊥c: 4.75
Thermal conductivity (300 K) [W/(mK)]	‖c: 69 [45] ⊥c: 60
Melting point [°C]	1,975 [46]
Melting point of Zn [°C]	420
Heat of formation [eV]	3.6
Cohesive energy [eV]	7.52 [47]

higher than that of silicon (2.6 × 10^{-6} K^{-1}). The stoichiometric width of ZnO is very narrow below 600°C and about 100 ppm at 1,000°C [53]. The thermodynamic properties of zinc oxide are summarized in Table 7.2.

7.2.4 *Applications of Zinc Oxide*

ZnO was used as a technical material since the Bronze Age especially for the production of brass [54]. Since the nineteenth century it was used in white paints, for rubber and porcelan enamel production.

Today, most of zinc oxide powder produced worldwide is used for rubber production, chemicals, paints, in agriculture and for ceramics [55].

Since about 1970 zinc oxide is also used for electronic devices: In displays ZnO powders are used as green phosphors [56–58]. With self-assembled ZnO nanowire micropatterns white cathodoluminescence was observed [59].

The piezoelectric properties ZnO [60] are used in surface-acoustic wave (SAW) devices and sensors [61–63] which can be applied as band pass filters in the telecommunications industry, primarily in mobile cell phones and base stations.

Acoustic wave sensors are advantageous due to its low costs, ruggedness, and a high sensitivity. Also in soft ferrites ($M_{\delta}Zn_{1-\delta}Fe_2O_4$, with M = Mg, Ni, Mn), used for resonant circuit applications, wide band transformers and for radio frequency interference (RFI) attenuation, ZnO is an essential component.

Today, the widest application of zinc oxide is in varistors, an electronic device with a highly nonlinear current-voltage curve which are used as voltage limiters [67]. The ZnO varistor, developed by Matsuoka in Japan at the beginning of the 1970s [68], consists of a polycrystalline matrix of ZnO (grain size some 10 μm) with additives of other metal oxides (for instance Bi_2O_3, Co_3O_4, Cr_2O_3, MgO etc.) with contents above ≈0.1 mol%. The current transport is governed by the inter-granular regions, which consist of depletion regions between ZnO grains, constituting electronic barriers for the current transport. Increasing the voltage at the varistor leads eventually to the breakdown of the barriers accompanied by a high current flow. While its current-voltage characteristic is comparable to a back-to-back Zener

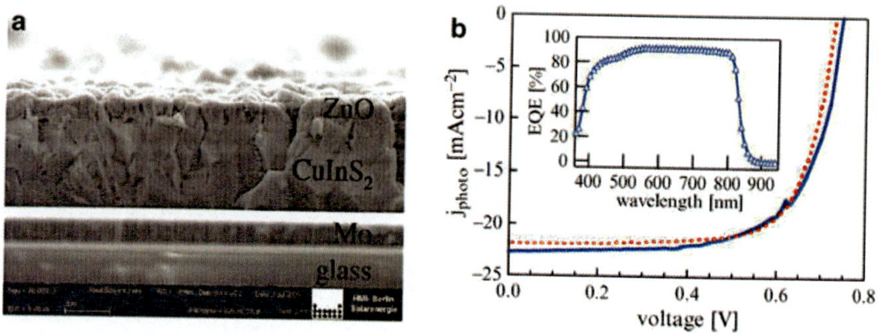

Fig. 7.5 (**a**) Scanning electron microscopy (SEM) picture of the layer stack of a magnetron-sputtered $CuInS_2$ solar cell, consisting of the Mo back contact, the 2 μm thick, compact absorber layer and the front contact consisting of a multilayer CdS/ZnO/ZnO:Al. (**b**) shows the I–V characteristics under AM1.5 illumination of a such a solar cell, while the *inset* displays the spectral quantum efficiency of the cell [64]. The molybdenum layer was prepared by DC magnetron sputtering, the $CuInS_2$ layer by a two-step reactive magnetron sputtering process and the ZnO/ZnO:Al layers by magnetron sputtering from ceramic targets [65]. For comparison the I–V curve (dotted line) of a thermally processed $CuInS_2$ solar cells is shown [66]

diode a varistor can handle much higher currents and energies and is a superior transient (response time of about ≈500 ps) and overvoltage protection device [69].

In varistors the high resistivity of the grain barriers is used; for another high-volume application the effect of the grain barriers on the electrical transport has to be minimized: transparent conductive electrodes, especially for thin film solar cells [18]. Highly conducting zinc oxide can be prepared by doping with group III elements (B, Al, In, Ga) up to carrier concentrations of 10^{21} cm^{-3} [70, 71]. As an example, Fig. 7.5 shows the SEM cross section of a thin film solar cell with a $CuInS_2$ absorber film and a ZnO/ZnO:Al window and contact layer as well as its I–V characteristics under illumination.

Recently, also ZnO nanowires are investigated as electrodes for extremely thin absorber thin film solar cells [72].

Emerging electronic applications of zinc oxide are UV sensitive photodiodes and transparent field effect transistors [17, 73].

7.3 Preparation of Zinc Oxide

7.3.1 Single Crystal Growth

The first synthetic ZnO single crystals were grown by Fritsch in 1935 by vapor transport at a temperature of about 1,450°C [4]. These crystals had a hexagonal cross section and grew preferentially along the c axis leading to pencil-like crystal habits. In the next decades this crystal growth technique was used by Scharowsky, Thomas and Lander and Bogner and Mollwo [74–76]. Other crystal growth

Table 7.3 Parameters of commercially available zinc oxide single crystals (ρ – resistivity, $N_{D,A}$ – donor/acceptor concentration, μ – Hall mobility, N_{dis} – dislocation density), taken from [83]

Company	Growth method	Growth parameters	Crystal size (mm)	ρ (Ω cm) N_D–N_A (cm^{-3}) μ (cm^2/V s)	N_{dis} (cm^{-2})	Reference
ZN Technology Inc. (formerly Eagle-Picher), Brea, USA	Seeded chemical vapor transport	1,100 °C	51	0.3 1×10^{17} 207		[84]
Cermet Inc., Atlanta, USA	Pressurized melt	2,000 °C	130	0.35 9×10^{16} 200 [85]		[82]
Mineral Ltd., Alexandrov, Russia	Hydrothermal	330 °C 50 MPa	51	500–1,000	<50	[86]
Tokyo Denpa Co., Tokyo, Japan	Hydrothermal	300. . .400 °C <100 MPa	51	390 8×10^{13} 200	<300	[79]

methods that have been used are hydrothermal growth [77–79] and solution growth from halogenide melts of PbF_2 or $ZnBr_2$ [80, 81]. Crystals grown from halogenide melts were not further investigated due to the high impurity content originating from the melt. Hydrothermal growth is quite efficient and yields large crystals of good crystalline quality. Due to the composition of the aqueous solution, which contains alkali metals, these ZnO crystals exhibit much higher resistivities compared to vapor phase grown crystals (see Table 7.3).

Czochralski growth of ZnO was not performed up to now, due to the difficulties with the high melting point and especially due to the high decomposition pressure of ZnO. Recently, larger ZnO crystals were grown by crucible techniques (cold and hot) which now have reached crystal sizes and qualities applicable for prospective electronic applications of ZnO [43, 82]. Photographs of hydrothermally grown undoped ZnO crystal bars and wafers are displayed in Fig. 7.6. Table 7.3 summarizes commercially available ZnO single crystals and its typical properties. Up to now only undoped single crystals are available, which poses a problem comparing electrical and optical properties of highly-doped polycrystalline films with that of single crystals.

7.3.2 *Thin Film Growth*

ZnO films can be deposited by a wide variety of thin film deposition methods [87]:

- Oxidation of evaporated metallic Zn films [4]
- Metal-organic chemical vapor deposition (MOCVD) [88, 89]
- Pulsed-laser deposition (PLD) [90]
- Electrochemical deposition [91]

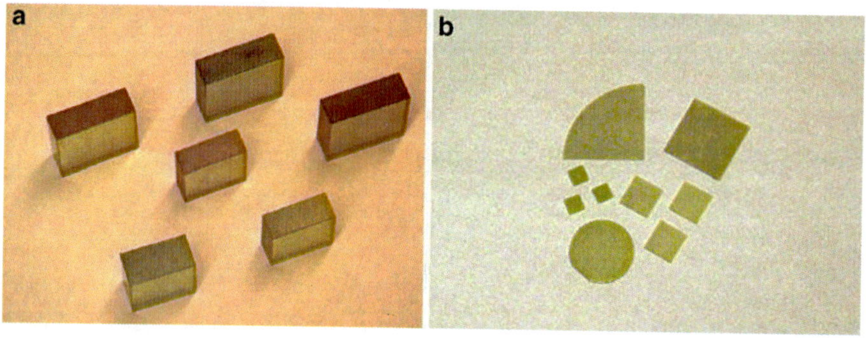

Fig. 7.6 Photographs of (**a**) hydrothermally grown undoped ZnO crystal bars (Mineral Ltd., Alexandrov, Russia) and (**b**) wafers of different sizes (5 × 5, 10 × 10, 20 × 20, ∅ 20 and a quarter of a ∅ 51 mm wafer, thickness 0.5 mm) prepared from the crystal bars in (**a**) (Crystec GmbH, Berlin, Germany)

- Chemical solution deposition [92–94]
- Reactive and nonreactive magnetron sputtering [87, 95]

In the following we will concentrate on magnetron sputtering since this technique is used for large-area deposition of doped and undoped ZnO films. This is due to the following unique features of magnetron sputtering:

- Low substrate temperatures due to plasma assistance
- Reactive deposition possible, i.e., preparation of oxides, nitrides, sulfides etc.
- High chemical reactivity
- High film density and good film adhesion
- Easily scalable to large areas (up to 20 m^2)

The first sputter-deposited ZnO films were prepared by diode sputtering around 1966 for piezoelectric applications [96, 97], which showed a very strong (001) texture.

7.3.2.1 Nonreactive Magnetron Sputtering

For this process a ceramic ZnO:dopant target is used. As dopant, aluminum is used mostly, but also boron, gallium or indium are suitable as dopants. Depending on the target preparation process the resistivity of the target can vary considerably. If the target is insulating, the plasma excitation has to be performed by an RF power supply, since with DC excitation no plasma can be ignited. For targets, which are sufficiently conductive (processed under slightly reducing conditions) DC power supplies can be used, which are significantly cheaper than RF power supplies.

During the deposition process a small amount of oxygen from the ZnO target is lost and pumped away. In order to compensate this oxygen loss, small oxygen

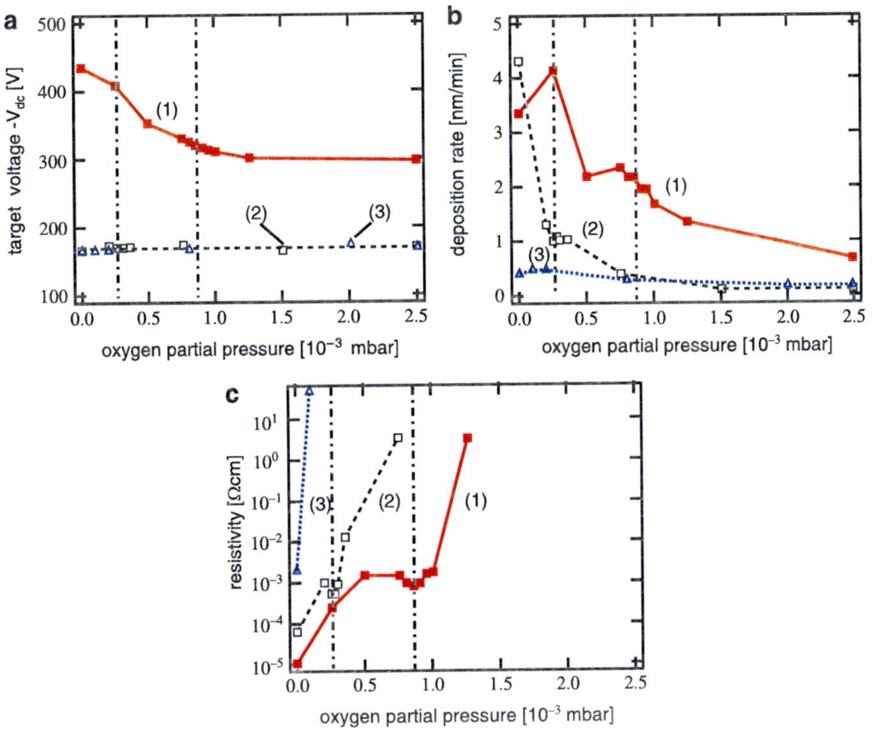

Fig. 7.7 Dependence of the discharge voltage – V_{dc} (**a**), the deposition rate R (**b**) and the resistivity ρ (**c**) on the oxygen partial pressure for three sputtering modes: (1) DC from a metallic target Zn:Al2wt%, (2) RF from a metallic target Zn:Al2wt%, (3) RF from a ceramic target ZnO: Al_2O_3(2 wt%). Sputtering conditions: P = 100 W, $p_{Ar+O2} = 0.5$ Pa, layer thickness ≈ 0.5 μm. The *vertical dot-dashed lines* mark the positions of the resistivity minima [98]

amounts are added to the sputtering gas, which is mostly argon. The relative portion of oxygen is typically between 0.1 and 1% and has to be adjusted carefully. Figure 7.7 shows the dependence of the discharge voltage, the deposition rate and the resistivity of ZnO:Al films on the oxygen partial pressure [98]. The deposition rate and the discharge voltage are independent of the oxygen partial pressure. This means that they are determined mainly by the bulk composition of the target (for instance ZnO:Al). The values of the discharge (target) voltages are different for DC and RF plasma excitation. The target voltage in an RF plasma is significantly lower than that in a DC discharge. This is caused by the fact, that the DC discharge is maintained essentially by secondary electrons emitted from the target surface through the Ar ion bombardment, which occurs in sufficient quantity only at higher (negative) potentials.

The resistivity of the ZnO:Al films is strongly dependent on the oxygen partial pressure p_{O2} (Fig. 7.7c, curve 3). At $p_{O2} = 0$ the resistivity is lowest. With increasing p_{O2} the resistivity increases steeply caused by a complete oxidation of the metals in the growing film (Zn and dopant Al), which leads to a phase mixture of two oxides, which is insulating. Sometimes it is observed that the lowest resistivity occurs only at an oxygen partial pressure slightly higher than zero. This behavior depends on the oxidation state of the ZnO targets, i.e., on its stoichiometry which can be adjusted by the target sintering temperature and atmosphere. The dependence of the resistivity on the stoichiometry of the ZnO films points to the decisive role of oxygen for the electrical and optical properties of ZnO (see Sect. 7.4).

Besides a low resistivity one also aims at obtaining a high optical transparency of the deposited ZnO films. Without oxygen during the sputter deposition the films exhibit occasionally a significant absorption leading to grayish or yellowish film colors. This is due to an insufficient oxidation of the films, leading to metal clusters or nanocrystals which absorbe light. Adding small amounts of oxygen or annealing such films in an oxygen containing atmosphere avoids the darkening of the films. Further increasing the oxygen partial pressure leads to a highly transparent, but also highly ohmic films.

As was shown above, two general dependences of film properties on the oxygen partial pressure have to be optimized for the deposition of TCO films:

1. High optical transparency, which increases with increasing oxygen partial pressure
2. Low resistivity, which occurs at lower oxgen partial pressures

The combination of both opposite property-parameter dependences brings about a (relatively) narrow process window, where low ohmic and highly transparent films can be prepared (see also Sect. 7.3.2.2 and the structure-phase diagram, Fig. 7.8).

7.3.2.2 Reactive Magnetron Sputtering

Sputtering of ZnO films from ceramic targets is a simple and reliable deposition method. However, it is not very well suited for technical applications since ceramic targets are expensive and mechanically fragile. Furthermore, ceramic targets are often too insulating to be sputtered by DC plasma excitation. Therefore, RF power supplies are used, which are much more expensive than DC power supplies.

An elegant way to avoid ceramic targets and expensive RF power supplies is the sputtering of metallic zinc or zinc alloy targets in argon/oxygen mixtures, the so-called reactive magnetron sputtering. This process is very attractive for the following reasons:

- Use of cheap metallic targets
- Use of DC power supplies
- Possibility to adjust the film stoichiometry through the reactive gas flow
- Higher deposition rates than for sputtering from ceramic targets

However, reactive sputtering often requires a careful process control since only within a narrow parameter window the desired layer properties can be obtained [99].

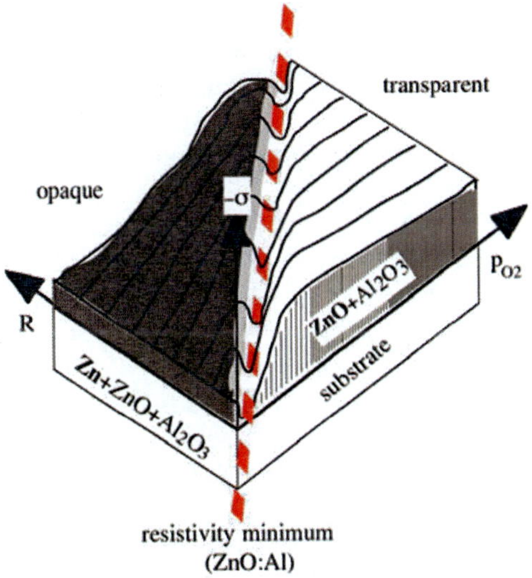

Fig. 7.8 Structure-phase zone model for the reactive magnetron sputtering of zinc oxide films [87]. The process parameters are the oxygen partial pressure p_{O2} and the deposition rate R (proportional to the discharge power). The compressive film stress (-σ) is depicted on the vertical axis. The *left dark triangular region* corresponds to opaque films while the *right triangle* represents the region of transparent films. The grain sizes and the phase composition are indicated in the two cross section areas: *Dark films* exhibit an X-ray amorphous or nanocrystalline structure with metal inclusions in an oxidic matrix, while the *transparent films* show a columnar structure with decreasing grain size for increasing oxygen partial pressure. The technological process window is coincident with the "stress valley" marked by the *dashed line*, which is linearly related to $R = f(p_{O2})$

The dependences of the discharge voltage and the resistivity versus the oxygen partial pressure for sputtering from different targets with dc and rf excitation are shown in Fig. 7.7 [98]. For dc sputtering from a metallic target (1) the discharge voltage exhibits a step like behavior at a specific oxygen partial pressure, accompanied by a decrease of the deposition rate (Fig. 7.7b). The decrease of the deposition rate can be explained by different surface states of the target in dependence on the oxygen partial pressure. With increasing oxygen portion in the sputtering atmosphere the target surface becomes oxidized, for zinc characterized by a higher secondary electron emission and a reduced sputtering rate. The change of the secondary electron emission for oxidized metal surfaces was investigated by Ellmer and Mientus [100] and Depla et al. [101].

It is this transition from the metallic to the oxidized target mode that defines the process window in reactive (magnetron) sputtering. The resistivity (Fig. 7.7c) shows a minimum right at this transition from the metallic to the oxidic target state. In the metallic mode (low oxygen partial pressure) the resistivity is low but the optical transmission too. Further increasing p_{O2} leads to transparent ZnO:Al films with low resistivity (see the dot-dashed lines in Fig. 7.7c). Still higher oxygen pressures do not improve the transparency of the films significantly but result in drastically higher resistivities. This is caused by the fact that the dopant (Al) as well as the host lattice metal atoms (Zn) are completely oxidized to ZnO and Al_2O_3, leading to a phase mixture of alumina and zinc oxide, which is insulating.

A similar picture for the behavior of the deposition rate and the resistivity is obtained for rf excitation (curve 2). However, the rf discharge voltage does not show a step like curve. This is caused by the fact that the rf discharge is nearly independent of the delivery of secondary electrons from the target surface [102].

In summary, the preparation of TCO films is a difficult balancing of structure and phase composition of polycrystalline thin semiconducting films. Though the dopants are added normally in such amounts that sufficiently high charge carrier concentrations ($\leq 10^{21}$ cm^{-3}) could be achieved, only part of the dopants are electrically active. Due to the interrelation between stoichiometry and structure a fine tuning is necessary in order to prepare low resistant and transparent films. From the results sketched above, a structure-phase zone model (Fig. 7.8) in analogy to the well known structure zone model of Thornton [103] for magnetron sputtering of metals can be established that relates the phase composition (Zn, ZnO, Al_2O_3), the mechanical stress, the transparency and the grain size with the deposition parameters oxygen partial pressure and deposition rate (discharge power). Films of lowest resistivities are obtained in a relatively narrow process window (see the dashed line in Fig. 7.8), where large strain free crystallites exist, which consist mainly of zinc oxide with the dopant (Al) on zinc lattice sites. A deviation from this region in the structure-phase diagram changes the structure, the composition and the electrical and optical properties of the films. The best films can be prepared in the "stress valley" extending from the lower left to the upper right. Similar diagrams can be constructed for the dependence of typical film properties on other preparation parameters such as sputtering pressure, substrate temperature, ion assistance during deposition etc.

Kluth et al. developed a modified Thornton-structure-zone model for the magnetron sputtering of ZnO:Al films from ceramic targets, see Fig. 7.9 [104]. Different to Thornton, who established his model for the magnetron sputtering of metallic films as a function of the argon sputtering pressure and the relative substrate temperature T/T_m, where T_m is the melting temperature of the metal, Kluth et al. used the absolute value of the substrate temperature and exchanged the pressure and the temperature axis. The best films with respect to low resistivities and a sufficient compactness were obtained in a medium range of sputtering pressure and substrate temperature, marked by (b) in Fig. 7.9. The microstructure, which develops after etching of such films in diluted HCl is advantageous for an efficient light trapping in a-Si:H thin film solar cells (see also [105]).

The influence of the different morphologies, depicted in Fig. 7.9, onto the resistivity and the etching rate of the ZnO:Al films is shown in Fig. 7.10 as a

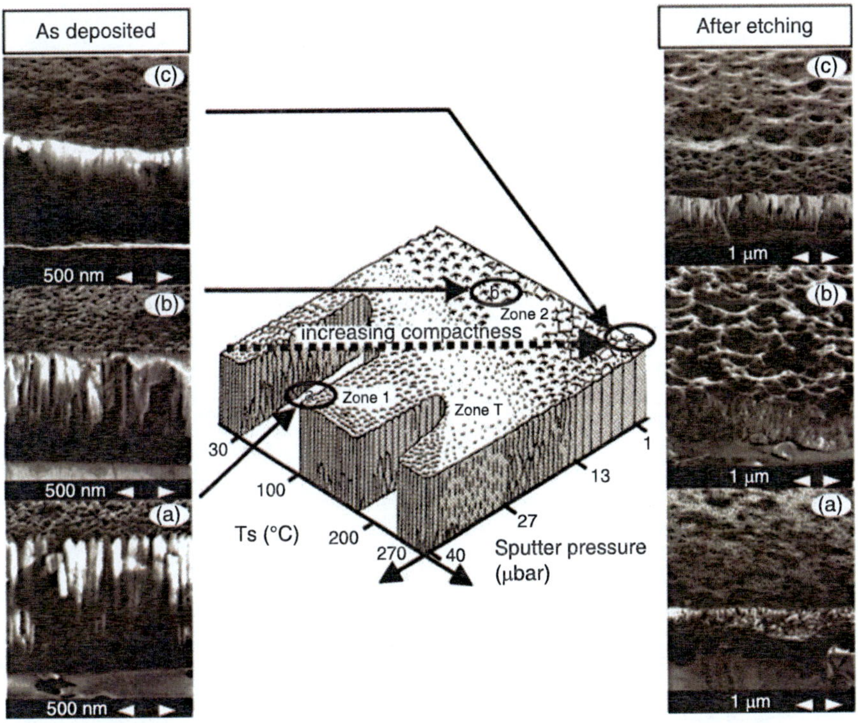

Fig. 7.9 Modified Thornton-structure-zone model for the RF magnetron sputtering of ZnO:Al films according to Kluth et al. [104]. The process parameters are the total sputtering pressure p_{Ar+O2} and the substrate temperature T_S. The SEM pictures show the film morphology in the as-deposited state (*left*) and after etching for 20 s in 0.5% HCl (*right*)

function of the total sputtering pressure. Depending on the substrate temperature and the plasma excitation (RF or DC) the resistivity starts to increase significantly for argon pressures between 0.7 and 2 Pa, accompanied by an increase of the etching rate. Therefore, the increased resistivity, mainly caused by a decreased mobility, is due to a more porous structure of the films, which can be inferred from the higher etching rate. This investigation points to the fact that the electrical properties of ZnO (and other TCO) films is determined not only by electrical parameters like carrier concentration and mobility but also by the morphology of the films. In many cases, the assumption of a compact film is not justified and the electrical (and optical) properties have to be described by effective medium theories, which was done for instance by Ederth et al. for spin-coated indium-tin oxide films (ITO) [106]. Also, the percolation ansatz has to be taken into account for porous films, as was done by Clarke et al. [107].

Recently an overview on magnetron sputtering of zinc oxide was given by Szyszka [95].

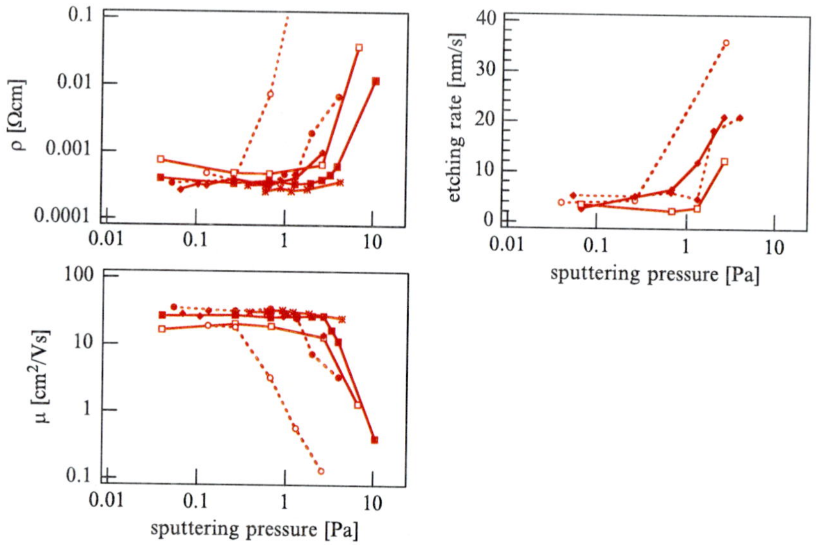

Fig. 7.10 Resistivity and mobility (*left*) and etching rate (*right*) of RF and DC magnetron sputtered ZnO:Al films as a function of the total sputtering pressure according to Kluth et al. [104]. The different symbols have the following meaning: (□) DC sputtering, 150°C; (▪) DC sputtering, 270°C; (◇) DC reactive sputtering, 270°C; (*) MF reactive sputtering; (○) RF sputtering, 150°C and (•) RF sputtering 270°C

7.4 Doping and Electrical Transport in ZnO Single Crystals

7.4.1 Intrinsic Doping of ZnO

As mentioned in the introduction intrinsic ZnO is an n-type semiconductor, which was already found by Fritsch in 1935 by Hall and Seebeck measurements [4]. Zinc oxide as a typical compound semiconductor can be made conductive both by intrinsic dopants (defects) as well as by extrinsic ones (foreign atoms). If zinc oxide single crystals are grown intrinsically, i.e., without intentional addition of extrinsic dopants, its resistivities are in the order of 0.1–10^3 Ω cm, depending on the preparation conditions (see Table 7.3) [4, 84, 108, 109]. By annealing in oxidizing or reducing atmospheres the carrier concentration can be adjusted in the range of about 10^{17}–10^{19} cm^{-3}. By compensation with group I elements (Li, Na, K) extremely high-resistive ZnO single crystals can be prepared [20, 110]. For instance, Kobiakov reported dark resistivities of up to $\approx 10^{12}$ Ω cm for ZnO crystals, which were compensated with lithium by annealing at 1,070 K in a Li_2CO_3 melt [111]. During hydrothermal growth of ZnO crystals alkali elements are added to the growth solution as “mineralizers,” in order to promote the crystal growth [112]. Therefore, these single crystals exhibit the lowest carrier concentration in the order of 10^{13} cm^{-3} (Table 7.3).

The low resistivity at room temperature, i.e., degenerate doping, which is required for the application of ZnO films as transparent electrode can be achieved in two different ways:

- Creation of intrinsic donors by lattice defects (for instance oxygen vacancies or zinc atoms on interstitial lattice sites) or
- Introduction of extrinsic dopants (either metals with oxidation number three on substitutional metal lattice sites or halogens with oxidation number minus one on oxygen lattice sites).

The first possibility can be realized during the deposition by carefully adjusting oxygen partial pressure and deposition rate (see Sect. 7.3). The other way is a reduction process of the oxide film after deposition, for instance by annealing in vacuum or in a hydrogen containing atmosphere [113]. However, it has been found that such intrinsically doped films exhibit properties, which are not well suited for application. First, the resistivity is only about 10^{-2} to 10^{-3} Ω cm. Furthermore, these films are not stable at ambient conditions (especially at higher temperatures) due to the reoxidation of the oxygen deficient or zinc rich films, which increases its resistivity significantly [71].

Though electrical properties of intrinsic zinc oxide were investigated since more than five decades, it was not clear until recently, which defect(s) constitute(s) the dominant donor(s) in intrinsic ZnO [83]. For a long time it was believed that the oxygen vacancy is the dominant donor, which was supported by the fact that an annealing of single crystals or thin films under reducing conditions increased the carrier concentration, while an oxidizing treatment decreased the carrier concentration. Recently, Look et al. performed an electrical transport study on pure, intrinsic ZnO single crystals in which crystallographic defects were generated by high-energy (MeV) electron irradiation [114]. Since the defect production was higher when the crystals were irradiated in (001) direction (Zn-face up) compared to electron bombardment along the (00-1) direction (O-face up), it was concluded that the zinc interstitials are the intrinsic shallow donors. Comprehensive density-functional calculations of defect formation energies and diffusion constants in ZnO, performed by Erhart et al. [31, 115, 116], support the results of Look et al. These authors assign the zinc vacancy and the oxygen interstitial as acceptors, and only zinc interstitials and oxygen vacancies have shallow energetic donor positions in the band gap. Depending on the growth atmosphere different defects are most likely: Under zinc-rich conditions the oxygen vacancy dominates, while the zinc vacancy and the oxygen interstitials are the most likely defects for an oxygen-rich atmosphere. Taking the lower energy of formation of the zinc interstitial, it should be the dominant donor in zinc oxide, thus explaining the results of Look et al. [114].

7.4.2 Extrinsic n-Type Doping of ZnO

Already in Sect. 7.2.4 the n-type doping of ZnO by the addition of group III dopants (B, Al, Ga, In) was noted, which was investigated for the first time in the 1950s.

Hauffe and coworkers [117] studied the relation between the electrical parameters and the addition of metal oxides to zinc oxide [118], which increased the conductivity of ZnO by orders of magnitude, for instance when the group III-oxides B_2O_3, Al_2O_3, In_2O_3 or Ga_2O_3 were added. It was assumed that the group III-dopant atoms were built in onto zinc lattice sites, spending the additional electron not required for the bonding to the conduction band according to the equation (M means metal) [119, 120]:

$$M_2O_3 \Leftrightarrow 2M_{Zn} + 2e + 2O_O + 1/2O_2, \tag{7.1}$$

leading to the following oxygen partial pressure dependence of the electron concentration [e]:

$$[e] = [M_{Zn}] \sim p_{O2}^{-1/8}. \tag{7.2}$$

Recently, by perturbed-angular correlation spectroscopy (PAC) using the radioactive probe $^{111}In/^{111}Cd$, it was confirmed, that indium as a shallow donor is located at substitutional zinc sites [121, 122]. In ZnO annealed in zinc vapor at 1,000°C, an In-defect complex was detected by PAC, which could be assigned to ^{111}In at Zn-interstitials at octahedral sites [121], thus confirming the theoretical calculations of the prevailing intrinsic defect in ZnO (see Sect. 7.4.1).

Also other oxides of metals with valence three or four (for instance Cr, Y, Ce, Ti, Zr, Hf) can be used to increase the conductivity of ZnO [118, 123]. The preparation of donor doped zinc oxide crystals at high oxygen partial pressures or annealing in an oxygen atmosphere leads to highly resistive crystals. One explanation for this effect is the formation of zinc vacancies, caused by the low enthalpy of formation under oxygen-rich conditions, which are acceptors [31, 116]. Another possibility is the oxidation of the dopant metal (for instance gallium) to the corresponding oxide, thus making the dopant atom (Ga) electrically inactive. The latter interpretation is supported by nuclear magnetic resonance (NMR) measurements of the group of Sleight who suggested $Ga_2^{3+}O_i^{2-}$ as double-electron traps, which can be viewed as a precursor of gallium oxide Ga_2O_3 [124]. Comprehensive investigations of the doping of ZnO thin films were performed by Minami et al. [70]. Besides the group III elements his group investigated elements like Si, Ge, Sn, Y, Sc, Ti, Zr, Hf, i.e., good oxid forming elements, as dopants. Also the halogens fluorine and chlorine were investigated as dopants in ZnO [125–127], however, these dopants were introduced by metal-organic chemical vapor deposition (MOCVD) or spray pyrolysis, but not by magnetron sputtering. All the n-type dopants constitute effective-mass-like (hydrogenic) donors with ionization energies around $E_D = 13.595$ eV $m^*/\varepsilon_O^2 = 66$ meV, which is confirmed by temperature-dependent Hall and conductivity measurements, see Table 7.4 [84, 128–131].

7.4.3 Hydrogen in Zinc Oxide

Already at the beginning of zinc oxide research, hydrogen was observed to be a shallow donor in ZnO. Mollwo [133], Thomas and Lander [75, 134] and Hutson

Table 7.4 Donors, its lattice position and ionization energy E_C–E_D in ZnO [83]

Element	Structure	E_C–E_D [meV]	References
Al	Al_{Zn}	65/53	[14, 131]
B	B_{Zn}		[117]
Cr	Cr_{Zn}		[117]
Ga	Ga_{Zn}	54.5	[14]
H	H_i	37/46	[131, 132]
In	In_{Zn}	63.2	[14]
Li	Li_i		[110]
O	V_O?	300	[131]
Zn	Zn_i	46	[108, 131]

[108] reported in the 1950s the doping effect of hydrogen. Unfortunately, this knowledge was forgotten (or ignored) for decades. This is an important fact, since hydrogen is present during almost all growth processes, introducing a background doping in single crystals and in films. The concentration of hydrogen in ZnO varies between 5×10^{16} cm^{-3} in single crystals grown by chemical transport and 1×10^{20} cm^{-3} in magnetron sputtered ZnO:Al films [135].

In 2000, hydrogen as a donor in ZnO was "rediscovered" by van de Walle by a density-functional theory (DFT) study [136], in which it was shown that under all circumstances, i.e., independently of the position of the Fermi level, hydrogen acts as a donor in ZnO (see also Fig. 7.4). Shortly after, it was unambiguously proved that hydrogen constitutes a shallow donor in zinc oxide [132]. The role of hydrogen in zinc oxide is peculiar, since it is a donor, independently of the position of the Fermi level, different from the behavior of H in silicon or other semiconductors. Some authors argued that (unintentionally introduced) hydrogen could have been the reason for the "natural" n-type behavior of ZnO. But experimental as well as DFT studies showed that the intrinsic n-type doping of ZnO is caused by intrinsic defects (see Sect. 7.4.1).

Only recently the intentional introduction of hydrogen as a donor during reactive magnetron sputtering of ZnO:Al was investigated [137]. Though a decrease of the resistivity from about 7 to 5×10^{-4} Ω cm was observed, it is not clear whether the hydrogen-doping is stable under illumination and at higher operating temperatures, for instance up to 85°C, which is a typical temperature that thin film solar cells can reach under operating conditions. The observed decrease of the resistivity of the ZnO:Al,H films is mainly caused by an increase of the carrier concentration by about a factor of two ($N \leq 7 \times 10^{20}$ cm^{-3}), while the mobility is decreased from 27 to 21 cm^2/V s. The built-in of hydrogen is directly proportional to the hydrogen gas flow F_{H2} and amounts to 2 at.% at $F_{H2} = 20$ sccm, corresponding to about 1.7×10^{21} H-atoms cm^{-3}, i.e., much higher than the electron concentration of 7×10^{20} cm^{-3} [137].

7.4.4 Extrinsic p-Type Doping of ZnO

Already at the beginning of ZnO research it was tried to make p-type zinc oxide. First experiments were performed by doping with group I elements like lithium and sodium [110, 138]. As already mentioned in Sect. 7.3.1, these dopants can be used to compensate the background donors. However, due to strong self-compensation effects it was not possible until recently, to prepare p-type ZnO [139]. Atomistic binding energy calculations by Binks and Grimes have shown that the built-in of Li^+ and Na^+ onto zinc lattice sites (Li_{Zn}, Na_{Zn}) is energetically favored, if interstitial oxygen is formed [140]. This explains the self-compensating mechanism, when alkali metals are added to ZnO, see also [141].

Since the end of the 1990s, with the stormy renaissance of the ZnO research, p-doping is again an important research topic, since the optional n- or p-doping is a prerequisite for most of the proposed optoelectronic applications of ZnO [13]. The new approaches to prepare p-type zinc oxide are focussed on group V elements (N, P, As), which should act as acceptors on oxygen lattice places. Compared to Li and Na, these acceptors exhibit lower ionization energies, which is promising for high hole concentrations (Fig. 7.11).

One of the first reports on p-type zinc oxide was given by Minegishi et al. [143], who deposited ZnO films by chemical vapor deposition from a ZnO source powder with H_2 as transport gas and NH_3 as dopant gas for doping with nitrogen. Under optimized growth conditions in a (very narrow) process window p-type ZnO films with a hole concentration of about 1.5×10^{16} cm^{-3} and a Hall mobility of 12 cm^2/V s were grown. Despite many reports on p-ZnO [144–148] and few papers on ZnO pn-junctions [142] a reliable p-doping process, that leads to high, reproducible, laterally homogeneous and long-term stable hole concentrations, is still missing. One severe problem seems to be that the grown films with acceptor doping exhibit laterally inhomogeneous n- an p-type regions, demonstrated for instance by Dadgar et al. for nitrogen-doped ZnO films on sapphire substrates with a GaN buffer layer [149]. Figure 7.12 shows locally resolved atomic-force microscope

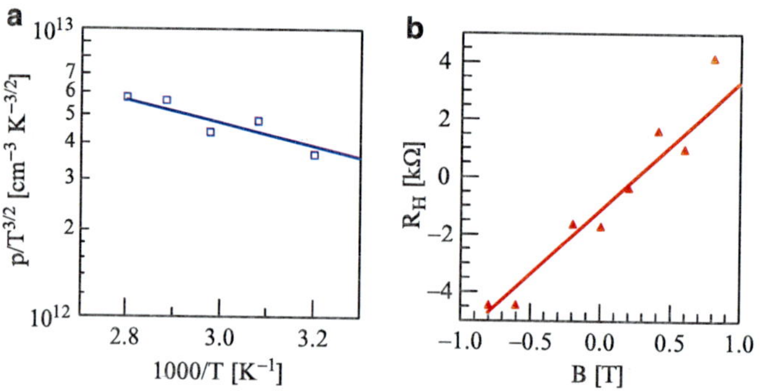

Fig. 7.11 **(a)** Temperature-dependent hole concentration p/T$^{3/2}$ of nitrogen-doped ZnO films, epitaxially grown on a $ScAlMgO_4$ (SCAM) substrate, which has a very low lattice mismatch relative to ZnO [142]. The nitrogen concentration in the film was about 2×10^{20} cm^{-3}. **(b)** Hall resistance versus the magnetic field from −1 to 1 T. The hole mobility decreases from 5 to 8 cm^2/V s when increasing the temperature from 300 to 350 K. Fitting the p/T$^{3/2}$ curve with the usual dependence under the assumption of an effective hole mass of $m_h = 0.9\ m_e$ yields an activation energy of the nitrogen donors of about 100 meV and a compensation ratio N_D/N_A of 0.8

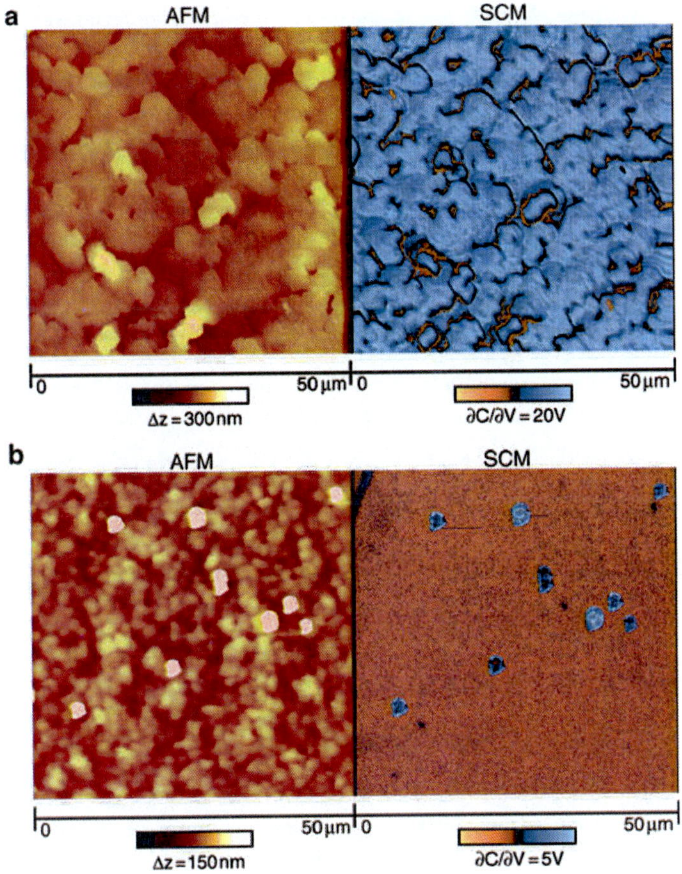

Fig. 7.12 Scanning microscopy images (AFM and SCM mode) of ZnO films grown on sapphire substrates on a GaN buffer layer. (a) Nitrogen doping (mainly n-type conduction), (**b**) co-doping by nitrogen and arsenic (dominant p-type conduction) [149]

(AFM) and scanning capacitance measurement (SCM) images of a nitrogen-doped and a nitrogen-arsenic co-doped ZnO film [149]. While the N-doped film shows p-conductivity only in small areas, the co-doped film exhibits a much more homogeneous p-type behavior.

Another nitrogen-doping method – implantation of nitrogen ions – was used by von Wenckstern et al. to produce $p^{+}n$ junctions in pressurized-melt grown ZnO single crystals [150]. The p-type conductivity of the N-implanted layer of a thickness of about 350 nm was proved by IV-curve measurements with Pd/ZnO contacts, which showed Schottky behavior for the unimplanted ZnO crystal and an ohmic IV curve for the N-implanted ZnO sample. By scanning capacitance measurements (SCM), the p-type character of the N^{+}-implanted ZnO surface was confirmed. The energetic position of acceptor states in the n-type region of the crystal was measured by deep-level-transient spectroscopy (DLTS). The authors

Table 7.5 Acceptors, its lattice position and ionization energy in zinc oxide [83]

Element	Structure	E_A–E_V [meV]	Type	References
Ag	Ag_{Zn}/Ag_i	200	Amphoteric	[151, 152]
As	As_{Zn}		Acceptor	[153]
Cu	Cu_{Zn}	380/190	Deep Acceptor	[152, 154]
Li	Li_{Zn}	800	Deep acceptor	[155] [14, 156]
	Li_{Zn}	280		[157]
N	N_O	110/165	Acceptor	[141, 158]
Na	Na_{Zn}	600	Deep acceptor	[156]
P	P_O	130	Acceptor	[159, 160]
Intrinsic	?	150	Acceptor	[150]

conclude that the level with $E_A - E_V \approx 150$ meV is an intrinsic acceptor, while another one with $E_A - E_V \approx 280$ meV is due to lithium [150], which is well known to be an acceptor in ZnO [110] (see also Table 7.5).

Overviews on the status of p-doping of zinc oxide can be found for instance in [139, 149]. Table 7.5 summarizes the reported acceptors in ZnO [83]. It can be seen that the acceptor ionization energies are much higher than that of the donors (see Table 7.4), explaining the difficulties to achieve high hole concentrations in ZnO.

7.4.5 *Carrier Transport in Single Crystalline ZnO*

For charge carrier scattering in semiconductors one can distinguish two general processes: Intrinsic scattering due to the band structure of the semiconductor, which is governed only by intrinsic material parameters (sometimes called lattice scattering) and extrinsic scattering caused by dopants and charge carriers. The intrinsic scattering process dominates for low carrier concentrations $N < 10^{16}$ cm^{-3} while the ionized impurity scattering is limiting the carrier transport for $N > 10^{18}$ cm^{-3}.

The experimental data and theoretical models on ionized impurity scattering were already reported in 2001 when estimating the mobility limit of highly-doped zinc oxide [71, 123, 161].

In the following a short summary, which is based on [162, 163], is given to lay the basis for the further discussion.

7.4.5.1 Lattice Scattering

In II-VI semiconductors this scattering process is determined by scattering at acoustic and polar-optical phonons as well as, in ZnO, by piezoelectric scattering from the phonon-induced electrical fields due to the absent inversion symmetry of the wurtzite lattice of ZnO [15, 164]. In the following an overview of these scattering processes is given. A more elaborated review was published recently in [162].

Optical mode scattering is due to the interaction of electrons with the electric field, induced by the lattice vibration polarization (polar longitudinal-optical phonons) occurring in polar semiconductors with partial ionic bonding. According to Devlin [165] the optical Hall mobility can be calculated by

$$\mu_{Hopt} = r_{Hopt}\phi \frac{e}{2\alpha\omega_0 m^*}\left[\exp\left(\frac{\hbar\omega_0}{kT}\right) - 1\right], \tag{7.3}$$

where the polaron coupling constant α is given by

$$\alpha = \left(\frac{1}{\varepsilon_\infty} - \frac{1}{\varepsilon_s}\right)\sqrt{\frac{m^* E_H}{m_e \hbar\omega_0}}. \tag{7.4}$$

ε_∞ and ε_s are the high frequency and the static dielectric constants and E_H is the first ionization energy of the hydrogen atom (13.595 eV). m* and m_e are the effective and the vacuum electron masses, while $\hbar\omega_0$ is the energy of the longitudinal optical phonon (73.1 meV). r_{Hopt} is the Hall coefficient factor for optical mode scattering and Φ is a slowly varying function of the temperature. Usually it is assumed that $r_{Hopt}\Phi$ is equal to 1 [129, 130, 165].

Acoustical mode scattering is a lattice deformation scattering process due to a local energetic shift of the band edges originating from acoustical phonons. According to Bardeen and Shockley [166] the acoustical lattice mode Hall mobility is:

$$\mu_{Hac} = r_{Hac}\frac{\sqrt{8\pi}\,\hbar^4 c_l e}{3E_1^2\sqrt{m^{*5}(kT)^3}} \tag{7.5}$$

where c_l is the averaged longitudinal elastic constant. E_1 is the deformation potential (energy shift of the conduction band per unit dilation). $r_{Hac} = 3\pi/8 = 1.178$ is the Hall coefficient for acoustic phonon scattering [167]. The deformation potential E_1 is not very well known for ZnO. In literature E_1 values scatter from 1.4 [168] to 31.4 eV [129]. For the calculation of the overall carrier scattering in ZnO this is not very important since other scattering mechanisms, especially polar-optical and piezoelectric scattering, are dominant [169].

Piezoelectric mode scattering occurs only in piezoelectric materials, i.e., in crystals without inversion symmetry, and is caused by the electric field associated with acoustical phonons. Zinc oxide exhibits very high electro-mechanical coupling coefficients P, exceeding that of quartz [5] [60], which is one of the piezoelectric materials used mostly. Zook [170] has calculated the piezoelectrically limited mobility as (see also Rode [169]):

$$\mu_{Hpie} = r_{Hpie}\frac{16\sqrt{2\pi}\hbar^2\varepsilon\varepsilon_0}{3eP^2_{\perp,\mathrm{II}}\sqrt{m^{*3}kT}}, \tag{7.6}$$

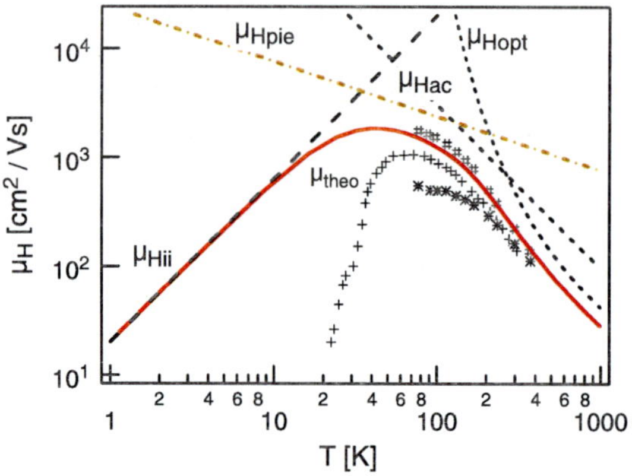

Fig. 7.13 Experimental Hall mobility data of Hagemark et al. [130, 171] in comparison to the theoretical lattice mobility μ_{theo}, calculated from the material data of zinc oxide (Table 7.7) by Wagner and Helbig [129]. The crystals were vapor-phase grown, not intentionally doped and had the following carrier densities at room temperature: + 3.1×10^{16} cm^{-3}, # 1×10^{16} cm^{-3}, * 2.5×10^{16} cm^{-3}. The contributions to the total mobility by the individual scattering processes: acoustic μ_{Hac} and polar-optical μ_{Hopt} phonon, piezoelectric μ_{Hpie} and ionized impurity scattering μ_{Hii} are also shown as *dashed* or *dotted lines*

with the Hall coefficient for piezoelectric mode scattering $r_{Hpie} = 45\pi/128 = 1.1045$. The averaged piezoelectric electro-mechanical coupling coefficients$P_{\perp,II}$ for electrical transport perpendicular or parallel to the c axis were calculated by [169, 170].

For most semiconductors, the piezoelectric electro-mechanical coupling coefficients P are of the order of 10^{-3} [172], while ZnO exhibits P values of about 0.2–0.4 (see Table 7.7).

Temperature-dependent mobility data of nominally undoped ZnO single crystals, reported by Hagemark et al. [130, 171], are displayed in Fig. 7.13. The room temperature mobility is in the range of 150–200 cm^2/V s, which is determined by the lattice scattering processes acoustic and polar-optical scattering (see also Fig. 7.15).The theoretical lattice scattering processes, taken from Wagner and Helbig [129] are shown for comparison. It can be seen that the experimental data are well described by theory (acoustical and polar-optical phonon as well as piezoelectric scattering) for temperatures above about 100 K. This means these crystals have reached the intrinsic limits given by the material parameters of ZnO (Table 7.7). For $T < 100$ K a significant deviation from theory, i.e., from ionized impurity scattering is observed. This is attributed by Li and Hagemark as well as by Wagner and Helbig to a (donor)-defect band conduction with a mobility of about 2 cm^2/V s [129, 171].

Hagemark and Chacka measured the temperature-dependent carrier concentrations of ZnO single crystals, which were doped by annealing in zinc vapor up to $N_D \approx 3 \times 10^{18}$ cm^{-3} and determined the donor activation energies E_D, which are plotted in Fig. 7.14a as a function of the cubic root of N_D. Despite some scattering

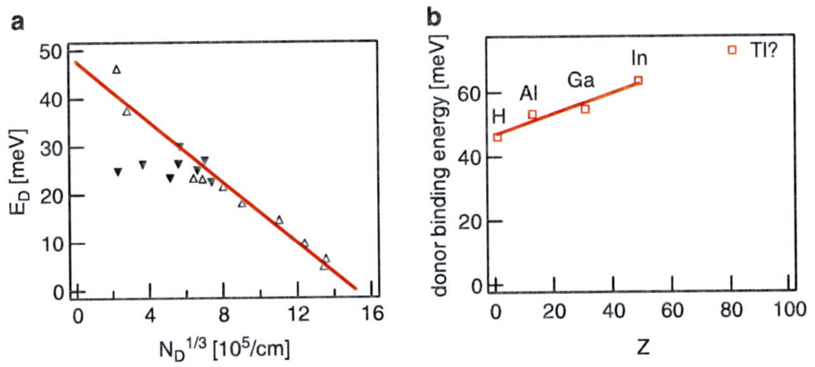

Fig. 7.14 (**a**) Donor activation energy E_D as a function of the cubic root of the carrier concentration [130]. Two sample sets with different diffusion temperatures were investigated: (Δ) – 1,000°C, (▼) – 800°C, while the Zn (dopant) source temperature was varied from 400 to 950°C. (**b**) Donor activation (binding) energy for different donors (hydrogen, aluminum, gallium and indium) depicted versus the atomic number Z (data taken from [14], see Table 7.4)

of the data points a linear correlation $E_D \sim E_{D0}–b{\cdot}N^{1/3}$ can be seen which reflects the fact that the ionization energy of the shallow donors decreases with increasing carrier concentration [130]. Extrapolating the donor activation energy to vanishing carrier density leads to a shallow donor ionization energy of about 45 meV, which compares well with the donor energies reported by others, given in Table 7.4.

The donor ionization energy vanishes at a carrier concentration of about 5×10^{18} cm^{-3}, which corresponds about to the density of states in the conduction band (see Table 7.7). This decrease of the ionization energy with increasing donor density is caused by the overlap of the wavefunctions of the donors, which was observed already by Pearson and Bardeen in doped silicon [173].

Meyer et al. [14] determined the donor binding energies from photoluminescence measurements of intrinsic ZnO single crystals with different donors, which are displayed in Fig. 7.14b versus the atomic number. It can be seen, that the binding (i.e., ionization) energy of a donor increases linearly with the atomic number of the donor. The donor with the largest binding energy was not attributed to a specific element by Meyer et al. Taking the linear dependence on the atomic number, it is most probably thallium, the last group III element, where also the other dopants come from, which occurs also as a contamination during crystal growth (see Fig. 7.14b). Closely inspecting Table 7.4 and Fig. 7.14b shows, that the model of the hydrogenic donor, where the outer shell electrons determine the binding energy is only an approximation. For donor atoms heavier than H, the inner shell electrons contribute significantly to the binding of the donors (see for instance [172]).

Figure 7.15 shows Hall mobilities and electron concentrations of intrinsic ZnO crystals from different authors [162]. The carrier density exhibits the typical thermal activation. The temperature dependence of the mobility is similar for the crystals with different carrier concentrations. With decreasing temperature,

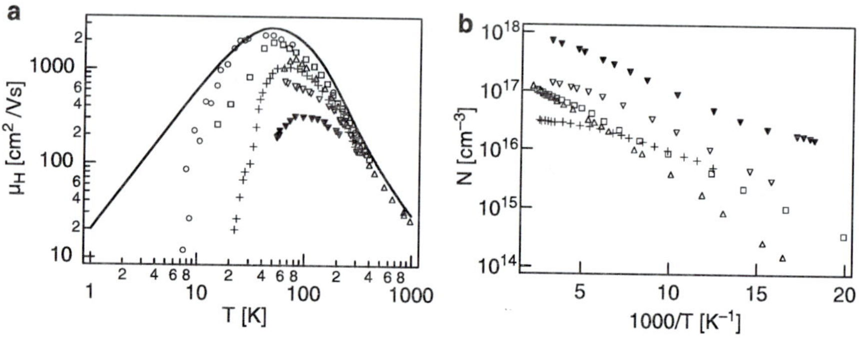

Fig. 7.15 (**a**) Hall mobility and (**b**) electron concentration of nominally undoped (*open symbols*) and doped ZnO single crystals (from [162]) reported by Hutson already in 1957 [108] (Δ – 7.6 × 10^{16} cm^{-3}; ∇ – 1.6 × 10^{17} cm^{-3}; ▼ – 1 × 10^{18} cm^{-3}). For comparison the mobilities given by Wagner and Helbig [129] (○ – 2.8 × 10^{16} cm^{-3}); Li and Hagemark [171] (+, 3.1 × 10^{16} cm^{-3}) and Look et al. (□ – 8 × 10^{16} cm^{-3}) are included. The *continuous line* is the theoretical fit given by Wagner and Helbig in 1974 (see also Fig. 7.14)

the mobility increases significantly, reaching values of 1,000–2,000 cm^2/V s around temperatures of 50 K. At even lower temperatures the mobility decreases steeply which is caused by neutral impurity scattering. At room temperature these (nearly) intrinsic crystals exhibit a mobility between 100 and 200 cm^2/V s, a limit which has not changed since more than 50 years (compare the data of Hutson from 1957).

7.4.5.2 Ionized Impurity Scattering

This scattering process is caused by ionized dopant atoms and dominates for carrier concentrations above about 5 × 10^{18} cm^{-3}. An analytical expression for the mobility μ_{ii} of degenerately doped semiconductors, taking into account the non-parabolicity of the conduction band, was given by Zawadzki [174] and refined by Pisarkiewicz et al. [175]:

$$\mu_{ii}^{ZP} = \frac{3(\varepsilon_r\varepsilon_0)^2 h^3}{Z^2 m^{*2} e^3}\frac{n}{N_i}\frac{1}{F_{ii}{}^{np}(\xi_d)} \quad \text{with} \quad \xi_d = (3\pi^2)^{1/3}\frac{\varepsilon_r\varepsilon_0 h^2 n^{1/3}}{m^* e^2}, \tag{7.7}$$

where the screening function F_{ii}^{np} is given by

$$F_{ii}^{np} = \left[1 + \frac{4\,\xi_{np}}{\xi_d}\left(1 - \frac{\xi_{np}}{8}\right)\right]\cdot \ln(1+\xi_d) - \frac{\xi_d}{1+\xi_d} - 2\,\xi_{np}\left(1 - \frac{5\,\xi_{np}}{16}\right), \tag{7.8}$$

with the parameter $\xi_{np} = 1 - m_0^*/m^*$, which describes the non-parabolicity of the conduction band (m^*, m_0^*-effective masses in the conduction band and at the conduction band edge, respectively). The non-parabolicity is usually described by the dependence $m^*/m_0^* = 1+2\beta(E - E_C)$, where β is the non-parabolicity parameter (Table 7.7) and E and E_C are the energies of the carriers in the conduction band and at its edge. The prefactor in (7.7) shows, that the ionized-impurity limited mobility depends as $\mu_{ii} \sim (\varepsilon_r/m^*)^2$ on the material constants of the semiconductor and as $\mu_{ii} \sim Z^{-2}$ on the charge of the dopants ($Z = 1$ for group III donors; $Z = 2$ for oxygen vacancies).

The theoretical model given above as well as the models of Conwell and Weisskopf [176, 177], Shockley [178], and Dingle [179] are based on the assumption of a statistically homogeneous distribution of scattering centers, i.e., dopants. However, this is no longer valid for extremely high dopant concentrations, where the dopants form clusters which lead, due to their higher charge, to lower mobilities ($\mu_{ii} \sim Z^{-2}$). This effect was already proposed in 1971 by Dakhovskii et al. [180]. Klaassen applied this cluster model to fit accurate measurements of mobilities in p- and n-type single crystalline silicon [181, 182]. He calculated cluster charges up to 2 for boron-doped and 3.5 for phosphorous-doped silicon at a carrier concentration of 10^{21} cm^{-3}. Recently, such clusters of zinc dopants were verified by atomically-resolved analysis in GaAs [183].

The carrier transport in highly-doped ZnO single crystals was investigated only occasionally and only for carrier concentrations below 10^{20} cm^{-3}. Higher electron concentrations were reported only for epitaxial or polycrystalline ZnO films (for a review see [71, 123, 190]). Masetti et al. [191] measured the mobility of silicon, degenerately doped by arsenic, phosphorus, and boron up to carrier concentrations of 5×10^{21} cm^{-3} and fitted their experimental values by the empirical curve

$$\mu^{Ma} = \mu_{min} + \frac{\mu_{max} - \mu_{min}}{1 + \left(n/n_{ref1}\right)^{\alpha_1}} - \frac{\mu_1}{1 + \left(n_{ref2}/n\right)^{\alpha_2}}. \tag{7.9}$$

This semi-empirical formula was used recently to fit the reported mobility data for single crystalline zinc oxide [71, 190], which is shown in Fig. 7.16. The fit parameters μ_{max}, μ_{min} and ($\mu_{min} - \mu_1$) describe the lattice mobility at low carrier concentrations, the mobility limited by ionized impurity scattering and the clustering mobility, discussed above (see Table 7.6). In the transition region from lattice to ionized scattering for $5 \times 10^{16} < N < 5 \times 10^{18}$ cm^{-3} a large scattering of the experimental ZnO data can be observed. Therefore, the data have been fitted in analogy to the silicon data, which exhibit a much higher accuracy [191]. However, the exact transition does not influence the conclusions much since we are interested predominantly in ionized impurity scattering in the region $N > 10^{19}$ cm^{-3}. It is interesting to note, that, though the lattice scattering mobilities of silicon are much higher, than that of the TCO materials, the μ_{min} values are comparable, pointing to the universal nature of the ionized impurity scattering process.

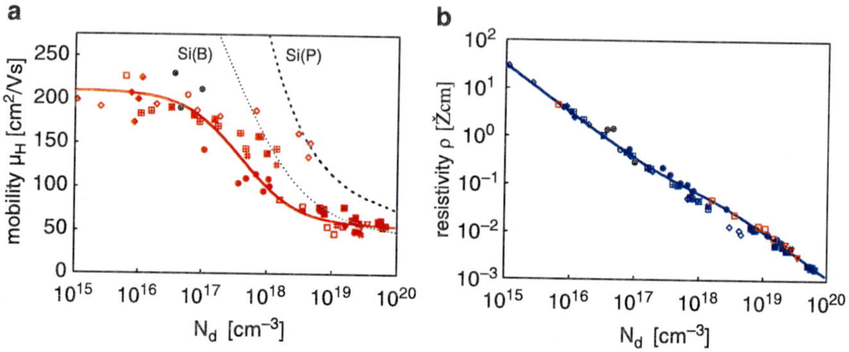

Fig. 7.16 (**a**) Hall mobility and (**b**) resistivity of undoped (*open symbols*) and indium-doped (*filled symbols*) ZnO single crystals as a function of the donor concentration N_d, measured at room temperature [162]. The data were reported by Hutson (⊞) [108], Rupprecht (■) [184], Thomas (●) [119], Baer (□) [185], Neville and Mead (▽) [186], Hausmann and Teuerle (#) [187], Wagner and Helbig (◆) [129], Hagemark and Chacka (⊗) [188], Utsch and Hausmann (◇) [189], Look et al. (○) [84], and von Wenckstern et al. (⊗) [131, 150]. The full lines in (a) and (b) were calculated with the semiempirical equation (7.9). In (a) the corresponding fit curves for boron- and phosphorous-doped silicon are shown as *dotted* and *dashed lines* for comparison. The fit parameters for (7.9) are summarized in Table 7.6

Table 7.6 Fit parameters for $\mu = f(N)$ (7.9) for phosphorous- and boron-doped silicon [191] and zinc oxide, indium oxide and tin oxide (μ_{max} – lattice mobility, μ_{min} – ionized impurity mobility, $\mu_{min}-\mu_1$ – clustering mobility)

Fit parameter	Si:P	Si:B	ZnO	ITO	SnO_2
μ_{max} [cm^2/V s]	1,414	470.5	210	210	250
μ_{min} [cm^2/V s]	68.5	44.9	55	55	50
$\mu_{min}-\mu_1$ [cm^2/V s]	12.4	15.9	5	5	10
n_{ref1} [10^{17} cm^{-3}]	0.92	2.23	4	15	20
α_1	0.711	0.719	1	1	1
n_{ref2} [10^{20} cm^{-3}]	3.41	6.1	6	20	6
α_2	1.98	2.0	2	2	2

7.4.5.3 Neutral Impurity Scattering

Neutral shallow-impurity scattering is often discussed in papers about transport in TCO films at room temperature [192, 193]. Since the shallow donors in TCO materials (for instance the group III elements in ZnO) exhibit ionization energies around about 50 meV [14] (see Table 7.4), the concentration of neutral donors at room temperature is very low, taking into account the further reduction of the ionization energy for degenerately doped semiconductors (see Fig. 7.14a). Therefore, in this chapter this scattering process is not taken further into consideration and mentioned only briefly.

The mobility due to neutral impurity scattering was first calculated by Erginsoy [194] who scaled the electron scattering at hydrogen atoms to a semiconductor by using its dielectric constant and carrier effective mass, which leads to:

$$\mu_n = \frac{m^* e^3}{A(T) 4\pi\varepsilon\varepsilon_0 \hbar^3 N_n}. \tag{7.10}$$

Here, A(T) is the generally temperature-dependent scattering cross-section factor and N_n is the density of neutral scattering centers. Erginsoy [194] selected a temperature-independent value for the scattering cross-section factor of A = 20, which is used mostly. Itoh took the correct e-H scattering cross section to describe low-temperature electron and hole mobilities in germanium [209]. The concentration of neutral impurities is given by $N_n = N_D – N_A – n(T)$, where N_D and N_A are the donor and acceptor concentrations, respectively.

Recently, Makino et al. calculated theoretical hole mobilities in ZnO for the above mentioned lattice-scattering processes [210]. For a hole concentration of 7×10^{17} cm^{-3} and an acceptor ionization energy of 100 meV, Makino et al. derived a hole mobility between 70 and 100 cm^2/V s for compensation ratios $K = N_D/N_A$ between 0 and 0.8. These values are much higher than the experimental hole mobilities $\mu_h < 10$ cm^2/V s, also reported by Tsukazaki et al. [142]. Therefore, Makino et al. come to the conclusion that defects like dislocations, space-charge regions etc. significantly decrease the mobility in heteroepitaxial p-type ZnO films [210] (see also Sect. 7.5.1.2).

Electrical properties of zinc oxide are summarized in Table 7.7. For comparison the corresponding data of two other TCO materials (In_2O_3 and SnO_2) and of silicon are given.

7.5 Electrical Transport in Thin Films

7.5.1 *Transport Models for Polycrystalline Films*

7.5.1.1 Grain Barrier Limited Transport

Polycrystalline films exhibit a vast amount of grain boundaries, which constitute crystallographically disturbed regions, leading to electronic defects in the band gap of semiconductors. These defects are charged by carriers from the interior of the grains. Depending on the type of the carriers (electrons or holes) and the type of the defects (electron trap or hole trap) charge balance causes depletion or accumulation zones around the barrier. In TCO films which are typically n-type a depletion zone is generated on both sides of a grain barrier accompanied by an energetic barrier of height Φ_b for the electrons. This is due to the electron trap character of the defects.

Table 7.7 Properties of zinc oxide and other TCO-materials in comparison to that of silicon (taken from [71])

Property (unit)	ZnO [162]	In_2O_3	SnO_2 [195]	Si [172, 196, 197]
Band gap E_g (300 K) [eV]	3.4 (dir)	2.7 (indir) 3.75 (dir)	3.6 (dir)	1.12 (indir) 4.18 (dir)
Pressure coefficient of E_g dEg/dp [meV/GPa]	23.5 [198]			−14.1(dir)
Temperature coefficient of E_g: dEg/dT [meV/K]	−0.3		−1.2	−0.47
Static dielectric constant ε_r	∥c: 8.75 ⊥ c: 7.8	≈9	∥c: 9.6 ⊥c: 13.5	11.9
Density of states effective electron mass m_n^*/m_e	0.28	0.35	0.275 (∥c: 0.23, ⊥c: 0.3)	0.337 [199]
Density of states effective hole mass m_p^*/m_e	0.59	0.6	1.0 [200]	0.55 [199]
Nonparabolicity parameter β [eV^{-1}]	0.29	0.18 [201]	0.96 [175]	0.27 [203]
	0.14 [201] 1.04 [202]			0.5 [204]
Eff. Conduction band dens. of states (300 K) N_C [cm^{-3}]	3.7×10^{18}	4.1×10^{18}	3.7×10^{18}	4.9×10^{18}
Exciton binding energy [meV]	59–65		130	14.3
Energy of the longitudinal optical phonon $\hbar\omega_0$ [meV]	73.1	41; 45; 51	34.1; 45.1; 86.8; 95.1	50.8; 51.9; 64.3
Deformation potential E_1 [eV]	31.4; 3.8; 18.9; 1.4	33.3 [206]	8 [207]	9.5 [172]
	3.5 [205]			
Longitudinal elastic constant c_l [GPa]	204.7; 207.2; 185.3; 204.5	159 [208]	326	156 [172]
Piezoelectric coefficients $P_\perp$; $P_\parallel$	0.21; 0.36	–	–	–
Extrinsic dopants	B,Al,Ga,In, Si, Ge, Sn, Y, Sc, Ti, Zr, Hf, F, Cl	Sn, Ti, Zr, F, Cl, Sb, Ge, Zn, Pb	Sb, (As, P) F, Cl	B,Al, Ga, In P,As,Sb

The carrier transport in polycrystalline silicon was first described comprehensively by Seto [211]. He assumed a δ-shaped density of electron trap states in the band gap which are completely filled. The origin of these trap states is not very clear in detail. For a ceramic ZnO varistor, Magnusson and Wiklund could show by photoelectron spectroscopy, that bismuth monolayers induce states, which are positioned 0.9 eV above the valence band maximum [212]. This suggests, that foreign atoms or other defects at grain boundaries induce these electron traps. A schematic band diagram according to Setos model is shown in Fig. 7.17.

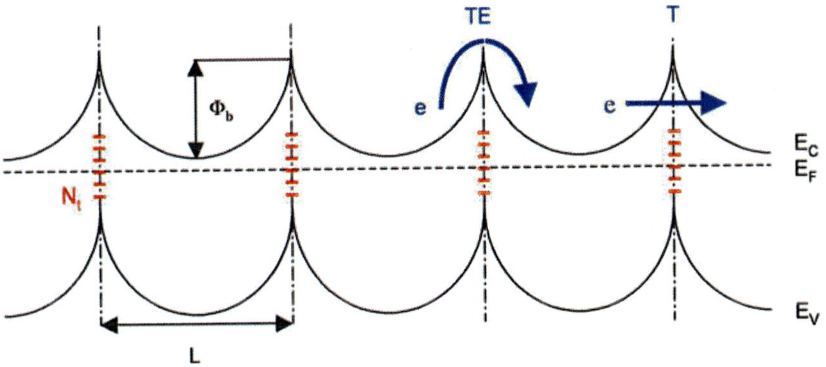

Fig. 7.17 Schematic band diagram of a linear row of grains of identical length L, doping N and with grain barriers of height Φ_b caused by a continuous distribution of electron trap states of density N_t [211]. Two different transport paths for electrons are indicated: *TE* thermionic emission across the barrier, *T* tunnelling through the barrier

An improved model was presented by Baccarani et al. [213] who considered a continuous energy distribution of trap states in the band gap. Additionally, these authors treated the possibility that the traps are only partially filled. Qualitatively, both models lead to the same conclusions. In the following we restrict our discussion to Setos model. The carrier transport across the grain barriers is described by the classical thermionic emission (see for instance Sze [196]) depicted by path "TE" in Fig. 7.17.

For very high carrier concentrations in the grains the depletion width is very narrow thus enabling quantum-mechanical tunneling of the barriers by the electrons, shown as path "T" in Fig. 7.17. It is interesting to note, that the potential drop over a grain barrier can be directly measured by scanning surface potential imaging with an atomic force microscope (AFM), which was done recently by Huey et al. on a commercial ZnO varistor [214].

The models of Seto and Baccarani et al. yield an effective mobility μ_{eff} dominated by thermionic emission across the grain barriers with an energetic height Φ_b:

$$\mu_{eff} = \mu_0 \exp(-\Phi_b/kT) \tag{7.11}$$

where Φ_b is the energetic barrier height at the grain boundary, T the sample temperature, and k is the Boltzmann constant, respectively. The prefactor μ_0 in (7.11) can be viewed as the mobility inside a grain given by [211]:

$$\mu_0 = \frac{eL}{\sqrt{2\pi m^* kT}}. \tag{7.12}$$

Depending on the doping concentration in the grains, two expressions for the barrier height can be derived:

$$\Phi_b = \frac{e^2 N_t^2}{8\varepsilon\varepsilon_0 N} \quad \text{for LN} > \text{N}_\text{t}, \tag{7.13a}$$

$$\Phi_b = \frac{e^2 L^2 N}{8\varepsilon\varepsilon_0} \quad \text{for LN} < \text{N}_\text{t}, \tag{7.13b}$$

where e is the elementary charge, N_t is the charge carrier trap density at the boundary, $\varepsilon\varepsilon_0$ is the static dielectric constant, N is the carrier density in the bulk of the grain and L the grain size. For LN > N_t the traps are only partially filled and hence the crystallites are completely depleted, while for LN < N_t only part of the grain is depleted and the traps are filled completely. The maximum barrier height Φ_{bmax} occurs for a doping concentration of $N(\Phi_{bmax}) = N_t/L$, accompanied by a minimum of the effective mobility according to (7.11).

For three situations (1–3), corresponding to low, medium and high carrier concentrations, the band structure at the grains is schematically shown in Fig. 7.18.

In the original models of Seto and Baccarani et al. [211, 213] only thermionic emission was taken into account. For very high carrier concentrations ($N > 10^{20}$ cm^{-3}) additional tunneling *through* the barriers takes place, which increases the current flow between the grains. If the thermionic model discussed above is applied for such high carrier concentrations lower barrier heights are calculated as expected from (7.13), which has to be kept in mind.

Concerning the prefactor μ_0, which is viewed as the intragrain mobility [216], in many cases too low values are extracted from the experiments [211, 213]. Here, additional carrier scattering processes take place in the grains, for instance point defect and dislocation scattering. Blom et al. performed a comprehensive analysis of the transport in undoped polycrystalline ZnO, which was deposited by RF magnetron sputtering in pure oxygen [217]. From van der Pauw and capacitance-voltage measurements they obtained the following transport parameters of the crystallites with lateral dimensions of 0.3 ± 0.1 μm: carrier concentration in the grains $N = 5 \times 10^{17}$ cm^{-3}, barrier height $\Phi_b = 1.2 \pm 0.3$ eV and the trap density at the grain boundaries $N_t = 5 \times 10^{12}$ cm^{-2} (see also Table 7.8).

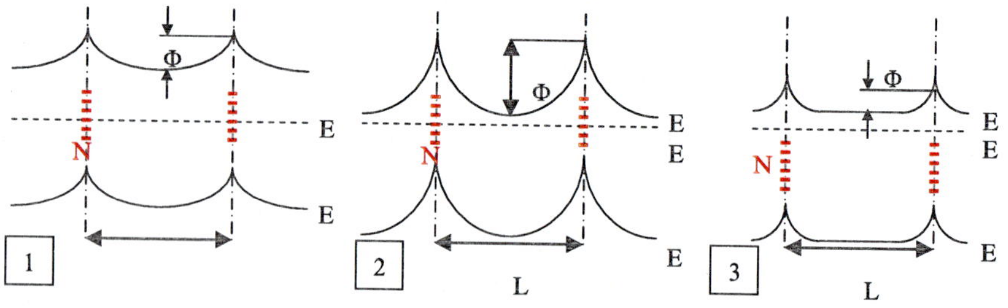

Fig. 7.18 Schematic band diagrams in the grains for different doping concentrations N in grains of identical size L (after Kamins [215]). The situations 1–3 correspond to low, medium and high carrier concentrations. The barrier height increases with increasing N up to a maximum at $N_{max} = N_t/L$. Further increasing N decreases the barrier height

Table 7.8 Trap densities N_t, deposition method and discharge voltage V_{dis} of epitaxial and polycrystalline TCO films [190]

Film:dopant	Growth	Method	Excitation (V_{dis} [V])	N_t (cm^{-2})	Source
ZnO:Al	Epitaxial	MS	RF (200)	1.3×10^{13}	[233]
ZnO:Al	On glass	RMS	DC (?)	$\approx 2 \times 10^{13}$	[261]
ZnO:Al	On glass	RMS	DC (425)	3×10^{13}	[229]
ZnO:Al	On glass	RMS	MF (340)	3×10^{13}	[230]
ZnO	On glass	MS*	RF	7×10^{12}	[123]
ZnO:Al	Epitaxial	PLD	–	5×10^{12}	[231]
ZnO	Epitaxial	PLD	–	1.5×10^{13}	[257]
In_2O_3:Sn	On glass	RMS/MS	DC (400) RF (250)	1.5×10^{12}	[190]
In_2O_3:Sn	On glass	MS	RF-diode (2000)	2.5×10^{13}	[262]
SnO_2:F,Cl	On glass	SP	–	4×10^{12}	[263]
$CdIn_2O_4$	On glass	RMS	DC (2000)	1.5×10^{13}	[175]

RMS, MS reactive magnetron sputtering, *PLD* pulsed laser deposition, *SP* spray pyrolysis, *DC, RF, MF* direct current, radio frequency, mid frequency excitation, *p-DC* pulsed DC excitation, * substrates arranged perpendicular to the target (reduced ion bombardment)

7.5.1.2 Defect-Limited Transport (Dislocation Scattering)

Besides by grain boundaries, discussed above, polycrystalline films are characterized by large densities of other crystallographic defects like point defects, dislocations or stacking faults (see for instance [218, 219]). While the mobility limitation by charged dislocations has been treated theoretically, the effect of point defects or stacking faults is unclear. Therefore, we will restrict the discussion to the role of dislocation scattering which is a natural scattering process in polycrystalline materials. However, this process is rarely used in explaining experimental data of carrier transport in polycrystalline semiconductors and especially transparent conducting oxides [193]. The main reason seems to be that the dislocation density of thin films is not measured.

Pödör [220] investigated bended Ge crystals with a dislocation density of about 10^7 cm^{-2} and could describe his results by (7.14), taking into account scattering by charged dislocations (see also Seeger [167]):

$$\mu_{disl} = \frac{30\sqrt{2\pi}\,(\varepsilon_r\varepsilon_0)^{3/2}a^2\sqrt{n}\,kT}{e^3f^2\sqrt{m^*}N_{disl}}, \tag{7.14}$$

where a is the distance between acceptor centers along the dislocation line, f is the occupation rate ($0 \leq f \leq 1$) of these acceptors and N_{disl} is the density of dislocations. For ZnO single crystals this scattering process can be neglected since the dislocation densities are very low ($N_{disl} < 300$ cm^{-2}, see Table 7.3).

7.5.2 Transport Data for Polycrystalline ZnO Films

Electrical parameters (resistivity and Hall mobility) of ZnO films deposited by a variety of deposition methods are shown in Fig. 7.19 versus the substrate temperature [71]. The lowest resistivity values reported are about 2×10^{-4} Ω cm, apart from two values of 1.4×10^{-4} Ω cm for a heteroepitaxially sputtered ZnO film [221] and a PLD film [222]. Most of the electrical measurements of heavily doped ($N > 10^{19}$ cm^{-3}) zinc oxide were obtained for polycrystalline thin films. Data on heavily doped single crystals or epitaxial layers are rare [184, 223]. The scattering of the resistivity and mobility data is large. The dopants used for the ZnO films: Al, B, Ga, In and Sn lead to comparable resistivities of the zinc oxide films [70]. However, aluminum is the dopant used by far mostly. As can be seen from Fig. 7.19a the resistivity data do not exhibit a correlation with the substrate temperature T_{sub}, while the mobility values tentatively show an increase with T_{sub}, which is understandable by the better crystalline quality of films prepared at higher substrate temperatures. In data series of the same research group the beneficial effect of an increased substrate temperature can be seen more clearly (see for instance Minami et al. [224]). When comparing the different deposition methods

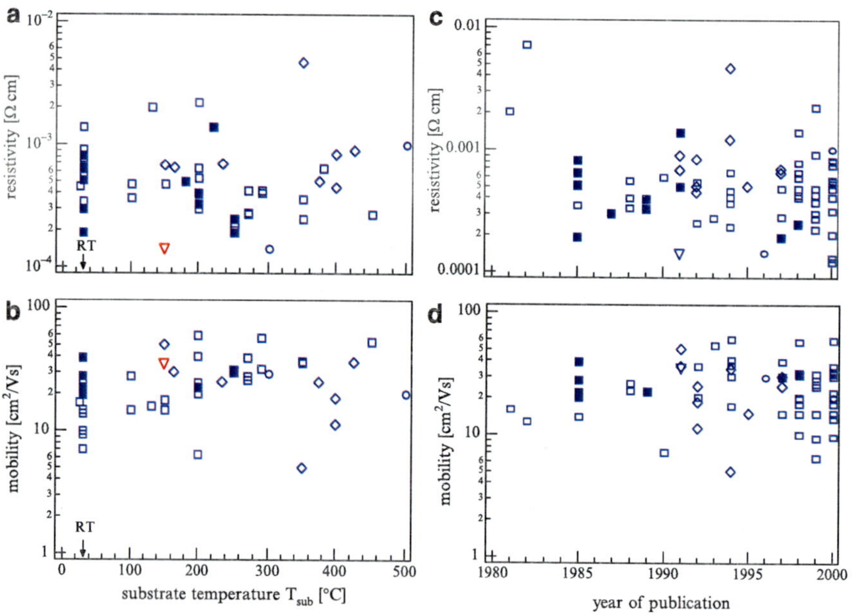

Fig. 7.19 Resistivity (**a**, **c**) and Hall mobility (**b**, **d**) of doped ZnO layers in dependence on the substrate temperature during film growth (**a**, **b**) and the publication year (**c**, **d**). The full symbols mark layers, which were prepared on substrates positioned perpendicular to the sputtering target. The point marked by a triangle belongs to an epitaxial ZnO film. The films were deposited by magnetron sputtering (5, 9,▽), MOCVD (◇) and pulsed laser deposition (○). Data taken from [87]. *RT* room temperature

one has to keep in mind that the growth is governed by different microscopic processes. While MOCVD relies on a pure thermal activation of the growth process, magnetron sputtering and PLD are characterized – as every plasma assisted process – by an additional energy input from the plasma into the growing film. Therefore, lower growth temperatures are possible in magnetron sputtering and pulsed-laser ablation deposition, which is also visible from Fig. 7.19a.

The fact, that the resistivity seems to be independent of such important experimental parameters like substrate temperature (and total sputtering pressure, not shown here) is certainly due to differences in the sputtering configurations of different groups (which are not always reported). Parameters like target-to-substrate distance, type of the used magnetron sputtering source (balanced or unbalanced magnetic field), deposition rate, arrangement of the substrate relative to the target (parallel or perpendicular) strongly influence the film properties. For example, the full symbols in Fig. 7.19 mark the electronic parameters of ZnO films, which were deposited onto substrates arranged perpendicularly relative to the target. This deposition configuration is characterized by a low deposition rate, a high ion-to-neutral ratio and reduces the bombardment of the films by species with high energies (reflected argon atoms or negative oxygen ions), thus leading to low resistivities even for room temperature deposition [225].

Obviously, the ion assistance of the film growth during magnetron sputtering allows the deposition of zinc oxide films with a high crystallographic and electronic quality even at room temperature [87, 226].

Highly conductive zinc oxide films have been prepared now for more than 30 years with different deposition methods and a lot of technological effort has been put into this research. Therefore, it is interesting to look for a trend of the electrical data of the ZnO films over the time (Fig. 7.19b). Surprisingly, since about 1985 the reported resistivity and mobility values have not changed significantly, despite the fact that a large number of groups have been working in this field. This suggests that there exists a physical limitation which prevents to reach resistivities lower than about 2×10^{-4} Ω cm and mobilities larger than about 50 cm^2/V s. This result is in qualitative agreement with a presentation of TCO-resistivity data over time, reported by Minami [123]. While Minami, in agreement with our assumption, suggests a lower resistivity limit of 1×10^{-4} Ω cm for tin-doped indium oxide (ITO) films, he supposed by a free-hand drawn line, that ZnO could reach resistivities below 1×10^{-4} Ω cm beyond the year 2000 (see his Fig. 7.1 in [123]). However, this "goal" has not been realized in the last 10 years, which supports the view of a physical limit of the mobility of ZnO films, and, more generally, for TCO films [161, 227]. Figure 7.20 shows the mobility of polycrystalline and epitaxial ZnO:Al films versus the carrier density [123, 228–231]. The mobility values for $N > 1 \times 10^{20}$ cm^{-3} exhibit a significant scattering from about 1 to 50 cm^2/V s. The highest μ-values of thin films reach the mobility-fit curve (7.9) for single crystalline ZnO. Though the epitaxial films exhibit a significantly better crystalline quality compared to films grown on glass or silicon the mobilities are almost comparable to the literature data reported for ZnO films on glass substrates. This points to the fact that for carrier concentrations above 1–3 $\times 10^{20}$ cm^{-3} the

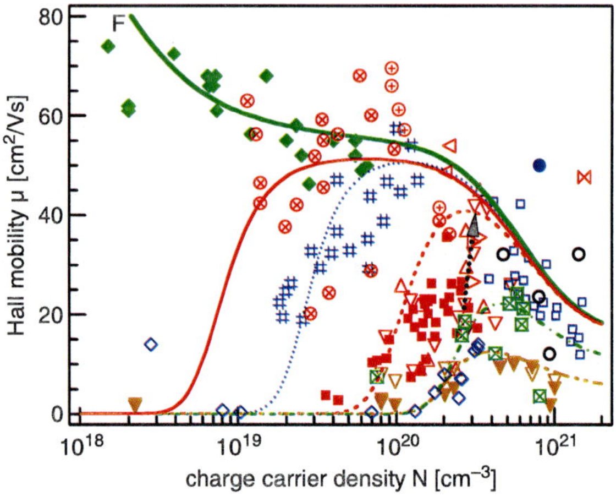

Fig. 7.20 Hall mobilities of undoped and doped zinc oxide thin films as a function of the carrier concentration (taken from [190]). Data from are shown for films deposited onto float glass (■) as well as sapphire substrates (Δ, ◁, ▷). For comparison mobility data of other groups were added, which have been measured for films deposited by magnetron sputtering and by pulsed laser deposition (PLD): Minami (□, #, ● – PLD) [202], Brehme et al. (▼, ▽) [229], Kon et al. (⊞) [230], Suzuki (○ – PLD) [222] and Lorenz et al. (⊗, ⊕ – PLD) [231]. The mobility values of ZnO single crystals (◆), already shown in Fig. 7.16 and the fit curve (F) according to (7.9), are also displayed. The thin film mobility data have been fitted by the combined ionized and grain boundary model (7.15), yielding the grain boundary trap densities summarized in Table 7.8

mobility is dominated by ionized impurity scattering as already pointed out by Bellingham, Minami and Ellmer [71, 123, 232]. Furthermore, the experimental dependence $\mu = f(N)$ for $N > 5 \times 10^{20}$ cm^{-3} can only be explained by taking into account the effect of impurity clustering (see [181–183]) and nonparabolicity of the conduction band of ZnO [71]. For carrier concentrations lower than about 3×10^{20} cm^{-3} the mobility in the epitaxial films decreases steeply, which is in qualitative agreement with data of Minami [228] for undoped ZnO films and of Brehme et al. [229] and Kon et al. [230] for ZnO:Al films.

This decrease of μ with decreasing N can be described by Setos model [211] for the effective mobility dominated by thermionic emission across grain barriers (see (7.11)). The experimental data for $N < 3 \times 10^{20}$ cm^{-3} have been fitted by the following expression:

$$\mu = \frac{\mu_{0s} \cdot \mu_{eff}}{\mu_{0s} + \mu_{eff}} \quad \text{with} \quad \mu_{eff} = \mu_{grain} \exp\left(-\frac{\Phi_b - \frac{\Delta\Phi_b^2}{2kT}}{kT}\right). \qquad (7.15a, b)$$

where μ_{0s} is the lattice scattering limited mobility of single crystalline ZnO (7.9) and μ_{eff} is the grain-barrier-limited mobility according to the Seto model (7.11).

These fits are shown as thin lines for the data of Minami [228] for nominally undoped polycrystalline ZnO films and for our epitaxial ZnO:Al films [233].

In order to fit the experimental data for ZnO and ZnO:Al films, different trap densities N_t at the grain boundaries had to be assumed: $N_{t(ZnO)} = 7 \times 10^{12}$ cm^{-2}; $N_{t\,(ZnO:Al)} = 1.3 \times 10^{13}$ cm^{-2}. A tentative explanation is a preferential built-in of aluminum or Al_2O_3 at grain boundaries leading to additional trap states in the band gap compared to undoped ZnO. These values can be compared to trap densities reported for other polycrystalline semiconductors: 3×10^{12} cm^{-2} (Si) [211], 1×10^{12} cm^{-2} (CdS) [234], 1.2×10^{13} cm^{-2} (FeS_2) [235]. Recently, also for gallium nitride films grown by MOCVD on sapphire substrates [236] and for GaAs layers [237] internal potential barriers (i.e., grain barriers) were reported to influence the carrier transport. That means, grain barriers or other structural inhomogeneities effect significantly the carrier transport in heteroepitaxially grown films, independent of the material. The dotted arrow in Fig. 7.20 shows the effect of a vacuum annealing at 500°C of the ZnO:Al films on sapphire substrates. While the carrier concentration does not change at all the mobility is increased by about 30–50%, moving the mobility values near to the general fit curve. This significant electrical change is not accompanied by a change of the structural parameters (not shown). This means that no recrystallization occurred. Instead it is plausible that point defects or dislocations have been annealed reducing the scattering at such centers.

This result can also explain that many mobility data from literature (see the data of [229, 230] in Fig. 7.20) have a comparable dependence on N as mentioned above, however its absolute values are lower by a factor of 2–4. Probably, such samples contain a large amount of point defects or dislocations which act as additional scattering centers (see Sect. 7.5.2.1).

Recent mobility data of Lorenz et al. [231] for Al- and Ga-doped ZnO films, deposited by pulsed laser deposition, were included in Fig. 7.20. These films exhibit mobilities that are comparable to mobility values reported for zinc oxide single crystals, pointing to a high structural quality of these PLD films. For $N < 5 \times 10^{19}$ cm^{-3} a tendency to lower mobilities (compared to single crystal values) can be observed, which means that the trap density is lower than in the other polycrystalline ZnO films.

7.5.2.1 Role of Defects on Transport in ZnO

Recently, heteroepitaxial ZnO films grown on GaN films on sapphire substrates were analyzed by transmission electron microscopy [218]. These films are characterized by a high stacking fault density in the range of 10^{18} cm^{-3} (see also Sect. 7.7). According to Gerthsen et al. [218] these stacking faults are mainly generated by the precipitation of Zn interstitials accompanied by the formation of oxygen vacancies in the vicinity of a stacking fault. Since oxygen vacancies exhibit a charge of $Z = 2$ its scattering power is higher than that of singly ionized dopands like the group III elements in ZnO (B^+, Al^+, Ga^+: $Z = 1$) leading to a reduced mobility. For polycrystalline films an even higher stacking fault and dislocation density is plausible. Sagalowicz and Fox [238] analyzed undoped polycrystalline ZnO films by TEM and found dislocation densities of around 10^{12} cm^{-2}, corresponding to a mean distance of 10 nm. Hence, the scattering of carriers at dislocations and oxygen vacancies would be even higher in such materials. Recent density-functional calculations [31] have shown that "point defect formation enthalpies in zinc oxide are very small" supporting the considerations given above. It is also noteworthy that ZnO exhibits a high radiation hardness, which is likely due to a rapid annealing of Frenkel pairs [116, 239], again caused by low defect formation energies, leading to high diffusion coefficients. Zinc oxide is a polar semiconductor, i.e., electrical charges are induced by stress along the c-axis. Its piezoelectric properties are also the reason for its use in surface acoustic wave (SAW) devices [240]. With this respect ZnO is similar to GaN [241]. This means intrinsic stress, due to lattice mismatch and/or to growth defects could induce charges at grain barriers or extended defects leading to additional scattering, a peculiar property of ZnO compared to ITO, which is not piezoelectric. Emtage used the same argument for a possible explanation of the fact that lowly-doped zinc oxide grain barriers exhibit highly nonlinear current-voltage characteristics, which are used in varistors [242, 243] (see also Sect. 7.2.4).

7.5.2.2 Carrier Concentration and Mobility Limits of Degenerately Doped Zinc Oxide

Figure 7.21 shows the temperature-dependent carrier densities and the Hall mobilities of ZnO:Al films prepared at different oxygen partial pressures, leading to carrier concentrations between 2×10^{19} and 7×10^{20} cm^{-3} [163]. The carrier concentrations are temperature-independent, that means, these films are degenerate semiconductors with a vanishing donor ionization energy (see Sect. 7.4.5.1). The temperature dependences of the mobilities exhibit a different behavior for different carrier concentrations. Films with the highest mobilities show a decrease of the mobility with increasing temperature, a behavior typical for metals [244]. This decrease of μ with T becomes less pronounced with decreasing mobility and for the ZnO:Al films with $N < 10^{20}$ cm^{-3} the mobility is independent of the temperature. Eventually, the film with the lowest carrier concentration (2×10^{19} cm^{-3}) shows a weak thermally activated mobility typical for grain-barrier limited transport [211, 245] (see Sect. 7.5.1.1). The metal-like mobility curves ($N > 10^{20}$ cm^{-3}) have been fitted by the following equations [163]:

$$\mu = \frac{\mu_{0m} \cdot \mu_{temp}}{\mu_{0m} + \mu_{temp}} \quad \text{with} \quad \mu_{temp} = \mu_{phonon}\left(\frac{T}{T_0}\right)^{-p}, \qquad (7.16a, b)$$

where μ_{0m} is the temperature-independent part of the total mobility, for instance due to ionized impurity scattering (see section 7.4.5.2) and μ_{temp} is the phonon activated metal-like mobility. μ_{phonon} is the temperature-dependent mobility at the reference temperature T_0, which is used to describe the metal-like scattering behavior of the degenerately doped films and p is an exponent which describes the temperature

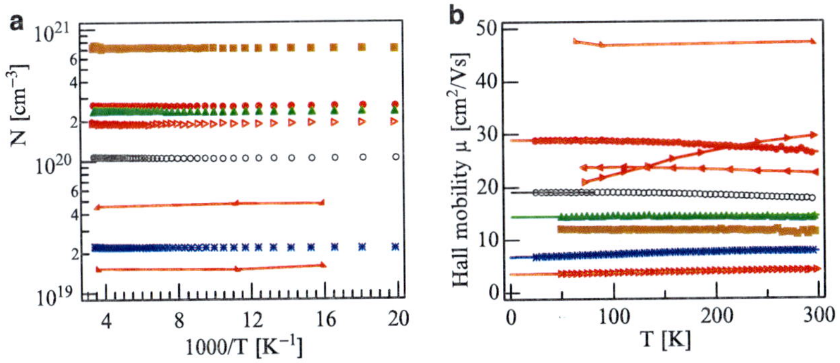

Fig. 7.21 (**a**) Carrier concentration and (**b**) mobility as a function of the temperature (50 K < T < 300 K) of ZnO:Al films prepared at different oxygen partial pressures and plasma excitation frequencies. The *thin lines* in (**b**) are the fits according to the models, described in the text. The films have the following carrier concentrations in [cm^{-3}]: (*) 2.2 × 10^{19}, (o) 1.1 × 10^{20}, (▷) 1.9 × 10^{20}, (▲) 2.4 × 10^{20}, (●) 2.6 × 10^{20}, (▪) 7 × 10^{20} [163]. For comparison data of In-doped ZnO single crystals (◣, ◢, [184]) and ZnO and ZnO:Al films (◀, ▶), reported by Young et al. [246], have been added

dependence. The extended Seto model (see Sect. 7.5.1.1) for grain barrier-limited transport of Werner is used to describe the μ(T) curves at lower carrier concentrations [245]. Again a temperature-independent term μ_{0s} has been added:

$$\mu = \frac{\mu_{0s} \cdot \mu_{eff}}{\mu_{0s} + \mu_{eff}} \quad \text{with} \quad \mu_{eff} = \mu_{grain} \exp\left(-\frac{\Phi_b - \frac{\Delta\Phi_b^2}{2kT}}{kT}\right). \qquad (7.17\text{a, b})$$

Here, μ_{grain} is the mobility inside the grain, Φ_b and $\Delta\Phi_b$ are the mean barrier height and its variation. The fitting curves are shown as lines in Fig. 7.21b. The temperature-independent mobility μ_{0m} varies from about 20 to 30 cm^2/V s, while the temperature-dependent part μ_{temp} spans the range of 220–310 cm^2/V s. The mobility curve for the two lowest carrier concentration were fitted by (7.17), yielding a mean barrier height of about 1 meV, an intra-grain mobility of about 140 cm^2/V s and a temperature-independent term of 30 cm^2/V s. The low barrier height of only 1 meV is consistent with the still high mobility at room temperature (see Fig. 7.21b). For carrier concentrations below 10^{19} cm^{-3}, it was not possible to measure Hall voltages.

7.5.2.3 Transport Data of Epitaxial Films on Different Substrates

Epitaxial ZnO films have been prepared since the end of the 1960s [97, 247, 248]. These early investigations were mainly devoted to highly resistive films for surface-acoustic wave (SAW) applications (see Sect. 7.2.4). Investigations of electrical properties of epitaxial ZnO films were reported only in the last 15 years. These ZnO

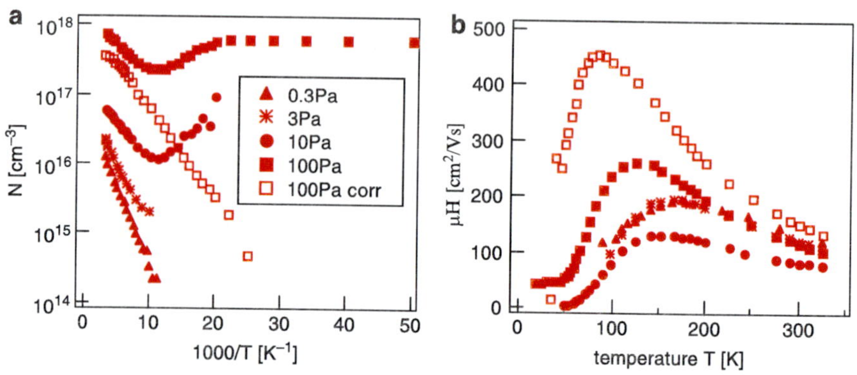

Fig. 7.22 **(a)** Electron concentration and **(b)** Hall mobility of nominally undoped ZnO layers grown on c-sapphire substrates by pulsed laser deposition (PLD) at different oxygen partial pressures by von Wenckstern et al. [254]. The open squares are corrected N and μ_H data of the film grown at $p_{O2} = 100$ Pa, which were calculated under the assumption that a degenerately doped layer at the interface to the sapphire substrate is formed

films were deposited onto different single crystalline substrates, especially sapphire (Al_2O_3) with c- and a-orientation, lithium niobate ($LiNbO_3$), periclase (MgO) and scandium-aluminum-magnesium spinel (SCAM–$ScAlMgO_4$) [221, 233, 249–253]. Some mobility data of epitaxial films on sapphire were shown already in Fig. 7.20. The deposition of epitaxial films has been performed by a number of deposition methods: magnetron sputtering, oxygen-radical-assisted evaporation or pulsed laser ablation deposition.

Epitaxial films grown by pulsed laser deposition (PLD) on sapphire substrates have been reported by different groups. Electrical data of such films are exemplarily shown in Fig. 7.22, which were deposited at four different oxygen partial pressures (0.3 to 100 Pa) [150]. While the films, deposited at low oxygen partial pressures exhibit electron concentrations below 2×10^{16} cm^{-3}, which decrease steeply at lower temperatures, the films, prepared at $p_{O2} \geq 10$ Pa show higher carrier densities, a smaller activation energy and a local minimum of N around measurement temperatures of 100 K. This is explained by von Wenckstern et al. by a two-layer transport model for the epitaxial films, which was already used by Look and Sizelove to describe the electrical transport in epitaxial GaN films [255]. Due to lattice defects, introduced at the heterointerface sapphire/film (see Sect. 7.5.2.1), a highly (degenerately) doped interface layer is formed, which takes over the current transport at low temperatures ($T \leq 100$ K). Correcting the measured data for the effect of the interface layer yields the data, shown in Fig. 7.22 by open squares [254]. The resulting mobility of the upper, nondegenerate ZnO layer is now higher, than the measured value. For the film deposited at 100 Pa oxygen partial pressure the mobility at room temperature is about 160 cm^2/V s ($N = 4 \times 10^{16}$ cm^{-3}), which is comparable to mobility values of single crystals (see Fig. 7.16). The films, which were prepared at lower oxygen partial pressures, exhibit significantly lower mobilities (at even lower electron densities), which is caused by the lower grain sizes (60–250 nm) of these films, compared to a grain size 2 μm for the film deposited at p_{O2} = 100 Pa. Tampo et al. also reported such degenerate interlayers, when growing heteroepitaxial ZnO films at 600°C on sapphire [256]. By introducing a ZnO buffer layer, deposited at a low temperature of 350°C with a thickness of 20–40 nm, this highly-doped interlayer could be avoided, leading to epitaxial ZnO films with the following electrical data: $N = 7.5 \times 10^{16}$ cm^{-3} and $\mu_H = 132$ cm^2/V s.

These results point to the strong effect of the lattice mismatch between ZnO and the substrate crystals as well as of the role of crystallografic defects.

The exotic spinel $ScAlMgO_4$ is especially suited for the heteroepitaxial growth of ZnO, since it has a very low lattice mismatch relative to ZnO ($\Delta a/a \approx 0.09\%$) [257]. It is interesting, to compare the mobility data of epitaxial ZnO films on a SCAM substrate to that of single crystalline ZnO, reviewed in Sect. 7.4.5.2. Figure 7.23 shows the data reported by Makino et al. [257] for an undoped ZnO film as a function of the temperature (a) as well as a function of the carrier concentration for Ga-doped ZnO films (b). The mobility value of the undoped ZnO/SCAM film is comparable to or even higher than that of bulk single crystalline ZnO with comparable carrier concentrations ($N < 10^{17}$ cm^{-3}). This holds both at room temperature, where polar-optical scattering dominates as well as at low temperatures around $T \approx 100$ K, where piezoelectric scattering is significant.

Makino et al. argue that the crystalline quality of their epitaxial films was better than that of ZnO single crystals, thus explaining the higher mobility. This reasoning rises doubts, since polar-optical and piezoelectric scattering processes depend only

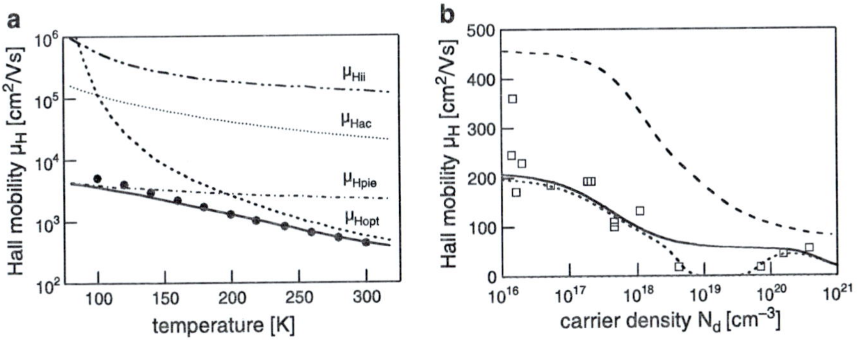

Fig. 7.23 (**a**) Hall mobility on a logarithmic scale as a function of the temperature for an undoped epitaxial ZnO layer and (**b**) Hall mobility of Ga-doped ZnO layers as a function of the carrier concentration (measured at room temperature). The ZnO films were grown epitaxially on lattice-matched $ScAlMgO_4$ (SCAM) by Makino et al. [257]. In (**a**) the calculated mobilities for acoustical, polar-optical, piezoelectric and ionized impurity scattering are shown, together with the total theoretical mobility. In (**b**) the *solid curve* is the fit curve (7.9) from Fig. 7.16, while the *dashed line* is the theoretical curve, calculated by Makino et al. [257]. The *dotted line* was calculated for transport across depletion regions at grain barriers (see Sect. 7.5.1.1), also present in epitaxial films [233]. Figure taken from [190]

on material parameters, i.e., on parameters, which were determined independently of electrical transport measurements (see Table 7.7), but not on the dopant concentration or the crystalline quality (see Sect. 7.4), Therefore, it is not plausible, that the mobility of the epitaxial films should be higher than that of good single crystals, see Fig. 7.15. This discrepancy becomes even more obvious from Fig. 7.23b, where the carrier concentration-dependent mobilities of ZnO:Ga films, also grown on SCAM by Makino et al., are shown. Though the experimental points scatter significantly, it are quite well described by the semi-empirical curve (7.9) (see Fig. 7.16), fitting the mobility data of ZnO single crystals reported from 1957 to 2005. On the other hand, the author's theoretical curve is higher by more than a factor of 2, caused by an unplausibly low value of the effective mass of ZnO of only $m^*/m_e = 0.24$ instead of 0.28 (see Table 7.7). In the carrier concentration region from 5×10^{18} to 8×10^{19} cm^{-3}, Makinos's mobility data exhibit a region, where the mobilities are almost zero, a fact not discussed by Makino et al. [257]. This "mobility hole" can be fitted by the model of transport across electrical grain barriers (dotted curve in Fig. 7.23b), which was explained in Sect. 7.5, and points to the fact, that even epitaxial films of high crystalline quality exhibit a high grain barrier trap density. The same was found for epitaxial films on sapphire substrates, which, due to the higher lattice mismatch, show a much lower crystal quality (see above) [233]. Figure 7.24 shows a compilation of mobility data versus the carrier concentration

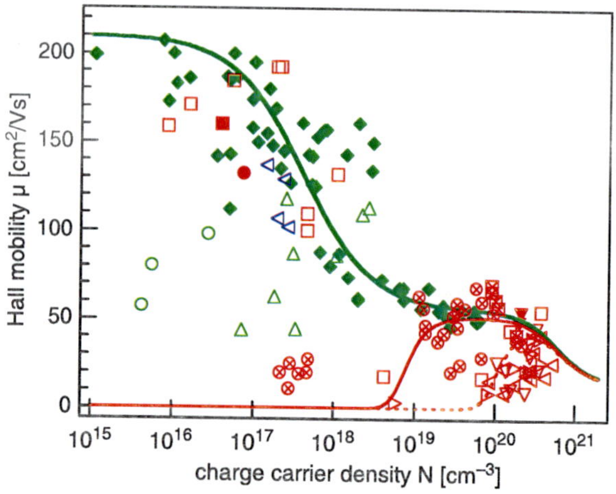

Fig. 7.24 Hall mobility of nominally undoped and doped ZnO layers epitaxially grown on different single crystalline substrates by pulsed laser deposition (PLD), magnetron sputtering and oxygen-radical assisted molecular-beam epitaxy (MBE): Ohtomo et al. (Δ, ○ – SCAM) [258], Tampo et al. (●) [256], Makino et al. (□) [257], Miyamoto et al. (◁) [259], von Wenckstern et al. (■) [260], and Lorenz et al. (⊗, ⊞ – PLD) [231]. The mobility values of ZnO single crystals (◆) are also displayed. The *thick line* is the semiempirical fit according to (7.9) to the mobility data reported for single crystalline ZnO (see Fig. 7.16). The *thin lines* are fits with the model of grain-barrier limited transport (see Sect. 7.5.1), from which the mean trap density N_t at grain barriers (see Table 7.8) can be estimated (according to [190])

of epitaxial ZnO layers, grown on different substrates and by different methods [190]. For comparison the data and the semiempirical fit curve (7.9) for single crystalline ZnO (c-ZnO) has been added. While some of the data of the epitaxial films do fit to the c-ZnO mobilities, other μ-values are significantly lower.

Though the crystalline quality of these epitaxial films is very good, it exhibit nevertheless grain boundaries with charged defects, inducing electrical grain barriers. Despite the significant scattering of the experimental data, the μ(N)-curves for the different data sets in Fig. 7.24 have been fitted by a combined scattering due to ionized impurities (7.9) and grain barrier limited transport (7.11), from which the electron trap densities N_t can be estimated. These trap densities are summarized in Table 7.8 for ZnO films and are compared with literature data and own measurements of ITO films [190]. As can be seen from Table 7.8 the charge trap densities of ZnO (and also other TCO) films prepared by different methods vary significantly, covering the range of 1.5×10^{12}–3×10^{13} cm^{-2}. These values are in the upper region compared to polycrystalline silicon (see for instance Seto [211]) or gallium arsenide, which is plausible taking into account that the ZnO (TCO) films were prepared with ion-assistance. Furthermore, in polycrystalline silicon the trap density can be reduced significantly by annealing or deposition in hydrogen-containing atmospheres, which is not viable for oxides due to the reducing effect of hydrogen. From Table 7.8 one can infer that the defect density is correlated with the discharge modus, especially with the discharge voltage. This means, the energy of the particles contributing to or assisting the film growth significantly influences the defect density. This can be seen both for ZnO as well as for ITO. However, for comparable deposition conditions in ZnO the defect density is higher than that of ITO, which seems to be caused by a lower defect formation energy in ZnO and/or self-annealing effects in ITO.

Recently, Yasui et al. [264, 265] used a biased grid in front of the magnetron sputtering source with a Zn target to reduce the energy of negative ions originating from the target surface (for instance O^-). They could show that negative grid voltages of 20–80 V improved the crystallographic film quality as well as the electronic properties significantly, supporting the argumentation presented above. Obviously, this influence of energetic particles on the ZnO properties occurs, though the defect formation energies of zinc oxide are quite high, compared to other semiconductors. While in silicon the vacancy formation energy is in the range of 11–22 eV, in ZnO ≈ 60 eV are necessary to create an oxygen vacancy [266, 267]. However, one has to keep in mind that in (reactive) magnetron sputtering, especially, when electronegative elements (oxygen, sulfur, fluorine etc.) are used, particle energies up to some hundred electron volts are possible. For instance negative ions (O^-, S^-, F^- etc.) are accelerated from the target surface towards the substrate, i.e., to the growing film, with an energy E_{ion}, which corresponds to the target voltage V_{dis} ($E_{ion} \approx q \cdot V_{dis}$, with q the ion charge) [268, 269]. These energies are more than sufficient to induce crystallographic defects.

As shown in [190], the grain barrier trap densities of ITO and ZnO are significantly different, covering the range of 1.5×10^{12}–3×10^{13} cm^{-2}. By optical and electrical analysis of ZnO:Ga and ITO films, Fujiwara and Kondo also concluded that the transport over grain boundaries is different for ZnO and ITO [201]. These differences cannot be explained by ionized impurity scattering, the process mostly used for the explanation of the mobilities for carrier concentrations above 10^{20} cm^{-3}. Taking into account the defect chemistry and the formation of planar defects in these oxides plausible reasons are given for the observed grain barrier densities. Especially remarkable is the piezoelectricity of ZnO along the c-axis, which distinguishes the polar ZnO from the cubic In_2O_3 lattice. One strategy to reduce the resistivity of zinc oxide further could be to find deposition processes or to use crystalline substrates which lead to ZnO films without a preferred texture along the polar c-axis.

7.6 Optical Properties of ZnO

Zinc oxide has an intrinsic band gap at room temperature of 3.37 eV [8], which is somewhat lower than that of the other TCO materials indium-tin oxide ($E_g = 3.75$ eV) and SnO_2 ($E_g = 3.6$ eV) and comparable to that of gallium nitride (GaN, $E_g = 3.3$ eV), which is used as a material for optoelectronic devices in the blue and ultraviolet spectral range. Since more than 10 years zinc oxide sees a renaissance as an optoelectronic material, due to its high exciton binding energy of 60 meV, allowing excitonic emission even at room temperature [271]. For comparison, the excitonic binding energy of GaN is smaller than 30 meV. Alloying ZnO with MgO or CdO can be used to shift the band gap to higher or lower values allowing a tuning of its optical emission [270].

The optical properties are characterized by the dielectric function DF: $\varepsilon = \varepsilon_1 + i\varepsilon_2$, which can be measured by transmission and reflection or by spectroscopic ellipsometry [272]. An overview of the real and imaginary parts of the DF is displayed in Fig. 7.25. It shows both the phonon optical modes in the infrared spectral range (most strongly the E_1–TO phonon) as well as the band-to-band transition at about 3.4 eV.

The number of papers, which report on optical properties of ZnO is too large to be listed here. Recent reviews can be found in [270, 271]. An important parameter for optoelectronic applications of ZnO is the band gap. A collection of band gap energies for bulk and thin film ZnO is given in Table 7.9 (data from [270]). The most accurate data for the optical transitions and the exciton binding energies have been determined by spectroscopic ellipsometry [273].

It is interesting to note, that the optical band gap values, measured on thin films are often 0.1–0.2 eV smaller than the lowest critical point optical transition energy.

This points to the influence of thin film effects, like impurities, strain and compositional variations, which cause band gap energy shifts and broadening of transitions. Zinc oxide is characterized, due to its dipole-allowed direct band-to-band transitions, by a strong exciton coupling to the light waves, leading to

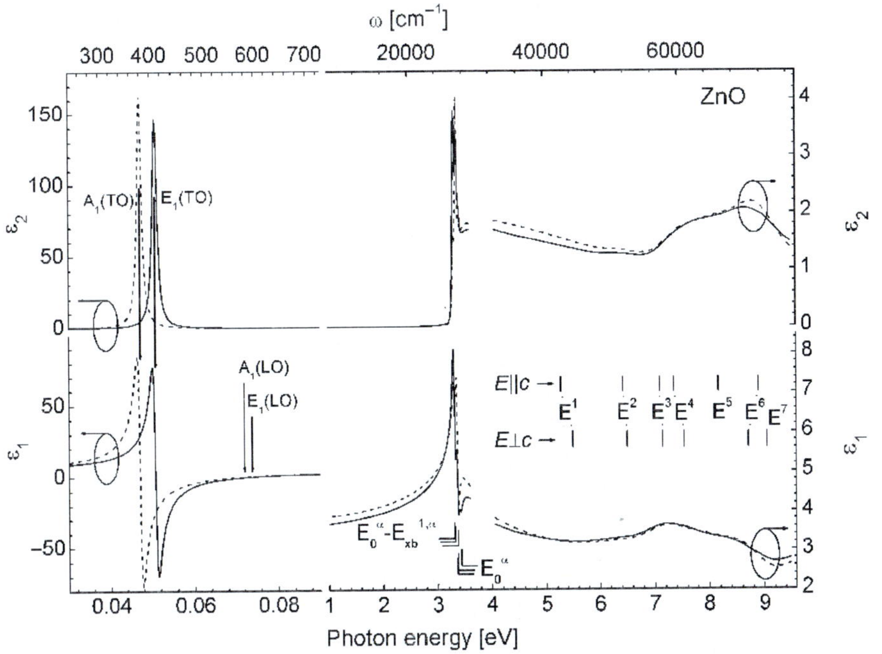

Fig. 7.25 Dielectric function $\varepsilon = \varepsilon_1 + i\varepsilon_2$ of single crystalline ZnO in the infrared (*left*) and NIR-VIS-UV (*right*) spectral range, measured by spectroscopic ellipsometry, taken from [270]. The *solid* (⊥c) and *dashed* (||c) *lines* mark the polarization direction of the light relative to the c-axis of the ZnO single crystal. The optical phonon modes and the transition energies are marked. Note the different energy and $\varepsilon_{1,2}$ scales in the IR and the UV-VIS, respectively

strong excitonic features in the absorption curves, even at room temperature [270, 277]. Typical transmission and absorption curves of two magnetron-sputtered ZnO:Al films are shown in Fig. 7.26. While the film with an optical band gap of 3.3 eV (□, ■), deposited at a high oxygen partial pressure, shows a high transmission down to a photon energy of 0.5 eV, the highly conductive ZnO:Al film (Δ, ▲), prepared at a low oxygen flow, exhibits a significantly larger optical gap of $E_{gopt} = 3.7$ eV and an absorption peak at about 0.9 eV, caused by the interaction of the light waves with the plasma excitation of the free electrons in the conduction band [284]. In Fig. 7.26b the absorption coefficient α as well as the quantity $(\alpha E_{ph})^2$, calculated from the data of Fig. 7.26a, are also displayed versus the photon energy. It can be seen, that the α(E) dependence corresponds to a direct semiconductor. The highly-doped ZnO film shows an extended absorption in the band gap, induced by the near-band gap donor states.

The shift of the absorption edge is due to the so-called Burstein-Moss effect [285, 286], which is due to the fact, that in degenerately doped semiconductors the Fermi level is shifted into the conduction band. Hence, the lowest conduction band states are already occupied by free electrons and an optical transition from the

Table 7.9 Band gap energy E_g, critical point energy E_0^A and free exciton binding energy E_{xb}^1 of ZnO single crystals and thin films (see also [270])

Author (year)	Sample (growth method)	E_g [eV]	E_0^A [eV]	E_{xb}^1 [meV]	[References] (optical method)
Yoshikawa (1997)	c-ZnO (CVT)		3.45	59	[274] (SE)
Jellison (1998)	c-ZnO (CVT, HT)		3.372/ 3.405	56/50	[275] (SE)
Ozaki (2003)	c-ZnO (CVT)		3.380	63.1	[276] (PR)
Muth (1999)	Epitaxial film (PLD)		3.40		[277] (T, R)
Rebien (2002)	Film (RMS)	3.37			[278] (SE)
Schmidt (2003)	Epitaxial films (PLD)		3.366	60	[272] (SE)
Srikant (1997)	Epitaxial films (PLD)	3.24–3.32			[279] (T, R)
Ohtomo (1998)	Epitaxial films (PLD)	3.29			[39] (T, R)
Tokumoto (2002)	Film (SP)	3.28			[280] (T, R)
Takeuchi (2003)	Epitaxial film (PLD)	3.28			[281] (T, R)
Misra (2004)	Epitaxial film (PLD)	3.28			[282] (T, R)
Zhao (2005)	FILM (PLD)	3.25–3.27			[283] (T, R)

CVT chemical vapor transport, *HT* hydrothermal growth, *RMS* reactive magnetron sputtering, *PLD* pulsed laser deposition, *SP* spray pyrolysis, *SE* spectroscopic ellipsometry, *T*, *R* transmission, reflection measurements, *PR* photoreflectance

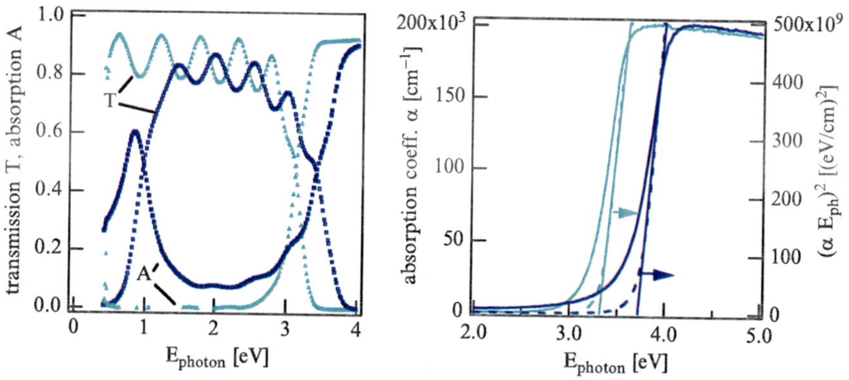

Fig. 7.26 (**a**) Transmission T (Δ, □) and absorption A (▲, ■, A = 1-R-T) versus the photon energy ($0.5 \leq E_{ph} \leq 4$ eV) of two ZnO:Al films deposited with different oxygen partial pressures. The film, deposited at $F_{O2 \approx 10\ sccm}$ (Δ, ▲) exhibits a high transmission from the IR to the bandgap energy $E_g = 3.32$ eV in the UV, while the highly conductive film, deposited at $F_{O2} = 6.1$ sccm (□, ■) shows an extended optical gap of $E_{gopt} = 3.72$ eV, accompanied by a high absorption below 1.3 eV, caused by absorption at free electrons, leading to the pronounced plasma absorption peak at 0.9 eV. (b) Absorption coefficient α versus the photon energy of the two films from (**a**). From the display of $(\alpha E_{ph})^2$ as a function of E_{ph}, the direct optical gaps, mentioned above, were derived by linear extrapolation to $\alpha = 0$ (*straight lines*)

valence band can occur only to unoccupied states higher in the conduction band. The Burstein-Moss shift is not the only effect which occurs in degenerately doped semiconductors. Due to the high density of free electrons the self energy is reduced,

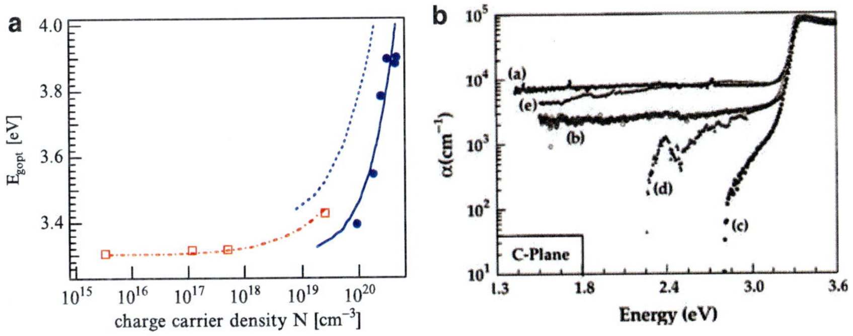

Fig. 7.27 (**a**) Optical band gap values E_{gopt} of ZnO films with different carrier concentrations (data taken from [288, 289]). The increase of the optical band gap for $N > 5\times10^{18}\ cm^{-3}$, due to the Burstein-Moss shift, can be clearly seen. The *lines* are theoretical dependences for E_{gopt}: *dot-dashed line*– fit of E_{gopt} taking into account the non-parabolicity of the ZnO conduction band (Ye et al.), *dashed line* – Burstein-Moss shift for a parabolic conduction band, *full line* – calculation of E_{gopt} by Sernelius et al. (see text). (**b**) Absorption coefficient (log scale) of epitaxial ZnO films on c-plane sapphire, which have been deposited by pulsed laser deposition (PLD) and annealed at different temperatures: a – as grown, 600°C, 10 mTorr O_2; b – 5 h at 800°C in air; c – 10 h at 800°C in air; d–2 h at 750°C, 1 mTorr O_2, e – 5 h at 750°C, 1 mTorr O_2 [279]

which leads to a band gap shrinking [287]. Sernelius et al. calculated the combined many-body and Burstein-Moss effect for highly-doped ZnO as a function of the carrier concentration N [287, 288]. Their theoretical dependence $E_g(N)$ is displayed in Fig. 7.27 together with experimental data from Jin et al. [284] as well as of Ye et al. [289]. It is interesting to note that the widening of the optical band gap is about the same for films with comparable carrier concentrations, independently if it were deposited by MOCVD (Ye et al.) or by reactive magnetron sputtering (Jin at al.). For films, which are doped up to carrier concentrations of $10^{21}\ cm^{-3}$, typical for transparent conductive electrodes, for instance in flat panel displays or thin film solar cells, the shift of the optical band gap reaches 0.6 eV. The old theory of the Burstein-Moss shift [285], which takes into consideration only the shift of the Fermi level into the conduction band, does not correctly describe the experimental data (see Fig. 7.27, dashed line). Taking into account also the shrinking of the ZnO band gap by many-body effects due to the high density of donors, the experimental results can be fitted very well [288].

In thin films another effect occurs: The formation of exponential band tails, which are manifested by a decreasing slope of the transmission edge in the UV spectral range (see Fig. 7.26). These band tails are caused by defects like donors, other point defects, dislocations or stacking faults [290]. Srikant and Clarke have demonstrated this effect for epitaxial ZnO films grown by pulsed laser deposition on sapphire substrates and silica [279]. Fitting the absorption edges by the expression:

$$\alpha(h\nu) = AE_0^{3/2}\exp\left(\frac{h\nu}{E_0}\right), \tag{7.18}$$

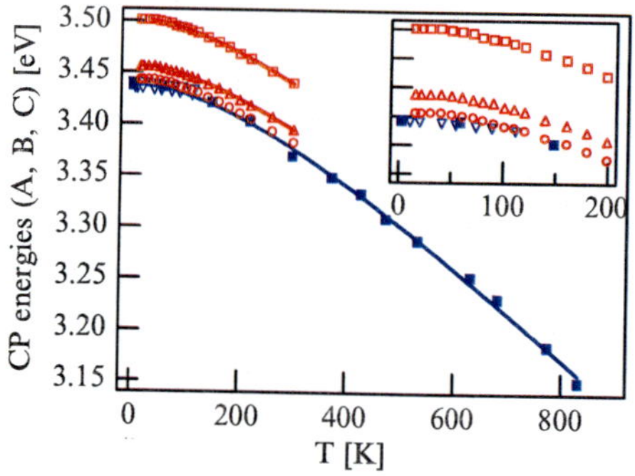

Fig. 7.28 Critical point energies $E_{A,B,C}$ (Δ, ○, □) of a ZnO single crystal and E_A of an epitaxial ZnO film on sapphire (■, ∇) as a function of the temperature [270, 276]. While the c-ZnO data were measured by photoreflectance spectroscopy, the epitaxial film was analyzed by photoluminescence measurements (∇) and spectral ellipsometry (■). The *lines* are fits with the Varshni expression (7.19). The *inset* is an extended view for T ≤ 200 K

where E_0 characterizes the width of the band tails, Srikant and Clarke determined the E_0 values for differently prepared films, which were in the range of 30–150 meV, the highest values measured for polycrystalline films on silica. This has to compared with E_0 = 10–25 meV for bulk single crystal GaAs [291]. Obviously, the higher defect density of the ZnO films leads to these extended band tails, displayed in Fig. 7.27b for ZnO/c-plane sapphire films [279]. Heating these films up to 10 h in air, decreases the band tail energy from 65 to 31 meV, due to the annealing of point defects.

The band gap of a semiconductor decreases with increasing temperature, which is caused by the increasing lattice constants, due to the thermal expansion. Figure 7.28 shows ZnO critical point energies $E_{A,B,C}$ by Bundesmann et al. [270] and Ozaki et al. [276], measured in the temperature range from 2 to 830 K. Despite the small difference in the absolute values, the temperature-dependence of both data sets is very similar. The transition from the B-valence band to the conduction band constitutes the lowest band gap in ZnO (E_B). By fitting the photoreflectance spectra, also the excitonic transition energies $E_{A,B,Cex}$ for the three valence bands can be determined, from which the exciton binding energies $E_{xb}{}^{A,B,C} = E_{A,B,C} - E_{A,B,Cex}$ were calculated (see Table 7.10, [276]). For comparison, data of other authors are given.

For the epitaxial ZnO films only one exciton binding energy is given, since the used analysis method (spectroscopic ellipsometry) allows no separation of the different valence bands.

The $E_g(T)$-curves are usually fitted by the empirical Varshni equation [294]:

$$E_{A,B,C}(T) = E_{A,B,C}(0) - \alpha_V T^2/(T + \theta_D), \tag{7.19}$$

which yields the fitting parameters, listed in Table 7.11. The parameter θ_D is thought to be closely related to the Debye temperature, but otherwise the physical background is unclear. Therefore, from a theoretical point of view, the analytical expression, proposed by Pässler [295]:

Table 7.10 Free exciton binding energies $E_{xb}^{A,B,C}$ determined for ZnO single crystals and an epitaxial film

Author (year)	Sample (method)	E_{xb}^{A} [meV]	E_{xb}^{B} [meV]	E_{xb}^{C} [meV]	Method (temperature)	References
Thomas (1960)	c-ZnO (CVT)		59		Absorption (4.2 K)	[292]
Liang (1968)	c-ZnO (CVT)	62.5	66.8	61.0	Absorption (4.2 K)	[293]
Ozaki (2003)	c-ZnO (CVT)	65	68	63	Photoreflectance (15–300 K)	[276]
Bundesmann (2008)	epi-ZnO (PLD)	60			Spectral ellipsometry (300 K)	[270]

Table 7.11 Fit parameters of the $E_{A,B,C}$ (T)-dependences according to the Varshni and the Pässler formulas (7.19, 7.20) for a ZnO single crystal and an epitaxial film [294, 295]

Author (year)	Sample (method)	$E_{A,B,C}(0)$ [eV]	α_V [meV/K]	θ_D [K]	α_P [meV/K]	θ_P [K]	p	References
Ozaki (2003)	c-ZnO (CVT)	3.455^{A}	0.65	660	0.3	215	2.8	[276]
		3.441^{B}						
		3.499^{C}						
Bundesmann (2008)	Epi-ZnO (PLD)	3.438^{B}	0.53	470	0.54	1,060	1.65	[270]

$$E_{A,B,C} = E_{A,B,C}(0) - \frac{\alpha_P \theta_P}{2}\left[\sqrt[p]{1 + \left(\frac{2T}{\theta_P}\right)^p} - 1\right]. \tag{7.20}$$

is better suited, since it accounts for the band gap shrinking in the electron-phonon spectral function. In the limit of $T \to \infty$ the band gap decreases linearly with T (compare Fig. 7.28), giving $d\,E_{A,B,C}/dT = -\alpha_P$. The exponent p in (7.20) describes the T-dependence of the electron-phonon spectral function and θ_P corresponds approximately to the phonon temperature. Ozaki et al. used both expressions for fitting, which yielded equally good agreement [276].

More detailed reviews on optical properties of ZnO were given recently by Bundesmann et al. [270], Özgür et al. [15] and Klingshirn [296, 297], the latter authors reporting also on photoluminescence and stimulated emission of ZnO and its compounds.

7.7 Structural and Morphological Properties of Zinc Oxide Films

Zinc oxide, deposited by plasma-assisted processes, i.e., magnetron sputtering or pulsed-laser deposition (even at room temperature), exhibits a very pronounced (001)-texture, i.e., the c-axis of almost all (hexagonal) zincite crystallites is oriented perpendicular to the substrate surface [298–300]. But also spray pyrolysis of indium-doped ZnO films leads to a preferred (001)-orientation of the films [301]. However, the extend of the texture, the grain size and the mechanical stress in the films depend significantly on the growth conditions, as shown for reactive magnetron sputtering in Fig. 7.29, which displays the (002)-diffraction peak intensities for different oxygen partial pressures during RF sputtering. With rising oxygen partial pressure the peak intensity decreases, the width of the peaks broadens and the peak location is shifted to smaller 2θ values compared to the reference peak of strain-free zincite powder.

These variations are due to:

- Different grain sizes (peak broadening, reduced intensity)
- Internal compressive strain (peak shift)

Similar variations have been found for dc reactive magnetron sputtering. The grain size and the strain derived from the (002)-peaks are shown in Fig. 7.29b in dependence on the relative oxygen flow (and hence of the oxygen partial pressure). Interestingly, the layers with the lowest resistivities (marked by an vertical arrow) exhibit the largest grains with the lowest strain. Tentatively, this can be explained by the substitutional built-in of the Al dopand atoms onto zinc lattice places, thus leading to the most efficient doping of the films. At higher (and) lower oxygen concentrations in the sputtering atmosphere, alumina (Al_2O_3) is formed, which segregates as a separate phase, thus introducing compressive strain in the films [302].

By pole figure measurements the preferential orientation of thin films can be quantified [303]. Typical pole figures of the ZnO (002)-reflex of a low resistant film (a) and a fully oxidized film (b) prepared by reactive RF sputtering from a Zn: Al2wt% target are shown in Fig. 7.30. The narrow peak around the tilt angle $\chi = 0°$ exhibits rotational symmetry, which means that the crystallites have their

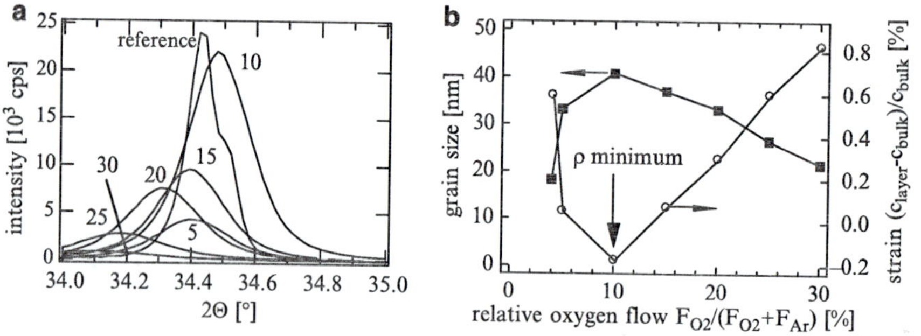

Fig. 7.29 (**a**) (002)-peak intensities of ZnO:Al films reactively sputtered by RF plasma excitation. The numbers at the peaks denote the relative oxygen flow $F_{O2}/(F_{O2} + F_{Ar})$ (in %). The reference peak was measured for a strain-free zincite powder. (**b**) Grain size and strain of the layers from (**a**) versus the relative oxygen flow $F_{O2}/(F_{O2} + F_{Ar})$ [299]. The *vertical arrow* marks the resistivity minimum which coincides with the strain minimum and the grain size maximum (compare also the structure-phase zone model, Fig. 7.8) [87]

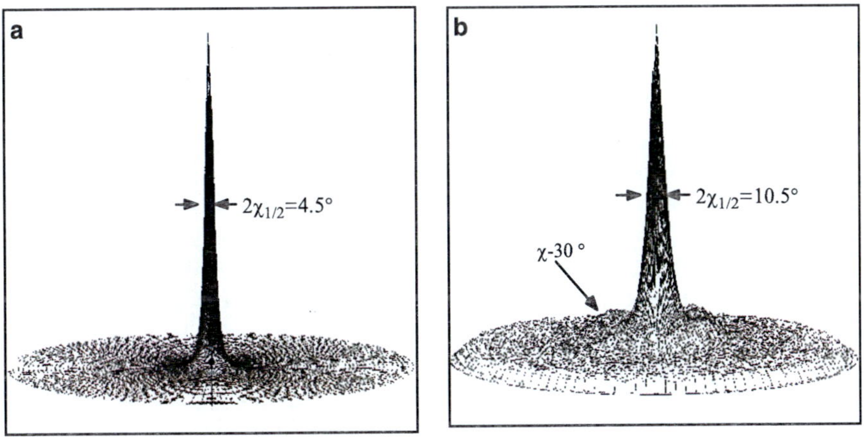

Fig. 7.30 Pole figures of the (002)-ZnO reflex in 3D-representation (in polar coordinates) as a measure of the texture of RF-magnetron sputtered ZnO films, (**a**) sputtered at $F_{O2}/F_{total} = 10\%$ (resistance minimum) and (**b**) deposited at $F_{O2}/F_{total} = 30\%$ (insulating films). Sputtering parameters: ZnAl2wt% target, $P_{rf} = 100$ W, $p_{total} = 0.3$ Pa [299]

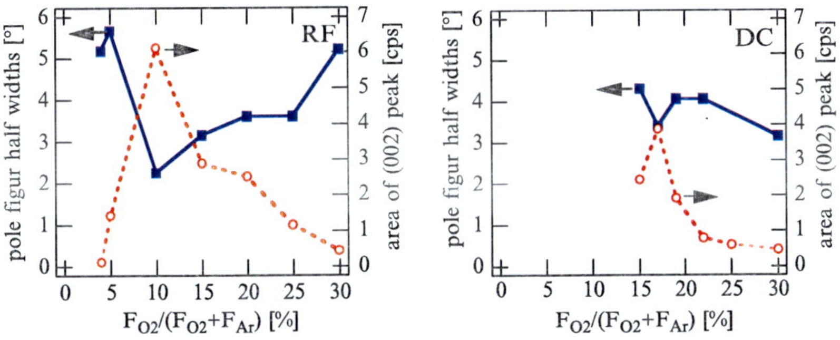

Fig. 7.31 Half widths (■) of the texture distributions and (002) peak area (o) in dependence on the oxygen flow (relative to the total flow) for DC and RF sputtered ZnO:Al films [299]

c-axis perpendicular to the substrate with a statistical mosaic spread around this axis. The fully oxidized film (Fig. 7.30b) displays besides the narrow peak at $\chi = 0°$ a lower intensity ring at $\chi \approx 30°$, which points to the fact that the crystallites are less textured in these films. The intensity ring at about 30° in Fig. 7.30 is caused by crystallites of the (101) orientation, since the angle between the (001) and the (101) planes of zinc oxide is 32°. This is in agreement with investigations of Kluth et al. [304]. The narrowest distributions are observed for the layers with the lowest resistances. This is displayed in Fig. 7.31, where the χ-halfwidths as a measure of the texture are shown in dependence on the oxygen gas flow relative to the total gas flow together with the areas of the (002)-peak. For an oxygen portion in the process gas of 10% (RF-sputtering) the halfwidth is 2.5°. Increasing or decreasing the oxygen partial pressure leads to an increase of the halfwidth. This behavior of the $\chi_{1/2}$-values is mirrored in the dependence of the (002)-peak areas. The most preferentially oriented films have the largest intensities of the (002)-line, as expected.

The texture, as quantified by pole figure measurements or from the (002)-peak area, depends on the deposition conditions, for instance on the oxygen partial pressure, the substrate temperature, the ion assistance or on the addition of dopants.

As a last example of the influence of the deposition conditions on structural properties of zinc oxide films the mechanical stress was measured for doped and undoped films. In microelectronics and thin film technology it is well known that mechanical stress influences not only the mechanical stability of thin film structures (adhesion) but also the electrical parameters. For instance, Ohler et al. could show, that the lattice mismatch at a heterojunction significantly influences the electronic band offsets [305]. It is clear, that different band offsets at a heterojunction lead to different electronic properties of a device made with such a heterojunction. Earlier investigations in our group of zinc oxide contact and window layers on $CuInS_2$ solar cells gave hints that excessive mechanical stress in the zinc oxide films decreases the efficiency of these cells [306]. In Fig. 7.32a the dependence of the ZnO film stress on the sputtering pressure is displayed. The stress was determined ex situ by measuring the lattice constant (XRD) and by evaluating the bending of a thin cantilever coated with the films. All films exhibit compressive stress (negative stress values) which decreases in its absolute value by increasing the sputtering pressure. The undoped zinc oxide films show significantly larger stress values compared to the doped zinc oxide. In difference to the behavior of sputtered metal films, which was extensively investigated by Hoffman and Thornton [307], no transition from compressive to tensile stress is observed for the zinc oxide films. Figure 7.32b shows the dependence of the stress on the excitation conditions of the magnetron discharge. Varying the ratio of DC-to-RF excitation, the stress can be

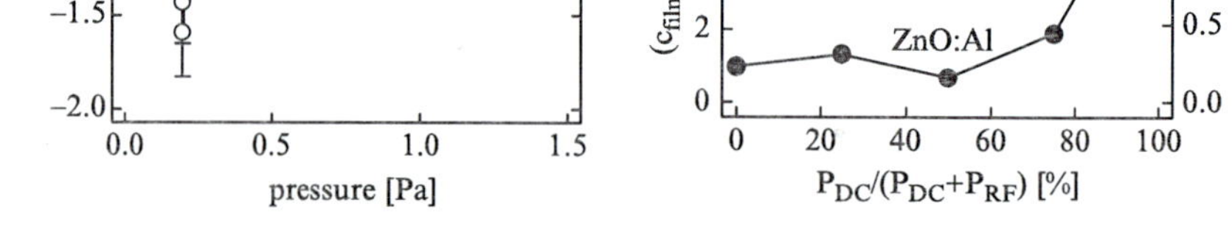

Fig. 7.32 (**a**) Dependence of the film stress σ_{film} on the sputtering pressure for ZnO and ZnO:Al films. Sputtering parameters: $P_{RF} = 30$ W, $d_{ZnO} = 120$ nm, $d_{ZnO:Al} = 350$ nm. (**b**) Strain (*left*) and stress (*right*) of ZnO and ZnO:Al films in dependence of the $P_{DC}/(P_{DC} + P_{RF})$-ratio. Sputtering parameters: $P_{DC} + P_{RF} = 80$ W, $p_{Ar} = 0.8$ Pa [308]

influenced significantly. In a pure DC magnetron discharge the stress is higher than for depositions with at least 50% RF excitation. The higher ion-to-neutral ratio for RF excitation improves the structural quality, in this case the stress state of the films [306]. It is interesting to note, that the grain size of the films does not change, when varying the $P_{dc}/(P_{dc} + P_{rf})$-ratio.

The morphology of ZnO films exhibits a wide variety, see also Fig. 7.9 above. In the following the morphological changes are illustrated for ZnO:Al films deposited onto sapphire and glass substrates. On sapphire, ZnO grows with an epitaxial orientation relative to the single crystalline substrate, as can be inferred from pole figure measurements [250, 309]. Nevertheless, the morphology of ZnO:Al films on sapphire, as displayed in Fig. 7.33, exhibits a columnar-type growth of crystallites. Obviously, the quite large lattice mismatch between ZnO and sapphire (<20%) leads to small-angle grain barriers, inducing the growth of columns. The size and the orientation of the columns depends on the deposition conditions, but also on the orientation of the sapphire substrate, compare Fig. 7.33. Some of the SEM pictures were taken from ZnO films, which were etched in diluted HCl. Due to the anisotropic etching, pyramidal structures have developed, which are different for the films on different sapphire substrates. These structures reveal not only the crystallite morphology but can be used for an efficient light coupling to solar cell absorber films, and are applied, for instance, in a–Si/μc–Si thin film solar cells [105]. It is interesting to note, that the ZnO crystallites on r-plane sapphire show an inclination of about 55° relative to the substrate normal, which is nearly equal to the angle of 53.5° between the (110) and the (012) planes in the wurtzite lattice, which corresponds to the orientation of the ZnO film on r-plane sapphire (Fig. 7.34).

A atomic-resolution analysis was performed by preparing transmission electron microscope cross sections (Fig. 7.34) of ZnO films grown on (001)- or c-plane oriented sapphire and on glass [309]. A clear difference can be inferred from these micrographs: The film on (001) sapphire shows only a slight columnar growth with c-axis orientation, while the film on glass exhibits a clear, distinct columnar grain structure.

At the interface substrate/film a bright thin layer can be seen. The origin of this layer is not clear at the moment. It could be due to the polishing process of the sapphire substrates or an ion bombardment during the initial states of the film growth.

The contrast variation seen in the cross sections of the films on both types of substrates indicates a high density of crystallographic defects, which is in agreement with the results of Gerthsen et al. [218, 310].

7.8 $Zn_{1-x}Me_xO$ Alloys

Zinc oxide can be alloyed with a variety of other elements to tailor the band gap [282, 311]. Earlier investigations of mixed crystals with ZnO were mainly devoted to catalytic powders [312], to varistors [67, 313] or to ZnO mixtures for medical applications.

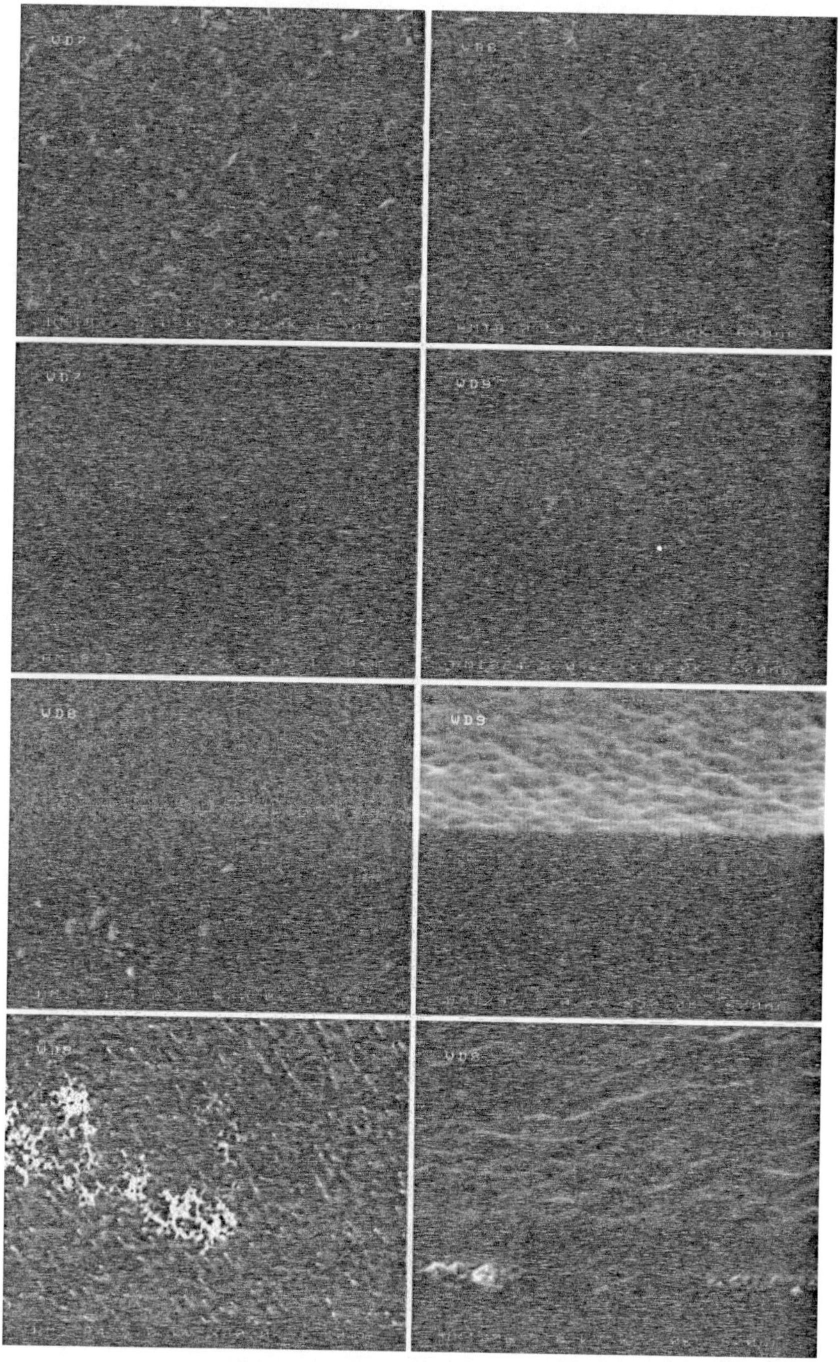

Fig. 7.33 Secondary electron micrographs (SEM) of epitaxial ZnO films on c-plane sapphire (*upper two rows*) and r-plane sapphire (*lower two rows*). On the *left side* the ZnO films can be seen in plain view, while the pictures on the *right* show the fractured sapphire/ZnO and glass/ZnO interfaces

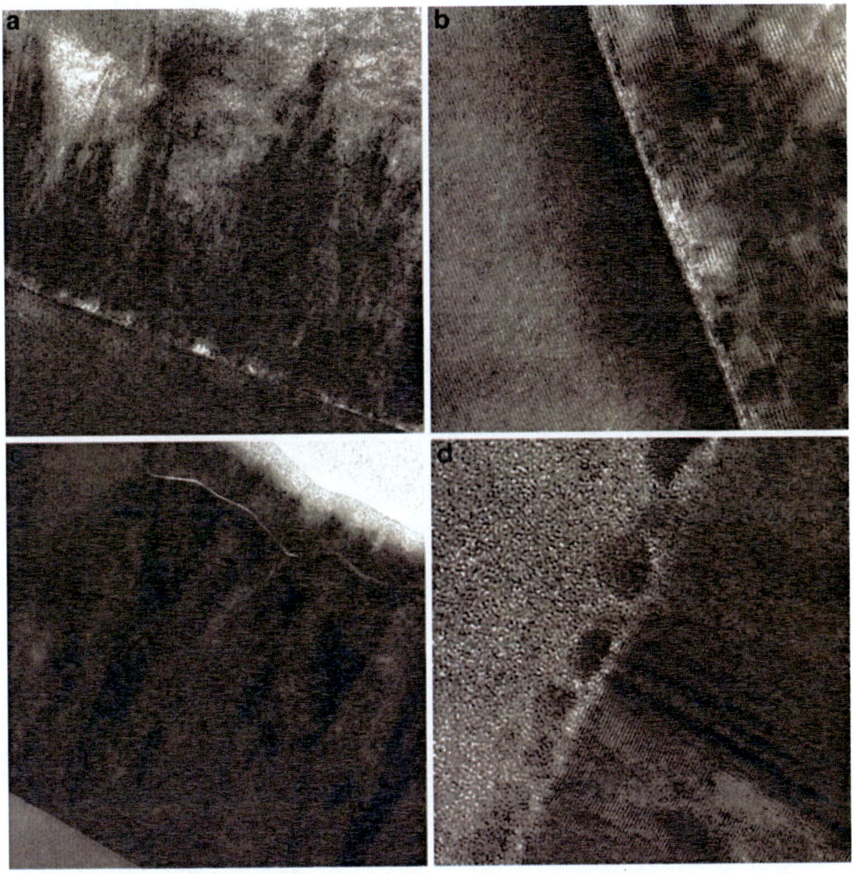

Fig. 7.34 Cross sectional transmission electron (XTEM) micrographs of epitaxial ZnO films on sapphire (*upper row*) and glass (*lower row*). On the *left side* the grain structure of the ZnO films can be seen, while the two pictures on the *right* show the interfaces sapphire/ZnO and glass/ZnO in atomar resolution

Table 7.12 Band gap energy E_g, electron affinity EA, heat of formation ΔH_f and the ionic radius of the metal of the earth alkali oxides in comparison to that of ZnO

Oxide	E_g [eV]	EA [eV]	ΔH_f [eV]	r_i (M^{2+}) [pm]
BeO	10.6	1.5	6.3	27
MgO	7.8	1.2	6.2	57
CaO	7.0	1.0	6.6	100
SrO	4.1	1.0	6.1	118
BaO	3.9	1.0	5.8	135
ZnO	3.4	1.7	3.6	60

For the optoelectronic use of ZnO alloys it is essential which oxides can be used to modify its intrinsic band gap. As can be inferred from the band gaps of the group II (earth-alkaline) oxides (Table 7.12), alloying of ZnO with Be, Mg and Ca could be used for extending the band gap. Barium oxide has a very low solubility in zinc oxide [314], and has been used occasionally for varistor applications, where a segregation layer between ZnO crystallites is essential for its function [313].

Magnesium is best suited, since its ionic radius is quite comparable to that of zinc. Beryllium was only rarely used up to now, due to its poisoning properties. Recently, Ryu et al. reported beryllium zinc oxide mixtures $Zn_{1-x}Be_xO$ up to $x = 0.8$, which allows a continuous band-gap tailoring from $E_g = 3.4$ to 10.6 eV [315].

Ohtomo et al. were the first, who prepared $Zn_{1-x}Mg_xO$ ($0 \leq x \leq 0.46$) films by pulsed laser deposition (PLD) onto (001) sapphire substrates at a substrate temperature of about 600°C [39]. Up to $x = 0.33$ the films showed a single-phase wurtzite structure. For larger Mg contents MgO precipitations were observed. Since the equilibrium solubility of MgO in ZnO is only 4 mol%, the films constitute a metastable wurtzite structure. The films were grown epitaxially with the c-axis perpendicular on the substrates and an in plane-orientation $Zn_{1-x}Mg_xO$ (1–10)||α-Al_2O_3(110), which is the same as for ZnO or GaN on sapphire. From optical transmission measurements, the band gaps of the $Zn_{1-x}Mg_xO$ films were calculated, which are displayed versus x in Fig. 7.35a. It can be seen that E_g increases linearly ($E_g = 3.32 + 2.2x$ [eV]) with x up to $x = 0.36$, which corresponds nearly to the solubility limit of MgO in these films. This means, that by alloying with magnesium the band gap of $Zn_{1-x}Mg_xO$ can be extended up to $E_g \approx 4.15$ eV, thus allowing to prepare electronic devices with band gap offsets up to 0.85 eV [39]. For comparison band gap values of other authors have been added in Fig. 7.35a. By photoluminescence (PL) analysis at a temperature of 4 K, Ohtomo et al. could show, that also ultra-violet emitting $Zn_{1-x}Mg_xO$ films were prepared with an emission energy up to $E_{PL} = 3.87$ eV.

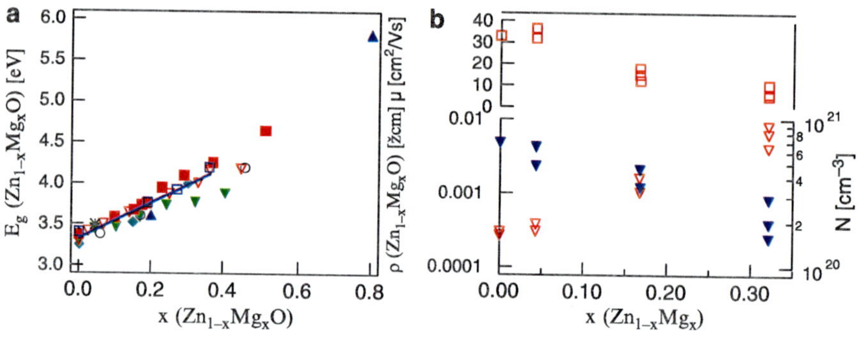

Fig. 7.35 (a) Band gap energies of $Zn_{1-x}Mg_xO$ films as a function of the magnesium content. The E_g data in (a) are from the following sources: ∇ – Ohtomo et al. [39], □ – Teng et al. [317], ◆ – Glatzel et al. [318], ■ – Schmidt-Grund et al. [319], ✳ – Ellmer and Vollweiler [233], ▲ – Kang et al. [320], ▼ – Matsubara et al. [321], ○ – Minemoto et al. [322]. The *line* is a linear fit to the data of Ohtomo et al. (see text). (**b**) Resistivity (∇), carrier concentration (▼) and mobility (□) of doped $Zn_{1-x}Mg_xO$ films as a function of the Mg content x [321]

Recently, Lau et al. prepared cubic $Zn_{1-x}Mg_xO$ films on MgO and $LaAlO_3$ single crystalline substrates by pulsed laser deposition (PLD) [316]. By doping with indium up to concentrations of 10 at.% the films could be made conductive ($\sigma > 100$ S/cm).

Matsubara et al. prepared aluminum-doped $Zn_{1-x}Mg_xO$:Al films by pulsed laser deposition (PLD) [321]. Up to a magnesium content of 15 at.% the resistivity of the films was better than 10^{-3} Ω cm, which means that such films can be used as transparent electrodes with an extended band gap. Increasing the Mg concentration up to $x = 0.33$ leads to a further increase of the resistivity, caused both by a decrease of mobility as well as of the carrier concentration. Resistivity, carrier concentration and mobility of the films of Matsubara et al. are displayed in Fig. 7.35b.

A prospective field for the application of $Zn_{1-x}Mg_xO$:Al films could be for window and contact layers in thin film solar cells, where a better band edge alignment is needed, for instance for wide-band gap chalcopyrit absorbers like $CuGaSe_2$ or $CuInS_2$ [323]. In such applications an upward shift of the conduction band edge is required, which can be fulfilled by $Zn_{1-x}Mg_xO$(:Al), as was shown recently by Rao et al. [324].

By *in situ* photoelectron spectroscopy Rao et al. measured the band alignment of ZnO and $Zn_{0.85}$ $Mg_{0.15}O$ films relative to the valence band of CdS. In this way it was shown that the wided band gap of $Zn_{0.85}$ $Mg_{0.15}O$ is mainly caused by an upward shift of $\Delta E_g = 0.3$ eV solely of the conduction band edge.

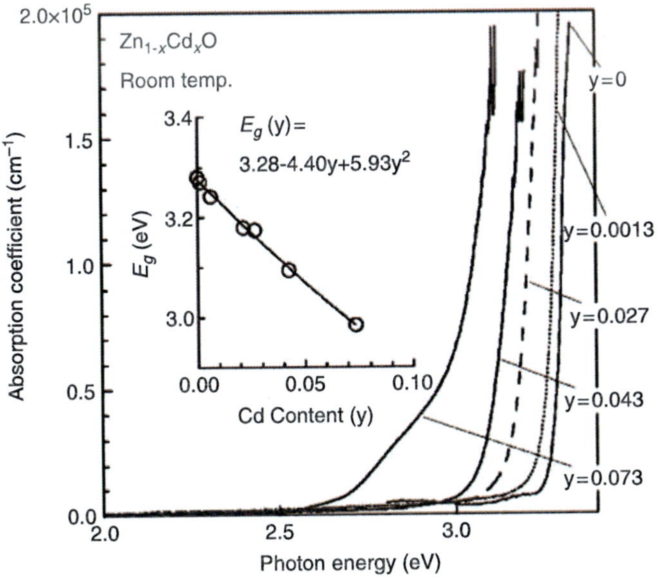

Fig. 7.36 Absorption edges of $Zn_{1-y}Cd_yO$ epilayers as a function of the cadmium content y, measured at room temperature. The *inset* shows the dependence of the band gap, derived from the absorption curves [311]

In order to decrease the band gap of zinc oxide cadmium can be mixed to ZnO. This has also been done by pulsed laser deposition by the group of Kawasaki [311], who already prepared $Zn_{1-x}Mg_xO$ films. Since CdO has a significantly higher vapor pressure compared to ZnO, the Cd concentration in the $Zn_{1-x}Mg_xO$ films was up to four times lower than in the target, when the deposition was performed at a substrate temperature of 750°C.

Due to the fact that CdO exhibits the rock salt crystal structure and the large ion radius of cadmium (74 pm), the solubility of CdO in ZnO in thermodynamical equilibrium is only 2%. On the other hand, Makino et al. and Misra et al. could show that under non-equilibrium conditions, like in pulsed-laser deposition, up to 8% Cd can be built-in into $Zn_{1-x}Cd_xO$ [282, 311]. The band gap of $Zn_{1-x}Cd_xO$ films could be reduced by about 0.4 eV for a cadmium content of $x = 0.08$ [282]; it depends linearly on the cadmium concentration (see Fig. 7.36).

Recently, Sadofev et al. used radical-source molecular beam epitaxy to deposit $Zn_{0.86}Mg_{0.14}O$ and $Zn_{0.84}Cd_{0.16}O$ films on single crystalline ZnO substrates [325]. These authors used significantly lower growth temperatures (350°C for ZnMgO and ≈150°C for ZnCdO), which could be the reason for the higher cadmium content in the $Zn_{1-x}Cd_xO$ films, compared to PLD films.

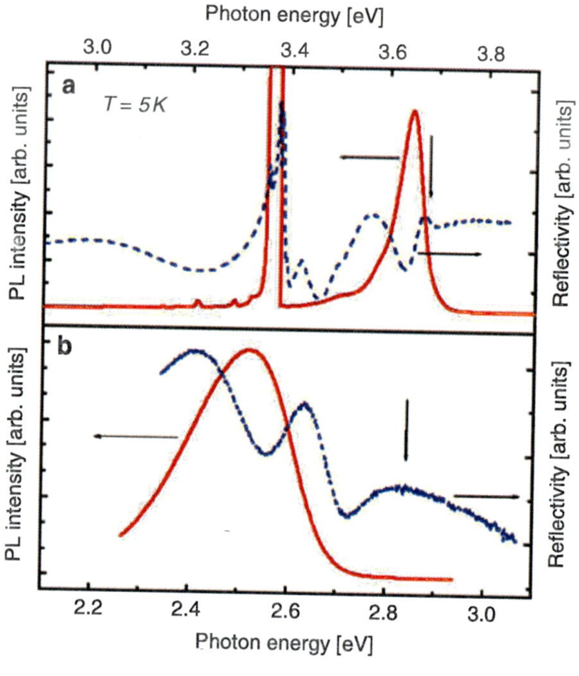

Fig. 7.37 Photoluminescence (*full lines*) and reflectivity at 5 K (*dashed lines*) spectra of (**a**) $Zn_{0.86}Mg_{0.14}O$ and (**b**) $Zn_{0.84}Cd_{0.16}O$ epitaxial films on ZnO single crystals. The *vertical arrows* mark the excitonic band gaps, calculated from the R spectra [325]

Though the crystalline quality of the $Zn_{1-x}(Mg)Cd_xO$ films on ZnO was much better (rocking curve widths as low as ≈20 arc sec) compared to heteroepitaxial films on sapphire (rocking curve widths ≈ 600 arc sec), the photoluminescence (PL) properties for both types of films were similar (see Fig. 7.37) [325]. From Fig. 7.37 it can be seen that the PL peak energy is shifted to lower energies compared to the excitonic band gaps, shown by vertical arrows, which is, according to the authors, caused by localization of the recombining carriers (higher for the $Zn_{1-x}Cd_xO$ film). Since neither the Stokes shift nor the width of the PL peaks is significantly different from films grown on sapphire, it is concluded by Sadofev et al., that the electronic localization is an intrinsic property of the $Zn_{1-x}(Mg)Cd_xO$ films [325].

7.9 Comparison with Other TCO Materials

Zinc oxide belongs, like cadmium, indium and tin oxide (CdO, In_2O_3, SnO_2), to the class of transparent conductive oxides which were investigated for a long time, beginning with the first reports on electrical properties of zinc oxide in 1912 [3], of cadmium oxide in 1971 [326], of indium oxide in 1954 [327], and of tin oxide in 1972 [328]. These oxides can be doped up to carrier concentrations of the order of 10^{21} cm^{-3} while still being transparent in the visible and near-infrared wavelength region. Some important data of these four TCO materials are summarized in Table 7.13. According to these data, cadmium oxide (CdO) exhibits the highest mobility and the lowest resistivity of 3×10^{-5} Ω cm. A disadvantage of CdO is the relatively low band gap of only 2.3 eV.

If one compares the main data of zinc oxide and indium (tin) oxide (ITO), it is not clear, why ITO exhibits a lower resistivity compared to ZnO. Hosono et al. argue, that indium oxide is characterized by the derivation of the conduction band from indium, i.e., metal s-states [329]. The isotropy of the s-like wave functions, Hosono et al. write, guarantees an overlap of the conduction band states, even if the order of the crystal lattice is disturbed or the structure is amorphous. This argument is supported by the fact, that ITO exhibits a quite high mobility even for amorphous films [330, 331]. On the other hand, the conduction band of ZnO is also derived from s-like metal (Zn) wave functions (see Sect. 7.2.2), hence the same reasoning should apply also for ZnO.

Obviously, further fundamental research on transport in TCO materials is needed, as already pointed out in the last review of the present author [71].

Of these oxides only In_2O_3, SnO_2 and ZnO are today of technological importance especially as transparent electrodes in flat panel displays, thin film solar cells low-emissivity architectural glass or heating windows [193]. While indium-tin oxide (ITO) is almost exclusively used for electrodes in flat panel displays, tin oxide and zinc oxide are used in increasing amounts in thin film solar cells, see for instance [18]. Though indium has the lowest availability in the earth crust, it is used

Table 7.13 Comparison of some important material data of the TCO materials CdO, In_2O_3, SnO_2, and ZnO

Property (unit)	CdO	In_2O_3	SnO_2	ZnO
Mineral name	–	–	cassiterite	zincite
Average amount of the metal in the earth's crust [ppm]	0.3	0.1	40	132
Band gap E_g (300 K) [eV]	2.3	2.7 (indir) 3.75 (dir)	3.6 (dir)	3.4 (dir)
Effective electron mass m^*/m_e	0.15/0.11	0.35	$\parallel c$: 0.23 $\perp c$: 0.3	0.28
Lattice mobility [cm^2/V s]	620 [332]	210	255	200
Minimum resistivity [Ω cm]	$\approx 3 \times 10^{-5}$ [332]	$\approx 1 \times 10^{-4}$	$\approx 3 \times 10^{-4}$	$\approx 2 \times 10^{-4}$

in flat panel almost exclusively, due to the fact that the resistivities are the lowest among the TCO (except CdO) materials and its good chemical etchability.

7.10 Research Opportunities

P-type ZnO. The strong research efforts in the last 10 years to develop ZnO as a future electronic material focused on the quest for p-type conducting zinc oxide [13, 17]. Though significant progress has been achieved, further intensive research is needed to overcome the difficulties with current p-type doping of ZnO:

- Higher carrier concentrations ($>10^{19}$ cm^{-3})
- Easier built-in of p-type dopants (N, P, or As)
- Persistence of the p-type conduction for long times [159]

Modulation doping. As explained in Sects. 7.4 and 7.5 the mobility of degenerately doped ZnO is limited by ionized impurity scattering. This physical limit is valid for a homogeneously doped semiconductor, where the doping function and the carrier transport occur in the same region of the semiconductor. However, since about 30 years it is well known that the mobility can be increased by the so-called modulation doping method, introduced by Dingle et al. [333] for GaAs/$Ga_{1-x}Al_xAs$ superlattice structures (for a review see [334]).

In a modulation doped structure the doping and the transport functions in a semiconducting material are separated by stacking lowly and heavily doped films with slightly different bandgaps, schematically depicted in Fig. 7.38. While the heavily doped films provide the charge carriers their transport occurs in the lowly doped layers of the stack which are not subjected to the ionized impurity scattering limitation of the mobility.

To our knowledge the modulation doping principle was applied only to lowly-doped semiconductors with carrier concentrations of about 10^{17} cm^{-3} and it is still open if this principle works also for very high carrier concentrations above

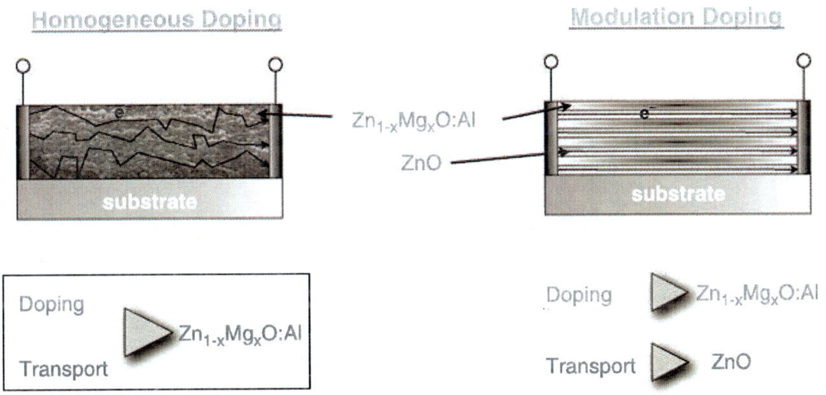

Fig. 7.38 Schematic model of the carrier transport in a homogeneously doped ZnO:Al film (*left*) and in a modulation-doped film, consisting of alternating doped $Zn_{1-x}Mg_xO$:Al and undoped ZnO films (*right*). The dots represent the ionized impurities in the films which act as scattering centers. The *zigzag lines* symbolize the scattering paths of electrons, while the *thin straight lines* are the electron path without scattering

10^{20} cm^{-3}. Rauf [335] was the first who reported very high mobilities in highly-doped ITO films which he claimed to be due to the modulation doping effect. Recently, Edahiro et al. reported the formation of a 2-dimensional electron gas at the interface of a ZnMnO(10 nm)/ZnO(30 nm), nominally undoped [336]. They observed Shubnikov-de Haas oscillations at T = 1.85 K for magnetic fields larger than 3 T, a strong indication of a 2D-electron gas.

In order that the modulation doping principle works, the thickness of the multilayers has to be in the range of nanometers, i.e., comparable to the Debye length in the doped semiconductor films $L_D = \sqrt{(\varepsilon\varepsilon_0 kT/e^2N)}$ in order to achieve a sufficient carrier transfer from the doping to the transport layer (see [337]). Cohen and Barnett [338] have performed simulations of the potential distributions of ZnO/ZnMgO/ZnMgO:Al multilayers and of the electron density distributions which are generated due to charge transfer from the doping layer (ZnMgO:Al) to the transport layer (ZnO). From these calculations the authors conclude that modulation-doped structures with this layer system are possible with mobilities as high as 145 $cm^2/V\,s$ at an average carrier density of 3.8×10^{18} cm^{-3}. The modulation-doping approach can be used for enhancing TCO properties, for instance for the design of transparent electronics (see Sect. 7.2.4). However, the resistivity of modulation-doped layers would be limited to 1.5×10^{-3} Ω cm since the transferred maximum electron sheet density is only about 10^{13} cm^{-2}. Recently, it was shown by Ohtomo and Hwang that a 2-dimensional electron gas exists also in highly-doped $Sr_{1-x}La_xTiO_3$ layers [339], making it plausible that it can be engineered also for ZnO or other TCO materials.

Transport in polycrystalline and amorphous films. In order to get further insight into the charge carrier transport mechanisms such as:

- Charge state of the donors in intrinsically and extrinsically doped ZnO

- Dopant clustering
- Band structure of the conduction band of heavily doped zinc oxide
- Formation of secondary (for instance Al_2O_3) and ternary (for instance gahnite, $ZnAl_2O_4$) phases in ZnO and its relation to doping, electrical and optical transport investigations of heavily doped zinc oxide single crystals and/or epitaxial films have to be performed.

A very exciting new question arose in the last years. In striking difference to covalently bonded semiconductors (Si, Ge, Se, C), where the amorphous phase exhibits a carrier mobility lower by two to three orders of magnitude (for instance amorphous silicon, a-Si:H) amorphous transparent oxides show an electron mobility comparable to that of the corresponding crystalline materials [32, 340]. Further investigations are necessary to clarify the transport processes in amorphous TCOs.

(Reactive) magnetron sputtering. As mentioned in Sect. 7.3 magnetron sputtering is a very promising large-area deposition technique, already used for the deposition of ZnO (and other TCOs) for thin-film solar cells or low ε glass. Nevertheless, there are still open questions which have to be solved:

- Stability of the reactive magnetron sputtering process
- Hysteresis problem for large sputtering targets
- Avoidance of negative oxygen ion bombardment of the growing film

References

1. W. Haidinger, "Handbuch der bestimmenden Mineralogie", (Wien, 1845).
2. P. Dunn, "Franklin and Sterling Hill: the World's Most Magnificent Mineral Deposits", (Excalibur Mineral Company, Peekskill, NY, 2004).
3. A. A. Somerville, Phys. Rev. I, 34(1912)311.
4. O. Fritsch, Ann. Phys., [5], 22(1935)375.
5. A. R. Hutson, Phys. Rev. Lett., 4(1960)505.
6. F. S. Hickernell, J. Appl. Phys., 44(1973)1061.
7. G. Heiland, E. Mollwo and F. Stöckmann, Solid State Phys., 8(1959)191.
8. W. Hirschwald, in "Current Topics in Material Science", Ed.: E. Kaldis (North-Holland, Amsterdam, 1981), p. 144.
9. K. L. Chopra and S. Major, Proc. Electrochem. Soc., 83(1983)346.
10. C. Klingshirn and H. Haug, Phys. Rep., 70(1981)315.
11. G. Fasol, Phys. Biol., 51(1995)925.
12. S. J. Pearton and C. Kuo, MRS Bull., 22(1997)17.
13. C. Klingshirn, M. Grundmann, A. Hoffmann, B. Meyer and A. Waag, Phys. J. 5(2006)33.
14. B. K. Meyer, J. Sann, D. M. Hofmann, C. Neumann and A. Zeuner, Semicond. Sci. Technol., 20(2005)S62.
15. U. Özgür, Y. I. Alivov, C. Liu, A. Teke, M. A. Reshchikov, S. Dogan, V. Avrutin, S.-J. Cho and H. Morkoc, J. Appl. Phys., 98(2005)041301.
16. "Zinc Oxide-A Material for Micro- and Optoelectronic Applications"; edited by N. H. Nickel and E. Terukov (Springer, Dordrecht, 2005).
17. "Zinc Oxide Bulk, Thin Films and Nanostructures"; edited by C. Jagadish and S. Pearton (Elsevier, Amsterdam, 2006).

18. "Transparent Conductive Zinc Oxide: Basics and Applications in Thin Film Solar Cells"; edited by K. Ellmer, A. Klein and B. Rech (Springer, Berlin, 2008).
19. W. L. Bragg, Phil. Mag., 39(1920)647.
20. G. Heiland, P. Kunstmann and H. Pfister, Z. Phys., 176(1963)485.
21. A. N. Mariano and R. E. Hanneman, J. Appl. Phys., 34(1963)384.
22. R. G. W. Wyckoff, "Crystal Structures", (Interscience, New York, 1963).
23. S. Desgreniers, Phys. Rev. B, 58(1998)14102.
24. S. O. Kucheyev, J. E. Bradby, J. S. Williams, C. Jagadish and M. V. Swain, Appl. Phys. Lett., 80(2002)956.
25. L. Gerward and J. S. Olsen, J. Synchrotron Radiat., 2(1995)233.
26. J. Serrano, A. H. Romero, F. J. Manjon, R. Lauck, M. Cardona and A. Rubio, Phys. Rev. B, 69(2004)094306.
27. E. Mollwo, in "Landoldt-Börnstein. Zahlenwerte und Funktionen aus Naturwiss. u. Technik. Neue Serie", Ed.: O. Madelung, M. Schulz and H. Weiss (Springer-Verlag, Berlin u.a., 1982), p. 35.
28. D. Vogel, P. Krüger and J. Pollmann, Phys. Rev. B, 54(1996)5495.
29. P. Schröer, P. Krüger and J. Pollmann, Phys. Rev. B, 47(1993)6971.
30. M. Usuda, N. Hamada, T. Kotani and M. van Schilfgaarde, Phys. Rev. B, 66(2002)125101.
31. P. Erhart, K. Albe and A. Klein, Phys. Rev. B, 73(2006)205203.
32. K. Nomura, H. Ohta, A. Takagi, T. Kamiya, M. Hirano and H. Hosono, Nature, 432(2004)488.
33. A. K. Kulkarni and S. A. Knickerbocker, J. Vac. Sci. Technol. A, 14(1996)1709.
34. S. A. Knickerbocker and A. K. Kulkarni, J. Vac. Sci. Technol. A, 14(1996)757.
35. H. Odaka, Y. Shigesato, T. Murakami and S. Iwata, Jpn. J. Appl. Phys., 40(2001)3231.
36. O. N. Mryasov and A. J. Freeman, Phys. Rev. B, 64(2001)233111/1.
37. J. Robertson, K. Xiong and S. J. Clark, Thin Solid Films, 496(2006)1.
38. U. Rau and M. Turcu, Mat. Res. Soc. Symp. Proc., 763(2003)335.
39. A. Ohtomo, M. Kawasaki, T. Koida, K. Masubuchi, H. Koinuma, Y. Sakurai, Y. Yoshida, T. Yasuda and Y. Segawa, Appl. Phys. Lett., 72(1998)2466.
40. V. A. Coleman, M. Buda, H. H. Tan, C. Jagadish, M. R. Phillips, K. Koike, S. Sasa, M. Inoue and M. Yano, Semicond. Sci. Technol., 21(2006)L25.
41. K. Ellmer and A. Klein, in "Transparent Conductive Zinc Oxide: Basics and Application in Thin Film Solar Cells", Ed.: K. Ellmer, A. Klein and B. Rech (Springer, Berlin, 2008), p. 32.
42. D. F. Anthrop and A. W. Searcy, J. Phys. Chem., 68(1964)2335.
43. D. Schulz, S. Ganschow, D. Klimm, M. Neubert, M. Roßberg, M. Schmidbauer and R. Fornari, J. Cryst. Growth, 296(2006)27.
44. H. A. Wriedt, Bull. Alloy Phase. Diagrams, 8(1987)166.
45. M. W. Wolf and J. J. Martin, Phys. Stat. Sol. (a), 17(1973)215.
46. I. Barin, "Thermochemical Data of Pure Substances", 3rd ed. (VCH, Weinheim, 1995).
47. J. E. Jaffe, J. A. Snyder, Z. Lin and A. C. Hess, Phys. Rev. B, 62(2000)1660.
48. N.-G. Vannerberg, Ark. Kemi, 14(1959)119.
49. F. Decremps, J. Zhang and R. C. Lieberman, Europhys. Lett., 51(2000)268.
50. A. Ashrafi and C. Jagadish, J. Appl. Phys., 102(2007)071101.
51. J. E. Jaffe and A. C. Hess, Phys. Rev. B, 48(1993)7903.
52. P. Erhart, N. Juslin, O. Goy, K. Nordlund, R. Müller and K. Albe, J. Phys. Condens. Matter, 18(2006)186585.
53. K. I. Hagemark and P. E. Toren, J. Electrochem. Soc., 122(1975)992.
54. H. E. Brown, "Zinc Oxide. Properties and Applications", (Int. Lead Zinc Research Org., New York, 1976).
55. J. Plachy, Zinc, http://minerals.usgs.gov/minerals/pubs/commodity/zinc/ (2003).
56. K. Vanheusdalen, C. H. Seager, W. L. Warren, D. R. Tallant and J. A. Voigt, Appl. Phys. Lett., 68(1996)403.
57. A. H. Kitai, Thin Solid Films, 445(2003)367.
58. T. Miyata, Y. Mochizuki and T. Minami, IEICE Trans. Electron., E88-C(2005)2065.

59. N. Saito, H. Haneda, T. Sekiguchi, N. Ohashi, I. Sakaguchi and K. Koumoto, Adv. Mater., 14(2002)418.
60. A. Ballato, in: Proc. 1996 IEEE Ultrasonics Symp., San Antonio, TX, USA 3-6 Nov. 1996, 1996 (IEEE), p. 575.
61. B. Drafts, Sensors, 17(2000)68.
62. J. W. Grate, S. J. Martin and R. W. White, Anal. Chem., 65(1993)987.
63. J. W. Grate, S. J. Martin and R. W. White, Anal. Chem., 65(1993)940.
64. S. Seeger, T. Unold and K. Ellmer, in: Proc. 22nd European Photovoltaic Solar Energy Conference, 3–7 September, 2007, Milano, Italy, 2007.
65. T. Unold, I. Sieber and K. Ellmer, Appl. Phys. Lett., 88(2006)213502.
66. K. Siemer, J. Klaer, I. Luck, J. Bruns, R. Klenk and D. Bräunig, Sol. Energy Mater. Sol. Cells, 67(2001)159.
67. D. R. Clarke, J. Am. Ceram. Soc., 82(1999)485.
68. M. Matsuoka, Jpn. J. Appl. Phys., 10(1971)736.
69. L. M. Levinson and H. R. Philipp, Am. Ceram. Soc. Bull., 65(1986)639.
70. T. Minami, H. Sato, H. Nanto and S. Takata, Jpn. J. Appl. Phys., 24(1985)L781.
71. K. Ellmer, J. Phys. D Appl. Phys., 34(2001)3097.
72. C. Levy-Clement, R. Tena-Zaera, M. A. Ryan, A. Katty and G. Hodes, Adv. Mater., 17(2005)1512.
73. J. F. Wager, Science, 300(2003)1245.
74. E. Scharowsky, Z. Phys., 135(1953)318.
75. D. G. Thomas and J. J. Lander, J. Chem. Phys., 25(1956)1136.
76. G. Bogner and E. Mollwo, J. Phys. Chem. Solids, 6(1958)136.
77. R. A. Laudise, E. D. Kolb and A. J. Caporaso, J. Am. Ceram. Soc., 47(1964)9.
78. D. Rykl and J. Bauer, Kristall und Technik, 3(1968)375.
79. K. Maeda, M. Sato, I. Niikura and T. Fukuda, Semicond. Sci. Technol., 20(2005)S49.
80. J. W. Nielsen and E. F. Dearborn, J. Phys. Chem., 64(1960)1762.
81. M. Hirose, Y. Furuya and I. Kubo, Jpn. J. Appl. Phys., 9(1970)726.
82. J. Nause and B. Nemeth, Semicond. Sci. Technol., 20(2005)S45.
83. I. Gur, N. A. Fromer, M. L. Geier and A. P. Alivisatos, Science, 310(2005)
84. D. C. Look, D. C. Reynolds, J. R. Sizelove, R. L. Jones, C. W. Litton, G. Cantwell and W. C. Harsch, Solid State Commun., 105(1998)399.
85. R. Khanna, K. Ip, K. K. Allums, K. Baik, C. R. Abernathy, S. J. Pearton, Y. W. Heo, D. P. Norton, F. Ren, S. Shojah-Ardalan and R. Wilkins, J. Electr. Mat., 34(2005)395.
86. E. V. Kortunova, P. P. Chvanski and N. G. Nikolaeva, J. Phys. IV France, 126(2005)39.
87. K. Ellmer, J. Phys. D Appl. Phys., 33(2000)R17.
88. J. Steinhauser, S. Faÿ, N. Oliveira, E. Vallat-Sauvain, and C. Ballif, Appl. Phys. Lett., 90(2007)142107.
89. S. Fay and A. Shah, in "Transparent Conductive Zinc Oxide: Basics and Application in Thin Film Solar Cells", Ed.: K. Ellmer, A. Klein and B. Rech (Springer, Berlin, 2007), p. 67.
90. M. Lorenz, in "Transparent Conductive Zinc Oxide: Basics and Application in Thin Film Solar Cells", Ed.: K. Ellmer, A. Klein and B. Rech (Springer, Berlin, 2007), p. 55.
91. T. Yoshida, K. Terada, D. Schlettwein, T. Oekermann, T. Sugiura and H. Minoura, Adv. Mater., 12(2000)1214.
92. I. Mora-Seró, F. Fabregat-Santiago, B. Denier, J. Bisquert, R. Tena-Zaera, J. Elias and C. Lévy-Clément, Appl. Phys. Lett., 89(2006)203117.
93. J. Tornow and K. Schwarzburg, J. Phys. Chem. C, 111(2007)8692.
94. T. Zhang, W. Dong, M. Keeter-Brewer, S. Konar, R. N. Njabon and Z. R. Tian, J. Am. Chem. Soc., 128(2006)10960.
95. B. Szyszka, in "Transparent Conductive Zinc Oxide: Basics and Application in Thin Film Solar Cells", Ed.: K. Ellmer, A. Klein and B. Rech (Springer, Berlin, 2008), p. 49.
96. G. A. Rozgonyi and W. J. Polito, Appl. Phys. Lett., 8(1966)220.
97. G. A. Rozgonyi and W. J. Polito, J. Vac. Sci. Technol., 6(1968)115.

98. K. Ellmer and R. Wendt, Surf. Coat. Technol., 93(1997)21.
99. C. May, R. Menner, J. Strümpfel, M. Oertel and B. Sprecher, Surf. Coat. Technol., 169–170 (2003)512.
100. K. Ellmer and R. Mientus, in: Proc. 4th Int. Symp. on Trends and New Appl. in Thin Films/ 11th Conf. on High Vacuum, Interfaces and Thin Films, Dresden, March, 1994 (DGM Informationsgesellschaft mbH, Oberursel), p. 131.
101. D. Depla, S. Heirwegh, S. Mahieu, J. Haemers and R. D. Gryse, J. Appl. Phys., 101(2007) 013301.
102. Y. Ohtsu, Y. Okuno and H. Fujita, J. Appl. Phys., 73(1993)2155.
103. J. A. Thornton, in "Annual Review of Material Science", Ed.: R. A. Huggins, R. H. Bube and R. W. Roberts (Annual Rev. Inc., Palo Alto, 1977), p. 239.
104. O. Kluth, G. Schöpe, J. Hüpkes, C. Agashe, J. Müller and B. Rech, Thin Solid Films, 442(2003)80.
105. J. Hüpkes, J. Müller and B. Rech, in "Transparent Conductive Zinc Oxide: Basics and Application in Thin Film Solar Cells", Ed.: K. Ellmer, A. Klein and B. Rech (Springer, Berlin, 2008), p. 56.
106. J. Ederth, G. A. Niklasson, A. Hultaker, P. Heszler, C. G. Granqvist, A. R. van Doorn, M. J. Jongerius and B. Burgard, J. Appl. Phys., 93(2003)984.
107. P. S. Clarke, J. W. Orton and A. J. Guest, Phys. Rev. B, 18(1978)1813.
108. A. R. Hutson, Phys. Rev., 108(1957)222.
109. E. Ziegler, A. Heinrich, H. Oppermann and G. Stöver, Phys. Stat. Sol. (a), 66(1981)635.
110. J. J. Lander, J. Phys. Chem. Solids, 15(1960)324.
111. I. B. Kobiakov, Solid State Commun., 35(1980)305.
112. L. N. Demyanets, D. V. Kostomarov and I. P. Kuzmina, Inorg. Mater., 38(2002)171.
113. O. Caporaletti, Solar Energy Mater., 7(1982)65.
114. D. C. Look, J. W. Hemsky and J. R. Sizelove, Phys. Rev. Lett., 82(1999)2552.
115. P. Erhart and K. Albe, Phys. Rev. B, 73(2006)115207.
116. P. Erhart and K. Albe, Appl. Phys. Lett., 88(2006)201918.
117. K. Hauffe and C. Seyferth, "Reaktionen in und an festen Stoffen", (Springer-Verlag, Berlin, 1966).
118. K. Hauffe and J. Block, Z. Phys. Chem., 196(1950)438.
119. D. G. Thomas, J. Phys. Chem. Solids, 9(1958)31.
120. F. A. Kröger, "The Chemistry of Imperfect Crystals. Imperfection Chemistry of Crystalline Solids", 2nd revised Ed. ed. (North-Holland, Amsterdam, 1973).
121. S. Deubler, J. Meier, R. Schiitz and W. Witthuhn, Nucl. Instrum. Methods Phys. Res. B, 63(1992)223.
122. T. Agne, M. Deicher, V. Koteski, H.-E. Mahnke, H. Wolf and T. Wichert, Hyperfine Interact., 159(2004)55.
123. T. Minami, MRS Bull., 25(2000)38.
124. N. Roberts, R.-P. Wang, A. W. Sleight and W. W. Warren, Phys. Rev. B, 57(1998)5734.
125. J. Hu and R. G. Gordon, in: Proc. Evolution of Thin-Film and Surface Structure, Boston, Massachusetts, USA, 1990 (MRS, Pittsburgh, PA), p. 457.
126. A. Sanchez-Juarez, A. Tiburcio-Silver, A. Ortiz, E. P. Zironi and J. Rickards, Thin Solid Films, 333(1998)196.
127. A. Guillen-Santiago, M. de la Olvera, A. Maldonado, A. Reyes and R. Asomoza, Phys. Stat. Sol. (a), 191(2002)499.
128. A. R. Hutson, J. Phys. Chem. Solids, 8(1959)467.
129. P. Wagner and R. Helbig, J. Phys. Chem. Solids, 35(1974)327.
130. K. I. Hagemark and L. C. Chacka, J. Solid State Chem., 15(1975)261.
131. H. von Wenckstern, S. Weinhold, G. Biehne, R. Pickenhain, H. Schmidt, H. Hochmuth and M. Grundmann, Adv. Solid State Phys., 45(2005)263.
132. D. M. Hofmann, A. Hofstaetter, F. Leiter, H. Zhou, F. Henecker, B. K. Meyer, S. O. Orlinskij, J. Schmidt and P. G. Baranov, Phys. Rev. Lett., 88(2002)045504.

133. E. Mollwo, Z. Phys., 138(1954)478.
134. D. G. Thomas and J. J. Lander, in "Halbleiter und Phosphore", Ed.: M. Schön and H. Welker (Vieweg, Braunschweig, 1958), p. 534.
135. N. Nickel, in "Zinc Oxide – A Material for Micro- and Optoelectronic Applications", Ed.: N. Nickel and E. Terukov (Springer, Dordrecht, 2005), p. 145.
136. C. G. van de Walle, Phys. Rev. Lett., 85(2000)1012.
137. F. Ruske, V. Sittinger, W. Werner, B. Szyszka, K.-U. v. Osten, K. Dietrich and R. Rix, Surf. Coat. Technol., 200(2005)236.
138. W. Jakubowski, Acta Phys. Pol. A, 37(1970)409.
139. D. C. Look, Semicond. Sci. Technol., 20(2005)S55.
140. D. J. Binks and R. W. Grimes, J. Am. Ceram. Soc., 76(1993)2370.
141. B. K. Meyer, Appl. Phys. A, 88(2007)119.
142. A. Tsukazaki, A. Ohtomo, T. Onuma, M. Ohtani, T. Makino, M. Sumiya, K. Ohtani, S. F. Chichibu, S. Fuke, Y. Segawa, H. Ohno, H. Koinuma and M. Kawasaki, Nat Mater., 4(2005)42.
143. K. Minegishi, Y. Koiwai, Y. Kikuchi, K. Yano, M. Kasuga and A. Shimizu, Jpn. J. Appl. Phys., 36(1997)L1453.
144. K.-K. Kim, H.-S. Kim, J.-H. Lim and S.-J. Park, Appl. Phys. Lett., 83(2003)63.
145. B. S. Li, Y. C. Liu, Z. Z. Zhi, D. Z. Shen, Y. M. Lu, J. Y. Zhang, X. W. Fan, R. X. Mu and D. O. Henderson, J. Mater. Res., 18(2003)8.
146. Y. R. Ryu, T. S. Lee and H. W. White, Appl. Phys. Lett., 83(2003)87.
147. Z.-Z. Ye, J.-G. Lu, H.-H. Chen, Y.-Z. Zhang, L. Wang, B.-H. Zhao and J.-Y. Huang, J. Cryst. Growth, 253(2003)258.
148. D. C. Look and B. Claflin, Phys. Stat. Sol. (b), 241(2004)624.
149. A. Dadgar, A. Krtschil, F. Bertram, S. Giemsch, T. Hempel, P. Veit, A. Diez, N. Oleynik, R. Clos, J. Christen and A. Krost, Superlatt. Microstr., 38(2005)245.
150. H. von Wenckstern, R. Pickenhain, H. Schmidt, M. Brandt, G. Biehne, M. Lorenz, M. Grundmann and G. Brauer, Appl. Phys. Lett., 89(2006)092122.
151. J. Fan and R. Freer, J. Appl. Phys., 77(1995)4795.
152. A. N. Gruzintsev, V. T. Volkov and E. E. Yakimov, Semicond., 37(2003)259.
153. U. Wahl, E. Rita, J. G. Correia, A. C. Marques, E. Alves and J. C. Soares, Phys. Rev. Lett., 95(2005)215503.
154. R. E. Dietz, H. Kamimura, M. D. Sturge and A. Yariv, Phys. Rev., 132(1963)1559.
155. O. F. Schirmer, J. Phys. Chem. Solids, 29(1968)1407.
156. D. Zwingel, J. Lumin., 5(1972)385.
157. O. Lopatiuk, L. Chernyak, A. Osinsky and J. Q. Xie, Appl. Phys. Lett., 87(2005)214110.
158. D. Pfisterer, J. Sann, D. M. Hofmann, M. Plana, A. Neumann, M. Lerch and B. K. Meyer, Phys. Stat. Sol. (a), 243(2005)R1.
159. B. Claflin, D. C. Look, S. J. Park and G. Cantwell, J. Cryst. Growth, 287(2006)16.
160. D.-K. Hwang, H.-S. Kim, J.-H. Lim, J.-Y. Oh, J.-H. Yang, S.-J. Park, K.-K. Kim, D. C. Look and Y. S. Park, Appl.Phys. Lett., 86(2005)151917.
161. M. Chen, Z. L. Pei, X. Wang, Y. H. Yu, X. H. Liu, C. Sun and L. S. Wen, J. Phys. D: Appl. Phys., 33(2000)2538.
162. K. Ellmer, in "Transparent Conductive Zinc Oxide: Basics and Application in Thin Film Solar Cells", Ed.: K. Ellmer, A. Klein and B. Rech (Springer, Berlin, 2008), p. 44.
163. K. Ellmer and R. Mientus, Thin Solid Films, 516(2008)5829.
164. D. L. Rode, Phys. Rev. B, 2(1970)4036.
165. S. S. Devlin, in "Physics and Chemistry of II-VI Compounds", Ed.: M. Aven and J. S. Prener (North-Holland, Amsterdam, 1967), p. 549.
166. J. Bardeen and W. Shockley, Phys. Rev., 80(1950)72.
167. K. Seeger, "Semiconductor Physics", (Springer, Berlin, 1991).
168. C. Solbrig, Z. Phys., 211(1968)429.

169. D. L. Rode, in "Semiconductors and Semimetals", Ed.: R. K. Willardson and A. C. Beer (Academic Press, New York, 1975), p. 1.
170. J. D. Zook, Phys. Rev., 136(1964)A869.
171. P. W. Li and K. I. Hagemark, J. Solid State Chem., 12(1975)371.
172. K. W. Böer, "Survey of Semiconductor Physics. Electrons and Other Particles in Bulk Semiconductors," (Van Nostrand Reinhold, New York, 1990).
173. G. L. Pearson and J. Bardeen, Phys. Rev., 75(1949)865.
174. W. Zawadzki, in "Handbook on Semiconductors", Ed.: T. S. Moss (North-Holland, Amsterdam, 1982), p. 713.
175. T. Pisarkiewicz, K. Zakrzewska and E. Leja, Thin Solid Films, 174(1989)217.
176. E. Conwell and V. F. Weisskopf, Phys. Rev., 77(1950)388.
177. E. Conwell and V. F. Weisskopf, Phys. Rev., 69(1946)258.
178. W. Shockley, "Electrons and Holes in Semiconductors", (van Nostrand, Toronto, 1950).
179. R. B. Dingle, Phil. Mag., 46(1955)831.
180. I. V. Dakhovskii, T. A. Polyanskaya, A. G. Samoilovich and Y. V. Shartsev, Sov. Phys. Semicond., 4(1971)1857.
181. D. B. M. Klaassen, Solid State Electron., 35(1992)961.
182. D. B. M. Klaassen, Solid State Electron., 35(1992)953.
183. P. Ebert, Z. Zhang, F. Kluge, M. Simon, Z. Zhang and K. Urban, Phys. Rev. Lett., 83(1999)757.
184. H. Rupprecht, J. Phys. Chem. Solids, 6(1958)144.
185. W. S. Baer, Phys. Rev., 154(1967)785.
186. R. C. Neville and C. A. Mead, J. Appl. Phys., 41(1970)3795.
187. A. Hausmann and W. Teuerle, Z. Phys., 257(1972)299.
188. K. I. Hagemark, J. Solid State Chem., 16(1976)293.
189. B. Utsch and A. Hausmann, Z. Phys. B, 21(1975)27.
190. K. Ellmer and R. Mientus, Thin Solid Films, 516(2008)4620.
191. G. Masetti, M. Severi and S. Solmi, IEEE Trans. Electron Devices., ED30(1983)764.
192. D. Mergel and Z. Qiao, J. Appl. Phys., 95(2004)5608.
193. H. L. Hartnagel, A. L. Dawar, A. K. Jain and C. Jagadish, "Semiconducting Transparent Thin Films", (Institute of Physics Publishing, Bristol, 1995).
194. C. Erginsoy, Phys. Rev., 79(1950)1013.
195. F. J. Schmitte, in "Landoldt-Börnstein. Zahlenwerte und Funktionen aus Naturwiss. u. Technik. Neue Serie", Ed.: O. Madelung, M. Schulz and H. Weiss (Springer, Berlin, 1983), p. 200.
196. S. M. Sze, "Physics of Semiconductor Devices", (John, New York, 1981).
197. "Handbook Series on Semiconductors. Vol.1: Si, Ge, C(diamond), GaAs, GaP, GaSb, InAs, InP, InSb"; edited by M. Levinshtein, S. Rumyantsev and M. Shur (World Scientific, Singapore, 1996).
198. S. J. Chen, Y. C. Liu, C. L. Shao, C. S. Xu, Y. X. Liu, L. Wang, B. B. Liu and G. T. Zou, J. Appl. Phys., 99(2006)066102.
199. A. L. Fahrenbruch and R. H. Bube, "Fundamentals of Solar Cells", (Academic Press, New York, 1983).
200. G. Sanon, R. Rup and A. Mansingh, Phys. Rev. B, 44(1991)5672.
201. H. Fujiwara and M. Kondo, Phys. Rev. B, 71(2005)075109.
202. T. Minami, H. Sato, K. Ohashi, T. Tomofuji and S. Takata, J. Cryst. Growth, 117(1992)370.
203. A. Dargys and J. Kundrotas, "Handbook of Physical Properties of Ge, Si, GaAs and InP", (Science Encycl. Publ., Vilnius, 1994).
204. A. Schenk, J. Appl. Phys., 79(1996)814.
205. A. Mang, K. Reimann and S. Rübenacke, Solid State Commun., 94(1995)251.
206. K. S. Ramaiah, V. S. Raja, A. K. Bhatnagar and R. D. Tomlinson, Semicond. Sci. Technol., 15(2000)676.
207. Z. M. Jarzebski and J. P. Marton, J. Electrochem. Soc., 123(1976)299C.

208. T. Wittkowski, J. Jorzick, H. Seitz, B. Schröder, K. Jung and B. Hillebrands, Thin Solid Films, 398–399(2001)465.
209. K. M. Itoh, W. Walukiewicz, H. D. Fuchs, J. W. Beeman, E. E. Haller, J. W. Farmer and V. I. Ozhogin, Phys. Rev. B, 50(1994)16995.
210. T. Makino, A. Tsukazaki, A. Ohtomo, M. Kawasaki and H. Koinuma, Jpn. J. Appl. Phys., 45(2006)6346.
211. J. Y. Seto, J. Appl. Phys., 46(1975)5247.
212. K. Magnusson and S. Wiklund, J. Appl. Phys., 76(1994)7405.
213. G. Baccarani, B. Ricco and G. Spadini, J. Appl. Phys., 49(1978)5565.
214. B. D. Huey, D. Lisjak and D. A. Bonnell, J. Am. Ceram. Soc., 82(1999)1941.
215. T. Kamins, "Polycrystalline Silicon for Integrated Circuit Applications", (Kluwer, Boston, 1988).
216. J. W. Orton and M. J. Powell, Rep. Prog. Phys., 43(1980)1263.
217. F. R. Blom, F. C. M. v. d. Pol, G. Bauhuis and T. J. A. Popma, Thin Solid Films, 204(1991) 365.
218. D. Gerthsen, D. Litvinov, T. Gruber, C. Kirchner and A. Waag, Appl. Phys. Lett., 81(2002) 3972.
219. V. A. Coleman, J. E. Bradby, C. Jagadish, P. Munroe, Y. W. Heo, S. J. Pearton, D. P. Norton, M. Inoue and M. Yano, Appl. Phys. Lett., 86(2005)203105.
220. B. Pödör, phys. stat. sol., 16(1966)K167.
221. Y. Igasaki and H. Saito, J. Appl. Phys., 69(1991)2190.
222. A. Suzuki, T. Matsushita, N. Wada, Y. Sakamoto and M. Okuda, Jpn. J. Appl. Phys., 35(1996)L56.
223. Y. Igasaki and H. Saito, J. Appl. Phys., 70(1991)3613.
224. T. Minami, H. Sato, H. Imamoto and S. Takata, Jpn. J. Appl. Phys., 31(1992)L257.
225. K. Tominaga, S. Iwamura, I. Fujita, Y. Shintani and O. Tada, Jpn. J. Appl. Phys., 21(1982)999.
226. J. E. Greene, S. A. Barnett, J.-E. Sundgren and A. Rockett, in "Ion Beam Assisted Film Growth", Ed.: T. Itoh (Elsevier, Amsterdam, 1989), p. 101.
227. K. Ellmer, M. Kanis and R. Mientus, in: Proc. 4th Int. Workshop on ZnO and Related Materials, Giessen, Germany, October 3–6, 2006.
228. T. Minami, S. Suzuki and T. Miyata, Mater. Res. Soc. Symp. Proc., 666(2001)F1.3.1.
229. S. Brehme, F. Fenske, W. Fuhs, E. Neubauer, M. Poschenrieder, B. Selle and I. Sieber, Thin Solid Films, 342(1999)167.
230. M. Kon, P. K. Song, Y. Shigesato, P. Frach, S. Ohno and K. Suzuki, Jpn. J. Appl. Phys., 42(2003)263.
231. M. Lorenz, E. M. Kaidashev, H. von Wenckstern, V. Riede, C. Bundesmann, D. Spemann, G. Benndorf, H. Hochmuth, A. Rahm, H.-C. Semmelhack and M. Grundmann, Solid State Electron., 47(2003)2205.
232. J. R. Bellingham, W. A. Phillips and C. J. Adkins, J. Mater. Sci. Lett., 11(1992)263.
233. K. Ellmer and G. Vollweiler, Thin Solid Films, 496(2006)104.
234. L. L. Kazmerski, W. B. Berry and C. W. Allen, J. Appl. Phys., 43(1972)3515.
235. J. Oertel, K. Ellmer, W. Bohne, J. Röhrich and H. Tributsch, J. Cryst. Growth, 198–199 (1999)1205.
236. H. Witte, A. Krtschil, E. Schrenk, K. Fluegge, A. Dadgar and A. Krost, J. Appl. Phys., 97(2005)043710.
237. V. Kazukauskas, G. Kühnel and W. Siegel, Appl. Phys. Lett., 70(1997)1751.
238. L. Sagalowicz and G. R. Fox, J. Mater. Res., 14(1999)1876.
239. C. Coskun, D. C. Look, G. C. Farlow and J. R. Sizelove, Semicond. Sci. Technol., 19(2004)752.
240. N. W. Emanetoglu, C. Gorla, Y. Liu, S. Liang and Y. Lu, Mater. Sci. Semicond. Proc., 2(1999)247.
241. J. J. Harris, K. J. Lee, J. B. Webb, H. Tang, I. Harrison, L. B. Flannery, T. S. Cheng and C. T. Foxon, Semicond. Sci. Technol., 15(2000)413.
242. P. R. Emtage, J. Appl. Phys., 48(1977)4372.

243. P. J. Clarke, J. Vac. Sci. Technol., 14(1977)141.
244. S. Kasap, C. Koughia, H. Ruda and R. Johanson, in "Handbook of Electronic and Photonic Materials", Ed.: S. Kasap and P. Capper (Springer, Berlin, 2006), p. 19.
245. J. H. Werner, in "Polycrystalline Semiconductors III-Physics and Technology", Ed.: H. P. Strunk, J. Werner, B. Fortin and O. Bonaud (Scitec Publ., Switzerland, Zug, Switzerland, 1993), p. 213.
246. D. L. Young, T. J. Coutts, V. I. Kaydanov, A. S. Gilmore and W. P. Mulligan, J. Vac. Sci. Technol. A, 18(2000)2978.
247. G. Galli and J. E. Coker, Appl. Phys. Lett., 16(1970)439.
248. F. Pizzarello, J. Appl. Phys., 43(1972)3627.
249. B. M. Ataev, A. M. Bagamadova, A. M. Djabrailov, V. V. Mamedov and R. A. Rabadanov, Thin Solid Films, 260(1995)19.
250. P. Fons, K. Iwata, A. Yamada, K. Matsubara, S. Niki, K. Nakahara, T. Tanabe and H. Takasu, Appl. Phys. Lett., 77(2000)1801.
251. P. Kuppusami, S. Fiechter and K. Ellmer, MRS Symp. Proc., 721(2002)93.
252. H. Kato, M. Sano, K. Miyamoto and T. Yao, J. Cryst. Growth, 237–239(2002)538.
253. E. M. Kaidashev, M. Lorenz, H. von Wenckstern, A. Rahm, H.-C. Semmelhack, K.-H. Han, G. Benndorf, C. Bundesmann, H. Hochmuth and M. Grundmann, Appl. Phys. Lett., 82(2003) 3901.
254. H. von Wenckstern, M. Brandt, H. Schmidt, G. Biehne, R. Pickenhain, H. Hochmuth, M. Lorenz and M. Grundmann, Appl. Phys. A, 88(2007)135.
255. D. C. Look and J. R. Sizelove, Phys. Rev. Lett., 82(1999)1237.
256. H. Tampo, A. Yamada, P. Fons, H. Shibata, K. Matsubara, K. Iwata, S. Niki, K. Nakahara and H. Takasu, Appl. Phys. Lett., 84(2004)4412.
257. T. Makino, Y. Segawa, A. Tsukazaki, A. Ohtomo and M. Kawasaki, Appl. Phys. Lett., 87(2005)022101.
258. A. Ohtomo, K. Tamura, K. Saikusa, K. Takahashi, T. Makino, Y. Segawa, H. Koinuma and M. Kawasaki, Appl. Phys. Lett., 75(1999)2635.
259. K. Miyamoto, M. Sano, H. Kato and T. Yao, J. Cryst. Growth, 265(2004)34.
260. H. von Wenckstern, H. Schmidt, C. Hanisch, M. Brandt, C. Czekalla, G. Benndorf, G. Biehne, A. Rahm, H. Hochmuth, M. Lorenz and M. Grundmann, Phys. Stat. Sol. (rrl), 4(2007)129.
261. S. Ghosh, A. Sarkar, S. Chaudhuri and A. K. Pal, Thin Solid Films, 205(1991)64.
262. J. Szczyrbowski, K. Schmalzbauer and H. Hoffmann, Thin Solid Films, 137(1986)169.
263. A. Messad, J. Bruneaux, H. Cachet and M. Froment, J. Mater. Sci., 29(1994)5095.
264. K. Yasui, A. Asano, M. Otsuji, H. Katagiri, A. Masuda, H. Nishiyama, Y. Inoue, M. Takata and T. Akahanea, Mater. Sci. Eng. B, 148(2008)26.
265. K. Yasui, N. V. Phuong, Y. Kuroki, M. Takata and T. Akahane, Jpn. J. Appl. Phys., 44(2005) 684.
266. D. Marton, K. J. Boyd and J. W. Rabalais, Chem. Phys. Lett., 283(1998)215.
267. K. Lorenz, E. Alvez, E. Wendler, O. Bilani, W. Wesch and M. Hayes, Appl. Phys. Lett., 87(2005)191904.
268. S. Mráz and J. M. Schneider, J. Appl. Phys., 100(2006)023503.
269. S. Mahieu and D. Depla, Appl. Phys. Lett., 90(2007)121117.
270. C. Bundesmann, R. Schmidt-Grund and M. Schubert, in "Transparent Conductive Zinc Oxide: Basics and Application in Thin Film Solar Cells", Ed.: K. Ellmer, A. Klein and B. Rech (Springer, Berlin, 2008), p. 46.
271. C. Klingshirn, H. Priller, M. Decker, J. Brückner, H. Kalt, R. Hauschild, J. Zeller, A. Waag, A. Bakin, H. Wehmann, K. Thonke, R. Sauer, R. Kling, F. Reuss and C. Kirchner, Adv. Solid State Phys., 45(2005)275.
272. R. Schmidt, B. Rheinländer, M. Schubert, D. Spemann, T. Butz, J. Lenzner, E. M. Kaidashev, M. Lorenz, A. Rahm, H. C. Semmelhack and M. Grundmann, Appl. Phys. Lett., 82(2003)2260.

273. M. Schubert, "Infrared Ellipsometry on Semiconductor Layer Structures", (Springer, Berlin, 2004).
274. H. Yoshikawa and S. Adachi, Jpn. J. Appl. Phys., 36(1997)6237.
275. G. E. Jellison and L. A. Boatner, Phys. Rev. B, 58(1998)3586.
276. S. Ozaki, T. Mishima and S. Adachi, Jpn. J. Appl. Phys., 42(2003)5465.
277. J. F. Muth, R. M. Kolbas, A. K. Sharma, S. Oktyabrsky and J. Narayan, J. Appl. Phys., 85(1999)7884.
278. M. Rebien, W. Henrion, M. Bär and C.-H. Fischer, Appl. Phys. Lett., 80(2002)3518.
279. V. Srikant and D. R. Clarke, J. Appl. Phys., 81(1997)6357.
280. M. S. Tokumoto, A. Smith, C. V. Santilli, S. H. Pulcinelli, A. F. Craievich, E. Elkaim, A. Traverse and V. Briois, Thin Solid Films, 416(2002)284.
281. I. Takeuchi, W. Yang, K.-S. Chang, M. A. Aronova, T. Venkatesan, R. D. Vispute and L. A. Bendersky, J. Appl. Phys., 94(2003)7336.
282. P. Misra, P. K. Sahoo, P. Tripathi, V. N. Kulkarni, R. V. Nandedkar and L. M. Kukreja, Appl. Phys. A, 78(2004)37.
283. J.-L. Zhao, X.-M. Lia, J.-M. Bian, W.-D. Yu and X.-D. Gao, J. Cryst. Growth, 276(2005)507.
284. Z.-C. Jin, I. Hamberg and C. G. Granqvist, J. Appl. Phys., 64(1988)5117.
285. E. Burstein, Phys. Rev., 93(1954)632.
286. T. S. Moss, Proc. Phys. Soc. London, Sec. B, 67(1954)775.
287. B. E. Sernelius, Phys. Rev. B, 36(1987)4878.
288. B. E. Sernelius, K.-F. Berggren, Z.-C. Jin, I. Hamberg and C. G. Granqvist, Phys. Rev. B, 37(1988)10244.
289. J. D. Ye, S. L. Gu, S. M. Zhu, S. M. Liu, Y. D. Zheng, R. Zhang and Y. Shi, Appl. Phys. Lett., 86(2005)192111.
290. D. Redfield, Phys. Rev., 130(1963)916.
291. J. I. Pankove, Phys. Rev., 140(1965)A2059.
292. D. G. Thomas, J. Phys. Chem. Solids, 15(1960)86.
293. W. Y. Liang and A. D. Yoffe, Phys. Rev. Lett., 20(1968)59.
294. Y. P. Varshni, Physica, 34(1967)149.
295. R. Pässler, Phys. Stat. Sol. (b), 216(1999)975.
296. C. Klingshirn, Phys. Stat. Sol. (b), 244(2007)3027.
297. C. Klingshirn, Chemphyschem, 8(2007)782.
298. H. Nanto, T. Minami, S. Shooji and S. Takata, J. Appl. Phys., 55(1984)1029.
299. K. Ellmer, K. Diesner, R. Wendt and S. Fiechter, in "Polycrystalline Semiconductors IV-Physics, Chemistry and Technology", Ed.: S. Pizzini, H. P. Strunk and J. H. Werner (Trans Tech, Zug, Switzerland, 1995), p. 541.
300. M. Lorenz, in "Transparent Conductive Zinc Oxide: Basics and Application in Thin Film Solar Cells", Ed.: K. Ellmer, A. Klein and B. Rech (Springer, Berlin, 2008), p. 55.
301. J. Nishino, T. Kawarada, S. Ohshio, H. Saitoh, K. Maruyama and K. Kamata, J. Mater. Sci. Lett., 16(1997)629.
302. I. Sieber, N. Wanderka, I. Urban, I. Dörfel, E. Schierhorn, F. Fenske and W. Fuhs, Thin Solid Films, 330(1998)108.
303. L. Lutterotti, D. Chateigner, S. Ferrari and J. Ricote, Thin Solid Films, 450(2004)34.
304. O. Kluth, B. Rech, L. Houben, S. Wieder, G. Schöpke, C. Beneking, H. Wagner, A. Löffl and H. W. Schock, Thin Solid Films, 351(1999)247.
305. C. Ohler, A. Förster, J. Moers, C. Daniels and H. Lüth, J. Phys. D Appl. Phys., 30(1997)1436.
306. R. Cebulla, R. Wendt and K. Ellmer, J. Appl. Phys., 83(1998)1087.
307. D. W. Hoffman and J. A. Thornton, J. Vac. Sci. Technol., 17(1980)380.
308. J. Hinze and K. Ellmer, J. Appl. Phys., 88(2000)2443.
309. P. Kuppusami, G. Vollweiler, D. Rafaja and K. Ellmer, Appl. Phys. A, 80(2005)183.
310. E. Müller, D. Litvinov, D. Gerthsen, C. Kirchner, A. Waag, N. Oleynik, A. Dadgar and A. Krost, in "Zinc Oxide – A Material for Micro- and Optoelectronic Applications", Ed.: N. Nickel and E. Terukov (Springer, Dordrecht, 2005), p. 99.

311. T. Makino, Y. Segawa, M. Kawasaki, A. Ohtomo, R. Shiroki, K. Tamura, T. Yasuda and H. Koinuma, Appl. Phys. Lett., 78(2001)1237.
312. P. Fontana and G. Gut, J. Cryst. Growth, 28(1975)165.
313. P. Singh, J. Mater. Sci. Lett., 9(1990)613.
314. K. Uematsu, T. Morimoto, Z. Kato, N. Uchida and K. Santo, J. Mater. Sci. Lett., 6(1987) 1285.
315. Y. R. Ryu, T. S. Lee, J. A. Lubguban, A. B. Corman, H. W. White, J. H. Leem, M. S. Han and Y. S. Park, Appl. Phys. Lett., 88(2006)052103.
316. C. H. Lau, L. Zhuang and K. H. Wong, Phys. Stat. Sol. (b), 244(2007)1533.
317. C. W. Teng, J. F. Muth, Ü. Özgür, M. J. Bergmann, H. O. Everitt, A. K. Sharma, C. Jin and J. Narayan, Appl. Phys. Lett., 76(2000)979.
318. T. Glatzel, D. Fuertes Marron, T. Schedel-Niedrig, S. Sadewasser and M. C. Lux-Steiner, Appl. Phys. Lett., 81(2002)2017.
319. R. Schmidt-Grund, M. Schubert, B. Rheinländer, D. Fritsch, H. Schmidt, E. M. Kaidashev, M. Lorenz, C. M. Herzinger and M. Grundmann, Thin Solid Films, 455-456(2004)500.
320. T. D. Kang, H. Lee, W.-I. Park and G.-C. Yi, Thin Solid Films, 455-456(2004)609.
321. K. Matsubara, H. Tampo, H. Shibata, A. Yamada, P. Fons, K. Iwata and S. Niki, Appl. Phys. Lett., 85(2004)1374.
322. T. Minemoto, T. Negami, s. Nishiwaki, H. Takakura and Y. Hamakawa, Thin Solid Films, 372(2000)173.
323. S. Siebentritt, Thin Solid Films, 403-404(2002)1.
324. G. V. Rao, F. Säuberlich and A. Klein, Appl. Phys. Lett., 87(2005)032101.
325. S. Sadofev, P. Schäfer, Y.-H. Fan, S. Blumstengel, F. Henneberger, D. Schulz and D. Klimm, Appl. Phys. Lett., 91(2007)201923.
326. F. P. Koffyberg, Can. J. Phys. , 49(1971)435.
327. G. Rupprecht, Z. Phys., 139(1954)504.
328. C. A. Vincent, J. Electrochem. Soc., 119(1972)515.
329. S. Narushima, M. Orita, M. Hirano and H. Hosono, Phys. Rev. B, 66(2002)035203.
330. J. R. Bellingham, W. A. Phillips and C. J. Adkins, J. Phys.: Condens. Matt., 2(1990)6207.
331. J. R. Bellingham, W. A. Phillips and C. J. Adkins, Thin Solid Films, 195(1991)23.
332. M. Yan, M. Lane, C. R. Kannewurf and R. P. H. Chang, Appl. Phys. Lett., 78(2001)2342.
333. R. Dingle, H.-L. Störmer, A. C. Gossard and W. Wiegmann, Appl. Phys. Lett., 33(1978)665.
334. J. J. Harris, J. A. Pals and R. Woltjer, Rep. Prog. Phys., 52(1989)1217.
335. I. A. Rauf, Mater. Lett., 18(1993)123.
336. T. Edahiro, N. Fujimura and T. Ito, J. Appl. Phys., 93(2003)7673.
337. A. A. Grinberg and S. Luryi, J. Appl. Phys., 61(1987)1181.
338. D. J. Cohen and S. A. Barnett, J. Appl. Phys., 98(2005)053705.
339. A. Ohtomo and H. Y. Hwang, Appl. Phys. Lett., 84(2004)1716.
340. K. Nomura, T. Kamiya, H. Ohta, K. Ueda, M. Hirano and H. Hosono, Appl. Phys. Lett., 85(2004)1993.

Chapter 8
Ternary and Multinary Materials: Crystal/ Defect Structure–Property Relationships

Thomas O. Mason, Steven P. Harvey, and Kenneth R. Poeppelmeier

8.1 Introduction

The best and most commercialized transparent conducting oxides (TCOs) are based on cations with filled d-shells, such at Zn^{2+}, Cd^{2+}, In^{3+}, and Sn^{4+}, which exhibit n-type character. We will refer to the parent oxides (ZnO, CdO, In_2O_3, and SnO_2) as the "basis" oxides for n-type TCO behavior. From their initial discovery circa 1900 [1] to the early 1990s, the basis oxides dominated the TCO literature. For example, the monograph by Hartnagel et al. [2] captured the "state of the field" at the time of its writing (ca. 1994); other than the basis oxides, only the spinel compounds, $CdIn_2O_4$ and Cd_2SnO_4, received attention. Ga_2O_3 is sometimes included as a "basis" oxide, and because of its significantly larger band gap, it is often considered to be an ultra-violet TCO [3]. We have excluded gallia as a "basis" oxide owing to (1) its relatively poor n-type conductivity even when doped [4], and (2) the tendency for Ga^{3+} to diminish the conductivity of n-type TCOs when substituted for other basis cations in large concentrations, e.g., In^{3+} (see below).

From the early 1990s, however, considerable work began to be reported on multi-cation systems, with several active groups: AT&T Bell Laboratories in the U.S. [5, 6], the Electron Device System Laboratory, Kanazawa Institute of Technology [7, 8] and the Materials and Structures Laboratory, Tokyo Institute of Technology [9, 10] in Japan, the Materials Research Center at Northwestern University [11, 12] and the National Renewable Energy Laboratory in the U.S. [13]. (These were just the "new materials" efforts; many other groups were active in TCO films and applications, and many other groups have since become active.) A search of ISI Web of Knowledge shows only 7 and 24 papers on "transparent conducting oxides" for the decades of the 1970s and 1980s, respectively, virtually

T.O. Mason (✉)
MCC Materials Science and Engineering, Northwestern University, 2220 Campus Dr 3037, EV3108 633 Clark Street, Evanston, IL 60208, USA
e-mail: t-mason@northwestern.edu

D.S. Ginley et al. (eds.), *Handbook of Transparent Conductors*,
DOI 10.1007/978-1-4419-1638-9_8, © Springer Science+Business Media, LLC 2010

Table 8.1 Some representative TCO compounds reported to date

Basis oxides	Binary compounds	Binary solid solutions
ZnO	Cd_2SnO_4	$Ga_{2-2x}In_{2x}O_3$
CdO	$CdIn_2O_4$	*Ternary compounds*
SnO_2	$CdSnO_3$	$Ga_{3-x}In_{5+x}Sn_2O_{16}$
In_2O_3	$ZnSnO_3$	*Ternary solid solutions*
(Ga_2O_3)	Zn_2SnO_4	$In_{1-x}Ga_{1+x}O_3(ZnO)_k$ ($k = 1, 2, 3$)
	$ZnGa_2O_4$	$Zn_{2-x}Sn_{1-x}In_{2x}O_4$
	$In_4Sn_3O_{12}$	$In_{2-2x}Sn_xZn_xO_3$
	$Zn_kIn_2O_{k+3}$ ($k = 2, 3, 7–15$ odd)	xCd_2SnO_4-$(1-x)CdIn_2O_4$
	$Ti_{(1-x)}M_xO_2$ (M = Nb,Ta)	$In_{2-2x}Cd_xSn_xO_3$
	$MgIn_2O_4$	$Cd_{1-x}Sn_{1-x}In_{2x}O_3$
	$12CaO{\cdot}7Al_2O_3$	xZn_2SnO_4-$(1-x)ZnGa_2O_4$
		$In_{4+x}Sn_{3-2x}Sb_xO_{12}$ ($x = 0–1.5$)

all of which were on the "basis" oxides; the decade of the 1990s saw 329 TCO publications, with a large portion being binary or higher compositions. The publication rate for the current decade should double that of the prior decade, with a large fraction of papers dealing with binary and higher TCOs. The rapid developments of the decade 1990–2000 are captured in the August, 2000 Materials Research Bulletin devoted to TCOs [13]. These include the discovery of p-type TCOs based upon the d^{10} cations, e.g., Cu^+ [14, 15]. (Although similar Ag^+ compounds meet the requirements as p-type TCOs, a p-type Ag-based TCO with reasonable conductivity has yet to be reported.)

Table 8.1 shows a list, not intended to be exhaustive, of TCO compounds and compositions known at present. The list of binary and ternary compounds and solid solutions is quite extensive, and is being added to on a regular basis. The current chapter is devoted to exploring "ternary and multinary" TCO materials. The "basis" oxides have been amply described in other chapters, as have some of the binary TCOs. We define our focus by first stating what we will *not* be covering. We then provide our definition of "ternary and multinary," and describe the focus of the chapter.

Our discussion will exclude "basis" oxides with simple intrinsic defect behavior (e.g., by oxygen vacancies) and/or by extrinsic defects (e.g., by aliovalent donor species). We refer to these systems as possessing conventional Brouwer (doping) diagrams. For example, we would consider Al-doped ZnO and Sb-doped SnO_2 as examples of conventional doping schemes. Representative Brouwer diagrams, log-log plots of defect concentration vs. doping level or vs. oxygen activity, are shown in Fig. 8.1. Although there may be two types of prevailing point defects, e.g., oxygen vacancies and donor species, these dominate separate electroneutrality regimes (e.g., $n = [V_O^{\bullet\bullet}]$ and $n = [D^{\bullet}]$, where D = donor) and do not extensively interact with one another except via point defect equilibrium.

On the other hand, we define "ternary and multinary" systems as:

- *Having complex point defect associates*, involving donors and acceptors in combination. In this way, even a doped unary TCO could be considered "ternary"

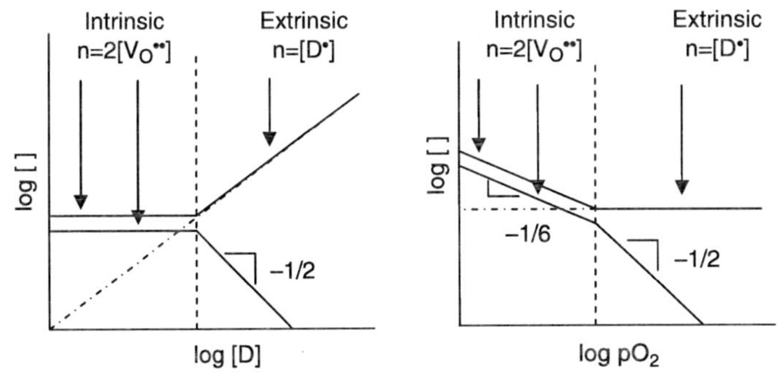

Fig. 8.1 Representative Brouwer diagrams for conventional doping schemes of "basis" TCOs, such as those of Al-doped ZnO and Sb-doped SnO_2

in character. A primary example involves the donor-oxygen interstitial associate in indium-tin oxide (ITO).
- *Having two or more cation sublattices.* This usually requires at least binary character, with ternary or higher character being induced by oxygen and/or donor-doping.

And of course, the most widely accepted definition of "ternary or multinary" behavior:

- *Existing in ternary or higher phase space*, i.e., true ternary or multinary cation compositions. We include the many isovalent substitution systems. For example, by hydrothermal synthesis (see below) an extended $CuAlO_2$-$CuGaO_2$ p-type solid solution can be synthesized in the Cu_2O-Al_2O_3-Ga_2O_3 system. Other systems exhibit co-doping, whereby extensive co-substitution is enabled by the isovalent nature of the substitution. A good example is the extended n-type solid solution of ZnO plus SnO_2 in In_2O_3 to levels far exceeding the solubility of either dopant separately (discussed below). In this case two In^{3+} cations are being replaced by a divalent (Zn^{2+}) cation and a tetravalent (Sn^{4+}) cation; acceptors and donors are present in near equal concentrations ($[Zn_{In}'] = [Sn_{In}^{\bullet}]$), hence the substitution is essentially "isovalent" in character. A similar mechanism is believed to take place in the $CdIn_2O_4$-Cd_2SnO_4 spinel solid solution (see below).

When dealing with ternary/multinary TCO systems, phase diagrams are important tools. They not only establish the stability ranges of the various equilibrium phases, but they also help resolve the doping mechanisms of the TCO materials, especially as related to their crystal/defect chemistry. Systematic composition-structure-property studies provide essential information about TCO behavior, with major implications for TCO fabrication/application in thin film form. This review revolves around four central themes related to composition–structure–property relationships in ternary/multinary TCOs. First, we consider the importance of *cation coordination* in TCO properties, both n-type (octahedral coordination) and p-type (linear coordination). Second, we emphasize the role of *defect complexation*, i.e., that donor-oxygen interstitial associates dominate the defect/transport properties of materials like indium-tin oxide (ITO) and copper aluminate ($CuAlO_2$), rendering them "ternary" in aspect. Third, we consider the origin and ramifications of *extended solid solubility* of TCO phases in ternary phase space, including isovalent substitutions/co-substitutions plus "self-doped" n-type materials, which appear to be inherently off-stoichiometric. Such extended solubility allows for tailoring of TCO properties, much as is done for compound semiconductors (band gap engineering). Finally, we consider *limiting factors* for ternary/multinary TCO behavior, including the influence of transport mechanisms (e.g., charge localization) and the role of isotropy/anisotropy. We also consider the limitations of structural randomness, with important ramifications for the emerging subfield of amorphous TCOs.

8.2 The Role of Cation Coordination

8.2.1 Octahedral Cation Coordination (n-Type TCOs)

As early as 1977, Shannon and coworkers [16] argued that "continuous edge sharing of Cd^{2+}, In^{3+} and Sn^{4+} octahedra is a necessary criterion for the formation of a transparent conductor." This is with reference to the best n-type TCOs. To test the Shannon criterion, the "best value" literature conductivity to date for each of the phases in the CdO-In_2O_3-SnO_2 system is plotted vs. the density (in #/cm^3) of octahedrally-coordinated cations in Fig. 8.2 [17]. The value for the distorted orthorhombic perovskite ($CdSnO_3$) is taken from two sources, one bulk and one thin film [18]. The value for orthorhombic Cd_2SnO_4 is for a bulk specimen, corrected for porosity [19]. The values for Cd_2SnO_4 spinel [20], Sn-doped bixbyite In_2O_3 [21], and In-doped rocksalt CdO [22] are for thin film materials. All these phases exist and are suitably donor-doped within Cd-In-Sn-O phase space shown in

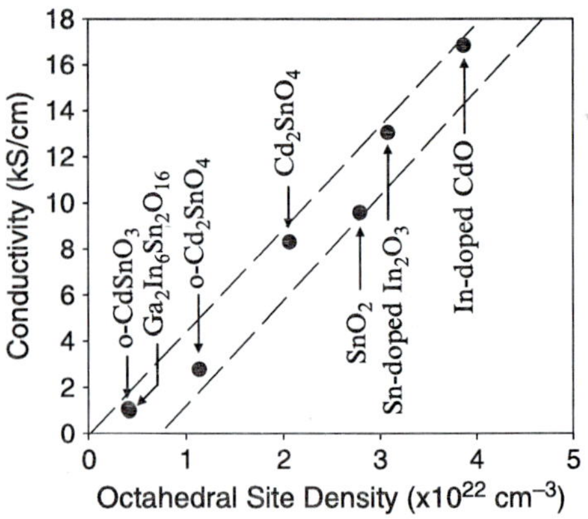

Fig. 8.2 Best reported conductivity values of TCO phases, plotted against the density (#/cm^3) of octahedrally coordinated cations (Cd, In, Sn). The o-$CdSnO_3$ and o-Cd_2SnO_4 are orthorhombic phases. The o-$CdSnO_3$ and $Ga_2In_6Sn_2O_{16}$ data points overlap [17]

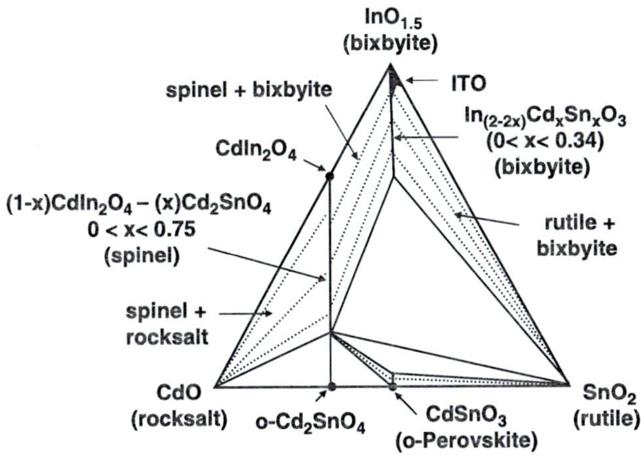

Fig. 8.3 Ternary phase diagram of $CdO-InO_{1.5}-SnO_2$ phase space at 1,175°C [19]

Fig. 8.3 [19]. There is no appropriate donor dopant for rutile SnO_2 in this system, however it is well known that SnO_2 can be donor-doped by pentavalent cations (e.g., Sb^{5+}) or monovalent anions (e.g., F^-); the value shown in Fig. 8.2 is from the compilation by Minami [23].

There is an excellent linear correlation of "best" or optimized conductivity vs. octahedral cation density in n-type TCOs. As pointed out by Gordon [24], the effective mass is very nearly the same ($m^* \sim 0.3$ m) in all the best n-type TCO materials, citing results for ZnO, SnO_2 and Cd_2SnO_4. This also holds for CdO [25]. This suggests that the trend in Fig. 8.2 has to do with the density of states and the corresponding carrier contents in the optimized materials, although mean free path may also play a role. In other words, the best TCOs are structures with the highest concentrations of octahedral species. By this criterion, the rocksalt structure is best (CdO) followed by bixbyite (In_2O_3) and rutile (SnO_2).

The other structures in Fig. 8.2 have a portion of their basis TCO cations in alternate sublattices/coordinations, thereby reducing their optimized conductivity. For example, the spinel structure has one-third of its cations in tetrahedral coordination (e.g., Cd_2SnO_4 is an essentially inverse spinel [26], i.e., $(Cd)[CdSn]O_4$, with the tetrahedral Cd in parentheses and the octahedral Sn plus remaining Cd in brackets). Similarly, only half of the cations in the orthorhombic perovskite structure (e.g., $o-CdSnO_3$) are in octahedral coordination.

The correlation extends to even more complicated crystal structures. Figure 8.2 shows a datum for the ternary compound, $Ga_2In_6Sn_2O_{16}$ [11]. This tetragonal structure can be described as an anion-deficient fluorite derivative (through removal of one-fifth of the anions), with cations in four-, six-, seven-, and eightfold coordination [27]. The datum in Fig. 8.2 was based upon the basis TCO cations (In^{3+}, Sn^{4+}) in octahedral coordination only.

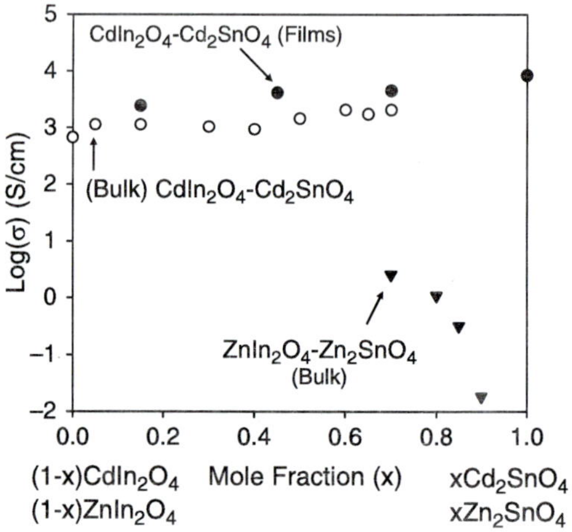

Fig. 8.4 Bulk and thin film conductivity values for $CdIn_2O_4$-Cd_2SnO_4 spinel solid solution, and the $ZnIn_2O_4$-Zn_2SnO_4 solid solution [28]

It should be pointed out that the datum for the orthorhombic perovskite, $CdSnO_3$, which has corner-shared rather than edge-shared octahedra, also obeys the trend line in Fig. 8.2. This would suggest that Shannon's criterion be modified as follows: *continuous edge- or corner-sharing of* Cd^{2+}, In^{3+} *and* Sn^{4+} *octahedra is a necessary criterion for the formation of an n-type transparent conductor.*

It is not immediately obvious why Zn^{2+}, which forms an outstanding TCO when in tetrahedral coordination (e.g., donor-doped ZnO), is excluded from the Shannon criterion. Figure 8.4 shows conductivity data for the $CdIn_2O_4$-Cd_2SnO_4 spinel solid solution (bulk and thin film) and for the $ZnIn_2O_4$-Zn_2SnO_4 solid solution (bulk specimens) vs. composition [28]. Note that $ZnIn_2O_4$ is not stable and that the $ZnIn_2O_4$-Zn_2SnO_4 solid solution is limited to the Zn_2SnO_4 end of the system. Nevertheless, the difference in optimized conductivity for the two systems is striking. Like cadmium stannate, zinc stannate is an essentially inverse spinel [29], with the Zn^{2+} divided between tetrahedral and octahedral sites. Its relatively insulating behavior in comparison with Cd_2SnO_4 argues that (1) Zn^{2+} is relatively ineffective as a TCO basis cation in octahedral coordination (whether owing to reduced carrier generation or mobility, or both) and (2) Zn^{2+} is similarly ineffective in a tetrahedral coordination different from that of ZnO. (The tetrahedral Zn-Zn distance is roughly 20% longer in Zn_2SnO_4 than it is in ZnO [28].) Therefore, insofar as optimized n-type TCOs with octahedral coordination are concerned, Zn^{2+} is effectively a diluent species, and does not contribute significantly to conductivity.

8.2.2 *Linear Cation Coordination (p-Type TCOs)*

It is well known that there is a strong localization of electron holes at the valence band edge of most oxides. This localization is due to the ionicity of metallic oxides, where the O-2p levels are generally far lower in energy than the valence orbitals of the metal species. This strong localization of holes translates into large effective masses and low mobilities vis-à-vis electrons. Hosono and co-workers [30, 31] designed a strategy to introduce covalency and induce the formation of an extended valence-band structure. Their strategy involved: (1) selecting cations with closed d-shells and energies comparable to that of the O-2p levels (e.g., $3d^{10}$ Cu^+ and $4d^{10}$ Ag^+), (2) placing them in linear coordination with oxygen (thus keeping the oxygen ligands at a distance, owing to strong repulsions between 2p electrons on the oxide ligands and the d^{10} cations), and (3) choosing structures with oxygen in tetrahedral coordination, resulting in reduced non-bonding character of the metal-oxygen bonds (all eight electrons around an oxygen are in σ, sp^3 hybrid orbitals), thus reducing the localization of the valence band edge.

A structure that satisfies these criteria for p-type TCOs is the delafossite structure (e.g., $CuAlO_2$). As shown in Fig. 8.5 [32], the ABO_2 delafossite structure (after the mineral, $CuFeO_2$) is composed of alternating layers of slightly distorted edge-shared $B^{3+}O_6$ octahedra sandwiching planes of close-packed A^+ cations (e.g., Ag^+, Cu^+) in linear or "dumbbell" coordination to oxygen anions in the adjacent BO_6 layers. Depending upon the stacking of the layers, one obtains either the rhombohedral 3R polytype (AaBbCcAaBbCc...) or the hexagonal 2H polytype (AaBbAaBb...).

There are considerable challenges in synthesizing the various delafossite compounds, especially the Ag-based materials, owing to the decomposition of the Ag-based precursor oxide. Table 8.2 lists the ABO_2 materials synthesized by a direct hydrothermal synthesis (HTS) route [33]. This procedure involves sealing starting

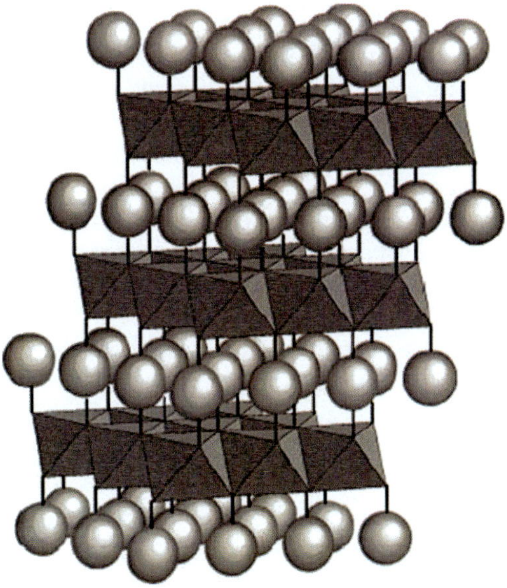

Fig. 8.5 Schematic representation of the delafossite structure, ABO_2 ($CuAlO_2$ in this case) for the "3R" polytype ($R\bar{3}m$) with AaBbCcAa... stacking along the c-axis [32]. The A^{1+} cation is in linear coordination (*circles* in figure), and the B^{3+} cation forms distorted BO_6 octahedra

Table 8.2 Delafossite compounds (ABO_2) synthesized by hydrothermal synthesis, where A is Cu or Ag, and B is Al, Ga, etc. [33]

	Al	Ga	Sc	In	Y	Eu	La
Cu	✓	✓	✓	×	×	×	×
Ag	✓	✓	✓	✓	×	×	×

materials plus controlled quantities of NaOH mineralizer in semi-permeable poly (tetrafluoroethylene) PTFE pouches. Loaded pouches are then placed in a PTFE-lined pressure vessel (Parr Instruments), which is backfilled with distilled water, sealed and heated to 150°C, followed by gradual incremental heating to the final temperature of 175–210°C, depending upon the delafossite in question. This temperature is held for 60 h, followed by cooling at 6°C/h to room temperature. Specific details regarding defect structure(s), doping, and physical properties are given elsewhere [32, 33]. However, two points need to be made about HTS of the delafossites. First, many more compounds and solid solutions are achievable by HTS than can be made by conventional solid-state synthesis (SSS). For example, only the $CuAlO_2$ and $CuGaO_2$ end members can be produced by SSS, whereas the entire $CuAlO_2$-$CuGaO_2$ solid solution can be synthesized by HTS [34], enabling us to more fully explore TCO behavior in ternary and multinary systems, e.g., the Cu_2O-Al_2O_3-Ga_2O_3 ternary system. Second, Ag-delafossites can be readily synthesized by HTS, whereas they are difficult to produce by standard SSS, owing to the decomposition of silver oxide mentioned previously. We report doping/defect character, physical properties, and limitations (in terms of doping and transport) in subsequent sections.

It should be stressed that the ABO_2 delafossite structure depicted in Fig. 8.5 represents the prototypical amphoteric TCO, owing to its unique crystal structure. Amphoteric character describes the ability of a given host to readily dope in both directions, both n-type and p-type. The delafossite structure allows for separation of the octahedral cation (B-site) manifold, which dominates conduction band behavior, and the linear cation (A-site) manifold, which dominates valence band behavior. This means that the B-sites can be donor-doped, leading to n-type character, or alternatively acceptor-doped, transferring the resulting holes to the A-sites (localized holes, as we show later), leading to p-type character. The prime example is $CuInO_2$, which can be both n-doped and p-doped to comparable (albeit low) levels of conductivity [35]. The unique character of $CuInO_2$ derives from its band structure [36], i.e., the presence of a lower energy conduction band minimum (CBM) combined with the fact that the transition from the valence band maximum to the CBM is direct-forbidden makes for electron stabilization; n-doping is therefore possible. The Ag-delafossites are also primarily p-type in character [37], however Sn-doped $AgInO_2$ shows strong n-type character, with relatively high conductivity [38, 39].

8.3 The Role of Defect Complexation

In the primary examples of n-type and p-type TCO behavior, i.e., indium-tin oxide (ITO) and copper aluminate ($CuAlO_2$), point defect complexes have been shown to exhibit a dominant role governing the doping/defect behavior and resulting transport properties. This suggests that other TCO systems should be investigated/considered with regard to potential point defect complex formation. Such complexation can have unexpected and unforeseen consequences: (1) larger defect populations than would otherwise be possible with isolated donors, whether intrinsic or extrinsic, (2) complex point defect (Brouwer) diagrams vs. doping level or oxygen partial pressure, (3) the imparting of ternary character to what would otherwise be considered binary systems, and even (4) the transformation of what should be a donor species into a complexed acceptor species.

8.3.1 *Point Defect Complexes in Indium-Tin Oxide*

The C-type rare-earth sesquioxide structure, otherwise known as bixbyite, has a body-centered lattice derived from that of fluorite (AO_2), but with one-fourth of the anions absent for charge balance ($AO_{1.5}$ or A_2O_3). These are often referred to as "structural vacancies," but in reality they are true anion interstitial positions. This situation is relatively unique compared to the other octahedral cation TCO structures, which have close-packed oxygen sublattices (e.g., rocksalt, rutile, spinel, etc.). Figure 8.6 shows the 8b and 24d cation positions in bixbyite (referred to as b and d positions for short). Each cation resides at the center of a distorted cube, with six corners occupied by oxygen anions and two corners vacant, which are the structural vacancies or available interstitial positions.

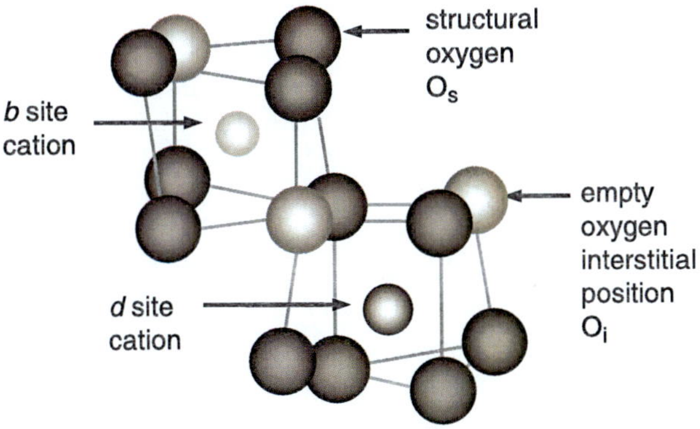

Fig. 8.6 Schematic representation of the b and d cation positions in the bixbyite crystal structure [41]

As early as 1982, Frank and Köstlin hypothesized the existence of 2:1 Sn-donor: O-interstitial associates in ITO [40]. These neutral associates decompose upon reduction according to:

$$(2Sn_{In}{}^{\bullet}O_i{}'')^x \longleftrightarrow {}^1\!/_2O_2(g) + 2\ Sn_{In}{}^{\bullet} + 2e' \tag{8.1}$$

leading to a unique $pO_2{}^{-1/8}$ dependence of the electron population at high pO_2 values, and a pO_2-independent regime at intermediate pO_2 values, once all reducible associates have been so reduced.

The Frank and Köstlin hypothesis was made on the basis of thin film specimens [40]. Later work confirmed the existence of Frank-Köstlin clusters in bulk specimens [41]. Combined Rietveld analysis of X-ray and neutron diffraction data from quenched bulk and nanocrystalline ITO specimens confirmed oxygen occupation of the structural vacancy sites in Fig. 8.6, i.e., oxygen interstitials. These sites were shown to be unoccupied for undoped In_2O_3 specimens. Furthermore, the oxygen interstitials were in approximate 2:1 ratio to Sn-donor concentration in ITO, as shown in Fig. 8.7. However, this does not prove that the two species are in proximity, i.e., complexed. However, EXAFS results for the first coordination shells at the In and Sn edges [41] showed that (1) the first-shell oxygen coordination at the Sn-edge was higher than 6 vs. precisely 6 for the first-shell oxygen coordination at the In-edge, and (2) the average first-shell Sn-O distance (2.05–2.08 Å) in ITO was significantly smaller than the average first-shell In-O distance (2.18–2.19 Å). The two values correspond nicely to the Sn-O distance in SnO_2 and the In-O distance in undoped In_2O_3.

These findings are nicely corroborated by Sn-Mossbauer spectroscopy studies of ITO [42], showing Sn^{4+} coordinated by six oxygens (isolated donors) but also seven or eight oxygens (Sn-O clusters). Hall effect studies confirmed that reductions in the electron population with oxidation treatments coincided with compensation of tin donors by Sn-O complex formation [43]. Combined X-ray diffraction and dielectric modeling of optical transmittance and reflectance spectra similarly confirm the existence of Sn-donors partially compensated by oxygen interstitials [44].

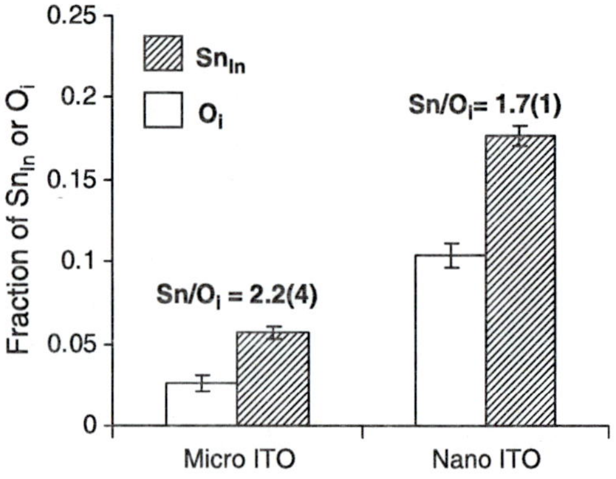

Fig. 8.7 Concentrations of Sn_{In} and O_i in bulk and nano ITO obtained from Rietveld analysis of TOF neutron diffraction data and X-ray diffraction data (all samples processed in air) [41]

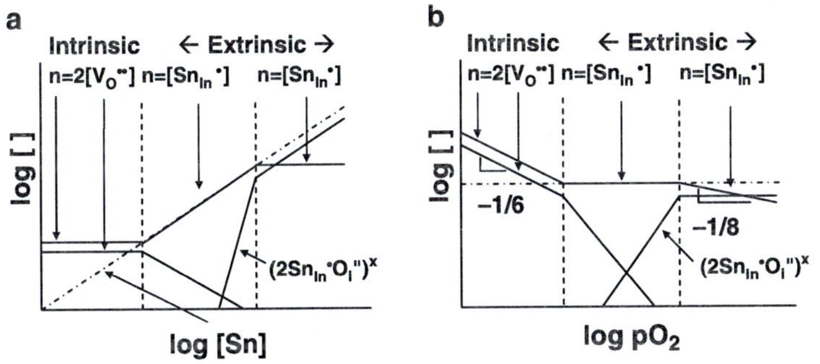

Fig. 8.8 Schematic Brouwer diagrams of point defect concentrations vs. (**a**) tin doping level, and (**b**) oxygen partial pressure

The implications for the defect structure of ITO are presented in Fig. 8.8, which shows Brouwer diagrams of point defect populations vs. Sn-doping level (Fig. 8.8a) and vs. oxygen partial pressure (Fig. 8.8b). These should be compared with the conventional Brouwer diagrams of Fig. 8.1. The presence of the 2:1 cluster imparts a second extrinsic point defect regime, in which $n = [Sn_{In}^{\bullet}]$, but where the concentration of free donors (and electrons) is directly tied to the population of defect clusters. Furthermore, a $pO_2^{-1/8}$-dependence of carrier content is imparted by the clusters, as specified by the reaction of (8.1).

There are several ramifications of the point defect character of ITO. First, it is believed that much higher doping levels can be achieved as a result of cluster formation, especially in thin films, where doping levels to tens of percents have been reported. This is a result of cluster stability; more tin (SnO_2) can be dissolved in the bixbyite lattice in conjunction with oxygen interstitial formation. Second, ITO is susceptible to two-step processing by (1) initial co-doping by Sn-donors complexed with O-interstitials, and (2) removal of the O-interstitials by reduction (see (8.1)). The typical 400–500°C reduction step for ITO films activates the otherwise compensated donors, resulting in large electron populations. Third, ITO is amenable to post-deposition surface dipole modification of ionization potential (work function) via surface oxidation [45]. The nature of the responsible surface species has yet to be conclusively established, but may be related to the bulk tendency to form oxygen interstitials. To conclude, ITO must therefore be considered to be a ternary "host-donor-compensating acceptor" system (In_2O_3-$Sn_{In}^{\bullet}$-O_i''), with the oxygen interstitials playing the role as acceptors.

8.3.2 Point Defect Complexes in $CuAlO_2$

Figure 8.9 shows three adjacent Cu-sites in the delafossite structure, with their linear coordination by oxygen, along with a trigonal oxygen interstitial defect site. Based upon Shannon's published radii [46], the size of this interstitial site is plotted vs. the

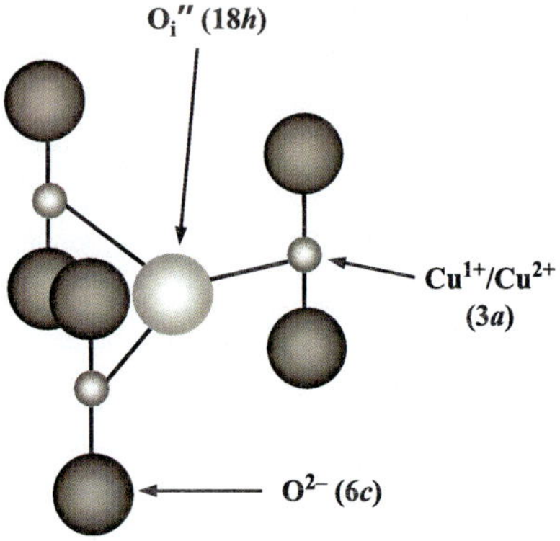

Fig. 8.9 Interstitial site (between three adjacent A sites) in the ABO_2 delafossite structure, shown here occupied by oxygen [47]

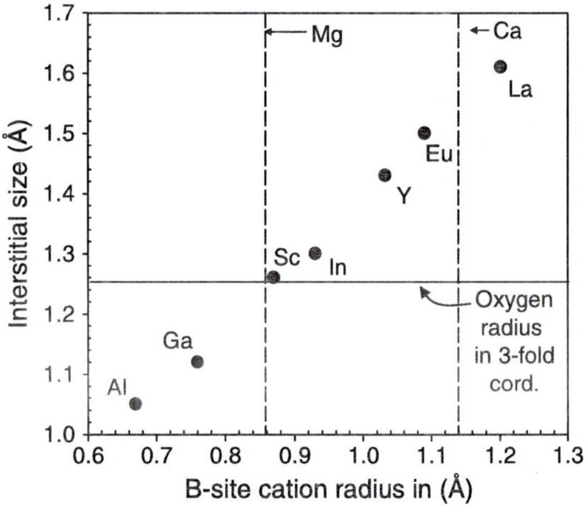

Fig. 8.10 Interstitial radius as a function of B-site cation in the delafossite compound ABO_2 [47]. The radius of oxygen in threefold coordination [46] is represented as the *horizontal line*. The size of Ca and Mg dopant ions in octahedral coordination [46] are shown as *vertical lines*

B-site cation radii (octahedral coordination) in Fig. 8.10 [47]. The horizontal line represents the oxygen anion radius for threefold coordination. Unlike delafossites with larger B-site cations (scandium and larger), which can readily accommodate isolated oxygen interstitials, the interstitial site in $CuAlO_2$ is ordinarily too small. This explains why $CuAlO_2$ exhibits relatively pO_2-independent electrical behavior [48]. Similarly, the B-site of $CuAlO_2$ is far too small for acceptor doping; Fig. 8.10 also shows the ionic radii of common B-site acceptor species, e.g., Mg^{2+} and Ca^{2+}. (Sr^{2+} and Ba^{2+} are still larger, beyond the scale of Fig. 8.10.) This likely explains why attempts to acceptor-dope $CuAlO_2$ have been relatively unsuccessful [49].

On the other hand, $CuAlO_2$ is known to exhibit the highest p-type conductivity of the undoped delafossites at elevated temperatures, e.g., 3.25 S/cm at 740°C (vs. 0.37 S/cm for $CuScO_2$ and 1.8×10^{-2} S/cm for $CuYO_2$) [32, 49]. The acceptor-doping mechanism remained unclear until it was demonstrated that $CuAlO_2$ exists

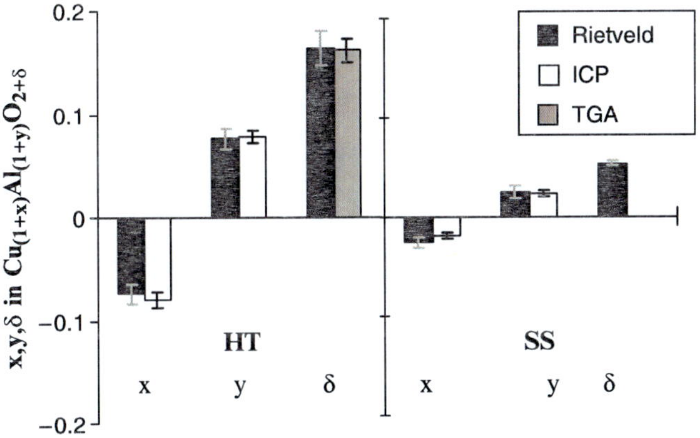

Fig. 8.11 Off-stoichiometric levels in $Cu_{(1+x)}Al_{(1+y)}O_{2+\delta}$ as determined from structure refinements and gravimetric results for hydrothermally synthesized (HT) and solid-state synthesized (SS) $CuAlO_2$. In both cases there is a Cu-deficiency and Al-excess that are nearly equivalent, which are approximately half of the oxygen excess [32]

with an excess of aluminum, i.e., aluminum anti-site defects ($[Al_{Cu}^{\bullet\bullet}]$) predominate [32]. However, such isolated defects would be donors rather than acceptors. Furthermore, aluminum prefers at least tetrahedral coordination by oxygen. Combined thermogravimetry (for oxygen content), chemical analysis (for cation content, via inductively-coupled plasma emission spectroscopy) and structural refinements (by Rietveld analysis of X-ray diffraction data) in Fig. 8.11 showed that Al-excess and Cu-deficiency are nearly equivalent, and amount to roughly half of the O-excess. Both conventional solid-state synthesis specimens and hydrothermal synthesis specimens (with larger Al-excess) were examined. Based upon these results, it was concluded that aluminum anti-site defects associate with two oxygen interstitials, i.e., $(Al_{Cu}^{\bullet\bullet}2O_i'')''$, thereby accomplishing roughly tetrahedral coordination of Al by oxygen – an oxygen above and below, and two oxygens in roughly trigonal (albeit distorted) interstitial sites. This also transforms an otherwise donor-species ($Al_{Cu}^{\bullet\bullet}$) into an acceptor species $(Al_{Cu}^{\bullet\bullet}2O_i'')''$, thereby accounting for the p-type character of $CuAlO_2$.

This behavior is most likely unique to $CuAlO_2$ and possibly $CuGaO_2$. This conclusion is based upon the plot of anti-site B-cation radius (in tetrahedral coordination) vs. B-site cation radius (in octahedral coordination) in Fig. 8.12. The horizontal line is the A-site Cu^+ radius (in linear coordination). This site is too small for trivalent cations larger than gallium. Furthermore, the situation of gallium may be borderline; reported conductivities of undoped $CuGaO_2$ tend to be quite small [37]. The situation in the Ag-delafossites is not clear. Although the A-site Ag^+ radius in linear coordination is considerably greater (0.81 Å [46]) than for Cu+, the larger trivalent cations tend to be incompatible with tetrahedral coordination.

What these findings suggest is that the stability range of $CuAlO_2$ exists to the Al-rich side of nominal stoichiometry, as shown schematically in Fig. 8.13.

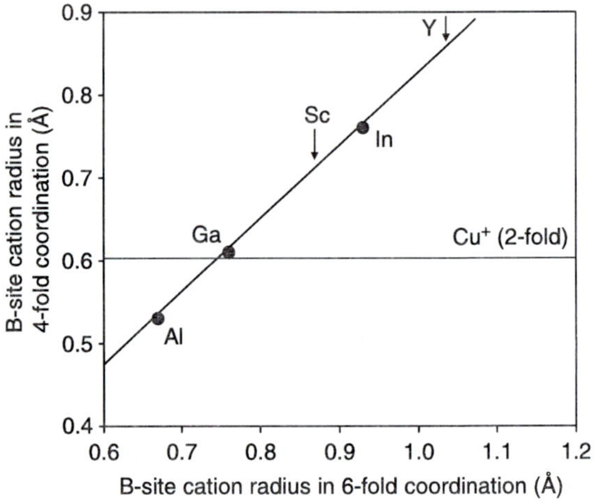

Fig. 8.12 Radius of delafossite (ABO_2) B-site cation in fourfold coordination vs. the radius in sixfold coordination. The *horizontal line* is the radius of Cu in twofold coordination. This copper site is too small for trivalent cations larger than gallium [47]

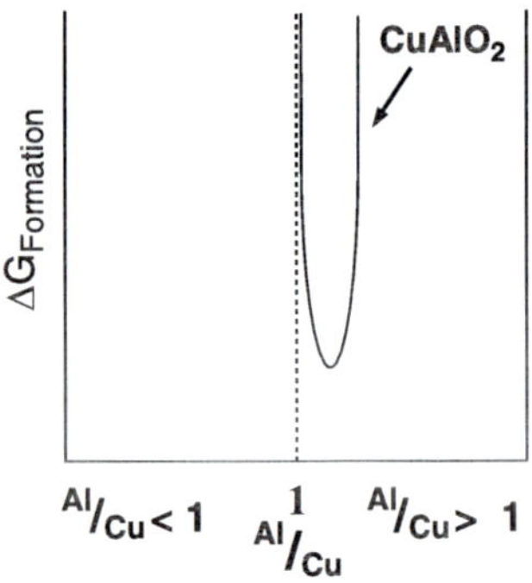

Fig. 8.13 Schematic free energy of formation for $CuAlO_2$, illustrating the tendency for $CuAlO_2$ to exist with a slight Al-excess

This conclusion is supported by experiments with intentionally off-stoichiometric mixtures of phases (Al-excess, Al-deficient); there was little detectable change in the hole content/conductivity of the majority delafossite phase [49]. A similar insensitivity to overall composition was demonstrated for $CuGaO_2$ [50]. This was interpreted in terms slight cation off-stoichiometry (i.e., copper vacancies as acceptors), but a slight Ga-excess (i.e., $(Ga_{Cu}^{\bullet\bullet}2O_i'')''$ associates) would also account for the much smaller hole contents/conductivities reported for $CuGaO_2$ [37].

Were it not for the localized character of transport (see below) and the tendency of the $(Al_{Cu}^{\bullet\bullet}2O_i'')''$ associates to trap holes, i.e., $(Al_{Cu}^{\bullet\bullet}2O_i''2Cu_{Cu}^{\bullet})^x$, with decreasing temperature, copper aluminate would be the ideal p-type TCO – self-doped, with large hole contents, and relatively insensitive to oxygen partial pressure during processing. However, localization limits mobility and trapping reduces the free hole content, such that hole content (and conductivity) is rather small at room temperature [32, 49]. Fortunately, other delafossites are more amenable to acceptor doping, whether by acceptor species, oxygen interstitials, or both. For example,

$CuScO_2$ can be acceptor-doped by Mg^{2+} on B-sites [49] and/or excess oxygen (interstitials) [32, 49], resulting in higher room temperature electrical conductivities [37].

8.3.3 Defect Complexation Summary

Point defect complex formation in TCOs is associated with "ternary" defect character – the host lattice, a dopant species, and a compensating (usually oxygen defect) species. Therefore, not every TCO will be susceptible to complex formation. When defect association occurs, however, it can result in enhanced doping levels, albeit with low doping efficiencies, owing to donor- or acceptor-compensation. For example, it is not uncommon to observe a linear doping regime ($n \propto [D^{\bullet}]$) followed by a doping-independent regime, i.e., $n = \text{constant} \neq f\,[D^{\bullet}]$, in highly donor-doped materials, e.g., Sn- and In-doped CdO thin films [51]. This is indicative of dopant compensation, which limits both carrier content (poor doping efficiency) and mobility (neutral impurity scattering). Only in special cases, e.g., ITO, will two-step processing result in elimination of the compensating defects, thereby "activating" the donor species and increasing the carrier concentration. The case of $CuAlO_2$ may be unique, whereby defect complexation transforms what would ordinarily be a donor, $Al_{Cu}^{\bullet\bullet}$, into an acceptor, $(Al_{Cu}^{\bullet\bullet}2O_i'')''$.

8.4 Extended Solid Solubility

The radii of d^{10} cations in octahedral coordination are fairly similar, e.g., 1.09 Å for Cd^{2+}, 0.88 Å for Zn^{2+}, 0.94 Å for In^{3+}, and 0.83 Å for Sn^{4+} [46]. It is therefore not surprising that there should be extended ranges of solid solubility in certain TCO phases. This is evidenced by straight lines in the ternary phase diagrams, corresponding to one of two "isovalent" substitution schemes (1) direct replacement of a given cation by a cation of identical charge, e.g., Ga^{3+} substituting for In^{3+}, or (2) co-substitution of a pair of cations by two cations whose average charge is the same as the host cation, e.g., Zn^{2+}/Sn^{4+} for two In^{3+} cations, or Cd^{2+}/Sn^{4+} for two In^{3+} cations.

There are significant ramifications of extended solid solubility on the stability, doping, and physical properties of TCO phases. First, substitutions can influence the stability ranges of TCO phases; compounds that are unstable in binary diagrams can be stabilized by a third component (i.e., these phases show up in ternary diagrams). Second, the defect structures (and carrier doping) of TCO phases can be significantly altered by extended solid solubility, especially in co-substitution phases, as we will show. Third, important TCO properties, such as optical band gap, can be tailored by adjusting composition. This is analogous to the well known "band gap" engineering of compound semiconductors through ternary alloying.

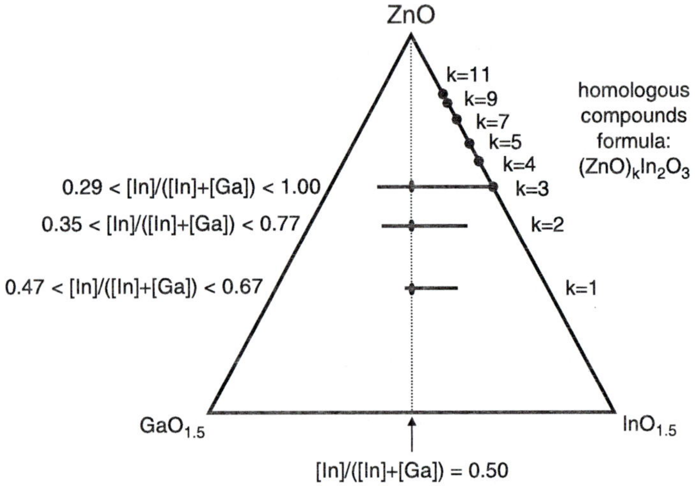

Fig. 8.14 Phase relationships in the ZnO-In_2O_3-Ga_2O_3 system at 1,400°C [52]

Figures 8.14 and 8.15 show the solid phase relationships in the ZnO-In_2O_3-Ga_2O_3 system at 1,400°C [52] and the ZnO-In_2O_3-SnO_2 at 1,275°C (adapted from [53, 54]). In each case the dots on the bounding ZnO-In_2O_3 binary refer to the $(ZnO)_kIn_2O_3$ homologous series of compounds; the binary phase relationships were established by Moriga et al. [55]. These compounds belong to space group $R\bar{3}m$ with $z = 3$ (the number of formula units per unit cell) when k is odd, and space group $P6_3/mmc$ with $z = 2$ when k is even. There are competing models for the precise structure. Cannard and Tilley [56] proposed a model in which k(0001) planes of ZnO (wurtzite structure) are sandwiched between two (111) planes of $InO_{1.5}$ (bixbyite structure). Kimizuka et al. [57] proposed that these compounds are isostructural with $(ZnO)_kLuFeO_3$, which has space group $R\bar{3}m$ for odd k values and space group $P6_3/mmc$ for even k values. In either case, In-based layers in octahedral coordination appear to control transport; the conductivity falls off dramatically as k (and the number of ZnO-planes) increases. Again, in the competition between octahedral coordination (In) and tetrahedral coordination (Zn), octahedral coordination wins out insofar as transport is concerned.

There are two interesting aspects of Fig. 8.14. First, the solubility ranges are horizontal lines, i.e., Ga^{3+} is clearly substituting for In^{3+}. Second, the $k = 1$ and $k = 2$ phases are stable in the ternary system at 1,400°C [52], whereas they are not stable in the ZnO-In_2O_3 binary [55]. This is most likely owing to the rather large size difference of Ga^{3+} (0.76 Å) vs. 0.94 Å for In^{3+} in octahedral coordination [46]. Clearly, the smaller gallium stabilizes the $k = 1$ and $k = 2$ phases.

The extended solid solubility ranges of the $k = 1$, $k = 2$, and $k = 3$ phases, with Ga^{3+} substituting for In^{3+}, allow for "band gap" engineering, i.e., the band gap varies significantly with the Ga/In ratio. This is not surprising, given the large difference in fundamental band gap for In_2O_3 (3.0 eV [58]) compared to Ga_2O_3

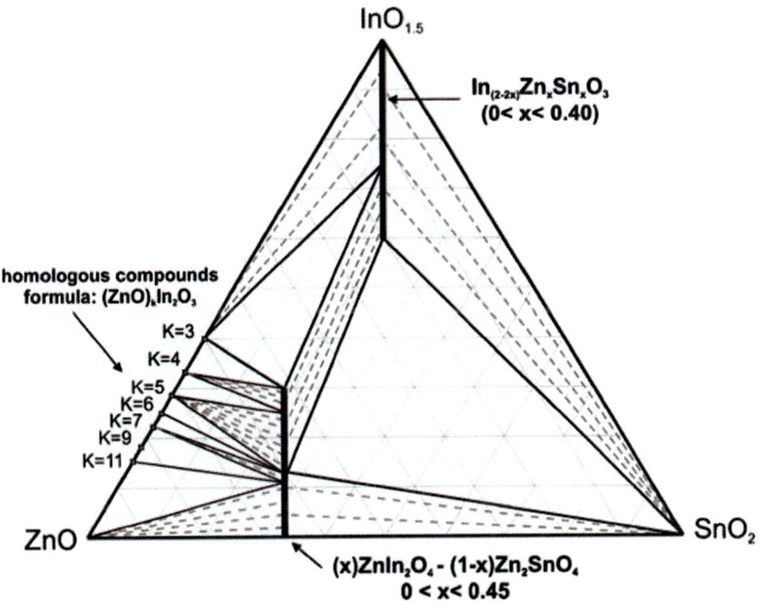

Fig. 8.15 Phase relationships in the In_2O_3-SnO_2-ZnO system at 1,275°C [53, 54]

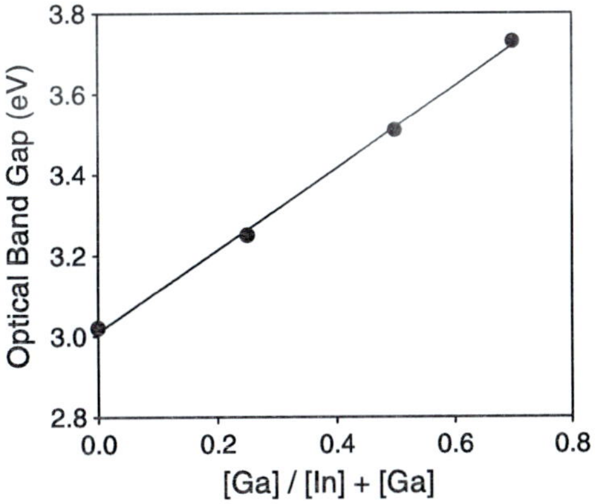

Fig. 8.16 Optical band gap vs. the gallium content of the k = 3 layered compound. $In_{1-x}Ga_{1+x}O_3(ZnO)_3$ [52]

(4.9 eV [59]). Figure 8.16 plots the optical band gap (from diffuse reflectance measurements) vs. the gallium content, [Ga]/([Ga]+[In]), for the k = 3 phase [52]. As expected, there is a large increase in band gap as indium is replaced by gallium. With relatively small changes in composition, the band gap can be "tuned" over a broad range.

The long vertical lines of solid solubility in both the CdO-In_2O_3-SnO_2 system (Fig. 8.3) and the ZnO-In_2O_3-SnO_2 system (Fig. 8.15) involve co-substitution. As described above, these substitutions are essentially isovalent in character, i.e., a 2+/4+ pair of cations replace a pair of 3+/3+ host cations. In the case of Cd/Sn, the average radius of the co-substituting cations is 0.96 Å (1.09 Å for Cd^{2+} and 0.83 Å for Sn^{4+}) vs. 0.94 Å for the In^{3+} they replace. Therefore, the substitution is not only isovalent, but is a nearly size-matched to the host. This enables the extended range of solid solubilities observed for both bixbyite and spinel phases. Similarly extended co-solubilities of Cu/Sn and Ni/Sn have been reported in In_2O_3, again attributable to the nearly size-matched character of such substations [60].

One ramification of this extended solid solubility is the potential for "band structure engineering." Kammler et al. [61] demonstrated that in lightly carrier-doped bulk $CdIn_2O_4$-Cd_2SnO_4 spinel solid solutions, the fundamental band gap decreased by as much as 0.3 eV with increasing Sn-content. In much more highly carrier-doped thin films, however, the effective band gap actually *increased* with Sn-content through donor-doping via the well-known Burstein-Moss shift. It was concluded that the fundamental band gap and Fermi level could be independently modified/tailored by cation composition, i.e., via solid solution/co-substitution, and carrier-doping, respectively.

What is surprising about both the bixbyite and spinel solid solutions is their predisposition for n-type character. In fact, to the authors' knowledge all attempts to dope them in the p-type direction have failed [62]. It is virtually impossible to prepare bulk or thin film Cd/Sn-doped bixbyites or spinels that are not highly conductive n-type TCOs. The same can be said of Zn/Sn-doped bixbyite, although the Zn/Sn-doped spinel is much less conductive, for the reasons outlined above.

If these phases involve truly isovalent substitutions they should be fully compensated, i.e., the concentrations of donors and acceptors should cancel. For example, in Cd/Sn co-doped In_2O_3, $[Cd_{In}'] = [Sn_{In}^{\bullet}]$, and the electron population would be set by the intrinsic electronic ionization constant (i.e., $n = p = K_i^{1/2}$). Instead, these materials are clearly donor-doped to high levels. We have suggested (and demonstrated) that both bixbyite and spinel phases have an inherent donor-excess character, i.e., $[Sn_{In}^{\bullet}] > [Cd_{In}']$, such that $n = [Sn_{In}^{\bullet}] - [Cd_{In}'] > 0$ under *all* conditions of preparation [63, 64]. In other words, the stability field actually exists to the donor-rich side of cation stoichiometry, much like the situation for $CuAlO_2$ represented in Fig. 8.13.

We have demonstrated recently a similar tendency for self-doping in the Zn/Sn co-doped bixbyite phase [65]. Intentionally Zn-excess and Sn-excess compositions (5% excess) were prepared to either side of the bixbyite line in Fig. 8.15. This resulted in X-ray detectable, but discontinuous, second phases (as per the phase diagram) to fix the majority bixbyite phase at its solubility limit to either side of nominal stoichiometry. The ratio of room temperature conductivities of as-prepared Sn-rich assemblages vs. Cd-rich assemblages is plotted vs. the extent of co-substitution in Fig. 8.17a. The ratio goes to zero as we approach the terminal extent of co-substitution in the phase diagram. However, the conductivity remains quite high and n-type character is maintained. In Fig. 8.17b we plot the ratio of reduced (4% H_2/96N_2, at 400°C for 4 h) vs. as-prepared conductivities of nominal compositions along the same line. Again, this ratio goes to zero as the terminal solid solubility is reached, and high conductivities are maintained. These observations indicate that

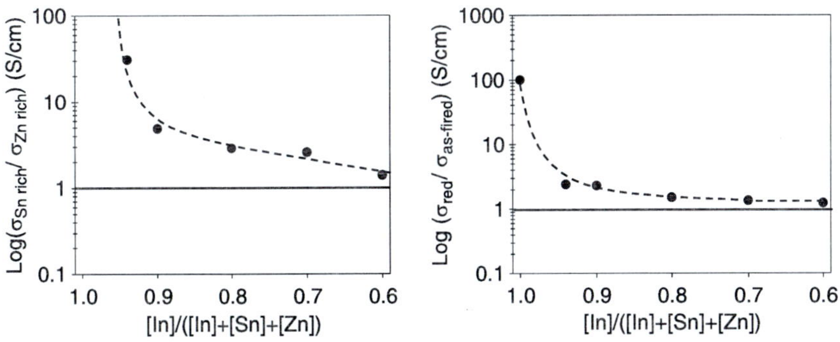

Fig. 8.17 (**a**) Ratio of room temperature conductivities of as-prepared Sn-rich vs. Zn-rich assemblages as a function of the extent of co-substitution. (**b**) Ratio of the reduced vs. as-prepared conductivities for specimens of nominal co-doping levels [65]

$n = [Sn_{In}^{\bullet}] - [Zn_{In}'] > 0$ under all conditions. At the terminal co-solubility, the donor-to-acceptor excess governs doping/conductivity to the extent that there is little detectable oxidation/reduction sensitivity. We refer to such robust compositions as inherently "self-doped." An important ramification of "self-doped" TCO compositions is their relative immunity to redox variations during the numerous processing steps required for fabrication of microelectronic and/or photovoltaic thin film devices.

8.5 Limiting Factors for TCOs

We now consider some important limiting factors for ternary/multinary TCO behavior – *charge localization* and how it limits the mobility of p-type TCOs, *crystalline anisotropy* and how it limits the transport of polycrystalline TCOs, and finally *structural disorder* and the limitations (or possibilities) for realizing amorphous TCOs.

The development of highly dopable, large-mobility p-type TCOs would be a significant breakthrough for a range of technological applications, including photovoltaics (for tandem cell design) and transparent electronics (in active devices). Unfortunately, as of this writing p-type oxide TCOs are either difficult to dope and/or possess low carrier mobilities. Doping limitations have already been considered for the delafossite class of p-type TCOs (see above) and elsewhere in the literature (see [37] and cited references). We now turn to the subject of *charge localization* and the origin of their large effective masses.

As discussed by Cann and co-workers [37], ABO_2 delafossite oxides can exhibit conductivities ranging from insulating to semi-metallic, depending upon the composition. The d^9 electron configuration of the monovalent A-cations, Pd and Pt, creates an unfilled d_z2-s hybrid orbital that is ultimately responsible for their

Table 8.3 Summary of the electrical properties of $CuBO_2$ at 740°C and $\log(pO_2) \sim -3.9$ [32, 49]

B=	Al	Sc	$Sc_{0.95}Mg_{0.05}$	Y	$Y_{0.95}Ca_{0.05}$
σ (S/cm)	3.75	0.37	2.28	0.018	0.26
μ ($cm^2/V/s$)	0.08	0.05	0.03	0.04	0.05
$p \times 10^{20}/cm^3$	3.02	0.45	4.48	0.026	0.33
E_h (eV)	0.14	0.22	0.25	0.23	0.28
μ_o ($Kcm^2/V/s$)	500	660	570	610	1,130

observed metallic conductivity [66–69]. Unfortunately, the Pd and Pt delafossites are optically opaque.

In contrast, the d^{10} electronic configuration of the Cu- and Ag-based delafossites leads to improved transparency, but with polaron formation. This is exhibited by small polaron hopping at elevated temperatures [32] and variable-range hopping at low temperatures [70, 71]. Band structure calculations agree with the experimental transport behavior, showing flat, non-disperse (large effective mass) valence bands in, e.g., $CuAlO_2$ [48].

Table 8.3 shows the derived transport parameters at 740°C for Cu-based delafossites, both undoped (B = Al, Y, Sc) and doped (B = Y, with Ca; B = Sc, with Mg) [32, 49]. At this temperature the conductivity is thermally activated, whereas the thermoelectric coefficient is not (or much less so) for each compound; therefore the mobility must also be thermally activated. The carrier content (and the concentration of localized carriers, i.e., Cu^{2+} on A-sites) can be calculated from the thermopower using existing small polaron models [32, 49]. From the mobility pre-exponential factors in ($\mu = (\mu_o/T)\exp(-E_h/kT)$, where μ_o = 500–1,000 Kcm^2/V s) and corresponding geometrical factors (e.g., hopping distances), reasonable lattice vibrational frequencies of 3–5 $\times$ 10^{12}/s can be obtained. The resulting mobilities are in the range of 0.01–0.1 cm^2/V s, also characteristic of small polaron conductors. This is to be compared with 10–75 cm^2/V s for the best n-type TCO materials.

Clearly, the strategies outlined above (1) cations with d-shell energies comparable to that of O-2p, (2) the use of linear coordination to keep oxygen ligands at a distance and (3) oxygen in tetrahedral coordination for reduced non-bonding character – do *not* result in delocalized holes in the Cu-delafossites. Activation energy decreases with B-cation radius (aluminum is the smallest), however there is no transition from localized to delocalized behavior. The situation in the Ag-delafossites remains unclear and is the subject of ongoing research. The Ag-based materials are a challenge to synthesize and are difficult to dope. Furthermore, they are hard to characterize, especially for transport properties vs. temperature (to test for thermally activated transport), given their susceptibility to thermal decomposition.

Some p-type success has been had with hybrid structures involving other anions. For example, higher hole mobilities have been be reported for oxychalcogenides (see Ref. [72] and references therein), which have alternating lanthanum oxide and copper chalcogenide layers. Mobilities in the 0.2–0.5 and 4–8 cm^2/V s range have been reported for doped LaCuOS and LaCuOSe, respectively. Even higher mobilities (~10 $cm^2/V\ s$) have been obtained when oxygen is completely replaced by other anions, e.g., doped BaCuSF [73]. Such mobilities are only a factor of ~5 lower than those of conventional n-type TCOs. The fact that the best p-type TCOs remain less conductive by 1–2 orders of magnitude than their n-type counterparts attests to how difficult they are to dope. To the authors' knowledge, p-type TCOs combining high mobility with large hole contents have yet to be realized.

A second limitation common to many p-type and certain n-type TCOs has to do with crystalline *anisotropy*. Most cost-effective fabrication methods result in the production of polycrystalline films. This is not a problem for highly degenerate, isotropic n-type materials. Coutts has argued that grain boundary scattering plays a relatively minor role at high doping levels in such cases; the electron mean free path (on the order of nanometers) is much less than typical grain sizes [74]. For less highly doped isotropic TCOs, however, grain boundary scattering will play an important role [23]. And this effect can be magnified for highly anisotropic/layered materials, where alignment of conducting planes at grain boundaries becomes important.

For example, Fig. 8.18 plots reported room temperature conductivities for bulk specimens of the layered $(ZnO)_kIn_2O_3$ homologous series of compounds vs. the octahedral site density of indium cations (from several sources [52, 55, 75]). Also shown for comparison is the slope of the line from Fig. 8.2, marked "3D n-type TCOs." The difference is striking, much more so than can be accounted for on the basis of porosity differences alone. Except for the k = 3 phase, whose conductivity lies intermediate between the 3D and 2D trend lines, the anisotropic phases have much lower conductivities than the essentially isotropic phases in Fig. 8.2. It should be noted that the k = 1 and k = 2 phases (not shown in Fig. 8.18) would be expected to have higher conductivities (owing to their larger concentrations of octahedral sites). However, the Ga^{3+} used to stabilize these structures has two deleterious effects (1) dilution of TCO-active cations (In^{3+}) and (2) reduction of carrier content (by binding oxygen, see below). What Fig. 8.18 suggests is that grain

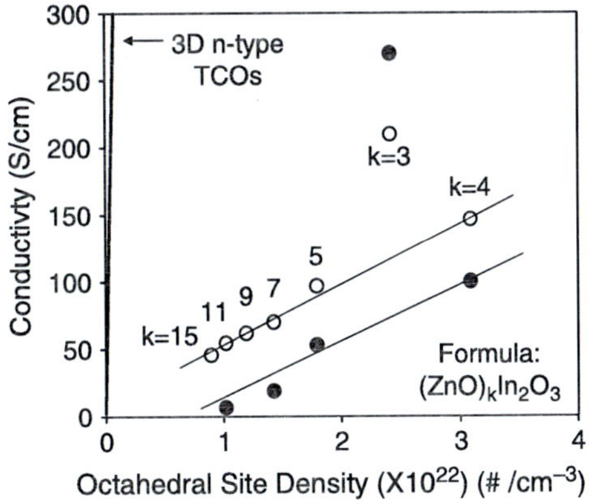

Fig. 8.18 Room temperature conductivity of homologous series of compounds $(ZnO)_kIn_2O_3$ vs. the octahedral site density, labeled with the appropriate k-number. The dark points are taken from Moriga et al. [55] and the light points from Kaga et al. [75]. The dark line near the y-axis is the slope from the plot in Fig. 8.2

boundary scattering needs to be taken into account when dealing with polycrystalline materials (bulk, thin films) of such highly anisotropic TCO materials. This does not, of course, preclude the attainment of high conductivity in highly textured or single crystal films, whose in-plane conductivity (parallel to their "conducting" planes) follows the trend line behavior of optimized n-type TCOs (Fig. 8.2).

The final limitation for TCO materials has to do with *structural disorder*. It has long been known that materials such as ITO can be deposited at low temperatures (see [76] and cited references). This results in amorphous thin films, but with electrical properties (carrier content, mobilities) not significantly less than for polycrystalline films (see below). We can evaluate the limitations/potential for achieving amorphous ternary/multinary TCOs, both p-type and n-type, by considering some of the factors already discussed in this chapter.

We began by emphasizing the important role of cation coordination in TCO transport. For most p-type TCOs studied to date, hole transport depends upon 2-D crystallographic layering, e.g., to stabilize monovalent Cu^+ in linear coordination as in the O-Cu-O dumbbell layers of the delafossite structure. Despite success in fabricating amorphous films of these materials, they do not show p-type electrical conduction [77]. This has been attributed to the inability to maintain the linear dumbbell coordination (and monovalent Cu^+) in the amorphous state [78]. This is not difficult to rationalize. Glasses typically require "network-formers" with a minimum cation coordination of 3 or higher; they cannot be made from concentrated "network modifiers" such as monovalent Cu+ (CN = 2). Furthermore, it would be unlikely for an occasional O-Cu-O dumbbell in the amorphous state to have a similar dumbbell at an appropriate distance, which is a necessary requirement for hole transport. These findings suggest that 3-D p-type TCOs must be developed as a first step toward amorphous p-type transparent conductors. Prototype oxides include p-doped NiO (cubic rocksalt structure) [79] and off-stoichiometric p-type $ZnRh_2O_4$ (cubic spinel structure), the latter of which has been successfully prepared as nanocrystalline or amorphous p-type films [80, 81].

On the other hand, considerable success has been achieved with amorphous films of the basis TCO n-type cations. An excellent case in point was the pioneering work of Phillips et al. [6], who grew films of Zn-In-Sn-O (ZITO) by pulsed laser deposition. For deposition temperatures below 450°C, their films were amorphous by X-ray diffraction. Nevertheless, large electron contents (up to $3.4 \times 10^{20}/cm^3$) and surprisingly high mobilities (up to 46 cm^2/V s) were obtained. The resulting best conductivity (2,500 S/cm) is only a factor of 4 less than the "optimized" crystalline bixbyite value in Fig. 8.2. This work has spawned the emerging field of "amorphous TCOs" involving binary systems (with representative citations among many that could be cited) such as ITO [82], ZTO or zinc-tin oxide [83], and IZO or indium-zinc oxide (see [84, 85] and references therein); and ternary systems such as ZITO (zinc-indium-tin oxide), of which Refs. [84, 85] are early representatives of a much larger body of work, and GIZO (gallium-indium-zinc oxide) [86].

We began this chapter by emphasizing the important role of cation coordination in TCO behavior. It was concluded that *continuous edge- or corner-sharing of Cd^{2+}, In^{3+} and Sn^{4+} octahedra is a necessary criterion for the formation of an n-type transparent conductor.* The observation can be made that several of the compounds in Fig. 8.2 have distorted octahedral environments. For example, the bixbyite structure has TCO basis cations in two different octahedral environments, with significant distortions of the ideal octahedron (see Fig. 8.6). The same can be said of the distorted perovskite structure (o-$CdSnO_3$) and even, to a lesser degree, of rutile (SnO_2). Nevertheless, these distortions do not appear to interfere with the conductivity vs. octahedral site density trend of Fig. 8.2. It is therefore not surprising that amorphous TCOs with the basis cations in octahedral (albeit distorted) coordination can be good transparent conductors.

This can be rationalized by consideration of electron mean free path. For lightly doped semiconductors with relatively large mean free paths, structural disorder would be expected to significantly reduce the electron mobility. However, for highly degenerate TCO materials, the mean free path of electrons is already on the order of nanometers [74]. When this becomes comparable to the scale of short-range order in amorphous TCOs, structurally disordered materials can be expected to exhibit good TCO properties – high mobilities and (if appropriately doped) correspondingly large electronic conductivities.

We can interpret the formation and "doping" of amorphous TCOs within the well-established "random-network" model of glasses. The best amorphous n-type TCOs to date are binary and ternary combinations of the Zn^{2+}, In^{3+} and Sn^{4+} basis cations, possibly in combination with Ga^{3+}. Unlike monovalent Cu^{+} (see above), which is a network-modifier, each of the n-type basis cations is a potential network-former. It can be posited that the majority cation (e.g., In^{3+} in a-ITO, a-IZO, a-ZITO, etc., where a = amorphous) sets the electroneutrality condition, i.e., a-In_2O_3 should be "undoped." Zn^{2+} on an In^{3+} site should be an acceptor and Sn^{4+} on an In^{3+} site should be a donor, with co-doping leading to compensation ([donors]=[acceptors]), as is the case for crystalline co-doped bixbyite. However, as we emphasized previously, the bixbyite structure is quite flexible with respect to oxygen non-stoichiometry, allowing for both oxygen vacancies and oxygen interstitials. This allows for even greater cation off-stoichiometry in the case of amorphous films.

For example, excess Zn^{2+} can be readily accommodated by ionic compensation. One can think of this as oxygen vacancies compensating Zn-acceptors ($[V_O^{\bullet\bullet}] = ½ [Zn_{In}']$). Alternatively, one can think of this as the local oxygen coordination being reduced around Zn^{2+} (which prefers tetrahedral coordination) so as to render it neutral insofar as electroneutrality is concerned. Note that Zn^{2+} in tetrahedral configuration, with Zn-Zn bond lengths comparable to those in crystalline ZnO, may be expected to contribute to n-type a-TCO behavior.

On the other hand, excess Sn^{4+} can be accommodated in association with oxygen interstitials, similar to the $(2Sn_{In}^{\bullet}O_i'')^x$ clusters in crystalline ITO. Alternatively, one can think of this as the local oxygen coordination being increased around Sn^{4+}, as is known to take place in crystalline ITO (see above). To the extent that some Sn^{4+}

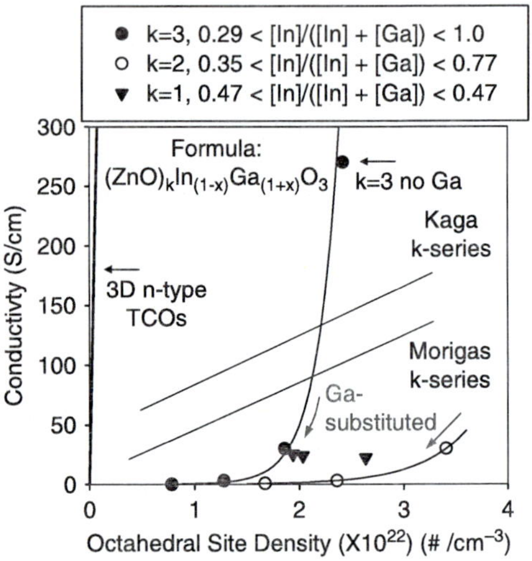

Fig. 8.19 Room temperature conductivity of homologous series of compounds with gallium doping on the indium site, $(ZnO)_kIn_{(1-x)}Ga_{(1+x)}O_3$, vs. the octahedral site density for compounds with k = 1,2, and 3. The data are taken from Moriga et al. [55]. The *trend lines* from Fig. 8.18 are shown for reference. The *dark line* near the y-axis is the slope from the plot in Fig. 8.2

remains in sixfold coordination, i.e., $n = [Sn_{In}^{\bullet}]$, and/or can be activated by reduction (as per (8.1)), n-type character will be achieved. In fact, regardless of the Sn/Zn ratio in ZITO, n-type character is consistently obtained in amorphous films. There is no evidence that Zn^{2+} acts as an acceptor imparting p-type character to amorphous films. This suggests that $n = [Sn_{In}^{\bullet}] - [Zn_{In}'] > 0$ under *all* conditions of preparation, similar to the situation in co-doped crystalline bixbyite.

Gallium appears to have a strong influence on the defect structure of both crystalline and amorphous n-type TCOs. Figure 8.19 is a replot of the data in Fig. 8.18 for the $(ZnO)_kIn_2O_3$ homologous series of compounds. Superimposed on the plot are data for Ga-substituted $(ZnO)_kIn_{2-x}Ga_xO_3$ k = 1, 2 and 3 phases. Conductivities fall well below the trend lines for the binary compositions. The curve for the Ga-doped k = 3 phase illustrates the marked reduction in conductivity with gallium substitution for indium. Similar reductions in carrier content and conductivity have been reported for amorphous Zn-In-Ga-O compostions [55]. These were attributed to the strong tendency of gallium to bind oxygen, decreasing oxygen vacancy donors and reducing carrier contents. This can be a disadvantage – for amorphous transparent electrodes – or an advantage – for amorphous semiconducting channel layers in field-effective transistors.

These are only preliminary observations regarding the "defect structure" of amorphous TCOs, based upon what is known about defects in their crystalline counterparts. In fact, little is known about the defect structures of amorphous materials, although some extrapolations may be made from amorphous silicon [87] and amorphous chalcogenides [88].

The advantages of amorphous TCOs are ease of processing and compatibility with polymer substrates (for low-temperature fabrication of flexible devices). The potential also exists for all-oxide/transparent devices such as diodes, near-UV light-emitting diodes, and thin film transistors [89].

8.6 Conclusions

Since the mid-1990s much transparent conductor research has been on ternary/multinary oxides and related materials. We define ternary/multinary TCOs as (a) having complex point defect associates, (b) having two or more cation sublattices, and/or (c) existing in ternary or higher phase space. In such systems, phase diagrams are helpful to establish stability ranges, to identify doping mechanisms, and to support systematic composition-structure-property studies. They are also useful in understanding differences between bulk and thin film behavior.

In this chapter we have considered the roles of (1) cation coordination, (2) point defect complexation, and (3) extended solid solubility on the structure–property relationships of both n-type and p-type ternary/multinary TCOs. We have also considered (4) limitations to TCO behavior, including charge localization, crystalline anisotropy, and structural disorder.

Cation coordination is of paramount importance in the very best n-type TCOs. These largely isotropic crystal types exhibit an excellent agreement between optimized transparent conductivity and the site density of basis cations (Cd^{2+}, In^{3+} and/or Sn^{4+}) in octahedral coordination. This allows us to modify the Shannon et al. criterion: "*continuous edge- or corner-sharing of* Cd^{2+}, In^{3+} *and* Sn^{4+} *octahedra is a necessary criterion for the formation of an n-type transparent conductor*." Note that Zn^{2+} is excluded from this criterion. Zn^{2+} behaves as a basis TCO cation in tetrahedral coordination (e.g., in donor-doped ZnO and possibly in amorphous TCOs), but in octahedral coordination it behaves as a diluent species, as does Ga^{3+}.

Cation coordination also plays an important role in achieving p-type behavior in transparent oxides. To overcome the ionicity of the metallic oxides, where the O-2p levels are generally far lower in energy than the valence orbitals of the metal species, d^{10} cations with O-2p-comparable energies (Cu+ and Ag+) can be placed in linear coordination by oxygen. The resulting structures, e.g., the layered ABO_2 delafossite structure with layers of edge-shared $B^{3+}O_6$ octahedra sandwiching planes of close-packed A^+ cations, produces the linear/dumbbell coordination of the A-cations (to keep the oxygen ligands at distance) and tetrahedral coordination of oxygen anions by cations (for reduced non-bonding character of the metal-oxygen bond), in an attempt to reduce the localization of the valence band edge. A wide range of ABO_2 (A = Cu, Ag) materials, many of which can only be fabricated by low temperature methods (e.g., hydrothermal synthesis) have now been made, most of which are p-type and transparent (to different degrees). At least one of the delafossites ($CuInO_2$) is amphoteric; the ability to alternatively acceptor-dope or donor-dope this material derives from the presence of two cation environments – an octahedral indium environment (for n-type doping/transport) and a linear copper environment (for p-type doping/transport).

Defect complexation may play a more important role governing the composition-defect structure–property relationships of ternary/multinary TCOs than previously thought. The dominant point defect species in ITO has been shown to be $(2Sn_{In}^{\bullet}O_i'')^x$ neutral associates, which can be reduced at intermediate temperatures (thereby activating Sn-donors), rather than oxygen vacancies. In $CuAlO_2$, another defect complex dominates the defect structure/doping. An inherent Al-excess is found to be associated with excess oxygen interstitials, i.e., $(Al_{Cu}^{\bullet\bullet}2O_i'')''$, which effectively transforms a donor defect into an acceptor species. As a result, $CuAlO_2$ is the very best *undoped* p-type delafossite at elevated temperatures. Except for the tendency of the associate to trap holes on cooling from elevated temperatures, $CuAlO_2$ would be the best *undoped* p-type delafossite at room temperature. Unfortunately, the B-site is too small to be acceptor-doped, so other delafossites (e.g., $CuScO_2$) are better candidates for acceptor doping in terms of room temperature conductivity.

A major result of phase diagram studies in ternary TCO systems is the observation of extended solid solubility ranges of important TCO phases. Two isovalent substitution schemes have been identified (1) direct replacement of a given cation by a cation of identical charge, e.g., Ga^{3+} substituting for In^{3+}, or (2) co-substitution of a pair of cations by two cations whose average charge is the same as the host cation, e.g., Zn^{2+}/Sn^{4+} for two In^{3+} cations, or Cd^{2+}/Sn^{4+} for two In^{3+} cations. These extended solubility ranges allow for composition control of band structure, i.e., "band structure engineering." Note that the effective band gap (i.e., the Fermi level in the n-type TCOs) can be independently varied by carrier-doping, e.g., through the well-known Burstein-Moss shift.

A second ramification of co-substitution is the observation that certain co-doped phases (e.g., bixbyite, spinel) are found to be inherently off-stoichiometric in favor of donors vs. acceptors, i.e., n = [donors] − [acceptors] > 0. This renders the co-doped TCO phases persistently n-type, regardless of preparation/treatment conditions. Such "self-doped" TCOs should be relatively immune to the redox steps employed in the fabrication of microelectronic and/or photovoltaic devices.

There are several limiting factors that can influence the behavior of TCOs. In the case of the p-type delafossites, charge localization severely limits their hole mobilities (owing to polaron formation), such that the best p-type TCOs remain orders of magnitude less conductive than their n-type counterparts. In the case of highly anisotropic or layered crystal structures, both n-type and p-type TCOs will suffer from grain boundary scattering in polycrystalline form (bulk or thin film). And structural disorder severely restricts the ability of layered crystal structures (most notably the p-type delafossites) to be p-type conductive in the amorphous state. On the other hand, such disorder is not a severe limitation for the largely isotropic n-type materials; when highly degenerate, the electron mean free path is comparable to the scale of short-range order in amorphous films.

References

1. F. Streintz, Ann. Phys. (Leipzig), 9, 854 (1902).
2. H. L. Hartnagel, A. L. Dawar, A. K. Jain, and C. Jagadish, Semiconducting Transparent Thin Films, Institute of Physics Publishing, Bristol (1995).

3. M. Orita, H. Ohta, M. Hirano, and H. Hosono, Appl. Phys. Lett., 77, 4166 (2000).
4. M. Orita, H. Hiramatsu, H. Ohta, M. Hirano, and H. Hosono, Thin Solid Films, 411, 134 (2002).
5. R. J. Cava, J. M. Phillips, J. Kwo, G. A. Thomas, R. B. van Dover, S. A. Carter, J. J. Krajewski, W. F. Peck, Jr., J. H. Marshall, and D. H. Rapkine, Appl. Phys. Lett., 64, 2071 (1994).
6. J. M. Phillips, R. J. Cava, G. A. Thomas, S. A. Carter, J. Kwo, T. Siegrist, J. J. Krajewski, J. H. Marshall, W. F. Peck, Jr., and D. H. Rapkine, Appl. Phys. Lett., 67, 2246 (1995).
7. T. Minami, H. Sonohara, T. Kakumu, and S. Takata, Jpn. J. Appl. Phys., 34, L971 (1995).
8. T. Minami, S. Takata, and T. Kakumu, J. Vac. Sci. Technol., A14, 1689 (1996).
9. H. Un'no, N. Hikuma, T. Omata, N. Ueda, T. Hashimoto, and H. Kawazoe, Jpn. J. Appl. Phys., Part 2 Lett., 32, L1260 (1993).
10. K. Yanagawa, Y. Ohki, T. Omata, H. Hosono, N. Ueda, and H. Kawazoe, Appl. Phys. Lett., 65, 406 (1994).
11. D. D. Edwards, T. O. Mason, F. Goutenoire, and K. R. Poeppelmeier, Appl. Phys. Lett., 70, 1706 (1997).
12. A. Wang, S. C. Cheng, M. P. Chudzik, T. J. Marks, R. P. H. Chang, and C. R. Kannewurf, Appl. Phys. Lett., 73, 327 (1998).
13. D. S. Ginley, and C. Bright, (eds) Mater. Res. Soc. Bull., 25, 8 (2000).
14. H. Kawazoe, M. Yasukawa, H. Hyodo, M. Kurita, H. Yanagi, and H. Hosono, Nature, 389, 939 (1997).
15. A. Kudo, H. Yanagi, H. Hosono, and H. Kawazoe, Appl. Phys. Lett., 73, 220 (1998).
16. R. D. Shannon, J. L. Gilson, and R. J. Bouchard, J. Phys. Chem. Solids, 38, 877 (1977).
17. B. J. Ingram, G. B. Gonzalez, D. R. Kammler, M. I. Bertoni, and T. O. Mason, J. Electroceram. 13, 167 (2004).
18. G. Haacke, Appl. Phys. Lett., 28, 622 (1976).
19. D. R. Kammler, B. J. Harder, N. W. Hrabe, N. M. McDonald, G. B. Gonzalez, D. A. Penake, and T. O. Mason, J. Am. Ceram. Soc., 85, 2345 (2002).
20. X. Wu, T. J. Coutts, and W. P. Mulligan, J. Vac. Sci. Technol. A, 15, 1057 (1997).
21. H. Ohta, M. Orita, M. Hirano, and H. Hosono, J. Appl. Phys., 91, 3547 (2002).
22. A. Wang, M. Lane, N. L. Edleman, A. W. Metz, M. A. Lane, R. Asahi, V. P. Dravid, C. R. Kannewurf, A. J. Freeman, and T. J. Marks, Proc. Natl. Acad. Sci. USA, 98, 7113 (2001).
23. T. Minami, Mater. Res. Soc. Bull., 25, 38 (2000).
24. R. S. Gordon, Mater. Res. Soc. Bull., 25, 52 (2000).
25. F. P. Koffyberg, Phys. Rev. B, 13, 4470 (1976).
26. D. Ko, K. R. Poeppelmeier, D. R. Kammler, G. B. Gonzalez, T. O. Mason, D. L. Williamson, D. L. Young, and T. J. Coutts, J. Solid State Chem., 163, 259 (2002).
27. D. D. Edwards, T. O. Mason, W. Sinkler, and L. D. Marks, J. Solid State Chem., 140, 242 (1998).
28. T. O. Mason, D. R. Kammler, B. J. Ingram, G. B. Gonzalez, D. L. Young, and T. J. Coutts, Thin Solid Films, 445, 186 (2003).
29. D. L. Young, D. L. Williamson, and T. J. Coutts, J. Appl. Phys., 91, 1464 (2002).
30. H. Kawazoe, H. Yanagi, K. Ueda, and H. Hosono, Mater. Res. Soc. Bull., 25, 28 (2000).
31. H. Yanagi, H. Kawazoe, A. Kudo, M. Yasukawa, and H. Hosono, J. Electroceram., 4, 427 (2000).
32. B. J. Ingram, G. B. Gonzalez, T. O. Mason, D. Y. Shahriari, A. Barnabe, D. Ko, and K. R. Poeppelmeier, Chem. Mater., 16, 5616 (2004).
33. W. C. Sheets, E. Mugnier, A. Barnabe, T. J. Marks, and K. R. Poeppelmeier, Chem. Mater., 18, 7 (2006).
34. D. Y. Shahriari, A. Barnabe, T. O. Mason, and K. R. Poeppelmeier, Inorg. Chem., 40, 5734 (2001).
35. H. Yanagi, T. Hase, S. Ibuki, K. Ueda, and H. Hosono, Appl. Phys. Lett., 78, 1583 (2001).
36. X. Nie, S.-H. Wei, and S. B. Zhang, Phys. Rev. Lett., 88, 066405 (2002).
37. M. A. Marquardt, N. A. Ashmore, and D. P. Cann, Thin Solid Films, 496, 146 (2006).

38. T. Otabe, K. Ueda, A. Kudoh, and H. Hosono, Appl. Phys. Lett., 71, 1036 (1998).
39. S. Ibuki, H. Yanagi, K. Ueda, H. Kawazoe, and H. Hosono, J. Appl. Phys., 88, 3067 (2000).
40. G. Frank, and H. Köstlin, Appl. Phys. A, 27, 197 (1082).
41. G. B. Gonzalez, T. O. Mason, J. P. Quintana, O. Warschkow, D. E. Ellis, J.-H. Hwang, and J. P. Hodges, J. Appl. Phys., 96, 3912 (2004).
42. N. Yamada, I. Yasui, Y. Shigesato, H. Li, Y. Ujihira, and K. Nomura, Jpn. J. Appl. Phys., 38, 2856 (1999).
43. N. Yamada, I. Yasui, Y. Shigesato, H. Li, Y. Ujihira, and K. Nomura, Jpn. J. Appl. Phys., 39, 4158 (2000).
44. D. Mergel, and Z. Qiao, J. Appl. Phys., 95, 5608 (2004).
45. S. P. Harvey, T. O. Mason, C. Körber, and A. Klein, Phys. Chem. Chem. Phys., 11, 3099 (2009).
46. R. D. Shannon, Acta Crystallogr. A, 32, 751 (1976).
47. B. J. Ingram, PhD dissertation, Northwestern University (2003).
48. B. J. Ingram, T. O. Mason, R. Asahi, K. T. Park, and A. J. Freeman, Phys. Rev. B, 64, 155114-1 (2001).
49. B. J. Ingram, B. J. Harder, N. W. Hrabe, T. O. Mason, and K. R. Poeppelmeier, Chem. Mater., 16, 5623 (2004).
50. N. A. Ashmore, and D. P. Cann, J. Mater. Sci., 40, 3891 (2005).
51. R. Asahi, J. R. Babcock, N. L. Edleman, D. R. Kammler, D. K., M. A. Lane, A. W. Metz, A. Wang, M. Yan, R. P. H. Chang, V. Dravid, A. J. Freeman, C. R. Kannewurf, T. J. Marks, T. O. Mason, and K. R. Poeppelmeier, Electrochem. Soc. Proc. 2001–10, Electrochemical Society, Pennington, NJ, (2001) p. 333.
52. T. Moriga, D. R. Kammler, and T. O. Mason, J. Am. Ceram. Soc., 82, 2705 (1999).
53. D. R. Kammler, D. D. Edwards, B. J. Ingram, T. O. Mason, G. B. Palmer, A. Ambrosini, and K. R. Poeppelmeier, Electrochem. Soc. Proc. Vol. 99–11, Electrochemical Society, Pennington, NJ (1999) p. 68.
54. S. P. Harvey, K. R. Poeppelmeier, and T. O. Mason, J. Am. Ceram. Soc., 91, 3683 (2008).
55. T. Moriga, D. D. Edwards, T. O. Mason, G. B. Palmer, K. R. Poeppelmeier, J. L. Schindler, C. R. Kannewurf, and I. Nakabayashi, J. Am. Ceram. Soc., 81, 1310 (1998).
56. P. J. Cannard, and R. J. D. Tilley, J. Solid State Chem., 73, 418 (1988).
57. N. Kimizuka, M. Isobe, and M. Nakamura, J. Solid State Chem., 116, 170 (1995).
58. L. F. J. Piper, A. Demasi, S. W. Cho, K. E. Smith, F. Fuchs, F. Bechstedt, C. Körber, A. Klein, D. J. Payne, and R. G. Edgell, Appl. Phys. Lett., 94, 022105 (2009).
59. H. H. Tippins, Phys. Rev., 140, A316 (1965).
60. L. Bizo, J. Choisnet, R. Retoux, and B. Raveau, Solid State Commun., 136, 163 (2005).
61. D. R. Kammler, T. O. Mason, D. L. Young, T. J. Coutts, D. Ko, K. R. Poeppelmeier, D. L. Williamson, J. Appl. Phys., 90, 5979 (2001).
62. A. Ambrosini, S. Malo, K. R. Poeppelmeier, M. A. Lane, C. R. Kannewurf, and T. O. Mason, Chem. Mater., 14, 58 (2002).
63. D. R. Kammler, T. O. Mason, and K. R. Poeppelmeier, J. Am. Ceram. Soc., 84, 1004 (2001).
64. D. R. Kammler, T. O. Mason, D. L. Young, and T. J. Coutts, J. Appl. Phys., 90, 3263 (2001).
65. S. P. Harvey, T. O. Mason, D. B. Buchholz, R. P. H. Chang, C. Korber, and A. Klein, J. Am. Ceram. Soc., 91, 467 (2008).
66. D. B. Rogers, R. D. Shannon, C. T. Prewett, and J. L. Gillson, Inorg. Chem. 10, 723 (1971).
67. M. Tanaka, M. Hasegawa, T. Higuchi, T. Tsukamoto, Y. Tezuka, S. Shin, and H. Takei, Physica B, 245, 157 (1998).
68. R. Seshadri, C. Felser, K. Thieme, and W. Tremel, Chem. Mater., 10, 2189 (1998).
69. M. Hasegawa, T. Higuchi, M. Tanaka, T. Tsukamoto, S. Shin, and H. Takei, Mater. Trans., 42, 961 (2001).
70. R. J. Cava, H. W. Zandbergen, A. P. Ramirez, H. Takagi, C. T. Chen, J. J. Krajewski, W. F. Peck, Jr., J. V. Waszczak, G. Meigs, R. S. Roth, and L. F. Schneemeyer, Solid State Chem., 104, 437 (1993).

71. N. Duan, A. W. Sleight, M. K. Jayaraj, and J. Tate, J. Appl. Phys. Lett., 77(9), 1 (2000).
72. K. Ueda, H. Hiramatsu, M. Hirano, T. Kamiya, and H. Hosono, Thin Solid Films, 496, 8 (2006).
73. H. Yanagi, S. Park, A. D. Draeseke, D. A. Keszler, and J. Tate, J. Solid State Chem., 175, 34 (2003).
74. T. J. Coutts, T. O. Mason, J. D. Perkins, and D. S. Ginley, Electrochem. Soc. Proc. Vol. 99–11, Electrochemical Society, Pennington, NJ (1999) p. 274.
75. H. Kaga, R. Asahi, and T. Tani, Jpn. J. Appl. Phys., 43, 3540 (2004).
76. B. Yaglioglu, Y.-J. Huang, H.-Y. Yeom, and D. C. Paine, Thin Solid Films, 496, 89 (2006).
77. S. Narushima, H. Mizoguchi, K. -I. Shimizu, K. Ueda, H. Ohta, M. Hirano, T. Kamiya, and H. Hosono, Adv. Mater., 15, 1409 (2003).
78. T. Kamiya, S. Narushima, H. Mizoguchi, K.-I. Shimizu, K. Ueda, H. Ohta, M. Hirano, and H. Hosono, Adv. Funct. Mater., 15, 968 (2005).
79. H. Sato, T. Minami, S. Takata, and T. Yamada, Thin Solid Films, 236, 27 (1993).
80. M. Dekkers, G. Rijnders, and D. H. A. Blank, Appl. Phys. Lett., 90, 021903 (2007).
81. S. Narushima, H. Mizoguchi, H. Ohta, M. Hirano, K. Shimizu, K. Ueda, T. Kamiya, and H. Hosono, in MRS Symp. Proc., Vol. 747, D. Ginley, S. Guha, S. Carter, S. A. Chambers, R. Droopad, H. Hosono, D. C. Paine, D. G. Schlom, and J. Tate, eds., Materials Research Society, Warrendale, PA, 2003, p. 235.
82. D. C. Paine, T. Whitson, D. Janiac, R. Beresford, C. Ow-Yang, and B. Lewis, J. Appl. Phys., 85, 8445 (1999).
83. J. D. Perkins, J. A. del Cueto, J. L. Alleman, C. Warmsingh, B. M. Keyes, L. M. Gedvilas, P. A. Parilla, B. To, D. W. Readey, and D. S. Ginley, Thin Solid Films, 411, 152 (2002).
84. T. Minami, S. Tsukuda, Y. Minamoto, and T. Miyata, J. Vac. Sci. Technol., A23, 1128 (2005).
85. T. Moriga, T. Okamoto, K. Hiruta, A. Fujiwara, and I. Nakabayashi, J. Solid State Chem., 155, 312 (2000).
86. M. Orita, H. Ohta, M. Hirano, S. Narushima, and H. Hosono, Philos. Mag. B, 81, 501 (2001).
87. G. Impellizzeri, S. Mirabella, F. Priolo, E. Napolitani, and A. Carnera, J. Appl. Phys., 99, 103510 (2006).
88. S. Narushima, M. Hiroki, K. Ueda, K. -I. Shimizu, T. Kamiya, M. Hirano, and H. Hosono, Philos. Mag., 84, 665 (2004).
89. T. Kamiya, and H. Hosono, Int. J. Appl. Ceram. Technol., 2, 285 (2005).

Chapter 9
Chemistry of Band Structure Engineering

Art Sleight

9.1 Introduction

The starting point for engineering a transparent conductor (TC) is usually a semiconductor with a band gap such that there is very little optical absorption over most or all the visible range. Introduction of carriers to produce electrical conductivity frequently produces some optical absorption in the visible region. A key to producing a good TC is high-mobility carriers so that significant electrical conductivity can be achieved with a small number of carriers, which can contribute to optical absorption. High mobility is readily achieved in s bands but not so easily achieved in other bands such as d bands. It is very common for the conduction band to be an s band. Thus, the engineering of n-type TCs is relatively easy. It is, however, rare that the valence band is an s band. Therefore, the engineering of good p-type TCs is very challenging.

The most common band structure situation for most known TCs is shown schematically in Fig. 9.1a where the band gap is about 3 eV. The donor level in a good n-type TC is typically so close to the conduction band that essentially all electrons are excited into the conduction band at room temperature. However, for p-type TCs the acceptor levels are typically far enough above the valence band so that only a fraction of these have produced holes in the valence band at room temperature. Large band gap materials (e.g., Al_2O_3 or MgO) are good for optical transparency, but they are generally not suitable as starting points for TCs. The donor levels and acceptor levels tend to be too far from the conduction and valence bands (Fig. 9.1b) [1, 2]. Thus, significant conductivity develops only well above room temperature.

Transition metal cations with open d-shells have generally not been considered to be good TC candidates due to relatively low carrier mobility in d bands and also

A. Sleight
Chemistry Department, Oregon State University, 339 Weniger Hall, Corvallis, OR 97331, USA
e-mail: sleight@oregonstate.edu

D.S. Ginley et al. (eds.), *Handbook of Transparent Conductors*,
DOI 10.1007/978-1-4419-1638-9_9,

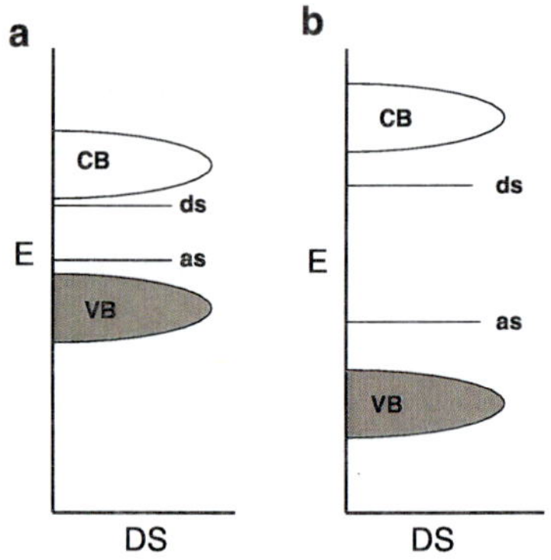

Fig. 9.1 Schematic diagrams of energy (E) vs. density of states (DS). (**a**) The typical situation for n-type and p-type TCOs where the band gap is about 3 eV. The donor states (ds) are very close to conduction band (CB), but the acceptor states (as) are not so close to the valence band (VB). (**b**) In wider band gap materials the donor and acceptor states are typically further from the conduction and valence bands, and significant conductivity is only obtained well above room temperature

due to the optical absorption that arises from d–d transitions. A special situation exists in the case of $4d^6$ and $5d^6$ configurations where the splitting between the t_{2g} and e_g orbitals can give compounds with classical semiconductor behavior. In this situation both the conduction band and the valence band are d bands. Doping of such materials can often produce either an n-type or p-type TC. The primary disadvantage of such materials is their small band gap; the splitting of t_{2g} and e_g levels is never enough to produce a material that is transparent throughout the entire visible region.

Very thin layers of metals, such as Ag, can be used as TCs, but such materials will not be considered here.

Although layout of this chapter might make it appear that this is a review of TCs, that is not the intent. The purpose is rather to indicate how a chemist might approach band structure engineering of TCs without the aid of calculations.

9.2 n-Type TCOs

Oxides are generally very stable and frequently have the band gaps suitable for TCs. Thus, transparent conducting oxides (TCOs) are the most common TCs. The conduction band in oxides is often an s band, and doping of carriers into the 4s, 5s, or 6s band in oxides is frequently easy to achieve. Given the high mobility expected for carriers in an s band, these oxides are excellent TCO candidates. For oxides where the conduction band is a 4s or 5s band, the band gaps are frequently suitable for TCOs. However, the band gap when the conduction band is a 6s band is too small for optical transparency throughout the entire visible region. Both Tl_2O_3 and PbO_2 are black compounds exhibiting metallic properties [3–5]. Band structure calculations have indicated that PbO_2 is indeed a semimetal [4]. However, stoichiometry issues confuse the situation, and it has been very recently suggested that stoichiometric PbO_2 would be a semiconductor with a band gap of 0.7 eV [5]. Although $ZnBi_2O_6$ and $MgBi_2O_6$ are n-type transparent conductors with conductivities as high as 0.4 S/cm, the band gaps are only about 1.8 eV [6].

The best example of a TCO with a 4s conduction band is ZnO. Substituting F for O injects carriers into the 4s band producing good TCO behavior. A variety of elements can be substituted for Zn to produce good TCO behavior. These include B, Al, Ga, In, Si, and Ge. To maximize the impact of these dopants, it is essential to use highly reducing conditions. Such conditions will give, for example, $Zn_{1-x}Al_xO{:}e_x$, where e is an electron in the conduction band. On heating such a material in air, oxidation to $Zn_{1-2y}Al_{2y}O_{1+y}$ readily occurs, and the carriers are lost.

Very high resistivity is normally found for ZnO equilibrated with air, even when shallow donors such as Al are present, because extra O in the ZnO lattice compensates these donors. On removing O from the ZnO lattice, two completely distinct processes occur. At low levels of reduction, such as typically occurs under vacuum synthesis of thin films, the conductivity of ZnO increases dramatically. When even more O is removed from the ZnO lattice by heating ZnO in vapors of elemental Zn at high temperatures, a Zn-rich zinc oxide is produced with a composition close to $Zn_{1.0003}O$ [7]. This Zn rich ZnO produced under these very harsh conditions has a deep red color. These mild and harsh reduction processes are completely different processes, as is apparent from the different colors they produce. The small amount of optical absorption caused by this mild reduction is that expected from conduction electrons. The harsh reduction process produces either Zn interstitials or O vacancies. The states created produce a red color, and they do not contribute significantly to conductivity because they lie deep within the band gap [7, 8]. The mild reduction of ZnO, on the other hand, only removes the extra O in the lattice that is compensating shallow donor states caused by impurities or other defects. There is no sign of a red color being produced during this mild reduction process. Conductivity actually drops somewhat as mildly reduced ZnO is further reduced to red ZnO, possibly because defects associated with red ZnO are scattering centers [8]. From experimental data, it is not yet possible to determine if the deep donor state in red ZnO is due to oxygen vacancies or Zn interstitials. Normally, an oxygen vacancy would be expected to be favored. However, in the case of Zn interstitials much depends on the energy associated with the Zn–Zn bonds that would exist. Current theory seems to be converging on oxygen vacancies as the dominant defect in Zn rich ZnO [9–13]. Thus, we might reformulate $Zn_{1.0003}O$ as $ZnO_{0.9997}$ to indicate the maximum oxygen vacancy level that has been achieved. Two electrons are trapped at each O vacancy. This is then a -U center, which is well established in cases such as BaO_{1-x}.

Another TCO with a 4s conduction band is Ga_2O_3. For β-Ga_2O_3 the band gap is 4.9 eV giving then good transparency into the UV. Good conductivity is reported for both crystals and films [14].

In the case where the conduction band is a 5s band, very good n-type TCOs can be engineered starting with CdO, In_2O_3, or SnO_2. More complex ternary oxides can be considered, but they seem to offer little advantage from the point of view of

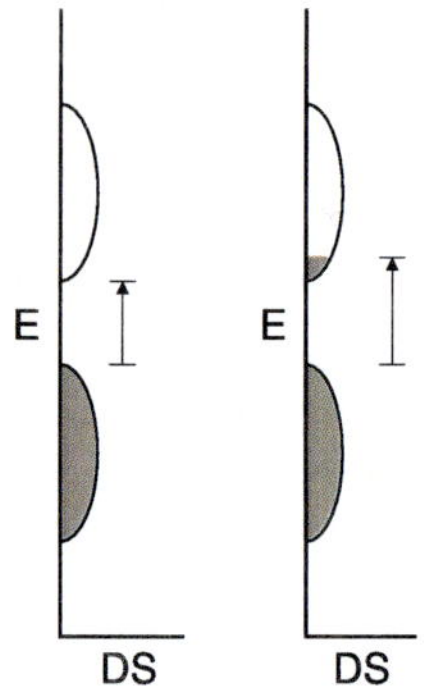

Fig. 9.2 Schematic diagrams of energy (E) vs. density of states (DS) showing how the slight filling of the conduction band effectively increases the energy of the lowest energy excitation

carrier mobility or optical transparency. Such materials may still be of interest, for example, to adjust band gaps to specific values or for increased stability. In 1977 Shannon et al. produced single crystals of $CdSnO_3$, Cd_2SnO_4, In_2TeO_6, and $CdIn_2O_4$, all of which showed high conductivity and high transparency throughout the visible region [15]. The high stability of Cd_2SnO_4 has attracted recent attention [16].

When a semiconductor is doped, its band gap can increase according to the Burstein-Moss effect (Fig. 9.2). In the case of ZnO, n-type doping has been shown to increase the optical band gap from 3.2 to 3.7 eV [17]. The band gap of CdO is too small for good optical transparency throughout the entire visible region. However, In doped CdO shows the expected increase conductivity and a significant increase in band gap. On going from $x = 0.0$ to $x = 0.05$ in $Cd_{1-x}In_xO$, the band gap increases from 2.82 to 3.11 eV [18].

It is also possible to produce n-type TCOs where the conduction band is a d band. Lower mobility is expected in this case. However, mobilities as high 7.6 cm^2/V s are reported for doped TiO_2 [19]. Although this is low compared to the values of greater than 50 cm^2/V s values obtained for some of the s band TCOs, this is sufficient to produce reasonable conductivities coupled with good transparency.

9.3 p-Type TCOs

Attempts to produce good p-type TCOs have met with several problems. Introduction of acceptor states is frequently difficult. When acceptor states are introduced, they have usually been far enough above the valence band to give a strongly temperature dependent conductivity over broad temperature ranges. This is in contrast to the common situation in n-type TCOs where the conductivity remains temperature independent over broad temperature ranges including room temperature. Finally, the carrier mobilities in p-type TCOs rarely exceed 5 cm^2/V s whereas values higher than 50 cm^2/V s are commonly reported for n-type TCOs.

For many oxides the valence band is comprised mainly of oxygen 2p states. Holes in this band should produce p-type conductivity, but there has been very little success in producing good TCOs based on such an approach. Actually, p-type conduction has very commonly been observed in oxides where the valence band is basically an oxygen 2p band, but this conduction is typically only found well above room temperature. Those studying oxygen ion conductivity in oxides frequently observe p-type conductivity, and this is usually regarded as a problem because it competes with the desired ionic conductivity.

Oxides differ from other materials in that simply equilibrating samples at varying temperatures and oxygen pressure can generally alter the carrier concentration. Low oxygen pressure tends to produce n-type materials with carrier concentrations increasing as oxygen pressure decreases. High oxygen pressure tends to produce p-type materials with the number of carriers increasing with increasing oxygen pressure. This is shown schematically in Fig. 9.3. For oxides that are good oxygen ion conductors there may be a central region where the ionic conductivity dominates over electronic conductivity. At the boundaries of the ionic and electronic regions, both ions and electrons can contribute significantly to conductivity. Because mobilities of ions are generally much lower than the mobilities of electrons, achieving oxygen ion conductivity in oxides that dominants over electronic conductivity requires high carrier concentrations achieved by doping to produce oxygen vacancies, e.g., $Zr_{1-2x}Y_{2x}O_{2-x}$. This intermediate region of dominant ionic

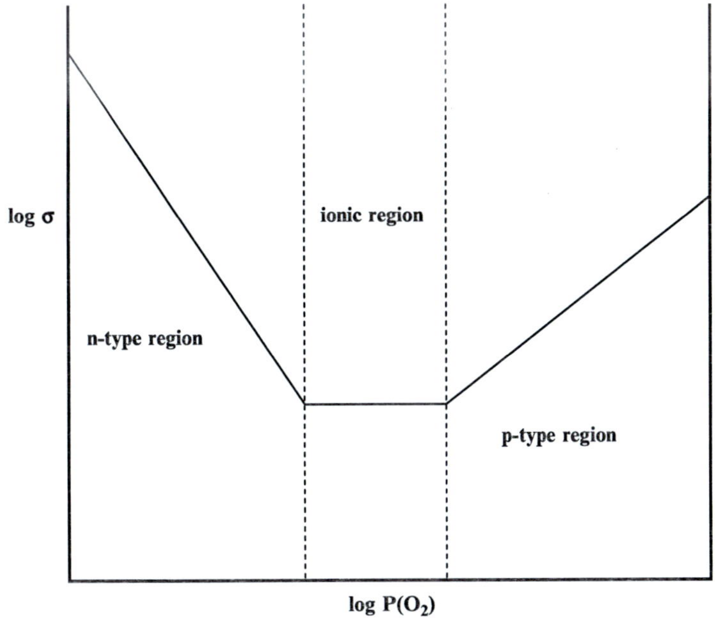

Fig. 9.3 Schematic of conductivity of an oxide as a function of oxygen pressure. The intermediate ionic region is frequently absent

conductivity shown in Fig. 9.3 is required for most applications of oxygen ion conductors. Where this region occurs as a function of oxygen pressure depends on the material. For example, n-type conductivity develops more readily with reducing oxygen pressure in ZrO_2 than in ThO_2. Conversely, the development of p-type conductivity develops more readily in ThO_2 than in ZrO_2 as oxygen pressure is increased. Thus, the ionic conducting zone in Fig. 9.3 shifts to the left on moving from doped ZrO_2 to doped ThO_2. From a practical point of view, doped ThO_2 is more useful as an oxygen sensor at very low oxygen pressures and doped ZrO_2 is more useful at high oxygen pressures. The reason for this difference between ZrO_2 and ThO_2 is the fact that Th^{4+} is more electropositive than Zr^{4+}. It should be noted that the doping used to produce oxygen ion conductivity actually facilitates the production of p-type carriers at high oxygen pressure through the reaction $Zr_{1-2x}Y_{2x}O_{2-x} + yO_2 \Rightarrow Zr_{1-2x}Y_{2x}O_{2-x+2y}$. The material with $y = 0.0$ has neither n- or p-type dopants and lies within the ionic region in Fig. 9.3. As y increases from 0.0 this material changes from an ionic conductor to a p-type electronic conductor. The oxygen introduced does not need to produce a defect such as an oxygen interstitial. It actually eliminates defects, the oxygen vacancies.

Experimental data illustrating the impacts of temperature and oxygen pressure on electrical conductivity and carrier type are shown in Figs. 9.4 and 9.5 for $Ba_{0.03}Sr_{0.97}TiO_3$ [20]. Thermoelectric power data show the change from n-type to p-type behavior as oxygen pressure is increased. There will always be a minimum in the plot of conductivity vs. oxygen pressure where the predominant carrier type switches from n-type carriers to p-type. At this minimum the n-type and p-type carriers contribute equally to the total electronic conductivity. The number of carriers is not fixed at fixed oxygen pressure. Oxides tend to lose oxygen with increasing temperature even when oxygen pressure is fixed. Thus, with increasing temperature the number of n-type carriers will increase and the number of p-type carriers will decrease. Therefore, both high temperature and low oxygen pressure

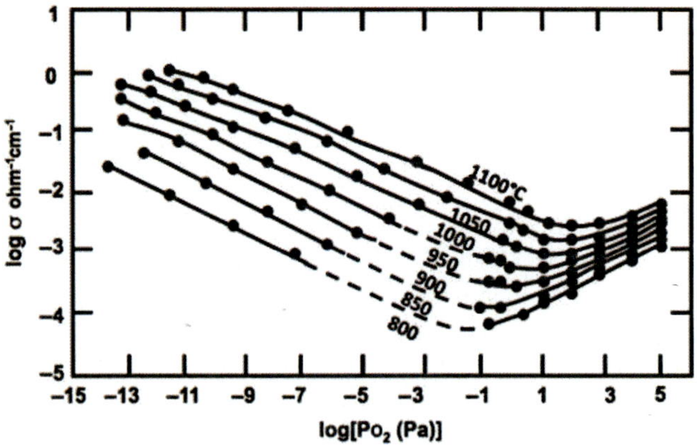

Fig. 9.4 Electrical conductivity (σ) of $Ba_{0.03}Sr_{0.97}TiO_3$ single crystal as a function of oxygen pressure (redrawn from ref. 20)

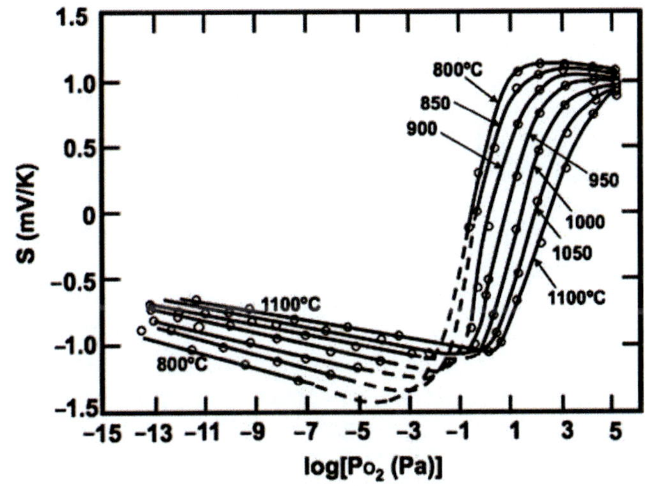

Fig. 9.5 Thermopower of $Ba_{0.03}Sr_{0.97}TiO_3$(S) as a function of oxygen pressure (redrawn from ref. 20)

enhance carrier concentration in the n-type region. Usually, this high carrier concentration can be retained on quenching to ambient conditions. High oxygen pressure and low temperatures enhance carrier concentration in the p-type region. For this reason it to common to anneal such materials under high oxygen pressure at low temperature after initial synthesis. The lowest feasible temperature is usually about 400°C for thin films and uncompacted powders. This temperature may be too low for equilibration in a reasonable time for mm size crystals or dense ceramics.

There are at least two reasons for the lack of symmetry between the n-type regions in Fig. 9.4. One is that the mobility of the n-type carriers is generally higher than that of the p-type carriers. This then contributes to the higher conductivity values observed in the n-type region. The other reason for the lack of symmetry between the two regions is the increased loss of oxygen with increasing temperature at fixed oxygen pressure. This causes the minimum in conductivity in the curves of Fig. 9.4 to move to the right with increasing temperature.

Examples of perovskite structure systems studied for oxygen ion conductivity are $CaTi_{1-2x}Al_{2x}O_{3-x}$, $CaTi_{1-2x}Ga_{2x}O_{3-x}$, $CaTi_{1-2x}Sc_{2x}O_{3-x}$, $La_{1-2x}Ca_{2x}AlO_{3-x}$, and $SrTi_{1-2x}Al_{2x}O_{3-x}$ [21]. In all cases, the oxygen vacancies start filling in at oxygen pressures as low as 10^{-6} atm. producing materials where p-type electronic conductivity dominates over oxygen ion conductivity. However, significant conductivity can only be observed well above room temperature. Despite the many studies of oxygen ion conductors where a problem of p-type electronic conductivity arose, these investigations have never produced a good p-type TCO candidate.

From a chemists perspective, a problem in producing good p-type TCOs based on holes in an O 2p band can be partially related to the size of O^{2-} vs. O^{1-}. A hole localized on an oxygen atom in an O 2p band is an O^{1-} species. Sizes of atoms or ions increase as electrons are added to them. Thus, Fe^{2+} is larger than Fe^{3+}. This effect is far from linear; the rate of increase increases as electrons are added [22]. For example, Re^{6+} is only slightly larger than Re^{7+} (0.55 Å vs. 0.53 Å). The size difference between Fe^{3+} and Fe^{2+} is considerably larger (0.64 Å vs. 0.78 Å). These size differences are relevant to the ease of delocalizing electrons. For example, ReO_3, which is a very good metal, can be considered in chemistry terms as a system where the nominal Re^{6+} valence state is degenerate with disproportionated $Re^{5+} + Re^{7+}$ valence state. This is because there is no energy barrier to move the conduction electron from one Re^{6+} to another Re^{6+}, conceptually producing then $Re^{5+} + Re^{7+}$. There are several factors that are important for such a degeneracy to occur; one is the relative size of the atoms in different oxidation states. The similarity of sizes of Re^{5+}, Re^{6+}, and Re^{7+} (0.58, 0.55, and 0.53 Å) is then one factor favoring degeneracy of the Re^{6+} state with the $Re^{5+} + Re^{7+}$ mixed valence state. At low temperatures a hole in an oxygen 2p band seems always to be localized on an oxygen atom producing a O^{1-} species among many O^{2-} species. The O^{1-} species is much smaller than the O^{2-} species, thus giving much shorter M–O distances. There is then a very considerable lattice relaxation at the O^{1-} site, which is presumably adjacent to the dopant site. This then places the energy of this O^{1-} state well above the valence band. A possible further consideration is that adjacent O^{1-} species will tend to bond to each other forming the $(O_2)^{2-}$ peroxides species, which could then further increase the energy of the acceptor level.

A rather exotic example of an O^{1-} species in an oxide occurs in the common mineral smoky quartz [23]. A common impurity in both natural and synthetic quartz is Al. This is generally compensated with H^+, e.g., $H_xSi_{1-x}Al_xO_2$ where x might be $\sim 10^{-4}$. On irradiation neutral H becomes separated from Al leaving an O^{1-} species attached to Al. No detectable conductivity is observed from this acceptor state because it is too far above the O 2p band. Such irradiated quartz crystals are gray or brown. Optical absorption due n-type carriers is mainly in the infrared region but typically tails into the visible red region, giving rise to a blue color for the TCO. Optical absorption associated with p-doped oxides is commonly much more spread out through the visible region, giving materials that are gray to brown.

9.3.1 ZnO

The low mobility of holes in an O 2p band is frequently attributed to nonbonding states at the top of this valence band. However, the states at the top of an O 2p band are not necessarily nonbonding. The 2p orbitals of O will not be fully involved in bonding if the coordination of O is less than 4 and there is no π bonding. This would, for example, be the situation in SnO_2, where the coordination of O is 3 and one of the three 2p orbitals is basically nonbonding leading then to nonbonding states at the top of the valence band. The coordination of O in ZnO is tetrahedral, and all three O 2p orbitals are fully and equally involved in strong covalent bonding to Zn. Thus, the states at the top of the valence band are not nonbonding. From this point of view, ZnO is a good candidate for a p-type TCO. The high covalency in ZnO is another favorable characteristic.

Historically, attempts to dope ZnO p-type have failed. Studies of Li-doped zinc oxide crystals indicate that Li acts as a donor when the doping is conducted under reducing conditions [24]. It was suggested that Li occupies an interstitial site in this situation, i.e., $ZnLi_xO{:}e_x$. When Li doping is conducted under oxidizing conditions, the Li apparently substitutes for Zn. If this doping occurs according to $Zn_{1-x}Li_xO{:}a_x$, acceptor states (a) are produced. If however the doping occurs according to $Zn_{1-2y}Li_{2y}O_{1-y}$, no donor or acceptor states are produced. It is clear that Li on the Zn site is very effective as an acceptor state to the extent that it can completely compensate for any donor states produced by defects or impurities. This is, in fact, an easy way to produce zinc oxide with nearly intrinsic properties. However, equilibration under impractically high oxygen partial pressures is required to produce p-type behavior [20]. In other words, extremely high oxygen pressure is required to oxidize $Zn_{1-2y}Li_{2y}O_{1-y}$ to $Zn_{1-x}Li_xO{:}a_x$. A study of Li-doped zinc oxide powders prepared in air also reported that the n-type conductivity decreased on Li doping and that p-type behavior did not occur [25]. The difficulty in doping zinc oxide p-type has been attributed to its work function [26]. This argument in chemical terms is that for oxides with a high work function holes in an oxygen 2p band will localize and combine on lattice oxygen converting 4O- to $2O_2$- and O_2, which then escapes the solid along with the holes. This does not preclude the possibility of preparing a metastable p-type zinc oxide. However, such a material would evolve O_2 on heating and the holes would be lost. Again, impractically high oxygen pressure would be required to prevent this reaction from occurring on heating. Recent synthesis of ZnO:Li films under nonequilibrium conditions including activated oxygen apparently has succeeded in obtaining p-type behavior [27].

It might seem that the best cation to substitute for Zn to produce p-type behavior would be Cu. However, the fact that Cu 3d states lie above the O 2p band precludes this possibility. Cu will substitute for Zn in ZnO, but the desired "hole" is firmly attached to Cu, which has become Cu^{2+} instead of the desired Cu^{1+}. It is very difficult to oxidize Ag^{1+} to Ag^{2+} whereas it is easy to oxidize Cu^{1+} to Cu^{2+}. Thus, substitution of Ag^{1+} for Zn might well be expected to produce p-type ZnO, and this seems to have been accomplished with conductivity as high as 0.03 S/cm [28].

There are many recent reports of p-type thin films of zinc oxide based on small substitutions of N for O [29–36]. Some of this work is based on codoping, such as doping with both N and Ga; however, most studies show no benefit of the codoping approach. Evidence for N doping of zinc oxide powders prepared from zinc nitrate has been obtained through EPR studies [37]. It has been determined that the N concentration in such powders can reach $ZnO_{0.997}N_{0.003}$, and iodometric titrations on such powders show evidence of acceptor states [38]. Zinc oxide powders prepared by decomposition of zinc peroxide also show evidence of acceptor states from iodometric titrations, suggesting that this is an oxygen-rich zinc oxide [38]. Acceptor level concentrations in N doped ZnO films and powders are always considerably lower than the N content. It is likely that the N is partially compensated with H. This H could readily be removed by heating at high temperatures, but such treatment will as well remove oxygen and holes.

Anions from the first row of the periodic table differ greatly in their chemistry from those in subsequent rows. Thus, the only anion that can be substituted for O^{2-} to produce n-type ZnO is F^{1-}, and the only anion expected to substitute for O^{2-} to produce p-type ZnO is N^{3-}. Doping of N into oxides is a challenge because it must occur under oxidizing conditions if it is to produce the desired acceptor level. Molecular N_2 is too inert to provide a good source of N. While NH_3 provides an active form of N, it also produces an undesirable highly reducing situation. The use of zinc nitrate to produce zinc oxide powders and the use of NO to produce ZnO films provides both a source of N and the required oxidizing conditions.

There have also been reports of p-type zinc oxide produced with dopants such as P, As, and Sb. Under equilibrium conditions significant substitution of such elements for oxygen is not expected. However, thin film synthesis conditions are typically far from equilibrium, expanding the compositions that can be prepared. For example, the amount of n-type dopants in thin films of zinc oxide can be much higher than allowed under equilibrium conditions.

9.3.2 NiO

Although d bands are not considered good candidates for high mobility carriers, they must be considered for p-type CTOs because there is little other choice. It is easy to produce p-type NiO, and the first p-type TCO thin films were NiO [39]. The properties of NiO:Li thin films have recently been optimized using a combinatorial approach. Conductivities as high as 1.42 S/cm were obtained on 40 nm thick films showing 80% transparency [40].

9.3.3 Delafossites

On proceeding across the 3d series from Ti to Cu, the 3d orbitals are filling and their energy is dropping to approach that of the O 2p band. At ZnO the filled 3d band has dropped below the O 2p band. But for Cu^{1+} oxides the filled 3d band lies just above the O 2p band. Thus, Cu^{1+} oxides will be p-type conductors if holes are introduced into this d band. It is well known that Cu_2O can readily be doped n- or p-type, and carrier mobilities are high in both cases. However, the band gap of Cu_2O is only 2.1 eV giving a deep red color. Ternary oxides of Cu^{1+} are relatively rare. When they can be made, they can have the larger band gaps desired for a TCO.

There is only one known structure for ternary oxides of Cu^{1+} where there is considerable flexibility for the countercation. This is the delafossite structure. The formula for this structure is $CuMO_2$ where countercation M can range in size from Al to La. This structure is shown in Fig. 9.6. The coordination of Cu^{1+} to O is

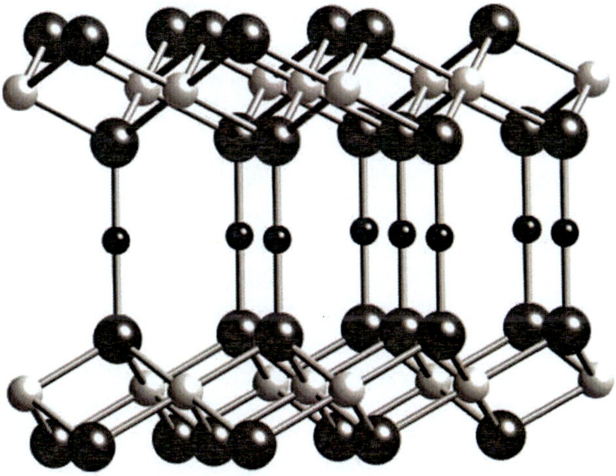

Fig. 9.6 The structure of $CuMO_2$ delafossite compounds where M atoms are *large dark spheres*, O atoms are *light spheres*, and Cu atoms are the *small black spheres*. The c axis is vertical

twofold linear, which is common for d^{10} cations in oxides. It is important to note that there are no Cu–O–Cu linkages in this structure. Thus, the covalency of the Cu–O bond will not lead to a pathway for hole carriers. The strong bonding pathway between Cu atoms is Cu–O–M–O–Cu. This would only be a good hole pathway if there were states on M that could mix strongly into the valence band, and this is not normally the situation. In Cu_2O there are extended –Cu–O–Cu–O- connections, and this may be the reason that hole mobilities as high as 100 cm^2/V s have been reported [41]. Hole mobilities in delafossites are less than 1 cm^2/V s.

It was noted by Jansen that the Cu–O distance in $CuMO_2$ delafossites decreases as the size of M and consequently the Cu–Cu distance increases [42]. Jansen rationalized this behavior for $CuMO_2$ delafossites on the basis of competition between Cu–O bonding and Cu–Cu bonding. Thus, as one bond becomes shorter and stronger, the other bond becomes longer and weaker. One would not normally expect any bonding interaction between $nd^{10}(n+1)s^0$ cations such as Cu^{1+}. However, strong 3d–4s hybridization occurs on Cu due to the linear O–Cu–O linkages. This effectively transfers some electron density from the filled d shell to the empty s shell. The Cu–Cu interaction then becomes weakly bonding, and it is this direct interaction that then provides the conduction mechanism for p-doped $CuMO_2$ compounds with the delafossite structure. However, the Cu–Cu bond appears too weak to explain the trend In Cu–O distances noted by Jansen. An alternative explanation has been offered [43]. The $s - d_{z^2}$ hybridization of the Cu cations necessary to form the twofold linear bonds to oxygen effectively polarizes the filled shell leading to an increase of the Cu–Cu repulsion. As the Cu^{1+} cations are forced closer by the smaller M cations, this polarization must decrease, leading to increased antibonding electron density for the Cu–O bond. Thus, the Cu–O bond becomes weaker and longer. This explanation was subsequently supported by calculations [44].

Because it is a direct Cu–Cu interaction that is responsible for the p-type conductivity in delafossites, one can expect improved conductivity as the Cu–Cu distance decreases, and this is observed experimentally [45]. There are two doping approaches for p-type $CuM^{3+}O_2$ delafossites. One is to substitute a divalent cation for M. This approach is frequently successful, but in the case of $CuAlO_2$ this approach has failed. Another approach is to insert O between Cu atoms in the Cu layer (Fig. 9.6). This approach succeeds if the Cu–Cu distance is large enough, and this occurs only when M is Sc or a larger cation. The conductivity increases with increasing temperature for all $CuMO_{2+x}$ delafossites studied. The mechanism for this conductivity appears, however, to be complicated. One indication is that a substantial deviation from linearity generally occurs for plots of log-conductivity vs. 1/T. The direct Cu–Cu interaction in the delafossite structure occurs only in planes perpendicular to the c axis (Fig. 9.6). Thus, one expects highly anisotropic conductivity. Measurements on oriented single crystals show that the conductivity perpendicular to the c axis can be 10^3 times higher than parallel to the c axis [46]. The activation energies are also significantly lower for the higher conductivity parallel to the c axis. A possible explanation is that the mobility is activated with different activation energies parallel and perpendicular to the c axis. An activated mobility model for $CuAlO_2$ has been advanced [47]. However, $^{63,65}Cu$ NQR studies for $CuYO_2$:Ca indicate a temperature independent mobility estimated to be 0.7 cm^2/V s with an acceptor level at least 250 meV above the valence band [48]. A possible explanation is then that the mobility is activated along the c axis but not parallel to it. A better understanding of the conductivity mechanism in delafossites awaits further transport and optical characterization on single crystals.

There is also a series of $AgMO_2$ compounds with the delafossite structure. However, there has been no success in introducing suitable acceptor states into such compounds. This is likely in part due to the fact that the Ag^{2+} state is very difficult to achieve relative to the Cu^{2+} state that is easy to achieve. Even annealing $AgMO_2$ delafossites under high oxygen pressures does not produce p-type materials. An important consideration is that the Ag–Ag distances are too short for O insertion in the known $AgMO_2$ delafossites. Because there are Ag–O bonds only along the c axis, the Ag–Ag distances in the Ag layers are determined only by the size of the M cations. As a consequence, the Ag–Ag distances in $AgScO_2$ are not significantly longer than the Cu–Cu distances in $CuScO_2$. However, Ag^{1+} has a diameter 0.4 Å larger than Cu^{1+} [22]. Oxygen can only barely find space to insert between Cu atoms in $CuScO_2$. There is not nearly enough space between Ag atoms in $AgScO_2$ for O insertion between Ag atoms. The M atom would need to be much larger than Sc for O insertion into a $AgMO_2$ delafossite, and attempts to prepare such compounds have failed.

9.3.4 *Rhodates*

It has been known since 1980 that the perovskite rhodates $RRhO_3$ where R is a rare earth atom such as Lu have band gaps of about 2.2 eV [49, 50]. Despite the open d shell, these compounds show classical semiconducting behavior where the valence band is 4d t_{2g} and the conduction band is 4d e_g. In the case of $LuRhO_3$, both n-type and p-type behaviors have been observed. Estimates of mobility are 2.5 cm^2/V s for holes and 1.0 cm^2/V s for electrons [50]. The conductivity is always activated with activation energies as low as 0.14 eV observed. The p-type semiconducting behavior of $LuRhO_3$ was used to construct a photochemical cell that produces simultaneously H_2, O_2, and electricity [49]. More recently the spinel $ZnRh_2O_4$ has been considered as a p-type TCO [51]. However, the band gap is again about 2.2 eV. Thus, such materials are not transparent in the blue region of the visible spectrum.

9.3.5 s Bands for p-Type TCOs

The Holy Grail for p-type TCOs could be considered the situation where the valence band is an s band. This leads us to the chemistry of s^2 cations such as Sn^{2+} and Bi^{3+}. In the case of Bi_2O_3 the 6s band from Bi^{3+} lies below the O 2p band (Fig. 9.7). Thus, holes introduced into Bi_2O_3 would be in an O 2p band, not the 6s band. Even in BiO_2, which is really $Bi^{3+}Bi^{5+}O_4$, the filled 6s band lies well below the O 2p band (Fig. 9.7). In both Bi_2O_3 and $Bi^{3+}Bi^{5+}O_4$ the valence band is O 2p, but the conduction band is Bi 6p for Bi_2O_3 and Bi 6s for $Bi^{3+}Bi^{5+}O_4$. Thus, the band gap is much smaller in $Bi^{3+}Bi^{5+}O_4$ than in Bi_2O_3. However, given the situation in $BaBiO_3$, an s-band p-type TCO might seem possible. If $BaBiO_3$ had the ideal perovskite structure, there would be a half-filled 6s band, and $BaBi^{4+}O_3$ would be

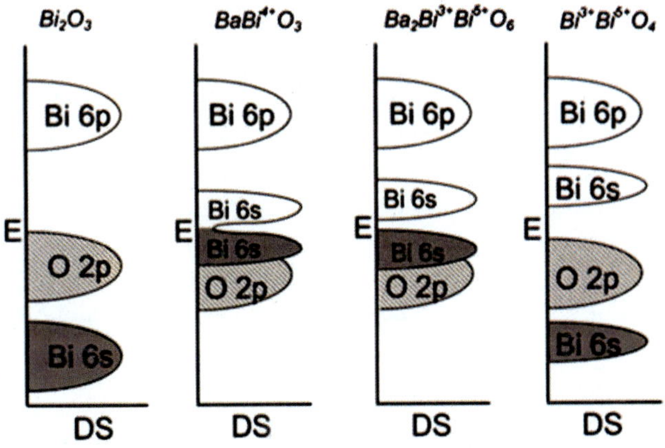

Fig. 9.7 Schematic diagrams of energy (E) vs. density of states (DS) showing how the Bi 6s band may be below or above the O 2p band. Bands *not shaded* are empty. Shown are both hypothetical metallic $BaBiO_3$ with a *half filled* 6s band and the actually disproportioned compound where the Bi 6s band has split into a *filled* band (Bi^{3+}) and an empty band (Bi^{5+}). The actual number of Bi states is much less than the number of O states

metallic (Fig. 9.7). However, the 6s band is split with a filled 6s band below the Fermi level and another empty 6s band above the Fermi level (Fig. 9.7). This compound is then actually $Ba_2Bi^{3+}Bi^{5+}O_6$ with considerably shorter Bi–O distances for Bi^{5+} relative to Bi^{3+}[52]. The situation shown in Fig. 9.7 with an unsplit 6s band does, however, pertain to the superconducting compositions of the type $BaPb_{1-x}$-Bi_xO_3. Thus, it is mainly the exact environment around Bi that determines whether this 6s band in oxides is above the O 2p band or below the O 2p band. A big issue here is the strong tendency of s^2 cations such as Bi^{3+} to undergo 6s–6p hybridization, which results in a very low symmetry environment for Bi^{3+}. This hybridization occurs because of the strong antibonding interaction between the filled 6s orbitals of Bi^{3+} and the filled 2p orbitals of O. The 6s–6p hybridization effectively pushes this antibonding electron density off to one side of Bi^{3+} where it becomes nonbonding with long Bi–O bonds on that side. This lone-pair distortion has occurred in Bi_2O_3 and $Bi^{3+}Bi^{5+}O_4$, pushing the filled Bi 6s band down in energy relative to the O 2p band. This lone-pair distortion for Bi^{3+} has not occurred in $Ba_2Bi^{3+}Bi^{5+}O_6$; thus, the 6s band is at much higher energy. The addition of Ba in the case of $Ba_2Bi^{3+}Bi^{5+}O_6$ also has an important impact. There are no Ba levels close to the Fermi level, but the presence of this highly electropositive element stabilizes the O 2p band driving the O 2p band down in energy, and the added ionicity with Ba narrows the O 2p band. Thus, the presence of Ba makes it easier for the Bi 6s band to be above the O 2p band. If one could manipulate the chemistry to obtain Bi^{3+} in a symmetric environment as in $BaBiO_3$, a filled 6s band above the O 2p band might be achieved as a starting point for the doping of holes. Attempts to produce this situation have failed. The very large Bi^{3+}–Bi^{3+} and Bi^{5+}–Bi^{5+} distances ($\sim$8 Å) in $Ba_2Bi^{3+}Bi^{5+}O_6$ means that there is very little mixing of the associated s orbitals. Thus, neither the introduction of electrons into the empty 6s band nor holes into the filled 6s band (Fig. 9.7) results in free carriers.

The situation might seem more promising for the 5s band because this filled band is always above the O 2p band in oxides of Sn^{2+}. Although p-type SnO can be prepared, it is not transparent throughout most of the visible region due to an apparent band gap of only 1.3 eV [53]. Ternary oxides of Sn^{2+} exist, and they are typically more stable than SnO and can have much larger band gaps. These oxides include $SnNb_2O_6$, $SnTa_2O_6$, Sn_2TiO_4, two polymorphs of $SnWO_4$, Sn_2WO_5, Sn_3WO_6, and various $Sn_{2-x}M_2O_{6+y}$ oxides and oxyflourides (M = Nb, Ta, Ti, or W) with the pyrochlore structure [54–59]. Despite some attempts, there has been no report of good p-type conductivity for these compounds with the larger band gaps.

A problem with the s band arising from the s^2 cations is that the top of the s band is dominated by nonbonding states. This is caused by the usual asymmetric environment favored by s^2 cations. Compounds of the formula $A_2M_2O_7$ with the pyrochlore structure might seem to offer the solution to obtaining s^2 cations in a highly symmetric environment. The compounds $Bi_2Ti_2O_7$, $Sn_2Nb_2O_7$, and $Sn_2Ta_2O_7$ have this structure [56, 60]. The A cation site in this structure is highly symmetric with an inversion center. However, structure analyses shows that these s^2 cations are strongly displaced off the ideal A site in a disordered fashion [56, 60].

In fact, they have the usual asymmetric bonding of s^2 cations. The Bi atoms in the pyrochlore $Bi_2Sn_2O_7$ are displaced off the ideal A cation site in an ordered way [61]. Calculated density of states for this compound indicates a Bi 6s band below the O 2p band, but they also show a strong Bi 6s component at the top of the valence band [62]. However, conductivity measurements on doped crystals of $Bi_2Sn_2O_7$ suggest that only n-type materials have been prepared [63].

9.4 Nonoxides

Compounds other than oxides can be considered TC candidates. Although sulfides generally have smaller band gaps than oxides, there are some sulfides with band gaps large enough for consideration. Furthermore, although sulfides are frequently not stable to air exposure, this is not always the case. For example, highly stable ZnS has a band gap of 3.7 eV, and both n-type and p-type behavior can be obtained. A carrier mobility of 120 cm^2/V s has been given for n-type ZnS [64], and a hole mobility as high as 86 cm^2/V s has been reported in a ZnS thin film [65], which is impressive for a p-type TC. Although significant solubility between ZnO and ZnS or ZnSe does not occur under equilibrium conditions, thin films of the type $ZnO_{1-x}S_x$ and $ZnO_{1-x}Se_x$ with a wide range of x have been reported. From a chemistry perspective the doping of ZnS with P or As should be much easier than the doping of ZnO with N. Another stable sulfide, $CuAlS_2$, has a band gap of 3.5 eV. Again, both n-type and p-type behaviors are reported with room temperature conductivities as high as 1 S/cm [66].

A series of compounds of the formula LaCuQO or BaCuQF where Q may be S, Se, or Te have been considered as p-type TCs [67–70]. These isostructural layered compounds have highly anisotropic properties. The CuQ sheets contain extended Cu–Q–Cu connections. The valence band may be considered as Cu 3d and the conduction band as Cu 4s, but the high covalency of the Cu–Q bond results in a significant contribution of Q states to both of these bands. Acceptor levels can be close enough to the valence band to give degenerate semiconductor behavior [68]. Band gaps as large as 3.1 eV and p-type conductivity as high as 167 S/cm have been reported [71]. A similar layered structure is also found for $BaCu_2S_2$, which can be a p-type TC with a band gap of 2.3 eV [72].

References

1. S.K. Tiku, F.A. Kröger, *J. Am. Ceram. Soc.* **63** (1980) 31.
2. D.R. Sempolinski, W.D. Kingery, H.L. Tuller, *J. Am. Ceram. Soc.* **63** (1980) 669.
3. A.W. Sleight, J.L. Gillson, B.L. Chamberland, *Mater. Res. Bull.* **5** (1970) 807–812.
4. M. Heinemann, H.J. Terpstra, C. Haas, R.A. de Groot, *Phys. Rev. B* **52** (1995) 11740.
5. D.J. Payne, R.G. Egdell, D.S.L. Law, P-A. Glans, T. Learmouth, K.E. Smith, J. Guo, A. Walsh, G.W. Watson, *J. Mater. Chem.* **17** (2007) 267.

6. H. Mizoguchi, N.S.P. Bhuvanesh, P.M. Woodward, *Chem. Commun.* **9** (2003) 1084.
7. R. Wang, A.W. Sleight, R.A. Platzer, J.A. Gardner, *J. Solid State Chem.* **122** (1996) 166.
8. A.W. Sleight, R. Wang, in *Solid-State Chemistry of Inorganic Materials*, MRS Symp. Proc. Vol. 453, edited by P.K. Davies, A.J. Jacobson, C.C. Torardi, T.A. Vanderah, pp. 323–330, Pittsburgh, PA, 1997.
9. A.F. Kohan, G. Ceder, D. Morgan, C.G. Van de Walle, *Phys. Rev. B* **61** (2000) 15019.
10. C.G. Van de Walle, *Physica B* **310** (2001) 899.
11. E.-C. Lee, Y.-S. Kim, Y.-G. Jin, K.J. Chang, *Phys. Rev. B* **64** (2001) 085120.
12. F. Oba, S.R. Nishitani, S. Isotani, H. Adachi, I. Tanaka, *J. Appl. Phys.* **90** (2001) 824.
13. A. Janotti, C.G. Van de Walle, *Appl. Phys. Lett.* **87** (2005) 122102.
14. M. Orita, H. Hiramatsu, H. Ohta, M. Hirano, H. Hosono, *Thin Film Solids* **411** (2002) 134.
15. R.D. Shannon, J.L. Gillson, R.J. Bouchard, *J. Phys. Chem. Solids* **38** (1977) 877.
16. D.S. Ginley, C. Bright, *Mater. Res. Soc. Bull.* **25** (2000) 15.
17. R. Wang, L.L.H. King, A.W. Sleight, *J. Mater. Res.* **11** (1996) 1659–1664.
18. A. Wang, J.R. Babcock, N.L. Edleman, A.W. Metz, M.A. Lane, R. Asahi, V.P. Dravid, C.R. Kannewurf, A.J. Freeman, T.J. Marks, *Proc. Natl. Acad. Sci. U.S.A.* **98** (2001) 7113.
19. M.A. Gillispie, M.F.A.M. van Hest, M.S. Dabney, J.D. Perkins, D.S. Ginley, *J. Appl. Phys.* **101** (2007) 033125.
20. C.M. Choi, H.L. Tuller, D. Goldschmidt, *Phys. Rev. B* **34** (1986) 6972.
21. S. Hashimoto, H. Kishimoto, H. Iwahara, *Solid State Ionics* **139** (2001) 179.
22. R.D. Shannon, *Acta Cryst.* **A32**, (1976) 751–767.
23. E. Fritsch, G.R. Rossnan, *Gems Gemol.* **24** (1988) 1.
24. J.J. Lander, *J. Phys. Chem. Solids* **15** (1960) 324.
25. K. Haufle, A.L. Vierk, *Z. Physik. Chem.* **196** (1950) 160.
26. S.B. Zhang, S.-Huai, A. Zunger, *J. Appl. Phys.* **83** (1998) 3192.
27. Y-J. Zenga, Z-Z. Ye, W-Z. Xu, L-L. Chen, D-Y. Li, L-P. Zhu, B-H. Zhao, Y-L. Hu, *J. Cryst. Growth* **283** (2005) 180.
28. B.D. Ahn, H.S. Kang, J.H. Kim, H.W. Chang, S.Y. Lee, *J. Appl. Phys.* **100** (2006) 093701.
29. D.C. Look, D.C. Reynolds, C.W. Litton, R.L. Jones, D.B. Eason, G. Cantwell, *Appl. Phys. Lett.* **81** (2002) 1830.
30. K. Minegishi, y. Koiwai, K.Kikuchi, *Jpn. J. Appl. Phys.* **36** (1997) L1453.
31. T.V. Butkhuzi, M.M. Sharvashidze, N.M. Gamkrelidze, Kh. V. Gelovani, T.G. Khulordava, N.P. Kekelidze, E.E. Kekelidze, *Semicond. Sci. Technol.* **16** (2001) 575.
32. M. Joseph, H. Tabata, H. Saeki, K. Ueda, T. Kawai, *Physica B* **302–303** (2001) 140.
33. A. Tsukazaki, H. Saito, K. Tamura, M. Ohtani, H. Koinuma, M. Sumiya, S. Fuke, T. Fukumura. M. Kawasaki, *Appl. Phys. Lett.* **81** (2002) 235.
34. X. Li, Y. Yan, T.A. Gessert, C. DeHart, C.L. Perkins, D. Young, T.J. Coults, *Electochem. Solid-State Lett.* **6** (2003) C56.
35. J. Lu, Z. Ye, L. Wang, J. Huang, B. Zhao, *Mater. Sci. Semicond. Process.* **5** (2003) 491.
36. J.M. Biam, X.M. Li, C.Y. Zhang, W.D. Yu, X.D. Gao, *Appl. Phys. Lett.* **85** (2004) 4070.
37. I.D. Campbell, *Faraday Trans. 1: Phys. Chem. Cond. Phases* **73** (1977) 487.
38. J. Li, R. Kykyneshi, J. Tate, A.W. Sleight, *Solid State Sciences* **9** (2007) 613–618.
39. H. Sato, T. Minami, S. Takata, T. Yamada, *Thin Solid Films* **236** (1993) 27.
40. U.S. Joshi, Y. Matsumoto, K. Itaka, M. Sumiya, H. Koinuma, *Appl. Surf. Sci.* **252** (2006) 2524.
41. S.J. Fonash, *Solar Cell Physics*, Academic Press (1981) 80.
42. M. Jansen, *Angew. Chem. Int. Ed. Engl.* **26** (1987) 1098–1110.
43. R. Nagarajan, N. Duan, M.K. Jayaraj, J. Li, K.A. Vanaja. A. Yokochi, A. Draeseke, J. Tate, A.W. Sleight. *Int. J. Inorg. Mater.* **3** (2001) 265–270.
44. H.C. Kandpal, R. Seshadri, *Solid State Sci.* **4** (2002) 1045–1052.
45. R. Nagarajan, A.D, Draeseke, A.W. Sleight, J. Tate, *J. Appl. Phys.* **89** (2001) 8022.
46. D.B. Rogers, R.D. Shannon, C.T. Prewitt, J.L. Gillson, *Inorg. Chem.* **10** (1971) 723.
47. B.J. Ingram, T.O. Mason, R. Asahi, K.T. Park, A.J. Freeman, *Phys. Rev.* **B64** (2001) 155114.

48. W.W. Warren, A. Rajabzadeh, T. Olheiser, J. Liu, J. Tate, M.K. Jayaraj, K.A. Vanaja, *Solid State Nucl. Mag. Reson.* **26** (2004) 209.
49. H.H. Kung, H.S. Jarrett, A.W. Sleight, A. Ferretti, *J. Appl. Phys.* **48** (1977) 2463–2469.
50. H.S. Jarrett, A.W. Sleight, H.H. Kung, J.L. Gillson, *J. Appl. Phys.* **51** (1980) 3916–3925.
51. H. Mizoguchi, M. Hirano, S. Fujitsu, T. Takeuchi, K. Ueda, H. Hosono, *Appl. Phys. Lett.* **80** (2002) 1207.
52. D.E. Cox, A.W. Sleight, *Acta Crystallogr.* **B35** (1979) 1–10.
53. Y. Ogo, H. Hiramatsu, K. Nomura, H. Yanagi, T. Kamiya, M. Hirano, H. Hosono, *Appl. Phys. Lett.* **93** (2008) 032113.
54. W. Jeitschko, A.W. Sleight, *Acta Crystallogr.* **B28** (1972) 3174.
55. W. Jeitschko, A.W. Sleight, *Acta Crystallogr.* **B30** (1974) 2088.
56. T. Birchall, A.W. Sleight, *J. Solid State Chem.* **13** (1975) 118–130.
57. H.Y. Chang, K.M. Ok, J.H. Kim, P.S. Halasyamani, M. Stoltzfus, P. Woodward, *Inorg. Chem.* **46** (2007) 7005.
58. Hiroshi Mizoguchi, A.W. Sleight, M.A. Subramanian, *Mat, Res. Bull.* **44** (2009) 1022.
59. N. Kumada, Y. Yonesaki, T. Takei, N. Kinomura, S. Wada, *Mater. Res. Bull.* **44** (2009) 1298.
60. I. Radosavljevic, J.S.O. Evans, A.W. Sleight, *J. Solid State Chem.* **136** (1998) 63.
61. I.R. Evans, J.A.K. Howard, J.S.O. Evans, *J. Mater. Chem.* **13** (2003) 2098.
62. A. Walsh, G.W. Watson, D.J. Payne, G. Atkinson, R.G. Egdell, *J. Mater. Chem.* **16** (2006) 3452–3458.
63. R.D. Shannon, J.D. Bierlein, J.L. Gillson, G.A. Jones, A.W. Sleight, *J. Phys. Chem. Solids* **41** (1980) 117–122.
64. S.J. Fonash, *Solar Cell Physics*, Academic Press (1981) 76.
65. L. Svob, C. Thiandoume, A. Lusson, M. Bouanani, Y. Marfaing, O. Gorochov, *Appl. Phys. Lett.* **76** (2000) 1695.
66. P.C. Donohue, J.D. Bierlein, J.E. Hanlon, H.S. Jarrett, *J. Electrochem. Soc.* **121** (1974) 829.
67. K. Ueda, S. Inoue, H. Hosono, N. Sarukura, M. Hirano, *Appl. Phys. Lett.* **78** (2001) 2333.
68. H. Hiramatsu, K. Ueda, H. Ohta, M. Hiranao, *Appl. Phys. Lett.* **82** (2003) 1048.
69. C-H. Park, D.A. Keszler, H. Yanagi, J. Tate, *Thin Fim Solids* **445** (2003) 288.
70. H. Yanagi, J. Tate, S. Park, C-H. Park, D.A. Keszler, *Appl. Phys. Lett.* **82** (2003) 2814.
71. R. Kykyneshi, D.H. McIntyre, J. Tate, C-H. Park, D.A. Keszler, *Solid State Sci.* **10** (2008) 921.
72. S. Park, D.A. Keszler, M.M. Valencia, R.L. Hoffman, J.P. Bender, J.F. Wager, *Appl. Phys. Lett.* **80** (2002) 4393.

Chapter 10
Non-conventional Materials

Hideo Hosono

This chapter describes new transparent and electro-conductive materials or p-type transparent oxides. Deep ultraviolet-transparent Ga_2O_3, electron-doped $12CaO{\cdot}7Al_2O_3$, and transparent p-type oxides are taken up. The main application expected for theses materials are not for transparent contact but for newly emerging areas such as transparent electronics or photonics.

10.1 Ultraviolet Transparent TCO: β-Ga_2O_3

Deep ultraviolet radiation is becoming an important light in various technological applications. Lithography in semiconductors, UV-sensors, anti-bacteriostatic activity and optical detection of DNA and nucleotides are typical examples. It is expected that TCO materials available for DUV have unique potential applications. Therefore DUV-transparent TCOs are needed for these applications. Conventional TCOs such as ITO, ZnO, and SnO_2 are opaque for DUV light (<300 nm) due to a band gap (~3 eV). β-Ga_2O_3 with a large band gap of ~4.8 eV is expected as a promising TCO material for this purpose.

10.1.1 Crystal Structure and Optical Properties

Although several polymorphs of Ga_2O_3 are known to exist, β-phase is stable under conventional conditions. Figure 10.1 shows the crystal structure and the crystallographic data [1] are summarized in Table 10.1. This crystal has monoclinic symmetry

H. Hosono
Frontier Research Center, Materials and Structures Laboratory, Tokyo Institute of Technology, 4259 Nagatsuta, Midori-ku, Yokohama 226-8503, Japan
e-mail: hosono@msl.titech.ac.jp

D.S. Ginley et al. (eds.), *Handbook of Transparent Conductors*,
DOI 10.1007/978-1-4419-1638-9_10, © Springer Science+Business Media, LLC 2010

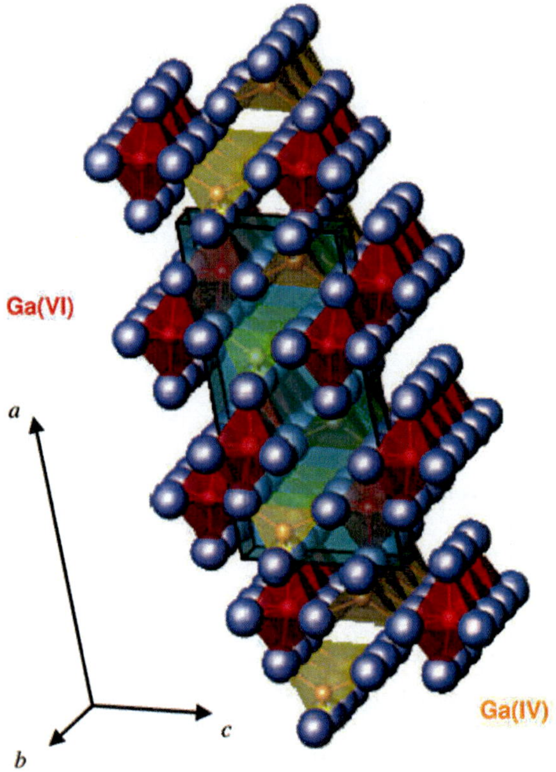

Fig. 10.1 Crystal structure of β-Ga_2O_3. Unit cell is shown by *lines*. There are octahedral and tetrahedral Ga^{3+} ions

Table 10.1 Fundamental properties of β-Ga_2O_3

Crystal symmetry	Monoclinic
Space group	$C2/m(C_{2h}{}^{3})$
Lattice constant	a = 1,223 pm
	b = 304 pm
	c = 580 pm
	β = 103.7°

and Ga^{3+} ions have octahedral and tetrahedral coordinations. The doubly connected straight chains of GaO_6 edge shared octahedral run along **b**, and these chains are connected by GaO_4 tetrahedra to one another. Anisotropy of the electrical and optical properties is expected between parallel (//**b**) and perpendicular (//**c**) directions to the chains.

10.1.2 Electron Transport and Optical Properties

Ueda et al. [2] examined the anisotropy of electron-transport and optical absorption using single crystals [3] grown by floating zone technique. The crystal is easily

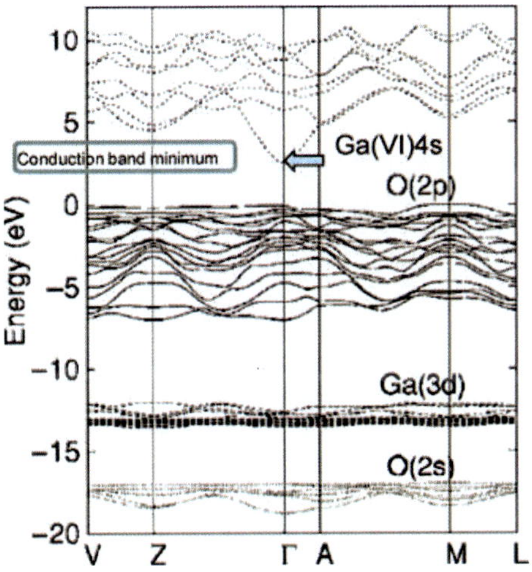

Fig. 10.2 Band structure for β-Ga_2O_3. The origin of the energy axis is the valence band maximum [4]. The calculation was carried out by DFT method employing GGA approximation. Γ-Z direction and Γ-A direction are almost parallel to b-axis and c-axis, respectively

cleaved along bc-planes. The band gap (4.9 eV) was evaluated to be larger by ~0.3 eV for **E//b** than that for **E//c**. Binet et al. [4] calculated the band structure by extended Huckel method. Figure 10.2 shows the energy dispersion curves in the conduction band along each the reciprocal axis (approximately the direction of each axis in real space is parallel to that of corresponding axis in reciprocal space). It is evident that the dispersion along the b-axis is much larger than that of the other axes and the octahedral Ga^{3+} 4s orbitals primarily constitute the bottom of the conduction band. This means that β-Ga_2O_3 is a one-dimensional electronic conductor and a chain composed of the edge-shared GaO_6 octahedra works as an efficient electron pathways. Recent density functional calculation [5] using the GGA approximation confirmed this view. The conduction band has a large dispersion around Γ-point (effective electron mass was evaluated to be 0.12 m_e), whereas the valence band is almost flat, indicating very large hole mass, and the valence band maximum is doubly degenerate at Γ and M. The observed difference in the energy gap is ascribed to the difference in these two points. The electron transport has a distinct anisotropy as predicted from the electronic structure; Hall mobility is 46 cm^2 $(V\ s)^{-1}$ for along **b** and 3 cm^2 $(V\ s)^{-1}$ along **c** in the single crystal [2].

Although transparent conductive single crystals of β-Ga_2O_3 are easy to grow by controlling the atmosphere during the growth, the fabrication of transparent conductive thin films needed further efforts. The critical parameters were choice of dopants and growth conditions. Orita et al. [6] fabricated the deep-UV transparent thin films by pulsed laser deposition on silica glass substrates at 880°C using Sn-doped Ga_2O_3 ceramic target. The resulting thin films were polycrystalline, having RT-conductivity of 1 S cm^{-1} (Hall mobility = 0.44 cm^2 $(V\ s)^{-1}$) and the internal transmittance of >50% for the 100-nm-thick film at the wavelength of 248 nm corresponding to the emission from KrF excimer laser. High temperature deposition was required to improve the optical transparency which is controlled by the formation of suboxides of Ga_2O_3. Lowering of the process temperature was realized by deposition on sapphire substrates [7]. Tin-doped Ga_2O_3 films were deposited on Al_2O_3(0001),

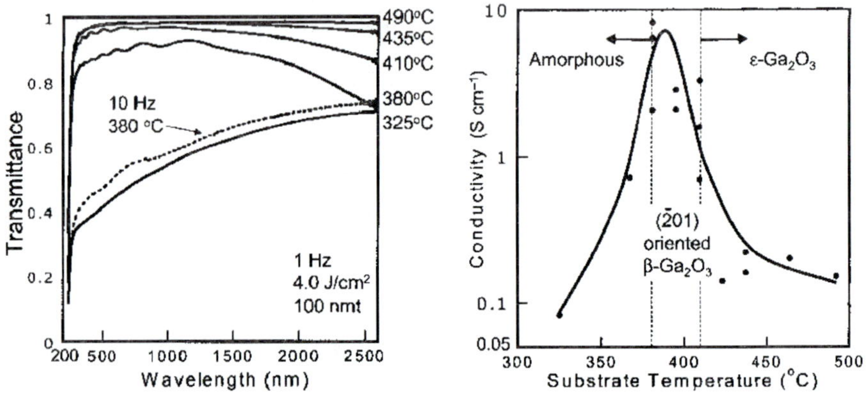

Fig. 10.3 Optical transmittance and dc conductivity at RT of β-Ga_2O_3:Sn thin films as a function of substrate temperature during pulsed laser deposition. Substrate: sapphire (0001) [7]

Table 10.2 Electron transport properties

Band gap (eV)	4.9(E//b), 4.6(E//c)
Hall mobility (cm^2 $(Vs)^{-1}$)	
For single crystal	46(//b), 2.6(//c)
For polycrystalline thin films	0.44
Effective electron mass (m_e)	0.12(//bc-plane)

(201)oriented β-phase films were grown at the temperature 380–435°C, and deep-UV transparent films with a conductivity as high as 8 S cm^{-1} were obtained at 380°C as shown in Fig. 10.3. Table 10.2 summarizes electron-transport properties of bulk single crystalline and polycrystalline thin films of β-Ga_2O_3.

The feature of the crystal structure of β-Ga_2O_3 was utilized for exploring new TCOs, i.e., it is expected that doubly connected chains of GaO_6 and GaO_4 tetrahedra connecting the GaO_6 chains work as electron pathways and electron-doping site, respectively. Cava et al. [8] reported the transparent conductivities of $GaInO_3$ with the same crystal structure as β-Ga_2O_3 by doping electrons through the introduction of oxygen vacancies, Sn doping for In, or Ge doping for Ga. The best conductivities measured on the bulk polycrystalline samples were 2–3 $\times$ 10^2 S cm^{-1} and the optical band gap was 3.3 eV. Their idea was to enhance the electron mobility by spatial separation of the electron pathway from the electron-doping site as realized in GaAs/AlGaAs high mobility transistors. Orita et al. [9] applied a similar idea to $InGaZnO_4$ (a band gap of 3.4 eV) which is composed of alternative stack of InO_2 layers and $GaZnO_4$ layers, but the maximum conductivity remained 5 $\times$ 10^2 S cm^{-1}.

10.1.3 Device Application

It is a straightforward idea to apply Ga_2O_3 thin films to deep-UV sensors. Kokubun et al. [10] fabricated the thin films on a-Al_2O_3(0001) by the sol-gel method and

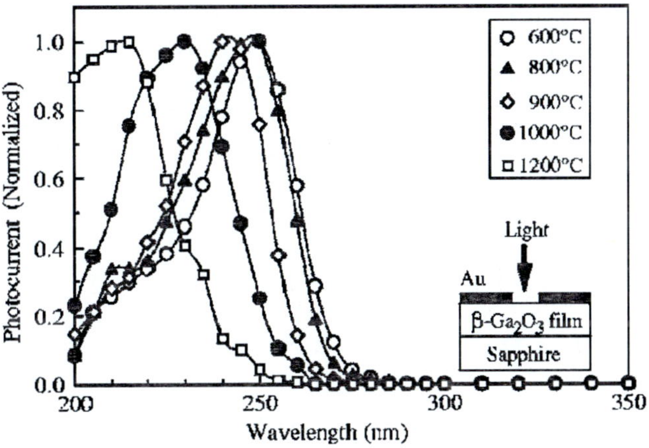

Fig. 10.4 Spectral photocurrent response of β-Ga_2O_3 sol-gel derived thin films prepared at various firing temperatures [10]

examined the spectral responses of the photoconductive detectors using planar geometry. Figure 10.4 shows the results, indicating the thin films have the sensitivity to deep UV-radiation with wavelength below 270 nm as expected.

Solar-blind TFTs are favorable for high sensitive UV-sensors because electric field-induced current modulation can be utilized through metal-insulator-semiconductor structures. The successful operation of oxide TFTs was restricted to transparent oxide semiconductors with the band gap $<$3.5 eV. Matsuzaki et al. [11] fabricated TFTs using β-Ga_2O_3 epitaxial thin films for n-channels on a-Al_2O_3(0001), reporting filed-induced current modulation (field effect mobility $= 5 \times 10^{-2}$ cm^2 $(V\ s)^{-1}$). The primary origin of lower field effect mobility to the that of Hall mobility is due to the presence of high tail state density below the conduction band bottom, that is, the Fermi level cannot raise to $>$ mobility edge by biasing. Although the performance is needed to be improved, this is the first demonstration of Ga_2O_3-TFTs.

10.2 Light Metal TCO: 12CaO·7Al_2O_3 with Built-in Nano-porous Structure

A wide variety of materials and crystal structure is a unique feature of oxides. Electro-conductive oxides are, however, restricted to transition metal or heavy metal cation-bearing materials. Representative light metal oxides which are most abundant in our terrestrial crust and environmentally benign, are classified as typical transparent insulators as describe in textbooks. They are the main constituents of traditional ceramics such as porcelain, cement, and glass, but are not considered to be a platform for realization of electro-active functionality. In fact, no one has succeeded in converting them to persistent electronic conductor until a report [12] in 2002 utilizing nano-structures built-in crystal structure of $12CaO{\cdot}7Al_2O_3$ (C12A7). Exploration for approach to conversion of light metal oxides to electro-active materials is a grand challenge in material science to resolve natural resource and environmental issues we are facing now. This section describes the electronic structure of C12A7 and various approaches to electron-doping along with unique properties of electron-doped C12A7.

10.2.1 Crystal Structure of $12CaO{\cdot}7Al_2O_3$

This compound can be easily prepared by a conventional solid-state reaction in air, but is not formed in O_2 or moisture-free atmosphere [13]. The crystal lattice of $12CaO{\cdot}7Al_2O_3$ (C12A7) with a micro porous structure has a cubic lattice constant of 1.199 nm, and the unit cell includes two molecules [14]. It is characterized by a positively charged lattice framework $[Ca_{24}Al_{28}O_{64}]^{4+}$ having 12 crystallographic cages with a per unit cell as shown in Fig. 10.5. Each cage with an inner diameter of ~0.44 nm and a formal charge of +1/3, is coordinated by eight cages to form a three-dimensional structure by sharing a monomolecular open mouse. This structure is a similar to a zeolite but the excess charge of the cage wall is reversed. Free oxygen ion O^{2-} is accommodated in 1/6 out of 12 cages in the unit cell to preserve electroneutrality in the stoichiometric state. The remaining two oxide ions (O^{2-}) are entrapped as counter anions in the cages in the stoichiometric composition and called *free oxygen ions* because the cage diameter is larger by ~50% than the diameter of O^{2-}. This fact is associated with the fast oxygen ion conduction in this material [15]. Monovalent anions, such as hydroxyl ions (OH^-) or halogen anions,

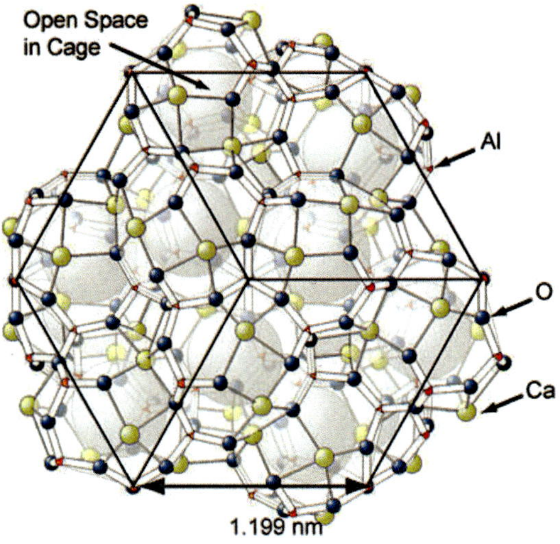

Fig. 10.5 Crystal structure of $12CaO{\cdot}7Al_2O_3$. Free O^{2-} is omitted for simplicity

X^-(X^- = F^- and Cl^-), are known to substitute for a part or all of the free oxygen ions to form derivatives $[Ca_{24}Al_{28}O_{64}]^{4+}\cdot 4(X^-)$. The replacement of a free O^{2-} ion by two monovalent anions likely stabilizes the structure through charge delocalization [16]. C12A7 crystals can be easily synthesized by conventional solid state reactions using $CaCO_3$ and $Al(OH)_3$ or Al_2O_3. Single crystals can be grown from the melt by using a floating-zone (FZ) method [17] or Czochralski (CZ) technique [18].

10.2.2 Electronic Structure and Our Approach

Figure 10.6 shows band structure of the stoichiometric C12A7 calculated by density functional method with GGA approximation using VASP code. Two major features are evident from the figure [19]. (1) The O2p levels of the free oxygen ions are located ~1 eV above the valence bands. (2) There are 2-types of conduction bands, i.e., one is a conventional conduction band which is primarily composed of Ca 4s levels, and is called framework conduction band(FCB). Another is a conduction band composed of three-dimensionally connected sub-nanometer-sized empty cages, which is not comprised of orbitals of specific ions constituting the cage wall but is constituted of a coupled "particle-in-a box", and is called "cage conduction band (CCB)" [19–21]. CCB, which is unique to C12A7, is located below by ~2 eV from the bottom of FCB and has moderate energy dispersion [22]. These features implied to us an idea that if the free oxygen ions are replaced by electrons, the resulting state may exhibit high electronic conduction; the energy

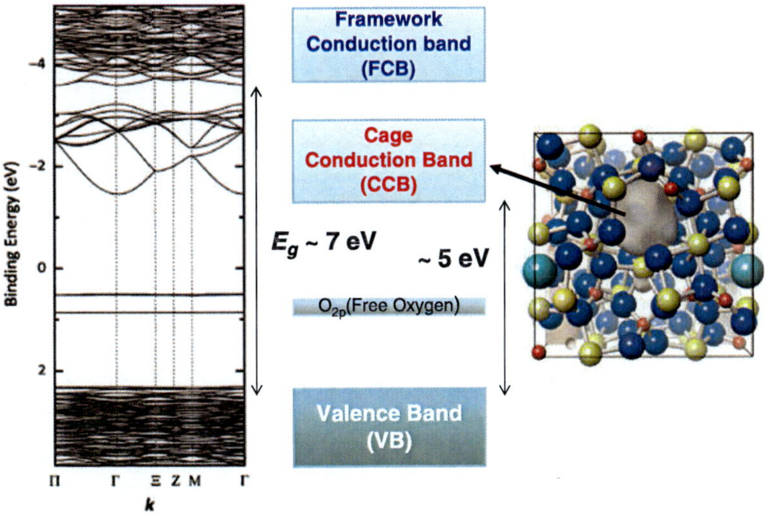

Fig. 10.6 Band structure of C12A7. Note that there are two types of conduction bands, and 2p levels of free O^{2-} ions are located above the framework valence band top [20]

level of FCB bottom is too high for electron-doping, whereas the location of CCB is marginal for successful electron-doping compared in the doping limit [23, 24] for other n-type semiconductors. Various methods to inject electrons to the CCB are described in the next section.

10.2.3 Electron Doping

10.2.3.1 Light-Induced Doping for C12A7:H$^-$

Hayashi et al. [12] demonstrated a process by which the transparent insulating oxide C12A7 can be converted into an electronic conductor. H$^-$ ions are incorporated into the sub-nanometer-sized cages of the oxides by a thermal treatment in a hydrogen atmosphere at 1,300°C; subsequent illumination of the resulting C12A7:H$^-$ with ultraviolet light ($\lambda < 350$ nm) results in a conductive state that persists after irradiation ceases. The photo-activated material exhibits moderate electrical conductivity (~0.5 S cm^{-1}) at room temperature, with visible light absorption losses of only 1% for 200-nm-thick films as shown in Fig. 10.7. The incorporation of H$^-$ anions in the cage may be understood through the following chemical reactions;

$$O^{2-}(\text{cage}) + H_2(\text{atmosphere}) \rightarrow OH^-(\text{cage}) + H^-(\text{cage}), \quad (10.1)$$

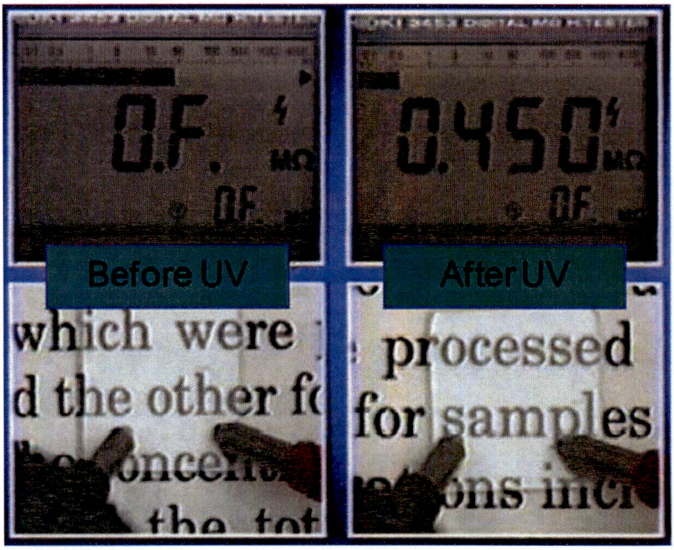

Fig. 10.7 Photos of C12A7:H$^-$ thin films (200 nm^t) before and after illumination with ultraviolet radiation

or

$$O^{2-}(\text{cage}) + H_2(\text{atmosphere}) \rightarrow 1/2O_2(\text{atmosphere}) + 2H^-(\text{cage}). \qquad (10.2)$$

Specimens C12A7: H^- look white in the polycrystalline powder or colorless and transparent for single crystals and is insulating (conductivity $<10^{-10}$ S cm^{-1} at RT). However, upon illumination with ultraviolet light (wavelength < 330 nm), two absorption bands centered at 2.8 and 0.4 eV are induced, and simultaneously, a drastic increase in the conductivity was observed along with the coloration. The conductivity increased from $<10^{-10}$ to 0.5 S cm^{-1}. The Seebeck coefficient of the conductive sample is negative (-360 μV K^{-1}), indicating n-type conduction. Figure 10.8 shows the temperature dependence of the conductivity (σ). Log σ is proportional not to T^{-1} but to $T^{-1/4}$ over a wide temperature range 300–50 K. This characteristic indicates that the conduction is not controlled by a simple thermal activation process but more probably by a variable-range hopping of carriers.

A possible mechanism for the photo-induced conversion of C12A7: H^- from an insulator to a conductor is proposed as summarized in Fig. 10.9: First, the UV-illumination induces an electron emission from the encaged H^- ions ($H^- \rightarrow H^0 + e^-$). Then, an empty cage electro-statistically captures the photo-induced electron to form an F^+-like center, i.e., an electron-trapped at the site of an oxygen vacancy. As the electron is weakly bound in the cage, the wave function would spread spatially. Consequently, electrons may migrate throughout the crystal by variable-range hopping. The resultant H^0 atoms rapidly diffuse throughout the cages and combine with the remaining the free O^{2-} ion to form OH^- releasing another electron. That is to say, an H^- ion works as a double photoelectron donor in C12A7.

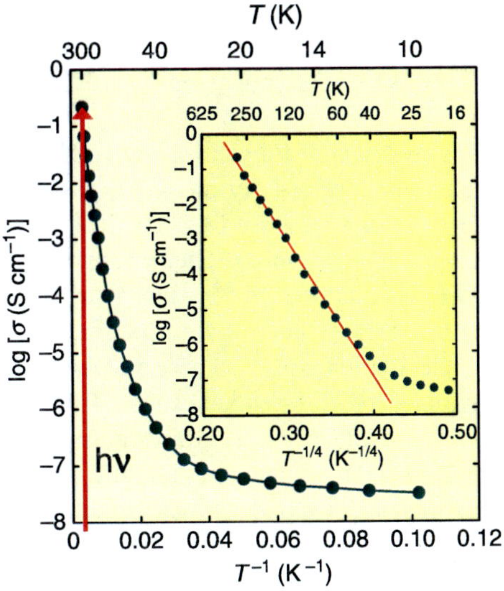

Fig. 10.8 Arrhenius plot of conductivity in UV-illuminated C12A7:H^- thin films. *Inset* is log σ vs. $T^{-1/4}$ plot [12]

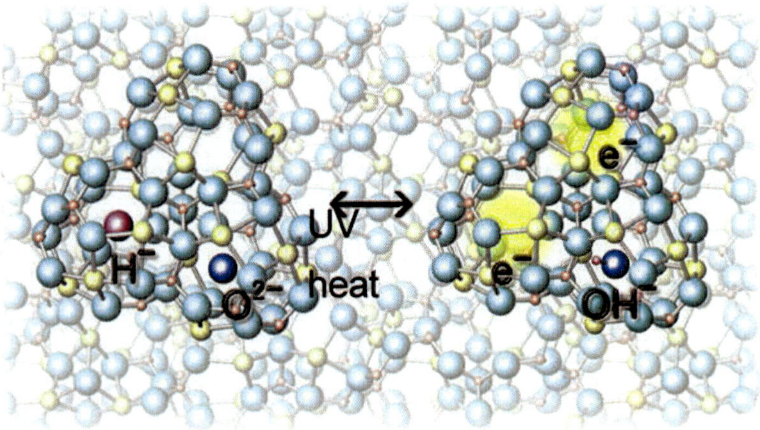

Fig. 10.9 Mechanism for insulator-conductor conversion in C12A7:H^- by UV-illumination. H^- works as a double photo-electron donor and electron generated moves to a neighboring empty cage via hopping

The validity of this model was justified by observation of H^0–F^+ pair generation in the sample illuminated at 4 K [25] and theoretical calculations [22].

The electron-doping from H^- is possible as well by an electron-beam irradiation in place of UV-radiation [26]. The quantum efficiency of conduction electron generation evaluated for 25 keV electron beam is 0.02. Since electron-beam can be focused to a level beyond the limit by UV-light, direct writing of electro-conductive pattering down to several ten nanometers is possible.

10.2.3.2 Doping by Chemical Reduction

Several chemical processes were developed for electron-doping via removal of the free O^{2-} ion.

Thermal annealing of C12A7 single-crystals in Ca metal vapor at 700°C led to the electrical conductivity of ~80 S cm^{-1} at 300 K [27]. However, the Ca-treatment needs a long duration (e.g. 10 days for a 0.4 mm thick sample) to extract the free oxygen ions because the upper annealing temperature is limited to ~700°C due to chemical reactions of Ca vapor with SiO_2 glass tubes. Further, the replacement may be incomplete due to the formation of a surface CaO layer that acts as a diffusion barrier of the free O^{2-}. These drawbacks were overcome by employing Ti metal [28] in place of Ca metal, because Ti metal forms stable nonstoichiometric oxides with high electronic conductivity over a wide chemical composition range ($1 < x < 2$ in TiO_{2-x}) and the out-diffusion of the free oxygen ions from C12A7 is not blocked even if a thick titanium oxide layer completely covers the sample surface. In addition, higher annealing temperatures up to ~1,300°C can be used utilizing an observation that Ti does not react with C12A7 up to this temperature.

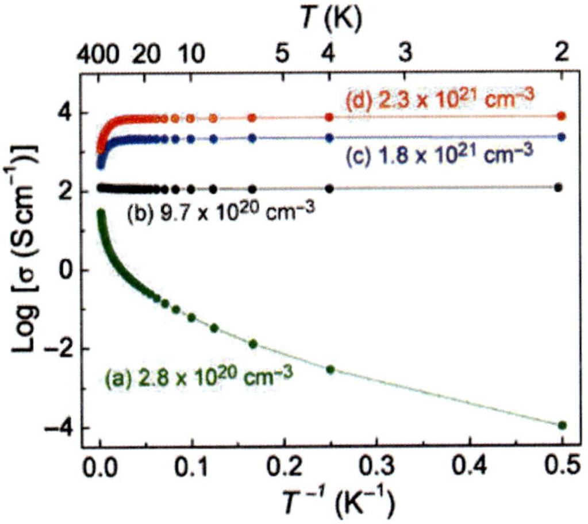

Fig. 10.10 Temperature dependence of conductivity in electron-doped C12A7 single crystals with different electron concentrations Ne [28]. Here the value of Ne was evaluated by analysis of dielectric function obtained from optical reflection spectra of the single crystals using Drude-Lorentz model [84]

Since an increase in the temperature from 700 to 1,100°C enhances the diffusion of the free oxygen ions in C12A7 by several orders of magnitude, the treatment time can be drastically shortened.

Figure 10.10 shows the Arrhenius plot of the electrical conductivities (σ) for the Ti-treated single-crystals with different electron concentrations (N_e). The N_e values were evaluated from the analysis of dielectric function obtained by the optical reflection spectra of the single crystals using Drude-Lorenz model. The Arrhenius plot has a negative slope when the electron concentration is low, indicating that a thermally-activated mechanism such as the hopping controls the migration of the electrons. On the other hand, the temperature dependence vanishes at $N_e \sim 1 \times 10^{21}$ cm^{-3} (N_{th}), demonstrating that the electron conduction is degenerated, and finally the slope becomes positive around room temperature at larger N_e values. The conductivity extrapolated to 0 K (σ_{0K}) as a function of N_e (Fig. 10.11) shows this systematic change more clearly: σ_{0K} starts to increase from zero to a finite value at $N_e = N_{th}$, confirming an Insulator-Metal (IM) transition and σ_{0K} super-linearly increases with N_e, showing a sharp increase in the drift mobility (from ~0.1 to 4 cm^2 V^{-1} s^{-1}) for $N_e > N_{th}$ (inset of Fig. 10.11).

These observations demonstrate that the insulating C12A7 is converted to metallic state by replacing more than half of the free oxygen ions by electrons. This IM transition accompanies a sharp increase in the drift mobility (μ_e), calculated from observed σ and Ne, from ~0.1 to 4 cm^2 V^{-1} s^{-1} as N_e increases. It should be noted that the doping and conduction mechanisms in C12A7 are much different from those in conventional semiconductors due to the unique crystal and electronic structures of C12A7.

In typical semiconductors such as phosphorus-doped Si [29], the IM transition occurs due to the percolative overlap of the donor wave functions as explained by Mott in terms of the critical donor concentration N_c, which follows the empirical relation $N_c^{1/3} \cdot R_c \sim 0.25$ (R_c denotes the radius of the electron orbital) [30]. When

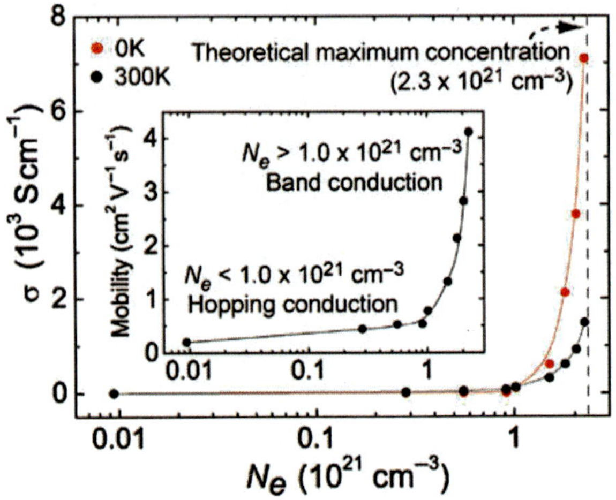

Fig. 10.11 Conductivity extrapolated to 0K and mobility as a function of carrier electron concentration

the N_{th} value of ~1 × 10^{21} cm^{-3} is put into this relation, obtained R_c value is of ~0.26 nm, which is close to the radius of the crystallographic cage embedded in C12A7 (~0.25 nm). Thus, it is feasible that the IM transition in C12A7 occurs via the percolation of the "donor" orbitals. However, this view cannot be directly applicable to C12A7 because even an "undoped" stoichiometric C12A7 crystal has a "donor wave function" in each empty cage as a component of the CCB. Although the "donor wave function" intrinsically exist, the electrons doped to the CCB do not contribute to the band conduction at $N_e < N_{th}$ at low temperatures because the electron is localized in a cage forming an F^+-like center [12, 25] due to the lattice relaxation as described above. However, a high electron doping may average out the deformations of the cage geometry throughout the crystal and it converts the localized F^+-like state to a delocalized state, invoking an IM transition.

This model for the IM transition is validated by MEM/Rietveld analysis and *ab initio* density functional theory (DFT) calculations [28]. Figure 10.12 displays the electron density maps in the (001) plane of stoichiometric C12A7 and metallic C12A7 obtained by the MEM/Rietveld analysis [31]. The electron densities of the Ca^{2+} ions on the cage wall of stoichiometric C12A7 have long tails like a dumbbell shape, indicating that the map is a superposition of the two deformed O^{2-}-encaging cages and the ten undeformed empty cages. The distances between the Ca^{2+} ions ($D_{Ca\text{-}Ca}$) in the deformed and empty cages obtained by the Rietveld analysis are 4.22 and 5.65 Å, respectively. Thus, the map in Fig. 10.12 provides solid evidence that the O^{2-}-encaging cage shrinks largely from the empty cage. On the other hand, the map in Fig. 10.12c indicates the cage structure of $[Ca_{24}Al_{28}O_{64}]^{4+}(4e^-)$ clearly differs from that of stoichiometric C12A7. Each atom has a spherical electron density, indicating that the local structural deformation is removed in $[Ca_{24}Al_{28}O_{64}]^{4+}(4e^-)$. That is, all the 12 cages are homogenized and $D_{Ca\text{-}Ca}$

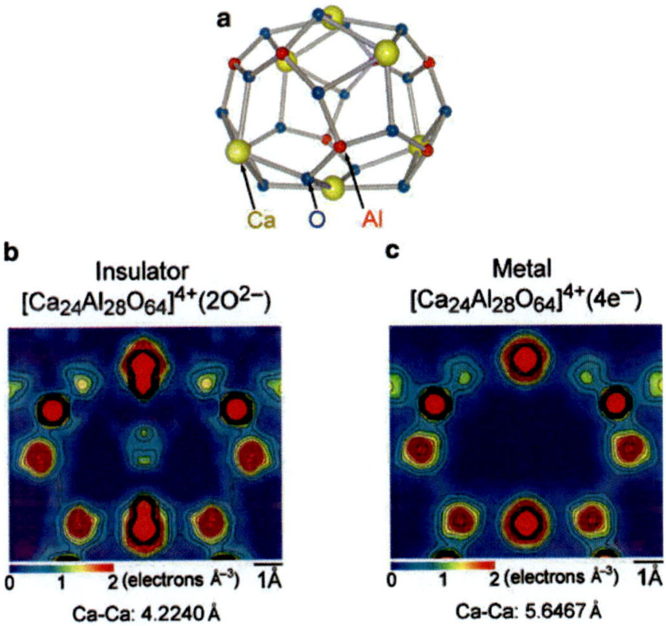

Fig. 10.12 Electron density map of cages (**a**) in insulating (**b**) and metallic conducting (**c**) C12A7determined by MEM-Rietvelt analysis of X-ray diffraction pattern. This density is on the (001) plane [31]. Note that electron density on Ca axially coordinated to an O^{2-} in (**b**) exhibits a distinct tail extending to the cage center. This means the Ca^{2+} position in O^{2-}-bearing cages with a distorted geometry is slightly different from that in O^{2-}-free cages with an undistorted geometry

(5.64 Å) becomes close to that of the empty cage, suggesting the electrons are distributed uniformly over the 12 cages within an accuracy of the MEM/Rietveld analysis. Figure 10.13 shows relaxed lattice structures calculated by DFT with the isosurfaces of the wave function ($|\Psi|^2$) of $2p$ level of the free oxygen ions and those of CCB bottom (isosurface state density) for (i) stoichiometric C12A7, (ii) $[Ca_{24}Al_{28}O_{64}]^{4+}(O^{2-})\cdot(2e^{-})$ and (iii) metallic C12A7. The DFT results are summarized in Fig. 10.13 as a schematic energy level diagram. Figure 10.13a(i) demonstrates that two free oxygen ions respectively occupy the cages at the center and the corners of the unit cell and the other cages are empty in the stoichiometric C12A7. The calculated $D_{Ca\text{-}Ca}$ of the O^{2-}-encaging cage is 4.27 Å, in good agreement with the observed value of 4.22 Å. Further, each empty cage has an s-like state, and the connection of these states forms the CCB even in the stoichiometric C12A7 (Fig. 10.5b(i)). On the other hand, the O^{2-}-encaging cage has no such CCB state because of the entrapping of O^{2-}, whose $2p$ energy level is located slightly above the valence band maximum of the cage framework (Fig. 10.13c(i)) Thus, stoichiometric C12A7 has no electron in the CCB and is a band insulator [22].

The DFT result for $[Ca_{24}Al_{28}O_{64}]^{4+}(O^{2-})\cdot(2e^{-})$, where a free oxygen ion is extracted from a cage and two electrons occupy two of the other cages in the unit

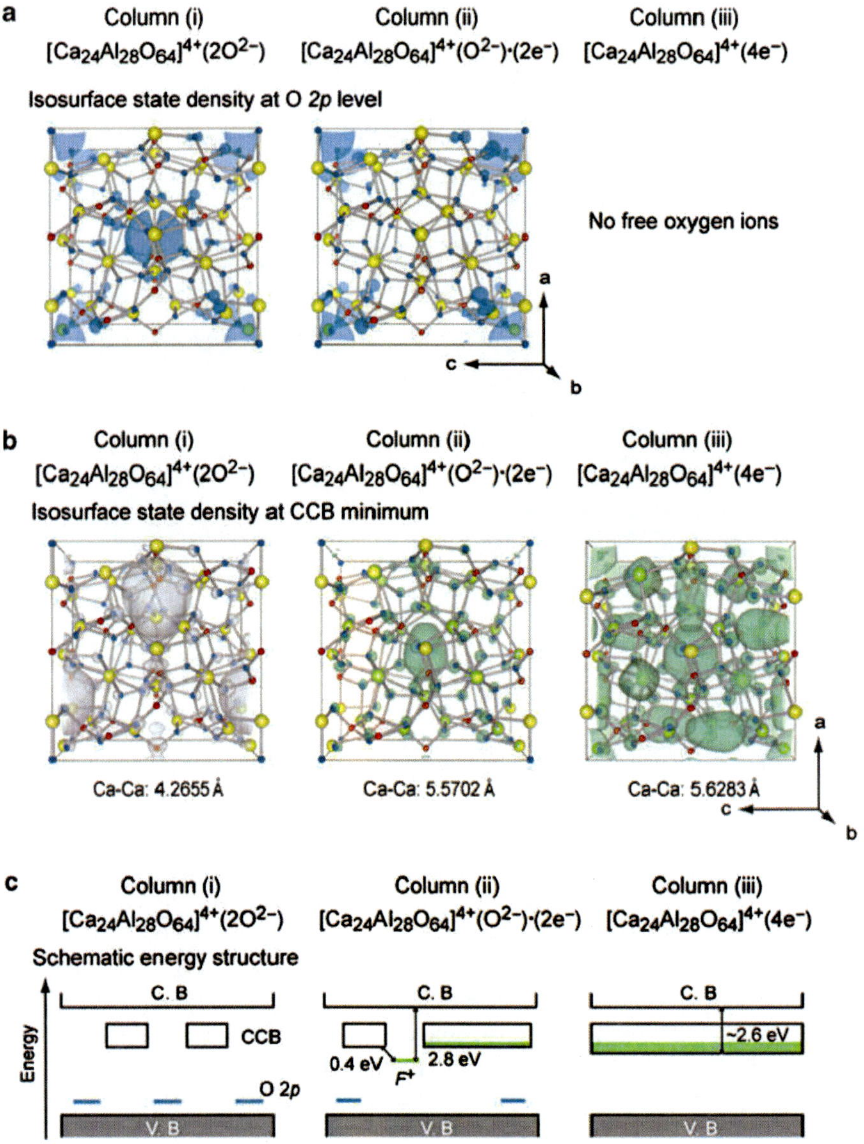

Fig. 10.13 Isosurface state density by DFT calculation on relaxed structure and schematic band structure of C12A7 [28]

cell, demonstrates that $D_{Ca\text{-}Ca}$ (5.57 Å) of the electron-encaging cage becomes larger compared to that of the O^{2-}-encaging cage (4.27 Å), but it is still smaller than that of the empty cage (5.77 Å). The wave function at the CCB minimum (Fig. 10.13b(ii)) implies that the electron is still localized in a specific cage, forming an F^+-like center (Fig. 10.13c(ii)).

Figure 10.13b(iii) is the relaxed lattice structure with the isosurface state density at the CCB minimum of $[Ca_{24}Al_{28}O_{64}]^{4+}(4e^{-})$, revealing that $D_{Ca\text{-}Ca}$ (5.63 Å) of the electron encaging cage further expands from that of $[Ca_{24}Al_{28}O_{64}]^{4+}(O^{2-})\cdot(2e^{-})$ and it becomes very close to that of the empty cage. It also indicates the isosurface state density in each cage is equivalent, indicating that the electrons are delocalized over the cages and consequently the cage conduction band has a rather large dispersion of ~2 eV.

Thus, the DFT calculation, consistent with the MEM/Rietveld analysis, qualitatively explains why the C12A7 shows the IM transition upon the electron-doping. That is, the successive reduction of deformation of the electron-encaging cage with increasing the electron concentration induces the IM transition, which accompanies the sharp increase in the electron mobility. It is noteworthy that simple ionic oxides such as MgO have not exhibited such a high conductive state although the F/F^{+}-like centers can be incorporated at high concentrations similar to the present case [32]. Thus, the three-dimensionally connected sub-nanometer-sized cage structure in C12A7 plays an essential role in the appearance of the metallic state.

10.2.3.3 Doping Via Melt Processes

The development of a simple synthesis process, which is compatible for mass production of C12A7:e^{-} is strongly required for practical applications. Possible candidate to satisfy the requirement is a direct synthesis of C12A7:e^{-} from the melt or glass state. Extensive studies regarding the relationship between melting condition and corresponding precipitated phases in CaO-Al_2O_3 system [33] verify that the presence of template anions to compensate the positive charge of the C12A7 framework is inevitable for the precipitation of the C12A7 phase. More specifically, the C12A7 phase is reproducibly prepared from the stoichiometric C12A7 melt in wet air [13], involving O^{2-} and/or OH^{-} ions that act as the template. On the contrary, the thermal treatment in inert gas atmospheres, where template anions are seemingly absent, hinders the formation of the C12A7 phase and only a mixture phase of C3A ($3CaO\cdot Al_2O_3$) and CA ($CaO\cdot Al_2O_3$) is formed. If we could precipitate the crystalline C12A7 under an appropriate reducing atmosphere, there is a possibility of direct synthesis of the C12A7:e^{-} from the melt or glass. The critical issue for the direct synthesis is to find a novel template anion species that can exist in the melt or glassy phase even when they are exposed to reducing atmospheres. With efforts in searching for such a template anion, successful results were obtained for the process via the reduced melt [34] or glass state [35] where C_2^{2-} ion most likely acts as a template for the C12A7 phase formation.

Figure 10.14 (upper) shows schematically a direct solidification process to prepare C12A7:e^{-} from the C12A7 melt. C12A7 powders of the stoichiometric mixture of CaO and Al_2O_3 are melted in a semi-airtight carbon crucible with a carbon cap at 1,600°C for 1 h in air. Heating the carbon crucible at 1,600°C produces a strongly reducing atmosphere (PO_2 = ~10^{-16} atm) inside the crucible. The crucible is slowly cooled down at a rate of ~400°C/h. X-ray diffraction

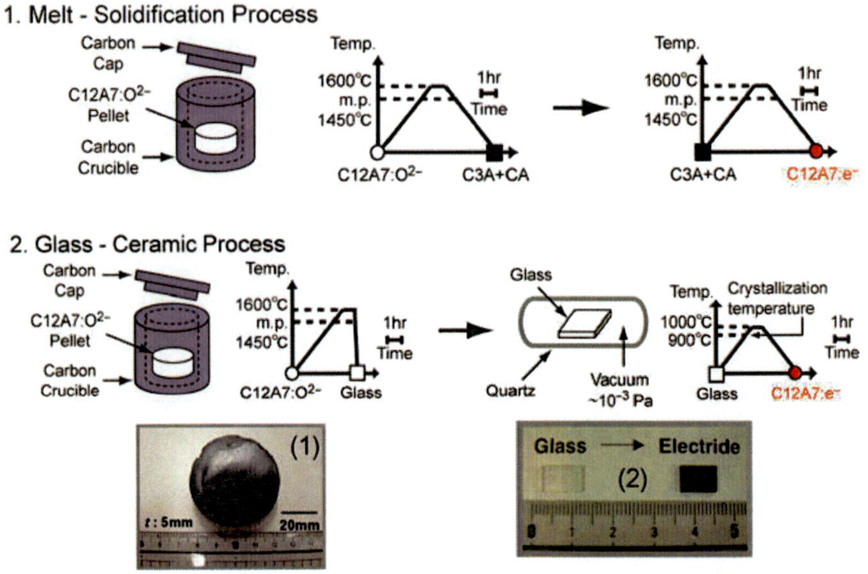

Fig. 10.14 Direct synthesis of electron-doped C12A7 from the melt and glass. Process procedures and photos of the resulting samples are shown [33]

measurements clarify that solidified products are mixtures of C3A and CA. Then, they are re-melted and solidified again according to the same temperature profile as that of the first run. The second run results in the formation of the C12A7 crystalline phase with a green color and the electrical conduction.

Raman spectra of the products after the first run and thermal desorption measurement in the cooling process in the second run suggest that C_2^{2-} ions are dissolved into the melt from the carbon crucible to compensate for the oxygen deficiency caused by the strongly reducing atmosphere. Furthermore, the ionic radius of C_2^{2-} (1.2 Å) is similar to O^{2-} (1.4 Å), which support that the C_2^{2-} anions serve as the template in a reducing atmosphere, instead of the free oxygen ions in an oxidizing atmosphere. Moreover, the C_2^{2-} Raman band is not observed in the C12A7:e$^-$ and CO gas is evolved in the cooling process, which implies that the encaged C_2^{2-} ions are only stable in the nucleation and/or initial stage of the crystallization and they are released from the cages, leaving electrons in the lattice during the cooling process. Thus, the electron formation processes in the cages may be expressed as follows:

$$C_2^{2-}(\text{melt}) \rightarrow C_2^{2-}(\text{cage}) \rightarrow 2\text{C (solid)} + 2e^-(\text{cage}), \tag{10.3}$$

and/or

$$C_2^{2-}(\text{cage}) + 2O^{2-}(\text{cage}) \rightarrow 2\text{CO (atmosphere)} + 6e^-(\text{cage}) \tag{10.4}$$

Figure 10.14 (bottom) shows schematically the glass-ceramics process for the preparation of C12A7:e-. Oxygen deficient glass, which is obtained by a rapid quench of the C12A7 melt in a carbon crucible placed in air, is crystallized in an evacuated silica tube by heating to 1,000°C, which is higher than the crystallization temperature (>900°C). The resultant C12A7 samples were green in color and exhibited rather high electrical conductivity, indicating that electrons are incorporated into the cages instead of free oxygen ions. Similar to the direct solidification process, C_2^{2-} ions, which are likely incorporated in the glass, are considered to act as the template. The resulting C12A7:e^- having 10^{19}–10^{20} cm^{-3} exhibits 0.1–1 S cm^{-1} at RT.

10.2.4 *Thin Film Fabrication of Electron-Doped C12A7*

Thin film fabrication is needed for device application of C12A7:e. Bulk formation processes described above is not applicable for thin film fabrication because the thickness of reaction layers formed by metal treatment is comparable to the conventional film thickness. Thus, different processes are required to be invented. Two processes to be described below have been reported for this purpose.

10.2.4.1 Ion Implantation

Ion implantation, a nonequilibrium physical process, has such advantages over conventional chemical doping processes that high concentration of ions can be introduced in a designated local area with an excellent controllability [29, 30], and thus it is widely used for the modification of thin films.

Miyakawa et al. [36, 37] conducted Ar^+ ion implantation into polycrystalline C12A7 thin films and found that the electron anions are incorporated into the cages as a result of kicking out the free oxygen ions from the cages through the nuclear collisions. The polycrystalline C12A7 thin films (~800 nm in thickness) were prepared on MgO(100) single crystal substrates by a pulsed laser deposition technique at RT and a subsequent annealing at 1,100°C for 1 h in air to crystallize the deposited amorphous films. Ion implantation was carried out at temperatures from RT to 600°C in a vacuum less than ~2×10^{-8} Torr. The Ar^+ fluence (accumulated dose) set to vary from 1×10^{16} to 1×10^{18} cm^{-2}. Moreover, the C12A7 framework is stable for the hot implantation, although a crystalline film is converted to amorphous phase if the implantation is performed at RT.

Figure 10.15 shows the electrical conductivities of 300 keV Ar^+-implanted films subjected to the implantation at high temperature of 600°C as a function of Ar^+-ion fluence. The electrical conductivity of the film increases with implanted fluence: films with fluences less than 1×10^{17} cm^{-2} are insulators, while those with 5×10^{17} to 1×10^{18} cm^{-2} exhibit the electrical conductivity of ~0.5 and 1 S cm^{-1}, respectively. The enhancement of the electrical conductivity is accompanied by the appearance

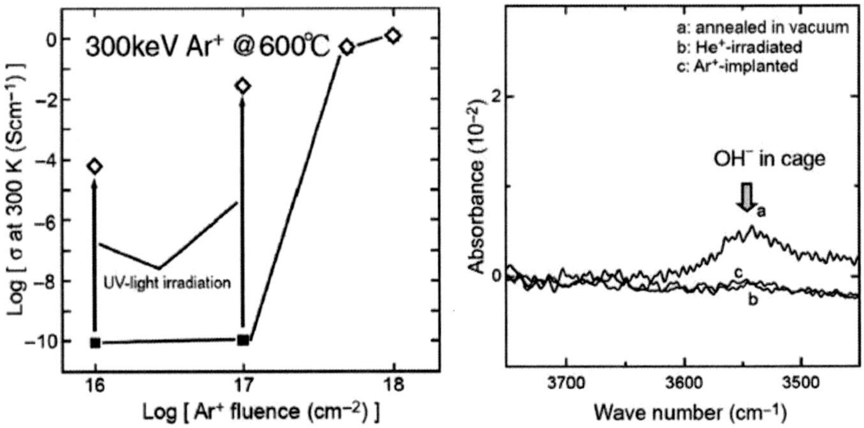

Fig. 10.15 Electrical conductivity in C12A7 thin films as a function of implanted Ar^+ fluence. *Right panel* shows infrared absorption spectra of OH^- in C12A7 thin films subjected to various treatments [37]

of optical absorption bands at 0.4 and 2.8 eV, indicating the F^+-like centers are simultaneously formed by the hot implantation. These results suggests that the free oxygen ions are kicked out from the cages by the collision between the free oxygen ions and Ar^+ ions, leaving electrons in the cages. As a result, the electrons are directly produced by Ar^+-implantation via the extrusion of free oxygen ions through the following reaction.

$$O^{2-}(\text{cage}) \rightarrow 1/2O_2(\text{atmosphere}) + 2e^-(\text{cage}) \quad (10.5)$$

It is noted in samples implanted with less than 1×10^{17} cm^{-2} Ar^+ ions that the conductivity and the optical absorption intensities increase after UV light irradiation. Since these phenomena are commonly observed in H^- incorporated C12A7, it suggests that H^- ions are created into the cages by the hot implantation of Ar^+ ions. It was consider that the formation of the H^- ions by Ar^+-implantation is caused by pre-existing OH^- groups that are introduced during post-annealing process for the crystallization of amorphous films in air. That is, electrons are generated by the extrusion of free oxygen ions through the reaction (10.5) and simultaneously protons (H^+) are created by the decomposition of pre-existing OH^- groups in the as-prepared film through the reaction (10.6).

$$OH^-(\text{cage}) \rightarrow O^{2-}(\text{cage}) + H^+(\text{attached to cage wall}) \quad (10.6)$$

Then, H^+ ions immediately react with the electrons in the cages, leading to the formation of H^- ions as follows;

$$2e^-(\text{cage}) + H^+(\text{attached to cage wall}) \rightarrow H^-(\text{cage}). \quad (10.7)$$

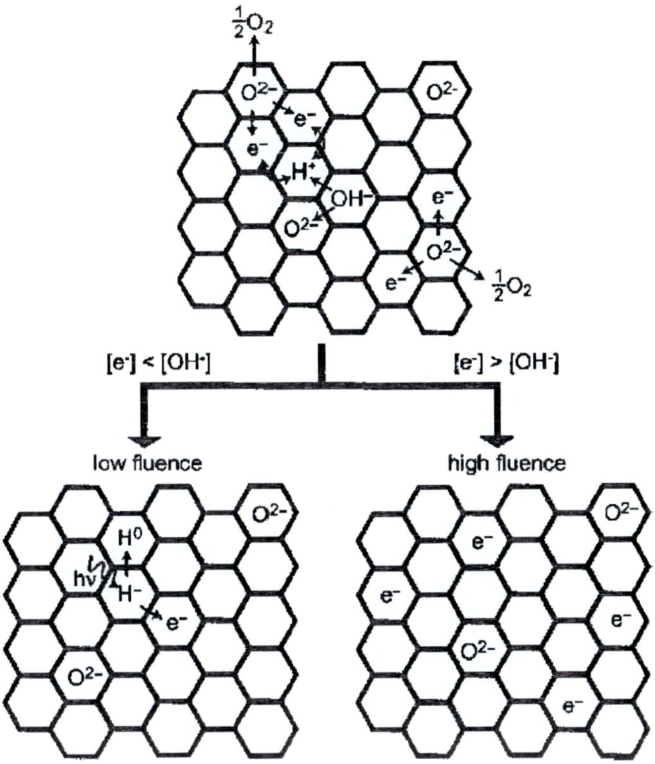

Fig. 10.16 Mechanism for electron carrier generation in C12A7 thin films by inert gas implantation. There are two independent processes, direct knock-on of free O^{2-} and photo-ionization of H^- which was created from OH^- in the cage by action of energetic ion [37]

Thus, H^- ions are formed in lower Ar^+-implanted samples, leading to the light-induced electronic conduction and coloration.

Figure 10.16 shows the schematic illustration of the electron and H^- formations by Ar^+-implantation as mentioned above. Ar^+-implantation produces the electrons and protons in the cages by nuclear collisions and electronic excitation effects through the reactions (10.5) and (10.6). When the number of the electrons produced by reaction (10.5) is less than that of protons, all of produced electrons are captured by H^+ ions through reaction (10.7) and H^- ions are formed in the cages. However, when the generated electron concentration exceeds the proton concentration, the excess electrons themselves are directly introduced into the cages.

Since higher concentration of the electrons are incorporated three-dimensionally in a designated narrow space, the synthesis of the C12A7:e^- thin films by the ion implantation paves a way to realize novel applications such as patterned electric wiring, electron injection electrode and cold electron emitter.

10.2.4.2 Chemical Reduction

Miyakawa et al. [38] found a method for fabrication of C12A7:e$^-$ thin films by using a deposition of oxygen-deficient amorphous C12A7 layer on crystalline C12A7 thin films. Figure 10.17 shows the fabrication processes. First, amorphous thin films of ~400 nm thick were deposited at RT by pulsed laser deposition (PLD) method using ceramic C12A7 targets. The partial O_2 pressure was ~1×10^{-3} Pa during the depositions, and the substrates employed were MgO(100). Polycrystalline (p-) C12A7 thin films were obtained by heating the resulting amorphous (a-) C12A7 thin films to 1,100°C in an ambient atmosphere. Next, a-C12A7 (~30 nm thick) layers were deposited on the p-C12A7 thin films at 700°C in an O_2 pressure of ~1×10^{-3} Pa by PLD. This deposited a-C12A7 layer is an oxygen-deficient and extracts free O^{2-} ions from the cages of p-C12A7 layer generating electrons in the cages under the condition. Last, the over-layered a-C12A7 layer was removed by chemical-mechanical polishing (CMP). By tuning the duration and temperature of post-annealing temperature, the electron concentration in C12A7 thin film can be increased up to ~2×10^{21} cm^{-3} (the theoretical maximum).

12SrO·7Al_2O_3 (S12A7) with the same crystal structure as C12A7 is known as a meta stable phase [39] Miyakawa et al. [40] applied this method to S12A7 and successfully obtained S12A7:e$^-$ thin films by tuning the temperature and atmosphere for the post-annealing of a-S12A7. Figure 10.18 shows the DC conductivities of resulting C12A7:e$^-$ and S12A7:e$^-$ thin films as a function of temperature. It is evident that both samples exhibit metallic conduction. The Hall mobility and

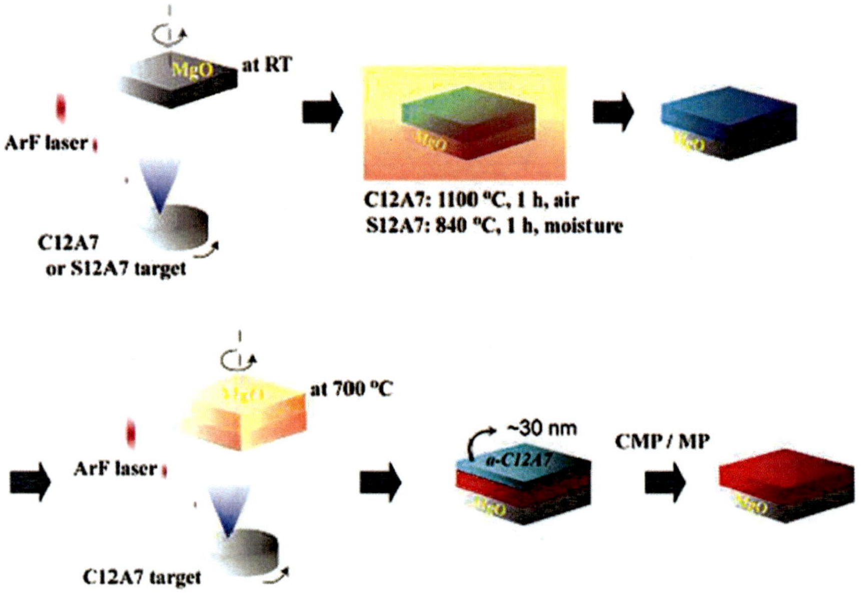

Fig. 10.17 Fabrication process of C12A7:e$^-$ and S12A7:e$^-$ thin films

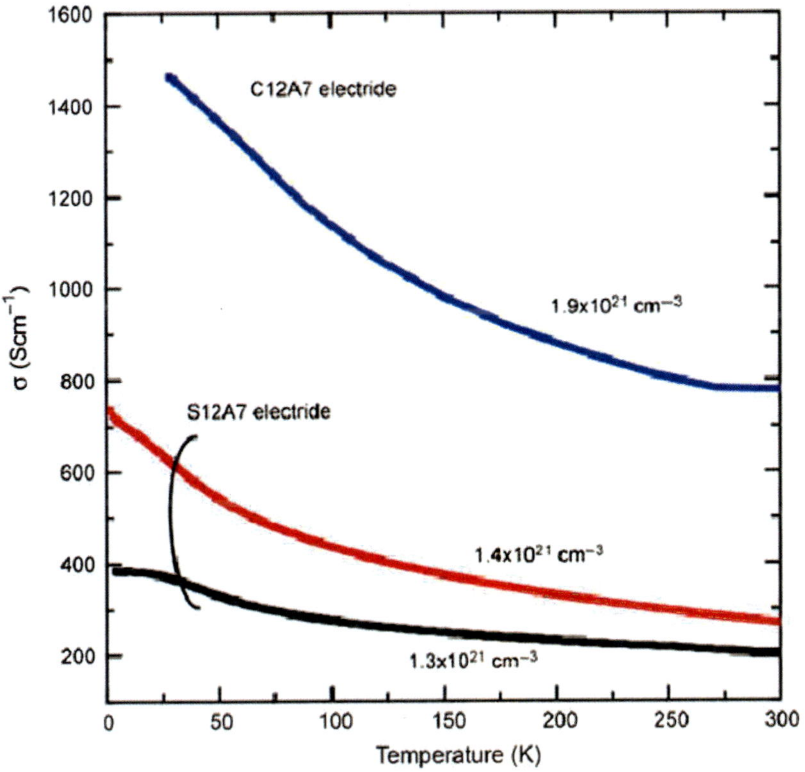

Fig. 10.18 Electrical conductivity in electron-doped $12SrO{\cdot}7Al_2O_3$ thin films. *Numbers in the panel* denote carrier electron concentrations measured by Hall effect. Data on electron-doped $12CaO{\cdot}7Al_2O_3$ thin film are shown for comparison [40]

carrier concentrations in the C12A7:e$^-$ film were evaluated to be 2.5 cm^2 (V s)$^{-1}$ and 1.3 × 10^{21} cm^{-3}, respectively. Similar values (1.0 cm^2 (V s)$^{-1}$ and 1.4 × 10^{21} cm^{-3}) were obtained for the S12A7:e$^-$ thin films. No carrier doping was observed in the same conditions without deposition of a-C12A7 on the pc-C12A7 or S12A7. Deposition of Au layers or Ti layers was examined for carrier generation. The former did not work, whereas the latter did. These observations imply that the role of a-C12A7 deposition on p-C12A7(S12A7) is to exact free O^{2-} accommodated in the cages and inject electrons to the cages.

10.2.5 Unique Properties of Electron-Doped C12A7 and Applications

C12A7:e$^-$ is an exceptional material that exhibits metal-insulator transition in light metal oxides. Their maximum conductivity at RT is 1,500 S cm^{-1}, which is higher by a factor of 2 than that of graphite and is comparable to metal manganese. The optical transmittance of thin films (100 nm thick) is >60% in the visible range. The conductivity and visible transmittance of metallic C12A7:e^- is obviously inferior to that of ITO compared as TCO, but C12A7:e^- has totally unique physical properties, i.e., very low work function and superconductivity.

10.2.5.1 Low Work Function

Toda et al. [21] evaluated work function (ϕ_{WF}) of the C12A7:e^- by the photoelectron yield spectroscopy (PYS) using bulk C12A7:e^- single crystals. A fresh surface of the sample was prepared by filing the surface with a diamond file in a preparation chamber attached to the measurement chamber under a vacuum of ~5×10^{-7} Pa before each photoelectron measurement. Figure 10.19a shows photoemission current (I_{PE}) as a function of V_{BIAS} at different excitation photon energies (hν). The I_{PE} rapidly increases with V_{BIAS} up to 10 V and levels off. I_{PE} is related to hν according to (10.8).

$$\ln I_{PE}T^{-2} = \ln F\left(\frac{h\nu - \varphi_{WF}}{k_B T}\right) + Z, \qquad (10.8)$$

where

$$F(x) = \int_0^{\infty} \frac{y}{1 + \exp(y - x)} dy.$$

Here k_B is the Boltzmann constant, T the temperature, and Z a constant. Figure 10.19b shows $(I_{PE})^{1/2}$–hν plots (the approximated Fowler plot) for samples A and B

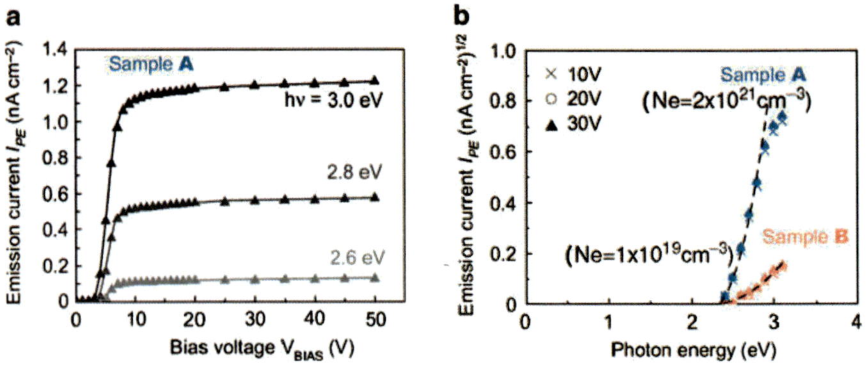

Fig. 10.19 Evaluation of work function (Φ_{WF}) in C12A7:e^- single crystals by photoelectron yield spectroscopy [21]. Two samples with different carrier electron concentrations were used for the measurements. (**a**) Photocurrent vs. bias voltage for different excited photon energy. (**b**) $I_{PE}^{1/2}$–hν plots. The intercept yields Φ_{WF} of 2.4 eV

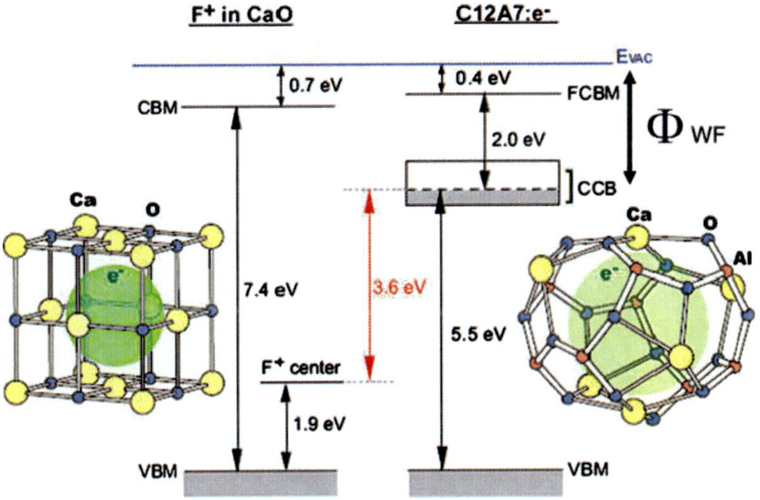

Fig. 10.20 Electronic energy levels of C12A7:e and F^+ center in CaO. Note that energy level of trapped electron at the site of oxygen vacancy is different by 3.6 eV between F^+ center in CaO and CCB in C12A7, whereas the energy gap and the location of conduction band minimum (CBM) and valence band maximum (VBM) are almost the same [21]

with different Ne at V_{BIAS} of 10, 20, and 30 V. The ϕ_{WF} value was determined to be 2.4 eV by obtaining the best-fit result to (10.8). It is seen that this ϕ_{WF} value gives fairly good agreement with the theory for all the V_{BIAS} especially at hν around the ϕ_{WF} value. These results indicate that the vacuum level (E_{VAC}) locates at ~2.4 eV above the E_F, ~0.4 eV above the FCBM, and ~7.9 eV above the VBM (see Fig. 10.20).

Here we consider the reason why C12A7:e$^-$ has a lower ϕ_{WF} than those of alkali and alkaline earth metals (Na = ~2.8 eV, K = ~2.3 eV, Rb = ~2.2 eV, Cs = ~2.1 eV, and Ca = ~2.9 eV) and why C12A7:e$^-$ is chemically stable regardless of the low ϕ_{WF}. Figure 10.20 compares the energy structures of F^+ center (an oxygen vacancy trapping an electron) in CaO and the CCB in C12A7:e$^-$. The CCB in C12A7:e$^-$ is composed of 3-dimensioanlly connected sub-nanometer-sized cages in some of which electrons are entrapped. An electron in the cage is octahedrally coordinated with 6 Ca^{2+} ions constituting the cage wall. Thus, the electron-trapped cage is an analog to the F^+ center in CaO except two points: (1) the separation between the electron and the Ca^{2+} is longer by ~50% than that in the F^+ center in CaO, and (2) an oxygen ion directly bonding with the Ca^{2+} in the lattice framework of C12A7:e$^-$ is connected one or two tetrahedrally-coordinated Al^{3+} ions. As a consequence, the net negative charge affecting at the cage center is smaller than that at the F^+ center in CaO. These differences cause a smaller Madelung potential for C12A7:e$^-$ than for CaO, resulting in the ~3.6 eV shallower electronic levels in the cages of C12A7:e$^-$ than at the oxygen vacancy site in CaO as shown in the figure. The chemical stability of C12A7:e$^-$ compared with the alkali and alkaline earth metal oxides is also understood from its crystal structure. When

C12A7:e$^-$ is exposed to an ambient atmosphere, the rigid structure of the lattice framework made of Ca–O and Al–O bonds prohibits in-diffusion of H_2O and O_2 molecules into a cage to react with the low work function electrons at around room temperature. This geometrical restriction is the primary cause for chemical and thermal stability of C12A7:e$^-$, providing an opportunity for practical application.

Recently, several applications of C12A7:e$^-$ have been reported., i.e., cold [41] and thermionic [42] electron emission sources, cathode materials in organic light-emitting diode (OLED) [43], and dielectric barrier in plasma panel displays (PDP) [44]. Better performance is obtained by using C12A7:e$^-$ for each application compared with the currently used materials. These are examples of straight forward applications based on unique properties of C12A7:e. An intriguing application has been found for a chemical reducing reagent available for various organic reactions in aqueous media [45].

10.3 Superconducting Transition [46]

In order to investigate whether the superconducting transition occurs in the metallic C12A7:e$^-$ or not, two samples exhibiting the highest conductivity in each crystal growth method (FZ and CZ) were used for measurements of electrical properties at low temperatures. Here C12A7 electride denotes that C12A7 in which most of free O^{2-} ions were replaced by electrons, i.e. an electron serves as an anion in this crystal. The electrical conductivities at 300 K of the FZ and CZ single-crystal electrides were ~810 and ~1,500 S cm^{-1}, respectively.

Figure 10.21 shows the temperature (T) dependence of the electrical resistivity (ρ) of two samples. A sharp drop of the resistivity is observed in the FZ single-crystal electride with an onset temperature of ~0.2 K. However, electrical resistivity of the metallic C12A7 electride prepared from the CZ single-crystal never becomes zero down to 90 mK, as shown in inset of the figure. Figure 10.21 shows that the onset temperature in the FZ sample decreases with the application of magnetic field, which strongly suggests that the FZ single-crystal C12A7 electride is a superconductor below a transition temperature of ~0.2 K. The superconductivity disappears by the magnetic field above a magnetic field of 30 mT. It is confirmed that FZ single-crystal electride with a conductivity of ~770 S cm^{-1} (sample B) also becomes superconductor with the onset temperature of ~0.19 K. The superconducting transition was further observed with the onset temperature of ~0.16 and ~0.4 K in epitaxial thin film electrides with N_e of 1.6×10^{21} and 2.0×10^{21} cm^{-3}, respectively. The variation of T_c in both single-crystal and thin film electrides suggests that the increase in N_e leads to an increase in T_c in the C12A7 electride. The critical magnetic field $H_c(0)$ is estimated to be ~19 mT for the bulk and ~33 mT for the thin film using a relation of $H_c(T) = H_c(0)[1-(T/T_c)^2]$. The low transition temperature and the small critical magnetic field imply that the C12A7:e$^-$ is a type-I superconductor. The perfect diamagnetism ($\chi \sim -1$) below ~0.20 K in the $\chi-T$ curve was also observed, confirming the bulk

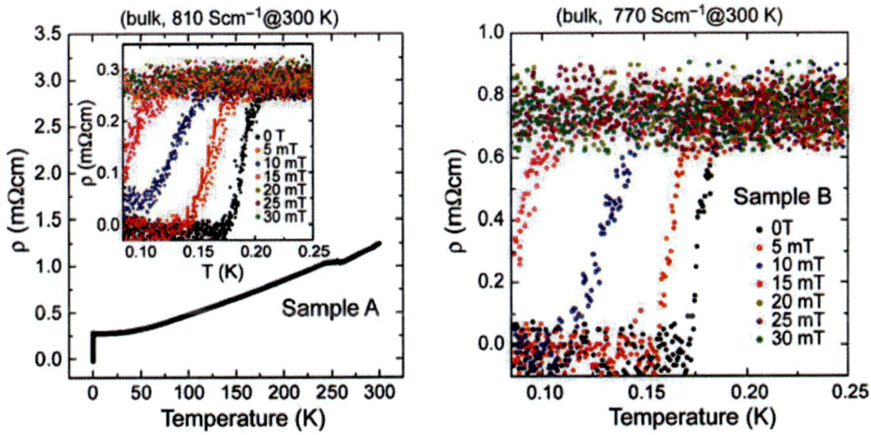

Fig. 10.21 Conductivity of heavy electron-doped C12A7 FZ-grown bulk single crystals. Zero resistivity at 0.2 K and its shift to lower temperature by applying magnetic field are clearly observed [46]

superconductivity in the C12A7 electride and 100% volume fraction of superconducting phase.

Here, the mechanism [47] for superconducting transition in C12A7 electride is described. In the Bardeen-Cooper-Schrieffer theory, T_c is expressed by McMillan's formula $T_c = [\Theta_D/1.45]\cdot\exp[-1.04\cdot(1+\lambda)/(\lambda-\mu^*\cdot(1+0.62\cdot\lambda))]$, where Θ_D is the Debye temperature, λ is the phonon-electron coupling constant and μ^* is the Coulomb pseudo-potential. λ is provided by $\lambda = N(0)\cdot V$, where $N(0)$ is the electron concentration at the Fermi energy, and V is phonon-mediated coupling constant for a single electron. Using the experimentally obtained value of $\Theta_D = 604$ K [48] from heat capacity and the assumption of μ^* as 0.2 in the low electron density system, we obtain $\lambda = \sim 0.45$, which is much larger than those of alkali metals as mentioned below. The observed increase in T_c with N_e is consistent with McMillan's formula, provided that N_e is correlated with $N(0)$.

Although the feature that the conduction electron in the electride is s-electron is similar to that of the alkali metal, superconducting transition is observed in C12A7 electride but not in alkali metals at ambient pressure except $T_c = 0.4$ mK of Li [49]. The difference is attributable to the magnitude of the electron phonon interaction. The anionic electron may interact strongly with the rigid lattice framework that is formed due to the strong covalent and ionic bonds among Ca^{2+}, Al^{3+} and oxygen ions. These facts possibly lead to the high values of λ ($\sim$0.45) and Θ_D (604 K), which are much higher than those of alkali metals such as Li ($\lambda = 0.38$, $\Theta_D = 420$ K), Na ($\lambda = 0.20$, $\Theta_D = 150$ K), and K ($\lambda = 0.14$, $\Theta_D = 100$ K). That is, the strong electron-phonon interactions likely resulted from the covalent and ionic bond nature of the oxide framework, is responsible for the emergence of the superconductivity in the C12A7 electride. Stable electrides with a rigid framework structure and with electrons populated in free space is a new category of superconductors.

10.4 p-Type TCOs

Although TCOs have been commercialized intensively as transparent metal for transparent window electrodes and interconnections, there is almost no application as transparent semiconductors. The absence of p-type TCO was the primary reason; many active functions of semiconductors come from intrinsic nature of pn-junction. No active electronic devices such as bipolar transistors and diodes can be fabricated without pn junctions. A breakthrough was the finding of the first p-type TCO, $CuAlO_2$ in 1997 by Kawazoe et al. [50] (Fig. 10.22), which triggered the development of a series of p-type TCOs and transparent pn junction devices such as UV light emitting diodes (LEDs). The achievement has significantly changed a view for TCOs and has opened a new frontier called "transparent oxide semiconductors (TOSs)". Therefore, we now consider that TOSs have the potential to develop new functionalities useful for novel optoelectronic devices that are hard to realize by current Si-based semiconductor technology. The discovery of the p-type TOS resulted from rational considerations regarding to the design of new TOSs based on knowledge about electronic structures that has accumulated experimentally and theoretically. The validity of this material design concept has been proven by discovery of a series of p-type TOSs [51]. These new TOSs led to transparent electronic devices such as UV LEDs and transparent thin film transistors (TFTs).

10.4.1 Materials Design of p-Type TCOs [52–54]

Many N-type TCO materials are known such as ZnO, In_2O_3, and SnO_2. Common features of N-type TCOs are (a) heavy metal cations with $(n-1)d^{10}ns^0$ electronic configuration ($n \geqq 4$, where n denotes the principal quantum number), and (b) the conduction band minimum (CBM) is located rather deep from the vacuum level. The former is required for realization of light electron mass at the CBM which

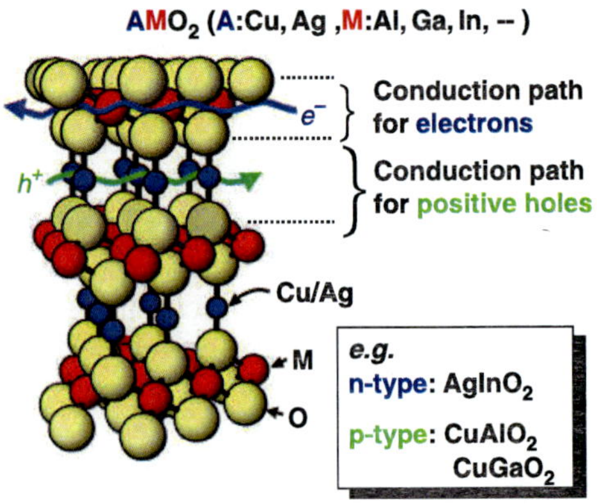

Fig. 10.22 Crystal structure of delafossite AMO_2

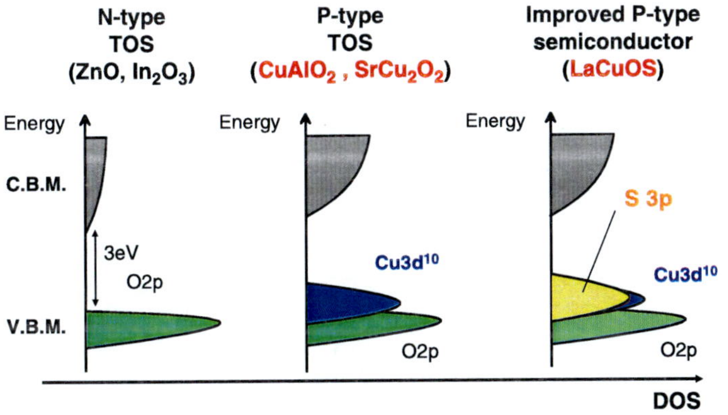

Fig. 10.23 Schematic energy diagram for N-type and P-type transparent semiconductors

works as electron pathway through large overlap between the orbitals of neighbouring metal cations, and the latter is needed for easiness of electron doping [52–54].

In contrast to n-type TOSs, it is much more difficult to attain good hole conduction because O 2p orbitals, which forms hole transport paths at the top of valence band of many oxides, are rather localized, therefore hole effective masses are rather large and the valence band maximum (VBM) levels measured from the vacuum level are deep. The former results in low hole mobilities and the latter causes difficulty in hole doping. These are the reasons why p-type oxides were not known for TCOs before 1997.

Considering the electronic structure discussed above, we expected that dispersion of VBM could be modified by (i) decreasing the nearest neighbouring oxygen–oxygen distance, (ii) using hybridization of metal orbitals that energy levels are close to those of O 2p (Fig. 10.23b), or (iii) employing more extended orbitals for anions (Fig. 10.23c). We first adopted the approach (ii), i.e., Cu^+-based oxides were chosen on the basis of the reason that the energy levels of Cu 3*d* are close to those of O 2p and the closed-shell configuration of Cu^+ $3d^{10}$ was expected not to give visible optical absorption due to *d*–*d* or O 2*p*–Cu 3*d* transitions, which therefore meets the requirements to obtain transparent materials. Following this guiding principle, we found a delafossite-type crystal, $CuAlO_2$, as the first p-type TOS in 1997 [51], followed by the findings of a series of p-type TOSs, $CuGaO_2$ [55] and $SrCu_2O_2$ [56].

The above consideration also gives an idea how to obtain a bipolar TOS in which either hole or electronic conduction is selected by intentional impurity doping. It suggested that electronic conduction can be imparted to p-type TOSs if heavy metals having largely extended *s* orbitals are incorporated to a delafossite oxide, which led to the finding of the first bipolar TOS, $CuInO_2$ [57] and fabrication of the first p/n homojunctions made of TOSs [58]. We also applied the p-type TOS $SrCu_2O_2$ to p/n hetero-junctions using ZnO for n-layer to fabricate near UV light-emitting diodes (LEDs) [59]. These achievements have demonstrated capability of TOSs for optoelectronic device applications, which opened "Transparent Oxide

Electronics" [60]. Very few papers on p-type TCOs have been reported to date. Detailed literature survey on this subject was reported as a review in 2005 [61].

10.4.2 Bipolar Transparent TCOs

Bipolar transparent TCOs are primarily important for realization of PN-junction devices such as CMOS and LEDs. Two TCO materials exhibiting bipolarity have been reported to date; $CuInO_2$ with delafossite structure and ZnO with wurtzite structure. For $CuInO_2$, hole and electron-doping were attained by replacing In^{3+} sites with Ca^{2+} and Sn^{4+}, respectively. As shown in Fig. 10.24. This material exhibits bipolarity exceptionally among delafossite family compounds. The optical band gap experimentally observed for 3.6 eV for $CuAlO_2$, 3.6 eV for $CuGaO_2$, and 3.9 eV for $CuInO_2$, respectively and bipolarity of $CuInO_2$ was hard to understand from these gaps. Since the valence band maximum (VBM) is primarily composed of Cu3d orbital and the conduction band minimum (CBM) is controlled by the trivalent cation (M), it is natural to expect that the location of VBM remains constant, whereas the CBM location varies with M. This idea apparently contradicts with the observed trend; $CuInO_2$ with the largest gap would have the highest CBM (should be most difficult for electron-doping). Nie et al. [62] explored this origin by band structure calculation and found a striking difference in minimum energy gap among them as a shown in Fig. 10.25. It is evident that $CuMO_2$ has an indirect-type material and all the transition from the Γ-point is exhibited. Thus, the observed gap corresponds to from the Γ-F-Γ in the VBM to the CBM (Γ-point). On the other hand, the easiness for electron-doping is determined by the location of CBM, which cannot be probed by optical transition due to the forbidden transition. Calculated VBM and CBM vary with M and this trend is the same as expected. However, the

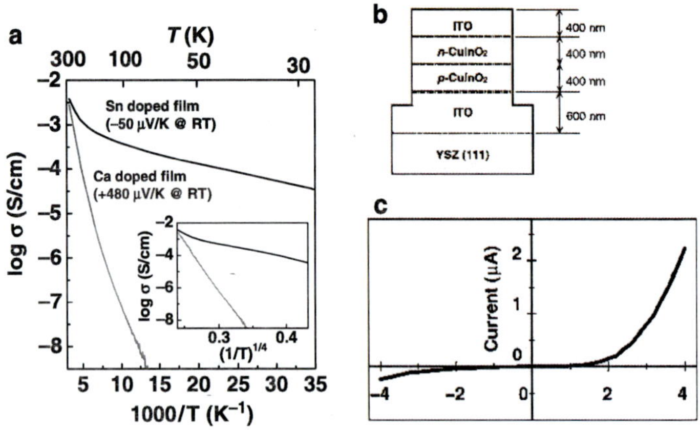

Fig. 10.24 DC conductivity in doped $CuInO_2$ showing bipolarity and PN-homo-junction diode [57]

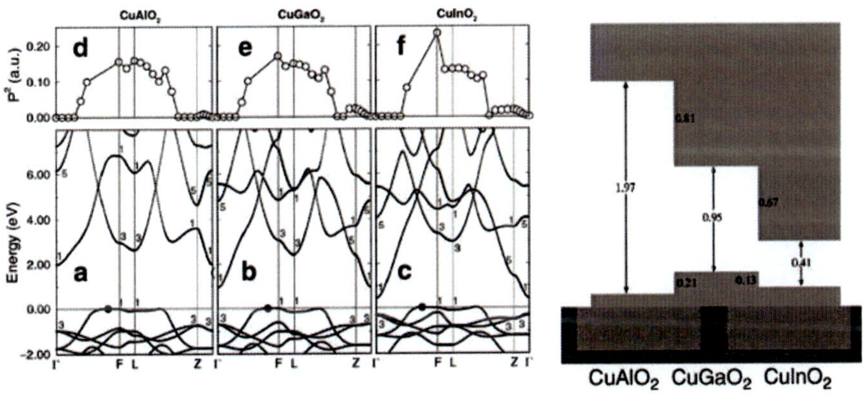

Fig. 10.25 Calculated band structures for delafossite compounds and band alignments. P^2 denotes transition probability for optical transitions [62]

transition for the energy gap responsible ~3.5 eV gap is strictly inhibited from the symmetry between the initial and final states only for $CuInO_2$.

Another bipolar TCO material is ZnO. Much effort has been denoted to realization of p-type ZnO including theoretical approach. Many papers [63] have been reported to date. Although very few papers have reported p-type ZnO, consensus of most of researchers associates with this subject has been not obtained except few papers, as far as the author knows. In 2005, Tsukazaki et al. [64] fabricated p-type ZnO by expitaxial growth on a lattice matched substrate (SCAM). They employed N^{3-} as the acceptor and doped it using temperature-modulation technique in pulsed laser deposition method. The resulting PN-homojunction exhibited blue emission, not an intrinsic excitonic emission, on injecting current. In 2006, Ryu et al. [65] reported p-type ZnO by doping As^{3-} instead of N^{3-}. They employed a hybrid beam method (PLD combined with K-cell in MBE) and deposited on deposited on conventional sapphire substrates. UV-excitonic emission at ~380 nm was first extracted by employing multiple quantum-well active layer and double hetero-structure shown in Fig. 10.26. Although the threshold voltage (3 eV) for this EL agreed with the band gap of ZnO, the quantum efficiency is still very low. Further technical breakthrough is needed to compete with the currently available UV-LED based on GaN. There appears to be recently reliable reports on P-doping in ZnO with As or P. Effectiveness of As or P as P-doping in ZnO was first suggested by one theoretical paper [66] in which As does not enter the lattice simply as As_O but a complex $As_{Zn}{-}2V_{Zn}$. More stable and reproducible P-doping is needed for practical application.

10.4.3 Layered Oxychalcogenides: Improved p-Type Performance

The first realized p-type TCOs are Cu^+-based oxides. However, the hole mobilities and concentrations are not enough for various applications. Thus, we then extended

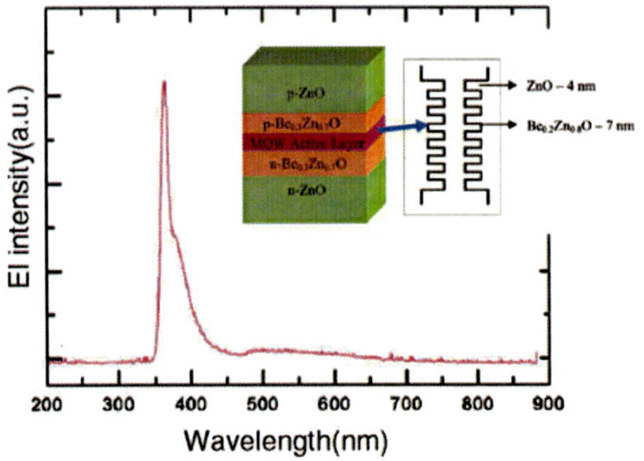

Fig. 10.26 Ultraviolet LED based on ZnO PN-homojunction. P-type ZnO is realized by As-doping [65]

the guiding principle to the approach (iii) "employing more spread orbitals for anions". Chalcogen ions were chosen to form largely hybridized VBM with Cu^+ $3d^{10}$ orbitals. To satisfy the condition to maintain large bandgaps, we chose layered oxychalcogenides, LnCuOCh (Ln = La, Ce, Pr, Nd; Ch = S, Se, Te) [67, 68].

Indeed, LnCuOCh exhibits intriguing opto-electronic properties, and it has been revealed that these properties are associated with their two-dimensional electronic structures. A large hole mobility (~8 $cm^2\ V^{-1}\ s^{-1}$) was obtained in nondoped LaCuOSe [69], and degenerate p-type conduction with moderately large hole mobility of ~4 $cm^2\ V^{-1}\ s^{-1}$ was attained in Mg-doped LaCuOSe (LaCuOSe: Mg). In addition, it was found that excitons were stable even at room temperature and sharp excitonic photoluminescence (PL) in blue-to-UV region were observed [70, 71]. This section describes these properties of the layered oxychalcogenides in relation to their electronic structures.

10.4.3.1 Crystal Structure and Epitaxial Film Fabrication: Reactive Solid-Phase Epitaxy

Figure 10.27 shows the crystal structure of LnCuOCh. It is composed of alternately stacked $(Ln_2O_2)^{2+}$ and $(Cu_2Ch_2)^{2-}$ layers along the <001> direction. Details about the synthesis and fabrication of epitaxial films were described in refs [72, 73]. It should be noted that epixial films were not obtained for LnCuOCh by a simple vapor phase deposition at high temperatures. Epitaxial films (Fig. 10.26) used for optical and electrical characterization in our study were fabricated by "Reactive Solid-Phase Epitaxy (R-SPE)" method [74]. In R-SPE, a very thin (~5 nm in thickness) and discontinuous Cu layer on (001) MgO surface was first deposited for an expitaxial template layer by pulsed laser deposition (PLD). Then a thick

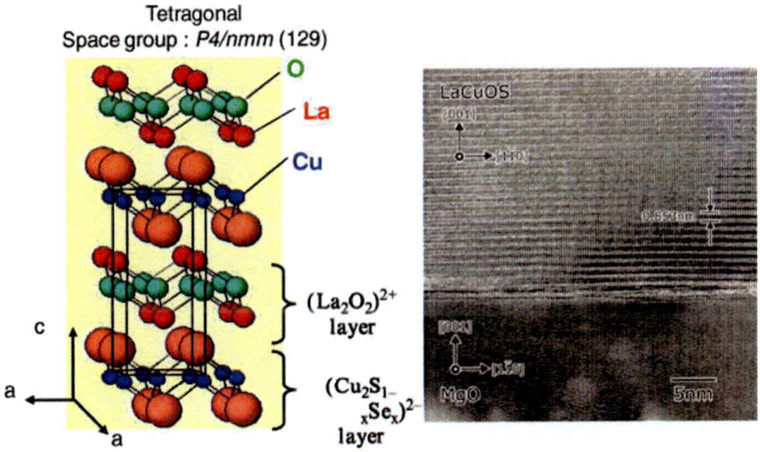

Fig. 10.27 Crystal structure of LaCuOCh and cross-sectional TEM photo of its epitaxial thin film grown on MgO (001) [73]

amorphous LnCuOCh layer was subsequently deposited at room temperature. The samples were taken out from the PLD chamber and sealed in an evacuated silica tube with LnCuOCh powder, followed by post-thermal annealing at ~1,000°C. This or similar procedures produced epitaxial films of a variety of LnCuOCh materials for a high-resolution TEM image of a LaCuOS film.

10.4.3.2 Carrier Transport, Light Emission and Excitonic Properties

All the LnCuOCh films exhibited p-type electrical conduction. Figure 10.28 shows temperature dependence of hole concentration (A) and mobility (B) in $LaCuS_{1-x}Se_x$ thin films estimated by Hall effect measurements. Mobility becomes larger with an increase in the Se content, which agrees with an expectation from the design concept. Since Se 4*p* orbitals have larger spatial spread than that of S 3*p* orbitals have, hybridization with Cu 3*d* becomes larger and thereby hole mobility increases as the Se content increases. The largest hole mobility $\sim 8.0\ cm^2\ V^{-1}\ s^{-1}$, which is comparable to that of p-type GaN:Mg, was obtained in LaCuOSe.

By contrast, LaCuOSe has distinct difference from GaN:Mg. Mg-doping to LaCuOSe increased the hole concentration up to $2 \times 10^{20}\ cm^{-3}$ and carrier transport changed to degenerate conduction [69]. We like to note that degenerate p-type conduction has not been attained in any type of wide bandgap semiconductors even including GaN:Mg. In addition, unlike conventional semiconductors, moderately large hole mobility ($\sim 4\ cm^2\ V^{-1}\ s^{-1}$) and intense PL are maintained even in the heavily-doped LaCuOSe:Mg although high-density carriers in conventional semiconductors generally reduce carrier mobility by charged-impurity scattering and exciton lifetime by increased Auger scattering and screening. Exciton binding

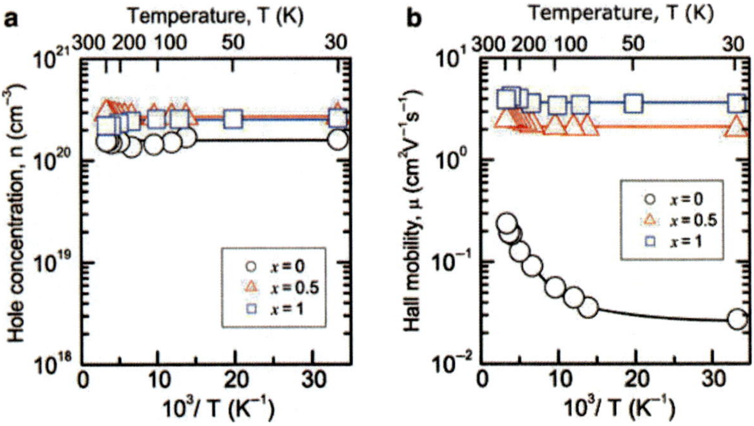

Fig. 10.28 Hole concentration and Hall mobility in $LaCuS_{1-x}Se_x$ thin films as a function of temperature [69]

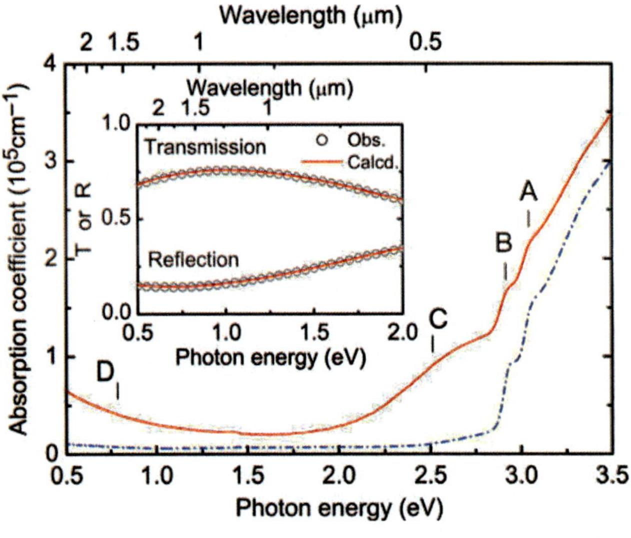

Fig. 10.29 Optical absorption spectra of LaCuOSe:Mg epitaxial thin films. Broken curve shows the undoped sample for comparison [75]. Absorption and reflection in the near infrared region are due to free carriers (hole). Fundamental transport parameters evaluated through the fitting by free absorption analysis using a Drude-type response are given in Table 10.3

energy estimated from temperature dependence of PL intensity was ~50 meV [72], which is much larger than the room-temperature thermal energy (~26 meV).

Recently, Hiramatsu et al. [75] successfully fabricated p-type metallic LaCuOSe epitaxial films with the hole concentration of 1.7×10^{21} cm^{-3}. These films exhibited distinct free carrier absorption as shown in Fig. 10.29, which has not observable so far for p-type TOSs. They extracted fundamental parameter associated with

Table 10.3 Fundamental parameters associated from hole-transport in p-type degenerated aCuOSe:Mg epitaxial thin films

Hole concentration (Hall effect, cm^{-3})	1.7 × 1,021
DC conductivity (300 K, S cm^{-1})	910
Mobility (Hall effect, cm^2 $(Vs)^{-1}$)	3.5
(Free carrier absorption)	4.6
Plasma frequency (s^{-1})	1.5×10^{15}
Momentum relaxation time (s)	4.2×10^{-15}
Effective mass (Hall effect, me)	1.6

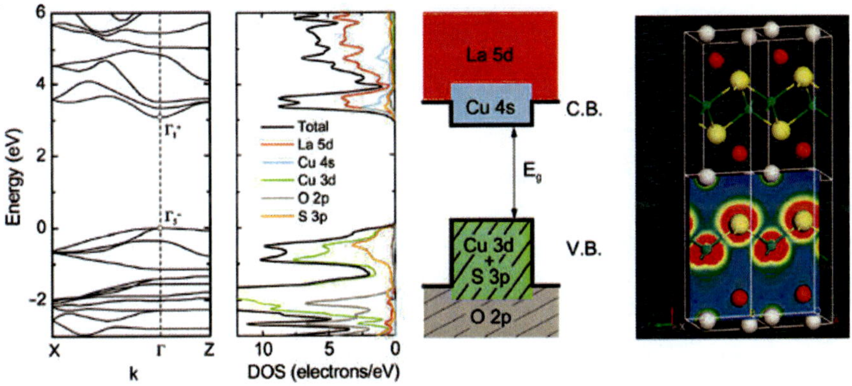

Fig. 10.30 Band structure and partial DOS of LaCuOS. Contour plot of electron density in the vicinity around valence band maximum showing that hole are well confined in the Cu-S hybridized orbitals [76]

hole-transport by analyzing the absorption as summarized in Table 10.3. The hole relaxation time is comparable to that (~6×10^{-15} s) of n-type degenerated TCOs such as ZnO:Ga but the mobility of the former is less by an order of magnitude than that of the latter. This large difference in mobility between the p-type TCO and n-type TCO originates from that in band effective mass.

10.4.3.3 Two-Dimensional Electronic Structure in LnCuOCh

The above unique properties are related with the layered crystal structure. Ab initio band calculations showed that the band structure is highly anisotropic and hole effective mass is smaller in the Γ-X direction than in the Γ-Z direction (Fig. 10.30) [76] which implies the electronic structure has highly two dimensional nature. Optical absorption spectra measured at 10 K (Fig. 10.31) showed a step-wise structure with sharp peaks just on the edges of the steps. Such a step-wise structure is similar to those observed in semiconductor artificial superlattices, reflecting two-dimensional DOS. Density plot of electrons in the vicinity of VBM (Fig. 10.30) shows that hole transport paths spread only in the $(Cu_2Ch_2)^{2-}$ layers and holes are

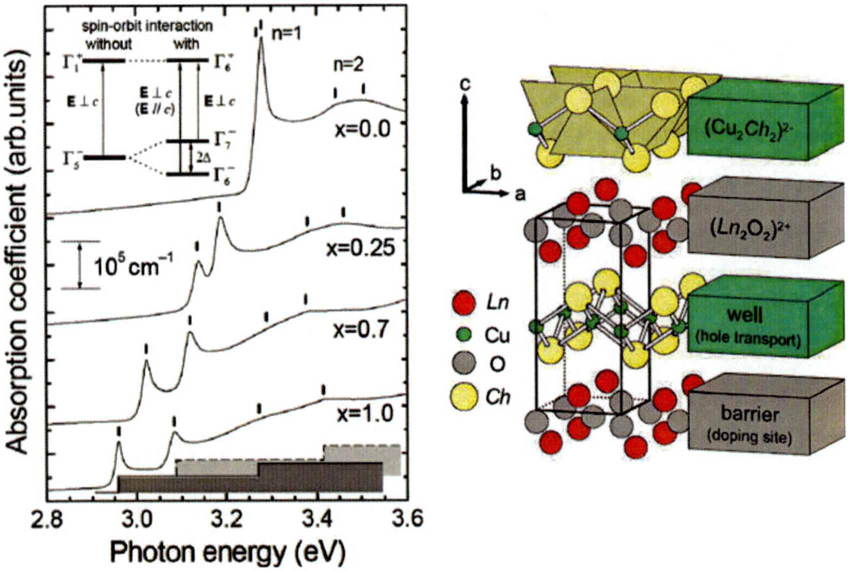

Fig. 10.31 Optical absorption spectra of LaCuOSSe epitaxial thin films at 10 K. A step-wise structure with sharp excitonic absorption split by spin-orbital coupling at the step edge was clearly observed. Such a spectrum is characteristic of two-dimensional electronic nature realized by confinement by exciton confined in CuCh layer working as a well [76]

confined two dimensionally, explaining the step-wise structure of the optical absorption spectra. It was also confirmed that the split of the sharp peaks comes from multi-level excitons split due to spin-orbit interaction in Ch ions [77–79].

Projected density of states (PDOS) showed that the VBM is mainly composed of hybridized orbitals of Cu^+ 3*d* and Ch *p* orbitals and the CBM of Cu 4s. Thus the bandgap is formed almost solely of the $(Cu_2Ch_2)^{2-}$ layer in LnCuOCh [80]. In contrast, the $(Ln_2O_2)^{2+}$ layer (except for Ln = Ce) has larger energy gap than the $(Cu_2Ch_2)^{2-}$ layer has (Fig. 10.29). This characteristic structure explains the moderately large hole mobility and intense PL in the heavily-doped LaCuOSe:Mg. In LaCuOSe:Mg, Mg ions are thought doped in the $(Ln_2O_2)^{2+}$ layer. Holes are generated from acceptor levels made from the Mg dopants, which are then transferred to the $(Cu_2Ch_2)^{2-}$ layer because it forms hole transport paths as seen in Fig. 10.29. This electronic structure spatially separates the conducting holes from the ionized acceptors (Mg^{2+} ions at Ln^{3+} sites). As a consequence, carrier scattering is effectively reduced. That is, modulation doping structure, which is realized in artificially in semiconductor superlattices and HEMT, is formed naturally in these layered oxychalcogenides.

Third-order nonlinear optical susceptibility $\chi^{(3)}$ [77] was measured to extract the photonic characteristics of RT-stable exciton. $\chi^{(3)}$ spectra showed peaks just at the excitonic absorption peaks. The maximum $\chi^{(3)}$ values are 4×10^{-9} esu for LaCuOS and $\sim 2 \times 10^{-9}$ esu for LaCuOSe, which are larger than that for ZnO films (1×10^{-9} esu). These results suggest that the large $\chi^{(3)}$ values are enhanced due to excitons in LaCuOCh. As exciton binding energy is larger for ZnO (~60 meV) than for LaCuOCh (~40–50 meV), these large $\chi^{(3)}$ values in LaCuOCh are attributable also to increases in the density of exciton states caused by exciton confinement in the $(Cu_2Ch_2)^{2-}$ layer.

It was of interest to note that the crystal structure of recently discovered iron-based high temperature superconductor LaFeOAs [81] is of the same as that of LaCuOCh; when a chalcogenide ion with −2 charge is replaced by a pnictide anion with −3 charge, Cu^+ with a $3d^{10}$ closed shell may be replaced by +2 charged 3d transition metal cations with unpaired spins. Then, one may expect that magnetic interaction between T_MPn layers intervening a nonmagnetic insulating LaO layer is reduced compared with a compound T_MPn and impurity doping layer is spatially separated from carrier transport layer. Thus we started to examine the electromagnetic properties in a series of LaTMPnO (TM: 3d transition metal) to explore magnetic semiconductors from 2004. In the course of the experiments, we found superconductivity in the Fe- and Ni-systems, Intensive studies are going on all over the world and the critical temperature reaches to 55 K and several unique features such as very high critical magnetic field up to ~200 T have been found to date. A small story of the discovery and recent progress of Iron-pnictide superconductors are reviewed in literature [82, 83].

10.5 Summary

This chapter described non-conventional transparent oxide semiconductors, i.e., UV-transparent TCO, Ga_2O_3, conversion of a light metal oxide $12CaO \cdot 7Al_2O_3$ to TCO, and p-type TCOs. These materials are not used in the industries yet, but have large potential for future application because their characteristics are complementary to those of currently used TCOs with respect to transparent regions, mother materials, and carrier polarity. The demand for transparent conductive oxides is expanding beyond the traditional needs such as a transparent contact material. A rapidly emerging trend is strong interest for transparent oxide electronics which covers application areas complementary to Si C-MOS technology. A unique feature of oxides is a wide material variety compared with so-called compound semiconductors whose crystal structure is restricted to diamond-type.

Here we focused on materials with nano-porous structure and layered structure. The electron trapped in the sub nanometer-sized cages may be regarded as an ultimate quantum dot. Thus, electron-trapped C12A7 corresponds to a three-dimensionally coupled quantum dot. A layered crystal structure is effective to enlarge the optical band utilizing the suppression of band dispersion. Transparent p-type $CuAlO_2$ with a band gap 3.1 eV may be taken as a material which contains a two-dimensional O-Cu-O dumbbell structure in a well-known p-type semiconductor Cu_2O with a band gap of 2.1 eV. The nature of valence band top is primarily determined by anion species. Thus, the layered structure composed of different anion species has a distinct band off-set at the valence band top. Such a band-off set makes delta-doping possible, which in turn can largely reduce the degradation of hole mobility arising from ion impurity scattering. Improved p-type transparent semiconductivity was realized in LaCuOCh by using this idea. Carrier and exciton confinement within a several atomic layers between heterointerfaces provides versatile opportunity to approach outstanding performance of electronic properties. An interesting example is a high performance thermoelectric power of $SiTiO_3/TiO_2$ heterointerfaces at which carrier electrons are confined near the interface. Extremely large carrier concentration at the interface yields Seebeck coefficient which is ~5 time s larger than that in bulk [85].

References

1. S. Geller, J. Chem. Phys. 33(1960) 676
2. N. Ueda, H. Hosono, R. Waseda, and H. Kawazoe, Appl. Phys. Lett. 70 (1997) 3561
3. N. Ueda, H. Hosono, R. Waseda, and H. Kawazoe, Appl. Phys. Lett. 71 (1997) 933
4. L. Binet, D. Gourier, and C. Minot, J. Solid State Chem. 113 (1994) 420
5. H. He, M.A. Blanco, and R. Pandey, Appl. Phys. Lett. 88 (2006) 261904
6. M. Orita, H. Ohta, M. Hirano, and H. Hosono, Appl. Phys. Lett. 77 (2000) 4166
7. M. Orita, H. Hiramatsu, H. Ohta, M. Hirano, and H. Hosono, Thin Solid Films 411 (2002) 134
8. R.J. Cava et al. Appl. Phys. Lett. 64 (1994) 2071
9. M. Orita, H. Sakai, M. Takeuchi, and Y. Yamauchi, Trans. Mater. Soc. Jpn 20 (1996) 573
10. Y. Kokubun, K. Miura, F. Endo, and S. Nakagomi, Appl. Phys. Lett. 90 (2007) 031912
11. K. Matsuzaki, H. Yanagi, T. Kamiya, H. Hiramatsu, K. Nomura, M. Hirano, and H. Hosono, Appl. Phys. Lett. 88 (2006) 092106
12. K. Hayashi, S. Matsuishi, T. Kamiya, M. Hirano, and H. Hosono, Nature 419 (2002) 462
13. R.W. Nurse, J.H. Welch, and A. Majumdar, Trans. Br. Ceram. Soc. 64 (1965) 323
14. H. Bartl, T. Scheller, N. Jb. Miner. Mh. 35 (1970) 547
15. M. Lacerda, J.T.S. Irvine, F.P. Glasser, and A.R. West, Nature 332 (1988) 525
16. J. Jeevaratnam, F.P. Glasser, and L.S. Dent Glasser, J. Am. Ceram. Soc. 47 (1964) 105
17. S. Watauchi, I. Tanaka, K. Hayashi, M. Hirano, and H. Hosono, J. Cryst. Growth 237 (2002) 801
18. K. Kurashige, Y. Toda, S. Matsuishi, K. Hayashi, M. Hirano, and H. Hosono, Cryst. Growth Des. 6 (2006) 1602
19. P.V. Sushiko, A.L. Shluger, K. Hayashi, M. Hirano, and H. Hosono, Phys. Rev. Lett. 91 (2003) 126401
20. P.V. Sushiko, A.L. Shluger, K. Hayashi, M. Hirano, and H. Hosono, Phys. Rev. B 73 (2006) 014101
21. Y. Toda, H. Yanagi, E. Ikenaga, J. Kim, M. Kubota, S. Ueda, T. Kamiya, M. Hirano, K. Kobayashi, and H. Hosono, Adv. Mater. 19 (2007) 3564
22. P.V. Sushiko, A.L. Shluger, M. Hirano, and H. Hosono, J. Am. Chem. Soc. 129 (2007) 942
23. A. Zunger, Appl. Phys. Lett. 83 (2003) 57
24. H. Hosono, T. Kamiya, and M. Hirano, Bull. Chem. Soc. Jpn. 79 (2006) 1
25. S. Matsuishi, K. Hayashi, M. Hirano, and H. Hosono, J. Am. Chem. Soc. 127 (2005) 12454
26. K. Hayashi, Y. Toda, T. Kamiya, M. Hirano, M. Yamanaka, I. Tanaka, T. Yamamoto, and H. Hosono, Appl. Phys. Lett. 86 (2005) 022109
27. S. Matsuishi, Y. Toda, M. Miyakawa, K. Hayashi, T. Kamiya, M. Hirano, I. Tanaka, and H. Hosono, Science 301 (2003) 626

28. S.W. Kim, S. Matsuishi, T. Nomura, Y. Kubota, M. Takata, K. Hayashi, T. Kamiya, M. Hirano, and H. Hoosno, Nano Lett. 7 (2007) 1138
29. N.F. Mott, Metal-insulator transition, Taylor & Francis, London (1990)
30. P.P. Edwards, M.J.J. Sienko, J. Am. Chem. Soc. 103 (1981) 2967
31. T. Nomura, K. Hayashi, Y. Kubota, T. Kamiya, M. Hirano, M. Takata, and H. Hosono, Chem. Lett. 36 (2007) 902
32. M. Tardio, R. Ramirez, R. Gonzalez, Y. Chen, and E. Alves, Nucl. Instr. Meth. B 191 (2001) 191
33. H.F.M. Taylor, Cement chemistry, 2nd edn, Thomas Telford Publishing, London (1997)
34. S.W. Kim, M. Miyakawa, K. Hayashi, T. Sakai, M. Hirano, and H. Hosono, J. Am. Chem. Soc. 127 (2005) 1370
35. S.W. Kim, Y. Toda, K. Hayashi, T. Sakai, M. Hirano, and H. Hosono, Chem. Mater. 18 (2006) 1938
36. M. Miyakawa, K. Hayashi, M. Hirano, Y. Toda, T. Kamiya, and H. Hosono, Adv. Mater. 15 (2003) 1100
37. M. Miyakawa, Y. Toda, K. Hyashi, T. Kamiya, M. Hirano, N. Matsunami, and H. Hosono, J. Appl. Phys. 97 (2005) 023510
38. M. Miyakawa, M. Hirano, T. Kamiya, and H. Hosono, Appl. Phys. Lett. 90 (2007) 182105
39. O. Yamaguchi, A. Narai, and K. Shimizu, J. Am. Ceram. Soc. 69 (1986) C36
40. M. Miyakawa, N. Ueda, T. Kamiya, M. Hirano, and H. Hosono, J. Ceram. Soc. Jpn. 115 (2007) 567
41. Y. Toda, S. Matsuishi, K. Hayashi, K. Ueda, T. Kamiya, M. Hirano, and H. Hosono, Adv. Mater. 16 (2004) 685
42. Y. Toda, S.W. Kim, K. Hayashi, M. Hirano, T. Kamiya, H. Hosono, T. Haraguchi, and H. Yasuda, Appl. Phys. Lett. 87 (2005) 254104
43. K. Kim, M. Kikuchi, M. Miyawaka, H. Yanagi, T. Kamiya, M. Hirano, and H. Hosono, J. Phys. Chem. C 111 (2007) 8403
44. M. Ono, S. Webster, S. Ito, and H. Hosono, Proc. SID'06 (2006) 1642
45. M. Buchammagari, Y. Toda, M. Hirano, H. Hosono, D. Takeuchi, and K. Osakada, Org. Lett. 9 (2007) 4287
46. M. Miyakawa, S.W. Kim, M. Hirano, Y. Kohama, H. Kawaji, T. Atake, H. Ikegami, K. Kono, H. Hosono, J. Am. Chem. Soc. 129 (2007) 7270
47. Y. Kohama, S.W. Kim, T. Tojo, H. Kawaji, T. Atake, S. Matsuishi, and H. Hosono, Phys. Rev. B 77 (2008) 092505
48. Y. Kohama, T. Tojo, H. kawaji, T. Atake, S. Matsuichi, and H. Hosono, Chem. Phys. Lett. 421 (2006) 558
49. J. Tuoriniemi, K. Juntunen-Nurmilaukas, J. Uusvuori, E. Pentti, A. Salmela, and A. Sebedash, Nature 447 (2007) 187
50. H. Kawazoe, M. Yasukawa, H. Hyodou, M. Kurita, H. Yanagi, and H. Hosono, Nature 389 (1997) 939
51. H. Kawazoe, H. Yanagi, K. Ueda, H. Hosono, MRS Bull. 25(8) (2000) 28
52. H. Yanagi, A. Kudo, K. Ueda, H. Hosono, and H. Kawazoe, J. Electroceram. 4 (2000) 407
53. T. Kamiya, H. Hosono, Semic. Sci. Technol. 20 (2005) 592
54. H. Hosono, Thin Solid Films 515 (2007) 600
55. K. Ueda, T. Hase, H. Yanagi, H. Kawazoe, H. Hosono, H. Ohta, M. Orita, and M. Hirano, J. Appl. Phys. 89 (2001) 1790
56. A. Kudo, H. Yanagi, H. Hosono, and H. Kawazoe, Appl. Phys. Lett. 73 (1998) 220
57. H. Yanagi, T. Hase, S. Ibuki, K. Ueda, and H. Hosono, Appl. Phys. Lett. 78 (2001) 1583
58. H. Yanagi, K. Ueda, H. Ohta, M. Orita, M. Hirano, and H. Hosono, Solid State Commun. 121 (2002) 13
59. H. Ohta, K. Kawamura, M. Orita, N. Sarukura, M. Hirano, and H. Hosono, Appl. Phys. Lett. 77 (2000) 475
60. H. Ohta and H. Hosono, Mater. Today 7 (2004) 42
61. A.N. Banerjee and K.K. Chattopadhyay, Prog. Cryst. Growth Charact. Mater. 50 (2005) 52

62. X. Nie, S.-H. Wei, and S.B. Zhang, Phys. Rev. Lett. 88 (2002) 066405
63. So many papers have been reported on p-ZnO to date. The following is a list of the papers surveyed to 2007: T.V. Butkhuzi, B.E. Tsekava, N.P. Kekelidze, E.G. Chikoidze, T.G. Khulordava, and M.M. Sharvashidze, J. Phys. D, Appl. Phys. 32 (1999) 2683, E.-C. Lee, Y.-S. Kim, Y.-G. Jin, and K.J. Chang, Phys. Rev. B 64 (2001) 085120, S.B. Zhang, S.-H. Wei, and A. Zunger, Phys. Rev. B 63 (2001) 075205, K. Nakahara, H. Takasu, P. Fons, A. Yamada, K. Iwata, K. Matsubara, R. Hunger, and S. Niki, J. Cryst. Growth 237–239 (2002) 503, A. Zeuner, H. Alves, D.M. Hofmann, B.K. Meyer, A. Hoffmann, U. Haboeck, M. Strassburg, and M. Dworzak, Phys. Stat. Sol. (b) 234 (2002) R7, T. Yamamoto, Thin Solid Films 420 (2002) 100, C.H. Park, S.B. Zhang, and S.-H. Wei, Phys. Rev. B 66 (2002) 073202, K.-K. Kim, H.-S. Kim, D.-K. Hwang, J.-H. Lim, and S.-J. Park, Appl. Phys. Lett. 83 (2003) 63, Y.R. Ryu, T.S. Lee, and H.W. White, Appl. Phys. Lett. 83 (2003) 87, C. Zhang, X. Li, J. Bian, W. Yu, and X. Gao, Solid State Commun. 132 (2004) 75, D.C. Look and B. Claflin, Phys. Stat. Sol. (b) 241 (2004) 624, Y.W. Heo, Y.W. Kwon, Y. Li, S.J. Pearton, and D.P. Norton, Appl. Phys. Lett. 84 (2004) 3474,Y.W. Heo, K. Ip, S.J. Pearton, and D.P. Norton, Phys. Stat. Sol. (a) 201 (2004) 1500, D.C. Look, G.M. Renlund, R.H. Burgener, and J.R. Sizelove, Appl. Phys. Lett. 85 (2004) 5269, A. Krtschil, A. Dadgar, N. Oleynik, J. Bläsing, A. Diez, and A. Krost, Appl. Phys. Lett. 87 (2005) 262105, D.C. Look, Semicond. Sci. Technol. 20 (2005) S55, E. Kaminska, A. Piotrowska, J. Kossut, R. Butkute, W. Dobrowolski, R. Lukasiewicz, A. Barcz, R. Jakiela, E. Dynowska, E. Przezdziecka, M. Aleskiewicz, P. Wojnar, and E. Kowalczyk, Phys. Stat. Sol. (c) 2 (2005) 1119, A. Krtschil, D.C. Look, Z.-Q. Fang, A. Dadgar, A. Diez, and A. Krost, Physica B 376/377 (2006) 703, Y. Cao, L. Miao, S. Tanemura, M. Tanemura, Y. Kuno, and Y. Hayashi, Appl. Phys. Lett. 88 (2006) 251116, B. Claffin, D.C. Look, S.J. Park, and G. Cantwell, J. Cryst. Growth 287 (2006) 16, H.S. Kang, B.D. Ahn, J.H. Kim, G.H. Kim, S.H. Lim, H.W. Chang, and S.Y. Lee, Appl. Phys. Lett. 88 (2006) 202108, J.G. Lu, Y.Z. Zhang, Z.Z. Ye, L.P. Zhu, L. Wang, B.H. Zhao, and Q.L. Liang, Appl. Phys. Lett. 88 (2006) 222114, Y. Yan, M.M. Al-Jassim, and S.-H. Wei, Appl. Phys. Lett. 89 (2006) 181912, J. Lee, J. Metson, P.J. Evans, R. Kinsey, and D. Bhattacharyya, Appl. Surf. Sci. 253 (2007) 4317, A. Allenic, W. Guo, Y. Chen, M.B. Katz, G. Zhao, Y. Che, Z. Hu, B. Liu, S.B. Zhang, and X. Pan, Adv. Mater. 19 (2007) 3333
64. A. Tsukazaki, A. Ohtomo, T. Onuma, M. Ohtani, T. Makino, M. Sumiya, K. Ohtani, S. F. Chichibu, S. Fuke, Y. Segawa, H. Ohno, H. Koinuma, and M. Kawasaki, Nat. Mater. 4 (2005) 42
65. Y. Ryu, T.-S. Lee, J. A. Lubguban, H. W. White, B.-J. Kim, Y.-S. Park, and C.-J. Youn, Appl. Phys. Lett. 88 (2006) 241108
66. S. Limpijumnong, S.B. Zhang, S.-H. Wei, and C.H. Park, Phys. Rev. Lett. 92 (2004) 155504
67. H. Hiramatsu, H. Kamioka, K. Ueda, H. Ohta, T. Kamiya, M. Hirano, and H. Hosono, Phys. Stat. Sol. (a) 203 (2006) 2800
68. K. Ueda, H. Hiramatsu, M. Hirano, T. Kamiya, and H. Hosono, Thin Solid Films 496 (2006) 8
69. H. Hiramatsu, K. Ueda, H. Ohta, M. Hirano, T. Kamiya, and H. Hosono, Appl. Phys. Lett. 82 (2003) 1048
70. K. Ueda, S. Inoue, H. Hosono, N. Sarukura, M. Hirano, Appl. Phys. Lett. 78 (2001) 2333
71. H. Hiramatsu, K. Ueda, H. Ohta, T. Kamiya, M. Hirano, and H. Hosono, Appl. Phys. Lett. 87 (2005) 211107
72. H. Hiramatsu, K. Ueda, K. Takafuji, H. Ohta, M. Hirano, T. Kamiya, H. Hosono. J. Appl. Phys. 94 (2003) 5805
73. H. Hiramatsu, H. Ohta, T. Suzuki, C. Honjo, Y. Ikuhara, K. Ueda, T. Kamiya, M. Hirano, and H. Hosono, Cryst Growth Des 4 (2004) 301
74. H. Ohta, K. Nomura, M. Orita, M. Hirano, K. Ueda, T. Suzuki, Y. Ikuhara, and H. Hosono, Adv. Funct. Mater. 13 (2003) 139
75. H. Hiramatsu, K. Ueda, H. Ohta, M. Hirano, M. Kikuchi, H. Yanagi, T. Kamiya, and H. Hosono, Appl. Phys. Lett. 91 (2007) 012104

76. K. Ueda, H. Hiramatsu, H. Ohta, M. Hirano, T. Kamiya, and H. Hosono, Phys. Rev. B69 (2004) 155305
77. H. Kamioka, H. Hiramatsu, H. Ohta, M. Hirano, K. Ueda, T. Kamiya, H. Hosono, Appl. Phys. Lett. 84 (2004) 879
78. H. Kamioka, H. Hiramatsu, K. Hayashi, M. Hirano, and H. Hosono, J. Photochem. Photobiol. A 166 (2004) 141
79. H. Kamioka, H. Hiramatsu, M. Hirano, K. Ueda, T. Kamiya, and H. Hosono, Opt. Lett. 29 (2004) 1659
80. S. Inoue, K. Ueda, and H. Hosono, Phys. Rev. B 64 (2001) 245211
81. Y. Kamihara, T. Watanabe, M. Hirano, and H. Hosono, J. Am. Chem. Soc.130 (2008) 3296
82. H. Hosono (2008) J. Phys. Soc. Jpn. 77SC:1
83. H. Hosono (2009) Physica C 469:314
84. S. Matsuishi, S.-W. Kim, To. Kamiya, M. Hirano, and H. Hosono, J. Phys. Chem. C 112 (2008) 4753
85. H. Ohta, S.-W. Kim, Y. Mune, T. Mizoguchi, K. Nomura, S. Ohta, T. Nomura, Y. Nakanishi, Y. Ikuhara, M. Hirano, H. Hosono, and K. Koumoto, Nature Mater. 6 (2007) 129

Chapter 11
Applications of Transparent Conductors to Solar Energy and Energy Efficiency

Claes G. Granqvist

11.1 Introduction

New energy sources and increased energy efficiency are important for development, and revolutionary advances in technology appear necessary even to maintain today's general standard of living and economic prosperity [1]. The problems ahead of us may seem daunting. For example, it has been stated that the warming and precipitation trends due to antropogenic, energy-related climate changes during the past 30 years already claim over 150,000 human lives annually [2, 3]. These climate changes are also expected to be accompanied by more common and/or extreme events such as heatwaves, heavy rainfall, and storms and coastal flooding; there are also fears that non-linear climate responses will lead to breakdown of ocean "conveyor belt" circulation, collapse of major ice sheets, and/or release of large quantities of methane at high latitudes thus intensifying global warming [4]. Adding to the precarious situation, the urgently needed advances in energy related technology must take place for an increasing population, whose growing concentration in mega-cities leads to "heat islands" which tend to aggravate the warming [5] and can increase the urban cooling load by up to 25% compared to the case of surrounding rural areas [6]. By 2050 there will be some ten billion people in the World. Energy must be available to them all, and it has to be clean. New technologies are necessary to accomplish this. Some of these technologies – mainly related to efficient solar energy utilization and to energy savings in the built environment – will be discussed in this Chapter.

Where do the transparent conductors (TCs) – oxides as well as others – fit into this scenario? A basic reason why they are important is their ability to be transparent in a limited and well-defined range, normally encompassing visible light in the

C.G. Granqvist
Department of Engineering Sciences, The Ångström Laboratory, Uppsala University, PO Box 534, SE-75121 Uppsala, Sweden
e-mail: claes-goran.granqvist@angstrom.uu.se

D.S. Ginley et al. (eds.), *Handbook of Transparent Conductors*,
DOI 10.1007/978-1-4419-1638-9_11,

$0.4 < \lambda < 0.7$ μm wavelength interval, while they can be reflecting in the infrared (IR) and absorbing the ultraviolet (UV). If the reflectance prevails in the range for thermal radiation, i.e., $3 < \lambda < 50$ μm at normal temperature, the emission of heat is impeded. If reflectance occurs at $0.7 < \lambda < 3$ μm, covering the IR part of the solar spectrum which carries about 50% of the solar energy, one can combine visible transmittance with rejection of a large part of the solar energy and at the same time have low thermal emittance. It is then evident that TCs have a number of diverse applications in the fields of solar energy utilization and energy efficiency. Other applications of TCs emerge from their electrical conductivity, which allows their use as current collectors in solar cells and as transparent electrodes for charging and discharging of electrochromic smart windows, etc. Indeed the TCs can be viewed as "solar energy materials", whose properties have been reviewed in detail recently [7], following a bird's eye perspective in some other recent articles [8–10]. The present Chapter can be viewed as an abbreviated version of ref. [7].

The applications of TCs hence can rely on their *spectral selectivity*, according to the discussion above. Other possible uses ensue from the *angular properties* of the radiation that surrounds us, specifically by the fact that one can take advantage of the Sun's passage over the vault of heaven to have different performances for midday and dawn or dusk. Still other applications emerge from the fact that *ambient radiation or human needs vary during the day and season*, so that solar energy and/or visible light ideally should be admitted or rejected as a function of time.

This Chapter is organized as follows: Sect. 11. 2 presents an overview of the radiation in our natural surroundings with the main object of defining the types of spectral and angular selectivity that are of interest for solar energy and energy efficiency. Also introduced are materials with variable optical properties, known as "chromogenic" materials [11], whose usefulness for energy savings in buildings is reviewed. Substates for TCs and antireflection coatings are mentioned too. Then follow discussions of TCs with tailored spectral selectivity, angular dependence, and temporal variability in Sects. 11.3, 11.4, and 11.5, respectively. Regarding spectral selectivity, most attention is devoted to architectural windows and glass façades which require TCs for admitting and reflecting visible light, solar energy, and thermal radiation in different wavelength regions. Special attention is devoted to noble-metal-based films, which are commonly used on today's windows, and *n*-doped wide band gap oxide semiconductor films with particular consideration of In_2O_3:Sn. Theoretical and experimental data are given for the spectral optical properties. The possible use of TCs to avoid or modify the condensation of water on windows is also touched upon, as are applications to "vacuum glazings", photovoltaic cells, and thermal solar collectors. Angular selectivity is given a shorter presentation with focus on obliquely deposited metal-based films having inclined columnar nanostructures. Materials and devices enabling temporal variability attract much current interest, and this part covers photochromic, thermochromic, electrochromic, and gasochromic options. In-depth discussions are given for thermochromic VO_2-based films and for oxide-based electrochromics; they can be regarded as TCs with temperature-dependent and electrical-charge-dependent transmittance, respectively. The Chapter is concluded with an outlook towards some future possible materials and applications of TCs in Sect. 11.6.

11.2 Foundations for Solar Energy Materials

When electromagnetic radiation impinges on a material one fraction can be transmitted, a second fraction is reflected, and a third fraction is absorbed. Energy conservation yields, at each wavelength, that

$$T(\lambda) + R(\lambda) + A(\lambda) = 1, \tag{11.1}$$

where T, R, and A denote transmittance, reflectance, and absorptance, respectively. Another fundamental relationship, also founded on energy conservation and referred to as Kirchhoff's law, is

$$A(\lambda) = E(\lambda) \tag{11.2}$$

with E being emittance, i.e., the fraction of the black-body radiation that is given off at a particular wavelength. Equation (11.2) is of practical relevance mainly for $\lambda > 3$ μm.

11.2.1 Spectral and Angular Selectivity of Ambient Radiation

Figure 11.1 introduces the electromagnetic radiation of our ambience in a unified manner [12]. The most fundamental property of this radiation stems from the fact that all matter sends out electromagnetic radiation, which is conveniently introduced by starting with the ideal black-body whose emitted spectrum – known as the Planck spectrum – is uniquely defined if the temperature is known. Planck's law is a consequence of the quantum nature of radiation. Fig. 11.1a depicts such spectra for four temperatures. The vertical scale denotes power per unit area and wavelength increment (hence the unit G W m^{-3}). The spectra are bell-shaped and confined to the $2 < \lambda < 100$ μm range. The peak in the spectrum is displaced toward shorter wavelength as the temperature goes up, which is called Wien's displacement law. The peak lies at about 10 μm for room temperature. Thermal radiation from a material is obtained by multiplying the Planck spectrum by a numerical factor – the emittance – which is less than unity. In general, the emittance is a function of λ.

Fig. 11.1b reproduces a solar spectrum for radiation just outside the Earth's atmosphere. The curve has a bell shape which defines the Sun's surface temperature (~6,000°C). One observes that the solar spectrum is limited to $0.25 < \lambda < 3$ μm, so there is hardly any overlap with spectra for thermal radiation. Hence it is possible to have surfaces with properties being entirely different with regard to thermal and

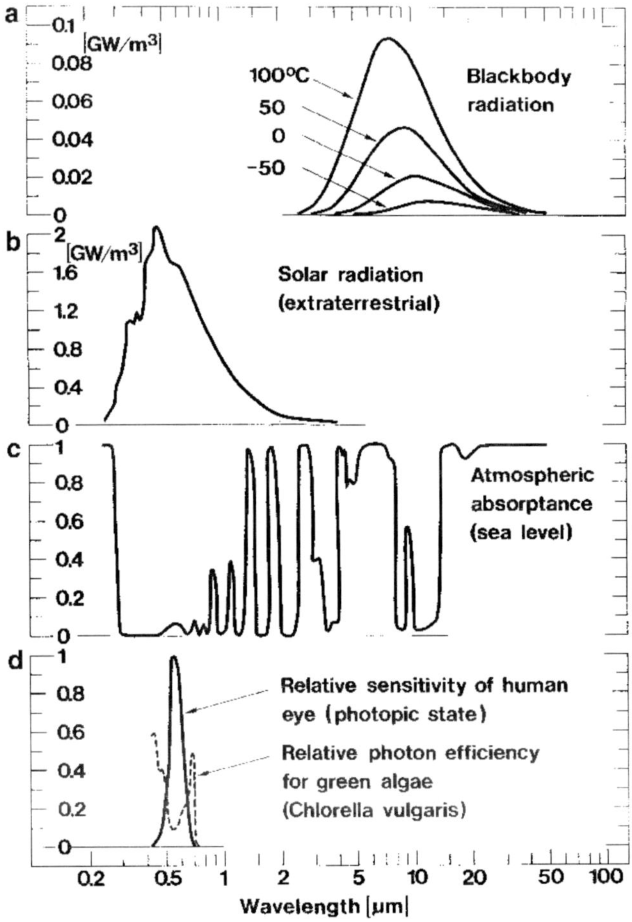

Fig. 11.1 Spectra for (**a**) black-body radiation pertaining to four temperatures, (**b**) solar radiation outside the Earth's atmosphere, (**c**) typical absorptance across the full atmospheric envelope, and (**d**) relative sensitivity of the human eye and relative photosynthetic efficiency for green algae. From Granqvist [12]

solar radiation. The integrated area under the curve gives the "solar constant" (1,353 $\pm$ 21 W m^{-2}); this is the largest possible power density on a surface oriented perpendicular to the Sun in the absence of atmospheric extinction.

Most solar energy conversion systems, taken with the widest meaning of this term, are located at ground level, and it is of obvious interest to consider to what extent atmospheric absorption influences solar irradiation and net thermal emission. Fig. 11.1c illustrates a typical absorption spectrum vertically across the full atmospheric envelope during clear weather. The spectrum is complicated and comprises bands of high absorption – caused mainly by water vapor, carbon dioxide, and ozone – as well as bands of high transparency. Evidently most of the solar energy can be transmitted to ground level, and only parts of the UV ($\lambda < 0.4$ μm) and IR ($\lambda > 0.7$ μm) radiation are strongly damped. The maximum power density perpendicular to the sun is about 1,000 W m^{-2}. Thermal radiation from a surface exposed to the clear sky is strongly absorbed except in the $8 < \lambda < 13$ μm range, called the main "atmospheric window", where the transmittance can be large provided that the humidity is not too high.

Fig. 11.1d illustrates two biological conditions of relevance for solar-energy-related applications. The solid curve shows the relative sensitivity of the human eye in its light-adapted (photopic) state; the bell-shaped graph extends across the $0.4 < \lambda < 0.7$ μm interval and has its peak at 0.555 μm. Clearly a large part of the solar energy comes as invisible IR radiation. The dashed curve indicates that photosynthesis in plants makes use of light with wavelengths in approximately the same range as those for the human eye, which is relevant for greenhouse applications.

It is frequently convenient to average the spectral data over the sensitivity of the eye, over a solar spectrum (with or without account for atmospheric effects), or over a black-body emittance spectrum (for a specific temperature) [13]. One then obtains luminous, solar, or thermal values pertinent to the respective optical property; they are denoted X_{lum}, X_{sol}, and X_{therm}, with X being T, R, A, or E.

Figure 11.1 emphasizes that ambient radiation is *spectrally selective*, i.e., confined to specific and usually well defined wavelength ranges. This property is of large importance with regard to TCs used for energy related applications, as discussed next in Sect. 11.2.2. Another type of selectivity – known as *angular selectivity* – originates from the fact that different angles may apply for different types of radiation; for example, solar radiation comes from a point far from the horizon during most of the day, whereas the visual contact between a person and his or her surroundings is often at near-horizontal lines-of-sight as further discussed in Sect. 11.2.3.

11.2.2 The Potential of Chromogenic Technologies

The ambient properties vary during the day or season and hence it is of interest to use "chromogenic" materials, capable of changing their optical properties [11]. The purpose can be to create energy efficiency by admitting solar energy if there is a heating demand or rejecting solar energy if there is a cooling demand. Considering such "smart" fenestration, there is also a "human dimension" related to comfort, implying a desire to experience good lighting, avoid discomfort though glare, and prevent thermal stress. Obviously the notion of "comfort" is a subjective one, even if quantification is possible to some extent [14]. Another aspect concerns skin protection (and, in general, materials protection) [15]. The key question here is the magnitude of the energy efficiency inherent in the ability of a chromogenic material to vary its optical properties. It is a complicated and multi-dimensional one as will become apparent.

The first line of reasoning is intutive and deals with a cooling situation [16]. Considering results consistent with existing "smart window" technology, and making reasonable assumptions about the occupancy of commercial buildings, it leads to a result that is best expressed as an analogy: for a given position a "smart window" is able to provide an *energy savings* of the same magnitude or larger than the *energy generation* that could have been accomplished with today's best solar cells.

Computer simulations of the energy use in buildings offer another, more elaborate, method to assess the energy savings inherent in chromogenics [17–20]. Specifically, calculations for a standard office module with well defined size, window area, lighting demand, occupancy, equipment and appliances, etc., showed that the energy savings potential was considerable for the cooling load [17, 18, 21]. The simulations were performed with climate data applicable to different locations within Europe. When using "smart windows" instead of conventional static solar control windows, the energy for space cooling, on an annual basis, could be reduced by as much as 40–50%. The amount of saved energy is obviously climate dependent. In moderately warm climates the number of days with very high outdoor temperature is relatively small but the energy required for balancing excessive solar energy inflow nevertheless is substantial, and an interesting result of the simulations was that the cooling power could be reduced so that air conditioning might be completely avoided when "smart" windows were used. Substantial energy savings appeared also in another recent study [19, 20] which also pointed at a number of important subjective assets of the technology [22].

The examples above were chosen to correspond to cooling of buildings. This choice is appropriate since the energy for air cooling is steadily rising and takes a growing share of the energy used in the built environment. The energy spent in buildings presently amounts to about 40% of the energy expenditure in the EU and corresponds to roughly 4% of the Gross National Product [5, 23]. As pointed out recently [24, 25], air conditioning and refrigeration becomes increasingly important for the peak electricity demand; it has risen by an average of 17% per year in the EU between 1995 and 2003 [26]. A further indication of the pressing demand for new and efficient cooling techniques is that more than half of the electricity consumption in households in some major Chinese cities is for air conditioning, refrigeration, and water cooling [27]. In an extreme climate, such as that of Kuwait [28], more than 75% of the electrical peak load is currently used for air conditioning.

11.2.3 Substrates for Transparent Conductors, and Their Antireflection Treatment

It is suitable to include some information on substrate materials for coatings of TCs. The most widely used substrate is glass. Float glass is a highly standardized product characterized by uniformity and flatness almost on the atomic scale. The middle

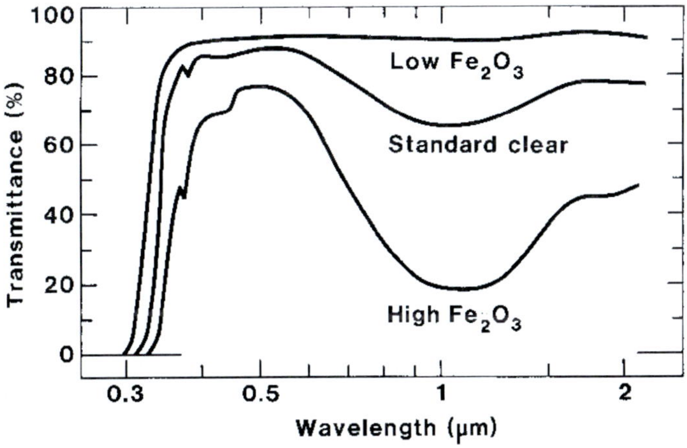

Fig. 11.2 Spectral transmittance for float glass with three amounts of Fe_2O_3. From Granqvist [29]

curve in Fig. 11.2 shows the spectral transmittance of 6-mm-thick standard float glass within the wavelength range relevant to solar radiation [29]. A characteristic absorption feature at $\lambda \approx 1$ μm, due to the presence of Fe oxide, limits T_{sol} to some noticeable extent. Special float glass is available with varying amounts of Fe oxide. Figure 11.2 shows that low Fe content leads to glass with very high T_{lum} and T_{sol}, and with substantial transmittance also in the UV [29]. Glass with a large Fe content limits T_{sol} while T_{lum} remains rather large. The reflectance of each interface between glass and air is about 4% in the $0.4 < \lambda < 2$ μm range, implying that the maximum transmittance for a glass pane is 92%. Glass is strongly absorbing for $\lambda > 3$ μm, and E_{therm} is as large as ~87%.

Polymers are also of much importance as substrates for TCs, and transparent polymers can, in principle, replace glass. Polyethylene terephthalate (PET) foil deserves special mention since it can be suspended between glass panes and used in practical window constructions. Another application of coated foil is in "window film" for upgrading the performance of existing fenestration. An alternative to PET is polyethylene napthalate (PEN) and flurocarbon plastics.

Laminated glass may be used for safety and other reasons. Such glass normally has a layer of polyvinyl butyral (PVB) sandwiched between two glass panes and bonded to the glass under heat and pressure. The laminate is an efficient absorber of UV light, and the transmittance at $\lambda < 0.38$ μm is almost zero for a layer thickness of the order of 1 mm.

Both for glass and polymers it is possible to modify the optical properties by incorporation of nanoparticles or ions. Noble metal particles give absorption at visible wavelengths, whereas particles with lower electron densities, such as heavily doped oxide semiconductors, display absorption peaks at longer wavelengths [30]. A particularly interesting situation, with rather high T_{lum} and moderately low T_{sol}, can be achieved with LaB_6 nanoparticles dispersed in PVB [31].

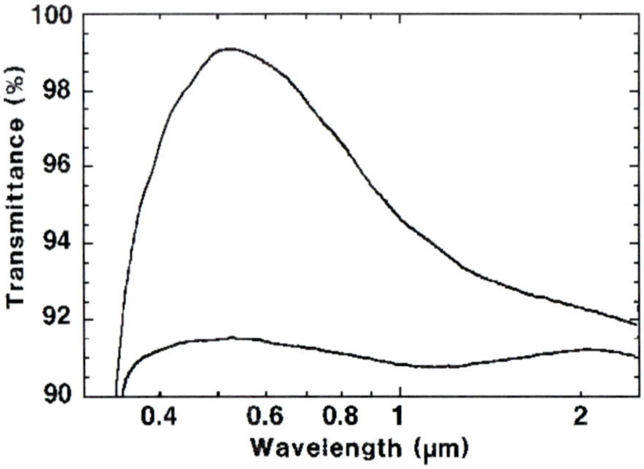

Fig. 11.3 Spectral transmittance of a glass plate coated on both sides by antireflecting thin films of porous silica. Replotted from Nostell et al. [32]

Antireflection (AR) coatings on glass and polymers are often of interest for boosting the performance of devices for solar energy and energy efficiency. For example, contemporary fenestration uses at least double glazing, and the overall value of T_{lum} and T_{sol} are limited by reflections at the various interfaces. To obtain high T_{lum} through glass one can apply a layer whose refractive index is close to the square root of the refractive index of the underlying substrate (i.e., 1.23) and with a thickness corresponding to a quarter wavelength for visible light (i.e., ~100 nm). By coating both sides of a glass pane with self-assembled silica nanoparticle films it is possible to obtain $T_{\mathrm{lum}} > 99\%$ in the middle of the luminous spectrum as illustrated in Fig. 11.3 [32]. Excellent AR properties are possible also with other types of coatings and with etched glass surfaces.

Simple AR treatments are able to decrease the reflectance within rather narrow wavelength ranges, but if broadband AR is required one can use multilayer films with carefully adjusted refractive indices and thicknesses; they can be designed using several different strategies. Normally an AR coating is efficient only for a specific range of incidence angles for the light, especially for the case of multilayer configurations, but wide-angle AR coatings are possible for narrow wavelength ranges. Recently AR treatments with a combination of graded index layers and layers having fixed refractive index have been investigated computationally as well as experimentally in order to specify coatings combining low reflectance in the full visible wavelength range and with low reflectance in a wide angular interval [33, 34].

11.3 Applications Based on Spectral Selectivity

Spectrally selective TCs can be used in different ways as discussed below. By far the largest applications are in architectural windows and glass façades which are dealt with in detail in Sects. 11.3.1 and 11.3.2, devoted to metal-based and oxide-based TCs, respectively. Section 11.3.3 then covers a number of diverse aspects of spectrally selective TCs and considers applications in "vacuum glazings", for preventing water from condensing on windows, self-cleaning and super-hydrophilicity, solar cells, and thermal solar collectors.

Architectural windows and glass façades are problematic from an energy perspective [35]. Their primary function – to provide unmitigated visual contact between indoors and outdoors, as well as day-lighting – must be kept in mind, though, so that window areas are not made too small. In a practical situation, there are often unwanted energy flows with too much thermal energy leaving or entering buildings via their windows, with concomitant demand for space heating and cooling. Present architectural trends are to increase the window areas, so the energy issue may become even more pressing in the future.

Thin films of TCs can alleviate the situation, and their ability to provide spectral selectivity is already used to a large extent for new windows. On some markets, the penetration of this technology is currently almost total, whereas the market share is smaller elsewhere. Given the obvious need to diminish the energy consumption on a global scale, the technology can only be expected to grow. The enormity of the market is obvious; the current production of float glass – used almost without exception in modern fenestration – is about four billion square meters per year [36]. Coating even a small portion of this with TCs makes architectural and automotive windows by far the largest field of applications for TCs, completely outsizing other presently much discussed applications such as displays and transparent electronics.

The radiation part of the heat transfer can be controlled by thin films with low E_{therm} which, in a multiply-glazed window, should face one of the confined air (or gas) spaces [29, 37]. In this way the heat transfer across a vertically mounted window can drop from ~3 to ~1.5 W m^{-2} K^{-1} for double glazing and from ~1.8 to ~1.0 W m^{-2} K^{-1} for triple glazing. Obviously, T_{lum} must be large for these films. The IR part of the solar radiation – which transmits energy through the glazing but is not needed for vision – can be stopped by a thin film having low transmittance at $0.7 < \lambda < 3$ μm.

These demands have led to two types of thin films, known as "low emittance coatings" ("low-*E*") characterized by

$$T(\lambda) = 1 \quad \text{for} \quad 0.4 < \lambda < 3\,\mu\text{m} \tag{11.3}$$

$$R(\lambda) = 1 \quad \text{for} \quad 3 < \lambda < 50\,\mu\text{m} \tag{11.4}$$

and "solar control coatings" characterized by

$$T(\lambda) = 1 \quad \text{for} \quad 0.4 < \lambda < 0.7\,\mu\text{m} \tag{11.5}$$

$$R(\lambda) = 1 \quad \text{for} \quad 0.7 < \lambda < 50\,\mu\text{m} \tag{11.6}$$

The spectral selectivity inherent in these relationships is of course idealized.

In order to go below the heat transfer values mentioned above it is necessary to diminish also non-radiative (conductive and convective) heat transfer, which is possible in evacuated “vacuum glazings”. A heat transfer coefficient less than 0.2 W m^{-2} K^{-1} has been estimated for a triple vacuum glazing [38]. The requirements of TCs in such devices are discussed briefly in Sect. 11.3.3.1 below.

11.3.1 Metal Based Thin Films for Energy Efficient Windows

11.3.1.1 Single-Layer Films

Metal layers can serve as TCs if they contain holes, are sufficiently thin, or combine these features. We first note that the properties of optically thick metal films with nanoholes (smaller that the wavelength) have attracted interest recently following the remarkable observation of extraordinary light transmission through metallic sub-wavelength hole arrays which does not obey classical optical theory [39, 40]. The effect is illustrated in Fig. 11.4 showing transmittance through Au films with the indicated nanostructures; the holes are 0.25 μm in diameter and have a lattice constant of ~2 μm. A transmittance peak centered at $\lambda \approx 615$ nm stands out clearly [41]. Peak transmission levels up to 39% have been reported recently [42]. Multiple transmission bands can be created with periodic arrays of nanohole apertures [43].

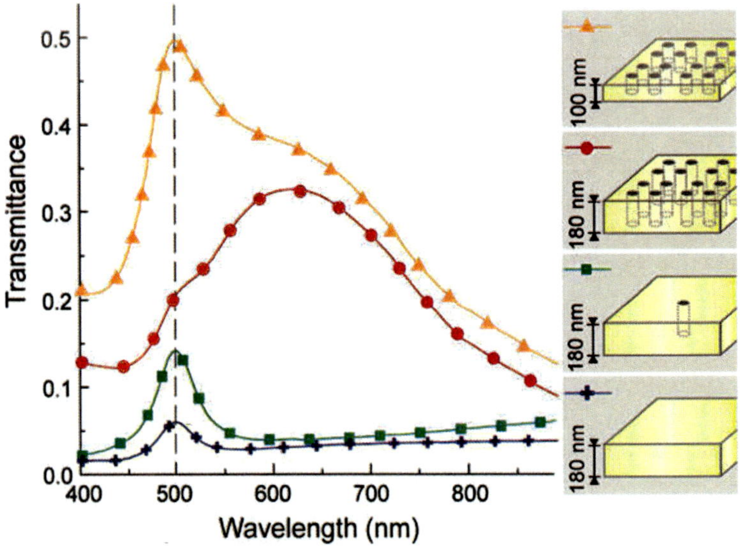

Fig. 11.4 Spectral wide-angle transmittance through Au films with the shown configurations. The sub-wavelength hole arrays are seen to lead to extraordinary transmittance in a band centered at a wavelength of 615 nm. From Gao et al. [41]

The transmittance may not be sufficient for most energy-related applications, but the phenomenon nevertheless deserves some attention here since it may affect T_{lum} and T_{sol} for a non-continuous metal film, as we return to shortly. An interpretation of the light transmission in terms of surface plasmon polaritons has been given strong support through near-field optical measurements in recent work [41, 42, 44]. Thermal radiation through nanoholes also shows unexpected and novel features, and the flux is not proportional to the fourth power of the temperature – as for Planck radiation (*cf.* Fig. 11.1a) – but rather to the eighth power [45].

The rest of this section is devoted to films that are thin enough to show some transparency. Very thin films of free-electron-like ("noble") metals have been used as transparent conductors and transparent heat reflectors at least since the early 1950s [12]. They can have properties to some extent resembling those in Eqs. (11.3)–(11.6) and hence provide a good starting point for development. In principle a very thin layer of Cu, Ag, or Au is the simplest way to accomplish a TC and transparent infrared reflector [46–48]. Alternative materials are TiN and ZrN [49, 50], as well as Pt [51] and Al [52].

Thin Ag films stand out as the superior option as a result of their low value of A_{lum}. The other metals exhibit some, usually unwanted, short-wavelength absorption at $\lambda < 0.5$ μm for Cu, Au, and the nitrides (hence their color), and at $\lambda < 0.8$ μm for Al. Chemical and mechanical ruggedness is an issue for the metal-based TCs – much more than for the oxide TCs to be discussed in Sect. 11.3.2 – and laboratory experiments as well as field tests show that they are only suited for application on surfaces facing cavities in double-glazed windows. The films are sometimes referred to as "soft coats" and are normally applied by high-rate magnetron sputtering.

The idealized performance of a metal-based thin film is readily computed using established techniques by representing it as a plane parallel slab of thickness t. The metal is uniquely characterized by two parameters – the optical constants or the complex dielectric function – pertinent to a uniform bulk-like material. Figure 11.5 shows computed results of T_{lum}, T_{sol}, R_{sol}, and E_{therm} as a function of t for Ag [47]. It is found that $t = 5$ nm yields an impressive set of data, specifically $T_{lum} = 85\%$, $T_{sol} = 74\%$, and $E_{therm} = 8\%$. A glass with such a thin film would constitute an excellent low-E glazing. Unfortunately, however, extremely thin noble metal films deposited onto glass cannot be represented as plane parallel slabs, as will be considered next.

When metals such as Cu, Ag, or Au are deposited onto glass or some other dielectric material the film goes through a series of rather well defined "growth stages" [48, 53]. This is illustrated in Fig. 11.6, where the pertinent structures and thickness scales are also shown. Initially, tiny metallic nuclei are formed at certain sites on the substrate. Continued deposition makes the nuclei grow via surface diffusion and direct impingement. Progressing coalescence growth leads to larger and more irregular islands. The growing film then passes through "large scale coalescence", meaning that an extended metallic network is formed. Subsequent deposition makes the voids between the metallic paths become smaller and increasingly regular. As noted above, the void structure can in itself be conducive to T_{sol},

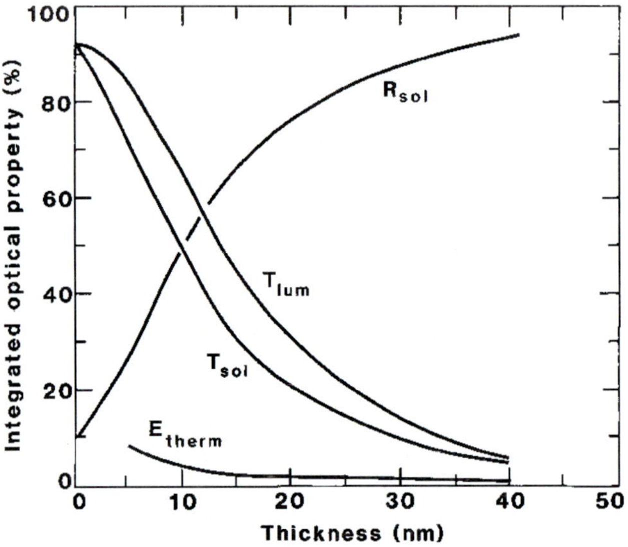

Fig. 11.5 Integrated optical properties as a function of thickness computed from an ideal plane-parallel Ag layer. Replotted from Valkonen and Karlsson [47]

Fig. 11.6 Survey over growth stages, structures, and thickness scales for thin Au films deposited onto glass by conventional evaporation (c.e.) and ion assisted evaporation (i.e.). After Smith et al. [48]

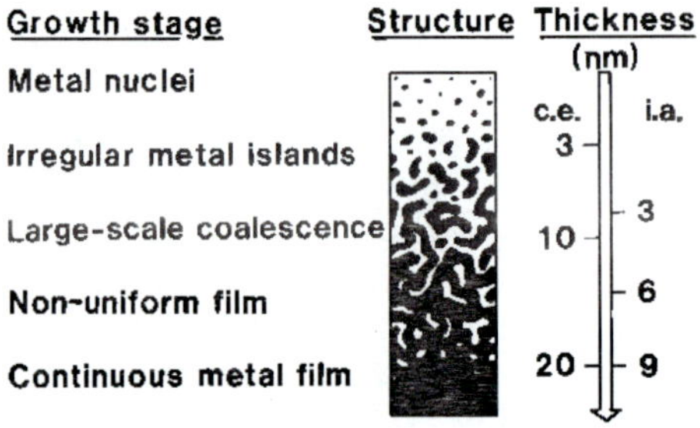

particularly in the near-infrared [48]. A close-to-uniform layer may ultimately be formed; its surface roughness can be as small as 0.4 nm for Au films sputter deposited onto float glass under optimized conditions [54]. The corresponding structures are illustrated in Fig. 11.6 with metal shown as black. Prior to large scale coalescence, the islands are ~10 nm in size.

It is now evident that a model of the film based on a plane-parallel slab can only be adopted for the later growth stages. However effective medium theories [55, 56] can be used to describe the optical properties of island structures as well as void structures [48], and theories can be used also for the cross-over stage at large scale coalescence [57] where the optical properties are gradually shifted from being associated with localized surface plasmons to extended plasmon-polaritons [58, 59].

The thickness scale appropriate for representing the growth stages depends on many parameters, such as deposited species, substrate material, presence of (artificially added) nucleation centers, substrate temperature, vacuum condition, electrical fields, etc. Figure 11.6 indicates two typical scales referring, respectively, to conventional evaporation (c.e.) and to ion-assisted evaporation (i.a.); the latter uses an optimized flow of energetic Ar^+ ions. Sputter deposition may be represented by an intermediate thickness scale. The most important conclusion of the discussion above is that noble metal films are reasonably uniform only at thicknesses above 10 nm, which is far beyond the optimal thickness for an idealized uniform film (*cf.* Fig. 11.5).

It is now possible to appreciate experimental data on spectral optical properties of noble metal films as a function of thickness. Figure 11.7 shows measured reflectance and transmittance for conventionally evaporated Ag films [46, 47]. At $t = 6$ nm one observes that R_{therm} is low, implying that E_{therm} is high. This can be reconciled with the occurrence of an island structure; if the layer were continuous then E_{therm} would have been low as inferred from Fig. 11.5. It appears that a continuous Ag layer is formed only at $t > 9$ nm. These films yield high IR reflectance combined with significant T_{lum} and hence are of interest for low-E windows. As expected, T_{lum} and T_{sol} decrease with increasing t.

The tendency for thin metal layers to form non-uniform structures at small thicknesses gives a significant limitation to their performance as TCs and transparent IR reflectors. If one prescribes high reflectance at long wavelengths, i.e., low E_{therm}, one is confined to $T_{\text{lum}} < 50\%$ and $T_{\text{sol}} < 40\%$ for evaporated films [46].

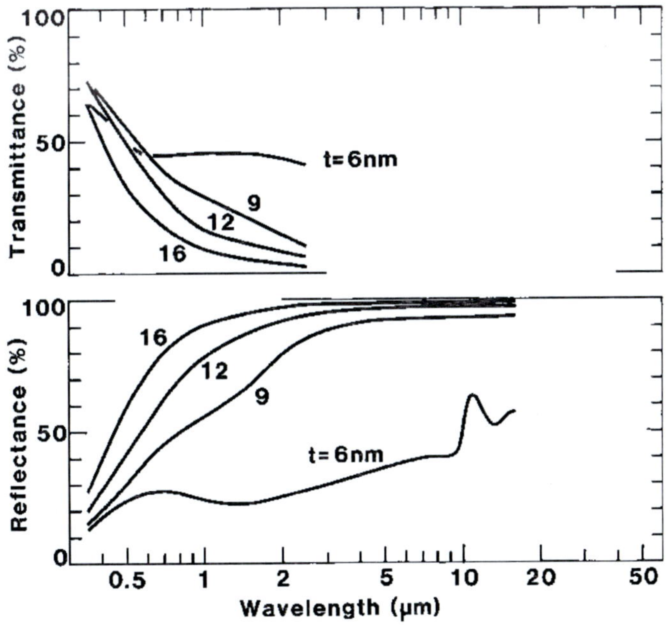

Fig. 11.7 Spectral normal transmittance and near-normal reflectance measured for Ag films on glass. The film thickness is denoted *t*. Replotted from Valkonen et al. [46]

Somewhat superior properties, with $T_{lum} \approx T_{sol} < 60\%$, can be achieved with ion assisted deposition [48] or special sputter technology [60]. If a higher transmittance is required – as is frequently the case for architectural windows and glass façades – one must use multi-layer coatings as considered next.

11.3.1.2 Multi-layer Films

The limitations in T_{lum} and T_{sol} for reasonably uniform noble metal films are largely due to reflectance at their surfaces, and consequently it is possible to increase the transmittance by adding layers that serve, effectively, to AR treat the metal. One is then led to dielectric/metal and dielectric/metal/dielectric multilayers. Dielectrics with high refractive indices – such as Bi_2O_3, In_2O_3, SnO_2, TiO_2, ZnO, and ZnS – give the largest enhancement. It is possible to optimize for low-*E* or for solar control trough proper selections of film thicknesses.

Figure 11.8 compiles measured transmittance spectra for TiO_2/Ag/TiO_2 films on glass [29]. These data are characteristic for what one can accomplish by using the three-layer design. It can be inferred that $T_{lum} > 80\%$ and $E_{therm} \ll 20\%$ are obtainable for all of the coatings. The dashed curve pertains to the coating with

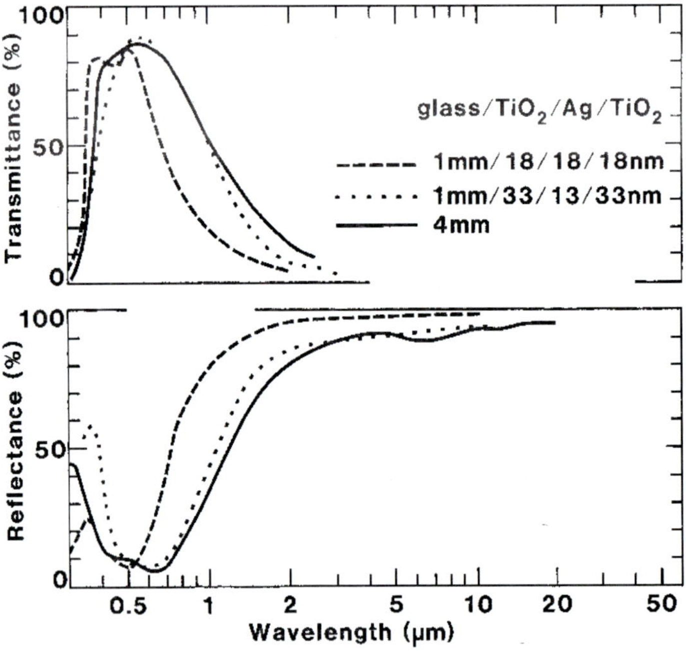

Fig. 11.8 Spectral normal transmittance and near-normal reflectance measured for TiO_2/Ag/TiO_2 coatings on glass. Film thicknesses are shown. *Solid curve* represents a commercial sample. From Granqvist [29]

maximum IR reflectance; it yields $T_{sol} \approx 50\%$ and $R_{sol} \approx 42\%$. The dotted curve corresponds to $T_{sol} \approx 67\%$ and $R_{sol} \approx 26\%$. The solid curve was measured for a commercial coating on float glass of the type represented by the middle curve in Fig. 11.2; this coating has $T_{lum} \approx 87\%$ and $T_{sol} \approx 72\%$. Work has been reported recently also on ZnS/Ag/ZnS [61, 62], ZnO/Ag/ZnO [63–65], and ZnO/Cu/ZnO [66, 67].

Figure 11.9 illustrates the joint development of electrical and optical properties in a three-layer film of ZnO(20 nm)/Ag/ZnO(20 nm) [65]. The sheet resistance $R_{\square}$ is given; it is defined by $R_{\square} = \rho/t$ where ρ is the resistivity. Clearly $t < 4$ nm yields poor electrical properties while $t > 8$ nm produces poor optical properties. The optimum is at $t \approx 6$ nm where one finds $R_{\square} \approx 3\ \Omega$.

A thin film design which in principle can give very low resistivity is In_2O_3:Sn/ Ag/In_2O_3:Sn [68–73]. In this case, all of the constituent layers are electrically conducting and a resistivity as low as $1.7 \times 10^{-4}\ \Omega$ cm, together with $T_{lum} > 80\%$, have been obtained on a polymer substrate [71]. Corrosion at the interface between Ag and In_2O_3:Sn can be diminished if the Ag is alloyed with Pd, Au, or Cu [73].

Coatings of the type TiO_2/TiN/TiO_2 have been investigated as well [74–77]. Figure 11.10 shows data from a recent study of a three-layer stack comprising TiO_2(35 nm)/TiN(22 nm)/TiO_2(35 nm) [77]. The top layer of TiO_2 can render this type of coating photocatalytically active and super-hydrophilic, as discussed further in Sect. 11.3.3.3 below. A very thin Ti interlayer between the TiN and the outer TiO_2 layer improved the heat insulating performance noticeably and also influenced the photocatalytic activity favorably [77].

The examples above have been selected to illustrate principles rather than today's (2007) commercial products. Ag-based thin films for energy efficient

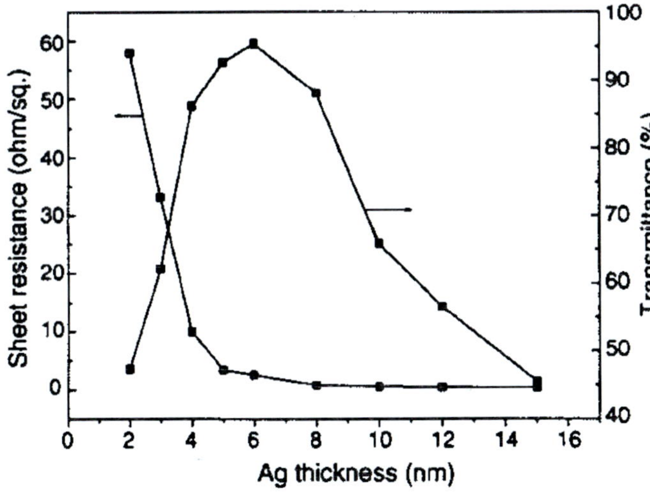

Fig. 11.9 Sheet resistance and maximum transmittance as a function of Ag thickness in a three-layer structure comprising ZnO(20 nm)/Ag/ZnO(20 nm). From Sahu et al. [65]

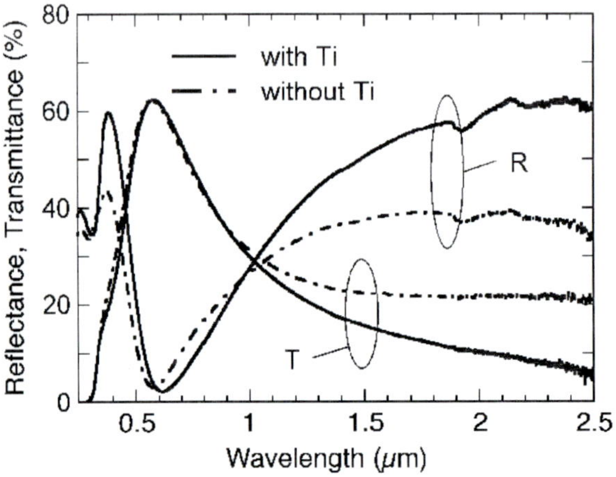

Fig. 11.10 Spectral normal transmittance and near normal reflectance for a TiO_2(35 nm)/TiN (22 nm)/TiO_2(35 nm) coating on glass. The effect of a thin interfacial layer of Ti is shown. Replotted from Okada et al. [77]

fenestration are now highly optimized, and a very large number of products with specified thermal, solar, and luminous (including color) properties are available on the market. Some of these products use multilayers that go far beyond the three-layer concept and may include as many as seven layers, with two metal films. It is remarkable that these TCs are prepared with thickness control approaching atomic precision in a high-performance production equipment handling glass sheets up to 30 m^2 in size. The technologies are normally proprietary and the reader is directed to vendors' technical publications for specific details, to the extent such are given.

11.3.2 Doped Oxide Based Thin Films for Energy Efficient Windows

11.3.2.1 Introduction

Suitably prepared metal oxide thin films have been used as TCs and transparent heat reflectors since the beginning of the 1900s, i.e., even longer than the noble metal based films discussed above [12, 78]. It appears that all of this early work was motivated by potential technical applications. The most widely used and studied *n*-type transparent conducting oxides (TCOs) are based on Zn, In, and Sn (and occasionally Cd). Doping is accomplished either by adding a higher-valent metal, by replacing some oxygen with fluorine, or by oxygen vacancies. The fundamental band gap is wide enough to allow very low magnitudes of A_{lum} and A_{sol}, and doping is feasible to a level high enough to render the materials IR reflecting and electrically conducting. The onset of reflectance lies in the near IR, implying that the oxides can have a small E_{therm} and thereby serve as low-E coatings. However, the material is less useful for solar control, the main reason being that the doping cannot be brought to a sufficiently high level, at least not without considerable difficulty. A specific advantage of the TCOs, compared to their noble metal based counterparts, is the chemical and mechanical stability which allows their use on glass surfaces exposed to air; the oxide films are sometimes referred to as "hard coats". Several useful reviews of TCOs have been published recently [78–84].

A resistivity of 10^{-4} Ω cm is often quoted as a number characterizing optimized films of In_2O_3:Sn (known as Indium Tin Oxide or ITO), ZnO:Al (AZO), and ZnO:Ga (GZO), even if lower resistivities were sometimes claimed already in the 1980s for ITO [85, 86]. Recent investigations of ITO films have demonstrated resistivities between 7.2 and 8×10^{-5} Ω cm and a free carrier density as high as 2.5×10^{21} cm^{-3} [87–95]. Films of AZO with $\rho \approx 8.5 \times 10^{-5}$ Ω cm [96] and of GZO with $\rho \approx 8.1 \times 10^{-5}$ Ω cm [97, 98] have been reported. It seems that pulsed laser deposition is particularly apt for making TCOs with exceptionally low resistivity. A large number of other investigations have stated resistivities between 1 and 6×10^{-4} Ω cm; reports from 2005 and later are listed elsewhere [7]. Attempts have been made to boost the properties of ITO by small additions of Ag [100–103]; such films could have lowered ρ, but T_{lum} decreased if the Ag content was excessive. A resistivity of 8.2×10^{-5} Ω cm was stated for the most conducting films; it had $T_{lum} \sim 50\%$ [102]. All resistivity data lie inside the intrinsic limits of the electrical conductivity of TCOs formulated some time ago [99].

Thin film TCO preparation is usually very sensitive to deposition conditions. In the case of ITO sputtering, for example, the properties of the target – such as its density – are important for the films [104, 105] as are the composition and conditions of the plasma [106, 107] and the distribution of the sputter gases [10]. High deposition rate is usually needed for practical manufacturing of ITO films; a rate of 650 nm min^{-1}, together with $\rho \sim 7.5 \times 10^{-4}$ Ω cm, have been reported recently for sputtering [95].

The critical nature of the thin film preparation conditions is evident from Fig. 11.11, which shows scanning electron micrographs of an ITO film deposited by DC magnetron sputtering onto a substrate at ambient temperature and subsequently annealed at 200°C for 1 h in air (panel a) and of such a film deposited onto a substrate kept at 200°C (panel b) [108]. The film morphologies are strikingly different although both films were treated at the same temperature.

SnO_2-based films are commonly prepared by spray pyrolysis or CVD and are then capable of yielding excellent properties as TCOs. These techniques normally operate at high temperature, but ozone-assisted deposition can decrease the substate temperature to ~100°C and yet give SnO_2:F films with $\rho \sim 10^{-3}$ Ω cm [109]. Concerning techniques operating at even lower temperature, SnO_2-based films with $\rho \sim 10^{-3}$ Ω cm have been prepared by sputtering [110, 111] and spin coating [112].

GZO deserves special attention since it is capable as serving as a TCO with UV transparency limited by a band gap as wide as ~5 eV [113, 114]. This opens possibilities to transmit radiation appropriate for photocatalysis, as further

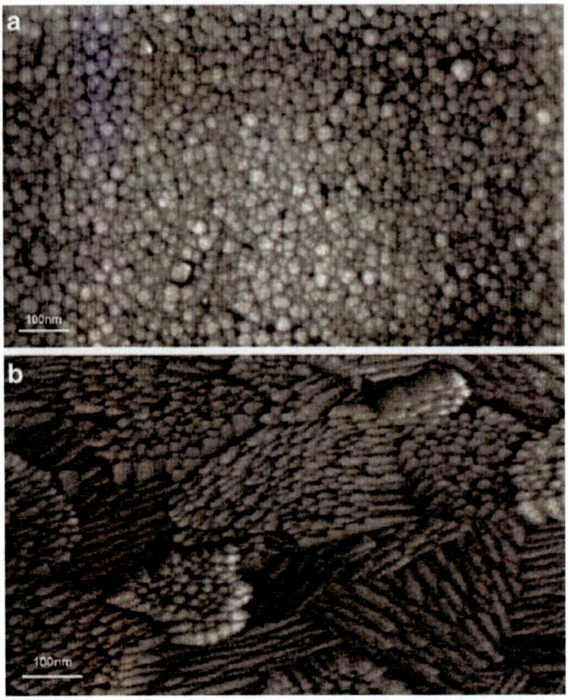

Fig. 11.11 Scanning electron micrographs for an ITO film sputter deposited at ambient temperature and annealing post treated at 200°C (**a**) and for a similar film sputter deposited onto a substrate kept at 200°C (**b**). *Horizontal bars* indicate the magnification. From Betz et al. [108]

discussed in Sect. 11.3.3.3. ZnO-based films are known to be sensitive to humidity ensuing from reactions of water molecules with oxygen vacancies. However the moisture resistance of GZO films can be very significantly improved if the "dopant" level is ~10%, i.e., much larger than in typical TCOs [115]. Special off-axis sputtering was used to avoid bombardment of energetic species and led to film growth with apparently random *c*-axis orientation.

Most work on TCOs has concerned films prepared on substrates heated to temperatures high enough that deposition onto standard plastics such as PET or PEN is out of the question. The difference in mechanical properties between TCO films and polymer substrates can also be an issue, and tensile failure may lead to crack patterns and thereby increased resistivity [116, 117]; surface activation by ion-assisted reactions can be useful for improving the adherence between TCO films and polymer substrates, though [118]. Nevertheless, despite a number of difficulties concerning the fabrication of TCOs on polymers, deposition at low substrate temperatures onto PET is feasible, and recent work on ITO yielded resistivities between 3 and $5 \times 10^{-4}\ \Omega$ cm [119–123], while In_2O_3-ZnO films have shown $\rho \sim 5 \times 10^{-4}\ \Omega$ cm [124]. ITO films have been prepared also on PEN [125, 126], polycarbonate (PC) [127], and on high-temperature stable flexible transparent polyimide [128]. GZO films on PEN yielded $\rho \sim 5 \times 10^{-4}$ Ω cm [129], and similar results have been reported for ITO on PC [130–133], poly methyl methacrylate [134] and polyethersulfone [135], IZO on PET and PC [136], and for In_2O_3-ZnO-SnO_2 on PET [137].

Still other alternatives for making *n*-type TCOs may appear in the future, such as TCOs based on TiO_2 [138–140]. Carbon nanotube layers is another emerging possibility, and the same may be true for $12CaO{\cdot}7Al_2O_3$ (a high-alumina cement). The latter two materials are discussed further in Sect. 11.6.1.

TCOs which are strongly *p*-doped, rather than *n*-doped as those discussed above, are of much current interest owing to their potential applications in optoelectronic devices [141, 142]. Future applications in energy-related technology are possible though still rather undeveloped.

11.3.2.2 Theoretical Considerations

The electromagnetic properties of doped oxides, above the Mott critical density [143], are conveniently discussed in terms of additive contributions to the complex dielectric function from valence electrons, free electrons, and phonons [144, 145]. They influence the optical properties of the oxides in different wavelength regions, as shown schematically in Fig. 11.12 pertaining to ITO [146]. Essentially, the heavy doping leads to *n*-type conduction and Sn^+ ions. Going from short to long wavelengths it is found that interband absorption prevails at $\lambda < 0.3$ μm due to excitations across the fundamental band gap denoted E_g. The band gap is shifted towards shorter wavelengths as a consequence of the free electrons entering states at the bottom of the conduction band of In_2O_3. This feature is conveniently described in terms of the Burstein-Moss effect [147, 148], but the underlying physics is considerably more complex than what can be captured in a rigid band model, and a quantitative theoretical description must include many-body representations of ion-ion and electron-electron scattering [149]. The theory has been developed in particular detail for ZnO-based materials and was found to agree well with experimental data for AZO [150, 151]. Recent work showed that this theory could account for band gap widening in GZOs (whereas it did not fit data for AZO) [152]. The band edge is of the Urbach type (i.e., logarithmic) [153]; there are several possibilities why this may be the case [146], one of them related to the effect of the Sn^+ impurities. From a practical point of view it is advantageous that the band gap widening is so strong that the "Urbach tail" at the band gap plays very little roles for the optical properties in the solar range and no role whatsoever in the luminous range.

Continuing the discussion around Fig. 11.12, the free electrons lead to onset of reflectance at a "plasma energy" denoted $\hbar\omega_p$, somewhat beyond 1 μm in the near IR. The electrons are scattered against several different kinds of centers; in the luminous wavelength range it seems that grain boundaries predominate, whereas ionized impurities, i.e., Sn^+, are important at longer wavelengths. These impurities become screened by the free electron plasma to an extent that increases gradually in

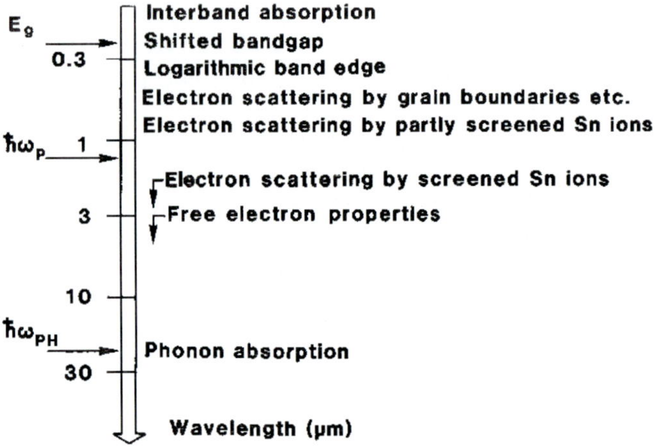

Fig. 11.12 Survey over fundamental physical processes and mechanisms for In_2O_3:Sn films. E_g is fundamental semiconductor band gap, $\hbar\omega_p$ is plasma energy, and $\hbar\omega_{PH}$ is phonon energy. From Hamberg and Granqvist [146]

the near IR and is fully developed in the thermal IR. This gradual onset is advantageous with regard to energy related applications and leads to larger transmittance than would have been the case if screening had been fully developed already at the plasma energy. Phonon effects come into play far into the thermal IR, indicated schematically by $\hbar\omega_{PH}$. They can be represented as Lorenzian oscillators added to the dielectric function. In practice, they are of little concerns since they are swamped by the effects of the free electron plasma.

It is now clear that the free electron plasma is of overriding importance in the wavelength range of relevance for solar energy applications and energy efficiency. Free electron theory is a very well understood subject capable of accounting for exchange and correlation [154, 155]. The theory can be quantitatively compared with experimental data, most expediently by use of the complex dynamic resistivity $\rho(\omega) = \rho_1(\omega) + i\rho_2(\omega)$ which is directly related to the optical constants and the dielectric function. This representation is suitable since $\rho_1 \sim \omega^s$, where s depends on the type of scattering that dominates [156]. Specifically, $s = -3/2$ for ionic impurities represented by a Coulomb potential, $s = -2$ or -1 for dislocations represented by a charged monopole or dipole line, respectively, $s = 1/2$ for a short-ranged neutral defect described by a δ-potential, $s = 1/2$ or $-1/2$ for acoustic phonons with deformation potential coupling or piezoelectric coupling, respectively, and $s = -1/2$ for optical phonons with polar coupling. Empirical data are consistent with $s = -3/2$, implying that point-like ionized impurities are dominating the free electron scattering beyond the plasma wavelength [146]. This result is important since it shows that the analyzed films are indeed optimized in the sense that their performance is largely governed by the type of scattering that is inherent in the doping mechanism, and the properties cannot be expected to be significantly improved by some conceivable post treatment such as annealing to improve the crystalline order. The frequency dependent scattering also shows that the classical Drude model [157] cannot be used for detailed quantitative work unless a frequency dependent mean scattering time or mean free path is invoked. The theoretical model outlined above is applicable not only to ITO films but also to nanoparticles of this material [158–161].

Having a quantitative theoretical model for the optical properties of ITO makes it possible to calculate the optical properties in detail [146] as discussed next. Figure 11.13 reports $R(\lambda)$ and $T(\lambda)$ in the full range covering luminous, solar, and thermal properties for an ITO film with $t = 0.2$ μm and backed by a substrate having a refractive index of 1.5. The latter value is appropriate for glass with low Fe content in the visible and near IR, or a foil of a polymer such as PET. In subsequent calculations the substrate was represented by the dielectric function for amorphous SiO_2 [162, 163], and the ITO was specified by a free electron density n_e in the

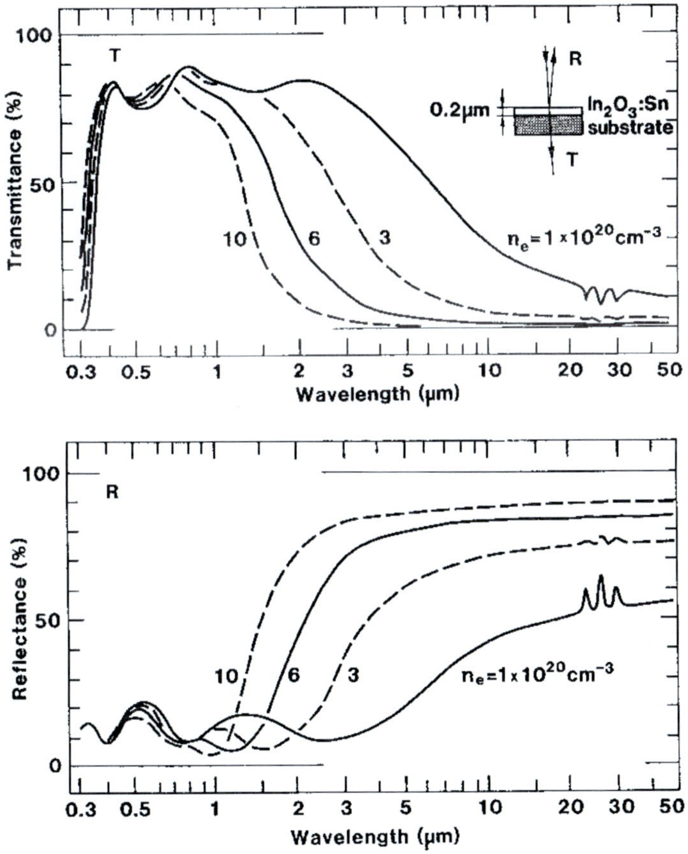

Fig. 11.13 Spectral normal transmittance T and reflectance R computed from a quantitative theoretical model for the optical properties of In_2O_3:Sn. The shown values of electron density n_e and film thickness were used. From Hamberg and Granqvist [146]

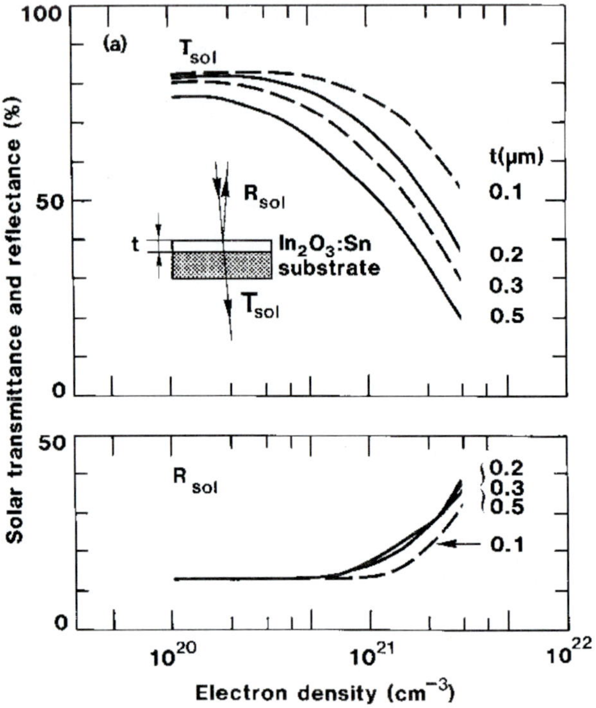

Fig. 11.14 Solar normal transmittance T_{sol} and reflectance R_{sol} as a function of electron density computed from a quantitative theoretical model for the optical properties of In_2O_3:Sn. Data are shown for four film thicknesses. From Hamberg and Granqvist [146]

$10^{20} \leq n_e \leq 3 \times 10^{21}$ cm^{-3} range. Figure 11.13 shows band gap widening, onset of reflectance at a wavelength which scales with n_e, and a gradual disappearance of phonon induced structure. It appears that the best low-E coating is obtained for $n_e \approx 6 \times 10^{20}$ cm^{-3}, and that one would expect to have a very good solar control coating at $n_e \approx 3 \times 10^{21}$ cm^{-3}. The latter value may be difficult to realize in practice, although data that are not too far off have been reported, as mentioned in Sect. 11.3.2.1.

The spectral data can be used to calculate T_{lum}, R_{lum}, T_{sol}, R_{sol}, and E_{therm} in order to assess the applicability of ITO films for energy efficient fenestration [146]; data are given in Figs. 11.14–11.16. Figure 11.14 shows T_{sol} and R_{sol} for ITO films with four thicknesses and $10^{20} \leq n_e \leq 3 \times 10^{21}$ cm^{-3}. It appears that T_{sol} decreases and R_{sol} increases as n_e goes up, which is an effect of the plasma wavelength gradually moving into the solar spectrum. At $n_e \leq 5 \times 10^{20}$ cm^{-3} one observes $R_{sol} \approx 13\%$ independently of t. Analogous data were obtained for T_{lum} and R_{lum}. At $n_e < 10^{21}$ cm^{-3} one finds $75 \leq T_{lum} \leq 85\%$ and $10 \leq R_{lum} \leq 20\%$ with the actual magnitudes being strongly influenced by optical interference. When n_e increases there is a tendency for R_{lum} to decrease while T_{lum} is more erratic. Figure 11.15 deals with E_{therm} and displays results for ITO backed by a

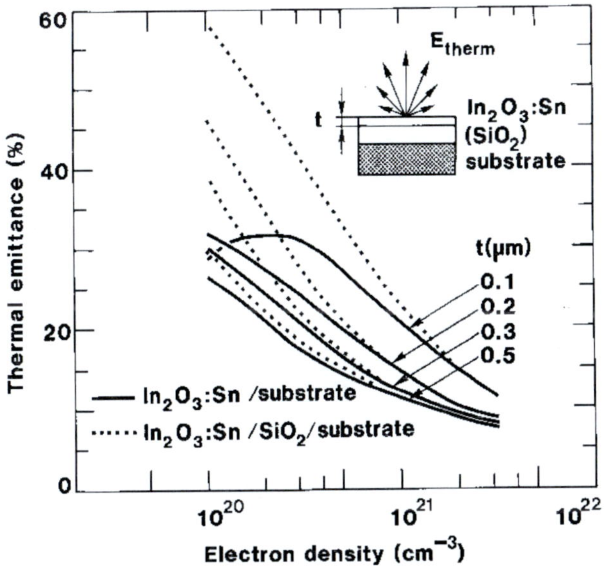

Fig. 11.15 Thermal emittance E_{therm} as a function of electron density computed from a quantitative theoretical model for the optical properties of In_2O_3:Sn. Data are shown for four values of the film thicknesses t. The substrate was taken to be non-emitting with a refractive index of 1.5 (*solid curves*) or specified by the dielectric function of amorphous SiO_2. From Hamberg and Granqvist [146]

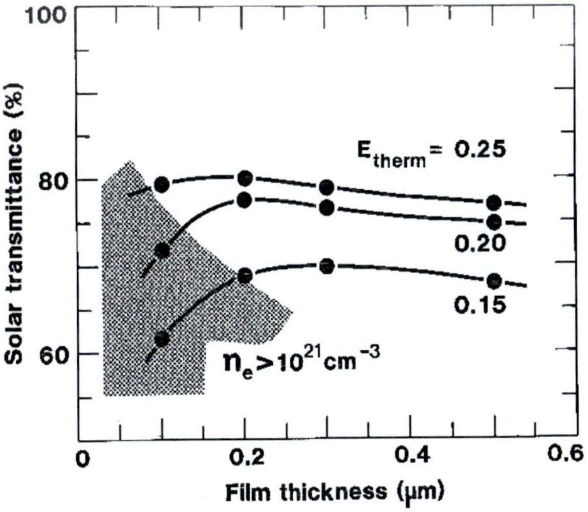

Fig. 11.16 Solar transmittance as a function of film thickness computed from a quantitative theoretical model for the optical properties of In_2O_3:Sn and specifying the substrate by the dielectric function of amorphous SiO_2. Data are shown for three values of the thermal emittance E_{therm}. *Shaded* region corresponds to electron density n_e larger than 10^{21} cm^{-3}. From Hamberg and Granqvist [146]

substrate with a refractive index 1.5 (solid curves) and a substrate simulating glass (dotted curves). Generally speaking E_{therm} drops when n_e and t increase, which is a consequence of the increased number of free electrons. It is found that substrate emission is important for $n_e \leq 5 \times 10^{20}$ cm^{-3} but negligible above this value as long as $t > 0.2$ μm; these latter films have $E_{therm} \leq 15\%$ for $n_e \approx 10^{21}$ cm^{-3}.

Figures 11.14 and 11.15 show that ITO films are useful for energy efficient windows. The actual performance of a window with such a film depends also on a number of other factors such as heat conduction and convection [37, 164], which largely fall outside the scope of the present discussion. However some general conclusions can be drawn. To that end it is appropriate to fix E_{therm} at a low magnitude. It is not meaningful to decrease it to the extreme [165], though, since non-radiative heat transfer always plays a role. Figure 11.16 shows computed results for E_{therm} set to 15, 20, and 25% and gives T_{sol} as a function of t. Requiring $E_{therm} = 15\%$ one cannot exceed $T_{sol} \approx 70\%$, which is undesirably low for many applications in architecture. However limiting the requirement to $E_{therm} = 20\%$ one can have $T_{sol} \approx 78\%$, and the solar transmittance can be marginally higher at a still larger E_{therm}. Optimum data for low-E applications can be obtained for $t = 0.2$ μm and $4 \leq n_e \leq 6 \times 10^{20}$ cm^{-3}; spectral data for such a film were shown in Fig. 11.13 above.

11.3.2.3 Some Empirical Results

The optimization of the TCO film in Fig. 11.16 did not consider T_{lum}, which obviously is a property of the greatest importance for a window. However here one can resort to AR coatings analogously with the case of noble metal based TCs. Figures 11.17 and 11.18 show, respectively, experimental data on $T(\lambda)$ in the luminous range and $R(\lambda)$ in the main part of the thermal range for a thin ITO film with and without an AR coating of AlO_xF_y or MgF_2 and backed by glass [146, 166]. Figure 11.17 illustrates two important features: (1) after AR treatment the transmittance of the coated glass is larger than for the uncoated glass and hence there is no grounds for the common claim that a low-E coating must limit the luminous performance, and (2) the interference effects are diminished which makes the AR treated glass less prone to show iridescence than the glass with only ITO. The AR coating degrades the thermal performance to some extent, as illustrated in Fig. 11.18 which shows $R(\lambda)$ at 45° incidence angle for s and p polarized light. The effect of the MgF_2 is not noticeable for s polarization, whereas the p-polarized reflectance displays a dip at $\lambda \approx 17$ μm consistent with the dielectric function of MgF_2. Considering the effect on E_{therm} one can conclude that this parameter is not increased by more than a few percent as a result of the AR treatment. Another AR coating for ITO is indium-tin oxyfluoride [167]; it is obviously convenient since it can be made from the same starting material as the ITO film itself. The same holds true for SnO_xF_y films as AR treatments for SnO_2-based films [168].

The discussion above has been confined to In_2O_3:Sn. However this is less of a restriction than one may at first think since detailed experimental and theoretical

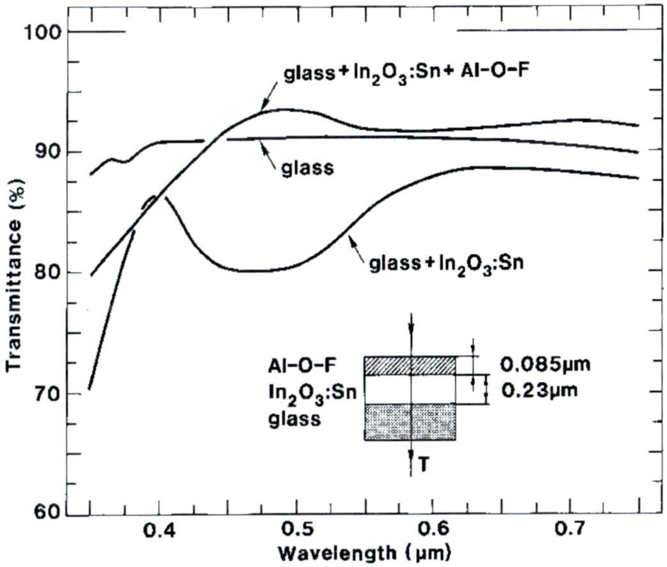

Fig. 11.17 Spectral normal transmittance of a glass substrate, after coating it with an In_2O_3:Sn film, and after a final deposition of AlO_xF_y. The experimental configuration is shown in the *inset*. From Hamberg and Granqvist [146]

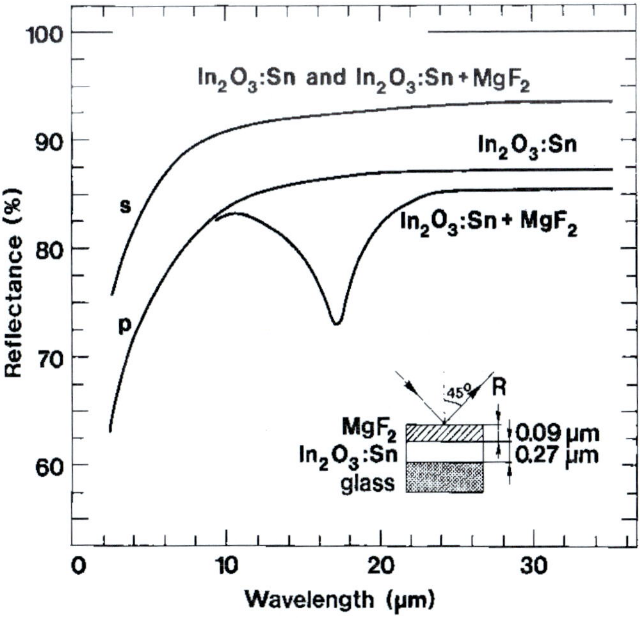

Fig. 11.18 Spectral reflectance of *s* and *p* polarized light incident at 45° angle towards a glass substrate, after coating it with an In_2O_3:Sn film, and after a final deposition of MgF_2. The experimental configuration is shown in the *inset*. From Hamberg and Granqvist [146]

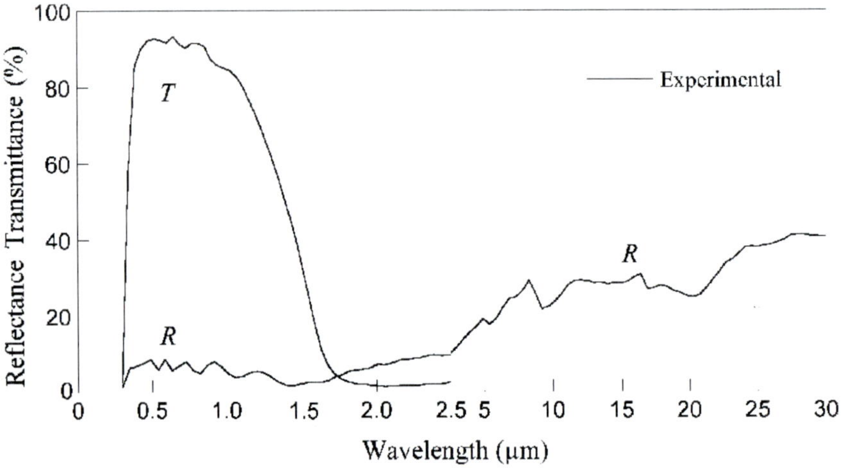

Fig. 11.19 Spectral normal transmittance *T* and near-normal reflectance *R* for an In_2O_3:Sn film made from sintered nanoparticles. Note the scale change at 2.5 μm. Replotted from Ederth et al. [158, 159]

studies – analogous to the one reported above – have been given for SnO_2:F and SnO_2:Sb [110] and for AZO [150, 169].

Thin film fabrication using nanoparticle deposition rather than atomistic, molecular, ion-based, or solution-based deposition may become of increasing importance in the future. A recent study [158–160, 170] specifically considered ITO films made from nanoparticles prepared by a wet chemical technique. A dispersion of these particles was spin coated and the ensuing film was annealed. Figure 11.19 shows some data on $R(\lambda)$ and $T(\lambda)$ for $\lambda < 30$ μm; they are consistent with $T_{\text{lum}} \sim 90\%$ and $E_{\text{therm}} \sim 70\%$. A detailed analysis of the structural, electrical, and optical properties showed that the individual nanoparticles had $\rho \sim 2 \times 10^{-4}\ \Omega$ cm – i.e., a resistivity comparable with that in the best films made by conventional techniques – but the film only had $\rho \sim 10^{-2}\ \Omega$ cm. Investigation of temperature-dependent resistivity gave clear evidence for fluctuation-induced tunneling [171] between μm-size clusters of internally connected nanoparticles. It is then obvious that a densification of the particulate ITO films is needed in order to boost ρ and produce a magnitude of E_{therm} that would be of interest for applications related to solar energy and energy efficiency.

11.3.3 Diverse Applications of Spectrally Selective Coatings

11.3.3.1 Low-Emittance Coatings for Vacuum Glazings

As pointed out above, exceptionally good thermal insulation can be achieved in "vacuum glazings". Essentially they comprise glass panes separated by tiny support pillars that do not significantly obstruct the view. The space between the panes is evacuated and permanently sealed. Low-*E* coatings are essential in order to obtain good thermal properties. This technology has been discussed in detail during the last several years [172] and vacuum glazings are available as commercial products. The heat transfer coefficient, not too close to the edges, can be ~0.4 W m^{-2} K^{-1} for double glazing [173] and ~0.2 W m^{-2} K^{-1} for triple glazing [38].

The ability to maintain good vacuum during extended periods of time is of obvious concern, and water vapor desorption during heating and carbon oxide desorption during solar irradiation have been investigated lately [174, 175]. Most of the desorption must emerge from the TCOs, specifically from SnO_2:F coatings, whose nanostructure therefore is critical for the successful use of "vacuum glazings".

11.3.3.2 Doped Tin Oxide Based Thin Films for Avoiding Radiative Cooling

The atmosphere can be transparent to thermal radiation in the $8 < \lambda < 13$ μm "window" range, as shown in Fig. 11.1c. This effect can lead to radiation cooling of surfaces oriented towards the sky [176, 177]. The resource for such cooling has been evaluated through detailed calculations showing that the cooling power for a black-body surface is 80–120 W m^{-2}, depending on the atmospheric condition, when the surface is at ambient temperature [178]. If nothing but radiation exchange is considered, the surface can drop to 14–26°C below the air temperature. In the presence of a non-radiative heat transfer equal to 1 W m^{-2} K^{-1}, the maximum temperature drop is expected to be 10–20°C. Practical tests have given clear evidence for radiative cooling of rigid plates in collector devices of different types [176], flexible foils [179, 180], and gases [176].

Radiative cooling is a well known phenomenon, although its dependence on the IR transmittance of the atmosphere has not always been clearly understood. The cooling manifests itself in water condensation during clear nights on surfaces exposed to the sky [181]. This is particularly noticeable on sky-facing windows and originates from the fact that the glass, which is a poor heat conductor and has limited heat capacity, can drop below the dew point before most other objects in the surrounding do so, implying that water condensation – with or without frost formation – takes place. Frost covered car windscreens are commonly seen in cold climates [182], and water condensation on architectural windows and glass façades is a growing concern owing to the introduction of glazed units with good thermal insulation [183]. A way to practically eliminate the condensation is to avoid radiative cooling, which can be done if the exposed glass surface has a coating with low emittance in the $8 < \lambda < 13$ μm interval. This is the case for the materials discussed in Sects. 11.3.1 and 11.3.2 above. If one further requires good chemical and mechanical durability, the favored material is doped SnO_2, normally prepared by spray pyrolysis onto hot, solidifying float glass. The "anti-condensation" property is compatible with a very thin top coating, for example an AR layer, as apparent from the discussion around Fig. 11.18.

Deicing and demisting are obviously possible by heating, using electrical power supplied to a TCO coating on glass. Electrical power can be used also for space heating, if there is a need for such, and for maintaining the impact strength of glass (for example in windows of airplane cockpits).

11.3.3.3 Comments on Photocatalytic Properties, Self-Cleaning, and Super-hydrophilicity of Transparent Conductors

The possibility of using TiO_2 films in conjunction with TCs as mentioned in Sect. 11.3.1.2, and the current interest in the wetting of glass surfaces discussed in Sect. 11.3.3.2, warrant some brief discussion of photocatalysis and related effects in TiO_2 based thin films. Photocatalytically active surfaces can cause breakdown of organic molecules as well as microorganisms adsorbed on these surfaces under the action of solar irradiation [184–186]. Thus hazardous or otherwise unpleasant pollutants can be decomposed into tasteless, odorless, or at least less toxic compounds. Well known photocatalysts are oxides of Cd, Ce, Fe, Sb, Sn, Ti, W, Zn, and Zr, with most work having been devoted to TiO_2 with anatase type crystal structure. The latter oxide works only with UV light owing to the width of the band gap. The photocatalytically active surfaces normally are comprised of nanocrystalline structural units, but polymer incorporated TiO_2 is another option [187].

Deodorization is possible, which in principle allows the use of natural light, rather than some energy-consuming air cleaning device, to purify indoor air. Recent work in this area has been reported for Volatile Organic Compounds (VOCs) as well as for a multitude of single gases [7]. Photocatalytic surfaces are usually backed by glass or metal, but coatings onto flexible PET, having a protective surface layer of SiO_2, has demonstrated photocatalytic elimination of trichloroethylene [188]. The photocatalytic layer can have a dual function and serve also as AR treatment of an underlying glass [189]. The photocatalysts tend to lose some of their efficiency over time, in particular under operation as air cleaners.

It may be an advantage to employ not only UV light but also the less energetic but more abundant visible light for photocatalysis, and dopants have been added to TiO_2 for that purpose in order to modify the spectral absorption range. There is much recent work in particular for nitrogen doping to form $TiO_{2-x}N_x$ [190–193]. An admixture of WO_3 was able to enhance the photocatalytic activity still further [194].

The photocatalytic surfaces exhibit photoinduced super-hydrophilicity, meaning that water droplets spread more or less evenly over the surface so that light scattering tends to be insignificant. Thus the water is not visible and, furthermore, drying-related contamination residues do not appear, which can be a considerable asset from a practical point of view. Most studies have been made on films coated onto glass substrates, but PC substrates have been studied recently too [195]. Self-cleaning by use of super-hydrophobicity – analogously with the features of the lotus leaves – represents another possibility for surfaces having nanoscale roughness

[196]. Transitions between super-hydrophobicity and super-hydrophilicity, depending on UV irradiation, have been reported for rough TiO_2-based surfaces [189].

11.3.3.4 Transparent Conductors for Solar Cells

Solar cells are used for converting solar radiation to electricity. Current collection takes place via electrodes; the one exposed to the sun can be comprised of metal strips, TCO, or a combination of these. For solar cells made of thin films it may be advantageous to have a scattering, sometimes called "milky", TCO which can make the average path length in the absorber longer than for a non-scattering film. This can be so both for a "substrate cell" with the TCO facing the sun or for a "superstrate cell" for which the solar radiation passes through the substrate before it reaches the TCO. Wet etching of sputter deposited AZO films has been studied several times for this purpose [197–199]. Figure 11.20 shows the development of roughness when a film is treated in 0.5% HCl [199]. The etched film has a roughness of the order of 1 μm, which is advantageous since maximum scattering occurs when the roughness is comparable with the wavelength of the light. The detailed deposition conditions are important for the roughness. This is illustrated in Fig. 11.21 which shows the haze – i.e., the diffuse transmittance divided by the total transmittance – at three wavelengths for AZO films made at different oxygen partial pressures in the sputter plasma [199]. Clearly there is an optimum pressure. The scattering is highest at the shortest wavelength.

In the rest of this section we scan through a range of different solar cell technologies and comment on the TCs that are frequently used; more details are given elsewhere [7]. Silicon is an excellent material for solar cells backed by glass or foils of metal or polymer. The Si can be of three types: crystalline, polycrystalline, or amorphous and hydrogenated. For single crystalline and polycrystalline cells it is conventional to have metallic "fingers" as current collectors, whereas TCO layers – normally based on SnO_2 – are used for thin film devices. A thin layer of TiO_2 can serve as a protection against hydrogen radical exposure inherent in some techniques for making amorphous silicon solar cells [200, 201].

Compound semiconductors of the III-V type – based on GaAs, (Al,Ga)As, InP, or (In,Ga)P – can show high efficiency as well as good resistance to ionizing radiation; they are costly, though, and mainly used for space applications. Among the II-VI compounds, CdTe and Cd(S,Te) are known for their robust manufacturability and are of interest for ground-based solar cells, although environmental concerns regarding large-scale use of a technology including Cd should not be overlooked. Regarding the I-III-IV_2 compounds, $CuInSe_2$, $Cu(In,Ga)Se_2$, and $Cu(In,Ga)(S,Se)_2$ are notable for their possibility to reach high photo-electric conversion efficiencies in thin film solar cells. Numerous other materials and approaches to solar cells are possible, and many other compound semiconductors may well turn out to have favorable properties. Some of those may find applications in thermo-photoelectricity, i.e., cells operating with thermal irradiation. Others can be used in tandem cells, encompassing superimposed cells with optical band gaps

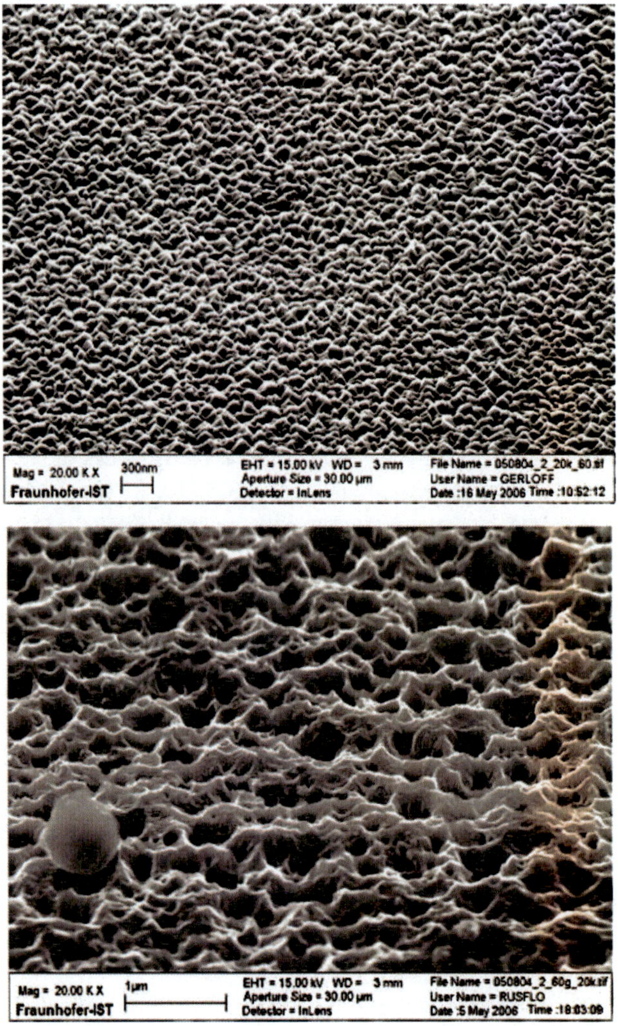

Fig. 11.20 Scanning electron micrographs for a sputter deposited ZnO:Al film before (*upper panel*) and after (*lower panel*) etching in 0.5% HCl. From Ruske et al. [199]

that are large on the exposed surface and gradually shrink towards the underlying substrate. The current collector is frequently a film of AZO, but other TCOs may work as well. In particular ZnO:B has been put forward recently as a superior material suitable for large-area deposition and characterized by great hardness [202, 203]. Deposition rates as large as $\sim$3 nm s^{-1} have been reported in work using vacuum arc plasma evaporation [203]. Recent work has shown that carbon nanotubes may replace the TCO in organic solar cells [204–207], as discussed further in Sect. 11.6.1. Alternatively, photo-electrochemical techniques can be employed to generate electricity, and studies of nano-crystalline dye cells are popular at present.

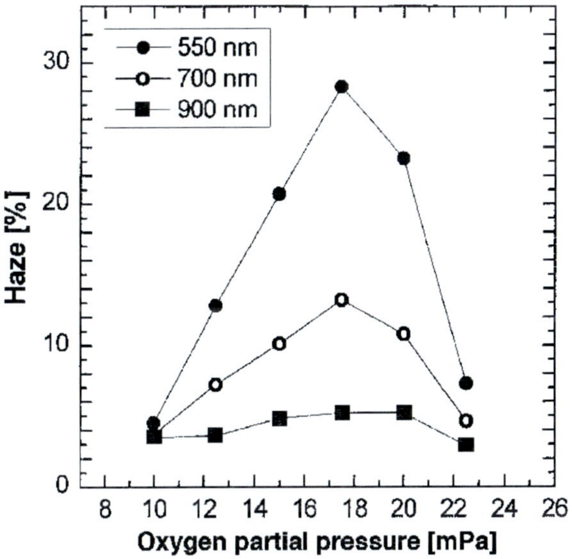

Fig. 11.21 Haze as a function of oxygen partial pressure in the sputter plasma for ZnO:Al films. Data were taken at three wavelengths. From Ruske et al. [199]

The TCO is normally SnO_2-based; the film is rough, which contributes to the ability to absorb the dye. TCs made from carbon nanotube layers have been demonstrated for such cells [208].

Transparent photovoltaic cells may be of interest for see-through applications. Thin-film devices with amorphous silicon as well as nano-crystalline dye cells can show some transparency. Another, more innovative, possibility can be realized by using superimposed *n*-type and *p*-type TCOs. Thus a $ZnO/CuAlO_2/ITO$ structure was able to demonstrate a photovoltaic effect [209].

In solar cell systems it is possible to make use of selectively reflecting mirrors which, ideally, reflect only radiation with energies above the band gap of a solar cell towards this cell. Such an arrangement prevents excessive heating of the solar cell and thereby maximizes its efficiency. The long wavelength radiation may be used to advantage in a solar thermal system. Selectively reflecting mirrors can be produced by using a TCO layer backed by metal, and the properties may be further optimized by having a dielectric layer between the TCO and the metal. A detailed study of such a system – comprising SnO_2:F, Al_2O_3, and Al – was recently carried out theoretically and experimentally [210].

11.3.3.5 Doped Oxide Based Thin Films for Solar Collectors

Solar collectors use photo-thermal conversion of solar energy to produce hot fluid or hot gas [211]. Efficient conversion requires that a solar absorbing surface is in thermal contact with the fluid or gas, and that thermal losses to the ambience are minimized. For most applications, the absorber should be positioned under a transparent cover, and a transparent thermal insulation material may be used as well. The back and sides of the solar collector should be well insulated. The most critical part for the photo-thermal conversion is the actual solar absorber surface, which obviously should have a large A_{sol}. The radiation heat losses should be minimized, and this requires low E_{therm}. TCOs can, in principle, be applied in two different ways: to serve as a heat reflector that brings radiation from the hot absorber surface back to the same, and as a top coating on a non-selective black absorber. The latter approach was attempted in some early work on solar absorbers based on black enameled surfaces [212], but it has not gained popularity.

11.4 Applications Based on Angular Selectivity

Pronounced angular properties can lead to energy efficiency as pointed out above. Section 11.4.1 treats general principles for obtaining angular selectivity, either with optical properties being symmetrical around the surface normal or with different optical properties on the two sides of the surface normal. Section 11.4.2 then presents experimental data focused on Cr based films prepared with physical vapor deposition arranged so that the incident flux has an oblique angle to the surface normal.

11.4.1 General Considerations

It may be advantageous to use windows having small transmittance for solar radiation coming from a point high up on the vault of heaven – thereby avoiding overheating – whereas the transmittance along a near-horizontal line-of-sight is high so that unmitigated visual contact can be maintained through the window. The situation is rather simple for vertical windows, at least in theory, and a TC coating will be able to bring advantages. A single metal film has the desired property to some extent. Spectral and angular selectivity hence can go hand in hand for vertical windows.

However the situation is more complicated for inclined windows such as those commonly used in atria and in transparent blinds; car windscreens is another common example. Then it is possible to use coatings including two or more metal layers to create an angular dependence qualitatively different from that in a coating with a single metal layer. Figure 11.22 illustrates this for the case of a five-layer configuration with two 12-nm-thick continuous Ag layers and three SiO_2 layers [29, 213]. When the SiO_2 thickness is 170 nm, T_{sol} goes from ~23% at normal incidence to as much as ~58% at 60° angle of incidence. When the SiO_2 thickness is 120 nm there is a monotonic decrease of T_{sol} with increasing angle, much as expected in

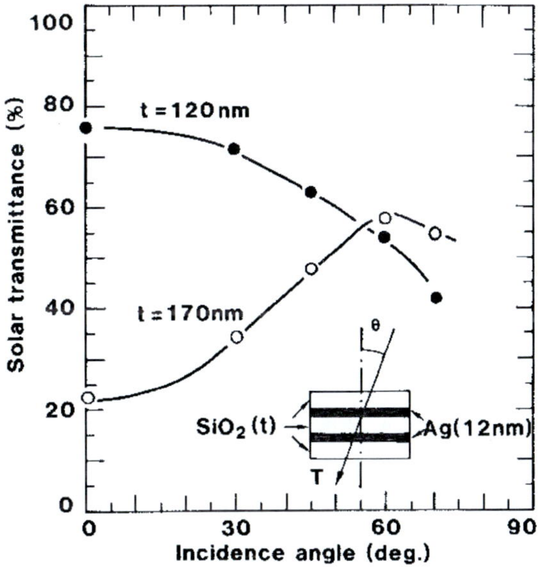

Fig. 11.22 Solar transmittance for thin films of SiO_2/Ag/SiO_2/Ag/SiO_2 as a function of SiO_2 thickness t and incidence angle θ. From Granqvist [29]

a coating with a single Ag layer. Clearly the metallic films render the coatings electrically conducting, i.e., they are TCs.

The coating type illustrated in Fig. 11.22 cannot be used universally. Considering, for example, a window inclined at 45°; one would like to have high transmittance at 45° on one side of the window's normal but low transmittance at 45° on the other side of the normal. This cannot be obtained with homogeneous thin films, but it is possible if the metal film has a nanostructure with directional selectivity, as considered next. In general, coatings of this type can exhibit different electrical properties in different directions [214]. The films are expected to show angular dependent values of E_{therm}, but this property does not seem to have been the subject of studies reported so far in the scientific literature.

Generally speaking, thin films can have a well defined nanostructure caused by the deposition conditions. In particular, oblique angle deposition using a directional flux of material can yield a columnar structure inclined towards the direction of this flux but not coinciding with it. The technique referred to above has been called "glancing angle deposition" (or "GLAD"). The film geometry has been described by an empirical "tangent rule", specifying that $\tan \beta = 1/2 \tan \alpha$, where β is the off-normal angle for the column orientation and α is the off-normal angle for the incident flux [215]; this "rule" is not generally obeyed, though.

The observation of columns is far from new [216], but it has been used in recent years – sometimes together with substrate rotation – to create thin films with a perplexing set of three-dimensional nanostructures, such as chevron-type or helical ones, with unusual optical, electrical, magnetic, mechanical, etc., properties; "sculptured films" is a term sometimes used. Essentially the limited mobility of the condensed atoms produces a "self-shadowing" effect which manifests itself in columnar film growth. Figure 11.23 illustrates this effect schematically from simulations of random depositions of two-dimensional hard discs ("atoms") onto a smooth substrate at the angle α [217]. The effect of different adatom mobilities was included in the simulation. Clearly the "film" develops inclined nanostructures with more or less "crystallinity" depending on the adatom mobility (i.e., "temperature"). Simulations for a large number of other situations have been carried out as well [7].

Another recently investigated method for making thin films with oblique columnar structures starts with granular films made by codeposition of metallic and dielectric materials. In particular, interest has been directed to Au-SiO_2 films. Heavy ion irradiation, with impingement from an off-normal direction, was able to change the shapes of the incipient gold particulates so that they became elongated in the direction of the ion beam [218].

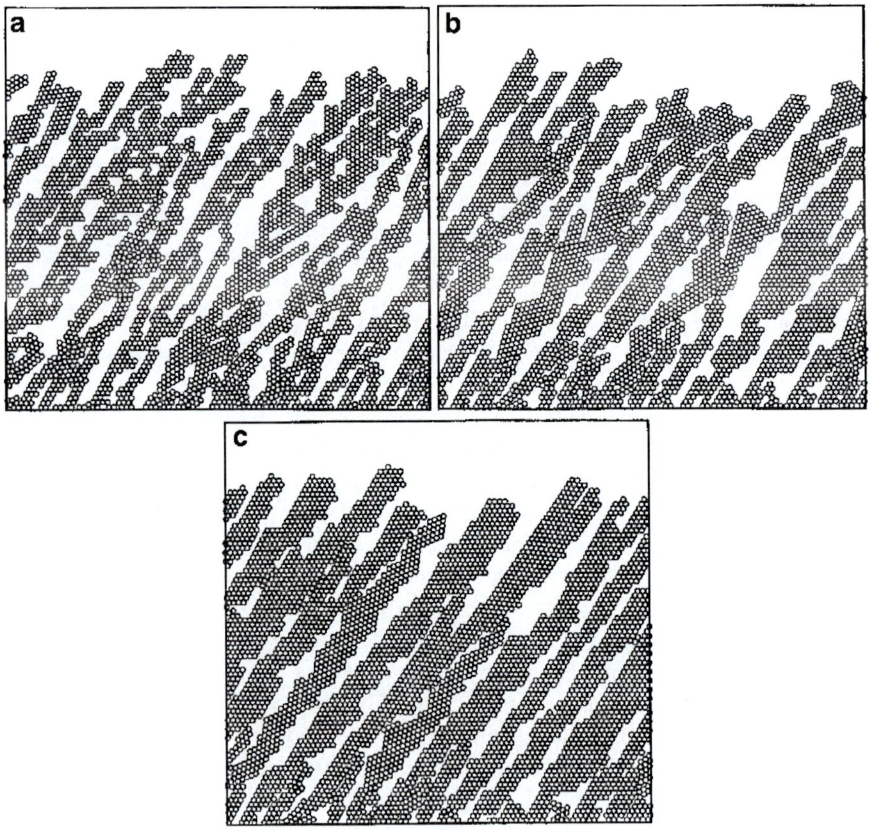

Fig. 11.23 Simulated deposits grown at an angle of incidence $\alpha = 50°$ and at low (**a**), medium (**b**), and high (**c**) adatom surface mobility. From Brett [217]

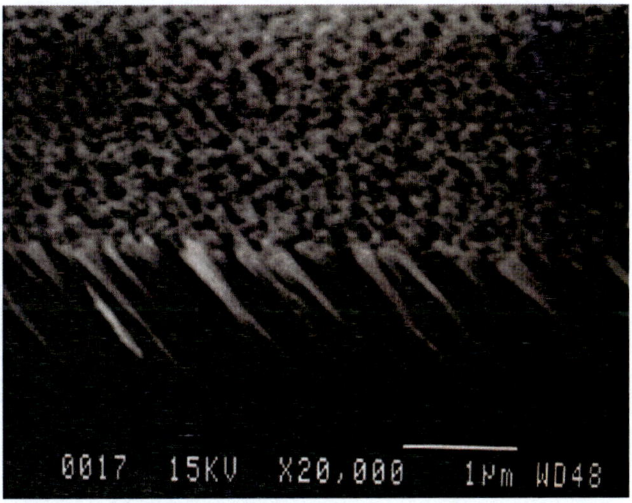

Fig. 11.24 Scanning electron micrograph of a Cr based film produced with oblique angle evaporation at $\alpha = 75°$. From Mbise et al. [221]

11.4.2 Data on Obliquely Deposited Cr Based Films

Films with oblique columnar nanostructures can show angular selectivity and be of interest for energy efficiency [219, 220]. Particularly detailed data are available for Cr based films made by oblique angle deposition [221–224]. Figure 11.24 shows a scanning electron micrograph of the cross section of a film prepared by evaporation with $\alpha = 75°$. A pronounced columnar structure is clearly seen with $\beta = 40 \pm 5°$; it is principle agreement with the simulation in Fig. 11.23.

Measurements of polarization dependent optical transmittance, $T_s(\theta, \lambda)$ and $T_p(\theta, \lambda)$, were performed on such films with the light beam incident at an angle θ to the film's normal and in the incidence plane (spanned by this normal, the direction of the evaporated flux, and the column direction). Data were taken for $-64° < \theta < +64°$, with positive angles taken to be on the same side of the normal as the deposition direction and the column orientation. It is necessary to consider s and p polarized light separately; the transmittance pertinent to unpolarized light is then the arithmetic mean of the two values. Figure 11.25 shows data for $T_s(0, \lambda)$, $T_s(\pm 55°, \lambda)$, $T_p(0, \lambda)$, and $T_p(\pm 55°, \lambda)$ for a film with $t = 45$ nm evaporated at $\alpha = 85°$. The curves display pronounced angular selectivity with the transmittance increasing monotonically with wavelength in the displayed solar range. It is found that $T_s(-55°, \lambda) = T_s(+55°, \lambda) < T_s(0, \lambda)$ and $T_p(-55°, \lambda) < T_p(0, \lambda) < T_s(+55°, \lambda)$, i.e., selectivity is confined to the p polarized component, as expected from theory [219]. Figure 11.26 illustrates $T_{s,\text{lum}}(\theta)$, $T_{p,\text{lum}}(\theta)$, and $T_{u,\text{lum}}(\theta)$, where u denotes unpolarized light, at $-64° < \theta < +64°$ for the same film as in Fig. 11.25. Expectedly, $T_{s,\text{lum}}(\theta)$ is symmetrical around $\theta = 0$, whereas $T_{p,\text{lum}}(\theta)$ is strongly angular selective. $T_{u,\text{lum}}(\theta)$ varies from $\sim$0.3 to $\sim$0.7 when θ goes from $-64°$ to 40°.

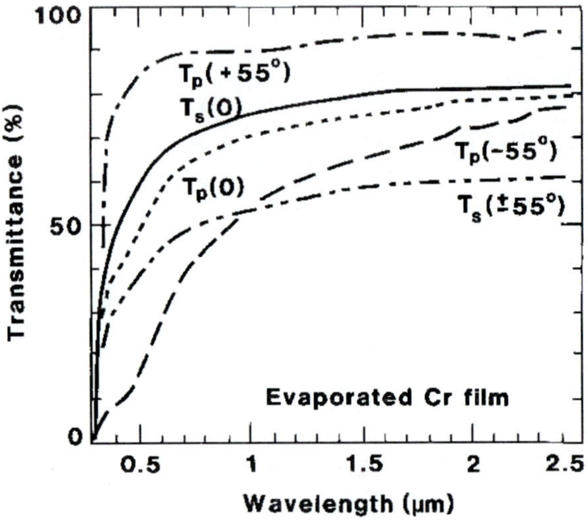

Fig. 11.25 Spectral transmittance T for a 45 nm thick Cr film made by oblique angle evaporation at $\alpha = 85°$. The *curves* refer to different incidence angles and states of polarization for the light. From Mbise et al. [221]

Fig. 11.26 Luminous transmittance as a function of incidence angle for the same Cr film as in Fig. 11.25. Data are given for s polarized, p polarized, and unpolarized light. From Mbise et al. [221]

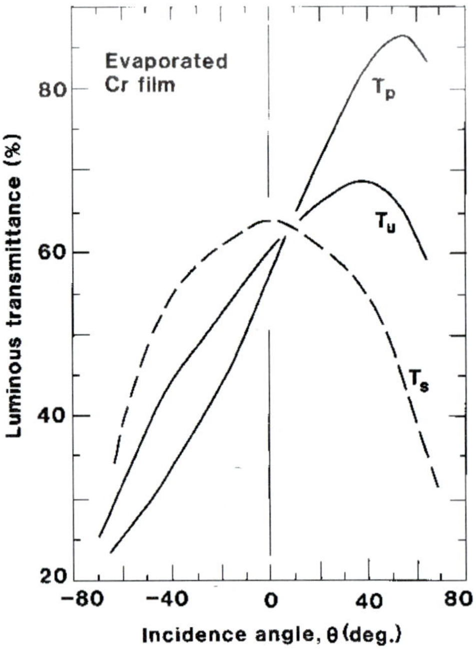

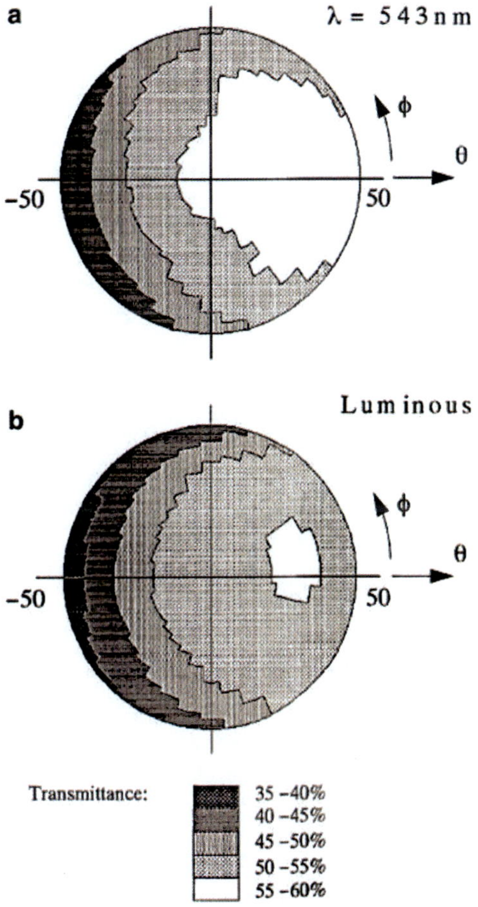

Fig. 11.27 Polar plot of fixed-wavelength and luminous transmittance through the same Cr film as in Fig. 11.25. Data are given as a function of incidence angle (θ) and azimuthal angle (φ). From Palmer et al. [224]

The maximum transmittance occurs at an angle θ_{max}. It was found that $\theta_{max} \approx \beta$, implying that the column orientation coincides with the direction for maximum transmittance.

Data were recorded not only with regard to the incidence plane but also for other directions. A polar representation of $T(\theta, \varphi, \lambda)$, where φ is the azimuthal angle, is then appropriate; $\varphi = 0$ and $\varphi = 180°$ correspond to the incidence plane. Measurements were taken with a laser giving light with $\lambda = 543$ nm and with a photodetector matched to the eye's sensitivity. Figure 11.27 shows data for the same film as above. The transmittance is illustrated by way of a grey scale with five 5%-wide intervals. The asymmetry between the left-hand and right-hand halves of the diagram provides striking evidence of angular selectivity for monochromatic as well as for luminous light, whereas the good agreement between the upper and lower halves of the diagrams indicates the symmetry expected from the geometrical features of the obliquely evaporated films. The lines representing $\varphi = 0$ and $\varphi = 180°$ can be directly compared with data in Fig. 11.26.

11.5 Applications Based on Temporal Variability (Chromogenics)

The variability of the weather conditions during the day and season makes it interesting to create energy efficiency by use of materials whose properties can be changed depending on external conditions, as discussed in Sect. 11.2.2, and windows and glass façades with variable properties have been the architects' dream for years [35, 225]. Such functionality is now becoming possible by exploiting chromogenic materials [11, 226], some of which can be classified as TCs. The exposition below gives an overview over chromogenics in Sect. 11.5.1 and then provides in-depth analyses of thermochromic VO_2-based films in Sect. 11.5.2 and on electrochromic materials and devices in Sect. 11.5.3, respectively.

11.5.1 Brief Survey of Chromogenic Materials and Devices

There are several categories of chromogenic materials. The most widely known are probably *photochromic* glasses [227] and polymers [228, 229] which have existed for decades; they darken under UV irradiation and bleach in the dark. Photochromism has been studied recently in thin films of several different oxides [7].

Another class of materials comprises *thermochromic* ones; their optical properties change reversibly when the temperature is altered [230, 231]. A number of transition metal oxides show strong thermochromism owing to metal-insulator transitions taking place at a well defined "critical" temperature τ_c. Materials based on VO_2 are of particular interest since τ_c lies in reasonable vicinity of a comfort temperature; thin films comprising this oxide are discussed in Sect. 11.5.2 below.

A problem with the photochromic and thermochromic materials with regard to many applications is that they do not allow operator control. This feature, however, is found among the *electrochromic* (EC) materials [232–235]. An EC device includes TCs for inserting or extracting charge in the EC materials whose optical properties thereby are changed. The device hence can be viewed as a variable-transmittance electrical conductor and infrared reflector. Considering electro-optical effects in general, there is a plethora of options that can be used more or less successfully in devices. A consideration of these technologies with regard to energy related applications must include aspects such as dynamical range in the modulation of T_{lum} and T_{sol}, the speed of the modulation in window-size apertures, open circuit memory (i.e., whether electrical power is needed to maintain a colored or bleached state), long term durability under operating conditions, transparency level in failure mode, device manufacturability, potential cost for mass fabrication, etc. Such considerations lead to a focus on inorganic EC devices incorporating oxide layers serving as anode and cathode in the same sense as in an electrical battery; they are discussed in detail in Sect. 11.5.3. EC devices can be made also with organic layers, rather than oxide films, but the former are prone to show degradation under UV irradiation and hence are less likely candidates for solar energy related applications; the various options are outlined elsewhere [7].

Among the numerous alternative technologies capable of electrically modulating the optical properties, special mention should be made of solution- and gel-based EC redox systems enabling optical absorption as a current is drawn through them and transparency in the absence of a current; such materials are successfully used in automatically dimming rear view mirrors for automobiles since almost 20 years [236]. The working substance in these devices is believed to be 5,10-dihydro-5,10-dimethylphenazine monomer. Rapid technological progress, as well as the proprietary nature of much basic information regarding several technologies for light modulation, makes a detailed comparison of various techniques difficult.

The oxides that display electrochromism are also "gasochromic", meaning that if their gaseous ambience is changed between oxidizing and reducing the optical transmittance is altered. Changing the gaseous ambience of certain metal hydrides modulates their optical properties, which can shift between transparent and reflecting with an intermediate strongly absorbing state. In principle a modulation of R_{sol} may be superior to a modulation of A_{sol} since heating is less pronounced in a practical fenestration-related application. Much work has been done recently on reflectance modulation and associated effects in hydrides based on rare earths and transition metals [7]. Electrochemical insertion and withdrawal of hydrogen is capable of yielding effects analogous to those shown under gas exposure [237].

11.5.2 Thermochromic Vanadium Dioxide Films

11.5.2.1 Basics

Thermochromic VO_2 films have been discussed for window applications since the early 1980s [238–240] and have received resurgent interest recently [241–243] among other things owing to the possibilities to achieve some energy efficiency of windows and glass façades through comparatively simple coatings.

VO_2 undergoes a first order structural transformation from a semiconducting, non-magnetic, and relatively IR transparent state with monoclinic structure below τ_c to a metallic, paramagnetic, and IR reflecting state with tetragonal structure above τ_c. The underlying physics of the metal-insulator transition has been debated in a sizeable current literature [7]. In geometrical terms, the main difference between the two states in VO_2 is that the monoclinic structure exhibits vanadium atoms that dimerize and twist with regard to their positions in the tetragonal structure. This transition can be highly repeatable in a thin film, whereas single crystals of VO_2 tend to disintegrate during structural transformations. The change of structure can happen in the picoseconds to femtoseconds range, but in a practical window application it is limited by the heat capacity of the glass. A window with a VO_2 coating has a T_{sol} that drops when heated to above τ_c, i.e., the window is in principle capable of automatically controlling the inflow of radiant energy to maintain a predetermined indoor temperature. A detailed investigation of the optical properties was presented recently for VO_2 films above and below τ_c [244].

In order to be of practical interest, τ_c must be close to a comfort temperature of about 20°C. Crystals of VO_2 are characterized by $\tau_c \approx 68°C$, which clearly is too high. However, other transition metal oxides, including many vanadium-oxide-based ones, have τ_cs that are much further off from a comfort temperature and are therefore less likely candidates as starting points for developing thermochromics as a means to accomplish energy efficiency [245]. The existence of a very large number of vanadium-oxide-based phases [246] – sometimes referred to as Magnéli phases and denoted V_nO_{2n-1} – makes it necessary to exercise accurate process control for thin film fabrication.

Figure 11.28a shows a typical example of data for a VO_2 film above and below τ_c [247]. T_{lum} is undesirably low for most applications to fenestration, but superior data with $T_{lum} \approx T_{sol}$ up to about 45% have been recorded recently for epitaxial films grown on sapphire (0001) [248]. The near IR optical properties are strongly temperature dependent – a feature that is further accentuated in the thermal IR – while the luminous properties are rather independent of temperature. However a careful study revealed that for $t < 50$ nm the low-temperature phase exhibits lower T_{lum} than the high-temperature phase, whereas the situation is reversed at larger

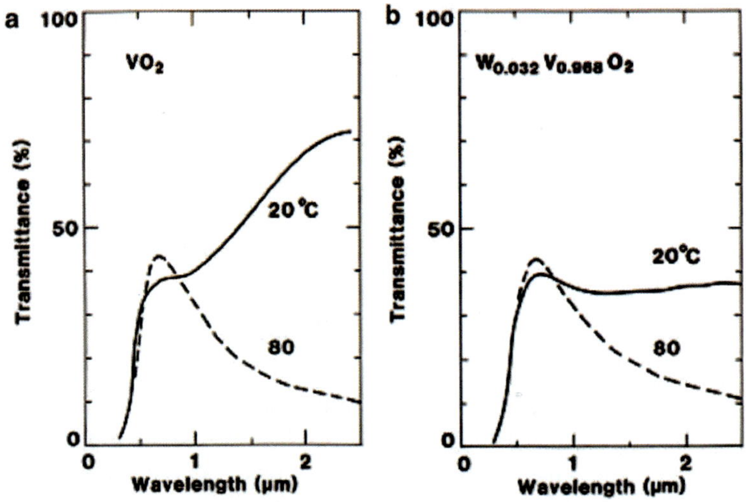

Fig. 11.28 Spectral transmittance of sputter deposited films of VO_2 (**a**) and $W_{0.032}V_{0.968}O_2$ (**b**) measured at the shown temperatures, i.e., below and above τ_c. From Sobhan et al. [247]

thicknesses [248]. The film in Fig. 11.28a was made by sputter deposition onto glass heated to ~400°C, but analogous results can be achieved with other thin film techniques involving similar substrate temperatures. Particularly facile techniques use thermal oxidation of metallic vanadium layers [249] or thermal reduction in vacuum of V_2O_5 [250]. The electrical conductivity is typically 10^4–10^5 Ω^{-1} m^{-1} well above τ_c independently of the details for making the film, but higher conductivities have been observed for epitaxially grown VO_2 layers [251]. The resistivity change is up to five orders of magnitude in bulk VO_2 but usually considerably smaller in thin films, and there is a direct correlation between the abruptness and amplitude of the transition and the crystallite size of the film [252, 253].

It should be clear from the discussion above that there are two main obstacles for practical uses of thermochromic films for energy efficient windows: (1) τ_c is too high and (2) T_{lum} is too low. Ways to alleviate these problems are considered next.

11.5.2.2 Towards Practical Applications in Energy Efficient Fenestration

It is possible to decrease τ_c in a VO_2 based material by replacing some of the vanadium by another metal. Tungsten addition to make films of $W_xV_{1-x}O_2$ has been investigated in several studies [238, 239, 241, 247, 254–258], and other dopants have been regarded too [7]. Figure 11.29 illustrates τ_c versus doping level x, showing that a few atomic percent of tungsten is all that is needed to reach the vicinity of a comfort temperature [247]. The relationship between τ_c and x is linear to within experimental uncertainties, which is consistent with literature data for bulk crystals [259, 260]. Figure 11.28b shows spectral transmittance data for a $W_xV_{1-x}O_2$ film with $x = 0.032$. The value of T_{lum} may be somewhat enhanced by the addition of tungsten, but the effect is not large. It was found recently that thin sputter deposited polycrystalline "buffer" layers of ZnO can improve the

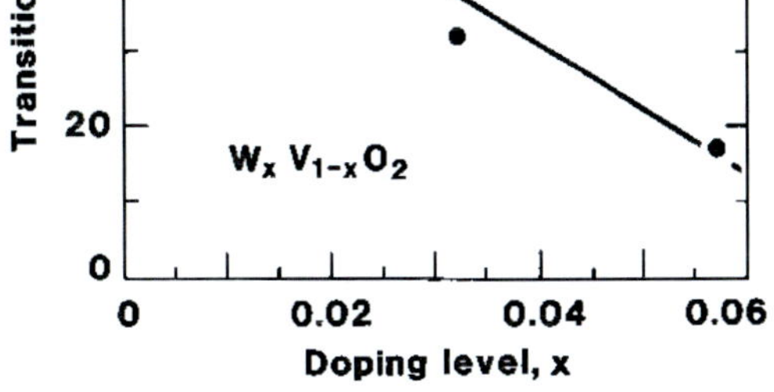

Fig. 11.29 Thermochromic transition temperature τ_c versus tungsten content in sputter deposited $W_xV_{1-x}O_2$ films. The *line* was drawn as a fit to the data points. From Sobhan et al. [247]

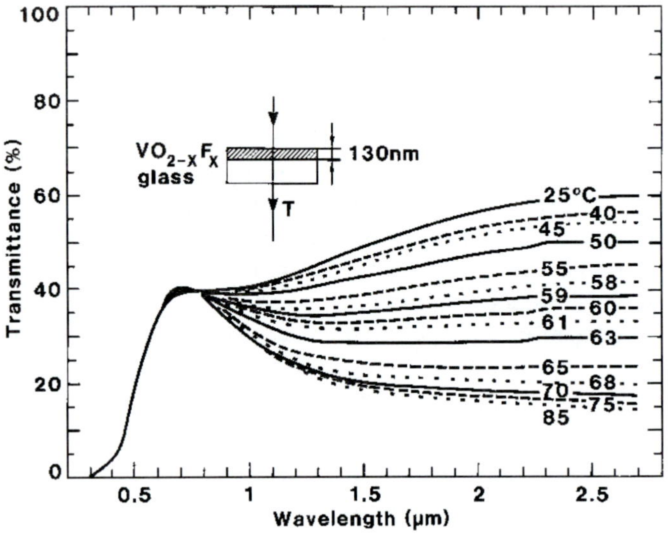

Fig. 11.30 Spectral transmittance for a sputter deposited film of VO_xF_y measured at the shown temperatures. From Khan et al. [265]

thermochromic performance of VO_2 and $W_xV_{1-x}O_2$ films [258]; adequately doped layers of this type may also contribute to give a low E_{therm} irrespectively of the temperature (*cf.* Sect. 11.3.2). VO_2 films on Si_3N_4-coated silica showed τ_cs as low as 34°C [261]. VO_2-SiO_2 composites made by different techniques have been reported to display enhanced T_{lum}s and broad metal-insulator transitions [262, 263]. VO_2 combines to some extent electrochromism and thermochromism, and τ_c can be modified by Li incorporation, though not to any larger degree [264].

Another way to influence τ_c is to replace some of the oxygen by fluorine [265, 266]. Figure 11.30 shows spectral transmittance data which are consistent with $T_{lum} \approx 28\%$ irrespective of temperature and T_{sol} ranging from 35% at 25°C to 28% at 70°C. The thermochromic transition is rather smooth around $\tau_c \approx 52$°C. The fluorination definitely enhances T_{lum} as apparent from Fig. 11.31 showing data for films of different thicknesses with and without fluorine. The effect is on the scale of several percent which is a clear improvement although T_{lum} may still be undesirably low for many applications.

Still another method to influence τ_c and T_{lum} is via mechanical stress induced by a suitable substrate, by an overlayer, or by a combination of both [240, 267]. Specifically, data recorded on glass coated with 50 nm of VO_2 and 110 nm of SnO_2 yielded $\tau_c \approx 49$°C together with $T_{lum} \approx 45\%$ and $T_{sol} \approx 53\%$ at 25°C, and $T_{sol} \approx 46\%$ at 80°C; the substrate with only the VO_2 layer had $T_{lum} \approx 35\%$. VO_2 films deposited on TiO_2 (001) had their τ_c decreased to about room temperature [268]. Metallic buffer layers are able to suppress τ_c to some extent as well [243]. Clearly a top layer can also serve as AR treatment, and a layer of TiO_2 on top of VO_2 was shown capable of yielding $T_{lum} \approx 49\%$ [269] whereas an optimized ZrO_2

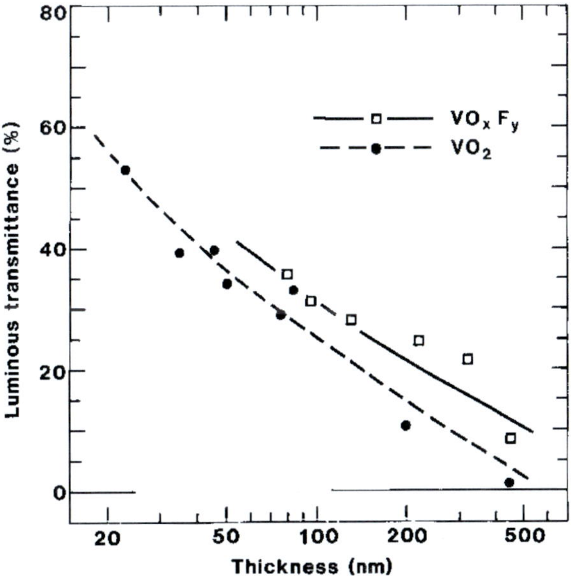

Fig. 11.31 Luminous transmittance versus thickness for sputter deposited films of VO_xF_y and VO_2. Curves were drawn for convenience. From Khan and Granqvist [266]

layer on VO_2 led to $T_{\text{lum}} \approx 50.5\%$ [270]. TiO_2 can also yield desirable multi-functionality and serve as a photocatalytic, self cleaning, and superhydrophilic layer, as discussed in Sect. 11.3.3.3.

11.5.3 Electrochromic Oxide Based Films and Devices

11.5.3.1 Introduction

Electrochromic materials were brought to public attention in the late 1960s and early 1970s through seminal work on WO_3 [271, 272]. Early applications were sought in information displays, but they did not stand up to the then rapidly developing liquid-crystal-based technology. The focus of much of the research on electrochromics was changed during the first part of the 1980s when it was realized that the technology may be capable of achieving energy efficiency in buildings [273]. The concept of a "smart window" was coined [274] and captured the attention of researchers as well as the general public. Considering the energy situation of the early 2000s, it is hardly surprising that "smart windows" technology experiences considerable market pull for applications in buildings and automobiles. EC materials and technology have been reviewed several times [232–235, 275].

EC technology is introduced in Fig. 11.32 which illustrates a standard device design and delineates basic features and operating principles [232]. The construction

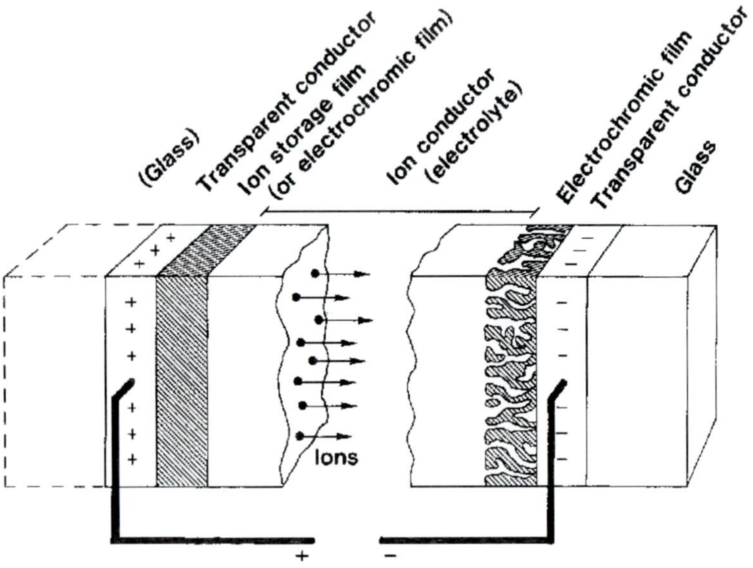

Fig. 11.32 Basic design of an electrochromic device, indicating transport of positive ions under the action of an electric field. From Granqvist [232]

comprises five superimposed layers on one substrate or between two substrates in a laminate configuration. The substrates are usually made of glass or polymer foil. The central part of the five-layer construction is a pure ion conductor (i.e., electrolyte) which can be organic (an adhesive polymer) or inorganic (often based on an oxide film). The ions should be small in order to be mobile; protons (H^+) or Li^+ are normally preferred, but Mg^{2+} has received some interest lately [276, 277]. Electrolytes based on PVB – i.e., a standard polymer for glass lamination – has been used in some recent work [278]. This ion conductor is in contact with an EC film (WO_3 being a typical example) capable of conducting electrons as well as ions. On the other side of the ion conductor is a film serving as ion storage, ideally with EC properties complementary to those of the first EC film (NiO being a typical example as further discussed below). This central three-layer structure is positioned between TCO films; the best material in terms of optical and electrical properties – or at least by far the most well-known one – is ITO but the other TCOs discussed in Sect. 11.3.2 can be used as well. SnO_2-based TCOs are well suited for window-size devices. Metal-based layers such as ZnS/Ag/ZnS [61, 62] and carbon nanotube layers [279] have been mentioned but not investigated in detail for EC applications.

When a voltage of about one volt is applied between the TCOs, ions are shuttled between the ion storage film and the EC film. The electrons extracted from or injected into the TCOs then alter the optical absorption. A reversal of the voltage, or short-circuiting, brings back the original properties. The coloration can be stopped at any intermediate level and consequently the device exhibits open-circuit memory. This means that the optical changes occur only when charge is moved, which is an important feature for an energy saving device. The voltage level is convenient for powering via photovoltaics [36]. Integrated devices with electrochromic WO_3 films and TiO_2-based dye sensitized solar cells (*cf.* Sect. 11.3.3.4) can provide "self powering" [280]. The EC devices are non-scattering, i.e., their light scattering level can remain below the one perceived by human eyes [281, 282]. If the film deposition is made with an oblique angle of the deposition flux, it is possible to create some angular selectivity [283]. A metallic pattern can be deposited to assist in the charge insertion and extraction. Additional layers can be used in order to diminish the reactivity between the electrolyte and the adjoining oxide layers [284, 285]. As pointed out above, WO_3 films have been used in the great majority of the device-oriented work, but recent research on Nb_2O_5-based devices has demonstrated multicolor capability [286]. It is clear that the EC device can be viewed as an electrical battery with its charging state manifested as optical absorption.

Recent work has been reported for EC oxides based W, Mo, Ir, Ti, V, Ni, and Nb. Some fifty to a hundred scientific and technical articles are currently published per year; those from 2002 and confined to thin films have been listed recently [7, 287]. The oxides can be divided into two main groups: those based on W, Mo, Ti, and Nb color under charge insertion and are called *cathodic* EC materials, whereas oxides of Ir and Ni color under charge extraction and are referred to as *anodic* EC materials; V oxide is of an intermediate nature and displays features of weak cathodic and anodic coloration in different wavelength regions. Among inorganic non-oxides, one notes that hexacyanometallates – i.e., compounds of the general type $M_k[M'(CN)_6]_l$, where M and M′ are transition metal ions with different valencies – can display pronounced anodic electrochromism [288].

A clear impression from the survey of the most recent literature referred to above is that much interest is currently focused on electrochromic WO_3 and NiO, and that sputtering and sol-gel deposition stand out as particularly relevant deposition techniques as a result of their capability of yielding mixed oxide films. These materials will be considered in some more detail next.

11.5.3.2 Films of Tungsten Oxide and Nickel Oxide

Electrochromic WO_3 is normally termed "amorphous", meaning that no clear features are discerned in X-ray diffraction. On an atomic scale, the structure comprises corner-sharing WO_6 octahedra with a spread in bond angles and bond lengths [232]. The bonding is predominantly ionic, though with a significant covalent contribution [289, 290]. The structural features regarding stoichiometric, substoichiometric, overstoichiometric, and ion intercalated amorphous WO_3 have been given a schematic though consistent interpretation recently [291]. Upon heating, the films transform in the sequence monoclinic → orthorhombic → hexagonal [232]. Electron-phonon interaction is strong so that structural rearrangement occurs easily, and doubly charged $(W\text{–}W)^{10+}$ complexes form [292] in accordance with the Anderson electron pairing mechanism. The valence band is largely composed of O 2*p* orbitals, while the conduction band derives mainly from

W 5*d* orbitals. The electronic structure was studied recently using chronopotentiometry and taking a rigid-band model to represent the effect of ion intercalation [293, 294]. Combining the chronoamperometry with electrical conductivity data it became apparent that the electronic states are localized far up in the conduction band [287, 295]. When ions are intercalated into the tungsten oxide, the charge compensating electrons enter these localized states. Hence the optical absorption underlying the EC effect would be best described as due to transitions between occupied and empty localized conduction band states. Concerning the specific optical absorption mechanism, polycrystalline tungsten oxide can be described in terms of polaron absorption [296, 297], which is consistent with recent temperature-dependent angle-resolved photoemission data for Na_xWO_3 [298]. However, the situation is less clear for the amorphous material, and a traditional polaronic model has been questioned in recent quantum chemical modeling [299]. The fact that optical absorption does not increase linearly with intercalation level has led to the notion of "site saturation" [300], whose extension to the case of transitions between three kinds of states – corresponding to W^{4+}, W^{5+}, and W^{6+} – has provided a new model and new insights into EC absorption in amorphous tungsten oxides up to high intercalation levels [287, 301]. A reflecting state is possible in sufficiently crystalline ion containing tungsten oxide; it can be represented by free electrons in the conduction band [302, 303].

Electrochromism in NiO is less well understood than in WO_3. NiO is an insulator with an onset of absorption at 3.1 eV and an absorption peak at 4.3 eV, which is consistent with recent first-principles calculations of the quasiparticle excitation spectrum [304]. Oxygen deficiency can lead to in-gap states about 0.5 eV below the Fermi level [305]. Generally speaking, however, there is no consensus regarding the detailed electronic structure of NiO despite decades of investigations. The knowledge of electrochromism in NiO films has advanced considerably recently particularly for sputter deposited films [306–309]. In overall terms, as-deposited films are characterized by oxygen excess, low density (high porosity), and small grains. The grain surface is richer in oxygen than the bulk. Independently of the coloration, there are always nickel atoms in valence states 2+ and 3+, but the number of Ni^{3+} ions increases upon coloration. A NiO structure was found by X-ray diffraction for films in all color conditions, while a presence of other nickel-based phases – such as $Ni(OH)_2$, Ni_2O_3, and NiOOH – was inferred from other data. This indicates that the coloration process is a surface phenomenon, most likely occurring in the outer parts of the grains. In the beginning of an electrochemical cycling sequence, a number of cycles are required for the charge capacity to reach its maximum value and stabilize. X-ray photoemission results were consistent with a transformation of the over-stoichiometric hydrated Ni oxide to hydroxide and oxy-hydroxide phases on the grain surfaces. The amount of oxygen in the films increases by $\sim$10% through intercalation of hydroxyl groups occurring simultaneously with intercalation of protons from the electrolyte. Electrochemical data [307] indicated crystallization of the hydroxide and oxy-hydroxide phases during the initial intercalation cycles. The electrochemical coloration occurs solely by proton intercalation in stabilized films, with the oxygen content in the films being equal under bleached and colored conditions. Proton extraction was consistent with the Bode reaction scheme [310, 311] originally developed for batteries. However, the Bode model does not explain the 2+ to 3+ valence change for nickel belonging to the NiO phase upon coloration, and an extension of the scheme was suggested to account for these results [309]. Because the coloration is a surface effect and the bulk remains NiO independently of the film being bleached or colored, one can argue that – after the transformation of the $Ni(OH)_2$ to NiOOH – the change of the NiO to Ni_2O_3 may take place at the interface between the NiO and the Ni oxy-hydroxide.

11.5.3.3 Flexible Electrochromic Foils

After having examined the basic properties of WO_3 and NiO films above, it is appropriate to turn to devices incorporating such films. Rigid, glass-based devices of this kind have been reported on in a number of studies [312–318] and recently work has been carried out also on flexible PET-based devices [319–322]. Some results from PET-based devices will be briefly discussed here. The main reason for using a flexible substrate is that roll-to-roll coating can be employed – rather than coating of large rigid substrates – which is a cost-cutting measure. Electrochromic WO_3 absorbs mainly in the red end of the visible spectrum whereas NiO absorbs mainly in the blue, and jointly they tend to give a neutral gray color in the dark state, a feature that is advantageous for architectural applications [35]. Sputter deposited NiO films display some unwanted absorption at short wavelengths, but this feature can be avoided by adding a second oxide to the nickel oxide. If the additive has a larger band gap than nickel oxide, such as MgO and Al_2O_3, the result will be a widened band gap for the mixed oxide and a lowered value of A_{lum} [306].

Successful device assembly requires that the EC film and counter electrode are appropriately precharged, which is a demanding task for manufacturing. However tungsten oxide can be efficiently precharged in H_2 during sputter deposition [323] and nickel oxide can be discharged by post treatment in ozone [324, 325]. Such foils were then laminated together by using a UV-transparent polymer electrolyte and roll-pressing. The edges of the double foil were sealed, electrical contacts were attached, and the EC device was ready for testing and use. Cycling between colored and bleached states took place using trapezoidal voltage pulses between -1.6 and $+1.5$ V, respectively; the precise voltage range was adjusted automatically to compensate for varying temperatures.

Figure 11.33 shows the transmittance at three different wavelengths for cycling with one full color-bleach cycle each 200 s [326]. The Ni oxide contained some V which had the effect of facilitating the film production by magnetron sputtering. The optical modulation is pronounced, especially for mid-luminous and red light. Most of the changes take place within a few tens of seconds after application of the pulse, but the coloration has not reached saturation even after several minutes. Figure 11.34 reports on the transmittance in fully colored and bleached states after cycling in the same way as in Fig. 11.33 [319]. The foils are seen to change between

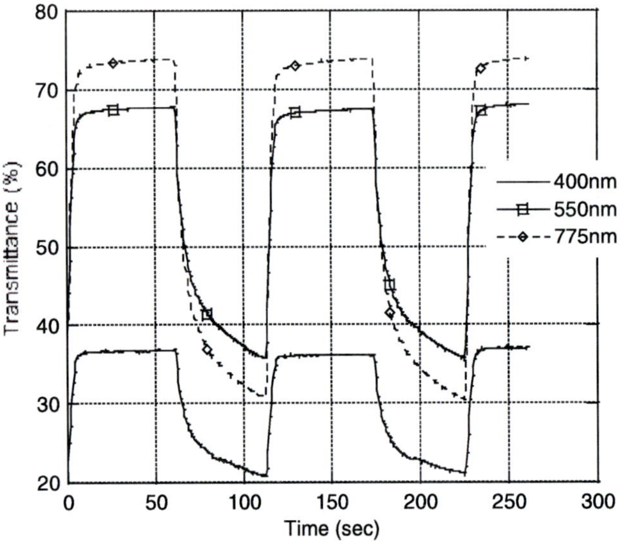

Fig. 11.33 Transmittance at three wavelengths versus time for electrochromic switching of a laminated device based on tungsten oxide and nickel oxide. From Azens [326]

Fig. 11.34 Transmittance at a wavelength of 550 nm versus number of color/bleach cycles for a laminated electrochromic foil device, based on tungsten oxide and nickel oxide, in colored and bleached states. From Azens et al. [319]

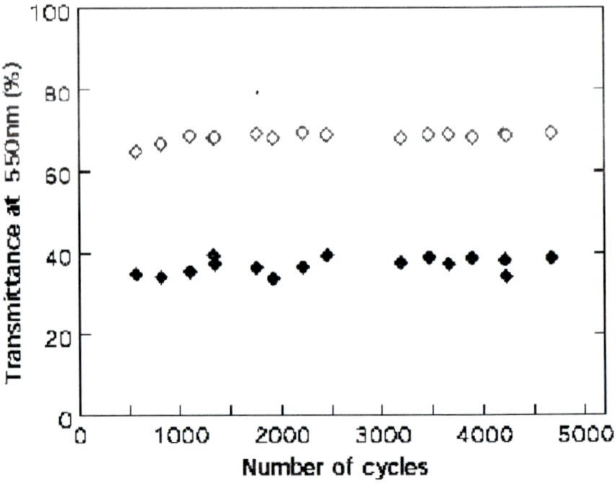

a high level of ~70% and a low level of 35–40% at a mid-luminous wavelength, which is in agreement with the data in Fig. 11.33. The span between colored and bleached states remains practically unchanged for 5,000 cycles. This by no means represents the maximum achievable number of cycles, though. Lower transmittance levels, down to 25% or less in the colored state, could be reached for longer coloration times. The devices have open-circuit memory, which is an asset since electrical power must be drawn only to effect changes in the optical properties.

The mid-luminous transmittance at for a device colored to a transmittance of ~20% increased only marginally after a time as long as 160 h [319], and hence there is no need for "refresh" charging to keep the EC foil dark during extended periods of time. This is a considerable advantage particularly for applications to energy savings in buildings and to vehicles.

Quality measures for EC devices are of obvious interest. Long-term testing is an option, ideally under some type of accelerated yet reliable conditions [327, 328], but there is an obvious need for other, faster techniques. One possibility may be offered by low-frequency current noise [329], which was investigated recently in a device of the type discussed above [330]. Figure 11.35 shows experimental current power spectral density S_i as a function of current I at a frequency f of 700 Hz during the final stage of discharging of a laminated EC device. It is found that $S_i(f) \sim I^2$. The most interesting feature of the data is that amplitude of S_i increased after cycling of the EC device so as to deliberately cause some degradation leading to irreversible coloration. The result is hardly surprising since current noise of $1/f$ type is a well known tool for providing information on corrosion processes [331, 332].

The work reported above demonstrated EC devices – in particular flexible EC foils – with sufficient optical modulation range, dynamics, and durability for making them interesting with regard to a number of energy-related and other applications, especially "smart" windows for uses in buildings and automobiles. Consumer products, such as visors for motorcycle helmets, may be another option

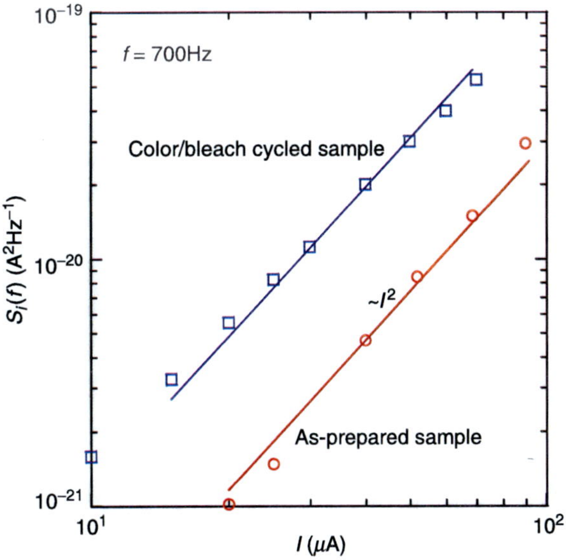

Fig. 11.35 Current noise power spectra, denoted $S_i(f)$ where $f = 700$ Hz, versus current I during the final stage of discharging of an electrochromic foil based on tungsten oxide and nickel oxide. Data are shown before and after harsh color/bleach cycling devised to give device degradation. The lines denote I^2 dependence. From Smulko et al. [330]

and the EC device then leads to diminished facial warming, which is an important issue with regard to comfort and hence riding safety [333].

The discussion of EC devices dealt with absorptance changes in order to modulate T_{lum} and T_{sol}. In principle a modulation of E_{therm} is possible, but this requires either that a crystalline WO_3 film is exposed to the ambience or that the layers in the device between an IR reflecting substrate and the ambience are IR transparent. This can be accomplished to some extent [334–337], but the relevance to solar energy and energy savings is uncertain.

11.6 Concluding Remarks and Futures Issues

11.6.1 Materials

We first consider whether there are new and promising TCs for energy related applications. One such possibility, it seems, is offered by carbon nanotube layers, especially single-walled ones consisting of one layer of the hexagonal graphite lattice rolled to form a seamless cylinder with a radius up to a few nanometers [338]. TCs comprise networks of "bundled" nanotubes; a typical example is shown in Fig. 11.36 [339]. Individual nanotubes can be as long as several centimeters. The nanotubes are of two types: metallic and semiconducting with conductivities comparable to those of copper and silicon, respectively. The individual nanotubes are separated by barriers leading to fluctuation-assisted electron transfer from one tube to the next [340, 341]. The optical transparency is high in the visible and infrared. Figure 11.37 displays sheet resistance as a function of transmittance at $\lambda = 0.55$ μm [339]; the curve represents a theoretical expression [339]. Quantitatively similar data have been reported in recent work on organic light-emitting

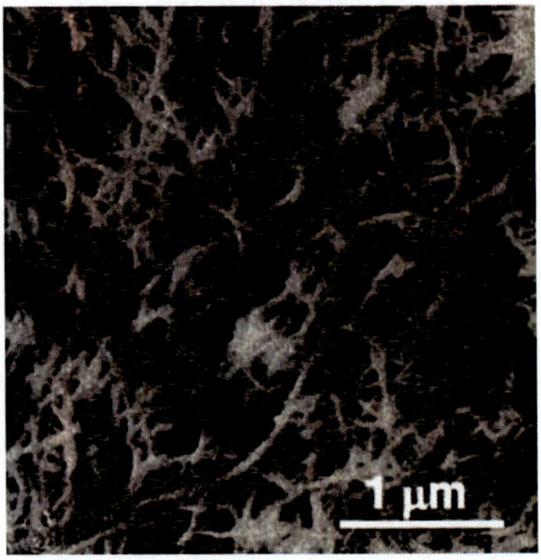

Fig. 11.36 Transmission electron micrograph of a film comprised of carbon nanotubes. From Zhou et al. [339]

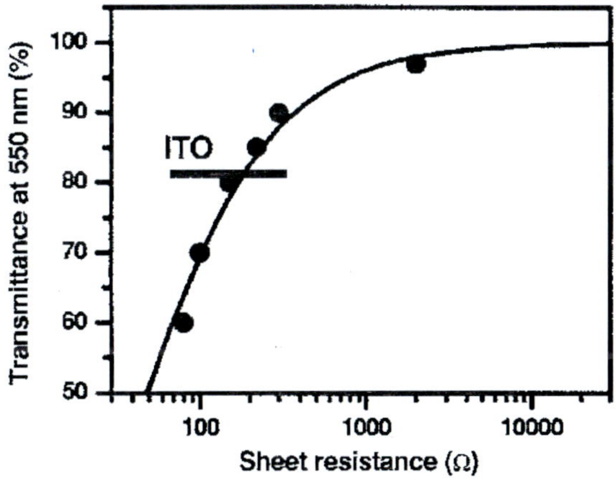

Fig. 11.37 Transmittance at a mid-luminous wavelength versus sheet resistance for carbon nanotube films as measured (*dots*) and calculated from theory (*curve*). The *horizontal bar* denotes typical data for ITO films. After Zhou et al. [339]

diodes with carbon nanotube electrodes [342]. It is apparent that the properties can not yet fully match those of the best TCOs, but advantages regarding cost can outweigh this depending on the application. In the future, nanotube layers with a higher fraction of the metallic variety, and perhaps means to diminish the intertube barriers, can lead to better performance. A few examples of uses of carbon nanotube layers in devices with a bearing of energy applications were noted above; they concern TCs for organic solar cells [204–208], for dye-sensitized solar cells [208], and for organic EC devices [279]. Chemical functionalization of carbon nanotubes is presently of much concern and may give new avenues to process and assemble tubes and to adjust their electrical and optical properties [343, 344]. Recently it has been discovered that cycloaddition functionalizations can preserve or control the conductivity of carbon nanotubes [345].

A newly discovered class of TCOs is based on $12CaO{\cdot}7Al_2O_3$ (or C12A7), i.e., a well known insulating oxide of much use in high-alumina cements [346]. It is characterized by a cage structure with two formula units (12 cages) per unit cell. Its empirical formula may be written $[Ca_{24}Al_{28}O_{64}]^{4+} + 2O^{2-}$; the free oxygen ions give charge neutrality and are located inside the cages. Hydrogen can be incorporated by a reaction that can be written O^{2-}(cage) + H_2(atm) → OH^- (cage) + H^-(cage). Following hydrogen incorporation the unit cell has two cages occupied by OH^-, two occupied by H^-, and eight empty ones. The hydrogen produces no apparent change in the optical and electrical properties of the material, but subsequent UV irradiation causes two optical absorption bands. The color is persistently changed from white to green and the conductivity is increased by as much as ten orders of magnitude by a process involving $H^0 \rightarrow H^+ + e^-$ [347]; the material can be described as an electride. Films of this material were made by pulsed laser deposition onto MgO followed by annealing at 800°C and hydrogen treatment at 1,200°C [347]. The electromagnetic properties have been subject to a number of detailed investigations, and it has been documented that the optical absorption is due to inter-cage and intra-cage electron transitions [348] and that electrical conductivity is due to variable range or polaron hopping of photoexcited electrons between neighboring cages [349–351]. The C12A7 type TCOs are very interesting for a variety of reasons: They are based solely on some of the most abundant elements and offer a new method to accomplish optical transmittance and electrical conductivity. Thus far the properties are not sufficiently good to permit applications to solar energy and energy efficiency, but the fundamental notions behind the electromagnetic properties of the electrides conceivably can be developed to more useful materials. Thin film manufacturing presently involves high temperature treatment, and alternative fabrication routes are needed.

11.6.2 Applications

Transparent conductors – given a generous definition – have a multitude of applications to solar energy and energy efficiency, as discussed above. These applications are in very different stages of development, ranging from established technology for example with regard to spectrally selective solar-control and low-emittance windows in buildings, to an early applications phase such as for some chromogenic materials allowing temporal variability of the transmittance, to an explorative phase such as for angular selective nanostructured materials.

Making rational use of energy in the built environment has a large potential for "passive" solar energy applications, thereby yielding large savings of fossil and nuclear fuels. In industrialized countries, typically 40% of the energy goes into heating, cooling, lighting, and ventilation of buildings, and to electrical appliances of various kinds, as noted in Sect. 11.2.2. It is obvious that improved materials could be used to alleviate the energy crunch, and coatings on low-cost substrates, such as glass and PET foil, are of particular importance. Today's coatings, which indeed are expected to remain with us for the forseeable future, are either based on ~0.02 μm thick silver or ~0.2 μm thick doped tin oxide. TC coatings on the exposed outer side of windows can almost eliminate water condensation which may otherwise occur under certain climatic conditions; this represents a new field of application which is compatible with self-cleaning induced by a very thin TiO_2-based outer layer. TC coatings may also be used on the inside of windows for buildings and vehicles, where they can serve as reflectors for thermal radiation and avoid what is sometimes referred to as "cold radiation", i.e., the coatings can reflect thermal radiation from a human body back to the same body.

The coatings mentioned above combine spectral selectivity with angular dependent properties but are usually only optimized with regard to light with normal incidence. Hence multilayer designs with angular performance tailored to a specific building site and orientation represent options for the future. Angular selectivity remains a possibility to be further explored.

Windows with adjustable optical properties have long been something of a Holy Grail in "high-tech" architecture. These windows can be developed using many different approaches involving photochromic, thermochromic, electrochromic, and gasochromic materials. Electrically regulated EC "smart windows" are, in the author's opinion, particularly interesting in view of their potential to give user-related operation; they can be considered as transparent conductors and infrared reflectors with variable transparency. Such windows make it possible to combine increased indoor comfort for the occupant of the building (less glare and thermal stress) with large energy efficiency (especially lowered air conditioning load in cooled spaces). Therefore, there is a strong incentive for their introduction. Durability, optical switching speed, and size constraints constitute an interrelated problem complex which is not fully solved today. However consensus is growing that a satisfactory solution indeed can be achieved. Cost is another major concern, but recent progress in manufacturing technology – such as roll-to-roll coating of polymer foil – open new avenues towards inexpensive products. It should be realized that electrochromism is an enabling technology of relevance for a vast number of applications so there are multiple roads towards devices eventually used in buildings. Contemplating the combined blessings of comfort and energy savings, the recent advances in technology, and the undeniable business opportunities, it may not be far fetched to envisage a gradual introduction of EC "smart windows" on a large scale within a few decades, perhaps leading to such windows being the norm rather than exception in the same way that fenestration with "static" surface coated glass is today the norm for a number of markets.

The EC technology may be combined with facilities to direct daylight deeply into buildings by use of light-guides. By equalizing the light level in a room, the eye – which tends to adjust to the brightest illumination – does not perceive deeper regions as disturbingly dark, and hence there is less need for artificial lighting. Similar notions, though employing a Venetian blind system, have been discussed in the past [352]. Generally speaking, the EC technology leads to new vista in daylighting, which is generally regarded as superior to artificial lighting by giving better task performance, improved visual comfort, and positive mood effects, especially if glare problems are eliminated. Particularly beneficial effects have been observed concerning student performance in daylit schools [353] and sales increases in daylit stores [354].

Solar cells for electricity generation are attracting much attention today. The cost of solar electricity is still high but there is little doubt that solar electricity will play an increasing role in off-grid situations and as building components offering special and desired architectural expressions. TCs are of great importance for many solar cells, and cheaper materials and deposition processes will remain of importance.

Solar-driven photocatalysis for air cleaning is a relatively unexplored field, but one with interesting possibilities to provide benign indoor conditions without excessive and energy intensive air handling. The future of this technology is still uncertain, though. Combinations with TCs are interesting and point towards multifunctionality as an important feature for future developments.

A concern with regard to applications is the availability of the raw materials needed for the different types of TCO coatings. This concern has become acute owing to the rapid price increase of indium during the first part of the 2000s.

The situation does not seem to be entirely clear today (2007), and on one hand it has been stated that the global availability of indium is ample and about as large as that of silver [355], but it has also been claimed that the availability is much too low to allow widespread deployment of indium-containing thin film solar cells [356]. Whatever view will turn out to be correct, resource availability is definitely an important aspect for large-scale applications to transparent conductors – as it is for any other large-scale implementation of new technology.

References

1. Smalley R.E., "Future Global Energy Prosperity: The Terawatt Challenge", MRS Bull. 30 (2005) 412–417.
2. Mc Michael A.J., Campbell-Lendrum D., Wilkinson P., Nicholls R., Hales S., Tanser F., Le Sueur D., Schlesinger M., and Andronova N., "Global Climate Change", in "Comparative Quantification of Health Risks: Global and Regional Burden of Disease due to Selected Major Risk Factors", edited by Ezzati M., Lopez A.D., Rodgers A., and Murray C.J.L. (World Health Organization, Geneva, Switzerland, 2004), Ch. 20, pp. 1543–1649.
3. Palz J.A., Campbell-Lendrum D., Holloway T., and Foley J.A., "Impact of Regional Climate Change on Human Health", Nature 438 (2005) 310–317.
4. Mitchell J.F.B., Lowe J., Wood R.A., and Vellinga M., "Extreme Events Due to Human-Induced Climate Change", Philos. Trans. Roy. Soc. A 364 (2006) 2117–2133.
5. Santamouris M., "Solar and Natural Resources for a Better Efficiency in the Built Environment", in "Solar Energy: The State of the Art", edited by Gordon J. (James & James, London, UK, 2001), Ch. 1, pp. 1–28.
6. Kolokotroni M., Zhang Y., and Watkins R., "The London Heat Island and Building Cooling Design", Solar Energy 81 (2007) 102–110.
7. Granqvist C.G., "Transparent Conductors as Solar Energy Materials: A Panoramic Review", Solar Energy Mater. Solar Cells 92 (2007) 1529–1598.
8. Granqvist C.G., "Solar Energy Materials", Adv. Mater. 15 (2003) 1789–1803.
9. Granqvist C.G., "Solar Energy Materials", in "Encyclopedia of Energy", editor-in-chief Cleveland C.J. (Elsevier, Oxford, UK, 2004), Vol. 3, pp. 845–858.
10. Granqvist C.G., "Solar Energy Materials", in "Kirk-Othmer Encyclopedia of Chemical Technology", 5th edition (Wiley, Hoboken, USA, 2006), Vol. 23, pp. 1–32.
11. Lampert C.M. and Granqvist C.G., editors, "Large-Area Chromogenics: Materials and Devices for Transmittance Control" (The International Society for Optical Engineering, Bellingham, USA, 1990), Vol. IS4.
12. Granqvist C.G., "Radiative Heating and Cooling with Spectrally Selective Surfaces", Appl. Opt. 20 (1981) 2606–2615.
13. Granqvist C.G., editor, "Materials Science for Solar Energy Conversion Systems" (Pergamon, Oxford, UK, 1991).
14. Fanger P.O., "Thermal Comfort: Analysis and Applications in Environmental Engineering" (Danish Technical Press, Copenhagen, Denmark, 1970).
15. Jelle B.P., Gustavsen A., Nilsen T.N., and Jacobsen T., "Solar Material Protection Factor (SMPF) and Solar Skin Protection Factor (SSPF) for Window Panes and Other Glass Structures in Buildings", Solar Energy Mater. Solar Cells 91 (2007) 342–354.
16. Azens A. and Granqvist C.G., "Electrochromic Smart Windows: Energy Efficiency and Device Aspects", J. Solid State Electrochem. 7 (2003) 64–68.
17. Roos A., Covalet D., Fanton X., Persson M.-L., Platzer W., Nielsen T.R., Wilson H.R., Zinzi M., Köhl M., Heck M., and Chevalier B., "Energy Performance of Switchable Glazing", in

"Proceedings of the International Conference on the Durability of Building Materials and Components" (Lyon, France, 2005), pp. 17–20.
18. Roos A., Persson M.-L., Platzer W., and Köhl M., "Energy Efficiency of Switchable Glazing in Office Buildings", in "Proceedings Glass Processing Days" (Tampere, Finland, 2005), pp. 566–569.
19. Lee E.S., DiBartolomeo D.L., and Selkowitz S.E., "Daylighting Control Performance of a Thin-Film Ceramic Electrochromic Window: Field Study Results", Energy Build. 38 (2006) 30–44.
20. Lee E.S., Selkowitz S.E., Clear R.D., DiBartolomeo D.L., Klems J.H., Fernandes L.L., Ward G.J., Inkarojrit V., and Yazdanian M., "Advancement of Electrochromic Windows" (California Energy Commission, PIER, 2006), CEC-500-2006-052, 87 pages.
21. Persson M.-L., "Windows of Opportunities: The Glated Area and Its Impact an the Energy Balanced of Buildings", PhD Thesis, Department of Engineering Sciences, Uppsala Universty , Uppsala, Sweden, 2006.
22. Clear R.D., Inkarojrit V., and Lee E.S., "Subject Responses to Electrochromic Windows", Energy Build. 38 (2006) 758–779.
23. Reinhart C.F., "Energy Efficient Solar Buildings", in "The Future for Renewable Energy 2: Prospects and Directions" (James & James, London, UK, 2002), Ch. 4, pp. 79–114.
24. Bojić M. and Yik F., "Cooling Energy Evaluation for High-Rise Residential Buildings in Hong Kong", Energy Build.37 (2005) 345–351.
25. Barnham K.W.J., Mazzer M., and Clive B., "Resolving the Energy Crisis: Nuclear or Photovoltaics?", Nat. Mater. 5 (2006) 161–164.
26. Jäger-Waldau A., editor, "REF-SYST Status Report 2004", EUR 21297 EN (JRC, Ispra, Italy, 2004).
27. Brockett D., Fridley D., Lin J.-M., and Jin J., "A Tale of Five Cities: The China Residential Energy Consumption Survey", in "Human and Social Dimensions of Energy Use: Understanding Markets and Demands" (ACEEE Summer Study on Building Energy Efficiency, 2002), pp. 8.29–8.40.
28. Darwich M.A., "Energy Efficient Air Conditioning: Case Study for Kuwait", Kuwait J. Sci. Eng. 32 (2005) 209–222.
29. Granqvist C.G., "Energy Efficient Windows: Present and Forthcoming Technology", in "Materials Science for Solar Energy Conversion Systems", edited by Granqvist C.G. (Pergamon, Oxford, UK, 1991), Ch. 5, pp. 106–167.
30. Smith G.B., Deller C.A., Swift P.D., Gentle A., Garrett P.D., and Fisher W.K., "Nanoparticle-Doped Polymer Foils for Use in Solar Control Glazing", J. Nanopart. Res. 4 (2002) 157–165.
31. Schelm S. and Smith G.B., "Dilute LaB_6 Nanoparticles in Polymer as Optimized Clear Solar Control Glazing", Appl. Phys. Lett. 82 (2003) 4346–4348.
32. Nostell P., Roos A., and Karlsson B., "Optical and Mechanical Properties of Sol-Gel Antireflective Films for Solar Energy Applications", Thin Solid Films 351 (1999) 170–175.
33. Janicki V., Gäbler D., Wilbrandt S., Leitel R., Stenzel O., Kaiser N., Lappschies M., Görtz B., Ristau D., Rickers C., and Vergöhl M., "Deposition and Spectral Performance of an Inhomogeneous Broadband Wide-Angular Antireflective Coating", Appl. Opt. 45 (2006) 7851–7857.
34. Tikhonravov A.V., Trubetskov M.K., Amotchkina T.V., Kokarev M.A., Kaiser N., Stenzel O., Wilbrandt S., and Gäbler D., "New Optimization Algorithm for the Synthesis of Rugate Optical Coatings", Appl. Opt. 45 (2006) 1515–1524.
35. Wigginton M., "Glass in Architecture" (Phaidon, London, UK, 1996).
36. Lampert C.M., "Large-Area Smart Glass and Integrated Photovoltaics", Solar Energy Mater. Solar Cells 76 (2003) 489–499.
37. Hollands K.G.T., Wright J.L., and Granqvist C.G., "Glazings and Coatings", in "Solar Energy: The State of the Art", edited by Gordon J. (James & James, London, UK, 2001), Ch. 2, pp. 29–107.

38. Manz H., Brunner S., and Wullschleger L., "Triple Vacuum Glazing: Heat Transfer and Basic Mechanical Design Constraints", Solar Energy 80 (2006) 1632–1642.
39. Ebbesen T.W., Lezec H.J., Ghaemi H.F., Thio T., and Wolff P.A., "Extraordinary Optical Transmission Through Sub-Wavelength Hole Arrays", Nature 391 (1998) 667–669.
40. Genet C. and Ebbesen T.W., "Light in Tiny Holes", Nature 445 (2007) 39–46.
41. Gao H., Henzie J., and Odom T.W., "Direct Evidence for Surface Plasmon-Mediated Enhanced Light Transmission through Metallic Nanohole Arrays", Nano Lett. 6 (2006) 2104–2108.
42. Lin L., Reeves R.J., and Blaikie R.J., "Surface-Plasmon-Enhanced Light Transmission through Planar Metallic Films", Phys. Rev. B 74 (2006) 155407-1–155407-6.
43. Ye Y.-H., Wang Z.-B., Yan D., and Zhang J.-Y., "Multiple Transmission Bands through Metal Films Perforated with Two Periodic Arrays of Apertures", Appl. Phys. Lett. 89 (2006) 221108-1–221108-3.
44. Degiron A., and Ebbesen T.W., "The Role of Localized Plasmon Modes in the Enhanced Transmission of Periodic Subwavelength Apertures", J. Opt. A Pure Appl. Opt. 7 (2005) S90–S96.
45. Prasher R., "Far Field Thermal Radiation Through Nanoholes and Apertures", Nano Lett. 6 (2006) 2135–2139.
46. Valkonen E., Karlsson B., and Ribbing C.-G., "Solar Optical Properties of Thin Films of Cu, Ag, Au, Cr, Fe, Co, Ni, and Al", Solar Energy 32 (1984) 211–222.
47. Valkonen E. and Karlsson B., "Optimization of Metal-Based Multilayers for Transparent Heat Mirrors", Energy Res. 11 (1987) 397–403.
48. Smith G.B., Niklasson G.A., Svensson J.S.E.M., and Granqvist C.G., "Noble-Metal-Based Transparent Infrared Reflectors: Experiments and Theoretical Analyses for Very Thin Gold Films", J. Appl. Phys. 59 (1986) 571–581.
49. Andersson K.E., Wahlström M.K., and Roos, A., "High Stability Titanium Nitride Based Solar Control Films", Thin Solid Films 214 (1992) 213–218.
50. Andersson K.E., Veszelei M., and Roos A., "Zirconium Nitride Based Transparent Heat Mirror Coatings: Preparation and Characterization", Solar Energy Mater. Solar Cells 32 (1994) 199–212.
51. Avrekh M., Thibadeau B.M., Monteiro O.R., and Brown I.G., "Transparent, Conducting, Metallic Thin Films", Rev. Sci. Instrum. 70 (1999) 4328–4330.
52. Kunz M., Niklasson G.A., and Granqvist C.G., "Optical and Electrical Properties of Sputter-Deposited Al Films Close to the Percolation Threshold", J. Appl. Phys. 64 (1988) 3740–3742.
53. Kaiser N., "Review of the Fundamentals of Thin-Film Growth", Appl. Opt. 41 (2002) 3053–3060.
54. Markert C., Lützenkirchen-Hecht D., and Frahm R., "Structural and Electrical Properties of Thin d.c. Magnetron-Sputtered Gold Films Deposited on Float Glass", Surf. Interface Anal. 38 (2006) 715–718.
55. Niklasson G.A., "Optical Properties of Inhomogeneous Two-Component Materials", in "Materials Science for Solar Energy Conversion Systems", edited by Granqvist C.G. (Pergamon, Oxford, UK, 1991), Ch. 2, pp. 7–43.
56. Niklasson G.A., Granqvist C.G., and Hunderi O., "Effective Medium Models for the Optical Properties of Inhomogeneous Materials", Appl. Opt. 20 (1981) 26–30.
57. Gadenne P., Yagil Y., and Deutscher G., "Transmittance and Reflectance *In Situ* Measurements of Semicontinuous Gold Films During Deposition", J. Appl. Phys. 66 (1989) 3019–3025.
58. Murray W.A., Astilean S., and Barnes W.L., "Transition from Localized Surface Plasmon Resonance to Extended Surface Plasmon-Polariton as Metallic Nanoparticles Merge to Form a Periodic Hole Array", Phys. Rev. B 69 (2004) 165407-1–165407-7.
59. Seal K., Genov D.A., Sarychev A.K., Noh H., Shalaev V.M., Ying Z.C., Zhang X., and Cao H., "Coexistence of Localized and Delocalized Surface Plasmon Modes in Percolating Metal Films", Phys. Rev. Lett. 97 (2006) 206103-1–206103-4.

60. Valkonen E. and Ribbing C.G., "Optical Selectivity of Thin Silver Films Prepared by rf-Assisted dc Magnetron Sputtering", Mater. Lett. 3 (1984) 29–32.
61. Leftheriotis G., Papaefthimiou S., and Yanoulis P., "Integrated Low-Emittance Electrochromic Devices Incorporating ZnS/Ag/ZnS Coatings as Transparent Conductors", Solar Energy Mater. Solar Cells 61 (2000) 107–112.
62. Papaefthimiou S., Leftheriotis G., and Yanoulis P., "Advanced Electrochromic Devices Based on WO_3 Thin Films", Electrochim. Acta 46 (2001) 2145–2150.
63. Sahu D.R. and Huang J.-L., "High Quality Transparent Conductive ZnO/Ag/ZnO Multilayer Films", Thin Solid Films 515 (2006) 876–879.
64. Sahu D.R., Chen C.Y., Lin S.Y., and Huang J.-L., "Effect of Substrate Temperature and Annealing Treatment on the Electrical and Optical Properties of Silver-Based Multilayer Coating Electrodes", Thin Solid Films 515 (2006) 932–935.
65. Sahu D.R., Lin S.-Y., and Huang J.-L., "ZnO/Ag/ZnO Multilayer Films for the Application of a Very Low Resistance Transparent Electrode", Appl. Surf. Sci. 252 (2006) 7509–7514.
66. Sahu D.R. and Huang J.-L., "Characteristics of ZnO-Cu-ZnO Multilayer Films on Copper Layer Properties", Appl. Surf. Sci. 253 (2006) 827–832.
67. Sahu D.R. and Huang J.-L., "Dependence of Film Thickness on the Electrical and Optical Properties of ZnO-Cu-ZnO Multilayers", Appl. Surf. Sci. 253 (2006) 915–918.
68. Bender M., Seelig W., Daube C., Frankenberger H., Ocker B., and Stollenwerk J., "Dependence of Film Composition and Thickness on Optical and Electrical Properties of ITO/Metal/ITO Multilayers", Thin Solid Films 326 (1998) 67–71.
69. Choi K.H., Kim J.Y., Lee Y.S., and Kim H.J., "ITO/Ag/ITO Multilayer Films for the Application of a Very Low Resistance Transparent Electrode", Thin Solid Films 341 (1999) 152–155.
70. Klöppel A., Meyer B., and Trube J., "Influence of Substrate Temperature and Sputtering Atmosphere on Electrical and Optical Properties of Double Silver Layer Systems", Thin Solid Films 392 (2001) 311–314.
71. Fahland M., Karlsson P., and Charton C., "Low Resistivity Transparent Electrodes for Displays on Polymer Substrates", Thin Solid Films 392 (2001) 334–337.
72. Bertran E., Corbella C., Vives M., Pinyol A., Person C., and Porqueras I., "RF Sputtering Deposition of Ag/ITO Coatings at Room Temperature", Solid State Ionics 165 (2003) 139–148.
73. Jung Y.S., Choi Y.W., Lee H.C., and Lee D.W., "Effects of Thermal Treatment on the Electrical and Optical Properties of Silver-Based Indium Tin Oxide/Metal/Indium Tin Oxide Structures", Thin Solid Films 440 (2003) 278–284.
74. Georgson M., Roos A., and Ribbing C.-G., "The Influence of Preparation Conditions on the Optical Properties of Titanium Nitride Based Solar Control Films", J. Vac. Sci. Technol. A 9 (1991) 2191–2195.
75. Zhao Q.-N. and Zhao X.-J., "Preparation and Characterization of TiO_2/TiN/TiO_2 Multi-Layer Solar Control Coatings Deposited by D.C. Reactive Magnetron Sputtering at Different Substrate Temperature", J. Wuhan Univ. Technol. 16 (2001) 9–12.
76. Jung M.J., Lee H.Y., Han J.G., Jung C.-K., Moon J.-S., and Boo J.-H., "High-Rate and Low-Temperature Synthesis of TiO_2, TiN, TiO_2/TiN/TiO_2 Thin Films and Study of Their Optical and Interfacial Characteristics", J. Vac. Sci. Technol. B 23 (2005) 1826–1831.
77. Okada M., Tazawa M., Jin P., Yamada Y., and Yoshimura K., "Fabrication of Photocatalytic Heat-Mirror with TiO_2/TiN/TiO_2 Stacked Layers", Vacuum 80 (2006) 732–735.
78. Granqvist C.G. and Hultåker A., "Transparent and Conducting ITO Films: New Developments and Applications", Thin Solid Films 411 (2002) 1–5.
79. Coutts T.J., Young D.L., and Li X., "Characterization of Transparent Conducting Oxides", MRS Bull. 25(8) (2000) 58–65.
80. Ginley D.S. and Bright C., "Transparent Conducting Oxides", MRS Bull. 25(8) (2000) 15–18.

81. Kawazoe H., Yanagi H., Ueda, K., and Hosono H., "Transparent *p*-Type Conducting Oxides: Design and Fabrication of *p-n* Heterojunctions", MRS Bull. 25(8) (2000) 28–36.
82. Minami T., "New *n*-Type Transparent Conducting Oxides", MRS Bull. 25(8) (2000) 38–44.
83. Minami T., "Transparent Conducting Oxide Semiconductors for Transparent Electrodes", Semicond. Sci. Technol. 20 (2005) S35–S44.
84. Edwards P.P., Porsch A., Jones M.O., Morgan D.V., and Perks R.M., "Basic Materials Physics of Transparent Conducting Oxides", Dalton Trans. (19) (2004) 2995–3002.
85. Nath P., Bunshah R.F., Basol B.M., and Staffsud O.M., "Electrical and Optical Properties of In_2O_3:Sn Films Prepared by Activated Reactive Evaporation", Thin Solid Films 72 (1980) 463–468.
86. Takaki S., Matsumoto K., and Suzuki, K., "Properties of Highly Conducting ITO Films Prepared by Ion Plating", Appl. Surf. Sci. 33/34 (1988) 919–925.
87. Adurodija F.O., Izumi H., Ishihara T., Yoshioka H., and Motoyama M., "Effect of Sn Doping on the Electronic Doping Mechanism of Indium-Tin-Oxide Films Grown by Pulsed Laser Deposition Coupled with Substrate Irradiation", J. Appl. Phys. 88 (2000) 4175–4180.
88. Adurodija F.O., Izumi H., Ishihara T., Yoshioka H., Motoyama M., and Murai K., "Pulsed Laser Deposition of Crystalline Indium Tin Oxide Films at Room Temperature by Substrate Irradiation", Jpn. J. Appl. Phys. 39 (2000) L377–L379.
89. Ohta H., Orita M., Hirano M., Tanji H., Kawazoe H., and Hosono H., "High Electrically Conductive Indium-Tin-Oxide Thin Films Epitaxially Grown on Yttria-Stabilized Zirconia (100) by Pulsed Laser Deposition", Appl. Phys. Lett. 76 (2000) 2740–2742.
90. Ohta H., Orita M., Hirano M., Tanji H., and Hosono H. (2002), "Surface Morphology and Crystal Quality of Low Resistive Indium Tin Oxide Grown on Yttria-Stabilized Zirconia", J. Appl. Phys. 91 (2002) 3547–3550.
91. Terzini E., Thilakan P., and Minarini C., "Properties of ITO Thin Films Deposited by RF Reactive Magnetron Sputtering at Elevated Substrate Temperature", Mater. Sci. Eng. B 77 (2000) 110–114.
92. Suzuki A., Matsushita T., Aoki T., Yoneyama Y., and Okuda M. (2001), "Pulsed Laser Deposition of Transparent Conducting Indium Tin Oxide Films in Magnetic Field Perpendicular to Plume", Jpn. J. Appl. Phys. 40 (2001) L401–L403.
93. Izumi H., Adurodija F.O., Kaneyoshi T., Ishihara T., Yoshioka H., and Motoyama M., "Electrical and Structural Properties of Indium Tin Oxide Films Prepared by Pulsed Laser Deposition", J. Appl. Phys. 91 (2002) 1213–1218.
94. Takaoka G.H., Yamazaki D., and Matsuo J., "High Quality ITO Film Formation by the Simultaneous Use of Cluster Ion Beam and Laser Irradiation", Mater. Chem. Phys. 74 (2002) 104–108.
95. Ohno S., Kawaguchi Y., Miyamura A., Sato Y., Song P.K., Yoshikawa M., Frach P., and Shigesato, Y., "High Rate Deposition of Tin-Doped Indium Oxide Films by Reactive Magnetron Sputtering with Unipolar Pulsing and Plasma Emission Feedback Systems", Sci. Technol. Adv. Mater. 7 (2006) 56–61.
96. Agura H., Suzuki A., Matsushita T., Aoki T., and Okuda M. (2003), "Low Resistivity Transparent Conducting Al-Doped ZnO Films Prepared by Pulsed Laser Deposition", Thin Solid Films 445 (2003) 263–267.
97. Park S.-M., Ikegami T., and Ebihara K., "Effects of Substrate Temperature on the Properties of Ga-Doped ZnO by Pulsed Laser Deposition", Thin Solid Films 513 (2006) 90–94.
98. Park S.-M., Ikegami T., Ebihara K., and Shin P.-K., "Structure and Properties of Transparent Conductive Doped ZnO Films by Pulsed Laser Deposition", Appl. Surf. Sci. 253 (2006) 1522–1527.
99. Bellingham J.R., Phillips W.A., and Adkins C.J., "Intrinsic Performance Limits in Transparent Conducting Oxides", J. Mater. Sci. Lett. 11 (1992) 263–265.
100. Suzuki T., Suzuki M., Sawada Y., and Matsushita J.-I., "DSC of Silver-Added Indium-Tin-Oxide (ITO) Transparent Conductive Materials", Thermochim. Acta 352–353 (2000) 87–90.

101. Hultåker A., Järrendahl K., Lu J., Granqvist C.G., and Niklasson G.A., "Electrical and Optical Properties of Sputter Deposited Tin Doped Indium Oxide Thin Films with Silver Additive", Thin Solid Films 392 (2001) 305–310.
102. Houng B., "Tin Doped Indium Oxide Transparent Conducting Thin Films Containing Silver Nanoparticles by Sol-Gel Technique", Appl. Phys. Lett. 87 (2005) 251922-1–251922-3.
103. Lu J., Hultåker A., Niklasson G.A., Granqvist C.G., and Olsson E., "Microstructure of Sputter Deposited Tin Doped Indium Oxide Films with Silver Additive", Thin Solid Films 479 (2005) 107–112.
104. Omata T., Kita M., Okada H., Otsuka-Yao-Matsuo S., Ono N., and Ikawa H., "Characterization of Indium-Tin Oxide Sputtering Targets Showing Various Densities of Nodule Formation", Thin Solid Films 503 (2006) 22–28.
105. Kim B.-C., Lee J.-H., Kim J.-J., and Ikegami T., "Rapid Rate Sintering of Nanocrystalline Indium Tin Oxide Ceramics: Particle Size Effect", Mater. Lett. 52 (2002) 114–119.
106. Rogozin A., Vinnichenko M., Shevchenko N., Kolitsch A., and Möller W., "Plasma Influence on the Properties and Structure of Indium Tin Oxide Films Produced by Reactive Middle Frequency Pulsed Magnetron Sputtering", Thin Solid Films 496 (2006) 197–204.
107. Iwase H., Hoshi Y., and Kameyama M., "Electrical Properties of Indium-Tin Oxide Films Deposited on Nonheated Substrates Using a Planar-Magnetron Sputtering System and a Facing-Targets Sputtering System", J. Vac. Sci. Technol. A 24 (2006) 65–69.
108. Betz U., Kharazzi Olsson M., Marthy J., Escolà M.F., and Atamny F., "Thin Films Engineering of Indium Tin Oxide: Large Area Flat Panel Displays Application", Surf. Coating Technol. 200 (2006) 5751–5759.
109. Bae J.W., Lee S.W., and Yeom G.Y., "Doped-Fluorine on Electrical and Optical Properties of Tin Oxide Films Grown by Ozone-Assisted Thermal CVD", J. Electrochem. Soc. 154 (2007) D34–D37.
110. Stjerna B., Olsson E., and Granqvist C.G., "Electrical and Optical Properties of Tin Oxide Films Doped with Oxygen Vacancies, F, Sb, or Mo", J. Appl. Phys. 76 (1994) 3798–3817.
111. Kim H. and Piqué A., "Transparent Conducting Sb-Doped SnO_2 Thin Films Grown by Pulsed-Laser Deposition", Appl. Phys. Lett. 84 (2004) 218–220.
112. Giraldi T.R., Ribeiro C., Escote M.T., Conti T.G., Chiquito A.J., Leite E.R., Longo E., and Varela J.A., "Deposition of Controlled Thickness Ultrathin SnO_2:Sb Films by Spin-Coating", J. Nanosci. Nanotechnol. 6 (2006) 3849–3853.
113. Hirata G.A., McKittrick, J., Cheeks T., Siqueiros J.M., Diaz J.A., Contreras O., and Lopez O.A., "Synthesis and Optoelectronic Characterization of Gallium Doped Zinc Oxide Transparent Electrodes", Thin Solid Films 288 (1996) 29–31.
114. Segura A., Sans J.A., Errandonea D., Martinez-García D., and Fages V. (2006), "High Conductivity of Ga-Doped Rock-Salt ZnO under Pressure: Hint on Deep-Ultraviolet-Transparent Conducting Oxides", Appl. Phys. Lett. 88 (2006) 011910-1–011910-3.
115. Nakagawara O., Kishimoto Y., Koshido Y., Yoshino Y., and Makino T., "Moisture-Resistant ZnO Transparent Conductive Films with Ga Heavy Doping", Appl. Phys. Lett. 89 (2006) 091904-1–091904-3.
116. Cairns D.R., Witte II R.P., Sparacin D.K., Sachsman S.M., Paine D.C., Crawford G.P., and Newton R.R., "Strain-Dependent Electrical Resistance of Tin-Doped Indium Oxide on Polymer Substrates", Appl. Phys. Lett. 76 (2000) 1425–1427.
117. Leterrier Y., Médico L., Månson J.-A.E., Betz U., Escolà M.F., Kharrazi Olsson M., and Atamny F., "Mechanical Integrity of Transparent Conductive Oxide Films for Flexible Polymer-Based Displays", Thin Solid Films 460 (2006) 156–166.
118. Cho J.S., Han S., Kim K.H., Beag Y.W., and Koh S.K., "Surface Modification of Polymers by Ion-Assisted Reaction", Thin Solid Films 445 (2003) 332–341.
119. Carcia P.F., McLean R.S., Reilly M.H., Li Z.G., Pillone L.J., and Messier R.F., "Low-Stress Indium-Tin-Oxide Thin Films rf Magnetron Sputtered on Polyester Substrates", Appl. Phys. Lett. 81 (2002) 1800–1802.

120. Wong F.L., Fung M.K., Tong S.W., Lee C.S., and Lee S.T., "Flexible Organic Light-Emitting Device Based on Magnetron Sputtered Indium-Tin-Oxide on Plastic Substrate", Thin Solid Films 466 (2004) 225–230.
121. Dekkers J.M., Rijnders G., and Blank D.H.A., "Role of Sn Doping in In_2O_3 Thin Films on Polymer Substrates by Pulsed-Laser Deposition at Room Temperature", Appl. Phys. Lett. 88 (2006) 151908-1–151908-3.
122. Kim J.H., Jeon K.A., Kim G.H., and Lee S.Y., "Electrical, Structural, and Optical Properties of ITO Thin Films Prepared at Room Temperature by Pulsed Laser Deposition", Appl. Surf. Sci. 252 (2006) 4834–4837.
123. Qiu Z., Shin D.-H., Nakada K., Murakami R.-I., and Yoon H.-K., "Influence of the SiO and SiON Layer on IZO Thin Films Deposited on PET by Inclination Opposite Target Type DC Magnetron Sputtering Method", Int. J. Mod. Phys. B 20 (2006) 3640–3645.
124. Fortunato E., Pimentel A., Gonçalves, A. Marques A., and R. Martins, "High Mobility Amorphous/Nanocrystalline Indium Zinc Oxide Deposited at Room Temperature", Thin Solid Films 502 (2006) 104–107.
125. Han H., Adams D., Mayer J.W., and Alford T.L., "Characterization of the Physical Properties of Indium Tin Oxide on Polyethylene Napthalate", J. Appl. Phys. 98 (2005) 083705-1–083705-8.
126. Han H., Mayer J.W., and Alford T.L., "Band Gap Shift in the Indium-Tin-Oxide Films on Polyethylene Napthalate after Thermal Annealing in Air", J. Appl. Phys. 100 (2006) 083715-1–083715-6.
127. Han Y.-G., Kim D.-H., Cho J.-S., Beag Y.-W., and Koh S.-K., "Study of the Substrate Treatment Effect on Initial Growth of Indium-Tin-Oxide Films on Polymer Substrate Using *in Situ* Conductance Measurement", Thin Solid Films 496 (2006) 58–63.
128. Cho S., Lee K., Lim H., and Ha C.-S., "Optical Spectra of Indium-Tin-Oxide Films Deposited on Flexible Colorless Polyimide Substrates", J. Korean Phys. Soc. 48 (2006) 468–471.
129. Fortunato E., Gonçalves A., Assunção V., Marques A., Águas H., Pereira L., Ferreira I., and Martins R., "Growth of ZnO:Ga Thin Films at Room Temperature on Polymeric Substrates: Thickness Dependence", Thin Solid Films 442 (2003) 121–126.
130. Kim J.S., Bae J.W., Kim H.J., Lee N.-E., Yeom G.Y., and Oh K.H., "Effects of Oxygen Radical on the Properties of Indium Tin Oxide Thin Films Deposited at Room Temperature by Oxygen Ion Beam Assisted Evaporation", Thin Solid Films 377–378 (2000) 103–108.
131. Kim H.J., Bae J.W., Kim J.S., Kim K.S., Jang Y.C., Yeom G.Y., and Lee N.-E., "Electrical, Optical, and Structural Characteristics of ITO Thin Films by Krypton and Oxygen Dual Ion-Beam Assisted Evaporation at Room Temperature", Thin Solid Films 377–378 (2000) 115–121.
132. Bae J.W., Kim J.S., and Yeom G.Y., "Indium-Tin-Oxide Thin Film Deposited by a Dual Ion Beam Assisted e-Beam Evaporation System", Nucl. Instrum. Methods Phys. Res. B 178 (2001) 311–314.
133. Kim D., "Low Temperature Deposition of ITO on Organic Films by Using Negative Ion Assisted Dual Magnetron Sputtering System", Vacuum 81 (2006) 279–284.
134. Kim D.-H., Park M.-R., Lee H.-J., and Lee G.-H., "Thickness Dependence of Electrical Properties of ITO Films Deposited on a Plastic Substrate by RF Magnetron Sputtering", Appl. Surf. Sci. 253 (2006) 409–411.
135. Jung S.K., Kim M.C., Sohn S.H., Park D.K., Lee S.H., and Park L.S., "Properties of Indium Tin Oxide on Polymer Films Deposited by Low-Frequency Magnetron Sputtering Method", Mol. Cryst. Liq. Cryst. 459 (2006) 167–177.
136. Park J.M., Kim J.J., Kim H.M., Kim J.H., Ryu S.W., Park S.H., and Ahn J.S., "Substrate Effects on the Characteristics of $(In_2O_3)_{1-x}(ZnO)_x$ Films, J. Korean Phys. Soc. 48 (2006) 1624–1627.
137. Woo B.-J., Hong J.-S., Kim S.-T., Kim H.-M., Park S.-H., Kim J.-J., and Ahn J.-S., "Effect of SiO_2 Buffer Layer on the Characteristics of In_2O_3-ZnO-SnO_2 Films Deposited on PET Substrates", J. Korean Phys. Soc. 48 (2006) 1579–1582.

138. Furubayashi Y., Hitosugi T., Yamamoto Y., Inaba K., Kinoda G., Hirose Y., Shimada T., and Hasegawa T., "A Transparent Metal: Nb-Doped Anatase TiO_2", Appl. Phys. Lett. 86 (2005) 252101-1–252101-3.
139. Furubayashi Y., Hitosugi T., Yamamoto Y., Hirose Y., Kinoda G., Inaba K., Shimada T., and Hasegawa T., "Novel Transparent Conducting Oxide: Anatase $Ti_{1-x}Nb_xO_2$", Thin Solid Films 496 (2006) 157–159.
140. Hitosugi T., Furubayashi Y., Ueda A., Itabashi K., Inaba K., Hirose Y., Kinoda G., Yamamoto Y., Shimada T., and Hasegawa T., "Ta-Doped Anatase TiO_2 Epitaxial Film as Transparent Conducting Oxide", Jpn. J. Appl. Phys. 34 (2005) L1063–L1065.
141. Ohta H., Nomura K., Hiramatsu H., Ueda K., Kamiya T., Hirano M., and Hosono H., "Frontier of Transparent Oxide Semiconductors", Solid-State Electron. 47 (2003) 2261–2267.
142. Ohta H. and Hosono H., "Transparent Oxide Optoelectronics", Mater. Today 7(6) (2004) 42–51.
143. Mott N.F., "Metal-Insulator Transitions", 2nd edition (Taylor & Francis, London, UK, 1990).
144. Grosse P., "Freie Elektronen in Festkörpern" (Springer, Berlin, Germany, 1979).
145. Mahan D., "Many Particle Physics" (Plenum, New York, USA, 1981).
146. Hamberg I. and Granqvist C.G., "Evaporated Sn-Doped In_2O_3 Films: Basic Optical Properties and Applications to Energy Efficient Windows", J. Appl. Phys. 60 (1986) R123–R159.
147. Burstein E., "Anomalous Optical Absorption Limit in InSb", Phys. Rev. 93 (1954) 632–633.
148. Moss T.S., "The Interpretation of the Properties of Indium Antimonide", Proc. Phys. Soc. Lond. B 67 (1954) 775–782.
149. Hamberg I., Granqvist, C.G., Berggren K.-F., Sernelius B.E., and Engström L., "Band-Gap Widening in Heavily Sn-Doped In_2O_3", Phys. Rev. B 30 (1984) 3240–3249.
150. Sernelius B.E., Berggren K.-F., Jin Z.-C, Hamberg I., and Granqvist C.G., "Band-Gap Tailoring of ZnO by Means of Heavy Al Doping", Phys. Rev. B 37 (1988) 10244–10248.
151. Jain A., Sagar P., and Mehra R.M., "Band Gap Widening and Narrowing in Moderately and Heavily Doped *n*-ZnO Films", Solid-State Electron. 50 (2006) 1420–1424.
152. Makino T., Sagawa Y., Yoshida S., Tsukazaki A., Ohtomo A., Kawasaki M., and Koinuma H., "Free-Carrier Effects on Zero- and One-Phonon Absorption of *n*-Type ZnO", Jpn. J. Appl. Phys. 44 (2005) 7275–7280.
153. Urbach F., "The Long-Wavelength Edge of Photographic Sensitivity and of the Electronic Absorption of Solids", Phys. Rev. 92 (1953) 1324.
154. Singwi K.S., Tosi M.P., Land R.H., and Sjölander A., "Electron Correlations at Metallic Densities", Phys. Rev. 176 (1968) 589–599.
155. Vashishta P. and Singwi K.S., "Electron Correlations at Metallic Densities. V", Phys. Rev. B 6 (1972) 875–887.
156. Gerlach E. and Grosse P., "Scattering of Free Electrons and Dynamical Conductivity", in "Festkörperprobleme/Advances in Solid State Physics", edited by Treusch J. (Vieweg, Braunschweig, Germany, 1977), Vol. XVII, pp. 157–193.
157. Wooten F., "Optical Properties of Solids" (Academic, New York, USA, 1972).
158. Ederth J., Heszler P., Hultåker A., Niklasson G.A., and Granqvist C.G., "Indium Tin Oxide Films Made from Nanoparticles: Models for the Optical and Electrical Properties", Thin Solid Films 445 (2003) 199–206.
159. Ederth J., Johnsson P., Niklasson G.A., Hoel A., Hultåker A., Heszler P., Granqvist C.G., van Doorn A.R., Jongerius M.J., and Burgard D., "Electrical and Optical Properties of Thin Films Consisting of Tin-Doped Indium Oxide Nanoparticles", Phys. Rev. B 68 (2003) 155410-1–155410-10.
160. Ederth J., Hultåker A., Niklasson G.A., Heszler P., van Doorn A.R., Jongerius M.J., Burgard D., and Granqvist C.G., "Thin Porous Indium Tin Oxide Nanoparticle Films: Effects of Annealing in Vacuum and Air", Appl. Phys. A 81 (2005) 1363–1368.
161. Solieman A. and Aegerter M.A., Modeling of Optical and Electrical Properties of In_2O_3:Sn Coatings Made by Various Techniques", Thin Solid Films 502 (2006) 205–211.

162. Eriksson T.S., Jiang S., and Granqvist C.G., "Dielectric Function of Sputter-Deposited Silicon Dioxide and Silicon Nitride Films in the Thermal Infrared", Appl. Opt. 24 (1985) 745–746.
163. Eriksson T.S. and Granqvist C.G., "Infrared Optical Properties of Silicon Oxynitride Films: Experimental Data and Theoretical Interpretation", J. Appl. Phys. 60 (1986) 2081–2091.
164. Rohsenow W.M., Hartnett J.P., and Cho Y.I., "Handbook of Heat Transfer", 3rd edition (McGraw-Hill, New York, USA, 1997).
165. Karlsson J. and Roos A., "Annual Energy Window Performance *vs.* Glazing Thermal Emittance: The Relevance of Very Low Emittance Values", Thin Solid Films 392 (2001) 345–348.
166. Hamberg I. and Granqvist C.G., "Color Properties of Transparent and Heat-Reflecting MgF_2-Coated Indium-Tin-Oxide", Appl. Opt. 22 (1983) 609–614.
167. Jiang S.-J., Jin Z.-C., and Granqvist C.G., "Low-Refractive-Index Indium-Tin-Oxyfluoride Thin Films Made by High-Rate Reactive dc Magnetron Sputtering", Appl. Opt. 27 (1988) 2847–2850.
168. Yin Z.Q., Stjerna B., and Granqvist C.G., "Antireflection Coatings of Sputter-Deposited SnO_xF_y and SnN_xF_y", Proc. Soc. Photo-Opt. Instrum. Eng. 1536 (1991) 149–157.
169. Jin Z.-C., Hamberg I., and Granqvist C.G., "Optical Properties of Sputter-Deposited ZnO:Al Thin Films", J. Appl. Phys. 64 (1988) 5117–5131.
170. Ederth J., Niklasson G.A., Hultåker A., Heszler P., Granqvist C.G., van Doorn A.R., Jongerius M.J., and Burgard D., "Characterization of Porous Indium Tin Oxide Thin Films Using Effective Medium Theory", J. Appl. Phys. 93 (2003) 984–988.
171. Sheng P., "Fluctuation-Induced Tunnelling Conduction in Disordered Materials", Phys. Rev. B 21 (1980) 2180–2195.
172. Collins R.E. and Simko T.M., "Current Status of the Science and Technology of Vacuum Glazing", Solar Energy 62 (1998) 189–213.
173. Griffiths P.W., di Leo M., Cartwright P., Eames P.C., Yianoulis P., Leftheriotis G., and Norton B., "Fabrication of Evacuated Glazing at Low Temperature", Solar Energy 63 (1998) 243–249.
174. Ng N., Collins R.E., and So L., "Thermal and Optical Evolution of Gas in Vacuum Glazing", Mater. Sci. Eng. B 119 (2005) 258–264.
175. Minaai T., Kumagai M., Nara A., and Tanemura S., "Study of the Outgassing Behavior of SnO_2:F Films on Glass in Vacuum under External Energy Excitation", Mater. Sci. Eng. B 119 (2005) 252–257.
176. Granqvist C.G. and Eriksson T.S., "Materials for Radiative Cooling to Low Temperatures", in "Materials Science for Solar Energy Conversion Systems", edited by Granqvist C.G. (Pergamon, Oxford, UK, 1991), Ch. 6, pp. 168–203.
177. Cucumo M., De Rosa A., and Marinelli V., "Experimental Testing of Correlations to Calculate the Atmospheric "Transparency Window" Emissivity Coefficient", Solar Energy 80 (2006) 1031–1038.
178. Eriksson T.S. and Granqvist C.G., "Radiative Cooling Computed for Model Atmospheres", Appl. Opt. 21 (1982) 4381–4388.
179. Nilsson T.M.J., Niklasson G.A., and Granqvist C.G., "A Solar Reflecting Material for Radiative Cooling Applications: ZnS Pigmented Polyethylene", Solar Energy Mater. Solar Cells 28 (1992) 175–193.
180. Mastai Y., Diamant Y., Aruna S.T., and Zaban A., "TiO_2 Nanocrystalline Pigmented Polyethylene Foils for Radiative Cooling Applications: Synthesis and Characterization", Langmuir 17 (2001) 7118–7123.
181. Nilsson T.M.J., Vargas W.E., Niklasson G.A., and Granqvist C.G., "Condensation of Water by Radiative Cooling", Renewable Energy 5 (1992) 310–317.
182. Hamberg I., Svensson J.S.E.M., Eriksson T.S., Granqvist C.G., Arrenius P., and Norin F., "Radiative Cooling and Frost Formation on Surfaces with Different Thermal Emittance: Theoretical Analysis and Practical Experience", Appl. Opt. 26 (1987) 2131–2136.

183. Werner A. and Roos A., "Condensation Tests on Glass Samples for Energy Efficient Windows", in "Conference Proceedings: EuroSun 2006, Glasgow, UK, 27–30 June", edited by Burek S., Hutchins M.G., Lockhart-Ball H., and Abrahamson S. (The Solar Energy Society, Abingdon, UK, 2006), 6 pages.
184. Fujishima A., Hashimoto K., and Watanabe T., "TiO_2 Photocatalysis" (BKC, Tokyo, Japan, 1999).
185. Kaneko M. and Okura I., editors, "Photocatalysis: Science and Technology" (Springer, Berlin, Germany, 2002).
186. Parkin I.P. and Palgrave R.G., "Self-Cleaning Coatings", J. Mater. Chem. 15 (2005) 1689–1695.
187. Paschoalino M.P., Kiwi J., and Jardim W.F., "Gas-Phase Photocatalytic Decontamination Using Polymer Supported TiO_2", Appl. Catal. B Environ. 68 (2006) 68–73.
188. Sánchez B., Coronado J.M., Candal R., Portela R., Tejedor I., Anderson M.A., Tompkins D., and Lee T., "Preparation of TiO_2 Coatings on PET Monoliths for the Photocatalytic Elimination of Trichloroethylene in the Gas Phase", Appl. Catal. B Environ. 66 (2006) 295–301.
189. Zhang X., Jin M., Liu Z., Nishimoto S., Saito H., Murakami T., and Fujishima A., "Preparation and Photocatalytic Wettability Conversion of TiO_2-Based Superhydrophobic Surfaces", Langmuir 22 (2006) 9477–9479.
190. Asahi R., Morikawa T., Ohwaki T., Aoki K., and Taga Y., "Visible-Light Photocatalysis in Nitrogen-Doped Titanium Oxides", Science 293 (2001) 296–271.
191. Lindgren T., Mwabora J.M., Avendaño E., Jonsson J., Hoel A., Granqvist C.G., and Lindquist S.-E., "Photoelectrochemical and Optical Properties of Nitrogen Doped Titanium Dioxide Films Prepared by Reactive DC Magnetron Sputtering", J. Phys. Chem. B 107 (2003) 5709–5716.
192. Mwabora J.M., Lindgren T., Avendaño E., Jaramillo T.F., Lu J., Lindquist S.-E., and Granqvist C.G., "Composition and Morphology of Photoelectrochemically Active $TiO_{2-x}N_x$ Thin Films Deposited by Reactive DC Magnetron Sputtering", J. Phys. Chem. B 108 (2004) 20193–20198.
193. Romualdo Torres G., Lindgren T., Lu J., Granqvist C.G., and Lindquist S.-E., "Photoelectrochemical Study of Nitrogen-Doped Titanium Dioxide for Water Oxidation", J. Phys. Chem. B 108 (2004) 5995–6003.
194. Gao B.-F., Ma Y., Cao Y.-A., Yang W.-S., and Yao J.-N., "Great Enhancement of Photocatalytic Activity of Nitrogen-Doped Titania by Coupling with Tungsten Oxide", J. Chem. Phys. 110 (2006) 14391–14397.
195. Takeuchi M., Yamasaki T., Tsujimaru K., and Anpo M., "Preparation of Crystalline TiO_2 Thin Film Photocatalysts on Polycarbonate Substrates by a RF-Magnetron Sputtering Deposition Method", Chem. Lett. 35 (2006) 904–905.
196. Yang C., Tartaglino U., and Persson B.N.J., "Influence of Surface Roughness on Superhydrophobicity", Phys. Rev. Lett. 97 (2006) 116103-1–116103-4.
197. Krč J., Zeman M., Kluth O., Smole F., and Topič M., "Effect of Roughness of ZnO:Al Films on Light Scattering in Hydrogenated Amorphous Silicon Solar Cells", Thin Solid Films 426 (2003) 296–304.
198. Hüpkes J., Rech B., Calnan S., Kluth O., Zastrow U., Siekmann H., and Wuttig M., "Material Study on Reactively Sputtered Zinc Oxide for Thin Film Silicon Solar Cells", Thin Solid Films 502 (2006) 286–291.
199. Ruske F., Jacobs C., Sittinger V., Szyszka B., and Werner W., "Large-Area ZnO:Al Films with Tailored Light Scattering Properties for Photovoltaic Applications", Thin Solid Films 515 (2007) 8695–8698
200. Kambe M., Sato K., Kobayashi D., Kurokawa Y., Miyajima S., Fukawa M., Taneda, N., Yamada A., and Konagai M., "TiO_2-Coated Transparent Conductive Oxide (SnO_2:F) Films Prepared by Atmospheric Pressure Chemical Vapor Deposition with High Durability against Atomic Hydrogen", Jpn. J. Appl. Phys. 45 (2006) L291–L293.

201. Natsuhara H., Matsumoto K., Yoshida N., Itoh T., Nonomura S., Fukawa M., and Sato K., "TiO_2 Thin Films as Protective Material for Transparent-Conducting Oxides Used in Si Thin Film Solar Cells", Solar Energy Mater. Solar Cells 90 (2006) 2867–2880.
202. Abduev A. Kh., Akhmedov A.K., and Asvarov A.Sh., "The Structural and Electrical Properties of Ga-Doped ZnO and Ga,B-Codoped ZnO Thin Films: The Effects of Additional Boron Impurity", Solar Energy Mater. Solar Cells 91 (2007) 258–260.
203. Miyata T., Honma Y., and Minami T., "Preparation of Transparent Conducting B-Doped ZnO Films by Vacuum Arc Plasma Evaporation", J. Vac. Sci. Technol. A 25 (2007) 1193–1197.
204. Du Pasquier A., Unalan H.E., Kanwal A., Miller S., and Chhowalla M., "Conducting and Transparent Single-Wall Carbon Nanotube Electrodes for Polymer-Fullerene Solar Cells", Appl. Phys. Lett. 87 (2005) 203511-1–203511-3.
205. Ulbricht R., Lee S.B., Jiang X., Inoue K., Zhang M., Fang S., Baughman R.H., and Zakhidov A.A., "Transparent Carbon Nanotube Sheets as 3-D Charge Collectors in Organic Solar Cells", Solar Energy Mater. Solar Cells 91 (2007) 416–419.
206. Rowell M.W., Topinka M.A., McGehee M.D., Prall H.-J., Dennler G., Sariciftci S., Hu L.-B., and Gruner G., "Organic Solar Cells with Carbon Nanotube Network Electrodes", Appl. Phys. Lett. 88 (2006) 233506-1–233506-3.
207. van de Lagemaat J., Barnes T.M., Rumbles G., Shaheen S.E., Coutts T.J., Weeks C., Levitsky I., Peltola J., and Glatkowski P., "Organic Solar Cells with Carbon Nanotubes Replacing In_2O_3:Sn as the Transparent Electrode", Appl. Phys. Lett. 88 (2006) 233503-1–233503-3.
208. Hino T., Ogawa Y., and Kuramoto N., "Dye-Sensitized Solar Cell with Single-Walled Carbon Nanotube Thin Film Prepared by an Electrolytic Micelle Disruption Method as the Counterelectrode", Fullerenes Nanotubes Carbon Nanostruct. 14 (2006) 607–619.
209. Tonooka K., Bando H., and Aiura Y., "Photovoltaic Effect Observed in Transparent *p-n* Heterojunctions Based on Oxide Semiconductors", Thin Solid Films 445 (2003) 327–331.
210. Mwamburi M., Wäckelgård E., Roos A., and Kivaisi R., "Polarization-Dependent Angular-Optical Reflectance in Solar-Selective $SnO_x/Al_2O_3/Al$ Reflector Surfaces", Appl. Opt. 41 (2002) 2428–2434.
211. Morrison G.L., "Solar Collectors", in "Solar Energy: The State of the Art", edited by Gordon J. (James & James, London, UK, 2001), Chs. 4 & 5, pp. 145–289.
212. Haitjema H. and Elich J., "The Physical Properties of Fluorine-Doped Tin Dioxide Films and the Influence of Ageing and Impurity Effects", Solar Energy Mater. 16 (1987) 79–90.
213. Mbise G., Smith G.B., Niklasson G.A., and Granqvist C.G., "Angular Selective Window Coatings: Theory and Experiment", Proc. Soc. Photo-Opt. Instrum. Eng. 1149 (1989) 179–199.
214. Otiti T., Niklasson G.A., Svedlindh P., and Granqvist C.G., "Anisotropic Optical, Magnetic, and Electrical Properties of Obliquely Evaporated Ni Films", Thin Solid Films 307 (1997) 245–249.
215. Leamy H.J., Gilmer G.H., and Dirks A.G., "The Microstructure of Vapor Deposited Thin Films", in "Current Topics in Materials Science", edited by Kaldis E. (North-Holland, Amsterdam, The Netherlands, 1980), Vol. 6, pp. 309–344.
216. Lakhtakia A. and Messier R., "Sculptured Thin Films: Nanoengineered Morphology and Optics" (SPIE Engr. Press, Bellingham, USA, 2004).
217. Brett M.J., "Simulation of Structural Transitions in Thin Films", J. Mater. Sci. 24 (1989) 623–626.
218. Lamarre J.-M., Yu Z., Harkati C., Roorda S., and Martinu L., Optical and Microstructural Properties of Nanocomposite Au/SiO_2 Films Containing Particles Deformed by Heavy Ion Irradiation", Thin Solid Films 479 (2005) 232–237.
219. Mbise G.W., Le Bellac D., Niklasson G.A., and Granqvist C.G., "Angular Selective Window Coatings: Theory and Experiments", J. Phys. D Appl. Phys. 30 (1997) 2103–2122.
220. Smith G.B., Dligatch S., Sullivan R., and Hutchins M.G., "Thin Film Angular Selective Glazing", Solar Energy 62 (1998) 229–244.

221. Mbise G.W., Niklasson G.A., and Granqvist C.G., "Obliquely Evaporated Cr Films with Large Angular Selectivity", J. Appl. Phys. 77 (1995) 2816–2818.
222. Mbise G.W., Niklasson G.A., Granqvist C.G., and Palmer S., "Angular-Selective Optical Transmittance through Obliquely Evaporated Cr Films: Experiments and Theory", J. Appl. Phys. 80 (1996) 5361–5364.
223. Le Bellac D., Niklasson G.A., and Granqvist C.G., "Angular-Selective Optical Transmittance of Anisotropic Inhomogeneous Cr-Based Films Made by Sputtering", J. Appl. Phys. 77 (1995) 6145–6151.
224. Palmer S., Mbise G.W., Niklasson G.A., and Granqvist C.G., "Angular Selective Properties of Thin Films: Measurement of Polar and Azimuthal Transmittance", Solar Energy Mater. Solar Cells 44 (1996) 397–403.
225. Campagno A., "Intelligente Glasfassaden/Intelligent Glass Façades", 5th edition (Birkhäuser, Basel, Switzerland, 2002).
226. Lampert C.M., "Chromogenic Smart Materials", Mater. Today 7(3) (2004) 28–35.
227. Hoffman H.J., "Photochromic Glass", in "Large-Area Chromogenics: Materials and Devices for Transmittance Control", edited by Lampert C.M. and Granqvist C.G. (The International Society for Optical Engineering, Bellingham, USA, 1990), Vol. IS4, pp. 86–101.
228. Chu N.Y.C., "Photochromic Plastics", in "Large-Area Chromogenics: Materials and Devices for Transmittance Control", edited by Lampert C.M. and Granqvist C.G. (The International Society for Optical Engineering, Bellingham, USA, 1990), Vol. IS4, pp. 102–120.
229. Minkin V.I., "Photo-, Thermo-, Solvato-, and Electrochromic Spiroheterocyclic Compounds", Chem. Rev. 104 (2004) 2751–2776.
230. Sone K. and Fukuda Y., "Inorganic Thermochromism" (Springer, Berlin, Germany, 1987), Springer Series on Inorganic Chemistry Concepts, Vol. 10.
231. Day J.H., "Science and Technology of Thermochromic Materials", in "Large-Area Chromogenics: Materials and Devices for Transmittance Control", edited by Lampert C.M. and Granqvist C.G. (The International Society for Optical Engineering, Bellingham, USA, 1990), Vol. IS4, pp. 122–141.
232. Granqvist C.G., "Handbook of Inorganic Electrochromic Materials" (Elsevier, Amsterdam, The Netherlands, 1995; reprinted 2002).
233. Monk P.M.S., Mortimer R.J., and Rosseinsky D.R., "Electrochromism: Fundamentals and Applications" (VCH, Weinheim, Germany, 1995).
234. Granqvist C.G., "Electrochromic Tungsten Oxide Films: Review of Progress 1993–1998", Solar Energy Mater. Solar Cells 60 (2000) 201–262.
235. Granqvist C.G., Avendaño E., and Azens A., "Advances in Electrochromic Materials and Devices: Survey of Some Recent Advances", Thin Solid Films 442 (2003) 201–211.
236. Byker H., "Electrochromics and Polymers", Electrochim. Acta 46 (2001) 2015–2022.
237. Notten P.H.L., Ouwerkerk M., Ledovskikh A., Senoh H., and Iwakura C., "Hydride Forming Electrode Materials Seen from a Kinetic Perspective", J. Alloys Compd. 356/357 (2003) 759–763.
238. Greenberg C.B., "Undoped and Doped VO_2 Films Grown from $VO(OC_3H_7)_3$", Thin Solid Films 110 (1983) 73–82.
239. Jorgenson G.V. and Lee J.C., "Doped Vanadium Oxide for Optical Switching Films", Solar Energy Mater. 14 (1986) 205–214.
240. Babulanam S.M., Eriksson T.S., Niklasson G.A., and Granqvist C.G., "Thermochromic VO_2 Films for Energy Efficient Windows", Solar Energy Mater. 16 (1987) 347–363.
241. Manning T.D., Parkin I.P., Pemble M.E., Sheel D., and Vernardou D., "Intelligent Window Coatings: Atmospheric Pressure Chemical Vapor Deposition of Tungsten-Doped Vanadium Dioxide", Chem. Mater. 16 (2004) 744–749.
242. Parkin I.P. and Manning T.D., "Intelligent Thermochromic Windows", J. Chem. Educ. 83 (2006) 393–400.
243. Miyazaki H. and Yasui I., "Effect of Buffer Layer on VO_x Film Fabrication by Reactive RF Sputtering", Appl. Surf. Sci. 252 (2006) 8367–8370.

244. Qazilbash M.M., Burch K.S., Whisler D., Shrekenhamer D., Chae B.G., Kim H.T., and Basov D.N., "Correlated Metallic State of Vanadium Dioxide", Phys. Rev. B 74 (2006) 205118-1–205118-5.
245. Jorgenson G.V. and Lee J.C., "Thermochromic Materials and Devices: Inorganic Systems", in "Large-Area Chromogenics: Materials and Devices for Transmittance Control", edited by Lampert C.M. and Granqvist C.G. (The International Society for Optical Engineering, Bellingham, USA, 1990), Vol. IS4, pp. 86–101.
246. Katzke H. and Schlögl R., "General Structural Relationships between Rutile-Type VO_2 and the Magnéli-Phases V_nO_{2n-1}", Z. Kristallogr. 218 (2003) 432–439.
247. Sobhan M.A., Kivaisi R.T., Stjerna B., and Granqvist C.G., "Thermochromism of Sputter Deposited $W_xV_{1-x}O_2$ Films", Solar Energy Mater. Solar Cells 44 (1996) 451–455.
248. Xu G., Jin P., Tazawa M., and Yoshimura K., "Thickness Dependence of Optical Properties of VO_2 Thin Films Epitaxially Grown on Sapphire (0001)", Appl. Surf. Sci. 244 (2005) 449–452.
249. Jiang S.-J., Ye C.-B., Khan M.S.R., and Granqvist C.G., "Evolution of Thermochromism During Oxidation of Evaporated Vanadium Films", Appl. Opt. 30 (1991) 847–851.
250. Yuan N.-Y., Li J.-H., and Lin C.-L., "Valence Reduction Process from Sol-Gel V_2O_5 to VO_2 Thin Films", Appl. Surf. Sci. 191 (2002) 176–180.
251. Rogers K.D., Coat J.A., and Lovell M.C., "Characterization of Epitaxially Grown Films of Vanadium Oxides", J. Appl. Phys. 70 (1991) 1412–1415.
252. Aliev R. A., Andreev V. N., Kapralova V. M., Klimov V. A., Sobolev A. I., and Shadrin E. B., "Effect of Grain Sizes on the Metal-Insulator Phase Transition in Vanadium Dioxide Polycrystalline Thin Films", Fiz. Tverd. Tela 48 (2006) 682–687 [English translation Phys. Solid State 48 (2006) 929–934].
253. Narayan J. and Bhosle V.M. (2006), "Phase Transition and Critical Issues in Structure-Property Correlations of Vanadium Oxide", J. Appl. Phys. 100 (2006) 103524-1–103524-6.
254. Jin P., Tazawa M., Ikeyama M., Tanemura S., Macák K., Wang X., Olafsson S., and Helmersson U., "Growth and Characterization of Epitaxial Films of Tungsten-Doped Vanadium Oxides on Sapphire (110) by Reactive Magnetron Sputtering", J. Vac. Sci. Technol. A 17 (1999) 1817–1821.
255. Jin P., Tazawa M., Yoshimura K., Igarashi K., Tanemura S., Macák K., and Helmersson U., "Epitaxial Growth of W-Doped VO_2/V_2O_3 Multilayer on α-$Al_2O_3(110)$ by Reactive Magnetron Sputtering", Thin Solid Films 375 (2000) 128–131.
256. Tazawa M., Jin P., and Tanemura S., "Optical Constants of $V_{1-x}W_xO_2$ Films", Appl. Opt. 37 (1998) 1858–1861.
257. Tazawa M., Jin P., Yoshimura K., Miki T., and Tanemura S., "New Material Design with $V_{1-x}W_xO_2$ Film for Sky Radiator to Obtain Temperature Stability", Solar Energy 64 (1998) 3–7.
258. Kato K., Song P.K., Odaka H., and Shigesato Y., "Study on Thermochromic VO_2 Films Grown on ZnO-Coated Glass Substrates for «Smart Windows»", Jpn. J. Appl. Phys. 42 (2003) 6523–6531.
259. Hörlin T., Niklewski T., and Nygren M., "Electrical and Magnetic Properties of $V_{1-x}W_xO_2$, $0 \leq x \leq 0.060$", Mater. Res. Bull. 7 (1972) 1515–1524.
260. Tang C., Georgopolous P., Fine M.E., Cohen J.B., Nygren M., Knapp G.S., and Aldred A., "Local Atomic and Electronic Arrangements in $W_xV_{1-x}O_2$", Phys. Rev. B 31 (1985) 1000–1011.
261. Wang H.-C., Yi X.-J., and Li Y., "Fabrication of VO_2 Films with Low Transition Temperature for Optical Switching Applications", Opt. Commun. 256 (2005) 305–309.
262. Lopez R., Boatener L.A., Haynes T.E., Haglund, Jr., R.F., and Feldman L.C., "Switchable Reflectivity on Silicon From a Composite VO_2-SiO_2 Protecting Layer", Appl. Phys. Lett. 85 (2004) 1410–1412.

263. Chen H.-K., Hung H.-C., Yang T.C.-K., and Wang S.-F., "The Preparation and Characterization of Transparent Nano-Sized Thermochromic VO_2-SiO_2 Films from the Sol-Gel Process", J. Non-Cryst. Solids 347 (2004) 138–143.
264. Khan M.S.R., Khan K.A., Estrada W., and Granqvist C.G., "Electrochromism and Thermochromism of Li_xVO_2 Thin Films", J. Appl. Phys. 69 (1991) 3231–3234.
265. Khan K.A., Niklasson G.A., and Granqvist C.G., "Optical Properties at the Metal-Insulator Transition in Thermochromic $VO_{2-x}F_x$ Thin Films", J. Appl. Phys. 64 (1988) 3327–3329.
266. Khan K.A. and Granqvist C.G., "Thermochromic Sputter-Deposited Vanadium Oxyfluoride Coatings with Low Luminous Absorptance", Appl. Phys. Lett. 55 (1989) 4–6.
267. Rakotoniaina J.C., Mokrani-Tamellin R., Gavarri J.R., Vacquier G., Casalot, A., and Calvarin G., "The Thermochromic Vanadium Dioxide: I. Role of Stresses and Substitution on Switching Properties", J. Solid State Chem. 103 (1993) 81–94.
268. Muraoka Y., Ueda Y., and Hiroi Z., "Large Modification of the Metal-Insulator Transition Temperature in Strained VO_2 Films Grown on TiO_2 Substrates", J. Phys. Chem. Solids 63 (2002) 965–967.
269. Jin P., Xu G., Tazawa M., and Yoshimura K. (2002), "A VO_2-Based Multifunctional Window with Highly Improved Luminous Transmittance", Jpn. J. Appl. Phys. 41 (2002) L278–L280.
270. Xu G., Jin P., Tazawa M., and Yoshimura K., "Optimization of Antireflection Coating for VO_2-Based Energy Efficient Window", Solar Energy Mater. Solar Cells 83 (2004) 29–37.
271. Deb S.K., "A Novel Electrophotographic System", Appl. Opt. Suppl. 3 (1969) 192–195.
272. Deb S.K., "Optical and Photoelectric Properties and Colour Centres in Thin Films of Tungsten Oxide", Philos. Mag. 27 (1973) 801–822.
273. Svensson J.S.E.M. and Granqvist C.G., "Electrochromic Tungsten Oxide Films for Energy Efficient Windows", Solar Energy Mater. 11 (1984) 29–34.
274. Svensson J.S.E.M. and Granqvist C.G., "Electrochromic Coatings for "Smart Windows", Solar Energy Mater. 12 (1985) 391–402.
275. Granqvist C.G., "Electrochromic Materials: Out of a Niche", Nat. Mater. 5 (2006) 89–90.
276. Sian T.S. and Reddy G.B., "Effect of Size and Valency of Intercalant Ions on Optical Properties of Polycrystalline MoO_3 Films", J. Electrochem. Soc. 152 (2005) A2323–A2326.
277. Sian T.S., Reddy G.B., and Shivaprasad S.M., "Effect of Microstructure and Stoichiometry on Absorption in Mg Intercalated MoO_3 Thin Films", Electrochem. Solid-State Lett. 9 (2005) A120–A122.
278. Kraft A., Rottman M., and Heckner K.-H., "Large-Area Electrochromic Glazing with Ion-Conducting PVB Interlayer and Two Complementary Electrodeposited Electrochromic Layers", Solar Energy Mater. Solar Cells 90 (2006) 469–476.
279. Wu Z.-C., Chen Z.-H., Du X., Logan J.M., Sippel J., Nikolou M., Kamaras K., Reynolds J.R., Tanner D.B., Hebard A.F., and Rinzler A.G., "Transparent, Conductive Carbon Nanotube Films", Science 305 (2004) 1273–1276.
280. Georg A., Georg A., and Opara Krašovec U., "Photoelectrochromic Window with Pt Catalyst", Thin Solid Films 502 (2006) 246–251.
281. Rönnow D., Kullman L., and Granqvist C.G., "Spectroscopic Light Scattering from Electrochromic Tungsten-Oxide-Based Films", J. Appl. Phys. 80 (1996) 423–430.
282. Lindström T., Kullman L., Rönnow D., Ribbing C.-G., and Granqvist C.G., "Electrochromic Control of Thin Film Light Scattering", J. Appl. Phys. 81 (1997) 1464–1469.
283. Le Bellac D., Azens A., and Granqvist C.G., "Angular Selective Transmittance through Electrochromic Tungsten Oxide Films Made by Oblique Angle Sputtering", Appl. Phys. Lett. 66 (1995) 1715–1715.
284. Azens A., Vaivars G., Veszelei M., Kullman L., and Granqvist C.G., "Electrochromic Devices Embodying W Oxide/Ni Oxide Tandem Films", J. Appl. Phys. 89 (2001) 7885–7887.

285. Yoo S.J., Lim J.W., and Sung Y.-E., "Improved Electrochromic Device with an Inorganic Solid Electrolyte Protective Layer", Solar Energy Mater. Solar Cells 90 (2006) 477–484.
286. Heusing S., Sun D.-L., Otero-Anaya J., and Aegerter M.A., "Grey, Brown and Blue Coloring Sol-Gel Electrochromic Devices", Thin Solid Films 502 (2006) 240–245.
287. Niklasson G.A. and Granqvist C.G., "Electrochromics for Smart Windows: Thin Films of Tungsten Oxide and Nickel Oxide, and Devices Based on These", J. Mater. Chem. 17 (2007) 127–156.
288. Agnihotry S.A., Singh P., Joshi A.G., Singh D.P., Sood K.N., and Shivaprasad S.M., "Electrodeposited Prussian Blue Films: Annealing Effects", Electrochim. Acta 51 (2006) 4291–4301.
289. Hjelm A., Granqvist C.G., and Wills J. M., "Electronic Structure and Optical Properties of WO_3, $LiWO_3$, $NaWO_3$, and HWO_3", Phys. Rev. B 54 (1996) 2436–2445.
290. Chatten R., Chadwick A.V., Rougier A., and Lindan P.J.D., "The Oxygen Vacancy in Crystal Phases of WO_3", J. Phys. Chem. B 109 (2005) 3146–3156.
291. Niklasson G.A., Berggren L., and Larsson A.-L., "Electrochromic Tungsten Oxide: The Role of Defects", Solar Energy Mater. Solar Cells 84 (2004) 315–328.
292. de Wijs G.A. and de Groot R.A., "Structure and Electronic Properties of Amorphous WO_3", Phys. Rev. B 60 (1999) 16463–16474.
293. Strømme M., Ahuja R., and Niklasson G.A., "New Probe of the Electronic Structure of Amorphous Materials", Phys. Rev. Lett. 93 (2004) 206403-1–206403-4.
294. Niklasson G.A., Berggren L., Jonsson A.K., Ahuja R., Skorodumova N.V., Backholm J., and Strømme M., "Electrochemical Studies of the Electron States of Disordered Electrochromic Oxides", Solar Energy Mater. Solar Cells 90 (2006) 385–394.
295. Berggren L., Ederth J., and Niklasson G.A., "Electrical Conductivity as a Function of Temperature in Amorphous Lithium Tungsten Oxide", Solar Energy Mater. Solar Cells 84 (2004) 329–336.
296. Larsson A.-L., Sernelius B.E., and Niklasson G.A., "Optical Absorption of Li-Intercalated Polycrystalline Tungsten Oxide Films: Comparison to Large Polaron Theory", Solid State Ionics 165 (2003) 35–41.
297. Ederth J., Hoel A., Niklasson G.A., and Granqvist C.G., "Small Polaron Formation in Porous WO_{3-x} Nanoparticle Films", J. Appl. Phys. 96 (2004) 5722–5726.
298. Raj S., Hashimoto D., Matsui H., Souma S., Sato T., Takahashi T., Ray S., Chakraborty A., Sarma D.D., Mahadevan P., McCarroll W.H., and Greenblatt M., "Angle-Resolved Photoemission Spectroscopy of the Metallic Sodium Tungsten Bronzes Na_xWO_3", Phys. Rev. B 72 (2005) 125125-1–125125-8.
299. Broclawik E., Góra A., Liguzinski P., Petelenz P., and Witek H.A., "Quantum Chemical Modeling of Electrochromism of Tungsten Oxide Films", J. Chem. Phys. 124 (2006) 054709-1–054709-11.
300. Denesuk M. and Uhlmann D.R., "Site-Saturation Model for the Optical Efficiency of Tungsten Oxide-Based Devices", J. Electrochem. Soc. 143 (1996) L186–L188.
301. Berggren L. and Niklasson G.A., "Optical Charge Transfer Absorption in Lithium-Intercalated Tungsten Oxide Thin Films", Appl. Phys. Lett. 88 (2006) 081906-1–081906-3.
302. Goldner R.B., Brofos A., Foley G., Goldner E.L., Haas T.E., Henderson W., Norton P., Ratnam B.A., Weris N., and Wong K.K., "Optical Frequencies Free Electron Scattering Studies on Electrochromic Materials for Variable Reflectivity Windows", Solar Energy Mater. 12 (1985) 403–410.
303. Svensson J.S.E.M. and Granqvist C.G., "Modulated Transmittance and Reflectance in Crystalline Electrochromic WO_3 Films", Appl. Phys. Lett. 45 (1984) 828–830.
304. Li J.-L., Riganese G.-M., and Louie S.G., "Quasiparticle Energy Bands of NiO in the *GW* Approximation", Phys. Rev. B 71 (2005) 193102-1–193102-4.
305. Nakajima N., Kato H., and Sakisaka Y., "Surface Metallic Nature Caused by an In-Gap State of Reduced NiO: A Photoemission Study", J. Electron Spectros. Relat. Phenomena 144–147 (2005) 873–875.

306. Avendaño E., Azens A., Niklasson G.A., and Granqvist C.G., "Electrochromism in Nickel Oxide Films Containing Mg, Al, Si, V, Zr, Nb, Ag, or Ta", Solar Energy Mater. Solar Cells 84 (2004) 337–350.
307. Avendaño E., Azens A., Niklasson G.A., and Granqvist C.G., "Proton Diffusion and Electrochromism in Hydrated NiO_y and $Ni_{1-x}V_xO_y$ Thin Films", J. Electrochem. Soc. 152 (2005) F203–F212.
308. Avendaño E., Kuzmin A., Purans J., Azens A., Niklasson G.A., and Granqvist C.G., "Changes in the Local Structure of Nanocrystalline Films of Hydrated Nickel Vanadium Oxide upon Ozone-Induced Coloration", Phys. Scr. T115 (2005) 464–466.
309. Avendaño, E., Rensmo H., Azens A., Sandell A., Niklasson G.A., Siegbahn H., and Granqvist C.G., "Coloration Mechanism in Proton Intercalated Electrochromic Hydrated NiO_y and $Ni_{1-x}V_xO_y$ Thin Films", J. Electrochem. Soc. 156 (2009) P132–P138.
310. Bode H., Dehmelt K., and Witte J., "Zur Kenntnis der Nickelhydroxidelektrode-I. Über das Nickel(II)-Hydroxidhydrat", Electrochim. Acta 11 (1966) 1079–1087.
311. Bode H., Dehmelt K., and Witte J., "Zur Kenntnis der Nickelhydroxidelektrode. II. Über die Oxydationsprodukten von Nickel(II)-Hydroxiden", Z. Anorg. Allg. Chem. 366 (1969) 1–21.
312. Mathew J.H.G., Sapers S.P., Cumbo M.J., O'Brien N.A., Sargent R.B., Raksha V.P., Ladaherne R.B., and Hichwa B.P., "Large Area Electrochromics for Architectural Applications", J. Non-Cryst. Solids 218 (1997) 342–346.
313. Lechner R. and Thomas L.K., "All Solid State Electrochromic Devices on Glass and Polymeric Foils", Solar Energy Mater. Solar Cells 54 (1998) 139–146.
314. Nagai J., McMeeking G.D., and Saitoh Y., "Durability of Electrochromic Glazing", Solar Energy Mater. Solar Cells 56 (1999) 309–319.
315. Karlsson J. and Roos A., "Angle-Resolved Optical Characterization of an Electrochromic Device", Solar Energy 68 (2000) 493–497.
316. Jonsson A.K., Larsson A.-L., Niklasson G.A., and Strømme M., "H^+ Conduction in Solid-State Electrochromic Devices Analyzed by Transient Current Measurements", J. Electrochem. Soc. 152 (2005) A377–A379.
317. Larsson A.-L. and Niklasson G.A., "Optical Properties of Electrochromic All-Solid-State Devices", Solar Energy Mater. Solar Cells 84 (2004) 351–360.
318. Subrahmanyam A., Kumar C.S., and Karuppasamy K.M., "A Note on Fast Protonic Solid State Electrochromic Device: $NiO_x/Ta_2O_5/WO_{3-x}$", Solar Energy Mater. Solar Cells 91 (2007) 62–66.
319. Azens A., Gustavsson G., Karmhag R., and Granqvist C.G., "Electrochromic Devices on Polyester Foil", Solid State Ionics 165 (2003) 1–5.
320. Azens A., Avendaño E., Backholm J., Berggren L., Gustavsson G., Karmhag R., Niklasson G.A., Roos A., and Granqvist C.G., "Flexible Foils with Electrochromic Coatings: Science, Technology, and Applications", Mater. Sci. Eng. B 119 (2005) 214–223.
321. Mecerreyes D., Marcilla R., Ochoteco E., Grande H., Pomposo J.A., Vergaz R., and Pena J.M.S., "A Simplified All-Polymer Flexible Electrochromic Device", Electrochim. Acta 49 (2004) 3555–3559.
322. Vergaz Benito R., Pena J.M.S., Gonzalo A.B., Ollero J.M., Vásquez C., Pomposo J.A., Grande H.-J., and Mecerreyes D., "Characterization of Novel All-Plastic Electrochromic Devices: Electro-Optic and Voltammetric Response", Opt. Eng. 43 (2004) 2967–2975.
323. Giri A.P. and Messier R., "Physical Structure and the Electrochromic Effect in Tungsten Oxide Films", Mater. Res. Soc. Symp. Proc. 24 (1984) 221–227.
324. Azens A., Kullman L., and Granqvist C.G., "Ozone Coloration of Ni Oxide and Cr Oxide Films", Solar Energy Mater. Solar Cells 76 (2003) 147–153.
325. Bardé F., Palacin M.R., Beaudoin B., and Tarascon J.-M., "Ozonation: A Unique Route to Prepare Nickel Oxyhydroxides: Synthesis Optimization and Reaction Mechanism Study", Chem. Mater. 17 (2005) 470–476.
326. Azens A., Private communication.

327. Lampert C.M., Agrawal A., Baertlien C., and Nagai J., "Durability Evaluation of Electrochromic Devices – An Industry Perspective", Solar Energy Mater. Solar Cells 56 (1999) 449–463.
328. Bell J.M. and Skryabin I.L., "Failure Modes of Sol-Gel Deposited Electrochromic Devices", Solar Energy Mater. Solar Cells 56 (1999) 437–448.
329. van Kampen N.G., "Stochastic Processes in Physics and Chemistry" (Elsevier, Amsterdam, The Netherlands, 1992).
330. Smulko J., Azens A., Kish L.B., and Granqvist C.G., "Low-Frequency Current Noise in Electrochromic Devices", Smart Mater. Struct. 17 (2) (2008) 025005.
331. Cottis R.A., "Sources of Electrochemical Noise in Corroding Systems", Elektrokhim. 42 (2006) 557–566 [English translation Russ. J. Electrochem. 42 (2006) 497–505].
332. Smulko J., "Methods of Electrochemical Noise Analysis for Investigation of Corrosion Processes", Fluctuation Noise Lett. 6 (2006) R1–R9.
333. Buyan M., Brühwiler P.A., Azens A., Gustavsson G., Karmhag R., and Granqvist C.G., "Facial Warming with Tinted Helmet Visors", Int. J. Ind. Ergon. 36 (2006) 11–16.
334. Larsson A.-L. and Niklasson G.A., "Infrared Emittance of All-Thin-Film Electrochromic Devices", Mater. Lett. 58 (2004) 2517–2520.
335. Beluze L., Morcrette M., Viana B., Badot J.C., Baffier N., and Tarascon J.M., "Infrared Electroactive Materials and Devices", J. Phys. Chem. Solids 67 (2006) 1330–1333.
336. Chandrasekhar P., Zay B.J., Birur G.C., Rawal S., Pierson E.A., Kauder L., and Swanson T., "Large, Switchable Electrochromism in the Visible through Far-Infrared in Conducting Polymer Devices", Adv. Funct. Mater. 12 (2002) 95–103.
337. Chandrasekhar P., Zay B.J., McQueeney T., Scara A., Ross D., Birur G.C., Haapanen S., Kauder L., Swanson T., and Douglas D., "Conducting Polymer (CP) Infrared Electrochromics in Spacecraft Thermal Control and Military Applications", Synth. Met. 135–136 (2003) 23–24.
338. Gruner G., "Carbon Nanotube Films for Transparent and Plastic Electronics", J. Mater. Chem. 16 (2006) 3533–3539.
339. Zhou Y.-X., Hu L.-B., and Grüner G., "A Method of Printing Carbon Nanotube Thin Films", Appl. Phys. Lett. 88 (2006) 123109-1–123109-3.
340. Stadermann M., Papadakis S.J., Falvo M.R., Novak J., Snow E., Fu Q., Liu J., Fridman Y., Boland J.J., Superfine R., and Washburn S., "Nanoscale Study of Conduction through Carbon Nanotube Networks", Phys. Rev. B 69 (2004) 201402-1–201402-3.
341. Goh S.K., Kaiser A.B., Lee S.W., Lee D.S., Yu H.Y., and Park Y.W., "Some Aspects of Conduction in Metallic Single-Wall Carbon Nanotubes", Curr. Appl. Phys. 6 (2006) 919–924.
342. Aguirre C.M., Auvray S., Pigeon S., Izquierdo R., Desjardins P., and Martel R., "Carbon Nanotube Sheets as Electrodes in Organic Light-Emitting Diodes", Appl. Phys. Lett. 88 (2006) 183104-1–183104-3.
343. Dyke C.A. and Tour J.M., "Covalent Functionalization of Single-Walled Carbon Nanotubes for Materials Applications", J. Phys. Chem. A 108 (2004) 11151–11159.
344. Lu X. and Chen Z.-F., "Curved Pi-Conjugation, Aromaticity, and the Related Chemistry of Small Fullerenes ($<C_{60}$) and Single-Walled Carbon Nanotubes", Chem. Rev. 195 (2005) 3643–3696.
345. Lee Y.-S. and Marzari N., "Cycloaddition Functionalizations to Preserve or Control the Conductance of Carbon Nanotubes", Phys. Rev. Lett. 97 (2006) 116801-1–116801-4.
346. Hayashi K., Matsuishi S., Kamiya T., Hirano M., and Hosono H., "Light-Induced Conversion of an Insulating Refractory Oxide into a Persistent Electronic Conductor", Nature 419 (2002) 462–465.
347. Sushko P.V., Shluger A.L., Hayashi K., Hirano M., and Hosono H., "Role of Hydrogen Atoms in the Photoinduced Formation of Stable Electron Centers in H-Doped $12CaO{\cdot}7Al_2O_3$", Phys. Rev. B 73 (2006) 045120-1–045120-13.

348. Toda Y., Miyakawa M., Hayashi K., Kamiya T., Hirano M., and Hosono H., "Thin Film Fabrication of Nano-Porous $12CaO{\cdot}7Al_2O_3$ Crystal and Its Conversion into Transparent Conductive Films by Light Illumination", Thin Solid Films 445 (2003) 309–312.
349. Sushko P.V., Shluger A.L., Hayashi K., Hirano M., and Hosono H., "Hopping and Optical Absorption of Electrons in Nano-Porous Crystal $12CaO{\cdot}7Al_2O_3$", Thin Solid Films 445 (2003) 161–167.
350. Sushko P.V., Shluger A.L., Hayashi K., Hirano M., and Hosono H., "Mechanisms of Oxygen Ion Diffusion in a Nanoporous Complex Oxide $12CaO{\cdot}7Al_2O_3$", Phys. Rev. B 73 (2006) 014101-1–014101-10.
351. Medvedeva J.E., Freeman A.J., Bertoni M.I., and Mason T.O., "Electronic Structure and Light-Induced Conductivity of a Transparent Refractory Oxide", Phys. Rev. Lett. 93 (2004) 016408-1–016408-4.
352. Beltrán, L.O., Lee E.S., and Selkowitz S.E., "Advanced Optical Daylighting Systems: Light Shelves and Light Pipes", J. Illum. Engr. Soc. 26(2) (1997) 91–106.
353. Heshong L., Wright R.L., and Okura S., "Daylighting Impacts on Human Performance in Schools", J. Illum. Eng. Soc 31 (2002) 101–114.
354. Heshong L., Wright R.L., and Okura S., "Daylighting Impact on Retail Sales Performance", J. Illum. Eng. 31 (2002) 21–25.
355. Schwartz-Schampera U. and Herzig P.M., "Indium: Geology, Mineralogy, and Economics" (Springer, Berlin, Germany, 2002).
356. Feltrin A. and Freundlich A., "Material Challenges for Terawatt Level Deployment of Photovoltaics", in "Conference Record of the 2006 IEEE Fourth World Conference on Photovoltaic Energy Conversion", Vol. 2, pp. 2469–2472.

Chapter 12
Nanostructured TCOs (ZnO, TiO_2, and Beyond)

Dana C. Olson and David S. Ginley

12.1 Other Applications of TCOs

Transparent conducting oxide (TCO) materials are traditionally used in applications as thin films, however there are a number of emerging technologies that require TCO materials in different form factors. This chapter will explore some three dimensional TCO nanostructures and their applications in hybrid photovoltaic devices. Additionally, we will look at the modification of TCO properties to allow for higher efficiency devices as well as novel materials, such as carbon nanotube assemblies, that might improve upon those properties. In many cases the nature of the TCO interface is constrained by the type of device. For example in hybrid organic–inorganic photovoltaic devices, poor exciton and charge carrier transport in the organic layer requires the distance to be short between the organic film and the TCO contact. In this case this distance is less than 10 nm necessitating a nanometer scale topology in a successful device. Similarly for solar cell and even display applications it is important to control the reflectance of the TCO surface through methods such as the etching facets in the surface as has been done in ZnO. This can, in fact, be accomplished simultaneously through the use of the TCO as the contact [1–8]. TCO nanofibers grown as a nanocarpet or incorporated in a composite can add conductivity without adversely affecting the optical properties [9–13]. In this chapter we review some of the more recent synthesis of nanostructured TCO materials and look at some of their emerging applications.

D.C. Olson (✉)
National Center for Photovoltaics, National Renewable Energy Laboratory, 1617 Cole Blvd, Golden, CO 80401-3393, USA
e-mail: dana.olson@nrel.gov

D.S. Ginley et al. (eds.), *Handbook of Transparent Conductors*,
DOI 10.1007/978-1-4419-1638-9_12, © Springer Science+Business Media, LLC 2010

12.1.1 Building Applications

Buildings represent one of the major energy users in the world. Not only that but they occupy a significant amount of area in populated regions. Nano-structured TCOs can have many applications in the building environment both on the outside and inside [14–25]. On the outside photocatalytic TiO_2 and other layers can provide active superhydrophylic layers that can help keep buildings clean and help remediate pollutants. Similarly on the inside of the building such layers can help indoor air quality, control static, and reduce cleaning loads. Thus surfaces coated with photocatalytic materials can be quite important. In addition there are a variety of devices that depend on nanostructured TCO materials that have tremendous impact on building energy efficiency and potentially building energy generation. Some of these are covered in other chapters but are covered here for completeness. These include nanostructured films for low-e windows, electrochromic windows, privacy coatings, wear resistant coatings, solid-state lighting/organic light emitting diodes (OLED) and building integrated solar. Thus in the complex environment of building systems, nano-structured TCO materials can play critical roles in remediating pollution, controlling air quality, controlling solar gain and loss and being integral to energy generation. TCO type materials used in these roles can be seamlessly integrated into the building in an integrated and architecturally important way.

12.1.2 Organic Photovoltaics

Photovoltaics have a long history with the first silicon solar cell being developed over 50 years ago. Historically, their utilization for large-scale energy production has been limited due to large fabrication, material and installation costs. A goal for more widespread application of photovoltaics would be to develop solar cells with moderate efficiency (15–20%) at a very low cost. Achieving this will require inexpensive materials for the active components, substrates and packaging, while allowing for low-temperature atmospheric processing techniques and high throughput manufacturing. Ideally these will be thin film devices that could be easily integrated in building or power production environments. Organic-based photovoltaics may have the potential to be one of the significant technologies helping to realize such goals. Organic photovoltaics have a theoretical efficiency equivalent to conventional semiconductor devices [26] and a cost structure derived from plastic processing [27]. As a result, organic photovoltaics offer the long-term potential as a PV technology that is economically viable for large-scale power generation.

Organic materials are inherently inexpensive and use low temperature processing techniques that most often take place at atmospheric pressure. This allows them to be deposited readily on plastic substrates through printing techniques in

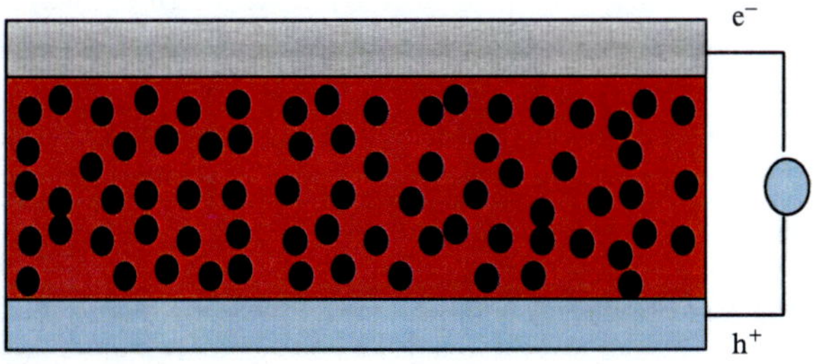

Fig. 12.1 A cartoon interpretation of the main components of a polymer/fullerene bulk heterojunction OPV device fabricated by spin coating from a single solution

a roll-to-roll process. Organic materials can be used in very thin film device architectures because their optical absorption coefficients are often very high. As a result, most of the incident light can be absorbed in only a few hundred nanometers of material. In addition, because of the short diffusion lengths for excitons (excited states) and carriers in the organics the length scale for the device architecture needs to be on the nanoscale. The processing techniques employed in the manufacturing of solution-processed organic devices require less capital investment, can be easily scaled up, and can be printed at higher speed than the manufacturing processes for current technologies [28]. Instead of using a planar heterojunction, the donor and acceptor materials can be combined and cast to form a blend through spin coating from a single solution. A blended architecture provides a large interfacial area between the donor and acceptor materials as illustrated in Fig. 12.1. This creates a "bulk heterojunction" where the intimate mixing of donor and acceptor species results in excellent exciton dissociation where any photogenerated exciton is within a diffusion length of an acceptor molecule leading to charge generation. Therefore, charges can be generated everywhere in the bulk, which dramatically increases the volume of the photoactive layer. If continuous pathways exist in each material from the bulk to the electrode, the efficiency can increase dramatically. Bulk heterojunction devices can be very cost effective as solution-processed devices. Initially, polymer/fullerene bulk heterojunction devices were fabricated and they demonstrated improved photovoltaic performance over equivalent bilayer devices [29]. The most successful polymer based photovoltaic cells to date are of this type, with efficiencies in excess of 5% being reported by numerous groups using a conjugated polymer and fullerene derivative [30–32].

The morphology of the polymer/fullerene blend plays a very important role in determining the device performance [33, 34]. After photoexcitation, excitons must be dissociated into charge carriers and in order for the charges to exit the device, there must be percolation through the acceptor and donor molecules to the cathode and anode, respectively. Without a percolation pathway, charges can be trapped leading to recombination and reduced efficiency. This also effectively limits the thickness of the device, reduces the amount of light absorbed, and decreases the efficiency. Defects in the network represent significant challenges to increased performance. Additionally, connectivity of the correct material with the corresponding electrode is not easily controlled. Presently, control of the blend morphology is influenced by the choice of the polymer (its chemistry and molecular weight), the solvent used, casting conditions, and thermal annealing either prior to or after deposition of the counter electrode.

The properties of the TCO used to contact the organic photovoltaic device are important for optimized performance. As in most TCO applications, the conductivity and transparency of the TCO need to be high to allow for low series resistance and efficient photon collection in the active layer of the organic photovoltaic. Such devices are typically less than 300 nm in thickness, which require TCO materials with very low surface roughness. Additionally, the work function of the TCO must be optimized to match the energy levels in the organic semiconductor for efficient charge collection. Also the surface chemistry of the TCO is critical for stability for the device and for the charge transfer at the interface.

Nanostructured TCO electrodes could help to enhance the performance of organic solar cell devices through increased charge collection area, or could serve as a replacement for the fullerene acceptor that is currently employed in polymer blend devices. As discussed below, the TCO-based electron acceptor could be incorporated into polymer-inorganic nanoparticle blend architectures or could be employed as an ordered array of nanorods grown normal to the substrate. Such architectures could enhance charge transport out of the active layer of the device compared to the more standard polymer-fullerene or polymer-nanoparticle blend devices [35].

12.1.3 Dye Sensitized Solar Cell

Since the realization of an efficient dye sensitized solar cell (DSSC) by Grätzel and coworkers in 1991 [36], efficient photovoltaic devices fabricated from solution-based techniques have been promising alternatives to traditional solar cell technologies. In a typical DSSC, a mesoporous titanium oxide (TiO_2) film is deposited onto a TCO-coated glass substrate from a colloidal solution of anatase TiO_2 nanoparticles with diameters of 15 nm. The 10 μm nanoparticle films are sintered at 450°C to form continuous electronic contact between the particles and the TCO electrode. A ruthenium dye is then absorbed onto the high surface area TiO_2 film and the device is filled with a liquid electrolyte and a platinum counter electrode completes the circuit. Such devices rely heavily on the properties of the nanocrystalline TiO_2 for electron transfer and transport to the thin film TCO electrode. Though to date it has been hard to improve on TiO_2, ZnO nanorods have been employed in DSSC devices with relative success due to enhanced charge transport properties [37]. Both are needed as electron transporters and both work best when they are near intrinsically doped.

12.2 TCO Requirements

12.2.1 *Work Function*

The work function of the TCO is of great importance for optimizing the injection/ extraction process in organic devices. The work function is tuned to minimize the barrier height for charge transfer to or from the corresponding molecular orbitals in the organic semiconductors. When the offset between the semiconductor and the TCO is minimized, ohmic contacts can be achieved.

The work function of untreated ITO is generally quoted at ~4.7 eV [38], where plasma cleaning of the TCO in O_2 generally increases the work function about 0.1–0.3 eV. While this is relatively adequate to inject and extract holes into widely used hole transport layers, problems with poor hole injection are experienced with the wider band gap host and dopant materials [39]. To aid in hole transfer at the ITO interface, the ITO can be modified with either a hole injection layer [39–45] or through chemical dipole modification [46, 47].

12.2.2 *Surface Chemistry*

Interface modification techniques have been shown to improve the properties of untreated ITO, but illustrates that at present the problem is actually twofold: one is the need to improve the anode work function and the other is to improve the anode surface chemistry so that this work function is actually operative in the device. The changes in surface chemistry can change the effective work function by up to a volt. TCOs with stabilized surfaces could significantly improve the contact and stability for the anode [48]. A sounder approach would be to develop a TCO material tailored to the needs of organic-based devices including the key properties of work function, material and interface stability, processability and optimized surface chemistry. Relatively few attempts have been made, however, to substitute improved TCOs for ITO. This omission is particularly curious given the repeated indications that acidic hole injection layers, such as polyethylenedioxythiophene doped with poly-styrene sulfonic acid (PEDOT:PSS), corrode ITO, and that In diffusion into the organic semiconductor layers causes accelerated device degradation [49, 50]. An In-free anode could, therefore, improve device efficiency (via the higher work function for hole injection), simplify the device structure (by removing the need for additional hole injection layers), and improve stability (via elimination of mobile In).

12.2.3 *Surface States*

Surface states on the oxide surface can be very important with respect to organic device performance. They can affect the charge transfer kinetics and recombination at the electrode interface. If located within the band gap they can significantly reduce the open circuit potential (photovoltage) of OPV devices, they can be a source for the decomposition of the oxide, and they can affect the molecular ordering of the polymer. Understanding surface states may allow for device optimization. Different oxides have demonstrably different wetting and charge transfer properties with respect to organics.

12.2.4 Organic Interface

As above the nature of the oxide/organic interface is critical to the performance of OPV and dye cell devices. The interface controls exciton decomposition, carrier recombination, molecular ordering in the organic semiconductor, and potentially the stability of the device. While you can create the appropriate nanostructure with a number of oxide systems much less is known about the how to optimize the interface. Additionally, It is important to control the wetting properties of the organic on the oxide. Nominally the oxides have hydroxylated surfaces with are normally not compatible with a typical polymeric molecule. Thus in some way the interface will have to be derivatized to get the right properties.

12.2.5 Etchability

The etchability of the oxide can be either desirable or detrimental to the fabrication and stability of the device. Etchability can aid in patterning the oxide and modifying the interface, however readily etchable oxides may create problems with the ultimate device stability. Much needs to be understood in this area, InSnOx is difficult to etch but the In diffuses into the organic, ZnO is much more etchable but may have better OPV properties.

12.2.6 Stability

The stability of the oxides may be critical for the ultimate device stability. Stability is a strong function of the solubility of the oxide, the surface stability (i.e. the ability to oxidize or reduce the surface), and the nature of the surface states. For most oxides this is not well understood and for the materials investigated for OPV and the dye sensitized solar cell the exact nature of the dye or organic interface is not well understood at present. Most of the best devices modify the surface to improve the device and stability properties.

12.2.7 Photochemistry

Many of the oxides being investigated (ZnO, TiO_2) can produce photoexcited holes or radicals upon the adsorption of light with energy greater than band gap. These holes are of sufficient potential to oxidize many of the polymers or dyes. Clearly this is not desirable and may ultimately require the use of charge scavengers or UV filters to prevent the excitation of the oxides. Fortunately most of the TCOs have large band gaps, mostly greater than 3.0 eV, and this represents a small portion of the solar spectrum. Again as above an appropriate surface or surface derivatizing species to block hole transfer from the oxide to the organic, which may help to eliminate this as a problem.

12.2.8 Transport

Transport of carriers is a critical role for any TCO and may be more critical in oxide/organic devices. In both the dye sensitized solar cells and the nanostructured OPV devices carriers need to diffuse a substantial distance from where they are created to where they are extracted by the back contact. Given that the highest performance devices to date employ oxides that have very low doping levels this implies that either they have to have very high mobilities or an efficient hopping type mechanism for carrier diffusion to minimize problems with series resistance. Currently, sufficient device models do not exist to be able to predict the best doping/mobility combination in such devices. This is an important area since the actual carrier diffusion lengths can be relatively long despite the nanostructured nature of the device.

12.3 The Hybrid OPV Example

12.3.1 Bulk Heterojunction Polymer/ZnO Nanoparticle Solar Cells

A cartoon illustrating the polymer/inorganic nanoparticle device is shown in Fig. 12.2, indicating device architectures similar to that of the polymer/fullerene bulk heterojunction devices. Photoinduced charge transfer has also been observed from a conjugated polymer donor to a number of metal oxide acceptors such as SnO_2, TiO_2, and ZnO [51, 52]. The main concern when choosing a donor–acceptor pair is obtaining a fast forward electron transfer with a much lower probability backward electron transfer process, which allows the charge to be extracted from the device prior to recombination of the charge carriers. This would effectively time

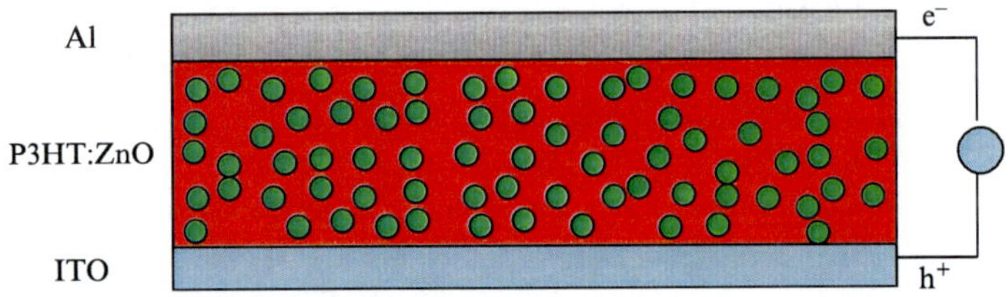

Fig. 12.2 A cartoon interpretation of a polymer/inorganic nanoparticle bulk heterojunction OPV device fabricated by spin coating from single solution

separate the two processes. This has been demonstrated for conjugated polymers and metal oxides, where the forward electron transfer rate is less than 100 fs, and the recombination rate, which includes both backward electron transfer and recombination, is greater than 1 ns [51]. Such electron transfer kinetics ensure that the exciton will be dissociated and the carriers will begin to diffuse away from the interface before any significant recombination occurs. While the electron mobility in nanoparticle ZnO can be rather high in pure nanoparticle films ($\mu_e \sim 10^{-2}$ cm^2 V^{-1} s^{-1}) [53], in a polymer/nanoparticle blend that value may be reduced to values similar to polymer/fullerene blends.

Devices based on composites of metal oxide semiconductor nanoparticles embedded in a conjugated polymer matrix have been developed and have demonstrated reasonable efficiencies approaching 2% [54]. Devices based on ZnO nanoparticles perform better than TiO_2-based bulk heterojunctions, which is presumably due to easier dispersion and crystallization of ZnO nanoparticles [54, 55]. These devices have demonstrated the feasibility of hybrid polymer/metal oxide composites for photovoltaic applications. Also, these devices perform comparably to the polymer/CdSe bulk heterojunction devices mentioned above, though the metal oxide acceptor has a large band gap (3.2 eV) and therefore does not contribute to light absorption in the device.

12.3.2 Ordered Polymer/ZnO Nanorod Array Solar Cells

Ordered organic/inorganic hybrid bulk heterojunction devices might be a means to combine a light absorbing conjugated polymer donor species with an electron acceptor (TCO) with control over the nanostructured morphology through the directed growth of a nanostructured metal oxide [27, 35]. Such a device would operate similar to a bulk heterojunction device based on a blend of donor and acceptor molecules, however the organized growth of the nanostructured oxide would determine the bulk heterojunction morphology in the composite and would be a stable defined nanostructure, as demonstrated in Fig. 12.3. This may be key in building flexible OPV devices on plastic substrates. Models developed to investigate the ideal morphology for ordered polymer/inorganic devices, indicate that the

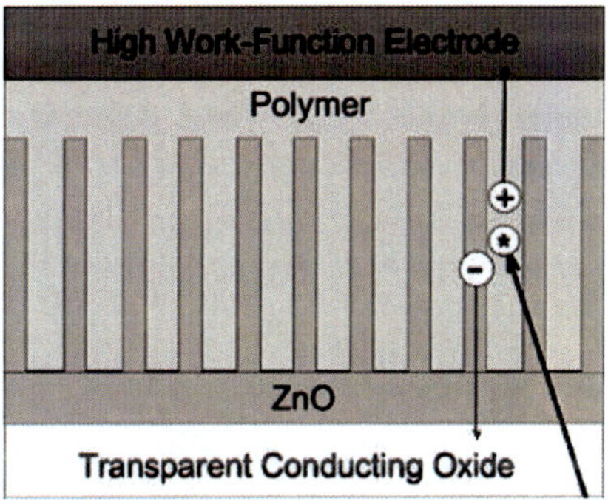

Fig. 12.3 Schematic diagram of an ordered bulk heterojunction polymer photovoltaic device

conjugated polymer should be intercalated into an inorganic electron acceptor made up of small diameter nanostructures oriented vertically off of the substrate [56]. A key advantage of this approach over the polymer/fullerene bulk heterojunction device structure is the potential for tailoring the morphology, chemistry, and transport of the donor/acceptor interface.

Systematic control of the donor–acceptor morphology might lead to better charge transport out of the device with connected pathways for charges to travel to the correct electrode, while maintaining efficient exciton dissociation in the composite. Devices made by incorporation of a conjugated polymer into the nanostructured oxide film may be more thermally stable over time than bulk heterojunction blend devices. Blends rely on phase separation to determine morphology and this can change at elevated temperatures such as those experienced during the day/night cycling of a PV device, while the morphology of a polymer/nanostructured oxide composite would not be expected to change to any significant extent.

A nanostructured oxide acceptor can be grown with a great deal of morphological control/self organization thanks to recent advances in growth of inorganic nanostructures. This allows for significant control in the spacing between the oxide nanostructures as well as their length or film thickness. In addition a variety of oxides can be grown in these morphologies and using ALD approaches can be conformally coated with alternate oxides or active interfacial agents to improve exciton dissociation, charge transfer and molecular ordering. Various metal oxide materials can be chosen, each with their own set of parameters and processes for controlled growth of nanostructures. Metal oxides can be grown with pore sizes and nanostructure spacing on the order 10 nm, which would allow for efficient exciton dissociation and charge transport out of the device [57–63]. Devices fabricated from uninterrupted phases of a polymer donor and a nanostructured oxide acceptor grown on a transparent conductive oxide would also have guaranteed percolation of both charge carrier materials as well as reduced recombination of carriers at the electrodes, due to only the acceptor being in contact with the cathode, and only the donor being in contact with the anode. This eliminates recombination due to charges not escaping the device as a result of isolated domains of a material in the film [35]. A hybrid conjugated polymer/nanostructured oxide device could also be made thicker due to the percolation of charge carrier through the donor/acceptor materials and to the increased carrier mobility of the metal oxide semiconductor as compared to electrons in a blend. Studies on dye-sensitized solar cells where carriers are efficiently transported through microns of near intrinsic TiO_2 clearly demonstrate the feasibility of this approach.

Due to the inverted nature of this device architecture as compared to the typical bulk heterojunction blend, a high work function metal electrode can be deposited to contact the polymer and extract holes [55, 64–66]. The use of electrodes with similar work functions results in a device where the current is dominated by a diffusion current rather than a drift current, which is the dominate mechanism in traditional inorganic photovoltaic devices utilizing a p–n junction. This potentially eliminates the oxygen sensitive Al, Ca or other low work-function electrodes from the device structure allowing for the use of metals like Ag, Au, etc., which may have the potential to increase device stability and lessen the strict requirements for device encapsulation.

Ordered hybrid polymer/metal oxide devices have been fabricated using a mesoporous TiO_2 acceptor species. The material exhibited a pore size of less than 10 nm, which is less than the typical exciton diffusion length in the conjugated polymer. Significant exciton dissociation (95%) was demonstrated by PL quenching data [57], which was anticipated with pores of this size. However, despite relatively efficient exciton dissociation and complete percolation of both species, reported efficiencies were estimated at less than 0.45% under AM 1.5 conditions [67, 68].

Based on optical absorption data, the conclusion was that the conjugated polymer species was not well ordered in contrast to what has been observed in annealed bulk heterojunction blend devices with P3HT and PCBM. This would lead to drastically reduced mobility in the P3HT and poor carrier transport properties [67]. When compared to annealed P3HT films, the absorption spectra of the intercalated P3HT was blue shifted with the shift increasing the deeper the polymer is infiltrated into the nanostructured TiO_2 film. A shoulder at 620 nm was also absent in the absorption spectra, which is often observed for P3HT with a large amount of intermolecular ordering [57]. Additional support for the polymer chains being disordered in the mesoporous TiO_2 stemmed from the decreased photoluminescence quenching observed at increased temperatures. This would also indicate a reduced exciton diffusion length, which is typically enhanced by molecular order.

It is likely that the reduced order in the P3HT component is related to the size and morphology of the interconnected pores in the nanostructured TiO_2 film. These pores are not vertical and are very small, leading to disordered polymer chains that coil upon themselves instead of ordered chains lying outstretched with longer conjugation lengths [57, 67]. When the polymer is more crystalline or ordered, it is expected that the charge mobility will be considerably higher, the absorption is red-shifted, and the exciton diffusion length is increased. The disordered nature of the polymer in these devices leads to a lower absorption coefficient and poor charge transport, thereby dramatically decreasing the short circuit current. Polymer films annealed for extended times at 170°C result in decreased hole mobility and exciton diffusion lengths. Films annealed for longer times have higher polymer optical densities in the mesoporous TiO_2 but demonstrate lower external quantum efficiencies [67]. It is thought that the oxide morphology is not ideal in this case, leading to poor intramolecular ordering. As a result, high efficiencies have not yet been realized using this mesoporous TiO_2 as an electron acceptor in an ordered hybrid solar cell.

Recently, the hole mobility of P3HT was improved by infiltration into small vertically oriented pores in anodic alumina [35, 69]. To characterize the electronic properties of the polymer in the nanoporous metal oxide, devices were prepared on fluorine doped tin oxide (FTO). The anodic alumina was deposited onto the FTO coated glass. Next, the polymer was intercalated into the porous metal oxide by annealing at 200°C for 1 min, and a gold counter electrode was thermally evaporated to complete the device. The hole mobility increased by more than an order of magnitude from 3×10^{-4} cm^2 V^{-1} s^{-1} for a P3HT thin film to 6×10^{-3} cm^2 V^{-1} s^{-1} in the charge transport direction, as estimated by fitting the space charge limited current vs. voltage data on the polymer/alumina devices [35, 69].

The polymer chain orientation can be determined through polarized optical transmission and reflection measurements [70]. It was observed that conjugated polymer chains prefer to lie in the plane of the substrate in neat films [70]. The same characterization technique was used to measure the polymer chain orientation of the P3HT in the porous anodic alumina. It was observed that the P3HT chains are partially aligned vertically along the sides of the oriented pores in anodic alumina film [69]. This is encouraging given that the charge transport direction in a photovoltaic device is in the vertical direction, normal to the substrate. Therefore, if such a structure could be made in a nanostructured oxide such as TiO_2 or ZnO, one might expect increased hole mobility and in turn efficiency in ordered hybrid polymer/metal oxide devices. If the polymer chains increase in order from the middle of the pore to the side of the pore, then the polymer's exciton diffusion length might be increased leading to enhanced photovoltaic device performance.

Another study efficiently infiltrated a number of different PPV-based polymers into a nanostructured TiO_2 film formed by the sintering of TiO_2 nanoparticles. This effort concluded that the limit in the short circuit current density of these devices is likely related to the photogeneration rate and the quality of the interfaces [68]. Neither increased polymer carrier mobility nor spectral response in the red to infrared showed dramatic improvement in device performance [68]. Ultimately, these devices were limited to a conversion efficiency of 0.4%, despite efforts to change the optical and electronic properties of the polymers investigated. Therefore, the current limitations of ordered hybrid polymer/nanostructured oxide devices are associated mainly with the interface between the organic and inorganic materials and the low electron mobility of the nanostructured oxide material. This is likely affected by the wetting of the polymer on the surface of the oxide as well as the complete filling of very small pores with ordered polymer chains. Additionally, the low electron mobilities in the sintered TiO_2 nanoparticle films, which have been estimated at 10^{-6} cm^2 V^{-1} s^{-1}, are much lower than the mobility in bulk heterojunction blend devices [68].

To increase the efficiencies of these ordered hybrid polymer/metal oxide bulk heterojunction devices, a promising route is to use a nanostructured oxide that is vertically aligned with respect to the substrate. Figure 12.3 illustrated what a composite device based on vertically aligned nanostructures might look like. This metal oxide morphology might allow ordering of the polymer chains in the vertical direction, thereby increasing the hole mobility compared to the mesoporous TiO_2 materials described above [68, 69]. Such an architecture also has the advantage of creating electron transport pathways connected to the negative electrode that possess very high electron mobility. For example, the electron mobility of ZnO nanorods has been measured as 10 cm^2 V^{-1} s^{-1} [37, 71], which is several orders of magnitude higher than is typically found in organic semiconductors or in sintered nanoparticle TiO_2 films.

Through a low-temperature chemical solution growth technique, single crystal ZnO nanorods can be grown that demonstrate high electron mobilities [63, 71]. The ZnO nanorods can be grown normal to the substrate in the 002 direction as seen in Fig. 12.4. However, the substrate must often first be coated with a nucleation layer for controlled growth to occur [60]. The length of the nanorods can be controlled by changing the growth time, solution concentration, temperature, and pH [62]. The nanorod diameter can be controlled by varying the grain size of the nucleation layer through the anneal conditions [72, 73]. The nanorod spacing can also be controlled through changes to the microstructure of the underlying nucleation layer and through the growth of more or less vertically oriented fibers [60, 61, 73]. Equally important in the optimization of such a device is that the ZnO nanorods,

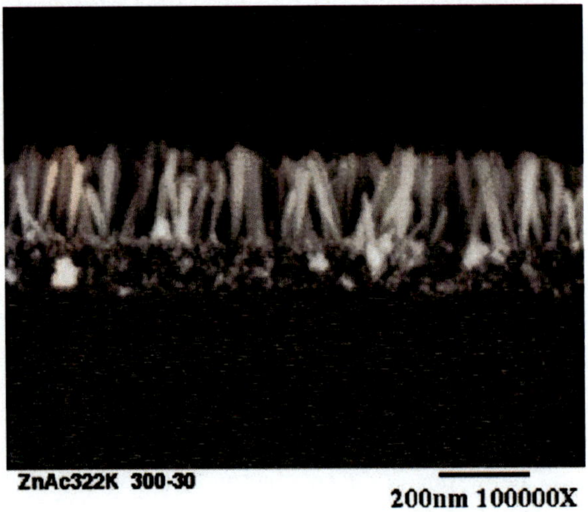

Fig. 12.4 Scanning electron microscope (SEM) images of ZnO fibers grown from an aqueous solution of zinc nitrate

based on a decomposed zinc acetate nucleation layer, make an ohmic contact to indium tin oxide (ITO), the electrode on which the fibers are grown [60].

Devices based on polymer/ZnO nanorod arrays have been prepared, however initial device efficiencies are lower than their polymer/fullerene blend counterparts [76–80]. The hybrid P3HT/ZnO nanorod devices have demonstrated dramatically improved conversion efficiency compared to planar (bilayer) hybrid devices [73], as seen in Fig. 12.5. The bilayer device of P3HT on the planar sol-gel ZnO layer exhibited a V_{OC} of 501 mV, a J_{SC} of 0.52 mA/cm^2, a *FF* of 55%, and an l of 0.15%. The performance of the nanostructured hybrid device was significantly improved with values for the V_{OC}, J_{SC}, *FF*, and l of 440 mV, 2.17 mA/cm^2, 56, and 0.53%, respectively. The increase in power conversion efficiency is due to enhanced J_{SC}, which is a result of increased interfacial area of the heterojunction between the ZnO nanorods and the P3HT. The performance is improved compared to previous attempts by other groups using nanostructured TiO_2, where J_{SC} values were much smaller [67, 68].

Additionally, the external quantum efficiency (EQE) spectra for the ITO/ZnO nanorods/P3HT/Ag devices are higher than the bilayer ITO/ZnO/P3HT/Ag are compared to the absorption spectrum for a corresponding nanostructured composite film as seen in Fig. 12.6. The photoaction spectra overlap with the absorption spectrum, however the action spectra for both devices indicate that much of the light absorbed is not converted to charge carriers. This is likely due to the large spacing between the nanorods used in this study (100 nm) and the short exciton diffusion length found in P3HT (3–5 nm) [74]. The bilayer device produced

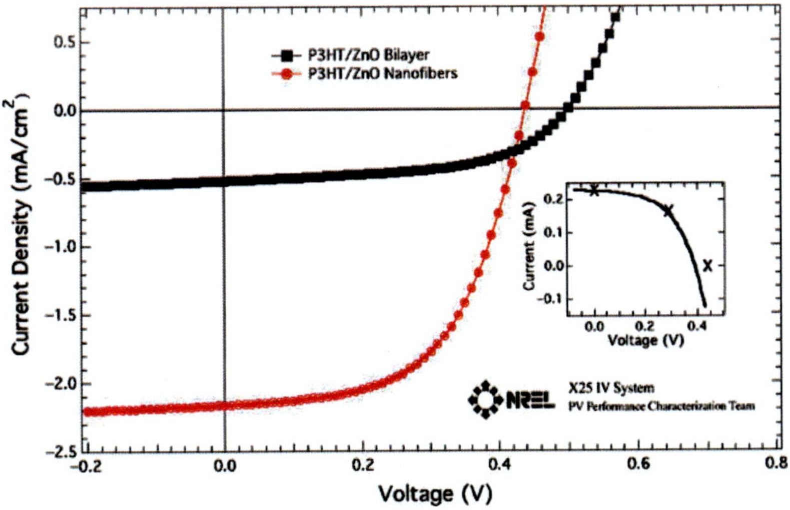

Fig. 12.5 Current density vs. voltage (J–V) for a typical ITO/ZnO/P3HT/Ag bilayer device (*black squares*) and ITO/ZnO nanorods/P3HT/Ag device (*black circles*) [73]. The devices were measured under AM 1.5 illumination with an intensity of 100 mW/cm^2, corrected for the spectral mismatch factor. *Inset* shows NREL certified measurement data

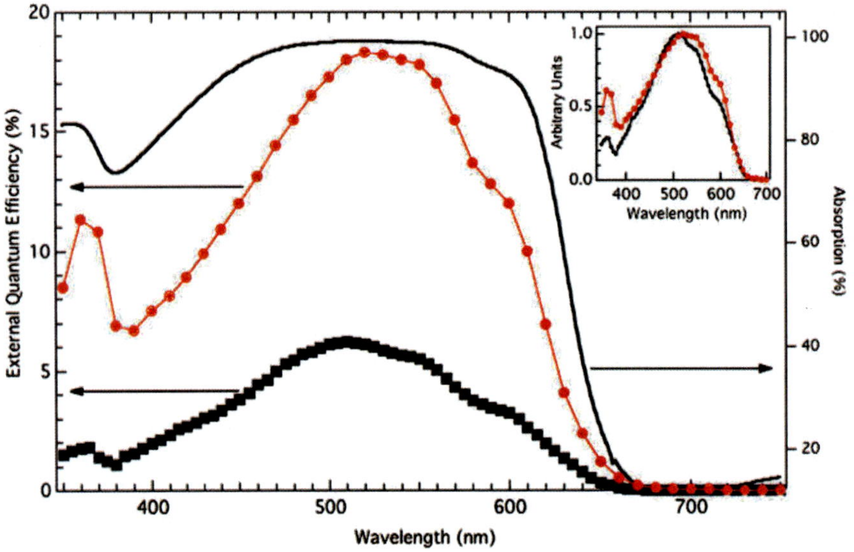

Fig. 12.6 External quantum efficiency of ITO/ZnO/P3HT/Ag bilayer device (*black squares*), ITO/ZnO fibers/P3HT/Ag nanostructured device (*black circles*) and UV–vis absorption spectrum of the nanostructured device (*smooth solid line*) [73]. *Inset* shows normalized spectra

a maximum EQE of 6.2% at 510 nm, corresponding to a J_{SC} of 0.52 mA/cm^2. The ZnO nanorod device achieved a maximum EQE of 18.3% at 520 nm corresponding to 2.1 mA/cm^2. The peak in both EQE spectra at 360 nm is the result of UV light absorption by the ZnO and subsequent hole transfer to the P3HT, which demonstrates clear contributions to the photocurrent from light absorption in both the P3HT and ZnO materials. Integration of the measured spectral response of the nanostructured device with the AM 1.5 solar spectrum (100 mW/cm^2) results in an estimated J_{SC} of 2.06 mA/cm^2, which agrees very well with the observed white light J_{SC} above. The normalized EQE spectra of both devices indicate that there is a small red shift in the photoresponse of the nanostructured device as seen in the inset of Fig. 12.6. This red shift in the EQE of polymer/ZnO nanorod device could be due to increased molecular order of the P3HT chains along the vertical surface of the nanorods.

There are several possible routes to improving the performance of the nanostructured devices. First, the V_{OC} is lower than expected from this donor–acceptor pair. Bulk heterojunction devices using a common donor polymer such as poly[2-methoxy-5-(3′,7′-dimethyloctyloxy)-1,4-phenylene vinylene] (MDMO-PPV) blended with PCBM as an acceptor have demonstrated similar V_{OC} values to hybrid blend devices using ZnO nanoparticles [54, 75]. Devices consisting of blends of P3HT and PCBM have yielded a V_{OC} of approximately 600 mV, therefore a similar V_{OC} might be expected in these devices. The reason the V_{OC} seen in the device is substantially lower than this is not clear, but possible explanations include the presence of mid-gap states on the surface of the ZnO that pin the Fermi level; or that the electron mobility of the ZnO nanofibers could actually be too high leading to space-charge build up, resulting in increased carrier recombination at the ZnO/P3HT interface and a reduced V_{OC}.

Secondly, the spacing between the ZnO nanofibers as grown is on the order of 100 nm. This is substantially larger than the typical exciton diffusion length in P3HT, which is presumed to be less than 10 nm. It is not clear how ordering of the P3HT chains inside the pores of the ZnO may affect this. This suggests that closer spacing of the fibers may yield a larger J_{SC} and thereby increase the J_{SC} and EQE to values similar to present P3HT/PCBM bulk heterojunction devices.

12.3.3 Inverted P3HT:PCBM Bulk Heterojunction Devices Based on ZnO

The inverted P3HT:PCBM bulk heterojunction device may prove invaluable for the development of devices with increased stability [58, 67–69, 82–89]. The inverted device geometry allows for the elimination of the low work function counter electrode required in traditional polymer/fullerene bulk heterojunction devices. Low work function metals have a large affinity for oxygen, resulting in the formation of an insulating oxide barrier. As a result, this may be a major degradation mechanism in these devices. An inverted device architecture also removes the need for the poly(3,4-ethylenedioxythiophene):poly(styrenesulfonate) (PEDOT:PSS) interfacial layer, which can introduce chemical and morphological instabilities with the ITO [76, 77]. Figure 12.7 compares the J–V curves under illumination for both the traditional ITO/PEDOT/P3HT:PCBM/Al device architecture fabricated in an argon atmosphere and the inverted ITO/ZnO/P3HT:PCBM/Ag device design fabricated in air and developed here. These devices prepared in the same manner, one in the glove box and one in air, clearly demonstrate that comparable performance is possible with the inverted device configuration. In fact, the performance of the inverted device is higher despite the device being processed in air, which historically has been strictly avoided in polymer/fullerene bulk heterojunction devices. While initial stability results have been inconclusive, the ability of inverted devices to be fabricated in air and still perform remarkably well is an indication that the configuration might lead to increased long-term device stability.

12.3.4 Increasing Charge Collection: P3HT:PCBM/ZnO Nanorod Devices

In order to dissociate more of the photogenerated excitons in the film and increase the J_{sc} of the cell, a blend of P3HT and PCBM can be intercalated into the ZnO nanorod or TiO_2 nanotube arrays [78–80]. The resulting device resembles the

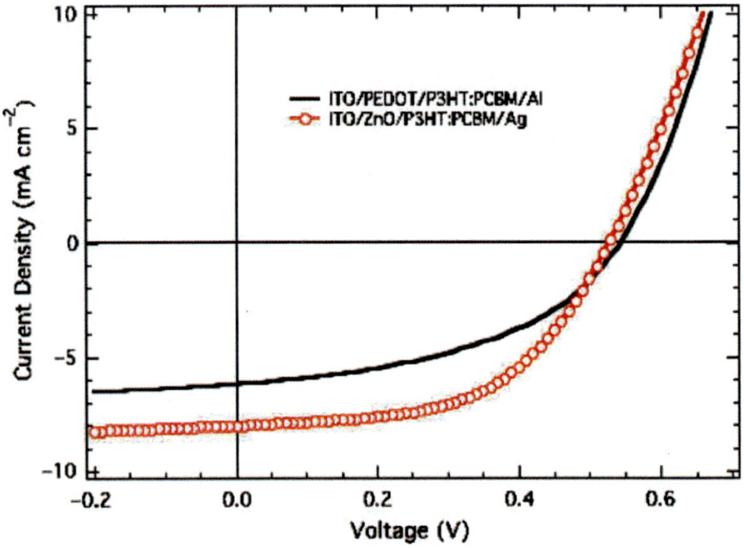

Fig. 12.7 Current density vs. voltage for typical ITO/PEDOT/P3HT:PCBM blend/Al device (*solid*) and ITO/ZnO/P3HT:PCBM blend/Ag device (*open circles*). The devices were measured under AM 1.5 illumination with an intensity of 100 mW/cm^2, corrected for the spectral mismatch factor

commonly studied polymer/fullerene device, except that the device structure is, again, inverted in terms of current collection relative to standard polymer/fullerene devices. ZnO nanorods also penetrate into the blend morphology. The performance of the ITO/ZnO nanorod/P3HT:PCBM/Ag device was significantly better than for similar devices without PCBM [78], demonstrating values for the V_{OC}, J_{SC}, FF and | of 475 mV, 10.0 mA/cm^2, 0.43, and 2.0%, respectively, as seen in Fig. 12.8. The incorporation of PCBM into the P3HT results in efficient exciton dissociation and increases the J_{SC} by nearly a factor of five over the device without PCBM as discussed above. Through optimization of the P3HT:PCBM ratio in the ZnO nanorod/blend device architecture the efficiency has been reported to increase to 2.7% [79], and when TiO_2 nanotube arrays are used as the electron transport material efficiencies of P3HT:$PC_{71}BM$ have been reported as high as 4.1% [80].

12.3.5 ZnMgO Alloys in Hybrid PV Cells

Few efforts in organic photovoltaic devices have significantly improved the open circuit voltage (V_{OC}) in polymer photovoltaic devices. Standard values for the V_{OC} in P3HT/PCBM bulk heterojunction devices are ~0.60 V, whereas the lowest energy photons absorbed by P3HT are approximately 1.9 eV. According to the conventional diode model yields for a material with a band gap of 1.9 eV, the

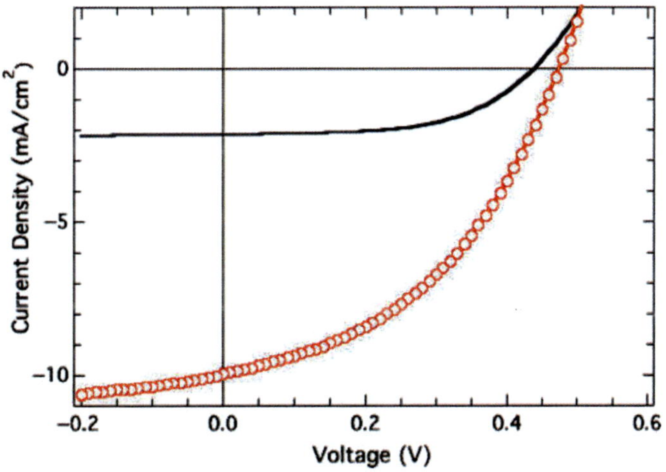

Fig. 12.8 Current density vs. voltage for typical ITO/ZnO/ZnO fibers/P3HT/Ag device (*solid*) and ITO/ZnO/ZnO fibers/P3HT:PCBM blend/Ag device (*open circles*) [78]. The devices were measured under AM 1.5 illumination with an intensity of 100 mW/cm^2, corrected for the spectral mismatch factor

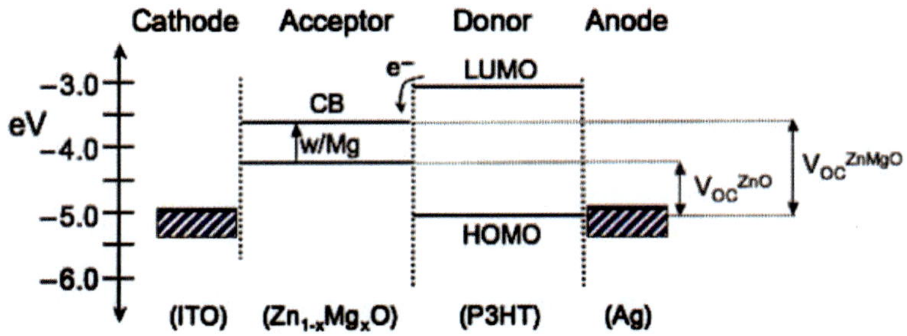

Fig. 12.9 Band diagram of ITO/$Zn_{1-x}Mg_xO$/P3HT/Ag hybrid polymer/metal oxide device [91]

maximum V_{OC} should be 1.5 V. Therefore 60% of the energy of an incoming photon is lost to thermalization.

The V_{OC} in an organic solar cell is fundamentally determined by the effective electronic band gap of the donor–acceptor system [81]. This is estimated by the offset between the conduction band or lowest unoccupied molecular orbital (LUMO) in the acceptor and the valence band or highest occupied molecular orbital (HOMO) of the donor as seen in Fig. 12.9. The V_{OC} measured when testing the device will be less than this value due, in part, to the binding energy of the exciton and carrier recombination either in the bulk or at the electrodes [82, 83]. The V_{OC} can also be reduced by chemical potential pinning to poorly optimized electrodes [84], by changing the work function of either electrode [82, 85], or by interface dipoles at the electrode [86]. Modifying the reduction potential of the electron acceptor (fullerene) with respect to the electron-donating polymer has demonstrated the relationship between the band offset and the V_{OC} [82, 87–89]. Similarly, changes in V_{OC} have been observed by varying the oxidation potential of the electron-donating polymer [90]. When the band offset is increased, the V_{OC} decreases.

The band offset in the donor–acceptor pair is important for controlling both the electron transfer kinetics as well as determining the maximum V_{OC}. The reduction of the donor–acceptor band offset, while maintaining sufficient driving force necessary for efficient electron transfer and slow recombination kinetics, should minimize the energy loss and allow for higher V_{OC}. Where the V_{OC} has not easily been increased using a given material set in polymer/fullerene systems, hybrid systems that rely on metal oxide electron acceptors are proving to be easily tailored for increased V_{OC} [91].

The band gap of ZnO can be modified through the substitution of divalent cations for Zn. Mg substitution has been demonstrated through pulsed laser deposition (PLD) experiments to increase the band gap [92], whereas Cd substitution has been shown to reduce it [93]. These appear to be true solid solution systems showing very small structural changes to the ZnO wurtzite structure. Additionally, the samples can be doped in the conventional fashion with Al, Ga or In giving rise to the potential for independently controlling the gap and doping level in the metal oxide [94]. Additionally, $Zn_{1-x}Mg_xO$ films have been fabricated through sol-gel deposition methods with Mg contents up to $x = 0.40$ [91]. A linear increase in the band gap from 3.2 to 4 eV was observed [91, 95] with no phase separation through annealing up to 700°C [96]. According to X-ray diffraction (XRD) data the films remain in the wurzite-type ZnO phase with increasing Mg content without any impurity phase until $x = 0.45$ when there is phase separation to include the cubic MgO phase [91]. The maximum Mg content is significantly larger than the thermodynamic limit of $x = 0.04$ [95]. Incorporation of Mg into the ZnO phase does decrease the crystallinity of the film, with the (100), (002), and (101) peaks becoming weaker and broadening with increasing Mg at the anneal temperatures employed here. There is not a dramatic shift in peak position, indicating that Mg is incorporated without significant strain on the lattice, which would be expected based on ionic radii of Zn^{2+} and Mg^{2+} (0.60 and 0.57 Å, respectively). Time resolved microwave conductivity measurements indicate that the photoconductivity of the $Zn_{1-x}Mg_xO$ films decreases with increasing Mg content [97].

Much of this increase in band gap should be the result of movement in the energy of the conduction band, which agrees with the common anion rule [98]. Hence, the V_{OC} could be increased by changing the conduction band level in such a film of $Zn_{1-x}Mg_xO$ as seen in Fig. 12.9. An increase in the oxide band gap could be of additional benefit as photoexcitation in the ZnO from ultraviolet radiation can result in hole transfer from the ZnO to the polymer. This often leads to degradation of the polymer due to the strong oxidizing potential of the ZnO. Increasing the band gap might reduce the amount of photoexcitation in the metal oxide, and thereby eliminate one of the degradation mechanisms in this cell design.

To observe the effects of varied band offsets on the V_{OC}, hybrid P3HT/$Zn_{1-x}Mg_xO$ bilayer devices were prepared with increasing Mg content. Bilayer devices demonstrate the concept, though nanostructured $Zn_{1-x}Mg_xO$ materials with increased surface area will ultimately be required for higher efficiency devices. The substitution of Mg into the ZnO acceptor results in a systematic increase in the V_{OC} from 500 mV for $Zn_{1.0}Mg_{0.0}O$ to greater than 900 mV for $Zn_{0.65}Mg_{0.35}O$, as seen in Fig. 12.10. When the Mg composition increased to x = 0.40, the V_{OC} drops dramatically. Similarly, the J_{SC} and I increase with increasing Mg content as well, though they begin to fall off above x = 0.25 as seen in Fig. 12.11. These observations might indicate that the band offset is not great enough to efficiently separate charges at that Mg substitution level, or that the series resistance in the device is too high to allow efficient transport of the charges out of the device.

The V_{OC} is observed to increase linearly with band gap of $Zn_{1-x}Mg_xO$ with a slope of 1.0 up to $Zn_{0.80}Mg_{0.20}O$, above which the slope decreases to 0.25. As the conduction band level moves closer to vacuum there is a corresponding increase in the V_{OC}, indicating that the increased band gap results in a shift of the conduction band of the $Zn_{1-x}Mg_xO$ acceptor.

The series resistance in these devices increases exponentially from 4.2 to 72 Ω cm^2 with increasing Mg content as seen in the inset in Fig. 4.9. This agrees with time resolved microwave conductivity (TRMC) measurements [97], which indicate that increasing Mg content results in an exponential decrease of over two orders of magnitude in the carrier mobility [97]. The carrier lifetime in the $Zn_{1-x}Mg_xO$ films remains constant throughout the series. These observations suggest that the Mg atoms do not act as trap defects but rather as scattering centers that reduce the carrier mobility. The increase in the series resistance should lead to

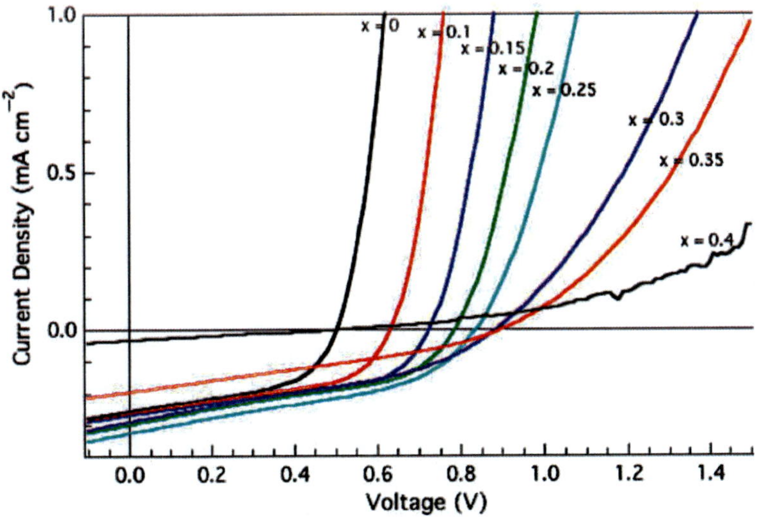

Fig. 12.10 Photovoltaic performance of synthesized P3HT/$Zn_{1-x}Mg_xO$ devices ($0 < x < 0.40$). Current density vs. voltage at various Mg contents was measured at an intensity of 100 mW cm^{-2}, corrected for spectral mismatch for an AM 1.5 spectrum [91]

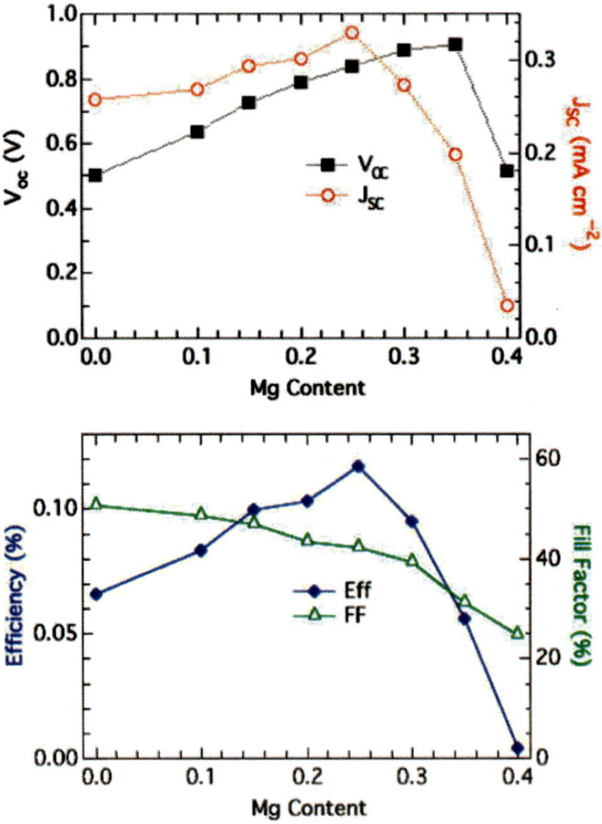

Fig. 12.11 Photovoltaic performance of synthesized P3HT/$Zn_{1-x}Mg_xO$ devices ($0 < x < 0.40$). *Top*, V_{OC} (*solid squares*) and J_{SC} (*open circles*) vs. Mg content. *Bottom*, Efficiency (*solid diamonds*) and *FF* (*open triangles*) vs. Mg content [91]

a decrease in the performance of the device. However despite the increase in series resistance, the J_{SC} actually increases by 25% up to $Zn_{0.75}Mg_{0.25}O$. This could be explained by several possible mechanisms, including an increase in surface roughness leading to increased interfacial area between polymer and oxide which will be discussed later, a change in the intrinsic carrier concentration in the $Zn_{1-x}Mg_xO$, an alteration of the chemistry of the interface leading to different exciton dissociation or carrier recombination rate, or a balancing of the electron and hole mobilities (time of flight mobility measurements for P3HT indicate a mobility of 4×10^{-4} $cm^2\ V^{-1}\ s^{-1}$) [99] that reduces recombination in the device [30].

Devices fabricated on $Zn_{0.75}Mg_{0.25}O$ using the Rieke P3HT exhibited a V_{OC} of 700 mV, a J_{SC} of 1.27 mA cm^{-2}, a *FF* of 56%, and a power conversion efficiency of 0.5%. Similar efficiencies were observed in hybrid P3HT/ZnO nanofiber composite devices, where the interfacial area was much larger. These results for P3HT/ $Zn_{1-x}Mg_xO$ indicate that moving the conduction band of the electron acceptor closer to vacuum yields a systematic increase of V_{OC} as anticipated. The ability to easily tune the band offsets in hybrid polymer/$Zn_{1-x}Mg_xO$ devices provides a tool for investigating the fundamental nature of the generation of photovoltage in these devices and provides a potential pathway to higher device efficiencies.

12.4 Dye-Sensitized Solar Cells Using Nanoparticle TiO_2

Though not an OPV device per say the dye-sensitized solar cell (DSSC) represents one of the most successful non-classical semiconductor devices to date. Simple enough to be made by school children it actually is a sophisticated device operating on the nanoscale. Figure 12.12 illustrates a typical schematic for a dye sensitized solar cell.

Here the nanostructured TiO_2 layer is on a TCO transparent contact (typically SnO_2) and has a photoactive dye adsorbed on it [100, 101]. Normally this layer is anatase, which has substantially better transport properties than rutile. Because of the nanostructure and therefore high surface area the dye (normally a Ru based coordination compound) has a high optical density. Upon excitation an electron is injected into the TiO_2 where it percolates to the back TCO contact [102–104]. The hole is captured by an Iodine/iodide couple and hops to the top electrode. Normally the TCO only has the intrinsic conductivity due to the defects formed during the deposition and anneal process. More recently it has been demonstrated the TiO_2 can be substitutionally doped by Nb or Ta [105–107].

12.4.1 3D Structure

Key to the functioning of the device is the very large surface are 3D structure of the oxide. As discussed above it is critical to get a sufficient optical density for the dye, it is key to have an intimate intermixing of the dye coated TCO and the electrolyte and it appears important for the transport of the electrons. TiO_2 while it is the gold standard of the DSSC is not unique, as cells have been fabricated with nanostructured ZnO similar to the OPV devices. With the advent of an increasing pallet of nanostructure oxides for TCO there may be additional materials with optimized properties [9–12, 108–111]. Figure 12.13 shows a nanorod array for alpha-Fe_2O_3 supporting the idea that many oxides binary and ternary can be grown in this kind of morphology potentially adding functionality to a variety of structures.

Included in this set is an increasing set of carbon nanotube based transparent conductive composite structures [112–116]. This opens a versatile new set of materials through combining nano-tubes single and multiwall with polymers in a composite. None-the-less it is hard to beat TiO_2 for cost in large area applications.

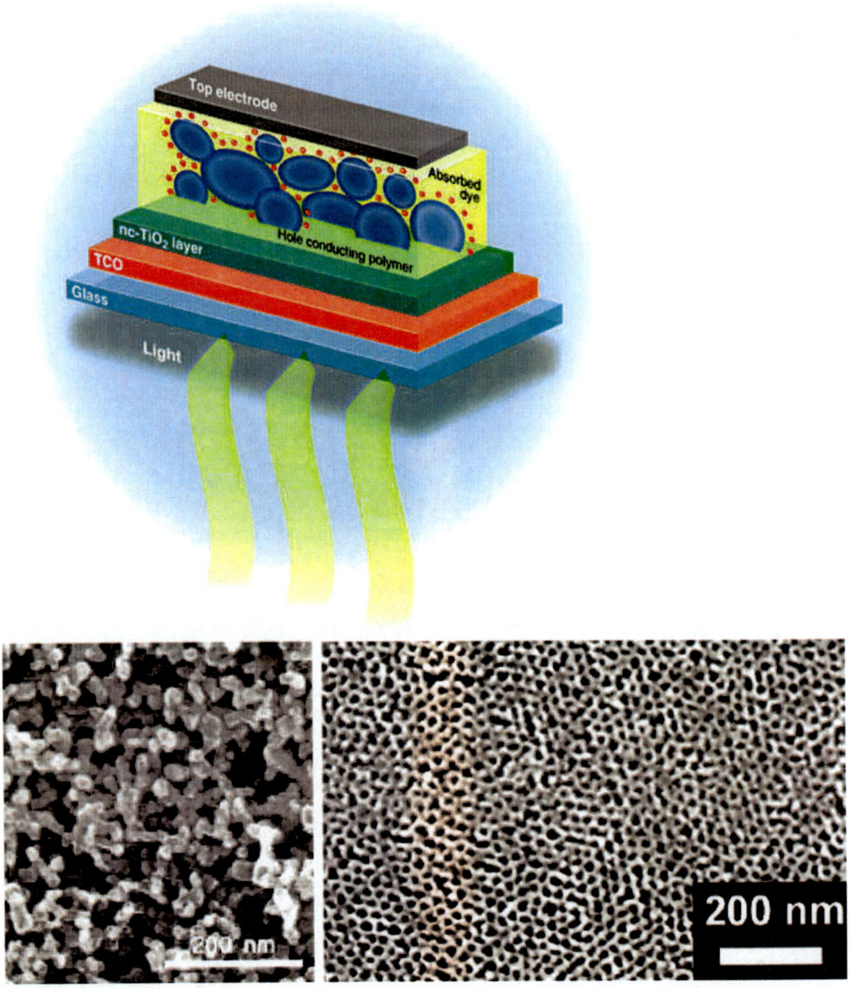

Fig. 12.12 Showing a schematic of a dye sensitized solar cell (*top*) and SEM images of a typical TiO_2 nanoporous electrode (*bottom, left*) and a mesoporous nanostructured electrode with 6 nm pores (*bottom, right*) [57]

12.4.2 Conductivity

To date the conductivity of the TiO_2 has been whatever weakly induced defect doping occurs in the sintering of the typical nano-particle slurries used to make the nanoporous electrode. While there is an effective percolation of electrons to the back TCO contact in fact anatase has historically been very difficult to dope, only showing a meta-stable conductivity under hydrogen reduction. Recently a number of laboratories have shown that Nb and Ta can substitutionally dope TiO_2 and that

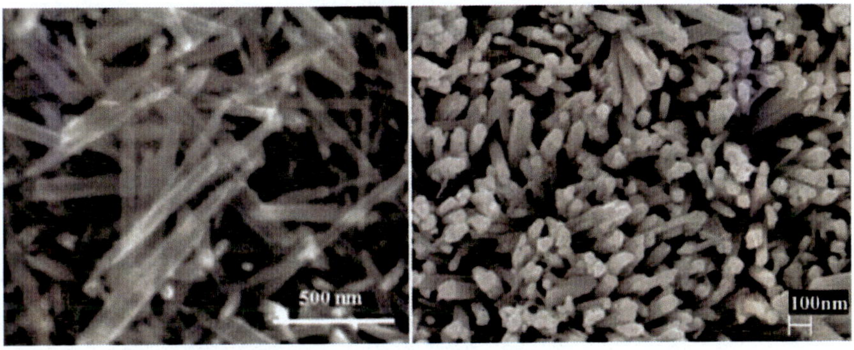

Fig. 12.13 SEM micrographs of α-Fe_2O_3 nanoparticle thin films consisting of oriented nanorods; parallel (*left*) and perpendicular (*right*) to the substrate [63]

epitaxial films can have conductivities as good as ITO [106–108]. This may be much too conductive for the DSSC, but as can be seen in the results on glass it appears that lower conductivities are readily attainable. Thus it may be possible to deliberately tune the conductivity of the TiO_2. This is also true for the ZnO nanorods where controlling the oxygen stoichiometry controls the conductivity over a wide range [4, 5, 117–119]. Another advantage of the ZnO potentially is that it can be doped with Al or Ga and can be substituted with Mg or Cd which can increase or decrease respectively the band gap [120–124]. This may be important for the solid state DSSC where the I/I^{3-} couple is replaced by an organic equivalent. To date this has not worked well but an optimized oxide may shorten diffusion lengths and control the band edges.

12.4.3 Transport

As discussed above transport in the TiO_2 is both critical and fortuitous as stated in the previous section the increasing interest in controlling the band edges and in doping TiO_2 potentially allows for much more deliberate control of the transport properties of the TiO_2 layer.

12.5 Next Generation Coatings

As alluded to in the buildings section above TCO layers may have multiple functions in a number of environments. As we have discussed above of many new applications based either on limited mobility of excitons or carriers in the materials or the need for very high surface area materials are leading to the development of controlled nanostructures for the TCO component. This is true in the case of TCO coatings for windows where in some cases such as the ZnO/Ag stacks used in low-e window the formation of an inherent nanostructure is critical to the device [4, 8, 125–128]. It is also true in electrochromic devices where the TCO layer may be involved as a contact as an optical layer and even as a Li storage layer [129–134]. All of which necessitate a controlled nanostructure. As the interior and exterior coating on buildings, automobiles, airplanes, etc. continue to become more sophisticated they will become multi-functional being involved in control of transparency, controlling static, having imbedded electronics and having photoactive surfaces that can catalyze the decomposition (oxidation) of pollutants both chemical and biological.

12.5.1 Superhydrophylic TiO_2

One of the best examples of this is superhydrophylic window and other coatings as has been enthusiastically endorsed in Japan and Europe. Most of the literature in this area is in the form of patents but there are a few review articles. Most of the focus has been on TiO_2 sputtered films with an anatase crystal structure. Thin nanostructured anatase films are both super hydrophilic (contact angle of zero) and highly photocatalytic generating oxygen radicals that are highly oxidizing. These coatings have been used for self-cleaning windows, anti-soiling coatings, anti-static coatings, and highly active photo-oxidation coatings in auto, appliance, fabric, and window applications [135–140]. Functionally they are similar to the TiO_2 coatings employed for the DSSC. Nominally the conductivity of these coatings was whatever was intrinsically formed during the sputtering or sol gel processing of the films. Recently, however, it has been shown that TiO_2 anatase may be substitutionally doped with Nb or Ta to produce highly conductive films in epitaxial materials but to produce moderately conductive films or glass or flexible substrates. This now provides an important new variable for controlling the properties of the highly active TiO_2 coatings [107, 108]. It has also been shown that ZnO and SnO_2 based films exhibit many of these same properties.

12.6 Novel Heterostructures

12.6.1 From OLEDs to Fuel Cells and Displays

It is increasingly clear in a variety of devices that simply having transparency and conductivity is not enough. In fact it is important to control the work function, the reflectivity, and the length scale of the TCO layer. This latter is critically important in a number of devices from solar cells, to OLEDs to fuel cells. This is due the nanoscale diffusion lengths of the excitons, carriers, and/or chemical diffusion in the device. It is only recently that we have begun to have synthetic control such as to define the nano-scale architecture of the TCO layer. Coupled to this is an increasing understanding of the ability to dope and to substitutionally modify the materials to control the conductivity well as the conduction and valance band positions. Thus increasingly, it appears possible to have large area TCO contacts optimize for the inherent transport in the device, to manage light absorption reflection, and to control the doping and band edge positions. This dramatically extends the potential applicability of TCOs to new organic/inorganic heterostructure devices and to become active elements in devices such as PEM fuel cells where transparency is not important but the rest of the properties are critical.

12.6.2 New Electro-Optical Devices

As discussed in the section above the increasing ability to control the morphology, structure and electronic properties of TCOs is leading to the potential for many new device structures. In addition this ability to integrate organic, biological and inorganic moieties leads to many new potential applications for sensors and active devices. Additional understanding of the surface chemistry of the TCOs will allow for an integration at the bonding level of active biological, catalyst and polymer electro-optic molecules leading to many new functionalities. In addition, our increasing ability to modify the TCO means that they now represent a basic building material for all oxide based electronics as well producing new high mobility transistors and spintronic materials. There is a plethora of opportunities that only 5 years ago were inconceivable.

12.7 New Nanostructured TCO Materials

12.7.1 Carbon Nanotube Composites

A possible alternative to ITO and PEDOT:PSS as transparent conducting electrodes are single-walled carbon nanotube (SWCNT) films [141–143]. This material shows high conductivities combined with high optical transmission and is solution processable. Additionally, SWCNT electrode materials could more easily allow for roll-to-roll processing on flexible substrates. Carbon nanotubes have previously been used in organic devices, such as a solar cell using multiwalled nanotubes as the hole collector [144]. This had very low short circuit current density (*Jsc*) and efficiency. Recently, a device was demonstrated using SWCNTs as the cathode in P3HT:PCBM bulk heterojunction device [145]. Unfortunately the active layer thickness in this device was 800 nm, which is much thicker than is necessary. This device was transport limited and resulted in a power-conversion efficiency of

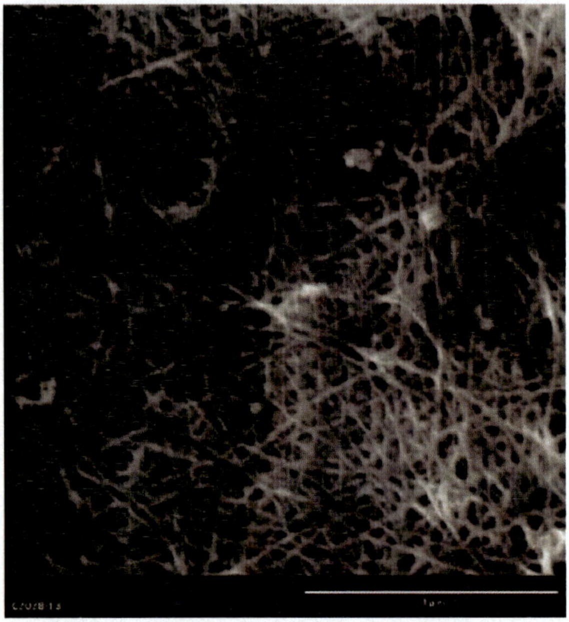

Fig. 12.14 SEM micrograph of carbon nanotube electrode materials

1%, however this clearly illustrates the promise of SWCNTs to replace ITO (Fig. 12.14).

More recent studies have employed SWCNT thin films and have used optimal P3HT:PCBM active layer thicknesses [146, 147]. In the first study, SWCNT films were spray coated onto glass through a shadow mask to form thin films. Devices with power conversion efficiencies of 1.5% were obtained by using SWCNT films with transparency of 70% at 650 nm and sheet resistance of 50 Ω/square [146]. A second study used a transfer printing method to deposit smooth SWCNT thin films onto poly(ethylene terephthalate) (PET), a flexible substrate [147]. Bulk heterojunction devices with efficiencies of 2.5% were obtained using SWCNT films with transparency of 85% in the visible and sheet resistance of 200 Ω/square. The flexibility of these devices was observed to be superior to devices fabricated on PET substrates coated with ITO.

References

1. Anna Selvan, J.A., Delahoy, A.E., Guo, S., & Li, Y.-M., A new light trapping TCO for nc-Si: H solar cells. *Solar Energy Materials & Solar Cells* 90(18–19), 3371–3376 (2006).
2. Groenen, R., Creatore, M., & van de Sanden, M.C.M., Dry etching of surface textured zinc oxide using a remote argon-hydrogen plasma. *Applied Surface Science* 241(3–4), 321–325 (2005).

3. Pan, Z.W., Dai, Z.R., & Wang, Z.L., Nanobelts of semiconducting oxides. *Science (Washington, DC, United States)* 291(5510), 1947–1949 (2001).
4. Shen, G., Bando, Y., & Lee, C.-J., Growth of self-organized hierarchical ZnO nanoarchitectures by a simple In/In2S3 controlled thermal evaporation process. *Journal of Physical Chemistry B* 109(21), 10779–10785 (2005).
5. Tornow, J. & Schwarzburg, K., Transient electrical response of dye-sensitized ZnO nanorod solar cells. *Journal of Physical Chemistry C* 111(24), 8692–8698 (2007).
6. Wang, H., Xie, C., Zeng, D., & Yang, Z., Controlled organization of ZnO building blocks into complex nanostructures. *Journal of Colloid and Interface Science* 297(2), 570–577 (2006).
7. Wei, Q., Meng, G., An, X., Hao, Y., & Zhang, L., Temperature-controlled growth of ZnO nanostructures: branched nanobelts and wide nanosheets. *Nanotechnology* 16(11), 2561–2566 (2005).
8. Zhu, F., Xue, Y.-m., Sun, J., Zhao, Y., & Geng, X.-h., Textured ZnO transparent conductive thin film. *Rengong Jingti Xuebao* 33(3), 419–421, 427 (2004).
9. Antonio, J.A.T. *et al.*, Application: US. US Patent No. 2005-121178, 2006078726 (20050504. 2006).
10. Chu, S.Z., Wada, K., Inoue, S., & Todoroki, S., Fabrication of oxide nanostructures on glass by aluminum anodization and sol-gel process. *Surface and Coatings Technology* 169–170, 190–194 (2003).
11. Spanier, J.E., One-dimensional semiconductor and oxide nanostructure. *Nanotubes and Nanofibers*, 199–232 (2006).
12. Ye, C., Wang, M., Hua, G., & Zhang, L., Application: CN. CN Patent No. 1009-4806, 1948127 (20051010. 2007).
13. Yoo, S., Akbar, S. A., & Sandhage, K.H., Ceram. Chemistry of Materials 17(19), 4825–4831 (2005).
14. Arbab, M., Shelestak, L.J., & Harris, C.S., Value-added flat-glass products for the building, transportation markets, part 2. *GlassResearcher* 14(2), 34–38 (2005).
15. Chen, H. & Zhou, Y., Application: CN. CN Patent No. 1009-955, 1888342 (20060705. 2007).
16. Doi, K., Application: JP. JP Patent No. 98-234737, 2000073276 (19980820. 2000).
17. Fukuwatari, I. & Mizuta, K., Application: JP. JP Patent No. 99-59670, 2000254043 (19990308. 2000).
18. Jonschker, G., Application: DE. DE Patent No. 2001-10116198, 10116198 (20010330. 2002).
19. Khan, M.I., Novel nanostructured solids for applications in environmental catalysis and gas sensing. *Abstracts of Papers, 229th ACS National Meeting, San Diego, CA, United States, March 13–17, 2005*, IEC-125 (2005).
20. Kitamura, A. & Hayakawa, M., Application: JP. JP Patent No. 97-31159, 10212139 (19970130. 1998).
21. Martens, J.A. *et al.*, Application: WO. WO Patent No. 2004-BE94, 2005000740 (20040628. 2005).
22. Sanderson, K.D., Mills, A., Hurst, S., Lepre, A., McKittrick, T., Rimmer, D., & Ye, L., Svc 46th Annual Technical Conference Proceedings, Publishers: Society of Vacuum Coaters, Albuquerque, Nm, Usa, 203–208 (2003).
23. Sugihara, S., Application: JP. JP Patent No. 98-287750, 2000103647 (19981009. 2000).
24. Sugihara, S., Application: JP. JP Patent No. 99-275283, 2001096168 (19990928. 2001).
25. Toshiaki, A. *et al.*, High performance photocatalytic coating on large area glass by sputtering. *International Congress on Glass, Proceedings, 20th, Kyoto, Japan, Sept. 27–Oct. 1, 2004*, 09 008/001-009 008/006 (2004).
26. Shockley, W. & Queisser, H.J., Detailed balance limit of efficiency of p–n junction solar cells. *Journal of Applied Physics* 32(3), 510–519 (1961).

27. Shaheen, S.E., Ginley, D.S., & Jabbour, G.E., Organic-based photovoltaics: toward low-cost power generation. *MRS Bulletin* 30, 10–15 (2005).
28. Shaheen, S.E., Radspinner, R., Peyghambarian, N., & Jabbour, G.E., Fabrication of bulk heterojunction plastic solar cells by screen printing. *Applied Physics Letters* 79(18), 2996–2998 (2001).
29. Yu, G., Gao, J., Hummelen, J.C., Wudl, F., & Heeger, A.J., Polymer photovoltaic cells: enhanced efficiencies via a network of internal donor–acceptor heterojunctions. *Science* 270, 1789–1791 (1995).
30. Li, G. *et al.*, High-efficiency solution processable polymer photovoltaic cells by self-organization of polymer blends. *Nature Materials* 4, 864–868 (2005).
31. Ma, W., Yang, C., Gong, X., Lee, K., & Heeger, A.J., Thermally stable, efficient polymer solar cells with nanoscale control of the interpenetrating network morphology. *Advanced Functional Materials* 15, 1617–1622 (2005).
32. Padinger, F., Rittberger, R.S., & Sariciftci, N.S., Effects of postproduction treatment on plastic solar cells. *Advanced Functional Materials* 13(1), 85–88 (2003).
33. Hoppe, H. *et al.*, Nanoscale morphology of conjugated polymer/fullerene-based bulk-heterojunction solar cells. *Advanced Functional Materials* 14(10), 1005–1011 (2004).
34. Shaheen, S.E. *et al.*, 2.5% efficient organic plastic solar cells. *Applied Physics Letters* 78(6), 841–843 (2001).
35. Coakley, K.M., Liu, Y., Goh, C., & McGehee, M.D., Ordered organic–inorganic bulk heterojunction photovoltaic cells. *MRS Bulletin* 30, 37–40 (2005).
36. O'Regan, B. & Grätzel, M., A low-cost, high-efficiency solar cell based on dye-sensitized colloidal TiO_2 films. *Nature* 353, 737–740 (1991).
37. Law, M., Greene, L.E., Johnson, J.C., Saykally, R., & Yang, P., Nanowire dye-sensitized solar cells. *Nature Materials* 4, 455–459 (2005).
38. Minami, T., Miyata, T., & Yamamoto, T., Work function of transparent conducting multicomponent oxide thin films prepared by magnetron sputtering. *Surface and Coatings Technology* 108–109 (1–3), 583–587 (1998).
39. Holmes, R.J. *et al.*, Efficient, deep-blue organic electrophosphorescence by guest charge trapping. *Applied Physics Letters* 83(18), 3818–3820 (2003).
40. Hains, A.W. & Marks, T.J. High-efficiency hole extraction/electron-blocking layer to replace poly(3,4-ethylenedioxythiophene): poly(styrene sulfonate) in bulk-heterojunction polymer solar cells. *Applied Physics Letters* 92, 023504 (2008).
41. Hains, A.W., Martinson, A.B.F., Irwin, M.D., Yan, H., & Marks, T.J., Bulk-heterojunction organic solar cells: interfacial engineering routes to increased open-circuit voltage and power conversion efficiency. *Polymeric Materials: Science and Engineering* 96, 814 (2007).
42. Irwin, M.D., Buchholz, D., Hains, A., Chang, R., & Marks, T., p-Type semiconducting nickel oxide as an efficiency-enhancing anode interfacial layer in polymer bulk-heterojunction solar cells. *Proceedings of the National Academy of Sciences of the United States of America* 105(8), 2783–2787 (2008).
43. Yan, H. *et al.*, High-brightness blue light-emitting polymer diodes via anode modification using a self-assembled monolayer. *Advanced Materials* 15(10), 835–838 (2003).
44. Yan, H., Huang, Q., Scott, B., & Marks, T.J., A polymer blend approach to fabricating the hole transport layer for polymer light-emitting diodes. *Applied Physics Letters* 84(19), 3873 (2004).
45. Yan, H. *et al.*, High-performance hole-transport layers for polymer light-emitting diodes. Implementation of organosiloxane cross-linking chemistry in polymeric electroluminescent devices. *Journal of American Chemical Society* 127, 3172–3183 (2005).
46. Hanson, E.L., Guo, J., Koch, N., Schwartz, J., & Bernasek, S.L., Advanced surface modification of indium tin oxide for improved charge injection in organic devices. *Journal of the American Chemical Society* 127(28), 10058–10062 (2005).
47. Osikowicz, W. *et al.*, Transparent low-work-function indium tin oxide electrode obtained by molecular scale interface engineering. *Applied Physics Letters* 85(9), 1616–1618 (2004).

48. Cui, J. *et al.*, Indium tin oxide alternatives – high work function transparent conducting oxides as anodes for organic light-emitting diodes. *Advanced Materials* 13(19), 1476–1480 (2001).
49. Lee, S.T., Gao, Z.Q., & Hung, L.S., Metal diffusion from electrodes in organic light-emitting diodes. *Applied Physics Letters* 75(10), 1404–1406 (1999).
50. Nguyen, T.P. & Ip, J., Study of indium tin oxide -4,4′-bis(4-dimethylaminostyryl) benzene interface: an X ray photoelectron spectroscopy investigation. *Applied Surface Science* 195 (1–4), 251–262 (2002).
51. Anderson, N.A., Hao, E., Ai, X., Hastings, G., & Lian, T., Ultrafast and long-lived photoinduced charge separation in MEH-PPV/nanoporous semiconductor thin film composites. *Chemical Physics Letters* 347, 304–310 (2001).
52. Savenije, T.J., Warman, J.M., & Goossens, A., Visible light sensitisation of titanium dioxide using a phenylene vinylene polymer. *Chemical Physics Letters* 287, 148–153 (1998).
53. Roest, A.L., Kelly, J.J., Vanmaekelbergh, D., & Meulenkamp, E.A., Staircase in electron mobility of a ZnO quantum dot assembly due to shell filling. *Physical Review Letters* 89(3), 036801 (2002).
54. Beek, W.J.E., Wienk, M.M., & Janssen, R.A.J., Efficient hybrid solar cells from zinc oxide nanoparticles and a conjugated polymer. *Advanced Materials* 16(12), 1009–1013 (2004).
55. Arango, A.C., Carter, S.A., & Brock, P.J., Charge transfer in photovoltaics consisting of interpenetrating networks of conjugated polymer and TiO2 nanoparticles. *Applied Physics Letters* 74(12), 1698–1700 (1999).
56. Kannan, B., Castelino, K., & Majumdar, A., Design of nanostructured heterojunction polymer photovoltaic devices. *Nano Letters* 3(12), 1729–1733 (2003).
57. Coakley, K.M., Liu, Y., McGehee, M.D., Frindell, K.L., & Stucky, G.D., Infiltrating semiconducing polymers into self-assembled mesoporous titania films for photovoltaic applications. *Advanced Functional Materials* 13(4), 301–306 (2003).
58. Gong, D. *et al.*, Titanium oxide nanotube arrays prepared by anodic oxidation. *Journal of Materials Research* 16(12), 3331–3334 (2001).
59. Gratzel, M., Photoelectrochemical cells. *Nature* 414, 338–344 (2001).
60. Greene, L.E. *et al.*, General route to vertical ZnO nanowire arrays using textured ZnO seeds. *Nano Letters* 5(7), 1231–1236 (2005).
61. Lee, Y.-J., Sounart, T.L., Scrymgeour, D.A., Voigt, J.A., & Hsu, J.W.P., Control of ZnO nanorod array alignment synthesized via seeded solution growth. *Journal of Crystal Growth* 304, 80–85 (2007).
62. Peterson, R.B., Fields, C.L., & Gregg, B., Epitaxial chemical deposition of ZnO nanocolumns from NaOH solutions. *Langmuir* 20(12), 5114–5118 (2004).
63. Vayssieres, L., Growth of arrayed nanorods and nanowires of ZnO from aqueous solutions. *Advanced Materials* 15(5), 464–466 (2003).
64. Breeze, A.J., Schlesinger, Z., Carter, S.A., & Brock, P.J., Charge transport in TiO_2/MEH-PPV polymer photovoltaics. *Physical Review B* 64(12), 125205 (2001).
65. Waldauf, C. *et al.*, Highly efficient inverted organic photovoltaics using solution based titanium oxide as electron selective contact. *Applied Physics Letters* 89(23), 233517 (2006).
66. White, M.S., Olson, D.C., Shaheen, S.E., Kopidakis, N., & Ginley, D.S., Inverted bulk-heterojunction organic photovoltaic device using a solution-derived ZnO underlayer. *Applied Physics Letters* 89, 143517 (2006).
67. Coakley, K.M. & McGehee, M.D., Photovoltaic cells made from conjugated polymers infiltrated into mesoporous titania. *Applied Physics Letters* 83(16), 3380–3382 (2003).
68. Ravirajan, P., Haque, S.A., Durrant, J.R., Bradley, D.D.C., & Nelson, J., The effect of polymer optoelectronic properties on the performance of multilayer hybrid polymer/TiO2 solar cells. *Advanced Functional Materials* 15(4), 609–618 (2005).
69. Coakley, K.M. *et al.*, Enhanced hole mobility in regioregular polythiophene infiltrated in straight nanopores. *Advanced Functional Materials* 15, 1927–1932 (2005).

70. McBranch, D., Campbell, I.H., & Smith, D.L., Optical determination of chain orientation in electroluminescent polymer films. *Applied Physics Letters* 66(10), 1175–1177 (1995).
71. Konenkamp, R., Dloczik, L., Ernst, K., & Olesch, C., Nano-structures for solar cells with extremely thin absorbers. *Physica E* 14, 219–223 (2002).
72. Ohyama, M., Kozuka, H., & Yoko, T., Sol-gel preparation of ZnO films with extremely preferred orientation along (002) plane from zinc acetate solution. *Thin Solid Films* 306, 78–85 (1997).
73. Olson, D.C., Shaheen, S.E., Collins, R.T., & Ginley, D.S. The effect of atmosphere and ZnO morphology on the performance of hybrid poly(3-hexylthiophene)/ZnO nanofiber photovoltaic devices. *Journal of Physical Chemistry C* 111, 16670–16678 (2007).
74. Kroeze, J.E., Savenije, T.J., Vermeulen, M.J.W., & Warman, J.M., Contactless determination of the photoconductivity action spectrum, exciton diffusion length, and charge separation efficiency in pholythiophene-sensitized TiO2 bilayers. *Journal of Physical Chemistry B* 107 (31), 7696–7705 (2003).
75. Beek, W.J.E., Wienk, M.M., Kemerink, M., Yang, X., & Janssen, R.A.J., Hybrid Zinc Oxide conjugated polymer bulk heterojunction solar cells. *Journal of Physical Chemistry B* 109(19), 9505–9516 (2005).
76. Greczynski, G. *et al.*, Photoelectron spectroscopy of thin films of PEDOT-PSS conjugated polymer blend: a mini-review and some new results. *Journal of Electron Spectroscopy and Related Phenomena* 121, 1–17 (2001).
77. Jong, M.P.d., IJzendoorn, L.J.v., & Voigt, M.J.A.d., Stability of the interface between indium-tin-oxide and poly(3,4-ethylenedioxythiophene)/poly(styrenesulfonate) in polymer light-emitting diodes. *Applied Physics Letters* 77(14), 2255–2257 (2000).
78. Olson, D.C., Piris, J., Collins, R.T., Shaheen, S.E., & Ginley, D.S., Hybrid photovoltaic devices of polymer and ZnO nanofiber composites. *Thin Solid Films* 496, 26–29 (2006).
79. Takanezawa, K., Hirota, K., Wei, Q.-S., Tajima, K., & Hashimoto, K., Efficienct charge collection with ZnO nanorod array in hybrid photovoltaic devices. *Journal of Physical Chemistry C* 111, 7218–7223 (2007).
80. Mor, G., Shankar, K., Paulose, M., Varghese, O., & Grimes, C.A., High efficiency double heterojunction polymer photovoltaic cells using highly ordered TiO[sub 2] nanotube arrays. *Applied Physics Letters* 91(15), 152111 (2007).
81. Gregg, B.A., The photoconversion mechanism of excitonic solar cells. *MRS Bulletin* 30, 20–22 (2005).
82. Brabec, C.J. *et al.*, Origin of the open circuit voltage of plastic solar cells. *Advanced Functional Materials* 11(5), 374–380 (2001).
83. Gregg, B.A. & Hanna, M.C., Comparing organic to inorganic photovoltaic cells: theory, experiment, and simulation. *Journal of Applied Physics* 93(6), 3605–3614 (2003).
84. Heller, C.M., Campbell, I.H., Smith, D.L., Barashkov, N.N., & Ferraris, J.P., Chemical potential pinning due to equilibrium electron transfer at metal/C60-doped polymer interfaces. *Journal of Applied Physics* 81(7), 3227–3231 (1997).
85. Frohne, H. *et al.*, Influence of the anodic work function on the performance of organic solar cells. *ChemPhysChem* 3(9), 795–799 (2002).
86. Mihailetchi, V.D., Koster, L.J.A., & Blom, P.W.M., Effect of metal electrodes on the performance of polymer:fullerene bulk heterojunction solar cells. *Applied Physics Letters* 85(6), 970–972 (2004).
87. Kooistra, F.B. *et al.*, Increasing the open circuit voltage of bulk-heterojunction solar cells by raising the LUMO level of the acceptor. *Organic Letters* 9, 551–554 (2007).
88. Lenes, M. *et al.*, Fullerene bisadducts for enhanced open-circuit voltages and efficiencies in polymer solar cells. *Advanced Materials* 20, 2116-+ (2008).
89. Ross, R.B. *et al.*, Endohedral fullerenes for organic photovoltaic devices. *Nature Materials* 8, 208–212 (2009).

90. Gadisa, A., Svensson, M., Andersson, M.R., & Inganäs, O., Correlation between oxidation potential and open-circuit voltage of composite solar cells based on blends of polythiophenes/fullerene derivative. *Applied Physics Letters* 84(9), 1609–1611 (2004).
91. Olson, D.C. *et al.*, Band-offset engineering for enhanced open-circuit voltage in polymer-oxide hybrid solar cells. *Advanced Functional Materials* 17(2), 264–269 (2007).
92. Ohtomo, A. *et al.*, $Mg_xMn_{1-x}O$ as a II-VI widegap semiconductor alloy. *Applied Physics Letters* 72(19), 2466–2468 (1998).
93. Chen, L. *et al.*, Influence of Ar/O_2 gas ratios on the crystal quality and band gap of $Zn_{1-x}Cd_xO$ thin films. *Journal of Crystal Growth* 274, 458–463 (2005).
94. Matsubara, K. *et al.*, Band-gap modified Al-doped $Zn_{1-x}Mg_xO$ transparent conducting films deposited by pulsed laser deposition. *Applied Physics Letters* 85(8), 1374–1376 (2004).
95. Zhang, X., Li, X.M., Chen, T.L., Bian, J.M., & Zhang, C.Y., Structural and optical properties of $Zn_{1-x}Mg_xO$ thin films deposited by ultrasonic spray pyrolysis. *Thin Solid Films* 492, 248–252 (2005).
96. Zhao, D. *et al.*, Photoluminescence properties of $Mg_xZn_{1-x}O$ alloy thin films fabricated by the sol-gel deposition method. *Journal of Applied Physics* 90(11), 5561–5563 (2001).
97. Piris, J. *et al.* The locus of free charge-carrier generation in solution-cast Zn1-xMgxO/Poly (3-hexylthiophene) Bilayers for photovoltaic applications. *Advanced Functional Materials* 17, 3849–3857 (2007).
98. Rao, G.V., Säuberlich, F., & Klein, A., Influence of Mg content on the band alignment at CdS/(Zn,Mg)O interfaces. *Applied Physics Letters* 87(3), 032101 (2005).
99. Kaneto, K., Hatae, K., Nagamatsu, S., Takashima, W., & Pandey, S.S., Photocarrier mobility in regioregular poly(3-hexylthiophene) studied by the time of flight method. *Japanese Journal of Applied Physics* 38, L1188-L1190 (1999).
100. Schropp, R.E.I., Solar cells based on nanocrystalline titania. *Nederlands Tijdschrift voor Natuurkunde* 63(5), 117–118 (1997).
101. Somani, P.R. *et al.*, Solid-state dye PV cells using inverse opal TiO2 films. *Solar Energy Materials & Solar Cells* 87(1–4), 513–519 (2005).
102. Kambili, A. *et al.*, Electron transport in the dye sensitized nanocrystalline cell. *Physica E: Low-Dimensional Systems & Nanostructures (Amsterdam, Netherlands)* 14(1–2), 203–209 (2002).
103. Schwarzburg, K., Ernstorfer, R., Felber, S., & Willig, F., Primary and final charge separation in the nano-structured dye-sensitized electrochemical solar cell. *Coordination Chemistry Reviews* 248(13–14), 1259–1270 (2004).
104. Shevalevskii, O.I., Tsvetkov, A.A., & Lim, K.S., Photosensitization of nanocrystalline layers of titanium dioxide with phthalocyanines in Gratsel-type solar cells. *Khimicheskaya Fizika* 21(2), 99–102 (2002).
105. Gillispie, M.A., van Hest, M.F.A.M., Dabney, M.S., Perkins, J.D., & Ginley, D.S., rf magnetron sputter deposition of transparent conducting Nb-doped TiO2 films on SrTiO3. *Journal of Applied Physics* 101(3), 033125/033121–033125/033124 (2007).
106. Hitosugi, T. *et al.*, Fabrication of TiO2-based transparent conducting oxide films on glass by pulsed laser deposition. *Japanese Journal of Applied Physics, Part 2: Letters & Express Letters* 46(1–3), L86–L88 (2007).
107. Hitosugi, T., Ueda, A., & Hasegawa, T., Nb-doped TiO2 as transparent conducting oxide on glass substrate. *Seramikkusu* 42(1), 32–36 (2007).
108. Khomutov, G.B. *et al.*, Interfacial nanofabrication strategies in development of new functional nanomaterials and planar supramolecular nanostructures for nanoelectronics and nanotechnology. *Microelectronic Engineering* 69(2–4), 373–383 (2003).
109. Lu, J.G., Chang, P., & Fan, Z., Quasi-one-dimensional metal oxide materials-Synthesis, properties and applications. *Materials Science & Engineering, R: Reports* R52 (1–3), 49–91 (2006).
110. Wang, X., Sun, X., Xu, J., & Li, Y., Nanowires and nanotubes of complex oxides. *Nanowires and Nanobelts* 2, 173–190 (2003).

111. Wang, Z.L. & Pan, Z., Nanobelts of semiconductive oxides: a structurally and morphologically controlled nanomaterials system. *International Journal of Nanoscience* 1(1), 41–51 (2002).
112. Behnam, A. *et al.*, Nanolithographic patterning of transparent, conductive single-walled carbon nanotube films by inductively coupled plasma reactive ion etching. *Journal of Vacuum Science & Technology, B: Microelectronics and Nanometer Structures – Processing, Measurement, and Phenomena* 25(2), 348–354 (2007).
113. Dettlaff-Weglikowska, U. *et al.*, Conducting and transparent SWNT/polymer composites. *Physica Status Solidi B: Basic Solid State Physics* 243(13), 3440–3444 (2006).
114. Kim, H.H. & Kim, H.J., Simple preparation and characteristics for transparent, conductive thin films of carbon nanotubes. *Journal of the Korean Physical Society* 50(2), L361-L364 (2007).
115. Lee, D.Y., Lee, W.J., Song, J.S., Koo, B.K., & Kim, H.J., Application: WO. WO Patent No. 2006-KR5135, 2007064164 (20061130. 2007).
116. Parekh, B.B., Fanchini, G., Eda, G., & Chhowalla, M., Improved conductivity of transparent single-wall carbon nanotube thin films via stable postdeposition functionalization. *Applied Physics Letters* 90(12), 121913/121911–121913/121913 (2007).
117. Gao, X., Li, X., & Yu, W., Synthesis and characterization of flowerlike ZnO nanostructures via an ethylenediamine-meditated solution route. *Journal of Solid State Chemistry* 178(4), 1139–1144 (2005).
118. Hughes, W.L. & Wang, Z.L., Controlled synthesis and manipulation of ZnO nanorings and nanobows. *Applied Physics Letters* 86(4), 043106/043101–043106/043103 (2005).
119. Kim, K.S. *et al.*, Photovoltaic properties of nano-particulate and nanorod array ZnO electrodes for dye-sensitized solar cell. *Bulletin of the Korean Chemical Society* 27(2), 295–298 (2006).
120. Ahn, B.D. *et al.*, Synthesis and analysis of Ag-doped ZnO. *Journal of Applied Physics* 100(9), 093701/093701–093701/093706 (2006).
121. Gopal, P. & Spaldin, N.A., Magnetic interactions in transition-metal-doped ZnO: An ab initio study. *Physical Review B: Condensed Matter and Materials Physics* 74(9), 094418/094411–094418/094419 (2006).
122. Hossain, M.A. *et al.*, Substitution of ZnO by CdO, CuO, CaO, Al_2O_3 and MgO in Mn-Zn ferrite. *Bangladesh Journal of Scientific and Industrial Research* 32(1), 156–160 (1997).
123. Inoue, K., Shoyama, M., Murayama, M., & Torii, Y., Chemical preparation and photoluminescence of partially MgO-substituted ZnO powders. *Journal of Materials Science* 41(4), 1269–1271 (2006).
124. Jin, X.-L., Lou, S.-Y., Kong, D.-G., Li, Y.-C., & Du, Z.-L., Investigation on the broadening of band gap of Wurtzite ZnO by Mg-doping. *Wuli Xuebao* 55(9), 4809–4815 (2006).
125. Haeussler, W., Villari, V., & Schicht, H., Application: EP. EP Patent No. 97-400822, 870601 (19970410. 1998).
126. Koss, V., Belkind, A., Memarzadeh, K., & Woollam, J.A., Determination of optical constants of silver layers in zinc oxide/silver/zinc oxide coatings using variable angle spectroscopic ellipsometry. *Solar Energy Materials* 19(1–2), 67–78 (1989).
127. Memarzadeh, K., Woollam, J.A., & Belkind, A., Ellipsometric study of zinc oxide/silver/zinc oxide optical coatings: determination of layer thicknesses and optical constants. *Proceedings of SPIE-The International Society for Optical Engineering* 823 (Opt. Mater. Technol. Energy Effic. Sol. Energy Conv. 6), 54–61 (1987).
128. Xu, J., Zhang, Z.-Y., Zhang, Y., Lin, B.-X., & Fu, Z.-X., Effect of Ag doping on optical and electrical properties of ZnO thin films. *Chinese Physics Letters* 22(8), 2031–2034 (2005).
129. Ahn, K.-S., Yoo, S.J., Kang, M.-S., Lee, J.-W., & Sung, Y.-E., Tandem dye-sensitized solar cell-powered electrochromic devices for the photovoltaic-powered smart window. *Journal of Power Sources* 168(2), 533–536 (2007).
130. Heckner, K.-H. & Kraft, A., Similarities between electrochromic windows and thin film batteries. *Solid State Ionics* 152–153, 899–905 (2002).

131. Kumasawa, K., Ogawa, H., & Okada, O., Application: JP. JP Patent No. 2005-291471, 2007101885 (20051004. 2007).
132. Limmer, S.J., Takahashi, K., & Cao, G., Electrochromic and transparent conducting oxide nanorods. *Proceedings of SPIE-The International Society for Optical Engineering* 5224 (Nanomaterials and Their Optical Applications), 25–32 (2003).
133. Vijayalakshmi, R., Jayachandran, M., & Sanjeeviraja, C., Electrochromic characterization of electrodeposited WO3 thin films. *Solid State Ionics: Trends in the New Millennium, Proceedings of the Asian Conference, 8th, Langkawi, Malaysia, Dec. 15–19, 2002*, 445–452 (2002).
134. Vijayalakshmi, R. & Sanjeeviraja, C., Impedance analysis and electrochromic properties of galvanostatically deposited tungsten oxide thin films. *Transactions of the SAEST* 40(2), 72–76 (2005).
135. Azzopardi, M.-J., Application: WO. WO Patent No. 2003-FR1219, 2003087005 (20030416. 2003).
136. Harris, C.S., Steffek, C.D., & Walck, S.D., Application: US. US Patent No. 2005-85328, 2006210810 (20050321. 2006).
137. Kiribayashi, T., Yasuda, N., Kawahara, T., & Maeda, T., Application: JP. JP Patent No. 2000-101093, 2001287970 (20000403. 2001).
138. Nakabayashi, A. & Ota, K., Application: JP. JP Patent No. 2002-271882, 2004106348 (20020918. 2004).
139. Paz, Y., Luo, Z., Rabenberg, L., & Heller, A., Photooxidative self-cleaning transparent titanium dioxide films on glass. *Journal of Materials Research* 10(11), 2842–2848 (1995).
140. Yoshimoto, T., Abe, S., Nakamura, T., Sakuma, E., & Abe, O., Application: JP. JP Patent No. 98-18430, 11207871 (19980130. 1999).
141. Blackburn, J.L. *et al.*, Transparent conductive single-walled carbon nanotube networks with precisely tunable ratios of semiconducting and metallic nanotubes. *ACS Nano* 2, 1266–1274 (2008).
142. Wu, Z. *et al.*, Transparent, conductive carbon nanotube films. *Science* 305, 1273–1276 (2004).
143. Zhang, M. *et al.*, Strong, Transparent, multifunctional, carbon natotube sheets. *Science* 309, 1215–1219 (2005).
144. Ago, H., Petritsch, K., Shaffer, M.S.P., Windle, A.H., & Friend, R.H., Composites of carbon nanotubes and conjugated polymers for photovoltaic devices. *Advanced Materials* 11(15), 1281–1285 (1999).
145. Pasquier, A.D., Unalan, H.E., Kanwal, A., Miller, S., & Chhowalla, M., Conducting and transparent single-wall carbon nanotube electrodes for polymer-fullerene solar cells. *Applied Physics Letters* 87, 203511 (2005).
146. Lagemaat, J.v.d. *et al.*, Organic solar cells with carbon nanotubes replacing In_2O_3:Sn as the transparent electrodes. *Applied Physics Letters* 88, 233503 (2006).
147. Rowell, M.W. *et al.*, Organic solar cells with carbon nanotube network electrodes. *Applied Physics Letters* 88, 233506 (2006).

Chapter 13
Transparent Amorphous Oxide Semiconductors for Flexible Electronics

Hideo Hosono

13.1 History of Amorphous Semiconductor

The most important feature of semiconductors is in the controllability of carrier concentration over several orders of magnitude. A unique advantage of amorphous materials over crystalline materials is capability of large-area deposition of uniform thin films at low temperatures. These process advantages make amorphous semiconductors extremely favorable for large-sized electronic devices fabricable on plastic substrates. Research on amorphous semiconductors started in 1950s to seek materials which can have both of these advantages [1, 2]. Figure 13.1 summarizes the history of amorphous semiconductors. The largest impact on electronics is the discovery of hydrogenated amorphous silicon (a-Si: H) in 1975 by Spear and LeComber [3]. This is the first material which can be used to control both carrier type and concentration by impurity doping as in crystalline Si, and it opened a new frontier called "Giant Micro-electronics" which means electronics based on circuits fabricated on a large area substrate. A thin film transistor (TFT) substrate a-Si: H on glass was also first reported by Lecomber et al. [4]. Although the field effect mobility remained a small value such as $\sim$0.5 cm^2 V s^{-1}, this TFT was enough to operate liquid crystal displays, which is a voltage-driven device and is easy to be fabricated on a glass substrate. Thus, a-Sigh-TFT met the requirement for the backplane of active LCD which was about to rise in 1980s and is now a fundamental building block of the circuit for active-matrix flat-panel displays.

Amorphous oxide semiconductors have a long history comparable to amorphous chalcogenides such as a-Se. The history of amorphous oxide semiconductors started in 1954 with a report of electronic conductive glasses containing a large amount of V_2O_5 in Nature by Stanworth et al. [5]. This report was a breakthrough in glass

H. Hosono
Frontier Research Center & Materials and Structures Laboratory, Tokyo Institute of Technology, 4259 Nagatsuta, Midori-ku, Yokohama 226-8503, Japan
e-mail: hosono@msl.titech.ac.jp

D.S. Ginley et al. (eds.), *Handbook of Transparent Conductors*,
DOI 10.1007/978-1-4419-1638-9_13, © Springer Science+Business Media, LLC 2010

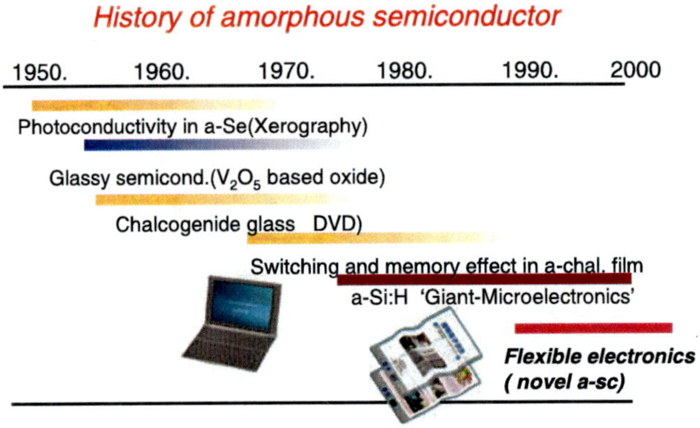

Fig. 13.1 History of amorphous semiconductors

science: it broke a widely believed recognition, "a glass is an insulator". Since then, a series of oxide glasses composed of a variable-valence transition metal oxide and glass-forming oxides such as P_2O_5 is named "glassy semiconductors" [6]. Electronic conduction in the glassy semiconductors is primarily controlled by variable-range hopping between transition metal cations with different valence states. Thus, the carrier drift mobility is very small such as 10^{-4} cm^2 V s^{-1} and is comparable to those of conventional chalcogenide glasses.

Recently, a new electronics is rapidly emerging for applications which cannot be fabricated by Si-C-MOS technology. This frontier named "flexible electronics" is characterized by electronic circuits fabricated on organic plastic (soft) substrates instead of inorganic (hard) glasses. This area was born to meet a strong demand for large-area light flexible displays because glass substrates, which are heavy and fragile, are obviously inconvenient. Amorphous semiconductors are preferable to crystalline semiconductors for flexible electronics. So far, organic semiconductors have been almost exclusively examined for such applications [7] but their performance and chemical/electrical instability are still insufficient for practical applications: e.g., field-effect mobilities of organic TFTs are insufficient to drive high-resolution, high-speed OLED displays and performance stability under electrical stress and an ambient atmosphere is rather poor.

A group at Tokyo Institute of Technology, Japan, started systematic material exploration of ionic amorphous oxide semiconductors in 1994 [8–10]. Figure 13.2 illustrates a location map of the various types of amorphous materials on the plane constituting of a chemical bonding nature axis and a band gap axis. One notice from the figure that conventional amorphous materials are composed of chemical bonds having high covalency and large band gap (transparent) are electrically insulating. It is seen that a transparent, ionic, amorphous semiconductor is an unexploited class of amorphous semiconductor. As an extremely high thermal quenching rate is needed to obtain an amorphous ionic materials such compared with a conventional oxide glass, physical deposition techniques from vapor phase on a substrate at room temperature (RT) are appropriate for this purpose.

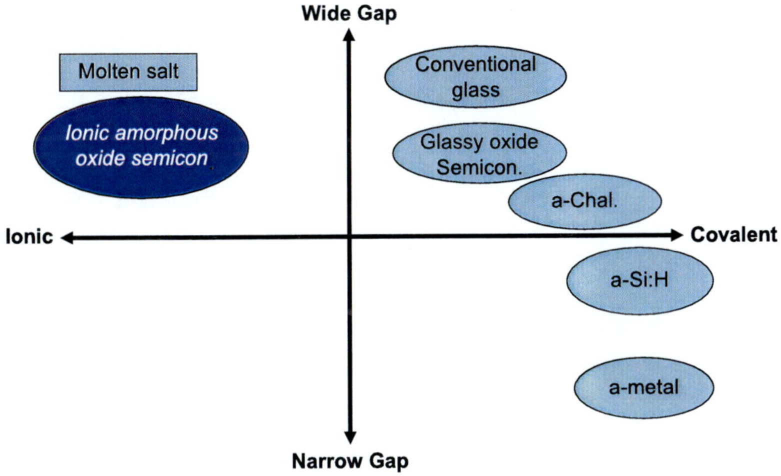

Fig. 13.2 Landscape of amorphous materials

13.2 Material Design for Ionic Amorphous Oxide Semiconductors with Large Electron Mobility [11–13]

In ionic materials, the nature of conduction band minimum (CBM) which works as an electron pathway totally differs from that of valence band maximum (VBM) which act as a hole pathway. The CBM in ionic oxides is primarily composed of vacant s-orbitals of a cation, and the contribution of oxygen 2p-orbitals, which are dominant at the VBM, is rather small. The spatial spread of this vacant s-orbital is so large that direct overlap between the s-orbitals of the neighboring cations is possible in heavy metal oxides, and therefore an effective mass of electron is small in these oxides. In fact, ionic oxides in which such a situation is realized, are called transparent conductive oxides (TCOs) represented by In_2O_3, SnO_2, and ZnO. What happens if these TCO materials become amorphous state? In an amorphous state, structural randomness concentrates on an energetically weak structural unit. In most amorphous materials structural randomness appears prominently as the bond angle distribution. *When the bond angle has a large distribution, how the effective mass (in other words, the transfer rate between neighboring cation s-orbitals) is modified for carrier electrons?* We considered the two cases for this question, covalent semiconductors and ionic semiconductors. In the former case, the magnitude of the overlap between the vacant orbitals of the neighboring atoms is very sensitive to the variation in the bond angle. As a consequence, rather deep localized states would be created at somewhat high concentrations and thereby the drift mobility would be largely degraded due to carrier scattering with these defects. On the other hand, the magnitude of the overlap in the latter case is critically varied by the choice of metal cations: when the spatial spread of the s-orbital is larger than the inter-cation distance, the magnitude should be insensitive to the bond angle distribution because the s-orbitals are isotropic in shape. As a consequence, we may anticipate that these ionic amorphous materials have large mobility comparable to that in the corresponding crystalline form. This is the amorphous material we have been seeking for.

In the case that the spatial spread of the metal s-orbital is small, such a favorable situation cannot be expected. The spatial spread of the s-orbital of a metal cation is primarily determined by the principal quantum number (n) and is modified by the charge state of the cation as shown in Fig. 13.3 [14]. Here, we take the value of n as a measure of the spatial spread of the metal cation s-orbital.

Thus candidates for transparent amorphous semiconductor having large electron mobilities comparable to those of the corresponding crystals are transparent oxides constituting of *post transition metal cations with an electronic configuration* $(n-1)d^{10}ns^{0}$*, where* $n \geq 5$. Note that a transition metal cation with an open shell structure is ruled out for candidates because they are not transparent due to absorptions arising from d-d transitions. In case of crystalline oxide semi-conducting oxides, this requirement is relaxed to be $n \geq 4$ as exemplified by ZnO, Zn^{2+} has the $(Ar)(3d)^{10}(4s)^{0}$ configuration, because crystalline materials have much regular and compact structures than amorphous oxides have.

Figure 13.4 illustrates the comparison in orbital drawings of Si and a post transition metal (PTM) oxide between crystalline and amorphous states. The drastic reduction of the electron mobility in the amorphous state from c-Si may be understood intuitively from the figure, whereas medium mobility in c-PTM oxides is reserved even in the amorphous state. In a sense, the situation of CBM in PTM oxides is similar to that in amorphous metal alloys in the aspect that metal orbitals dominantly constitute the electron pathways. The conductivity of amorphous/liquid metal alloys remains at slightly lower level compared with the corresponding

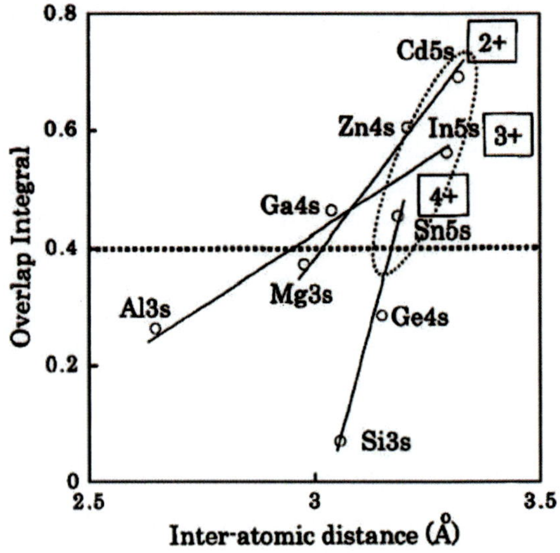

Fig. 13.3 Calculated values of the overlap integral between the ns orbital functions of metals with $(n-1)d^{10}ns^{0}$ electronic configuration versus the interatomic separations in the crystals of simple metal oxides. Slater-type orbital functions were used for the calculation. The crystals exhibiting electronic conductivity have an overlap integral larger than the threshold of ~0.4, which is indicated by a *dotted line*. A *dotted circle* encloses s-orbitals of $n = 5$

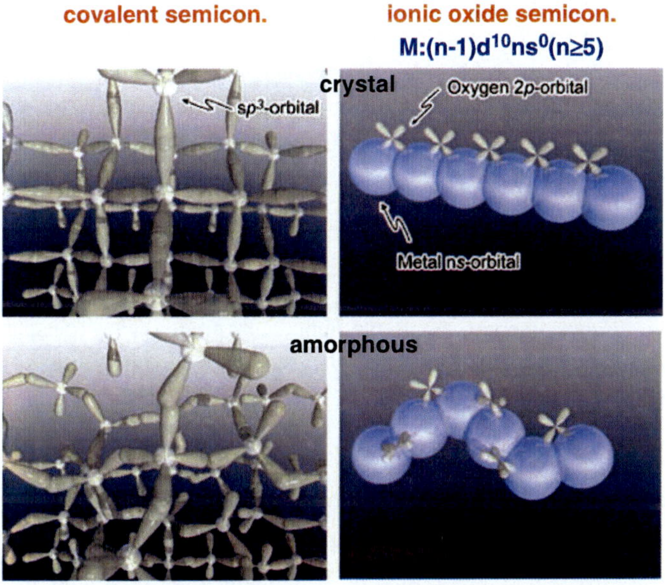

Fig. 13.4 Schematic orbital drawing of electron pathway (conduction band bottom) in conventional compound semiconductor and ionic oxide semiconductors

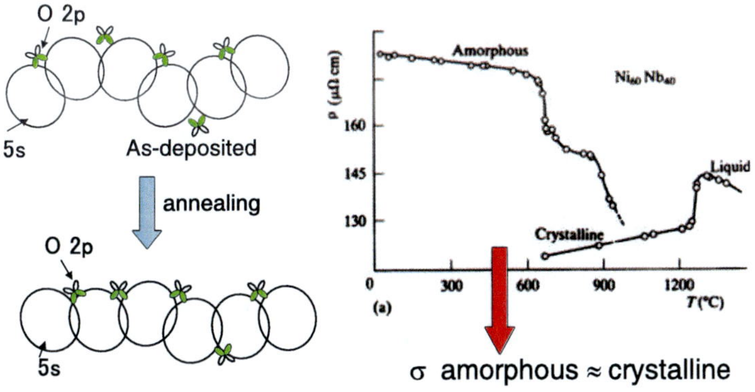

Fig. 13.5 Similarity between ionic oxides and metals. (*Right*) comparison in resistivity among liquid, amorphous, and crystalline metals [2]. (*Left*) Expected change in wavefunctions at CBM of IAOS with annealing. The electron mobility is expected to be enhanced by annealing as in the case of a-metals

crystalline phases as illustrated in Fig. 13.5 [1, 2]. The structural of amorphous metal is modeled by dense random packing of metal sphere and occupation of metalloid at the interstitial position. We may consider that in a-PTM oxides the vacant ns-orbitals of the metal cations work as metal atoms in a-metal.

Electron dopability in semiconductors is determined by the thermodynamic stability of doped electron, expressed by the magnitude of electron affinity, and easiness of formation of the counter defect, i.e., electron-capturing defect, in the hosts [15]. If a higher valence state of candidate PTM cations is not so stable, electron doped is captured by the low valence state. Tin ion, which meets the criteria as the PTM cation, has this tendency, i.e., Sn^{4+} is a cation suitable as a-TOS but low valence state Sn^{2+} is easily formed compared with In^{3+} with the same electron configuration. This valence stability of PTMC is reflected in the optimization of fabrication processes [16–20]

As for stability of doped electron, electron-neutrality level of the candidate metal oxides is close or above the CBM due to a large energy dispersion of CBM reflecting a large overlap in ns-orbitals between the neighboring metal cations [11, 21]. As a consequence, the above hypothesis predicts that transparent amorphous oxides following the hypothesis is being capable of electron-doping and have a large electron mobility comparable to the corresponding crystalline phases.

13.3 Example of Transparent TAOS

A series of transparent amorphous oxide semiconductors were found, following the working hypothesis described in Sect. 13.2, which includes $CdO\text{-}PbO_2$ [22], $AgSbO_3$ [23], $InGaZnO_3$ [14]. Although a variety of amorphous TCOs and TOSs in multi-component oxides have been reported to date, each system exhibiting a large mobility (>10 $cm^2V\ s)^{-1}$ includes at least In^{3+} or Sn^{4+} as a PTM cation, exemplified as In_2O_3-ZnO (IZO) [24, 25], In_2O_3-Ga_2O_3 [26], In_2O_3-SnO_2 [27], and SnO_2-ZnO [28]. A comprehensive review up to the end of 2006 was published in the form of a book by Wager et al. [29].

Here the electronic structure of these materials is explained by taking a-2CdO-GeO_2 [19, 20] as a typical example. Note that only Cd^{2+} with a $[Kr]4d^{10}5s^0$ configuration meets the above criteria. The amorphous thin films of this material were fabricated by r.f. sputtering on a glass substrate in an O_2-bearing Ar atmosphere without intentional heating of the substrate. The amorphous structure of the resulting thin films was confirmed by XRD measurements and TEM observation. The sample is optically transparent (Tauc gap = 3.4 eV) and electrically insulating ($\sigma_{300K} = 10^{-9}$ S cm^{-1}). The conductivity was increased from 10^{-9} to 10^2 S cm^{-1} by ion implantation of appropriate cations such as H^+ [30] and Ti^+ at RT as shown in Fig. 13.6. The activation energy of the electrical conductivity continuously decreases from almost half of the energy gap to ~0 as the implanted dose increases, indicating that the Fermi level is controllable from a mid-gap the position to CBM by doping. The Hall mobility in the degenerate state (electron carrier concentration: ~10^{20} cm^{-3}) is 10–12 cm^2 V s^{-1}, which is slightly smaller than that (~15 cm^2 V s^{-1}) in the crystalline Cd_2GeO_4. Electron carrier generation by hydrogen implantation is a common feature in TAOS and may be understood by comparing the most stable charge state of H at the bottom of the conduction band minimum [21, 31]. Recently this feature has been applied to the fabrication of co-planar-type TFT structure based on a-In_2O_3-Ga_2O_3-ZnO (a-IGZO) channel [32–34].

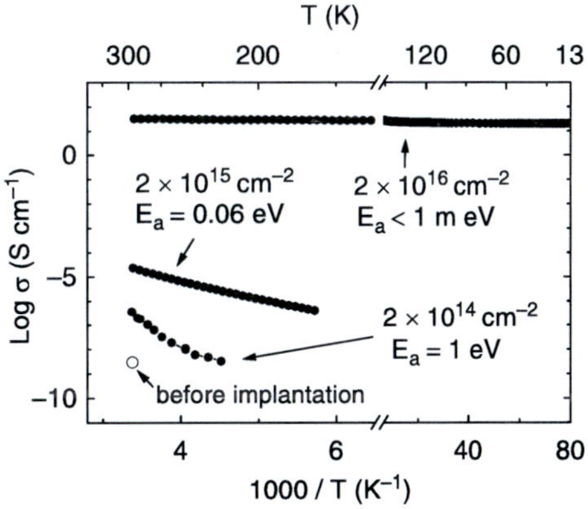

Fig. 13.6 DC conductivity in a-2CdO-1GeO_2 thin films before and after implantation with H^+ ion to 2×10^{14}–2×10^{16} cm^{-2} as a function of temperature. Activation energy of conduction is described in the figure [30]

The free carrier absorption appearing in the infrared region can be described by a simple Drude formula using a relaxation time [35–37]. The effective mass evaluated was 0.33 m_e, which agreed with that estimated from the Burstein-Moss shift (band gap widening due to electron filling at the CBM). Note that the mean free path is larger by an order of magnitude than the chemical bond length. This is the reason why Hall voltage measurements, which yield Hall sign anomaly for conventional amorphous semiconductors [38], gives the reasonable result for a-PTM oxides as in crystalline semiconductors.

Next, electronic structure near the CBM was examined by combining photoelectron spectroscopy with calculations on the structural model determined by radial distribution function assisted by reverse Monte-Carlo simulation [39]. Figure 13.7 shows the observed and calculated density of states (DOS). It is evident from the figure that the CBM is primarily composed of Cd 5s orbitals and the contribution of Ge 4s is much smaller than that of Cd 5s. This means that the orbital mainly participating to the electron pathway is not Ge 4s but Cd 5s, and the GeO_2 component primarily works as a stabilizer of amorphous structure. The close overlap between neighboring Cd 5s orbitals at the CBM is clearly observed in the electron density counter map in Fig. 13.8. Figure 13.9 shows the connectivity of the neighboring Cd ions, visualizing that the 5s orbitals are well overlapped with each other. It is evident that 3D-network of electron pathways is formed by percolation. These findings demonstrate the validity of the working hypothesis for exploring transparent ionic amorphous oxide semiconductors

13.4 Unique Electron Transport Properties of TAOS

Ionic amorphous oxide semiconductors have several common properties which are not seen in conventional amorphous semiconductors. First is their large electron mobility such as 10–40 cm^2 V s^{-1}, which is higher by 1–2 orders of

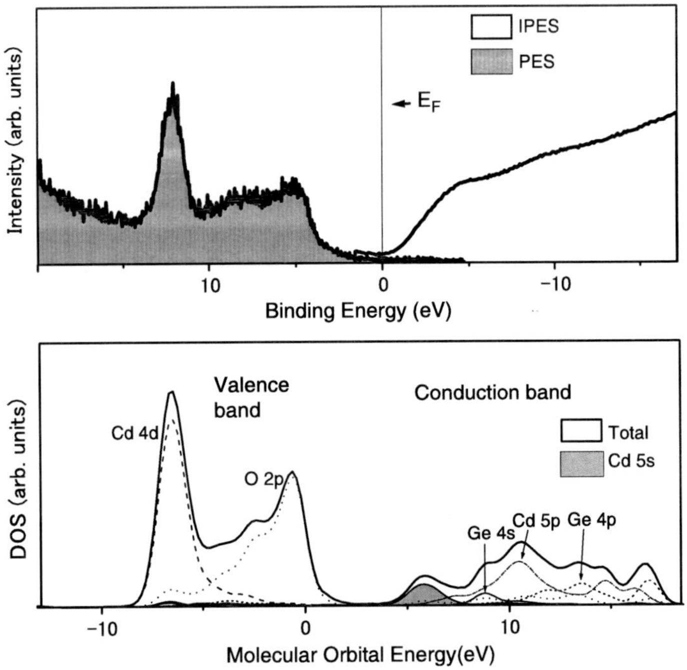

Fig. 13.7 Density of States in a-2CdO-1GeO$_2$. (*Top*) observed by ultraviolet and inverse photoemission measurements, (*bottom*) calculated [39]

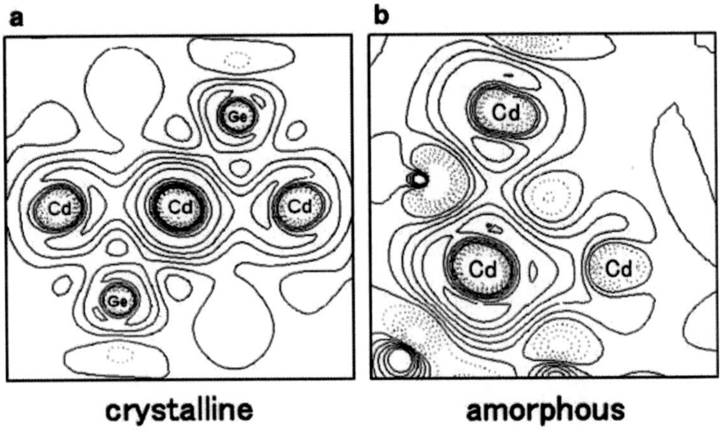

Fig. 13.8 Counter map of calculated wavefunction at the conduction band minimum in 2CdO-1GeO$_2$ [39]

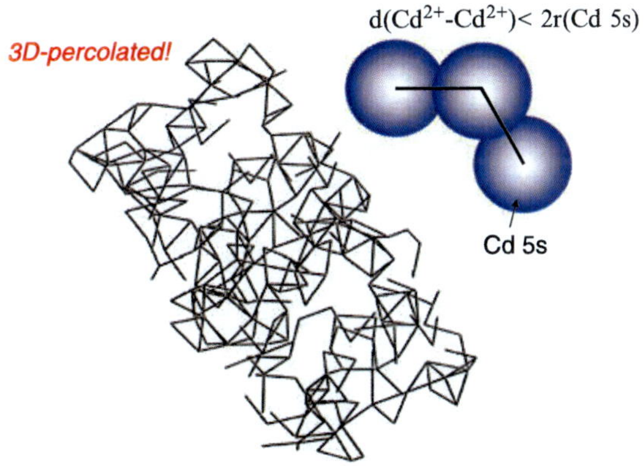

Fig. 13.9 Connectivity of Cd 5s orbitals in a-2CdO-1GeO_2. The atomic position was determined by Reverse Monte Carlo method. When the neighboring Cd-Cd separation is within the sum of two Cd^{2+} 5s Slater orbital radii, a pair of Cd ion is connected by a *line* [39]

magnitude than that in a-Si:H or by several orders of magnitude than that in glassy semiconductors or amorphous chalcogenides. Second is that a degenerate state can be realized by doping. This is totally different from the other amorphous semiconductors. For instance, c-Si is easily changed to the degenerate state by carrier doping (approximately several 100 ppm), but no such a state has been attained in a-Si:H [1–3] to date. That is, conduction takes place by hopping or percolation between tail states in conventional amorphous semiconductors. This is the reason why drift mobility in the amorphous state is so small compared with that in the crystalline state. On the other hand, in a-PTM oxides Fermi level can exceed the mobility gap easily by carrier doping, leading to band conduction. It is considered that this striking difference originates from that in chemical bonding nature between the materials, i.e., strong ionic bonding with spherical potential is much favorable to form a shallow tail state having small density of state.

The last unique feature is that electron carriers are primarily injected by chemical doping. Substitutional doping, which is effective in crystalline semiconductors and a-Si:H, is very inefficient. Each ion is primarily connected by ionic bonding and has a large flexibility in a coordination structure compared with the tetrahedral unit in a-Si:H. Thus, doping (or substitution) of aliovalent ions does not yield charge carriers if neutrality of the total formal charge of the constituent ions is maintained. Therefore, an effective doping way is to alter stoichiometry of oxygen ion by controlling the oxygen gas pressure during deposition processes or ion implantation of cations with a low electron affinity. Both processes modify the charge valence of cations and anions, leading to the injection of carrier electrons to the CBM. This situation is similar to that in the organic semiconductors such as polyacetylene for which carrier doping is carried out by chemical doping using such as PF_6(hole-doping) and I_3(electron-doping). Table 13.1 summarizes the bonding nature and carrier transport properties of various classes of amorphous semiconductors.

Table 13.1 Comparison in various amorphous semiconductors

Material System	Chemical bond	Mechanism	Hall voltage sign	Mobility $cm^2/(Vs)$	Example
Tetrahedral	covalent	hopping	abnormal	~1	Si:H
Chalcogenide	covalent	hopping	abnormal	$>10^{-3}$	Tl_2Se-As_2Se_3
Oxides(glass semiconductors)	covalent+Ionic	hopping		$\sim 10^{-4}$	V_2O_5-P_2O_5
(Ionic amorphous oxide semicondutors)	covalent	Band conduction	normal	10~30	In-Ga-Zn-O

13.5 TAOS in the System In_2O_3-Ga_2O_3-ZnO and Their Application to Flexible Transistor

13.5.1 Transport Properties

There are so many composition varieties for transparent amorphous oxide semiconductors (TAOS) as described in Sect. 13.3. Among them amorphous In_2O_3-Ga_2O_3-ZnO (IGZO) has been extensively studied as the semi-conducting channel layer of transparent TFTs since the first report [40] in late 2004. Figure 13.10a summarizes electrical properties (Hall mobility μ_{Hall} and carrier concentration N_e) for the films in the In_2O_3–Ga_2O_3–ZnO system. All the films were deposited at the same condition: i.e., on SiO_2 glass substrate at RT and oxygen partial pressure of 1.0 Pa. Although pure In_2O_3 and ZnO films exhibited large Hall mobilities of ~34 and ~19 $cm^2\ V^{-1}\ s^{-1}$, respectively, they were crystalline even if the films were deposited at RT (Fig. 13.10b). Moreover, it is not easy to control carrier concentration down to $<10^{17}\ cm^{-3}$ in these films without compensation doping. Pure Ga_2O_3 forms amorphous films but carrier doping, i.e., formation of a shallow oxygen vacancy, was very hard irrespective of deposition condition examined [41].Thus, these end-member materials in this ternary system are not appropriate because of local non-uniformity due to grain boundaries, no stable amorphous phase and/or the difficulty in carrier generation. As known in glass science, incorporation of aliovalent and different size cations is effective to enhance amorphization, and it is much favorable to introduce network forming cations [42]. Indeed, stable amorphous phases are formed in the binary systems of In_2O_3–Ga_2O_3 (a-IGO) and ZnO–Ga_2O_3 (a-GZO), and in the ternary system of In_2O_3–Ga_2O_3–ZnO (a-IGZO).

Both the Hall mobility and the carrier concentration rapidly decrease with increasing the Ga^{3+} ion content. Hall motilities in the a-IGO films decreased from ~25 $cm^2\ V^{-1}\ s^{-1}$ at $N_e \sim 10^{20}\ cm^{-3}$ to ~1 $cm^2\ V^{-1}\ s^{-1}$ at $\sim 10^{18}\ cm^{-3}$ as the Ga^{3+} ion content increased from 30 to 50%. However, we should note that the Hall mobility values here are not maximum potential of these materials because Hall mobility largely depends on carrier concentration in TAOS due to the presence of

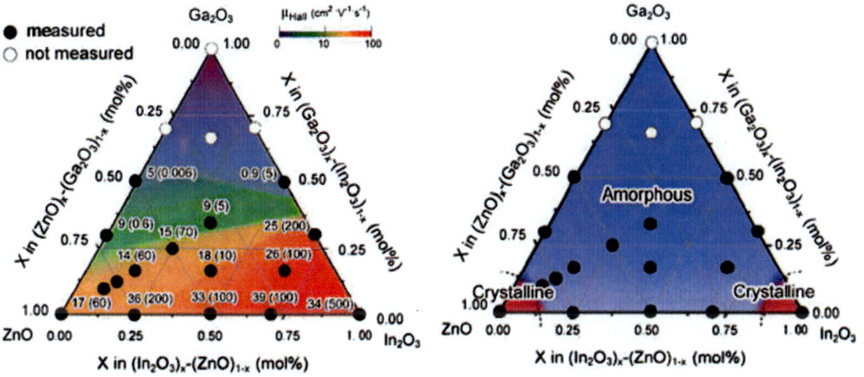

Fig. 13.10 The amorphous formation region (*right*) and Hall mobility and carrier electron concentrations (*left*) in the In_2O_3-Ga_2O_3-ZnO system. The thin film was deposited on a glass substrate by a pulsed laser deposition ($P_{O2} = 1$ Pa). Numbers in the parenthesis denote carrier electron concentration ($\times 10^{18}$ cm^{-3})

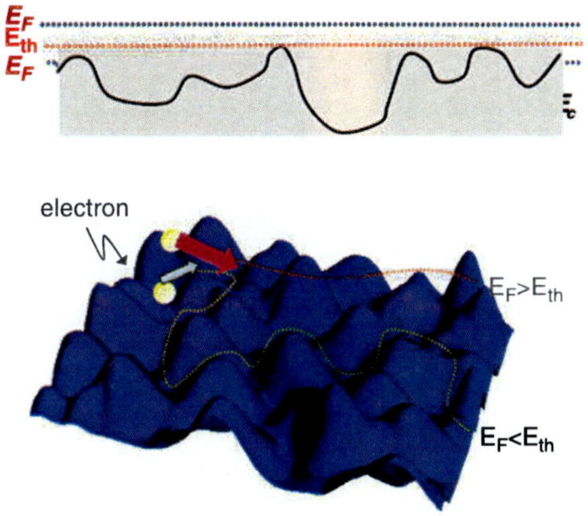

Fig. 13.11 Schematic illustration of near the conduction band bottom. *Arrows* denote conduction paths of electron for the cases of $E_F > E_{th}$ or $E_F < E_{th}$

potential barriers arising from structural randomness, as seen in Fig. 13.11: Carrier mobility strongly depends on carrier concentration, and large mobilities are obtained at carrier concentrations larger than a threshold value (e.g., $\sim 10^{18}$ cm^{-3} for a-IGZO). However, introduction of high-density carriers (e.g., $>10^{20}$ cm^{-3}) became very difficult in the larger Ga content films. This result indicates that large mobility is not easily obtained in the a-IGZO films with large Ga contents if one tries to dope carriers by impurity doping or introducing oxygen vacancies. This feature is unfavorable for TCO application, but it would not be disadvantage for semiconductor device applications because the difficulty in the carrier doping by oxygen vacancy suggests better controllability and stability of carrier concentration, especially at low concentrations. Even if high-density doping is difficult by choosing deposition condition, it is still possible to induce high-density carriers by external electric field if TFT structures are employed, which may make it possible to utilize the potential large mobilities that may be available at large carrier concentrations.

Hall mobilities larger than 10 cm^2 V^{-1} s^{-1} are obtained also in the In_2O_3–ZnO (a-IZO) and the In_2O_3–Ga_2O_3–ZnO (a-IGZO) systems. However, the controllability and stability of carrier concentrations are not satisfactory in a-IZO films [43]. Figure 13.12 shows controllability of the carrier concentration of the a-IGZO (nominal chemical composition in atomic ratio was In:Ga:Zn = 1:1:1) and a-IZO (In:Zn = 2:3) films, in which carrier concentration is plotted against oxygen partial pressure during the film deposition. The carrier concentration was well controlled from $<10^{15}$ to 10^{20} cm^{-3} by varying oxygen pressures from 0.1 to 7 Pa for a-IGZO. In contrast, it is hard to reduce carrier concentration down to 10^{17} cm^{-3} for a-IZO. It was not impossible to further reduce carrier concentration if oxygen pressure was further increased. However, the properties of such films were not stable: excess carriers were easily generated in the a-IZO films even if they are kept in ambient air or even by photolithography processes to fabricate TFTs.. We also like to note that similar effects to suppress unintentional carrier generation were confirmed also using Al^{3+} ions instead of Ga^{3+} ions by keeping reasonable mobilities of ~10 cm^2 V^{-1} s^{-1} (Fig. 13.13).

13.5.2 Electronic Structure [44]

The local coordination structures in a-IGZO thin films by XAFS measurements and constructed a structure model with assistance of ab initio calculations. It was found that the short-range ordering structures, i.e., the nearest-neighbor distances, in the

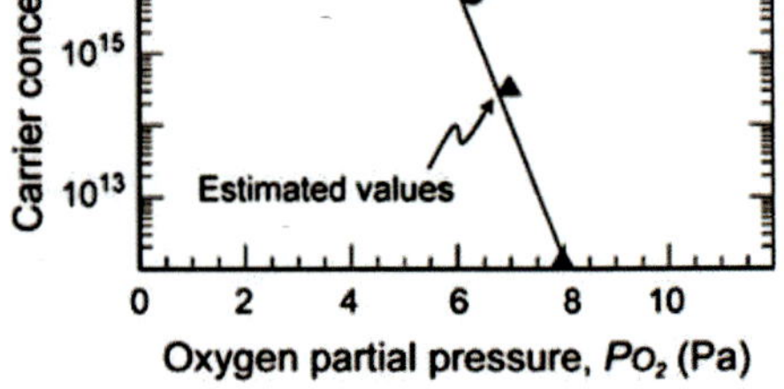

Fig. 13.12 Comparison in dependence of carrier concentration on oxygen partial pressure (P_{O2}) between a-IGZO and a-IZO films [13]. *Closed triangles* show values estimated by supposing a drift mobility to be 10 cm^2 V^{-1} s^{-1}

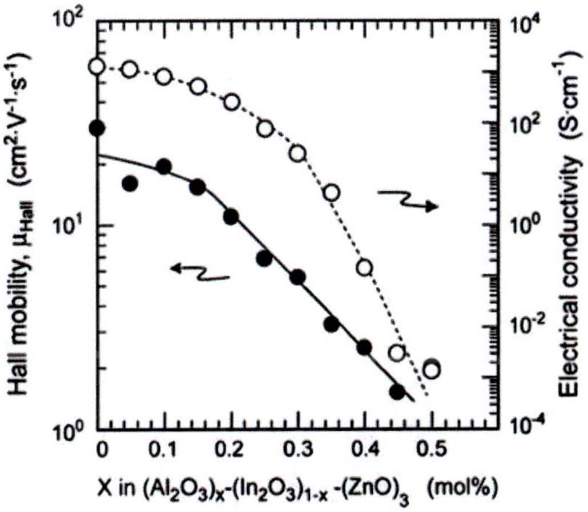

Fig. 13.13 Hall mobility and electrical conductivity as functions of Al content in IAZO films. IAZO films were deposited at RT in $P_{O2} = 1.0$ Pa [43]

amorphous phase were similar to those in the crystalline phase. Coordination structures of the ions are basically similar to those in crystal phase, i.e., but average coordination numbers are reduced in a-IGZO from IGZO crystal, giving a smaller density. Some of In–O polyhedra maintain the edge-sharing network structures like in IGZO crystal, but the other In–O polyhedra are connected by corner-sharing. Ab initio variable-cell relaxation on the relaxed amorphous structure of the 84 atoms MD cell was performed to obtain a quantum-mechanically stable structure. The relaxation calculations were performed by the pseudo-potential and plane wave method at the local-density approximation (LDA) level using the code VASP ("LDA-relaxed structure"). Figure 13.14 shows the calculated local structures compatible with XAFS analysis and wave function at the CBM. It is evident that the CBM is composed of 3d-percolated In 5s-orbitals. This result substantiates the validity of the material design concept described in Sect. 13.2 because only In^{3+} meets the criterion among the three cations. Although a-IGZO is sometimes classified to ZnO-based transparent semiconductors, it is clear that In-based semiconductor is a correct classification. The calculation revealed that the In ions form an extended conduction band and the small band effective mass of ~0.2 m_e (experimentally determined value is ~0.3 m_e [36] from the analysis of free carrier absorption) is formed through the 5s orbitals of the In ions separated by 0.32–0.40 nm in the a-IGZO structure. This separation is somewhat large, but similar to that in In_2O_3 crystal.

13.5.3 TFT Application

TFTs are fundamental building blocks for state-of-the-art microelectronics such as flat-panel displays and system-on-glass [45]. Furthermore, fabricating

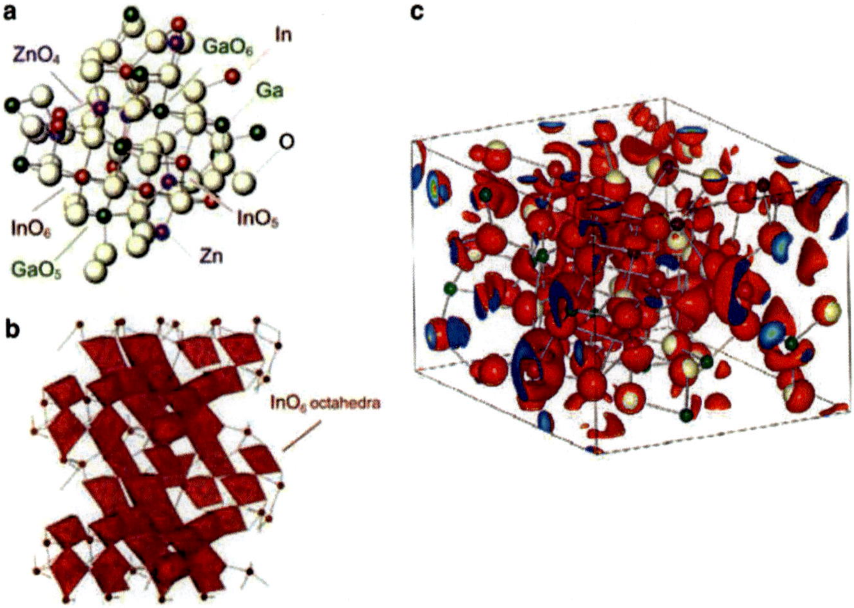

Fig. 13.14 (**a**) LDA-relaxed a-IGZO structure containing ($InGaZnO_4$) 12 atoms. "MOn" (M = metal cation, n = integer) indicate MO_n polyhedra. (**b**) Polyhedral view of (**a**). In ions are shown by the *red spheres* and In–O polyhedra by the *red polyhedra*. *Straight lines* show In–In distances smaller than double of the Slater's radius of In^{3+} 5s orbital (0.15 nm). (**c**) The *Red surfaces* show isosurfaces of the norm of the conduction band bottom wavefunction $|\psi|^2$. The *blue-to-red planes* show cross-sections of the $|\psi|^2$ at the edge planes of the cell [44]

low-temperature TFTs will allow flexible large-area electronic devices to be developed. These devices are flexible, lightweight, shock resistant, and potentially affordable, which are inevitable for large, economic, high-resolution displays, wearable computers, and paper displays. Further, when combined with "transparent circuit technology", TFTs can integrate display functions even on windshields of automobiles.

Amorphous(a−) Si:H and organic semiconductors are most extensively investigated for flexible electronics and have demonstrated the ability to fabricate flexible solar cells and TFTs. However, device performance is limited by the low mobilities of the channel materials, e.g., field effect mobilities, μ_{eff}, are <1 cm^2 V^{-1} s^{-1} for a-Si:H and ~0.5 cm^2 V^{-1} s^{-1} for pentacene TFTs. In addition, Si-based devices are of less interest for transparent circuits since they are not transparent due to the narrow band gap.

In contrast, degenerate band conduction and large mobility (>10 cm^2 V^{-1} s^{-1}) are possible in amorphous oxide semiconductors (AOSs) that are composed of post-transition metal cations as described in the previous section. In this section, low temperature fabrication and performances of transparent TFTs fabricated using a TAOS in an In-Ga-Zn-O system, a-IGZO, for an active n-channel layer. A-IGZO films were prepared by a pulsed laser deposition (PLD) with KrF excimer laser [40, 43] or a conventional sputtering techniques [46–53] using a polycrystalline target of $InGaZnO_4$ or an appropriate compound in this ternary system target at room temperature in an oxygen atmosphere (oxygen pressure P_{O2}). Since the resulting thin film is amorphous, thin film fabrication processes are not critical compared with process parameters such as partial oxygen pressure during deposition and post annealing treatments. These characteristics provide a large allowance in device fabrication process.

Figure 13.15 shows top-gate flexible transparent TFTs using the a-IGZO film as an n-channel active layer (the channel length = 50 μm, and width = 200 μm on 0.2-mm-thick polyethylene terephthalate (PET) films) [13]. A 140-nm-thick Y_2O_3 layer was used for the gate insulator and ITO (Sn:10%) was used for source, drain, and gate transparent electrodes. The performance of the flexible transparent TFTs was measured in air at room temperature. Figure 13.15a shows typical source-to-drain current (I_{DS}) – voltage (V_{DS}) characteristics of the virgin device. The current I_{DS} markedly increases as V_{DS} increases at a positive gate bias (V_{GS}), indicating that the channel is n-type. The I_{DS} reaches ~0.06 mA at V_{GS} of 10 V. The $I_{DS} - V_{DS}$ characteristics exhibit a clear pinch-off and current saturation, confirming that the TTFT operation follows the standard field-effect transistor theory. The saturation regime field-effect mobility (μ_{sat}) is obtained from the I_{DS}-V_{DS} curves in the saturation region using the equation $I_{DS} = \left(\frac{C_i \cdot \mu_{eff} \cdot W}{2L}\right) \cdot (V_{GS} - V_{th})^2$, where C_i and V_{th} denote the gate capacitance and threshold gate voltage, respectively. The estimated μ_{sat} value is ~4.9 $cm^2\ V^{-1}\ s^{-1}$, which is much larger than those for organic and a-Si:H TFTs. Similarly, a linear regime field-effect mobility (μ_{lin}) estimated from the I_{DS}-V_{DS} curve in linear region $\left(I_{DS} = \left(\frac{C_i \cdot \mu_{eff} \cdot W}{L}\right) \cdot V_{DS} \cdot \left(V_{GS} - V_{th} - \frac{1}{2} V_{DS}\right)\right)$ is ~4.8 $cm^2\ V^{-1}\ s^{-1}$ at V_{DS} = 2.5 V, which is consistent with the μ_{sat} value.

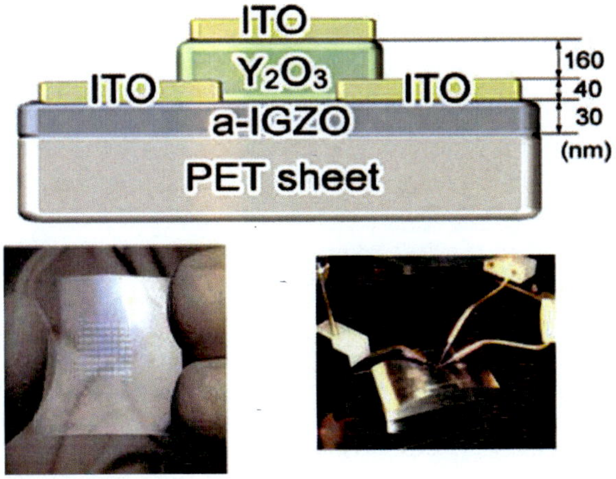

Fig. 13.15 a-IGZO-TFT fabricated on a PET substrate at RT [40]

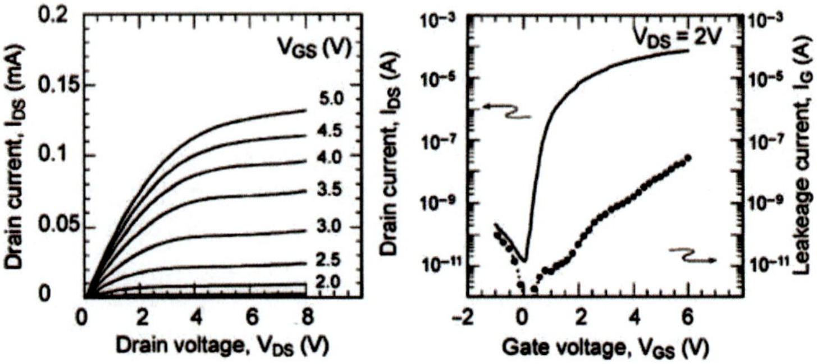

Fig. 13.16 Performance of a-IGZO TFT fabricated on a glass substrate [13, 43]. $\mu = 13\text{cm}^2 (\text{V s})^{-1}$ and $I_{on}/I_{off} \sim 10^7$

The threshold gate voltage is positive ($V_{th} \sim +1.6$ V), showing that the TTFT operates in the enhancement mode (normally-off characteristics).

The TTFT is curved with a surface curvature radius (R) of 30 mm (corresponding to tensile strain of ~0.3% in the TFTs) as shown in the figure. The TTFT after the bending exhibits characteristics such as $\mu_{sat} \sim 4.3$ cm^2 V^{-1} s^{-1} and an on-to-off current ratio ~10^3. We would like to stress that the TFT performance is largely unaffected by bending, but a slight decrease is observed in the saturation current. After the initial bending, the TFT characteristics are stable and reproducible during and after repetitive bending. The TFT is stable at temperatures up to 120°C, but is broken at higher temperatures probably due to the softening of the PET substrate.

Next, TFTs were fabricated on glass substrates [43]. Figure 13.16 shows a typical drain–source current (I_{DS}) – voltage (V_{DS}) characteristic measured at various gate biases (V_{GS}) at RT in air. The saturation regime field-effect mobility (μ_{sat}) was estimated to be ~10 cm^2 V^{-1} s^{-1}, which is almost the same as that obtained from the linear region ($\mu_{lin} \sim 9$ cm^2 V^{-1} s^{-1} at $V_{DS} = 2$ V). The transfer characteristic (I_{DS} versus V_{GS}) showed a low off current of less than 10^{-10} A . I_{DS} was modulated by approximately six orders of magnitude by the applied V_{GS} of ~6 V. The TFT performances equivalent or better than the above were reported by conventional sputtering [46–53].

13.6 Amorphous TCOs

13.6.1 Why Amorphous TCOs

Amorphous TCOs have distinct advantages in fabrication processes over crystalline TCOs; large-sized and homogeneous thin films can be fabricated at low temperatures.

These unique features make them essentially favorable as transparent conductive thin films on plastic substrates for flexible applications.

The stable amorphous formation of TCO materials containing a single cation by conventional physical vapor deposition processes is restricted to In_2O_3. ZnO does not form amorphous state, whereas Ga_2O_3 forms amorphous phase but is insulating. Amorphous In_2O_3 and In_2O_3:Sn exhibit similar properties including conductivity and optical transparency. The fundamental properties of these materials were examined. Bellingham et al. [54] measured the correlation the oxygen stoichiometry and the conductivity, suggesting that oxygen deficiency leads to the carrier electron generation like the corresponding crystals.

As described in Sect. 13.2, n-type TCOs based on the metal with $(n - 1)d^{10}ns^0$ electronic configuration ($n \geq 5$) exhibit a large mobility comparable to that in the crystal. Thus, we may expect that CdO, SnO_2, and In_2O_3-based amorphous oxides are candidate materials for a-TCOs. A decisive difference between a-TCO and TAOS is carrier controllability; TAOS is required to have high controllability of carrier concentration without serious degradation of mobility, whereas a-TCOs need higher conductivity and high transparency.

13.6.2 Advantage of Binary Oxides

Although the a-TCOs are inferior slightly to c-TCOs in the respect to the minimal resistivity, they have several advantages over the crystalline materials; formation of homogeneous thin films with a very smooth surface at low temperatures. The formation of a-In_2O_3 and a-In_2O_3:Sn is so sensitive to the total and water vapor pressure during sputtering processes [55] that the process window to obtain microcrystal-free amorphous thin films is rather narrow. In addition, they are not so stable and easily crystallize at ~200°C. Incomplete formation of a-In_2O_3 and a-In_2O_3:Sn films has for a drawback in the practical applications, i.e., residual micro-crystals remain as particles after wet-etching and work as the origin for cracking under bending.

These drawbacks were largely resolved by using binary oxides, especially In_2O_3-ZnO (IZO) system. Amorphous thin films with rather low resistivity are easily obtained in the composition of 60–90 In % by conventional sputtering processes. The resulting thin films have no residual microcrystals and higher crystallization temperature, >500°C. Thus, a-IZO thin films deposited on organic substrates are used as flexible TCOs. Figure 13.17 summarizes conductivity data on amorphous In_2O_3, In_2O_3:Sn, IZO and crystalline ITO fabricated by sputtering using Ar gas with different O_2 contents [56]. The minimal resistivity of $6 \times 10^{-4}\ \Omega$ cm is obtained for a-IZO, which is higher by a factor of 2–3 than that for c-ITO.

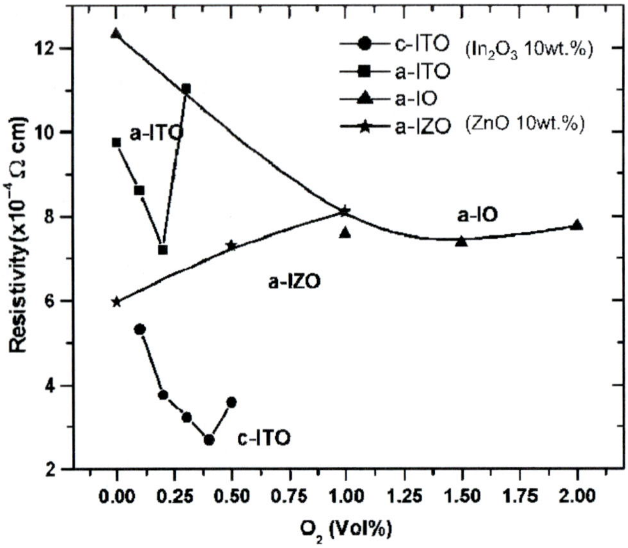

Fig. 13.17 Resistivity in various a-TCO thin films as a function of O_2% in the sputtering Ar gas [56]

13.6.3 General Electron Transport Properties

Impurity ions which are known as effective donors in the crystalline TCOs do not work in the amorphous TCOs. Carrier electrons generation is performed by chemical (interstitial) doping or via the formation of oxygen ion vacancy. Ion implantation of electropositive cations is a typical example of chemical doping as exemplified in Fig. 13.6. In practically important a-IZO thin films, electron carriers are produced by the latter. What is the relationship between mobility and carrier concentration in a-IZO with different In contents? Recently, Leenheer et al. [57] examined the optical and electron-transport properties of a-IZO thin films as a function of In-content or process parameters and obtained the results of Fig. 13.18. It is evident that Hall mobility does not depend on the In-content but on the carrier concentrations (N_e) and that N_e is primarily determined by the oxygen content in the sputtering gas. The conduction mechanism is switched from thermally activated type in the $N_e < 10^{19}$ cm^{-3} to the degenerate-type above this N_e. Further increase of N_e reduces the mobility by impurity scattering. It is considered from the results obtained for a-IGZO that the CBM of a-IZO is primarily composed of In 5s orbitals as well. Electron transport properties are roughly determined by whether In 5s-orbitals percolates three-dimensionally or not. The atomic fraction of In more than 60% is amply higher than the percolation threshold. Thus, Electron transport of a-IZO is insensitive to the In content within the examined range.

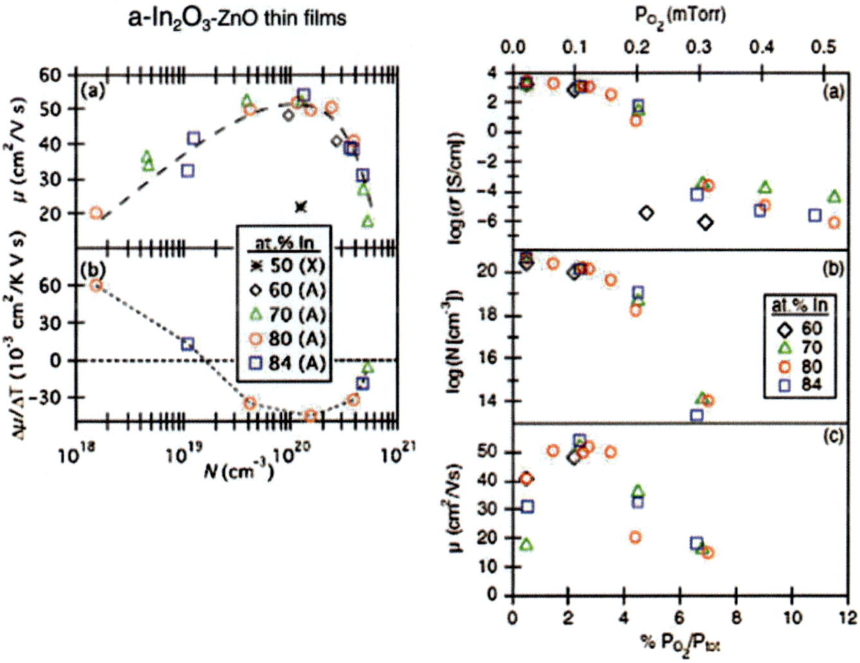

Fig. 13.18 Correlation between electron-transport properties and carrier concentration (*left*) or partial oxygen pressure during deposition (*right*) [57]. *Asterisk* denotes polycrystalline sample

13.6.4 Nano-fabrication [58, 59]

Etching rate of a-TCO is much larger than c-TCO, generally. Here an example of nanofabrication of Bragg gratings utilizing this feature and laser crystallization with spatially interfered pulsed UV laser light is introduced. Heating to >200°C is needed for crystallization of a-ITO. This temperature is so high that most plastics are deformed. This difficulty can be overcome by crystallization using pulsed UV-laser such as a KrF excimer laser (light wavelength 248 nm, pulse duration ~20 ns). Crystallization was observed at a threshold laser energy >40 mJ cm^{-2} $pulse^{-1}$ and resistivity was reduced from 6×10^{-4} to 2×10^{-4} Ω upon crystallization. This change is due primarily to the increase of carrier concentration which originates from the activation of Sn^{4+} upon crystallization.

Excimer laser irradiation through a phase mask yields spatially periodic modulated light intensity pattern. If the light intensity is tuned, this periodic pattern can be transferred to the periodic pattern of crystal-amorphous state. A dilute acid selectively etched the amorphous parts, leaving the periodic patterns of c-ITO portion on the substrates. Figure 13.19 shows the SEM photo of the resulting c-ITO pattern along with a photo exhibiting diffraction by the ITO Bragg gratings.

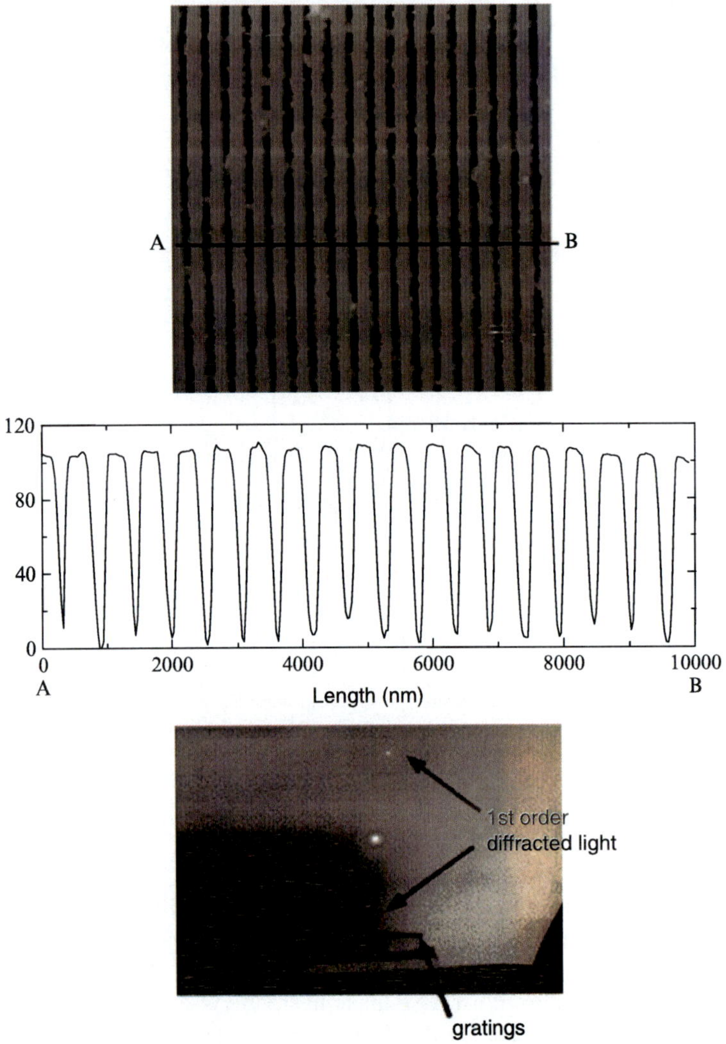

Fig. 13.19 SEM photomicrograph (*upper*) and surface profile (*middle*) of an ITO grating fabricated by excimer laser irradiation though a phase mask and subsequent acid-etching [58, 59]. The *bottom* demonstrates diffraction of light by the ITO-gratings

13.7 TFT Device Application

Recently research and development of display application of TAOS-TFT are rapidly emerging toward practical application. Main drivers in this area are major display related companies and SID (Society for Information and Displays) meetings are now the main forum of presentation including exhibitions. So many papers on

TAOS-TFTs have been published and the U. Oregon group led by Wager successfully summarized them in the form of book "Transparent Electronics" (Springer, 2008 [29]) along with devices and circuits. Here recent advances in display applications using TAOS-TFT arrays are reviewed so as to avoid the overlap with the above book.

13.7.1 Unique Features of TAOS-TFTs

TAOS-TFTs have three unique characteristics originating from the intrinsic nature of TAOS compared with other TFTs First is high field mobility >10 $cm^2\ V\ s^{-1}$. Second is easy fabrication at low temperature using conventional DC sputtering. The last is a large process allowance in fabrication. For instance, TFTs fabricated at unoptimized conditions exhibits poor performance, but the TFT performance can be much improved to that prepared under the optimized condition just by annealing at an appropriate temperature far below the crystallization temperature of TAOS [16–20, 44, 46–53]. Figure 13.20 is an example of a-IZGO TFTs showing effectiveness of post annealing to improve the TFT performance. The annealing temperature is 250–300°C which is much lower than the crystallization temperature (>500°C). No distinct structural change around each metal cation was noted before and after annealing. Pronounced annealing effects are observed for a-ZnO-SnO_2 TFTs [60, 61].

Figure 13.21 shows the performance histograms of a-IGZO TFTs which were fabricated on a glass substrate by a conventional sputtering and subsequently annealed. About 100 TFTs were fabricated from a 1 cm × 1 cm area of a-IGZO thin film. The TFT exhibits excellent uniformity and high average performance. The value of μ_{sat} resides within the range of 0.5 $cm^2\ V\ s^{-1}$ and σ is 0.11 $cm^2\ V\ s^{-1}$

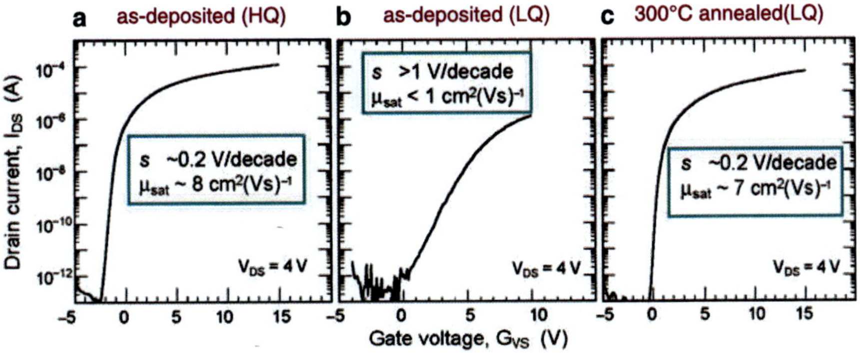

Fig. 13.20 Effect of thermal annealing on the IGZO-TFT performance. HQ and LQ denote high quality and low quality samples, respectively. The sample quality was controlled by tuning the deposition condition [19]

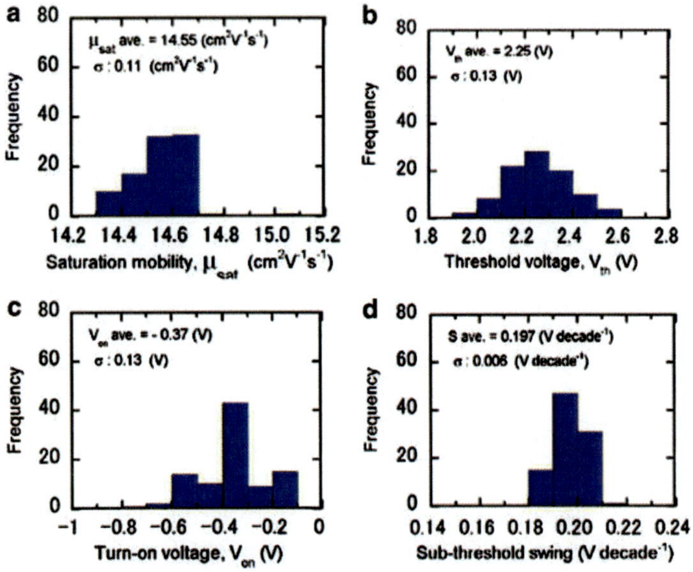

Fig. 13.21 Histogram of device performance in 97 a-IGZO-TFTs fabricated on a 1 cm × 1 cm area. The average value and the standard deviation are shown [62]

(0.76% of the average value), demonstrating the excellent uniformity of the *a*-IGZO TFTs [62] It strongly suggests that the *a*-IGZO TFTs essentially have a good short-range uniformity and are advantageous in integrated circuits and large area

13.7.2 Novel Display Structure

An innovative e-paper display structure called "front drive" type was recently proposed by M. Ito et al. of Toppan Printing [63, 64]. Alignment of TFT array to color filter array is a troublesome process in the display assembling because a-Si:H is non-transparent and variation of substrate dimension with aging. Their idea to escape this difficulty was to directly deposit the TAOS-TFT arrays on the color filter arrays utilizing the low temperature process and optical transparency simultaneously. This is a first demonstration of a device structure benefited from optical transparency of TAOS. Figure 13.22 shows the front-drive structure applied to an electronic paper based on E-ink imaging film. Taking advantage of transparency of TAOS, the TFT and the color filter arrays can be positioned at the viewing side of displays. This display structure is applicable to other displays, facilitating the alignment.

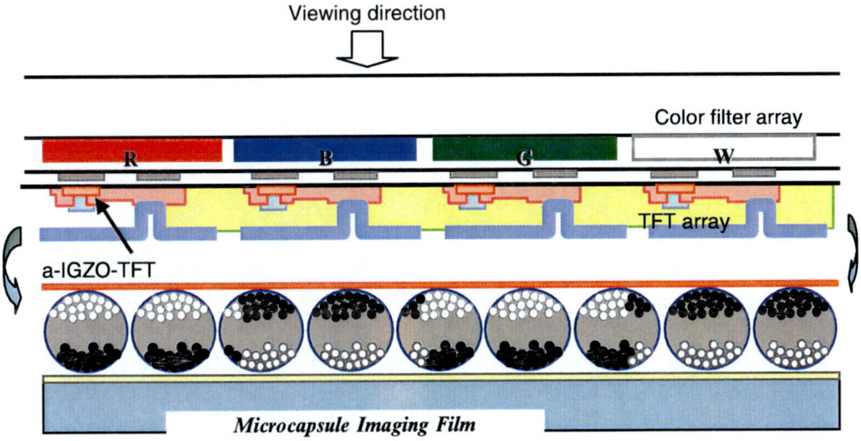

Fig. 13.22 Front drive structure applied for electrophoresis type e-paper [63, 64]. A TFT array of TAOS can directly fabricated on the color filter array using low temperature process compatibility and visible transparency of TAOS

13.7.3 Driving Backplane of OLED and LCD Panels

Today conventional active-matrix (AM) flat panel displays are based on amorphous or polycrystalline silicon thin-film transistor (TFT) technology. Limitation of the amorphous silicon (a-Si:H) include visible light sensitivity and a low field-effect mobility, which reduce the pixel aperture ratio and driving ability for some applications.

A typical example is an organic light emitting diode (OLED) which drives by current. Since the luminous intensity of OLEDs is proportional to the flow current, higher mobility TFTs are required. Although polycrystalline silicon TFTs have a larger field-effect mobility, it's uniformity over large area is not acceptable for a high yield manufacturing. Over last several years, there has been a great interest in thin-film transistors made of transparent oxide semiconductors. This is mainly due to metal oxide semiconductor thin-film transistors' unique advantages such as visible light transparency, a large-area uniform deposition at low temperature, and a high carrier mobility. However, conventional metal oxide semiconductor such as zinc oxide (ZnO) is polycrystalline in nature even at room temperature. The grain boundaries of such polycrystalline metal oxide could affect device properties, uniformity and stability over large area.

Recently, OLED and LED panels driven by a-IGZO-TFT backplane have been presented in SID Annual Meeting and SID-sponsored display meetings. In SID'07 LG electronics presented a 4-in. size AM full color OLED using this backplane [65]. This is a first demonstration of OLED-based on an oxide semiconductor TFTs. Subsequently, they presented a flexible OLED at IMID'07 by depositing a-IGZO on a thin stainless plate [66]. Figure 13.23 shows the photo of flexible a-IGZO TFT-array and OLED. SID'07 was a memorial year for TAOS-TFTs because a technical session on oxide TFT was launched for the first time at this conference which is the largest and most important in the display area. In SID'08 oxide TFTs were highlighted as the backplane of LCD as well as OLED. Samsung SDI demonstrated a 12.1″ WXGA AMOLED prototype display fabricated by an a-IGZO TFT backplane as shown Fig. 13.24 [67]. They successfully fabricated a-IGZO THF array by the same lithographic process as a-Si:H TFTs. TAOS-TFTs are now attractive as a backplane of next generation liquid crystal displays (LCDs). Larger size and higher frame frequency operation are requirements for next generation LCDs. When the display size is increased to >60 in., the number of TFTs required becomes much more to keep the same pixel density of the current size, In such a case "on"-resistance of TFT arrays cannot be negligibly small. Furthermore, frame frequency doubling (60–120 Hz) is now rapidly spreading. Thus, the performance of a-Si:H TFTs is insufficient to meet these requirements as shown in Fig. 13.25 This is a reason why, a-IGZO TFTs are examined for LCD. Samsung Electronics and SAIT presented a 15-in. AM-LCD panel using a-IGZO backplane [68].

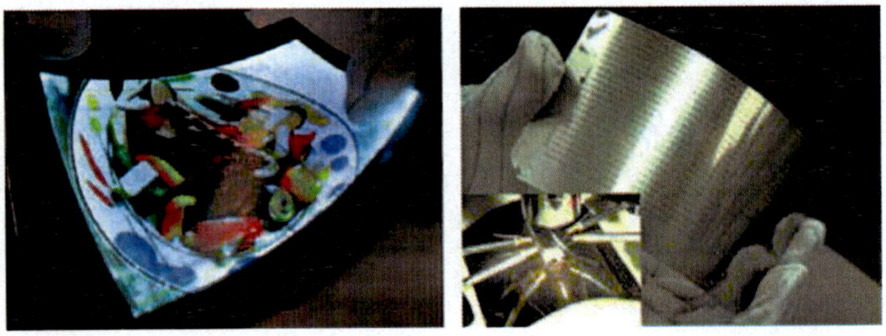

Fig. 13.23 Flexible a-IGZO TFTs fabricated on a stainless steel plate and 4 in. AM-OLED display using this backplane [65, 66]

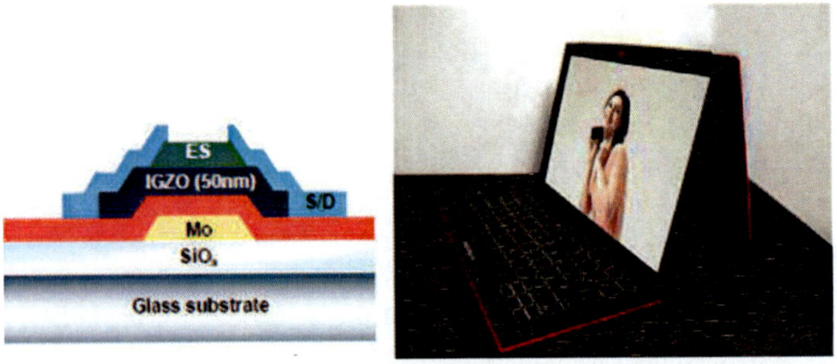

Fig. 13.24 Schematic cross-section of a-IGZO-TFT used for 12 in. AM-OLED panel. ES: etch stopper layer. This TFT can be fabricated by the almost same lithographic process as a-Si:H TFTs which are currently used for LCD backplane [66]

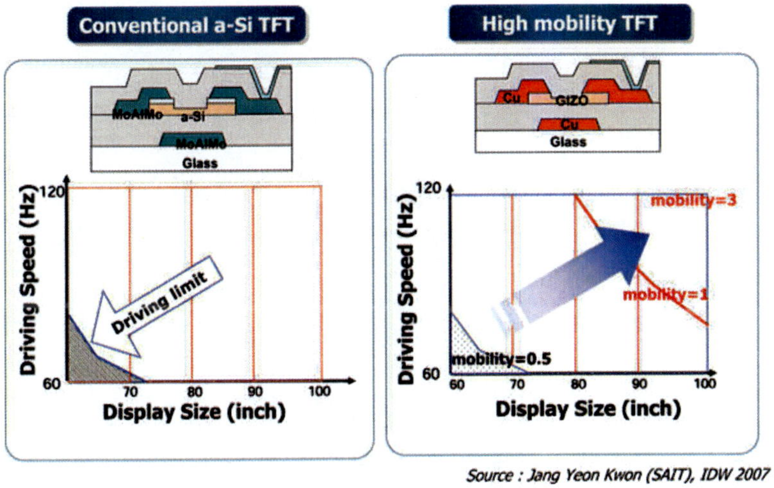

Fig. 13.25 Trend in AM-LCD panels and demanding TFT performance

13.8 Future Challenge and Opportunity

Demand for TAOS and TCO is rapidly increasing to meet the requirements from display electronics and flexible electronics. The advantages of these materials over crystalline materials are to be depositable at low temperature and performance comparable to the corresponding polycrystals. The application of TAOS-TFTs to 3D stacking memory was reported based on excellent operation of short channel length less than 100 nm for a-IGZO-TFTs [69]. It is of interest to note that the field effect mobility remains a value of ~10 cm^2 V s^{-1} even the channel length is reduced less than 100 nm.

Fabrication of TAOS-TFTs at low temperature by a non-vacuum process such as spin-coating and ink-jet-printing is one of the important challenges toward printable circuits on flexible substrates [70, 71]. A group at Oregon State University and Hewlett-Packard fabricated TFTs based on a-In_2O_3-ZnO (IZO) by wet process and subsequent annealing at ~400°C, reporting the field effect mobility of 7–16 cm^2 V s^{-1} [70]. Recently, Ryu et al. [72] of Samsung Advanced Institute Technology developed active matrix type organic LED display with a solution processed a-IZO TFT (mobility 0.9 cm^2 V s^{-1} and threshold voltage ~2.2 eV after annealing at 350°C) backplane.

Viewed from semiconductors, there have been no report on *p-type* amorphous oxide semiconductors with a mobility >1 cm^2 V s^{-1} to date. Figure 13.26 shows the PN-diode fabricated on plastic substrates using N-type a-IGZO and P-type a-ZnO-Rh_2O_3. The current rectifying ratio is good (~10^3) [73, 74]. However, p-channel TFTs did not work well using this p-type amorphous oxide. If a PN-junction with good performance and stability is fabricated by amorphous

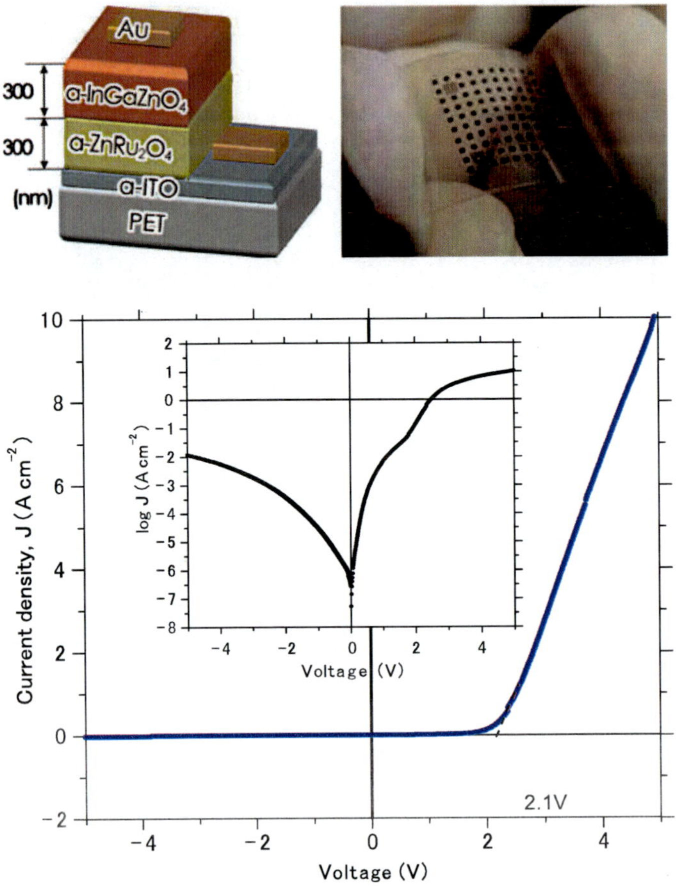

Fig. 13.26 All oxide PN-diode fabricated on a plastic substrate and its current-voltage characteristics [73, 74]

oxide semiconductors on plastic sheets by non-vacuum processes, this would be a breakthrough for solar cells as well as flexible electronics. This is a truly worth challenging subject in this area.

Recently, Ogo et al. [75] reported a p-channel oxide TFT with a field effect mobility of ~1 cm^2 V s^{-1} as shown in Fig. 13.27. The active layer employed is SnO which exhibits p type conduction. The Field effect mobility is ~1/2 of that of Hall effect mobility. Valence band maximum (VBM) of conventional oxides is primarily composed of energetically isolated O 2p-level. This is the main reason why p-type conduction is hard to realize. Oxides of a low valence state post transition metal cations such as Sn^{2+} and Pb^{2+} has a VBM composed of PTM ns-orbitals. This situation is almost reverse to the n-type PTM oxides. Thus, we may expect amorphous p-type oxides would be realized in these material system.

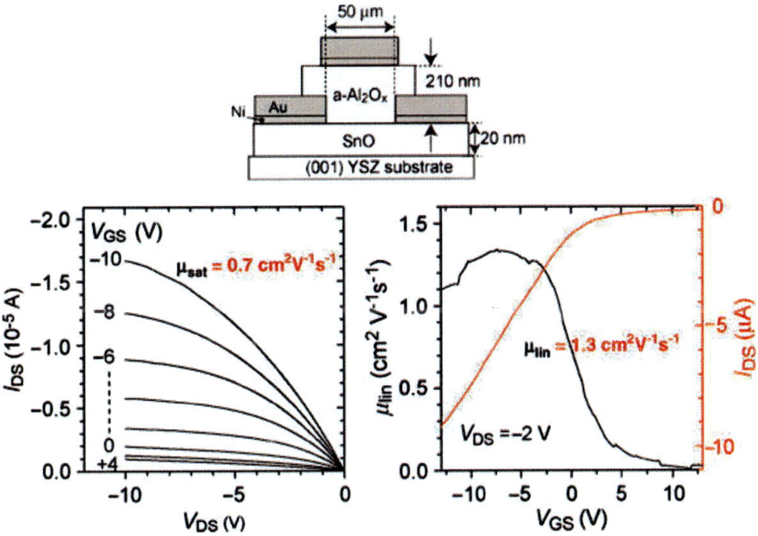

Fig. 13.27 Output characteristics of p-channel TFT using crystalline SnO as active layer [75]

References

1. N.F. Mott and E.A. Davis, *Electronic Processes in Non-Crystalline Materials*, Clarendon Press, Oxford (1979).
2. S.R. Elliott, *Physics of Amorphous Materials*, 2nd Edn. Wiley, New York (1990).
3. W.E. Spear and P.G. LeComber, Solid State Commun. **17**, 1193(1975).
4. P.G. LeComber, W.E. Spear, and A. Gaith, Electron. Lett. **15**, 179(1979).
5. E.P. Denton, H. Rawson, and J.E. Stanworth, Nature **173**, 1030(1954).
6. I.G. Austin and N.F. Mott, Adv. Phys. **18**, 41(1969).
7. C.D. Dimitrakopoulos and D.J. Mascaro, IBM J. Res. Dev. **45**, 11(2001).
8. H. Hosono, T. Kamiya, and M. Hirano, Bull. Chem. Soc. Jpn. **79**, 1(2006).
9. H. Ohta and H. Hosono, Mater. Today **4**, 42(2004).
10. T. Kamiya and H. Hosono, Semicond. Sci. Technol. **20**, S92(2005)
11. H. Hosono, M. Yasukawa, and H. Kawazoe, J. Non-Cryst. Solids **203**, 338(1996).
12. H. Hosono, N. Kikuchi, N. Ueda, and H. Kawazoe, J. Non-Cryst. Solids **198–200**, 165(1996).
13. H. Hosono, J. Non-Cryst. Solids **352**, 851(2006).
14. M. Orita, H. Ohta, M. Hirano, S. Narushima, and H. Hosono, Philos. Mag. B **81**, 501(2001).
15. A. Zunger, Appl. Phys. Lett. **83**, 57(2003).
16. P. Gorn et al., Appl. Phys. Lett. **9**, 063502(2007).
17. E.M.C. Fortunate et al., Appl. Phys. Lett. **92**, 222103(2008).
18. H.Q. Ching et al., J. Non-Cryst. Solids **354**, 2826(2008).
19. H. Hosono, K. Nomura, Y. Ogo, T. Uruga, and T. Kamiya, J. Non-Cryst. Solids **354**, 2796 (2008).
20. Y. Ogo et al., Phys. Stat. Sol. Phys. Stat. Solidi (a) **205**, 1920(2008).
21. J. Robertson, J. Non-Cryst. Solids **354**, 2791(2008).
22. H. Hosono, Y. Yamashita, N. Ueda, H. Kawazoe, and K. Shimizu, Appl. Phys. Lett. **68**, 661 (1996).
23. M. Yasukawa, H. Hosono, N. Ueda, and H. Kawazoe, Jpn. J. Appl. Phys. **34**, L281(1996).

24. E. Fortunate, A. Pimentel, A. Goncalves, A. Marques, and R. Martins, Jpn. J. Appl. Phys. **502**, 104(2006).
25. B. Yaglioglu,Y.-J. Huang, H.-Y. Yeom, and D.C. Paine, Jpn. J. Appl. Phys. **496**, 89(2006).
26. B. Kumar, H. Gong, and R. Akkipeddi, J. Appl. Phys. **98**, 073703(2005).
27. K. Utsumo, H. Iigusa, R. Tokumaru, P.K. Song, and Y. Shigesato, Thin Solid Films, **445**, 229 (2003).
28. T. Minami, Semicond. Sci. Technol. **20**, S35(2005).
29. J. Wager, D.A. Keszler, and R.E. Presley, Transparent Electronics, Springer, New York (2008).
30. S. Narushima, H. Hosono, J. Jisun, T. Yoko, and K. Shimakawa, J. Non-Cryst. Solids **274**, 313 (2000).
31. C.G. Van de Walle and J. Neugebauer, Nature **423**, 626(2003).
32. R. Hayashi, SID'08 Digest 621(2008).
33. A. Sato et al., Appl. Phys. Lett. **94**, 133502(2009).
34. J. Park et al., Appl. Phys. Lett. **93**, 053501(2008).
35. K. Shimakawa, S. Narushima, H. Hosono, and H. Kawazoe, Philos. Mag. Lett. **79**, 755(1999).
36. A. Takagi, K. Nomura, H. Ohta, H. Yanagi, T. Kamiya, M. Hirano, and H. Hosono, Thin Solid Films **486**, 38(2005).
37. M. K. Jayaraj, K.J. Saji, K. Nomura, T. Kamiya, and H. Hosono, J. Vac. Sci. Technol. B **26**, 495(2008).
38. L. Friedman, J. Non-Cryst. Solids **6**, 329(1971).
39. S. Narushima, M. Orita, M. Hirano, and H. Hosono, Phys. Rev. B **66**, 035203(2002).
40. K. Nomura, H. Ohta, A. Takagi, T. Kamiya, M. Hirano, and H. Hosono, Nature **432**, 488 (2004).
41. M. Orita, H. Hiramatsu, H. Ohta, M. Hirano, and H. Hosono, Thin Solid Films **411**, 134(2002).
42. R. Doremus, *Glass Science*, Wiley, New York (1994).
43. K. Nomura, A. Takagi, T. Kamiya, H. Ohta, M. Hirano, and H. Hosono, Jpn. J. Appl. Phys. **45**, 4303(2006).
44. K. Nomura, T. Kamiya, H. Ohta, T. Uruga, M. Hirano, and H. Hosono, Phys. Rev. B **75**, 035212(2007).
45. D.K. Schroder, *Semiconductor Materials and Device Characterization*, Wiley, New York (2006).
46. H. Yabuta, M. Sano, K. Abe, T. Aiba, T. Den, H. Kumomi, K. Nomura, T. Kamiya, and H. Hosono, Appl. Phys. Lett. **89**, 112123(2006).
47. H. Kumomi, K. Nomura, T. Kamiya, and H. Hosono, Thin Solid Films **516**, 1516(2008).
48. J.K. Jeong, J.H. Jeong, H.W. Yang, J-S. Park, Y-G. Mo, and H.D. Kim, Appl. Phys. Lett. **91**, 113505(2007).
49. H.Q. Chiang, B.R. McFarlane, D. Hong, R.E. Presley, and J.F. Wager, J. Non-Cryst. Solids **354**, 2826(2008).
50. W. Lim, S.H. Kim, Y-L. Wang, J.W. Lee, D.P. Norton, S.J. Pearton, F. Ren, and I.I. Kravchenko, J. Electrochem. Soc. **155**, H383(2008).
51. J.H. Jeong, H.W. Yang, J-S. Park, J.K. Jeong, Y-G. Mo, H.D. Kim, J. Song, and C.S. Hwang, Electrochem. Solid State Lett. **11**, H157(2008).
52. P.F. Carcia, R.S. McLean, M.H. Reily, and G. Nunes, Appl. Phys. Lett. **82**, 1117(2003).
53. R. Martins, P. Barquinha, A. Pimentel, L. Pereira, and E. Fortunate, Phys. Stat. Solidi **A202**, R95(2005).
54. J.R. Bellingham, A.P. Mackenzie, and W.A. Phillips, App. Phys. Lett. **58**, 2506(1991).
55. P.K. Song, Y. Shigesato, M. Kamei, and I. Yasui, Jpn. J. Appl. Phys. **38**, 2921(1999).
56. B. Yaglioglu, Y-J. Huang, H-Y. Yeom, and D.C. Paine, Thin Solid Films, **496**, 89(2006).
57. A.J. Leenheer, J.D. Perkins, M.F.A.M. Van Hest, J.J. Berry, R.P. O'Hayre, and D.S. Ginley, Phys. Rev. B **77**, 115215(2008).
58. H. Hosono, M. Kurita, and H. Kawazoe, Jpn. J. Appl. Phys. **37**, L1119(1998).
59. H. Hosono, M. Kurita, and H. Kawazoe, Thin Solid Films **351**, 137(1999).

60. P. Gorren et al., Appl. Phys. Lett. **90**, 063502(2007).
61. P. Gorren et al., Adv. Mater. **17**, 590(2005).
62. R. Hayashi et.al. J. SID **15**, 915(2007).
63. M. Ito et al., IEICE Trans. Electron. **E90-C**, 2105(2007).
64. M. Ito et al., J. Non-Cryst. Solids **354**, 2777(2008).
65. H.-N. Lee et al., J. SID **16/2**, 265(2008).
66. M.-C. Sung et al., IMID'07 Digest 133(2007).
67. J.K. Jeong et al., SID Digest'08 1(2008).
68. K.-S. Son et al., SID Digest'08 633(2008).
69. I. Sing et al., IEEE Electron Device Lett. **29**, 549(2008).
70. D-H. Lee, Y-J. Chang, G.S. Herman, and C-H. Chang, Adv. Mater. **19**, 843(2007).
71. Y. Sun and J.A. Rogers, Adv. Mater. **189**, 1897(2007).
72. M.K. Ryu et al., SID'09 Digest 188(2009).
73. S. Narushima, K. Ueda, H. Mizoguchi, H. Ohta, M. Hirano, K. Shimizu, T. Kamiya, and H. Hosono, Adv. Mater. **15**, 1409(2003).
74. T. Kamiya, S. Narushima, H. Mizoguchi, K. Shimizu, H. Ohta, M. Hirano, and H. Hosono, Adv. Funct. Mater. **15**, 968(2005).
75. Y. Ogo et al., Appl. Phys. Lett. **93**, 032113(2008).

Chapter 14
Junctions

Hiromichi Ohta

14.1 Introduction

The aim of this chapter is an introduction of *p–n* junction diode applications of transparent oxide semiconductors (TOSs). *p–n* junctions of TOSs are of great importance in order to realize transparent electronic circuits since *p–n* junctions are basic building block of electronic circuits and functional properties of commercial semiconductor electronics are attributable to such junctions. The built-in electric field, which is formed at a *p–n* junction region of a semiconductor, leads to various electronic and optoelectronic functions of semiconductor electronics. In this chapter, the general physics, the fabrication and the measurement of TOS *p–n* junctions will be described. It concludes with a discussion of several optoelectronic devices, e.g., transparent current rectifier, photovoltaic and light-emitting diode (LED) composed of *p–n* junctions of TOSs.

14.2 *p–n* Junction Diodes

When *n*-type and *p*-type semiconductors are joined together (*p–n* junction, Fig. 14.1), an interaction takes place between the two types of semiconductor at the junction as a result of the positive holes in one semiconductor and the free electrons in the other semiconductor. When a *p–n* junction is formed, some of the free electrons from the *n*-type semiconductor diffuse across the junction and recombine with positive holes in the *p*-type semiconductor; similarly, some of the positive holes in the *p*-type semiconductor diffuse cross the junction and recombine

H. Ohta
Graduate School of Engineering, Nagoya University, Furo-cho, Chikusa, Nagoya, 464-8603, Japan
e-mail: h-ohta@apchem.nagoya-u.ac.jp

D.S. Ginley et al. (eds.), *Handbook of Transparent Conductors*,
DOI 10.1007/978-1-4419-1638-9_14, © Springer Science+Business Media, LLC 2010

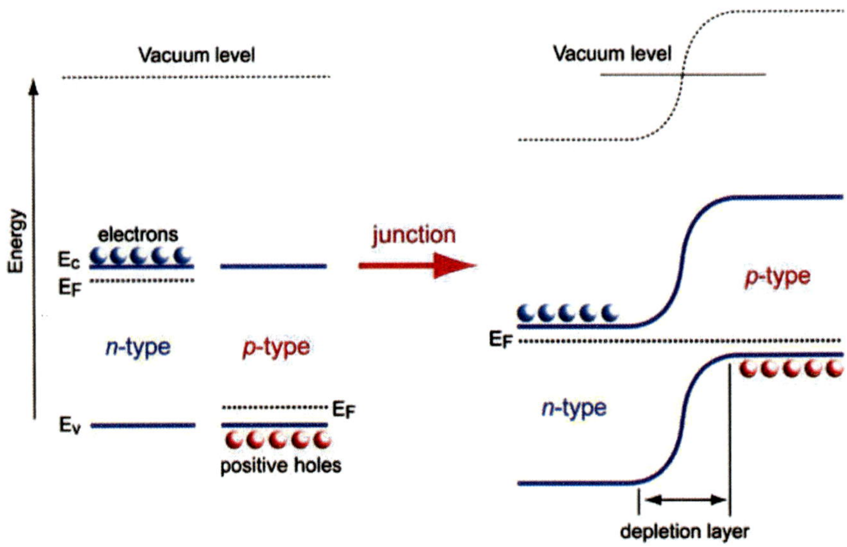

Fig. 14.1 Schematic energy diagram of *p*-type and *n*-type semiconductors before junction formation (*left*) and of the *p–n* junction (*right*)

with free electrons in the *n*-type semiconductor. In thermal equilibrium this charge transfer leads to a narrow space-charge region (depletion layer). Various functions of a *p–n* junction diode, i.e., current rectification, photovoltaic effect, and light-emission originate by the depletion layer.

Generally, *p–n* junction diodes are fabricated by several processes as shown schematically in Fig. 14.2. First, a multilayer, which is composed of both *p*- and *n*-type semiconductors, is fabricated on an appropriate substrate by vapor phase methods such as MBE, CVD or PLD. Then, photoresist is coated on the surface of the multilayer. After the exposure through a photomask and the development of the photoresist the multilayer is etched using a reactive gas such as CF_4 to fabricate a mesa structure. After that, the remaining photoresist is removed by O_3 ashing. Then, the surface is covered by photoresist again to fabricate contact electrode. Finally, appropriate contact electrodes are deposited on the surface.

A *p–n* junction diode is basically characterized by an asymmetric current–voltage (*I–V*) characteristics. While the *I–V* measurement of a Si-diode is generally performed under dark conditions, those of TOS-diode can be performed under room light illumination (Fig. 14.3) because *p–n* junction diode composed of TOSs is basically insensitive to visible light illumination. When an electric power source is connected across a *p–n* junction, the amount of current flow is determined by the polarity of the applied voltage. In the forward-bias region, current rises rapidly when the applied voltage is greater than the built-in electric field (~bandgap of the semiconductor, e.g., 2.6 V for In-Ga-N *p–n* junction diode). The current in the reverse-bias region is usually much lower. This current–voltage behavior of *p–n*

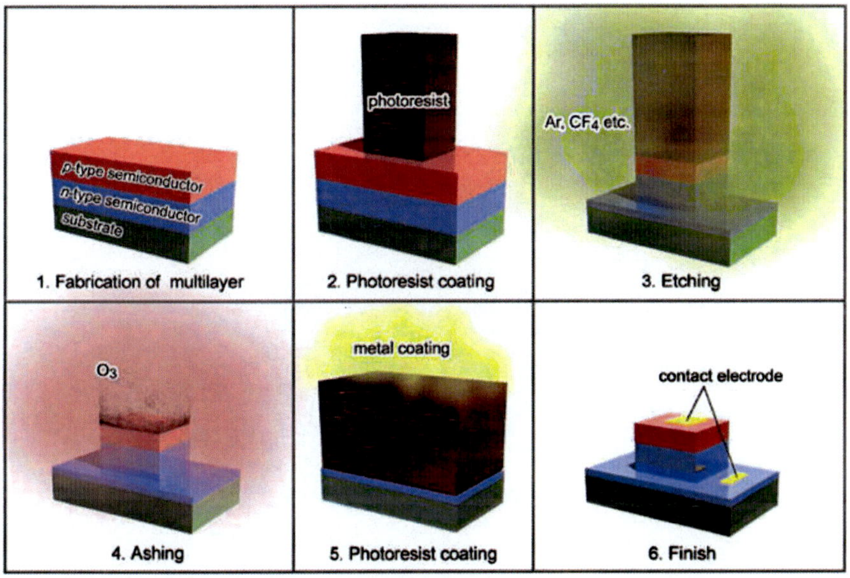

Fig. 14.2 Typical fabrication procedure of a *p–n* junction diode. *Step 1*: Multilayer fabrication. A multilayer, which is composed of both *p*-type and *n*-type semiconductor film, is grown on a substrate by an appropriate growth method, like PLD. *Step 2*: Photoresist coating. The multilayer is processed by a photolithography technique to fabricate a mesa structure having a small surface area. Then, photoresist is coated on the multilayer surface followed by exposure to UV-light through the photomask. The photoresist is developed and rinsed. *Step 3*: Etching. The multilayer is etched by an ion beam etching method using sputter gas such as Ar. *Step 4*: Ashing. Photoresist is removed by ozone ashing. *Step 5*: Photoresist coating. Photolithography process is performed again. Finally, metal coated multilayer is rinsed with acetone to remove photoresist completely

junction is applied for electric current rectification. Typical *p–n* junction diodes composed of TOSs are summarized in Table 14.1.

14.3 Transparent Diodes

In 1999, Kudo et al. first demonstrated a transparent $p = n$ heterojunction diode (diode factor = 1.62), composed of polycrystalline *p*-type $SrCu_2O_2$ (SCO) and *n*-type ZnO, which was deposited by PLD on a glass substrate (Fig. 14.4) [1]. Thin films of highly-conducting ITO were deposited by magnetron-sputtering on commercial glass substrates of size 13 × 13 mm^2. Polyimide thin films were spin-coated on the ITO for electrical insulation and four rectangular holes (2 mm × 2 mm) were fabricated in the polyimide layers by a photolithography technique. SCO layer was deposited into the four holes by a reactive evaporation technique at 350°C. Next, *n*-type ZnO was deposited by magnetron sputtering onto the SCO at

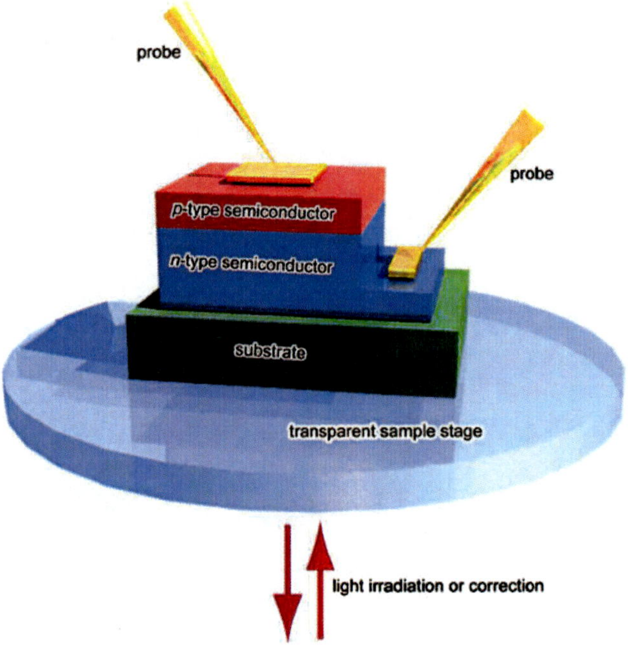

Fig. 14.3 *I–V* measurement of *p–n* junction diode

250°C to fabricate the *p–n* heterojunctions. Thickness and electron density in the *n*-ZnO layer were 300–1,000 nm and 5×10^{18} cm^{-3}, respectively. Finally, n^+-type ZnO with the conductivity of 1×10^2 S cm^{-1} was deposited on *n*-type ZnO as a transparent metallic electrode (Fig. 14.4a). The junction exhibits nonlinear and rectifying *I–V* characteristics (Fig. 14.4b). A small leakage current was observed in the reverse bias region, but the ratio of forward current to the reverse current is larger than 80 in the applied voltage range of -1.5 to $+1.5$ V. The probability of observing a rectifying characteristic of the junctions was 1/10 on the average. As described later, this combination is applicable to fabricate UV-light emitting diodes when the layers are heteroepitaxially grown on the (111)-face of a single-crystalline YSZ substrate.

In 2002, Yanagi et al. reported a transparent *p–n* homojunction diode using bipolar doped $CuInO_2$ with the delafossite structure [2]. The *p–n* $CuInO_2$ homojunction sandwiched by metallic ITO films as *n*-type and *p*-type contacts, exhibited rectifying characteristics with a turn-on voltage of ~1.8 V, keeping an optical transmission of 60% in the visible light region.

Comparable current rectification ratios of *p–n* heterojunction diode composed of *p*-type $CuYO_2$: Ca with delafossite structure and *n*-type ZnO was reported by Hoffman et al. in 2001 [3]. They also used ZnO as active *n*-type semiconductor, no light emission from the diode was observed most likely due to its poor crystalline quality of the films, which were grown on a glass substrate.

Table 14.1 Data of *p*–*n* junction diodes reported in the literature

Year	First author	Technique	*p*-Type layer	*n*-Type layer	Substrate	Light emission	References
1999	Kudo (Tokyo Inst. Tech.)	PLD	$SrCu_2O_2$	ZnO	SiO_2 glass	No	[1]
2000	Ohta (Tokyo Inst. Tech.)	PLD	$SrCu_2O_2$	ZnO	YSZ	382 nm	[12]
2001	Hoffman (Oregon State Univ.)	RF magnet. sputt.	$CuYO_2$	ZnO	Glass	No	[3]
2002	Yanagi (Tokyo Inst. Tech.)	PLD	$CuInO_2$	$CuInO_2$	YSZ	No	[2]
2003	Ohta (Tokyo Inst. Tech.)	PLD	NiO	ZnO	YSZ	No	[9]
	Narushima (Tokyo Inst. Tech.)	RF sputt.	a-ZnO·Rh_2O_3	a-In-Ga-Zn-O	PET film	No	[4]
	Alivov (Inst. Microelectronics Tech.)	CVD	AlGaN	ZnO	6H-SiC	389 nm	[13]
2005	Tsukazaki (Tohoku Univ.)	PLD	ZnO	ZnO	$ScAlMgO_4$	440 nm	[15]
	Hirmatsu (Tokyo Inst. Tech.)	PLD	LaCuOSe	a-In-Ga-Zn-O	MgO	430 nm	[14]
	Hwang (Gwangju Inst. Tech.)	RF magnet. sputt.	ZnO	GaN	Sapphire	409 nm	[16]
2006	Ryu (MOXtronnics)	PLD/MBE hybrid	BeZnO	BeZnO	Sapphire	363 nm	[17]

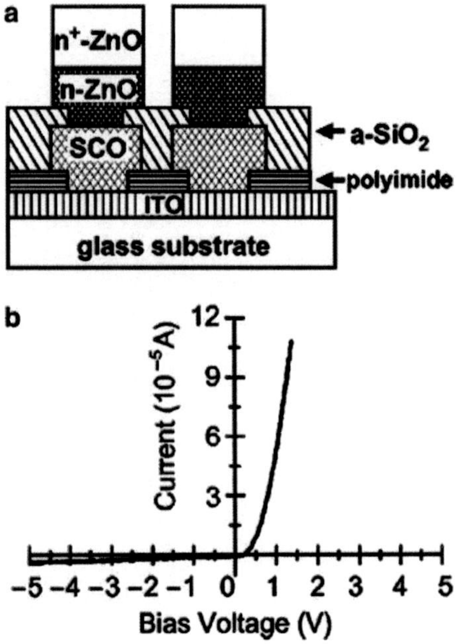

Fig. 14.4 (**a**) Schematic device structure and (**b**) typical *I–V* characteristic of a *p*-$SrCu_2O_2$/*n*-ZnO heterojunction diode *Source*: A. Kudo, H. Yanagi, K. Ueda, H. Hosono, H. Kawazoe, and Y. Yano, Appl. Phys. Lett. 75 (1999) 2851–2853

Thus, the interface between *p*-type and *n*-type semiconductors is essential for improving the current rectification behavior, since crystallographic imperfections at the interface generate leakage currents. In 2003, Narushima et al. demonstrated an all-amorphous *p–n* junction diode composed of *p*-type $ZnORh_2O_3$ and *n*-type In-Ga-Zn-O layers [4]. Since the heterointerface between the two amorphous layers was atomically flat, the diode exhibited a very clear *I–V* characteristics.

14.4 Transparent Photovoltaic Devices

When a photon is absorbed at a *p–n* junction, an electron-hole pair is generated in the absorber layer. The generated electrons and holes are quickly separated into *n*-type semiconductor and *p*-type semiconductor, respectively, by the built-in electric field of the junction. This phenomenon is the so-called photovoltaic effect. Although *p–n* junction diodes from Si (bandgap ~1.1 eV) are widely applied as solar cells, *p–n* junction diode of TOSs may be useful as active devices for ultraviolet (UV) detectors, since the bandgap of TOSs is greater than the maximum energy of visible light (3.1 eV).

UV-radiation that reaches the surface of the Earth is 280–400 nm in wavelengths (UV-A and UV-B), and plays a harmful role that may cause skin cancer. To prevent skin cancer due to the UV-radiation, the development of mobile UV-detectors that are mainly composed of *p–n* junctions of wide-gap semiconductors are required today. Although several UV-detectors have recently been developed using *p–n* junction and Schottky junction diodes of wide-gap semiconductors such as GaN [5], ZnSe [6], ZnS [7] and diamond [8] systems, TOSs are much more preferable for the fabrication of UV-detectors, because TOSs are optically transparent in visible and near UV-light region, environmental friendly and thermally and chemically stable.

Ohta et al. reported an UV-detector which was composed of *p*-type NiO: Li and *n*-type ZnO (Fig. 14.5a) [9]. Transparent trilayered films of ZnO/NiO/ITO were heteroepitaxially grown onto a (111) YSZ substrate by PLD combined with solid-phase epitaxy (SPE). This epitaxial structure is displayed in the cross-sectional HRTEM image (Fig. 14.5b). The resulting trilayered films were dry-etched to form a mesa structure 300 μm square. Rectifying *I–V* characteristics were measured with a forward threshold voltage of ~1 V (Fig. 14.5c). The ideality factor of the diode obtained from the dark current density versus voltage (*J–V*) characteristics was ~2, indicating that carrier recombination in the depletion layer dominates carrier transport. Although the photoresponsivity was fairly weak at zero-bias voltage, it was enhanced up to ~0.3 A W^{-1} through the application of a reverse bias of −6 V

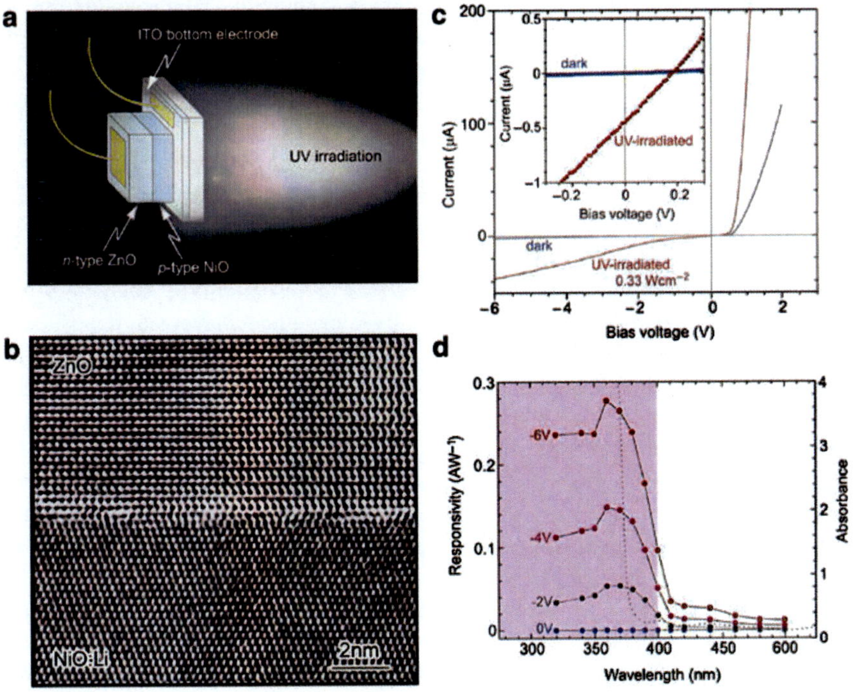

Fig. 14.5 (**a**) Schematic illustration of an UV-detector device composed of a *p*-NiO:Li/*n*-ZnO heterojunction diode. (**b**) Cross-sectional HRTEM image of the interface between NiO and ZnO. (**c**) Typical *I–V* characteristics of the *p–n* heterojunction diode. (**d**) Spectral response of the diode at several reverse bias voltages
Source: H. Ohta, M. Hirano, K. Nakahara, H. Maruta, T. Tanabe, M. Kamiya, T. Kamiya, and H. Hosono, Appl. Phys. Lett. 83 (2003) 1029–1031

under the irradiation of 360 nm light, a value comparable to that of commercial devices (Fig. 14.5d).

14.5 Light-Emitting Diodes

When in light-emitting semiconductors, i.e., GaAs, GaN and ZnSe a *p–n* region has been formed, the *p–n* junction diode can emits electroluminescence light (EL) under forward-bias when the applied voltage is greater than the built-in electric field ($E_{\text{built-in}} \approx$ bandgap).

Among TOSs, ZnO is the most promising candidate as active layer to fabricate ultraviolet emitting diodes (UV-LED). ZnO has been known to emit visible light efficiently when excited with either electron beams or UV-light. The combination of this feature together with *n*-type conductivity has been commercialized in the form of green phosphors for vacuum fluorescence devices and cathode ray tubes (CRTs). ZnO is also promising for use as an UV-emitting phosphor (emission wavelength, $\lambda \sim 380$ nm) at room temperature because the free excitons (a complex of an electron and a positive hole) have a sufficiently large binding energy (59 meV) to be stable even at room temperature. Laser operation excited with electron beams [10] or UV-light [11] have been reported.

Since *p*-type doping of ZnO is very difficult, ZnO-based LEDs were fabricated by combining *n*-type ZnO with a *p*-type semiconductor other than ZnO. In 2000, Ohta et al. first reported an ultraviolet light-emitting diode (UV-LED) operating at room temperature using a *p–n* heterojunction composed of *p*-$SrCu_2O_2$ and *n*-ZnO (Fig. 14.6a) [12]. A (111)-face of YSZ single-crystalline plate with an extremely flat surface was used as the substrate. An ITO epitaxial layer, used as the transparent bottom contact, was grown on the substrate by PLD. Then *n*-type ZnO and *p*-type $SrCu_2O_2$ (SCO) layers were deposited. Finally, a metallic Ni layer was deposited on top of the multilayered film as the top contact. Cross-sectional HRTEM observation reveals that no reaction layer can be seen at the interface (Fig. 14.6b). A rather sharp electroluminescence (EL) band centered at 382 nm was generated when a forward bias voltage exceeding the turn-on voltage of 3 V was applied to the junction. The emission may be attributed to a transition associated with the electron-hole plasma of ZnO. Increased EL intensity with increased injection current is observed without significant change in the band shape or peak energy (Fig. 14.6c). The estimated EL efficiency was less than 10^{-3}%.

Efficient EL emission from *n*-ZnO/*p*-AlGaN heterojunction LED was reported by Alivov et al. in 2003 [13]. Hydride vapor phase epitaxy (HVPE) was used to grow *p*-type AlGaN, while chemical vapor deposition (CVD) was used to produce the *n*-type ZnO layers (Fig. 14.7a). The built-in potential of this combination is ~3.2 eV (Fig. 14.7b). Intense ultraviolet emission with a peak wavelength near 389 nm was observed when the diode is forward biased; this emission is found to be stable at temperatures up to 500 K and shown to originate from recombination within the ZnO.

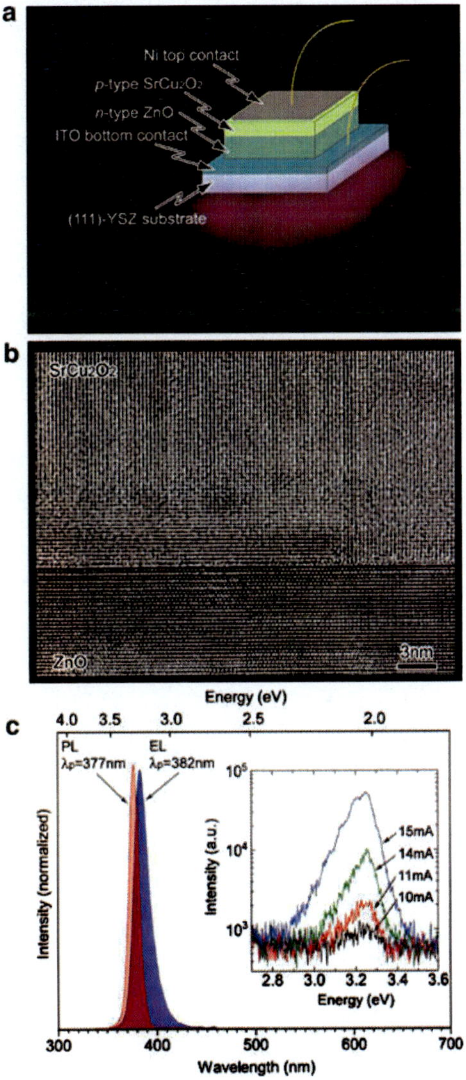

Fig. 14.6 (**a**) Schematic illustration of a LED device structure. The film thicknesses of the ZnO and $SrCu_2O_2$ layers are 500 and 200 nm, respectively. ITO and metallic Ni are used as bottom and top electrodes, respectively. (**b**) Cross-sectional HRTEM image near the *p–n* heterojunction region. The electron beam is injected along $[11\bar{2}0]$ ZnO. A very abrupt heterointerface can be seen. (**c**) Electroluminescence spectra of the LED *Source*: H. Ohta, K. Kawamura, M. Orita, M. Hirano, N. Sarukura, and H. Hosono, Appl. Phys. Lett. 77 (2000) 475–477

In addition, a very unique heterojunction LED was reported by Hiramatsu et al. in 2005 [14]. They reported fabrication of blue LED composed of new TCO materials of *p*-type La-Cu-O-Se and *n*-type amorphous In-Ga-Zn-O. The diode exhibited a relatively sharp blue electroluminescence (EL) that peaked at $\lambda{\sim}430$ nm at room temperature when a forward bias voltage above 8 V was applied (Fig. 14.8). The wavelength and bandwidth of the EL band agreed well with those of the excitonic photoluminescence band in La-Cu-O-Se, which indicates that the EL band originates from the exciton in La-Cu-O-Se. This experiment strongly suggests that layered compounds, *Ln*-CuO-*Ch* (*Ln* = lanthanide, *Ch* = chalcogen),

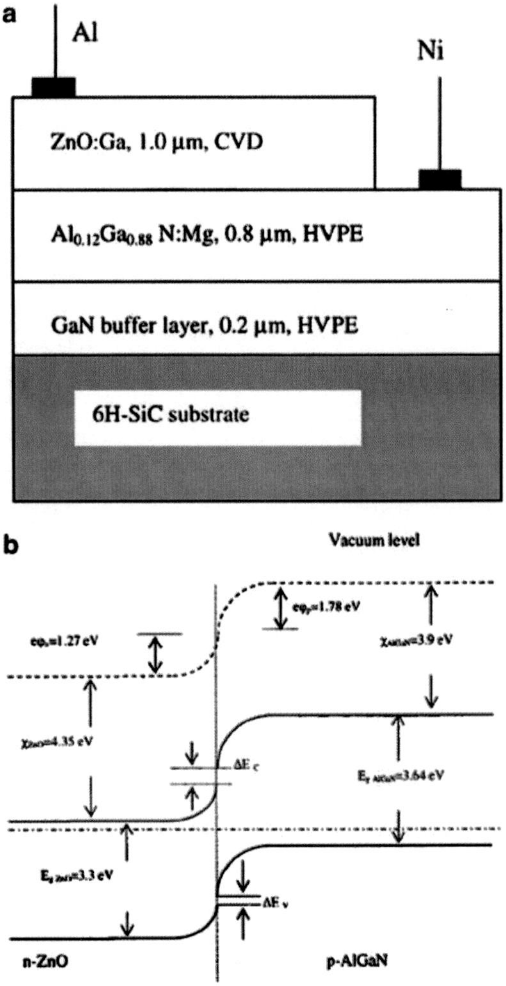

Fig. 14.7 (**a**) Schematic structure of a *p*-type AlGaN/*n*-type ZnO heterojunction diode. (**b**) Schematic band diagram of the diode *Source*: Y. I. Alivov, E. V. Kalinina, A. E. Cherenkov, D. C. Look, B. M. Ataev, A. K. Omaev, M. V. Chukichev, and D. M. Bagnall, Appl. Phys. Lett. 83 (2003) 4719–4721

are promising as the light-emitting layer in optoelectronic devices that operate in the blue-ultraviolet region.

After the publication of several reports on the realization of *p*-type ZnO in 2000, many researchers quickly tried to fabricate *p*-type ZnO. Although many reports on fabrication and characterization of *p*-type doped ZnO film have been published in 2000–2004, there was no report about EL emission from *p*–*n* ZnO homojunction LED most likely due to the poor reproducibility of *p*-type ZnO.

In 2005, Tsukazaki et al. first reported a *p*–*n* ZnO homojunction blue-LED (Fig. 14.9a) [15]. All the ZnO layers were deposited by the PLD method. An *n*-type ZnO: Ga ($n_e = 2 \times 10^{18}$ cm^{-3}) bottom contact layer was deposited on

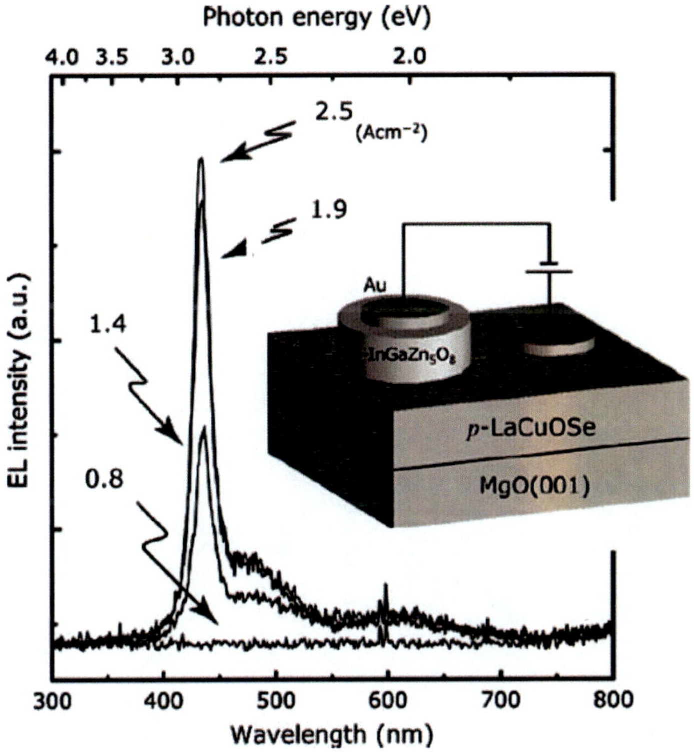

Fig. 14.8 Electroluminescence spectra of the *p*-type LaCuOSe/*n*-type In-Ga-Zn-O (amorphous) heterojunction LED
Source: H. Hiramatsu, K. Ueda, H, Ohta, T. Kamiya, M. Hirano, and H. Hosono, Appl. Phys. Lett. 87 (2005) 211107

the ZnO-buffered (0001)-face of $ScAlMgO_4$ (SCAM) single-crystalline substrate. After the undoped ZnO layer was deposited, highly N-doped *p*-type ZnO (n_h $\sim 10^{16}$ cm^{-3}) was grown by repeated temperature modulation (RTM) technique. Photograph of the device taken with feeding a direct forward bias current of 5 mA in dark (Fig. 14.9b lower) demonstrates blue light emission through the electrode. EL spectra show luminescence from violet to green regions with multi-reflection interference fringes (Fig. 14.9c upper). The higher-energy side peak around 430 nm in the EL spectrum matches well with the photoluminescence (PL) spectrum of a *p*-type ZnO film (Fig. 14.9c lower). The EL from active *i*-ZnO is partly absorbed in the *p*-type layer because of a slight redshift of the absorption edge for this layer.

In 2005, Hwang et al. reported a violet light-emitting diode composed of *p*-type ZnO and *n*-type GaN [16]. The LED structure consisted of a P-doped *p*-type ZnO film with a positive hole concentration of 6.7×10^{17} cm^{-3} and a Si-doped *n*-type GaN film with an electron concentration of 1.1×10^{18} cm^{-3} (Fig. 14.10a). The *I–V* characteristic of the LED showed a threshold voltage of 5.4 V (Fig. 14.10b) and an

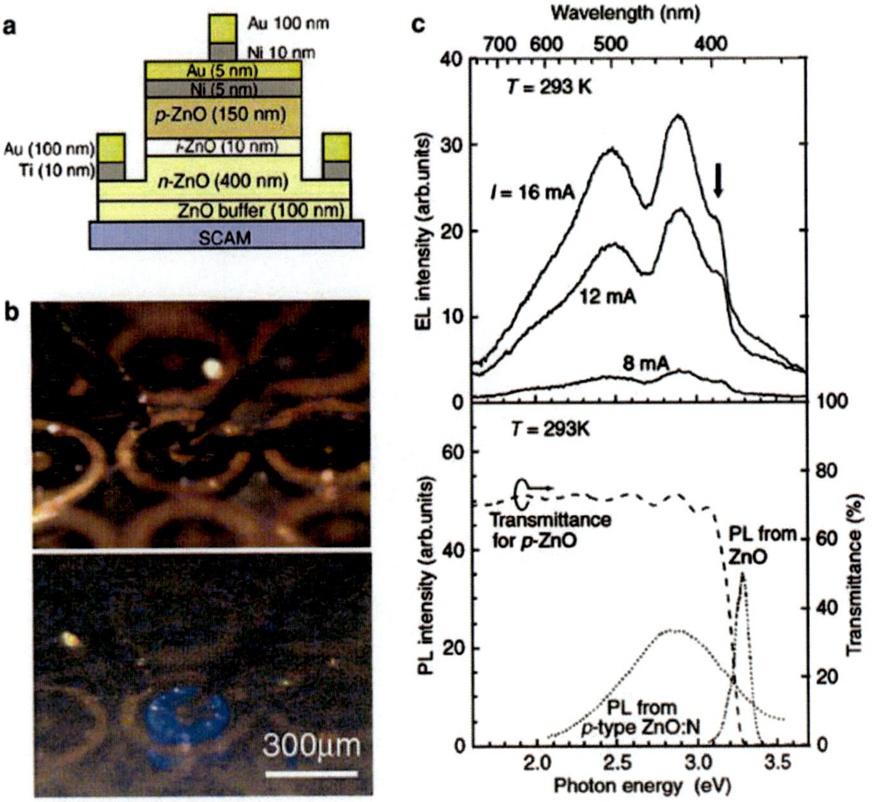

Fig. 14.9 (**a**) Schematic structure of a ZnO *p–n* homojunction LED. (**b**) Photographs of the LED. (**c**) Electroluminescence spectra of the LED
Source: A. Tsukazaki, A. Ohtomo, T. Onuma, M. Ohtani, T. Makino, M. Sumiya, K. Ohtani, S. F. Chichibu, S. Fuke, Y. Segawa, H. Ohno, H. Kuinuma, and M. Kawasaki, Nat. Mater. 4 (2005) 42–46

EL emission at 409 nm (Fig. 14.10c) at room temperature. The EL emission peak at 409 nm was attributed to the optical bandgap of *p*-type ZnO which was reduced as the result of the band offset at the interface between *p*-type ZnO and *n*-type GaN.

In 2006, Ryu et al. reported bright *p–n* homojunction UV-LED composed of ZnO and ZnO-based multiple quantum well (MQW) [17]. The active layer is composed of seven quantum wells (QWs) for which undoped $Be_{0.2}Zn_{0.8}O$ and ZnO form barrier and well layers, respectively, as illustrated schematically in Fig. 14.11a. The thickness of each BeZnO barrier layer is about 7 nm, and each ZnO well layer is about 4 nm. The *p*-type ZnO and BeZnO layers are formed with As as the acceptor dopant, while the *n*-type ZnO and BeZnO layers were formed with Ga as the dopant. The positive hole concentrations in *p*-type layers of the ZnO-based UV-LEDs are in the

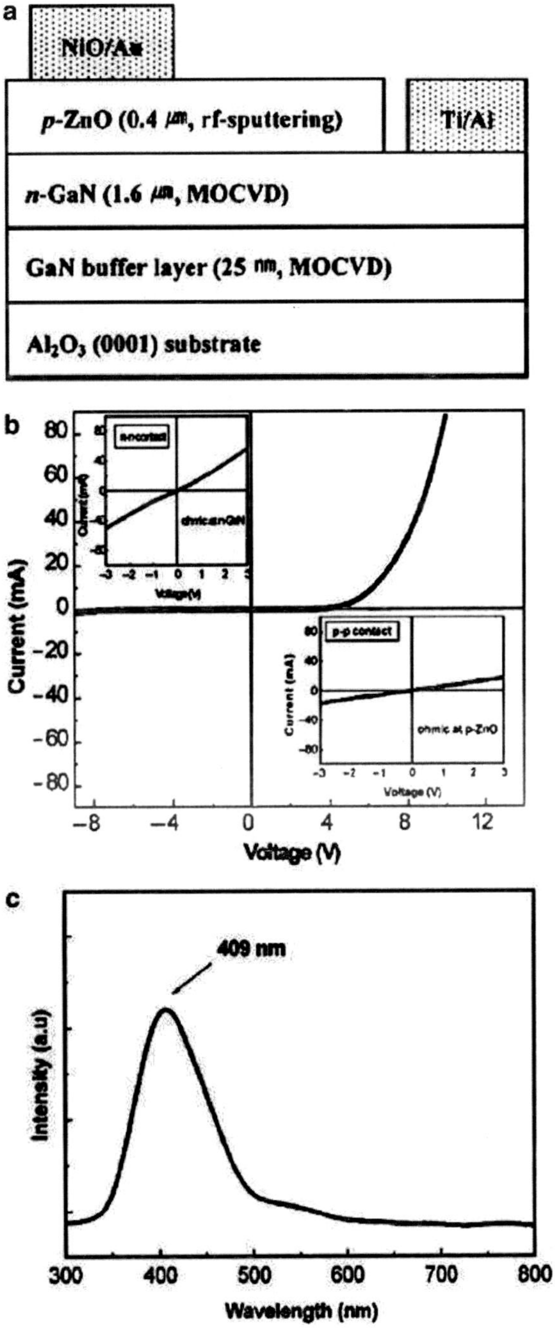

Fig. 14.10 (**a**) Schematic structure of a *p*-type ZnO/*n*-type GaN heterojunction LED. (**b**) *I*–*V* characteristic of the LED. (**c**) Electroluminescence spectrum of the LED
Source: D. Hwang, S. Kang, J. Lim, E. Yang, J. Oh, J. Yang, and S. Park, Appl. Phys. Lett. 86 (2005) 222101

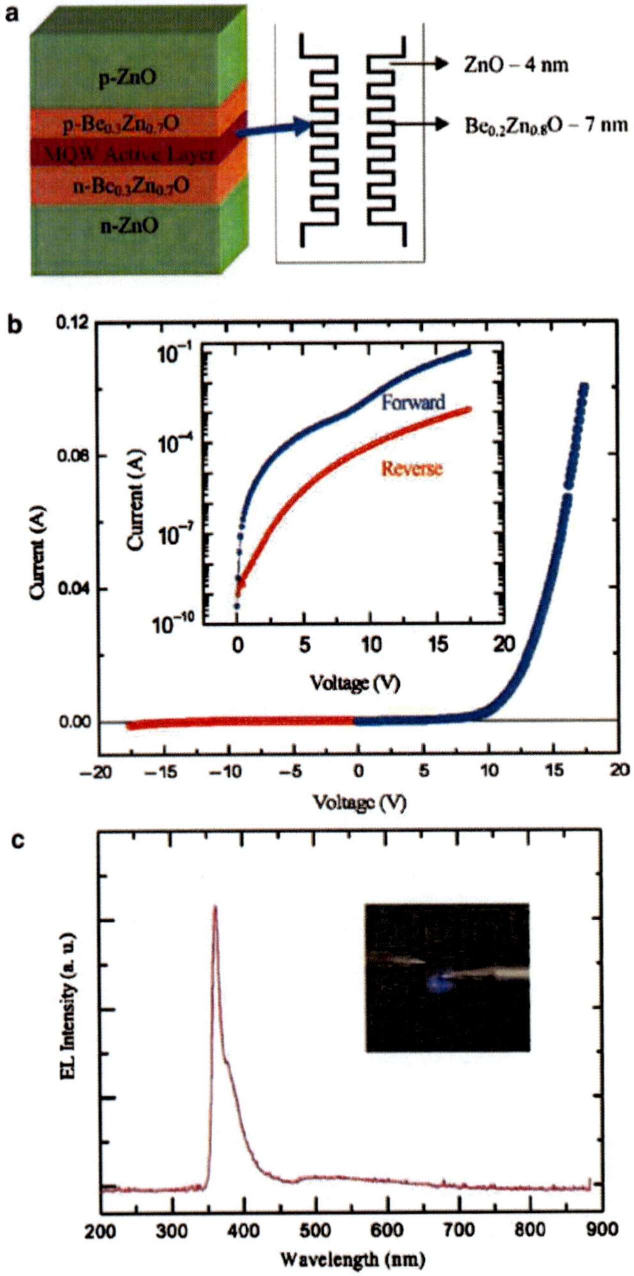

Fig. 14.11 (**a**) Schematic illustration of the structure of the ZnO-based UV-LED device that employs a BeZnO/ZnO active layer comprised of MQWs. (**b**) *I*–*V* characteristics for the *p*–*n* junction LED. (**c**) The spectral curve from the LED with 50 mA continuous injection current *Source*: Y. Ryu, T-S. Lee, J. A. Lubguban, H. W. White, B-J. Kim, Y-S. Park, and C-J. Youn, Appl. Phys. Lett. 88 (2006) 241108

mid-10^{17} cm^{-3} range for ZnO and in the upper 10^{16} cm^{-3} range for BeZnO, while the electron concentrations in *n*-type layers are in the mid-10^{18} cm^{-3} and low 10^{17} cm^{-3} ranges for ZnO and BeZnO, respectively, as determined by Hall-effect measurements. All ZnO and BeZnO layers in the devices were deposited by the hybrid beam deposition (HBD) growth method. The *I–V* characteristic of the LED showed a threshold voltage of ~10 V (Fig. 14.11b) and an intense EL emission at ~360 nm (Fig. 14.11c) at room temperature. The purple illumination (the inset) clearly demonstrates strong EL emission of the LED.

Furthermore, in 2007, Ryu et al. fabricated UV-laser diodes based on ZnO/BeZnO films (Fig. 14.12) [18]. The devices have *p–n* heterojunction structures with a MQW active layer sandwiched between guide-confinement layers. The MQW active layer comprises undoped ZnO and BeZnO, while the two guide-confinement layers were As-doped *p*-type ZnO/BeZnO and Ga-doped *n*-type BeZnO/ZnO films, respectively. Exciton-related lasing was observed by optically pumping the MQWs (Fig. 14.12b bottom). Figure 14.12a shows the EL spectra measured at RT with continuous current injection values of 10 and 60 mA for a ZnO/BeZnO-based MQW device. As the injected current was increased, the rate at which the P_n band (exciton–exciton collision) grew was much higher than for the EX (exciton) peak – it grew superlinearly, indicating stimulated emission. Such evolution in the behavior for excitonic emissions is very similar to that observed in the PL spectrum. With higher injected current densities, several sharp peaks appear in the neighborhood of the P_n band, as depicted in Fig. 14.12b upper. The lasing mechanism is inelastic exciton–exciton collision.

14.6 Summary

In this chapter, the general physics, the fabrication and the measurement of TOS *p–n* junctions were reviewed using several optoelectronic devices, e.g., transparent current rectifier, photovoltaic and light-emitting diode (LED) composed of *p–n* junctions of TOSs as examples. Although *p*-type doping of TOSs is extremely difficult, the discovery of several *p*-type TOSs together with the development of high-quality epitaxial film growth methods has led to optoelectronic devices using TOSs. To realize practical devices, the development of new TOS materials, especially *p*-type TOS, is still a critical issue as represented by *p*-type ZnO. Another important issue is the development of a transparent amorphous oxide semiconductor (TAOS) to realize flexible optoelectronic devices because amorphous thin films can be fabricated on plastics near room temperature. Efforts to realize the device application of TOSs make the TOS world more exciting and fruitful.

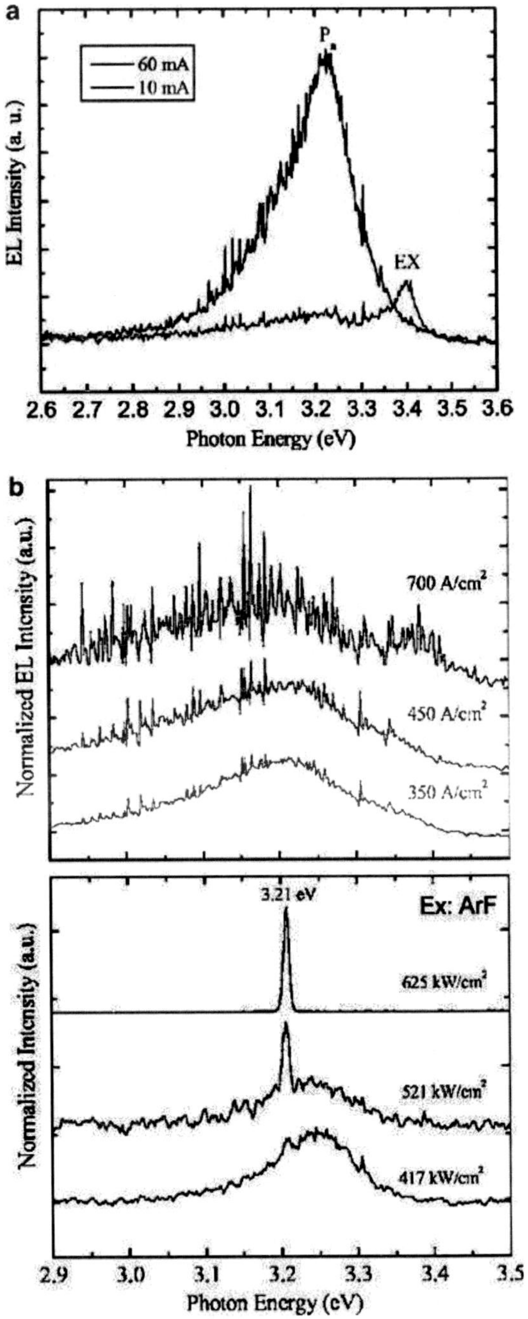

Fig. 14.12 (**a**) Electroluminescence (EL) spectra measured at RT with continuous current injection values of 10 and 60 mA for a ZnO/BeZnO-based MQW device. The device area used for electrical pumping was $300 \times 600\ \mu m^2$. (**b**) EL spectra (*upper*) measured at RT in pulse current injection mode (10% duty cycle). Normalized PL spectra measured at RT of a ZnO/BeZnO superlattice are also shown (*bottom*). Sharp Fabry-Pérot type EL oscillations become more prominent with increasing current density, indicating lasing action *Source*: Y. R. Ryu, J. A. Lubguban, T. S. Lee, H. W. White, T. S. Jeong, C. J. Youn, and B. J. Kim, Appl. Phys. Lett. 90 (2007) 131115

References

1. "Fabrication of transparent *p–n* heterojunction thin film diodes based entirely on oxide semiconductors", A. Kudo, H. Yanagi, K. Ueda, H. Hosono, H. Kawazoe, and Y. Yano, *Appl. Phys. Lett.* **75** (1999) 2851–2853
2. "Fabrication of all oxide transparent *p–n* homojunction using bipolar $CuInO_2$ semiconducting oxide with delafossite structure", H. Yanagi, K. Ueda, H. Ohta, M. Orita, M. Hirano, and H. Hosono, *Solid State Commun.* **121** (2002) 15–18
3. "Electrical characterization of transparent *p-i-n* heterojunction diodes", R. L. Hoffman, J. F. Wager, M. K. Jayaraj, and J. Tate, *J. Appl. Phys.* **90** (2001) 5763–5767
4. "A *p*-type amorphous oxide semiconductor and room temperature fabrication of amorphous oxide *p–n* heterojunction diodes", S. Narushima, H. Mizoguchi, K. Shimizu, K. Ueda, H. Ohta, M. Hirano, T. Kamiya, and H. Hosono, *Adv. Mater.* **15** (2003) 1409–1413
5. "High-responsivity photoconductive ultraviolet sensors based on insulating single-crystal GaN epilayers", M. Asif Khan, J. N. Kuznia, D. T. Olson, J. M. Van Hove, M. Blasingame, and L. F. Reitz, *Appl. Phys. Lett.* **60** (1992) 2917–2919
6. "Time response analysis of ZnSe-based Schottky barrier photodetectors", E. Monroy, F. Vigue, F. Calle, J. I. Izpura, E. Munoz, and J.-P. Faurie, *Appl. Phys. Lett.* **77** (2000) 2761–2763
7. "Molecular-beam-epitaxy-grown ZnMgS ultraviolet photodetectors", I. K. Sou, M. C. W. Wu, T. Sun, K. S. Wong, and G. K. L. Wong, *Appl. Phys. Lett.* **78** (2001) 1811–1813
8. "Thin film diamond photodiode for ultraviolet light detection", M. D. Whitfield, S. S. Chan, and R. B. Jackman, *Appl. Phys. Lett.* **68** (1996) 290–292
9. "Fabrication and photoresponse of a *pn*-heterojunction diode composed of transparent oxide semiconductors, *p*-NiO and *n*-ZnO", H. Ohta, M. Hirano, K. Nakahara, H. Maruta, T. Tanabe, M. Kamiya, T. Kamiya, and H. Hosono, *Appl. Phys. Lett.* **83** (2003) 1029–1031
10. "Ultraviolet ZnO laser pumped by an electron beam", F. H. Nicoll, *Appl. Phys. Lett.* **9** (1966) 13–15
11. "Optically pumped ultraviolet lasing from ZnO", D. C. Reynolds, D. C. Look, and B. Jogai, *Solid State Commun.* **99** (1996) 873–875
12. "Current injection emission from a transparent *p–n* junction composed of *p*-$SrCu_2O_2$/*n*-ZnO", H. Ohta, K. Kawamura, M. Orita, M. Hirano, N. Sarukura, and H. Hosono, *Appl. Phys. Lett.* **77** (2000) 475–477
13. "Fabrication and characterization of *n*-ZnO/*p*-AlGaN heterojunction light-emitting diodes on 6H-SiC substrates", Y. I. Alivov, E. V. Kalinina, A. E. Cherenkov, D. C. Look, B. M. Ataev, A. K. Omaev, M. V. Chukichev, and D. M. Bagnall, *Appl. Phys. Lett.* **83** (2003) 4719–4721
14. "Excitonic blue luminescence from *p*-LaCuOSe/*n*-$InGaZn_5O_8$ light-emitting diode at room temperature", H. Hiramatsu, K. Ueda, H, Ohta, T. Kamiya, M. Hirano, and H. Hosono, *Appl. Phys. Lett.* **87** (2005) 211107
15. "Repeated temperature modulation epitaxy for p-type doping and light-emitting diode based on ZnO", A. Tsukazaki, A. Ohtomo, T. Onuma, M. Ohtani, T. Makino, M. Sumiya, K. Ohtani, S. F. Chichibu, S. Fuke, Y. Segawa, H. Ohno, H. Kuinuma, and M. Kawasaki, *Nat. Mater.* **4** (2005) 42–46
16. "*p*-ZnO/*n*-GaN heterostructure ZnO light-emitting diodes", D. Hwang, S. Kang, J. Lim, E. Yang, J. Oh, J. Yang, and S. Park, *Appl. Phys. Lett.* **86** (2005) 222101
17. "Next generation of oxide photonic devices: ZnO-based ultraviolet light-emitting diodes", Y. Ryu, T-S. Lee, J. A. Lubguban, H. W. White, B-J. Kim, Y-S. Park, and C-J. Youn, *Appl. Phys. Lett.* **88** (2006) 241108
18. "Excitonic ultraviolet lasing in ZnO-based light-emitting devices", Y. R. Ryu, J. A. Lubguban, T. S. Lee, H. W. White, T. S. Jeong, C. J. Youn, and B. J. Kim, *Appl. Phys. Lett.* **90** (2007) 131115

Chapter 15
Process Technology and Industrial Processes

Mamoru Mizuhashi

A quick review of the deposition methods that achieved sufficient low resistivities and comparison of these methods in terms of the capability for large area mass production will be made. At the later half our experience on two pilot runs by evaporation and magnetron sputtering machines will be described in detail.

15.1 Challenges to Lower Resistivities

It was 1966 when the indium oxide doped with tin, later called as ITO, was recognized as an outstanding low resistivity transparent conductor by the work of R. Groth [1]. He and his colleagues later reported $2 \times 10^{-4}\ \Omega$ cm in 1968 [2] by spray pyrolysis from an acidic alcoholic solution of $SnCl_4$ and $InCl_3$ onto 5 mm thick borosilicate glass substrates at 500°C range. The optimal doping level of tin was 2 at.%. This triggered many extensive R & D activities all over the world toward the liquid crystal displays (LCDs), especially in the US and Japan. According to this movement we could have many findings and in-depth understandings of the ITO materials. These research works provided firm position of ITO in the LCD devices.

During the very beginning of LCD development the transparent thin film electrode on both sides of liquid crystal layer was tin oxide doped with antimony, SnO_2:Sb. Its resistivity was in some lower $10^{-3}\ \Omega$ cm at best. The first commercial LCD was applied to the calculator where the required sheet resistance was of the order of hundred ohms with a thickness less than 40 nm so as not to be seen to unaided human eyes. This means the resistivities less than $1 \times 10^{-3}\ \Omega$ cm as an essential demand from the device side. Based on this market demand sheet glass

M. Mizuhashi
School of Science and Engineering, Aoyama-Gakuin University, 5-10-1 Fuchinobe, Sagamihara, Kanagawa 229-8558, Japan
e-mail: dsatelmn@ss.iij4u.or.jp

D.S. Ginley et al. (eds.), *Handbook of Transparent Conductors*,
DOI 10.1007/978-1-4419-1638-9_15,

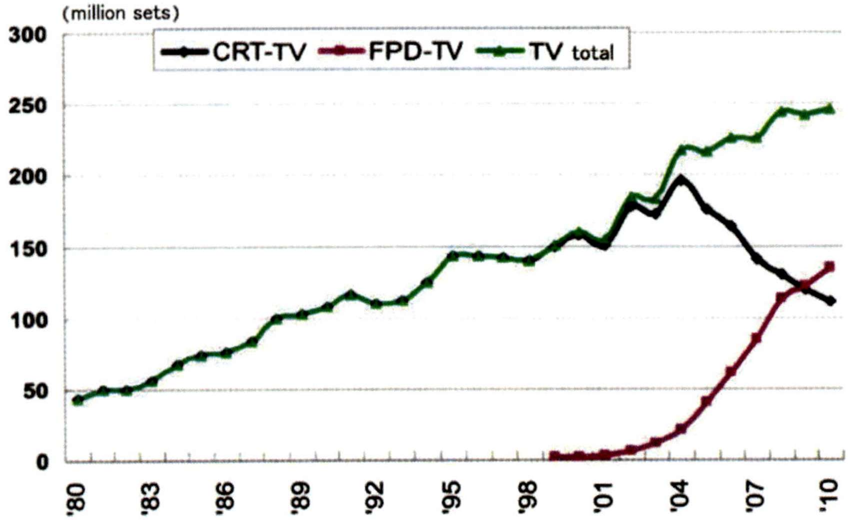

Fig. 15.1 Market development of flat panel display in comparison to the traditional CRT

companies got deeply involved into the development of the production technologies of high quality ITO coating from early 1980s. However, the large area ITO coating process was rapidly absorbed into the production process of the LCD panel manufacturers to improve the over-all production yield.

The position of the LCD monitors in the world display market has long been the second. It was 2003 when the flat panel LCD took over the first position from CRTs as a summation of built-in LCD in the laptop PC and stand-alone LCD monitors and 2004 as stand-alone LCD monitors. Figure 15.1 shows the world-wide competition between CRT-TV and FPD-TV since 1980 with some future prospect [3]. We can see that there will be crossing at 2009 in terms of the number of television sets.

In 1972 D. B. Fraser and H. D. Cook showed that the resistivity of the same level could be attained by the dc diode sputtering with a sintered oxide target. It was sufficiently conducting to allow use of the method [4]. The minimum resistivity of $1.77 \times 10^{-4}\ \Omega$ cm was observed at around 9 mol% addition of tin oxide. Deposition was carried out on a Corning 7059 substrate heated at 500°C. This paper also indicated that the target has to be cleaned by argon bombardment prior to the deposition (importance of pre-sputtering to eliminate hydrated surface layer on the target).

Next reported were the CVD and thermal evaporation methods in 1975. The CVD-ITO by J. Kane, H. P. Schweizer and W. Kern [5] reported $2.2 \times 10^{-4}\ \Omega$ cm at around 8 at.% doping, similar to that of the spray pyrolysis. The vapor source material was β-diketone synthesized from the reaction of hydrolyzed $InCl_3$ and dipivaloyl methane. Vacuum evaporation from the oxide mixture by J. C. C. Fan and F. J. Bachnar also provided the resistivity of the same level, $2 \times 10^{-4}\ \Omega$ cm at some slightly lower doping levels of 4–6 mol% [6].

As a result, all the main methods of thin film formation revealed their potential capabilities for the low resistivity ITO deposition until 1980. There have been several breakthroughs in the electrical properties down to 10^{-5} Ω cm up until 2007, i.e. activate reactive evaporation by P. Nath, R. F. Bunshah, B. M. Basol and O. M. Stuffsud [7, 8], magnetron sputtering by S. Ray, R. Banerjee, N. Basu, A. K. Batabyal and A. K. Barna [9], biased magnetron sputtering by S. Maniv, C. J. Miner and W. D. Westwood [10], high magnetic field magnetron sputtering by S. Ishibashi and H. Higuchi [11], zone-confining arrangement by I. A. Rauf [12]. However some of them were not successfully reproduced by the others, suggesting that there are still some critical conditions or parameters not explicitly mentioned in such papers. Now they are widely known as slow deposition rate, appropriate growth temperature and contaminant-free vacuum to form higher crystallinity where highest free carrier creation and transportation can be realized.

15.2 Comparison of Methods for the Uniform Large Area Deposition

Large area deposition technology available at around 1980 was the thermal evaporation, based on the long experiences in the area of sheet glass coating, so-called heat reflecting glass, manufactured by Owens-Ford Glass Company in 1970s.

The dc diode sputtering was also used in a small scale production in some area of electronic device production. However, it has drawbacks of very slow (several tens nm/min) deposition rate and low efficiency process due to enormous heat release at the surfaces of target and substrate when compared with the electron-beam evaporation.

The magnetron sputtering process now prevailing worldwide was in the stage just after the completion of development and yet needed actual experiences of production. See original patents [13, 14]. This new technology is characterized by the high rate, low heat generation and capability of low-pressure operation due to the adoption of magnetic field, and hence considered to be suited for the large area deposition.

The pyrolytic spray deposition had also been frequently used for the glass coating. However, this method often results in a coarse appearance of irregular reflection patches because droplets in the spray mist vaporize in the very vicinity of hot glass surface. Thus such products would not be acceptable as a electronic component from the cosmetic point of view.

The CVD process was thought to be more acceptable for the deposition of oxide films than the spray pyrolysis or diode sputtering. The process is capable of operation for the production at higher speeds with better uniformity. Indeed the CVD has grown to be a very sophisticated high speed process for the deposition of semiconductor crystalline materials such as Si or GaAs. However it does not provide satisfactory uniformity over entire area of glass substrates of a few to

Table 15.1 Summary of characteristic features of four main thin film deposition methods

	Spray pyrolysis	CVD	Evaporation	Diode sputtering	Magnetron sputtering
Machine cost	Low	Medium	High	High	High
Depo. rate	Very fast	Fast	Slow	Very slow	Slow
Material	Chloride	β-diketon	Oxide/metal	Oxide/metal	Oxide/metal
Material cost	Low	Very high	Med. to high	Med. to high	Med. to high
Ito quality	Yes	Yes	Yes	Yes	Yes
Large area	Yes	Yes	Yes	Yes	Yes
Cosmetic	Unacceptable	Good	Very good	Very good	Very good
Eco-measure	Heavy	Heavy	Medium	Medium	Medium

several meter sizes. In addition one of its fatal drawback is the high cost of raw materials as a vapor source like β-diketone. The second is that it often requires some kind of neutralization treatment of the exhaust gases to make them harmless. The biggest issue was the high cost of vapor source material.

Points of discussion is summarized in Table 15.1, suggesting that evaporation and sputtering are good candidates.

15.3 Two-Source Evaporation

In the case of vacuum evaporation large area deposition can be done either passing a large (wide) substrate over plural vapor sources in line across the width of the substrate or rotating the substrate over a single source placed in an off-axis arrangement, ordinarily with using at least one deposition mask in between. The substrate size is rather limited in the turn-around method although it can achieve precise and uniform deposition over entire substrate, say within 1% deviation, and hence suited for the production of precise optical elements such as interference filters and cold mirrors.

In the deposition using in-line sources, two electron beam (EB) sources in our case, we used a powder mixture of In_2O_3 and SnO_2. The point of primary importance was that we have to keep the evaporation rate of the two sources at a predetermined value during the deposition. This is a matter of monitoring and control. However, the point of anxiety was whether the film composition was automatically uniform and constant as long as the deposition rate was kept constant. There is a substantial difference in their vapor pressures at evaporation temperatures, possibly resulting in a gradual change in the film composition. There is also a possibility of forming volatile (or nonvolatile) compound through chemical reactions at evaporation temperatures. Fortunately In_2O_3 and SnO_2 as well as their electron-beam-reduced sub-oxides sublimed instantaneously when heated and soon reach a stationary evaporation state. Thus preliminary experiments showed no compositional deviation during the deposition period of about 30 min, the time necessary to deposit roughly 100 nm thick ITO film over a 1 m wide by 2 m long glass substrate, except in a very early stage where tin oxide rich composition was found to exist. In order to avoid the composition deviation it is sufficient to provide shielding means by which the substrate is not exposed to vapor at the very beginning of evaporation. From this result we decided to use the sintered oxide mixture instead of powder mixture.

In the deposition configuration we chose the substrate was put in a stainless metal frame and linearly transported across the deposition. Thus our configuration provides the thickness distribution of axial symmetry.

It is necessary for the development of technology for the large area deposition to know the spatial distribution of evaporated vapor. This enables us to calculate thickness distribution and optimum configuration among the two sources and substrate, then to establish the method of deposition rate control for the reproducibility. Finally it is necessary to examine various deposition conditions so as to get the electrical and optical properties required.

15.3.1 Spatial Distribution of Evaporated Substance

The EB evaporators used were the ULVAC model EGM-35 EB guns with a 10 kW (10 kV × 1,000 mA) power supply. This is a 270° deflection type. The source material was a sintered block of the size of 35 mmϕ by 20 mm high.

Figure 15.2 shows a typical thickness distribution in nanometer unit at each point with a 2 cm interval along both x- and y-axes directions on a square feet sheet glass placed 15 cm above the evaporant surface. Deposition was carried out at 5×10^{-5}

X -14	-12	-10	-8	-6	-4	-2	0	2	4	6	8	10	12	14	Y
400,	420,	450,	510,	320,	67,	60,	440,	570,	590,	525,	520,	470,	460,	0	14
420,	440,	520,	500,	420,	210,	260,	620,	650,	660,	595,	575,	530,	490,	430	12
420,	500,	530,	590,	610,	670,	700,	710,	705,	720,	636,	600,	590,	525,	450	10
460,	520,	570,	660,	710,	740,	770,	770,	760,	730,	642,	630,	620,	565,	490	8
500,	570,	640,	710,	760,	810,	840,	840,	850,	790,	665,	655,	630,	590,	540	6
510,	540,	650,	720,	760,	830,	860,	840,	850,	830,	750,	670,	640,	600,	540	4
510,	530,	620,	700,	760,	810,	860,	840,	800,	830,	750,	670,	635,	590,	530	2
480,	530,	640,	700,	750,	790,	800,	820,	800,	780,	730,	680,	640,	600,	530	0
450,	510,	610,	660,	710,	760,	780,	790,	790,	760,	740,	660,	630,	560,	510	-2
430,	520,	580,	620,	660,	730,	760,	760,	780,	740,	720,	640,	610,	530,	450	-4
410,	510,	540,	600,	610,	680,	700,	700,	680,	690,	660,	610,	560,	460,	420	-6
360,	410,	450,	530,	560,	580,	590,	610,	635,	610,	560,	530,	470,	420,	380	-8
310,	330,	420,	430,	510,	540,	535,	570,	560,	550,	510,	450,	420,	380,	350	-10
0,	270,	360,	400,	430,	480,	480,	510,	500,	480,	470,	400,	380,	320,	0	-12
0,	0,	260,	330,	380,	420,	430,	410,	420,	0,	370,	280,	340,	0,	0	-14

Fig. 15.2 Thickness distribution of an ITO film on glass placed above 15 cm from an EB source

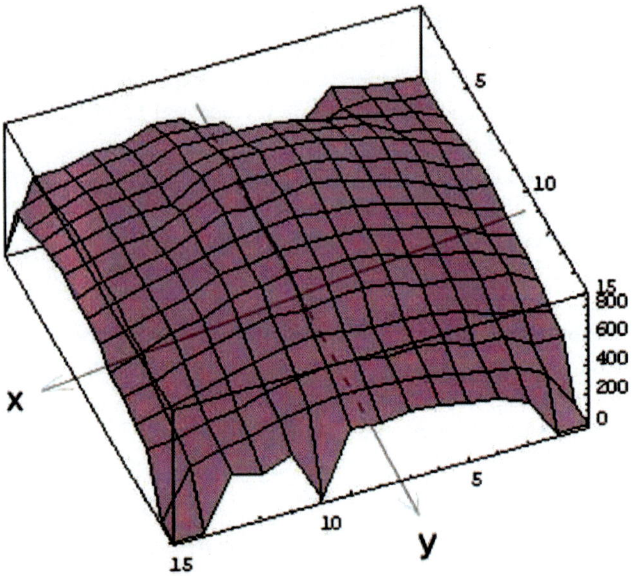

Fig. 15.3 Resultant 3D map of the thickness distribution

torr without introducing reactive gas. Note that in this numerical map the thickest position deviates by about 2 cm in the positive y-direction. The thickness contour is approximately concentric and not anisotropic, as shown in Fig. 15.3. Thinner areas seen along the edge reflect the shadowing effect of the instruments such as quartz crystal monitor.

This 2 cm deviation is explained as a result of the formation of virtual evaporation source above the evaporant surface. This is formed by the intense electron beam focused on the evaporant surface. Focused electron beam digs an inclined erosion pit, resulting in an inclined vaporization. The virtual source is a spatial point from which the evaporation apparently takes place and said to coincide with the position of most intensive part of plasma consisting of ionized hot particles of evaporant and electrons as shown in the insert of Fig. 15.4.

This deviation was not correctly expressed in terms of deflection angle and distance to the substrate. It was found that the deviation was almost the same when the source-substrate distance was 45 cm. Thus this can be explained by assuming a virtual evaporation source as mentioned above.

Eliminating unevenness from the concentric contour we examined to fit this curve with the curves of the cosine law.

$$\frac{d(\theta)}{d_0} = h^{-\alpha}\cos^n\theta$$

$$\theta = \tan^{-1}\frac{\sqrt{X^2+Y^2}}{h}$$

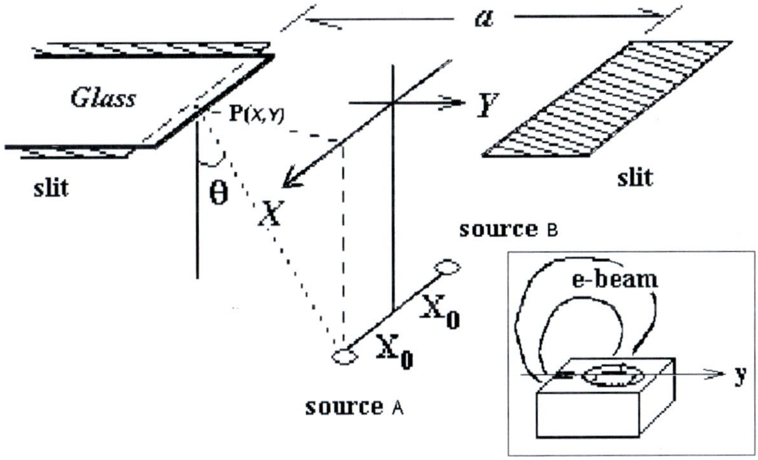

Fig. 15.4 Coordinate of the two-source evaporation system with respect to the substrate

where d_0 the film thickness just above the virtual source, h the source-substrate distance, (X, Y) the coordinate on the substrate. We got $a = 2.05 \pm 0.05$ and $n = 4.65 \pm 0.01$ finally as a result of fitting trials.

15.3.2 Calculation of the Thickness Distribution

In the configuration depicted in Fig. 15.4 the glass substrate is transported in a plane situated at h above the line connecting two EB sources A and B.

The substrate is coated when it moves across the slit. Putting the speed of transportation v, slit width $\boldsymbol{a}$, source distance $2x_0$ and evaporation rate C (an arbitrary constant in the calculation) the deposition rate $\Re$at a point (X, Y) on the substrate is given by

$$\Re(X, h, t) = \Re(-X, h, t) = \frac{C}{h^{2.05}}(\cos^{4.65}\theta_1 + \cos^{4.65}\theta_2)$$

$$where \quad \theta_{\frac{1}{2}} = \tan^{-1}\frac{\sqrt{(X_0 \mp X)^2 + (Y_0 - Y)^2}}{h}$$

$$Y = vt$$

An important assumption here is that the two sources are completely equivalent and that the evaporation rate is kept constant during the substrate transportation across the slit. The resultant thickness d after completion of deposition at the other end of slit is expressed as an integration of $\Re$ with time.

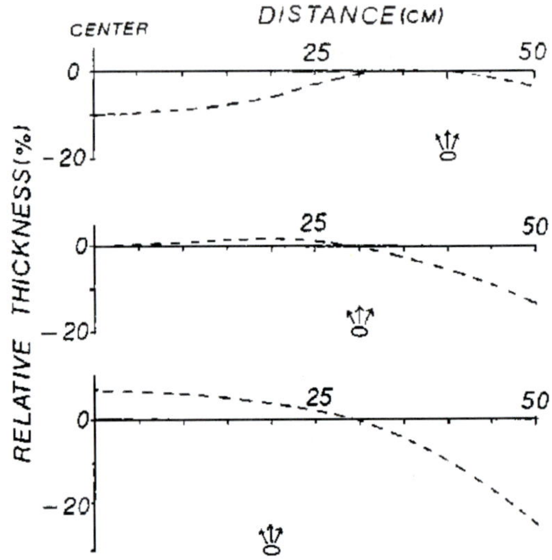

Fig. 15.5 Calculated thickness distribution across a 1 m-wide glass substrate

$$d = \int_{-\frac{a/2}{v}}^{+\frac{a/2}{v}} \Re(X, h, t)dt$$

Some results of calculation are shown in Fig. 15.5. Note that this figure shows only the half side of the distribution because of its axial symmetry. The best separation between two sources is somewhere around 60 cm when the distance to the substrate is 60 cm, whereby we can have a thickness distribution of about $\pm 7\%$. Too close and too apart configurations give a center-thick and a edge-thick distribution, respectively.

More detailed calculations led to the optimum configuration in term of the configuration ratio.

$$2X_0/h = 1.08$$

It is evident from Fig. 15.5 that further improvement in the uniformity is possible if we put the third source at the center, or third and fourth sources at both sides.

15.3.3 *Large Area Deposition from Two Point Sources*

Figure 15.6 gives an experimental result carried out on a 1 m by 2 m glass substrate. It is important to establish stable and equal evaporation rate of the two EB sources before we begin the transportation of substrate. In order to achieve this we used two quartz rate monitors to watch each source. Then we found that there is a suitable

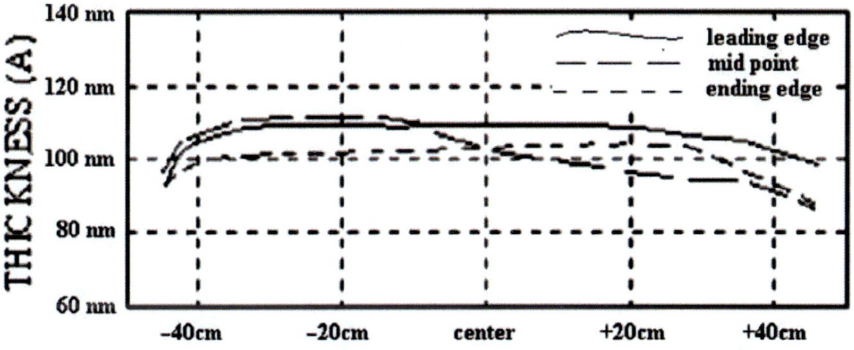

Fig. 15.6 Experimental result on thickness distributions on a 1 m by 2 m glass substrate

position to place them, i.e. the position of monitor head should be above the x axis in the coordinate system of Fig. 15.4. Otherwise, i.e. when it is above the y axis, the actual rate increases gradually with time even the monitoring value is kept constant, because of the formation of an oblique v-shaped pit declined in the y direction. In other words the monitor sees the change of the evaporation surface when it is above the y axis, while it sees much smaller change when above the x axis. The monitor controls the output power of an EB gun to maintain the predetermined rate. However, it should be mentioned that a co-operative action of the two monitors was not possible at that time around year 1980. The evaporation amount is about 10 g for each source. Time to complete the deposition onto 1 m by 2 m sheet glass is roughly 30 min with an evaporation rate of 5 Å/s at the monitor. This low rate operation was necessary to avoid splashes of evaporant. Splashes even if they are seemingly weak undoubtedly reach the substrate surface to stick, then each particle constitutes defect on the film.

As seen from the figure, at the leading edge the thickness distribution is almost symmetric because of careful rate adjustment of the two sources before the deposition. It lost the symmetric shape in the +X side at the midpoint of 100 cm from the leading edge, due to the rate imbalance. Then we decreased the rate of the EB source on the −X side, resulting again in a almost symmetric distribution at the rear end of substrate.

This experiment tells us that the wide-width in-line deposition from a linear array of point sources is rather difficult from the cooperative control point of view.

15.4 Sputter Deposition

The most prominent success in the sputter deposition technology would be the invention of planar magnetron sputtering by Chapin in 1974 [14]. In combination with the invention of Corbani [13] the planar magnetron sputtering technology succeeded in establishing its position worldwide. Greatest progress was the

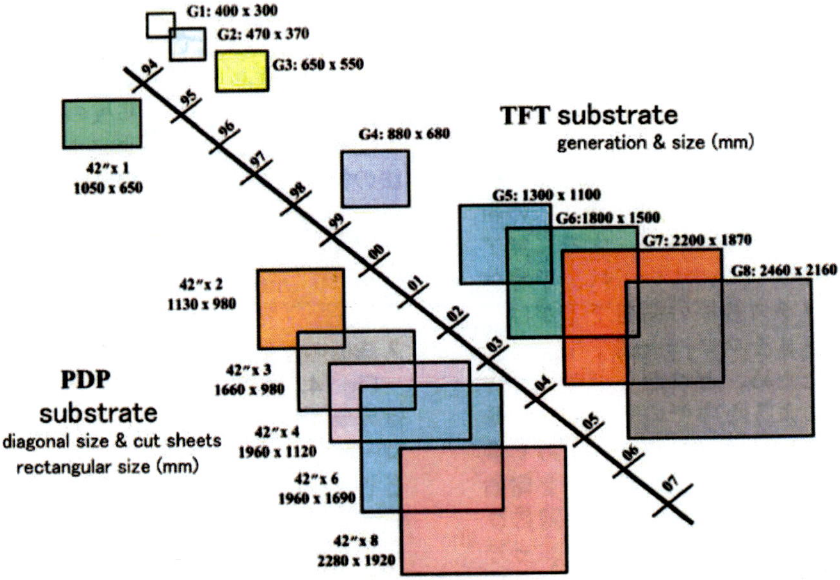

Fig. 15.7 Historic expansion of LCD and PDP substrate sizes from G1 to G8 and from 42″ × 1 to 42″ × 8, respectively. For example, 42″ × 4 on the PDP stream means that four sheets of glass with 42″ diagonal can be obtained after a series of long production processes

deposition rate of several hundred nm/min, roughly one to two orders of magnitude higher than the diode sputtering of old generation as in the claim No. 1 and 8 of the patent [14], describing *"Apparatus comprising a cathode defining a substantially planar sputtering surface magnet means to form an endless arch over a closed loop erosion region, an anode, and means for connecting said cathode and said anode to a source of electrical potential*", and "*A high rate sputtering apparatus for coating a substrate with material sputtered from a substantially planar sputtering surface in a low pressure environment, comprising magnet means, arc, a first electrode, a second electrode, and means for connecting said first and second electrodes to a source of electric potential*". In addition it has a high potential for the large area in-line deposition if longer targets become available.

The in-line sputter machine to date is so-called G8 (eighth generation in term of liquid crystal display), and capable of coating the size of 2,400 by 2,200 mm. See the FPD's expansion history in Fig. 15.7 [3].

The ITO deposition process has been taken into the LCD panel manufacturers from the coating service companies early 1990s. The time almost coincides with the beginning of TFT (thin film transistor) epoch. They called at that time the size 400 × 300 as G1 and the size 470 × 370 as G2. It has been said in mid-1990s that some of LCD manufacturers constructed two kinds of sputter-coating line, an in-line coater for the front glass of the LCD panel and many batch coaters for the rear glass. The ITO coating on the front glass was formed on the color filter array. A number of batch coaters were for the rear glass with the TFT array. This was due to the fact that a very serious caution was necessary in treating the TFT surface. Now such a difficult situation was overcome, and very recently it is said that a huge, unified vacuum system consisting of two in-line coaters is in serious examination for the production.

The thickness uniformity is always a deep concern for the vacuum engineers in the manufacturing division. However this issue is not so serious in the discussion of machine specifications to be guaranteed, because the machine price will goes up considerably if this specification becomes steeper. Improvement in the uniformity is rendered to the engineers in the production sector, ordinarily.

One thing to be mentioned here in relation to the uniformity is the width of the peripheral zone without coating. The peripheral zone width has increased as the substrate generation proceeded. It was 10 mm till G4, 15 mm till G7 and will be 25 mm from G8 on every peripheral. The bigger and thinner the glass substrate, the broader the peripheral zone to hold glass sheet safely and firmly.

The style of in-line coater is also changing. The conveyer system had a return path outside the in-line chambers till G4. The system adopted so-called vacuum return formulation from G5 in which the conveyer chain is not exposed to air during its return to the coater's entrance position, as shown in Fig. 15.8.

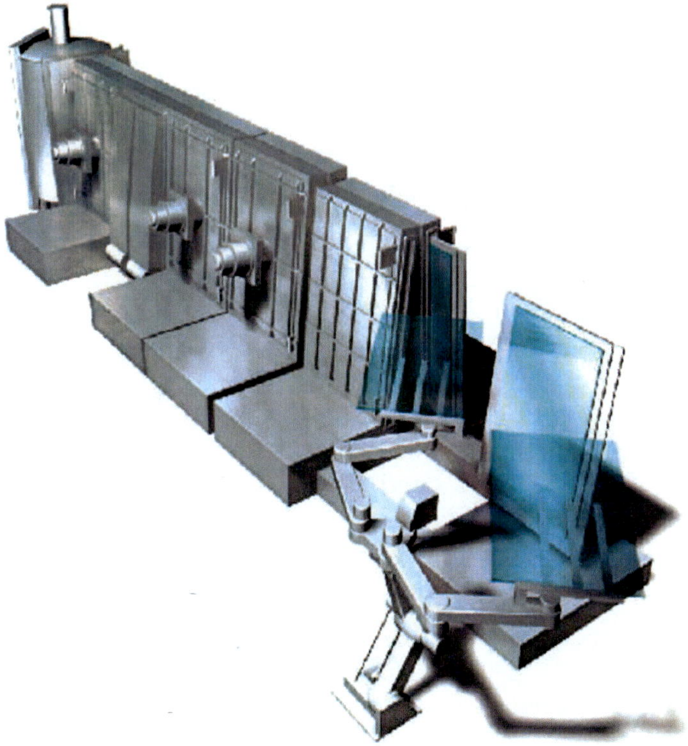

Fig. 15.8 Bird's eye view of a typical G8 sputter machine (courtesy of ULVAC corporation). A cylindrical (diameter >2.5 m) vacuum return system is depicted at the far end

15.5 Transparent Electrode for Large Area Solar Cells

The control of conduction types on the amorphous Si (a-Si) by phosphorous and boron found by Prof. W. Spear at Dundee in 1975 has opened up the way to the electric power generation with large area flat panel solar cells as well as the way to thin film transistors now indispensable for the LCDs. There are two types of cell structure, one called substrate-type and the other superstrate-type, distinguishing the location of glass substrate with respect to the incident light, as shown in Fig. 15.9.

In making substrate-type we first prepare the electrode which is not necessarily transparent and hence metal layer or metal foil is used as a conducting substrate. On this electrode a-Si:H layers of n, i, p type and finally a transparent conductor (TCO) layer are formed. On the other hand in the case of superstrate we have to put first the TCO, then p, i, n, and finally reflective metal electrode layers. The solar cell structure is finally laminate-enclosed with the other sheet glass or polymer film such as highly durable fluoropolymers. Note that the sunlight passes through the cell structure from p to n in both cases.

Apart from the very active research on the semiconductors and device structures, the other main issue has been an artificial formation of efficient textures on top of the TCO film surface in the superstrate-type. However, first of all, the TCO should be of low resistivity to conduct a current with minimum loss and of high transmittance over the sunlight spectrum to inlet the sunlight as much as possible. These two requirements are fully satisfied by In_2O_3:Sn (ITO), but it has a serious drawback in the plasma durability. The ITO surface suffers from very strong chemical reduction when exposed to the $SiH_4 + H_2$ gas plasma. Reduction takes place within a few

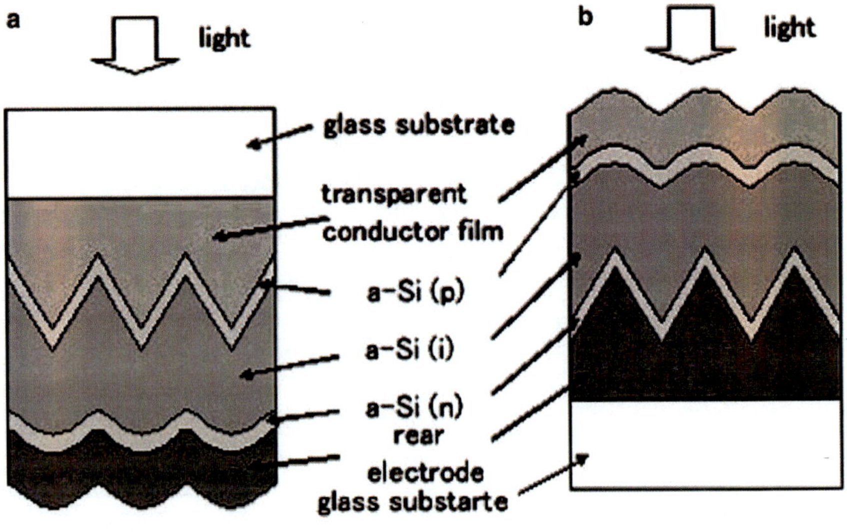

Fig. 15.9 Two types of thin film solar cell structure: (**a**) superstrate, (**b**) substrate

seconds, resulting in an absorption of a few to 10%, depending on the plasma excitation strength. This is the reason why the ITO has never been accepted for the solar cell.

The size of substrate in the thin film solar cell production is normally more than 1 m in dimension to achieve sufficient power output of a few to several hundreds watts at several tens volts. Thin film solar cell module is expected to provide solutions against the cost and the ecological issues with using TCO materials other than ITO.

15.5.1 Properties Requested to Textured TCOs

The texture should be such that it nicely scatter the sunlight so as to increase the path length of light inside the cell. The texture, however, does not mean the steep and rugged surface because it makes the smooth and stable junction formation difficult.

One of such smooth, coherent texture was reported on CVD deposited SnO_2:F film in 1990 [15]. Also the chemical stability of ZnO on top of the textured SnO_2:F has been clarified to be better than SnO_2:F bare surface in 1994 [16]. Figure 15.10 shows the type "U" texture reported in [15]. It consists of regular, smooth surface of microcrystals with a good coherence and is recognized as a standard texture for a-Si solar cells. U texture has been referred in many reports including papers on ZnO based TCO to compare with theirs.

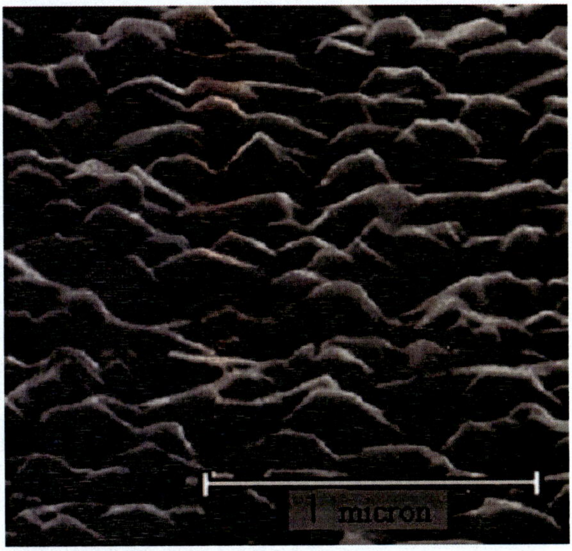

Fig. 15.10 Type "U" texture prepared by CVD

The market demand on the TCO coated substrates for use in solar cells is comprises of three main features, i.e. (1) resistivity as low as possible to minimize the loss due to Joule heat, (2) absorption of sunlight (covering from near UV to near infrared) as low as possible to get better conversion efficiency, (3) a proper texture to confine as much sunlight as possible in the junction region of the cell [17]. The second requirement is now recognized as more important because the light utilization ratio of near infrared/visible is increasing in a-Si:H/mc-Si:H tandem structures recently available on the market. This means that TCOs of higher mobilities are better because of lower absorption in the infrared.

The effect of texture can be explained in more detail as shown in Fig. 15.11. The first is the effect of reflection reduction (or suppression of reflection) provided by the rugged structure with a characteristic length of somewhat smaller than the quarter wavelength of average sunlight. This gives rise to a graded refractive index layer which serves as an anti-reflection (AR) layer at the interface (TCO/a-Si). Taking into account the refractive index 2.0 of tin oxide and that of a-Si c.a. 4.0, this material combination well satisfy the amplitude condition for the AR coating.

The second is strong scattering power to extend the traveling length of light in the i-layer where photoelectrons and holes are generated. Multiple passage of light increases the electron-hole pair generation. Thus generated electrons and holes must be separated as soon as possible. The separation is driven by the electric field between p- and n-layers. Accordingly the separation efficiency goes up as the i-layer thickness decreases. This trade-off relation suggests that there is an optimum thickness of the i-layer. Multiple scattering of light helps semiconductor to generate more electron-hole pairs so that its thickness can be reduced and the separation field is strengthened. Thickness reduction may also serve as saving the Si raw materials.

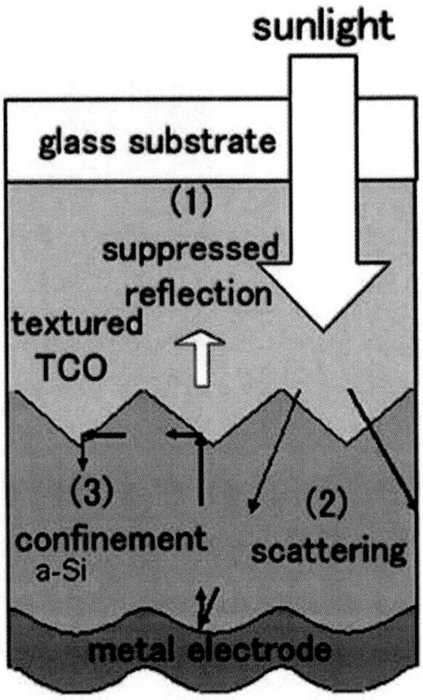

Fig. 15.11 Three functions of the TCO texture

The third is light confinement. The sunlight yet not absorbed in the photoactive i-layer suffers from diffuse scattering and back again into the active i-layer where this remainder sunlight takes part in the electric power generation. The higher the diffuse back scattering and the broader the wavelength region covered by the texture, its contribution to the power generation is enhanced. Note again that the formation of high quality junction cannot be achieved unless the texture has gentle curvatures at its ridges, peaks and valley bottoms [17].

15.5.2 The Present State of the Art of SnO_2:F TCO

Tin oxide conductor has a long history of development. The first patent using antimony as a dopant was granted in 1950 to Corning Glass Works. They called this materials as "NESA". Consideration of defect semiconductor with this material led to examination of various elements as a dopant, resulted in SnO_2:F materials in 1960s. Blue hue observed at higher Sb concentrations were avoided and the resistivity was further improved. Tin oxide thin films found a market, so called reach-in doors, a see-through door of refrigerator for customers to look into and pick items up to buy. A dense amorphous SiO_2 underlayer was found to be efficient for the reproducible production because it blocks the outdiffusion of alkali from soda lime silicate glass [18, 19].

The method of making SnO_2:F TCO is the atmospheric pressure CVD (APCVD). A new development after the type "U" has been reported recently [20]. The basic concept of this texture is to introduce some new class of structures to enhance light scattering at longer wavelengths. This has to be achieved without losing the scattering characteristics in the shorter wavelengths that the type U structure has. Resultant TCO has two levels of dimensional hierarchy, namely, small grains are growing on the large grain, which is something like "many baby tortoises on mother tortoise".

Figure 15.12 shows SEM images of some typical type "HU" textures recently developed [20]. It is obvious from the figure that the sizes of small and large grains are almost the same among these images. This was achieved by a proper adjustment of the CVD parameters. Accordingly it is possible to enhance the scattering in the longer wavelengths as well as in the shorter wavelengths.

Figure 15.13 shows the scattering intensities as a function of wavelength. The scattering intensity is expressed in terms of haze, the difference of the integrated transmission and the direct transmission. It is clear that the long wavelength scattering increases with the increase in the height of the primary grains (big and coarse grain) and that also the short wavelength scattering increases with the sculpture depth of the secondary grains. Table 15.2 summarizes the optical, electrical and geometric properties. This new generation TCOs is expected to enhance the conversion efficiency of the solar cells consisting of microcrystalline Si such as a-Si:H/mc-Si:H stack whose sensitivity is extended to wavelength region more than 800 nm.

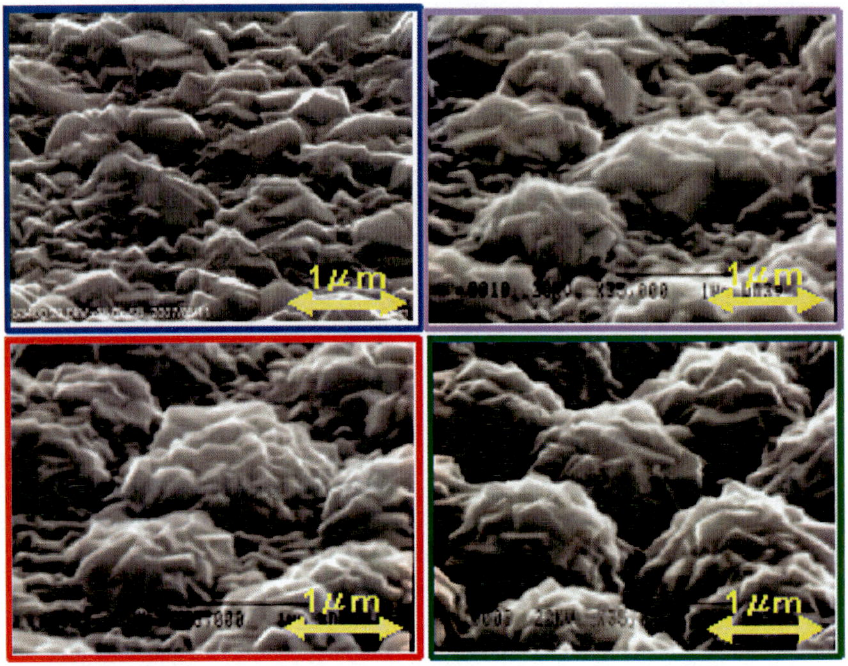

Fig. 15.12 TCO's features with different ruggedness

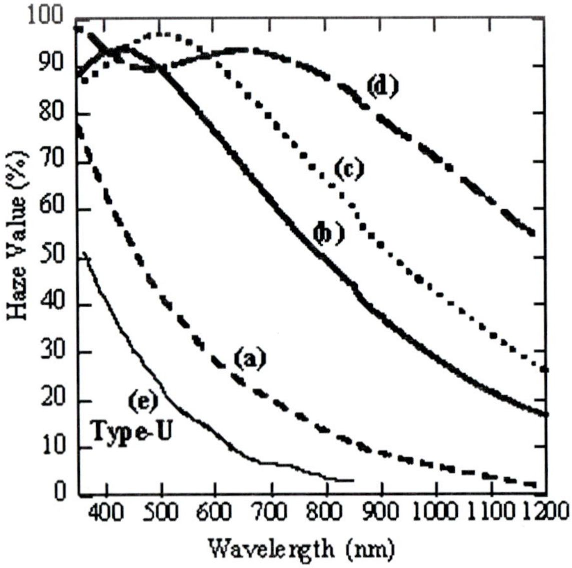

Fig. 15.13 Scattering spectra of the type "HU" TCOs with thicknesses around 1 μm

Table 15.2 Physical properties of the type "HU" TCOs (about 1 mm thick)

Sample name	Haze at 550 nm (%)	Transmission at 550 nm (%)	Transmission at 800 nm (%)	Sheet resistance (Ω/sq)	rms Roughness (nm)
(a)	35	89	88	10	65
(b)	84	89	88	9	109
(c)	95	88	88	8	122
(d)	90	89	88	10	150

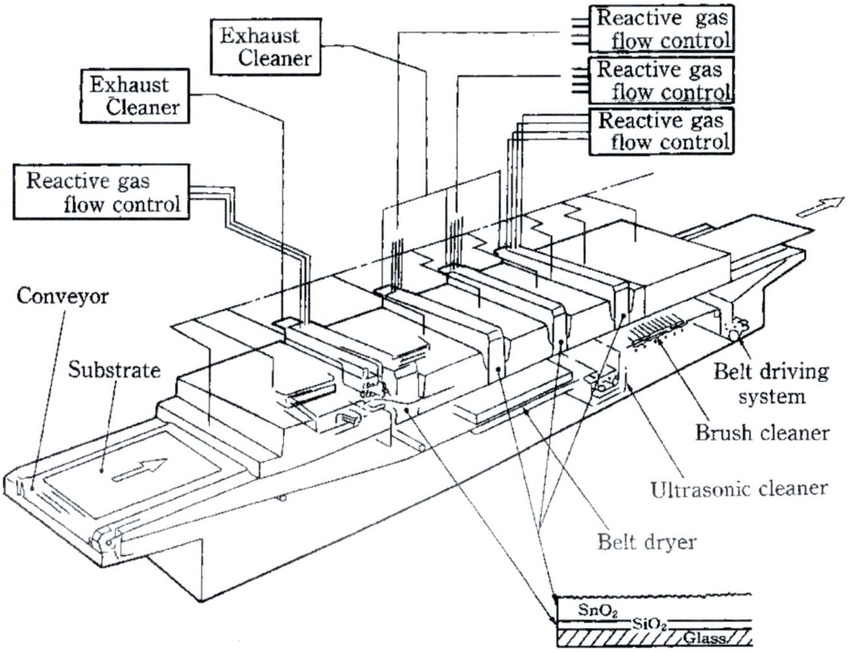

Fig. 15.14 Schematic view of an APCVD machine

15.5.3 CVD Deposition of SnO_2:F TCOs

A typical CVD machine to deposit tin oxide TCO is depicted in Fig. 15.14. Glass substrates are put onto the belt conveyer and transported into the furnace zones to be heated to a CVD temperature at around 500°C. Then the substrate is subjected to the CVD gases to deposit the silica barrier, then tin oxide films. Gas injection zone can accommodate multistep injectors to provide enough conductivity and texture. The CVD gases for SiO_2 deposition are monosilane and nitrogen, and $SnCl_4$, methanol, HF, H_2O and nitrogen for SnO_2:F deposition. These gases are mixed each other just before entering the injection zone. Flow of the component gases are precisely controlled by mass flow meters [21]. Uniform injection in the form of laminar flow and uniform aspiration of exhaust gases in and around each injector are of prime importance for the uniform deposition.

The glass substrates are cooled down after the deposition, then taken out, whereas the belt conveyer turns back to the furnace entrance, passing through the cleaning station underneath the furnace to remove all powdery and flaky deposits adhered on it.

15.5.4 *Present State of the Art of ZnO TCOs*

The first R&D activity on ZnO has reported in 1979 on the spray pyrolysis. Low resistivity ZnO in the range of 10^{-4} Ω cm was reported with using the RF magnetron sputtering in 1982 [22]. Thermally stable ZnO:Al was found in 1983, which is important from the practical application point of view. Common resistivity of ZnO: Al at present by the dc magnetron sputtering is about 3×10^{-4} Ω cm when deposited on heated (~300°C) substrate to thicknesses more than 400 nm. Thinner films show higher resistivities depending very much on the thickness. In addition, in contrast to the case of ITO, the resistivity is sensitive to the oxygen partial pressure and hence sensitive to the location on the substrate or to the position in the vacuum chamber. Optimization at the substrate center resulted in higher resistivities in the peripheral zone, particularly distinctive on large substrates [23].

ZnO is chemically fragile, in particular to acids. The film disappears even in a weak acidic solution, for example, a solution of 0.5% HCl. This has been a serious neck when considering LCD applications. In the LCD production, thin sheet glasses with TCO film are immersed in such an acidic etchant to generate precise patterns on the substrate. Precise etch is almost impossible for ZnO films due to unintentional excess etching at the pattern edges.

However ZnO is regarded as the most possible candidate to replace SnO_2 materials. It also has a high potential to realize lower resistivities comparable to the ITO. It has several other obvious advantages, for example, the Clarke number, i.e. Zn is far more rich than In. This situation encouraged worldwide development activities of this material toward the thin film solar cells. Fortunately low sheet resistance, transparency and texture in particular are seemingly achieved when it is thick enough.

In the long history of research, many elements has been found to be efficient as dopants. They are typically B and Ga, denoted as BZO and GZO [24–26]. There are many methods for the production of ZnO transparent conductors. Texture can be formed during the deposition or by selective etching after deposition. There are two main streams of deposition: one is the sputtering [27, 28] and the other is low pressure CVD [29].

Reactive mid-frequency sputtering with ZnO:Al target [30] clarified that the deposition rate rose remarkably up to more than 50 nm · m/min (~30% increase) when the substrate temperature range was at 200–150°C, due probably to the increase in the sticking probability of Zn atoms. The resistivity under the optimum gas pressure was 264 mΩ cm, not so bad. The results were brought into a project aiming at the solar cell module production, where the texture was formed by chemical etching [31]. Thorough understanding of the magnetron sputtered film properties and optimization concept were reported from the same group [32]. Similar results have been reported in the laboratory scale in 2002 and 2003 [33, 34].

Low pressure CVD (LPCVD) has also been a good candidate for the large area deposition [35]. State of the art of LPCVD ZnO:B describes that the carrier mobility can be raised to 35 cm^2/V s if the carrier density is controlled lower than (3–4) $\times$ 10^{19} cm^{-3}, resulting in a high transparency and a haze level of 50–70% at a thickness of 5 μm, which is considered to be very favorable as an as-grown texture [29].

Needless to say, highly reproducible production requires almost perfect cleaning of the substrates. Then it is of critical importance to maintain the cleanliness throughout the coating process. There are many traps that are often overlooked.

Acknowledgement The author would like to express sincere thanks to Dr. T. Oyama, Mr. Ishikawa, Mr. T. Haranoh, Dr. Takaki, Prof. Y. Shigesato Dr. K. Suzuki and Mr. T. Iwamatsu for supporting him on this subject.

References

1. R. Groth, Phys. Stat. Solidi., 14(1966)69
2. H.J.J. van Boort and R. Groth, Philips Tech. Rev., 29(1968)17
3. H. Ishikawa, Report Res. Lab. Asahi Glass, 57(2007)37
4. D.B. Fraser and H.D. Cook, J. Electrochem. Soc., 119(1972)1368
5. J. Kane, H.P. Schweizer and W. Kern, Thin Solid Films, 29(1975)155
6. J.C.C. Fan and F.J. Bachnar, J. Electrochem. Soc., 122(1975)1719
7. P. Nath and R.F. Bunshah, Thin Solid Films, 69(1980)63
8. P. Nath, R.F. Bunshah, B.M. Basol and O.M. Stuffsud, Thin Solid Films, 72(1980)463
9. S. Ray, R. Banerjee, N. Basu, A.K. Batabyal and A.K. Barna, J. Appl. Phys., 54(1983)3497
10. S. Maniv, C.J. Miner and W.D. Westwood, J. Vac. Sci. Technol., A1(1983)1370
11. S. Ishibashi and H. Higuchi, Proc. 1st Intn'l Symp. on Sputtering and Plasma (ISSP91), pp 153
12. I.A. Rauf, J. Appl. Phys., 79(1996)4057
13. Corbani patent, USP3,878,085
14. Chapin patent, USP4,166,018
15. K. Sato, Y. Gotoh, Y. Hayashi, K. Adachi and H. Nishimura, Report Res. Lab. Asahi Glass, 40 (1990)233
16. T. Ikeda, K. Sato, Y. Hayashi, Y. Wakayama, K. Adachi and H. Nishimura, Solar Energy Matr. Solar Cells, 34(1994)379
17. T. Oyama, "Developments of Transparent Conductive Films III – Present Status of ITO and its Substitute Material Developments", edited by Tadatsugu Minami, CMC Publishing, Tokyo, March 2008 (Japanese), pp 143
18. M. Mizuhashi, Glastechnische Ber., 56K(1983)1040
19. M. Mizuhashi, Y. Gotoh and K. Adachi, Report Res. Lab. Asahi Glass, 36(1986)1
20. N. Taneda, T. Oyama and K. Sato, Tech. Digest PVSEC-17, pp 309 as 5O-B6-01
21. K. Sato, Y. Gotoh, Y. Wakayama, Y. Hayashi, K. Adachi and H. Nishimura, Report Res. Lab. Asahi Glass, 42(1992)129

22. T. Minami, H. Nanto and S. Takata, Appl. Phys. Lett., 41(1982)958
23. D. Horwat and A. Billard, Thin Solid Films, 515(2007)5444
24. H. Ogura, A. Suzuki, T. Matsushita, T. Aoki and M. Okuda, Thin Solid Films, 445(2003)263
25. T. Nakada, Y. Ohkubo, N. Murakami and A. Kunioka, Jpn. J. Appl. Phys., 34(1995)3623
26. T. Minami, H. Nanto and S. Takada, Jpn. J. Appl. Phys., 24(1985)781
27. F. Ruske, C. Jacobs, V. Sittinger, B. Szyszka and W. Werner, Thin Solid Films, 515(2007) 8695
28. W. Beyer, J. Hüpkes and H. Stiebig, Thin Solid Films, 516(2007)147
29. S. Fay, J. Steinhauser, N. Oliveila, E. Vallat-Sauvain and C. Ballif, Thin Solid Films, 515 (2007)8558
30. B. Szyszka, V. Sittinger, X. Jiang, R.J. Hong, W. Werner, A. Pflug, M. Ruske and A. Lopp, Thin Solid Films, 442(2003)179
31. J. Müller, G. Schöpe, O. Kluth, B. Recht, V. Sittinger, B. Szyszka, R. Geyer, P. Lechner, H. Schade, M. Ruske, G. Dittmar and H.-P. Bochem, Thin Solid Films, 442(2003)158
32. C. Agashe, O. Kluth, G. Schöpe, H. Siekmann, J. Hüpkes and B. Rech, Thin Solid Films, 442 (2003)167
33. M. Kon, P.K. Song, Y. Shigesato, P. Frach, A. Mizukami and K. Suzuki, Jpn. J. Appl. Phys., 41(2002)814
34. M. Kon, P.K. Song, Y. Shigesato, P. Frach, S. Ohno and K. Suzuki, Jpn. J. Appl. Phys., 42 (2003)263
35. J. Meier, U. Kroll, S. Dibail, S. Golay, S. Fay, J. Dubail and A. Shah, Proc. 28th IEEE PV Specialist Conf., pp 746

Index